Heat Treatment of Metals

Volume 1

Contents of Volume 2

Heat Treatment of Metals

Volume 1

J. L. SMITH MS, PhD
G. M. RUSSEL M Tech
S. C. BHATIA BE

CBS PUBLISHERS & DISTRIBUTORS PVT. LTD.
NEW DELHI • BANGALORE • PUNE • COCHIN • CHENNAI (INDIA)

ISBN : 978-81-239-1644-6

First Edition : 2008
Reprint : 2009, 2012

Published by Satish Kumar Jain and produced by V.K. Jain for
CBS Publishers & Distributors Pvt. Ltd.,
CBS Plaza, 4819/XI Prahlad Street, 24 Ansari Road, Daryaganj,
New Delhi - 110002, India. • Website: www.cbspd.com
e-mail: delhi@cbspd.com, cbspubs@airtelmail.in
Ph.: 23289259, 23266861, 23266867 • Fax: 011-23243014

Branches:

- ***Bengaluru:*** Seema House, 2975, 17th Cross, K.R. Road,
  Bansankari 2nd Stage, Bengaluru - 560070 Ph.: +91-80-26771678/79
  • E-mail: cbsbng@gmail.com, bangalore@cbspd.com
- ***Pune:*** Bhuruk Prestige, Sr. No. 52/12/2+1+3/2,
  Narhe, Haveli (Near Katraj-Dehu Road By-pass), Pune - 411051
  Ph.: +91-20-64704058/59, 32342277 • E-mail: pune@cbspd.com
- ***Kochi:*** 36/14, Kalluvilakam, Lissie Hospital Road,
  Kochi - 682018, Kerala • Ph.: +91-484-4059061-65
  Fax: +91-484-4059065 • E-mail: cochin@cbspd.com
- ***Chennai:*** 20, West Park Road, Shenoy Nagar, Chennai - 600030
  Ph.: +91-44-26260666, 26208620 E-mail: chennai@cbspd.com

Printed at :
India Binding House, Noida (UP)

Preface

Heat Treatment is the controlled heating and cooling of metals to alter their physical and mechanical properties without changing the product shape. Heat treatment is sometimes done inadvertently due to manufacturing processes that either heat or cool the metal such as welding or forming.

Heat Treatment is often associated with increasing the strength of material, but it can also be used to alter certain manufacturability objectives such as improve machining, improve formability, restore ductility after a cold working operation. Thus it is a very enabling manufacturing process that cannot only help other manufacturing process, but can also improve product performance by increasing strength or other desirable characteristics.

Steels are particularly suitable for heat treatment, since they respond well to heat treatment and the commercial use of steels exceeds that of any other material. Steels are heat treated for one of the following reasons: (i) softening; (ii) hardening; and (iii) material modification.

The reference book is intended for practising engineers—metallurgical, mechanical, mining, material science and persons working in steel plants, rolling mills, automobile and mechanical industries.

This volume dealing with basic concepts of metallurgy and theory of alloys, heat treatment processes, chemical heat treatment of steel and surface hardening of steel is the first volume of heat treatment of metals.

This reference book is an effort to put together various new techniques of heat treatment of metals. Section I deals with basic concepts of metallurgy and theory of alloys. Chapter 1 is devoted to identification of metals and alloys. Various metal identification tests such as magnetic test, observation test, file hardness test, chemical reaction tests, spart tests have been discussed in detail. Chapter 2 is confined to nature of metals and alloys. Chapter 3 focuses on metallography which is the study of the microstructure of metals and alloys by making use of both the microscope and the metallograph to make photomicrographs. Chapter 4 deals explicitly with iron carbon and alloy system. A study of iron, carbon alloy is valuable not only because it helps explain the properties of steel but also because it provides a means of understanding solid state reactions in general.

Section II describes various heat treatment processes. Heat treatment is an important operation in the manufacturing process of any machine parts and tools. Only by heat treatment it is possible to impart the high mechanical properties to steel required for the normal operation of modern machinery and tools. Heat treatment is usually classified into four typical groups. The first group comprises first-order or recrystallisation which is employed to relieve internal stresses, reduce hardness and to increase the ductility of strain-hardened metal. The second group, second-order or full annealing, which involves phase recrystallisation, consists of heating alloys above the temperatures required for phase transformation. This is followed by slow cooling. In the third group, hardening alloys are heated above the phase transformation temperature and are then rapidly cooled (quenched) to retain at room temperature, either a phase stable at high temperatures or various stages of decomposition or transformation of this phase. The fourth group, tempering involves the reheating of a hardened alloy (steel) to a temperature below that required for phase transformation so as to bring it nearer to an equilibrium state. Chapter 5 provides information on basic concepts of heat treatment of steel. Various basic concepts of heat treatment of steel such as—annealing, normalising, quenching, tempering, etc., have been discussed briefly. Chapter 6 explains the theory of heat treatment of steel. The theory of heat treatment of steel is understood as the analysis of the processes of structure formation and the particularities of structural state of alloys. Chapter 7 is confined to stress-relief heat treating steel which is used to relieve stresses that remain locked in a structure as a consequence of a manufacturing sequence. Chapter 8 concentrates on annealing of steel. Annealing is a generic term denoting a treatment that consists of heating to and holding at a suitable temperature followed by cooling at an appropriate rate, primarily for softening of metallic materials. Chapter 9 is devoted to normalising of steel which is a heat treating process that is often considered from both thermal and microstructural standpoints. Chapter 10 and 11 elucidates the hardening and tempering of steel. Hardening is a heat treating process in which steel is heated to a temperature above the critical point held at this temperature and then quenched; where as tempering of steel is a process in which previously hardened or normalised steel is heated to a temperature below the transformation range and cooled at a suitable rate, primarily to increase ductility and toughness. Chapter 12 provides information on austempering of steel. Austempering is the isothermal transformation of a ferrous alloy at a temperature below that of pearlite formation and above that of martensite formation. Chapter 13 focuses on martempering of steel. Martempering is a term used to describe an elevated-temperature quenching procedure aimed at reducing cracks, distortion or residual stresses. It is not a tempering procedure, as the name implies and is more properly termed marquenching. Hardness is defined as resistance to plastic deformation or penetration and hardenability is defined as the ease with which hardness may be attained by quenching. Considering this Chapter 14 emphasises on hardenability of steel. Chapter 15 throws light on heat treatment of ultrahigh-strength steels, which are heat treated by the use of equipment and techniques similar to those employed for heat treating constructional alloy steels. The ultrahigh-strength steels ordinarily are quenched and tempered to specified hardness, but for critical applications it may be necessary to pull tensile specimens to ensure that a required combination of strength and ductility has been achieved. Chapter 16 elucidates on quenching of steel and is the rapid cooling of the steel from a suitable elevated temperature. This generally is accomplished by immersion in water, oil, polymer solution or salt, although forced air is some times used.

Section III describes the chemical heat treatment of steel. In chemical heat treatment, parts are usually placed in a medium rich in the desired elements which penetrates the metal by diffusion. Chapter 17 is devoted to basic concepts of chemical heat treatment of steel. Theory of chemical treatment along with types of diffusion movements of atoms in metals is briefly discussed. Chapter 18 highlights carburising of steel. The main purpose of the carburising process is to obtain a hard and wear-resistant surface on the machine parts of enrichment of the surface layer with carbon to a concentration from 0.75 to 1.2 per cent and subsequent quenching. Steel which has been carburised and quenched (case hardened) has a high fatigue limit. Chapter 19 concentrates on cyaniding and carbonitriding of steel. Various methods of cyaniding and carbonitriding—such as liquid cyaniding, pack cyaniding, absorption of carbon and nitrogen are briefly discussed. Nitriding is the process of saturating the surface of steel with nitrogen by holding it for a prolonged period at a temperature from 480 to 650°C in an atmosphere of ammonia. Nitriding increases the hardness of the surface to a very high degree and also increases the wear resistance, fatigue limit and resistance to corrosion in such media as the atmosphere, water and steam etc. Keeping this in mind Chapter 20 focuses on nitriding of steel. Various methods of nitriding and their advantages and disadvantages are discussed in detail. Chapter 21 deals with diffusion coatings and metallic concentration. Saturation of the surface of steel by aluminium, chromium, beryllium and other elements is called diffusion coating. Diffusion coating of steel by various methods and also by silicon is done by embedding the parts in the corresponding powdered mixtures. The rate of diffusion depends to a considerable extent upon the composition of the steel.

Section IV deals with surface hardening of steel. Aim of surface hardening is to increase the hardness only on the surface while the core does not show any structural change. Surface hardening is advantageous over chemical heat treatment as it requires appreciably less time. Chapter 22 is an overview of surface hardening of steel. Various surface hardening methods are briefly discussed. Chapter 23 emphasises on flame hardening which is a heat treating process in which a thin surface shell of a steel part is rapidly heated to a temperature above the critical point of the steel. Various methods of flame hardening along with operating procedures and control and safety precautions are discussed. Chapter 24 is confine to induction hardening by which we can achieve a considerable increase in surface hardness and only those steels are used having carbon content of 0.30 to 0.50 per cent. Procedures of induction hardening along with their advantages and disadvantages are discussed. Chapter 25 highlights the electron beam hardening which is a selective hardening process in which the surface of a hardenable ferrous alloys is heated rapidly above the transformation temperature of the alloy by direct impingement of an accelerated stream of electrons. Chapter 26 is an indepth study on laser surface transformation hardening. In recent years, industrial lasers have become available for metal working uses, including surface hardening. In laser surface transformation hardening, thermal energy is generated by absorption of the laser radiation at the surface transformation hardening including size of the laser spot, power density, processing speed, thermal properties of the material and laser hardenability of the material.

This reference book is throughout supplemented with diagrams and tables. The treatment of all topics is in a congent, lucid style aimed at enabling the reader to grasp the information quickly and easily. The references and index which are given in the end will prove useful for the readers.

Smith and Russel are consultants in material science and metallurgy and have been involved with heat treatment industry for over two decades. They have also published various research papers.

J. L. SMITH
G. M. RUSSEL
S. C. BHATIA

Contents

SECTION III

SECTION IV

Contents in Detail

SECTION I

SECTION II

SECTION III

SECTION IV

SECTION I

Basic Concepts of Metallurgy and Theory of Alloys

Chapter 1

Identification of Metals and Alloys

INTRODUCTION

During the construction stages and fabrication of parts metals have been known to become mixed. Sometimes the craftsman will notice the difference and make the correct replacement. In other instances correct metal was used but incorrectly oriented with the direction of stress.

A metal is composed of a single metallic element or group of elements. An alloy is made up of two or more elements, one of which must be a metallic element.

A metal is generally considered to have the following characteristics:

1. Crystalline when solid.
2. Good conductor of heat and electricity.
3. Can be deformed with an application of stress in a plastic manner.
4. Will reflect light when polished

These characteristics are based on general conditions. Some special types of preparation may produce other results. An element is the smallest divisible part of a substance that may be made without changing the substance. It cannot be separated into substances different than itself by ordinary chemical means. There are a limited number of elements that individually or in combination make up or compose every material substance on earth.

METAL IDENTIFICATION TESTS

Complete identification of metals can be complex and time consuming. There are several simple methods that help a person separate common metals. A few of these will be discussed.

Magnetic Test

The magnetic test is a fast, simple test to determine if a particular metal is attracted to a magnet. One of the most widely used metals is steel. It is composed of a large amount of the element iron and a small amount of the element carbon. Steel is strongly attracted to a magnet. Nickel and cobalt also have good magnetic properties. Most of the other metals are not attracted by a magnet. Examples of these non-magnetic metals are copper, aluminium, tin, zinc, chromium, manganese, magnesium and titanium.

One of the most notorious examples of the magnetic test is found during the process of differentiating between two types of stainless steel, as shown in Fig. 1.1.

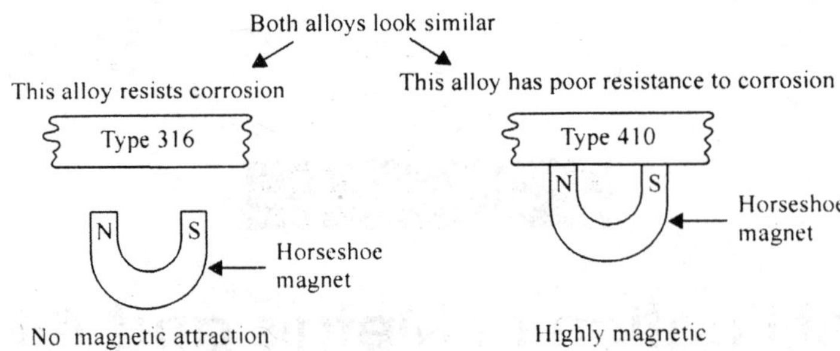

Fig. 1.1. Identification of two different types of stainless steel.

Visual Observation Test

With a little experience this test will provide good dividends. When trying to identify an alloy, one should use known samples as a means of reference. For example, if a type 316 stainless steel had been available as a comparison standard in the magnetic test, the colour could have been a deciding factor. Type 316 seems to have a white appearance, while type 410 is more bluish-white.

One very positive identification in steel can be made by visually observing the surface. Some steels, such as the structural grades, are finished at the steel mill by hot rolling to final size. After hot rolling, the steel becomes covered with a layer of mill scale or iron oxide. This makes the surface slightly rough and irregular. This is acceptable for I beams, angle shapes, plate stock, and other components used for bridges or buildings. If the finished surface is important, then cold-rolled steel should be specified. This stock will usually be delivered with a heavy grease coating, when clean, cold-rolled steel will be smooth and shiny. Cold-rolled steel also has better machining characteristics than hot-rolled steel.

Visual appearance is also valuable in determining certain mechanical properties of fractured sections. Reference to this will be made later in a discussion of the ductile and brittle fracture of stressed specimens.

File Hardness Test

A file test is a very simple method for estimating the relative hardness of an unknown piece of metal. Obtain a known sample and compare the cutting action of the file on it with that on the unknown piece. A good example of the use of a file test is in the fabrication shop. Assume that a bar ½ inch in diameter is needed to be welded in a structure as a brace. Both tool steel and structural steel were in the shop. A quick method for identifying the two would be the file test shown in Fig. 1.2.

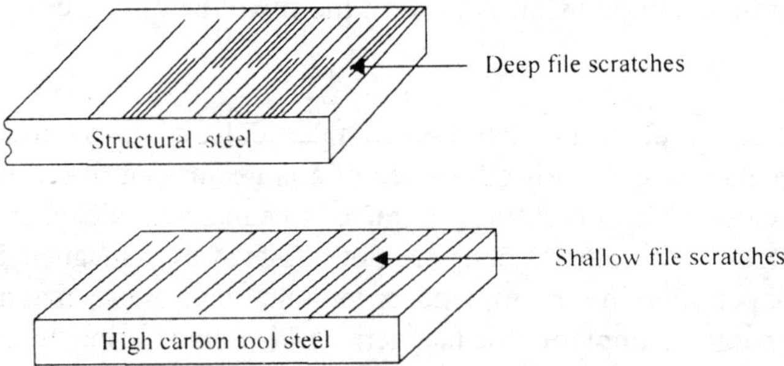

Fig. 1.2. The results of a file test to separate tool steel from structural steel.Both sections are hot rolled mill stock.

Surface Reflectivity Test

With a little experience it is possible to identify alloys by their ability to reflect light. The reflectivity can be compared in various stages, such as in the mill condition (with equal treatment) or in the highly polished stage. For example, reflectivity might aid in separating aluminium and magnesium. Aluminium is 1.25 times more reflective than magnesium and aluminium alloys are generally more reflective than magnesium alloys. In separating lead-tin alloys, the higher the tin content, the greater the reflectively. Tin is 1.21 times more reflective than lead as cast. After exposure to air for a long period of time, tin then becomes more reflective than lead, due to the thick oxidation layer on lead. Here, reflectivity is a visual test.

Weight Per Volume Test

Alloys can be separated by comparing their weight per volume to that of known samples. The test is conducted by obtaining a small sample that will fit in a small graduate. The 25 millilitre (ml) or 100 ml graduate is probably the most useful. Weigh the metal and place it in the graduate partially filled with water. Then divide the weight of the metal by the volume of water displaced. Compare the weight per volume of the known sample with the unknown, as shown in Fig. 1.3.

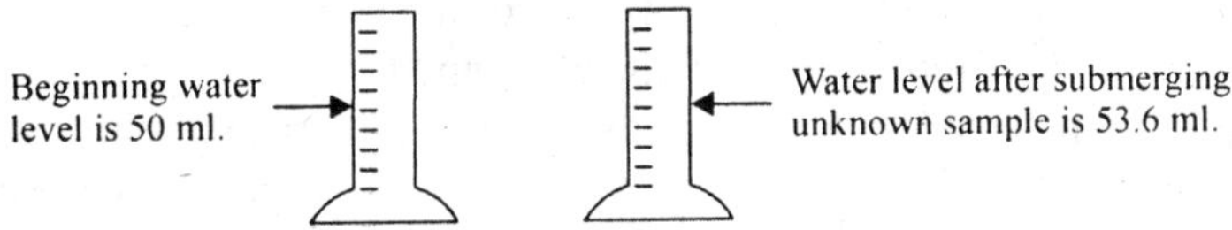

Question: Do both samples have the same weight per mass?

Known sample   Unknown sample

Given: Weight per mass for this problem will be density in g/cm^3. (1 cm^3 = 1 ml of water at 4°C). Density is specific gravity.

Formula: $$\text{Density} = \frac{\text{Weight of metal (g)}}{\text{ml of displaced water}}$$

Data: Known sample weighs 183.4 grams and displaces 28.00 ml of water

$$\text{Density} = \frac{\text{Weight (g)}}{\text{ml (water)}} = \frac{183.4}{28.0}$$

Density = 6.55 (of known sample)

Unknown sample weighs 23.57 grams and displaces 3.60 ml of water

$$\text{Density} = \frac{\text{Weight (g)}}{\text{ml (water)}} = \frac{23.57}{3.60}$$

Density = 6.55 (of unknown sample)

Therefore both samples have the same weight per mass.

Fig. 1.3. Procedure to determine weight per volume (or density in g/cm^3).

The result in the Fig. 1.3 example does not mean the metals are identical, but the specific gravity rating indicates some similarity. Common metals range from a low specific gravity (weight per equal volume of water) of about 1.74 for magnesium to about 19.3 for tungsten.

Chemical Reaction Test

Testing the reaction of a metal to certain acids can be simple or it can be complex, depending upon the type of alloy and the amount of alloying elements present. The mixing of acids can be hazardous and should be done by an experienced person. A few selected tests are listed in Table 1.1.

Table 1.1. Typical procedures for chemical reaction tests.

Objective	*Procedure*
Estimate carbon content of carbon steel	Grind surface smooth, place 3 to 4 drops to 50 per cent nitric acid and alcohol on cold surface. After 1 minute transfer fluid to filter paper. The darkness of the stain will determine amount of carbon.
Test for manganese	Grind surface smooth, place 3 to 4 drops of 50 per cent nitric acid and alcohol on the surface. Add enough sodium bismuthate reagent until a brown cluster of particles appears. Wait for 3 minutes and add a drop of nitric acid. If a pink colour appears, there is manganese in the metal.
Separate 300 stainless from monel	Monel shows an immediate reaction to concentrated nitric acid and usually becomes blue-green.

Spark Test

The spark test can be useful in separating alloys that contain known alloying elements. It is especially helpful in determining wide variations in carbon content, such as separating mild steel, which has good weldability, from carbon tool steel, which should not be welded by ordinary procedures. Spark testing is based on the theory that heat generated by friction between the alloy and the grinding wheel and oxygen in the air cause the particles to burn. The technician analyses the burning pattern and relates it to patterns from past experience and reference charts. In the analysis the technician is concerned with volume, length from wheel to burn-out, colour and explosions along the stream or at the end of the spark pattern. Sparks from iron carbon alloys develop a very distinctive appearance. The spark pattern is easy to photograph in colour by using colour film of at least ASA 160 (tungsten) and setting the shutter speed no slower than 1/30 seconds. With a little skill, rapid identification can be made with an inexpensive portable grinder (Fig. 1.4).

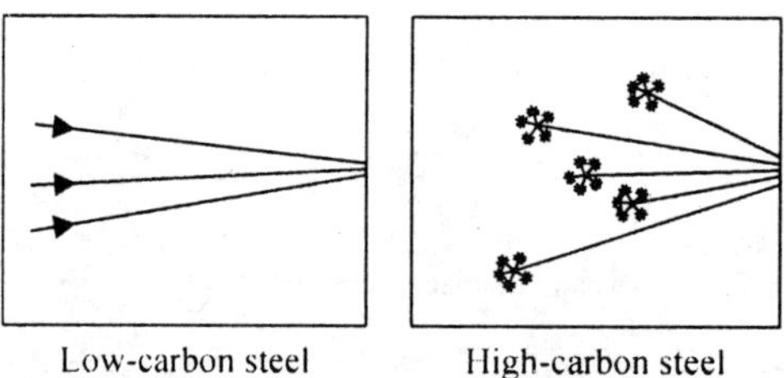

Fig. 1.4. Spark patterns used to help separate low carbon steel from high carbon steel.

Cast iron and the alloy steel may also be separated with experience and characteristic spark patterns. (Fig. 1.5).

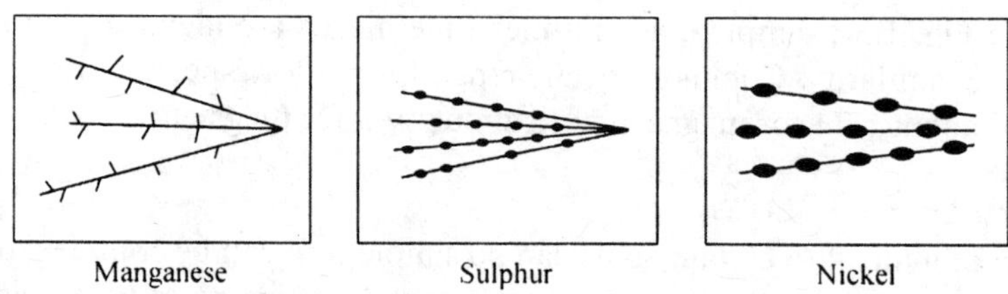

Fig. 1.5. Spark patterns from a few selected steels.

CLASSIFICATIONS OF METALS

Much has been done to standardise operations in the metals industry. Non-profit societies have established controls in the manufacturing, design, purchase and construction of metals.

Many criteria can be used in classifying metals. Typical systems include the method of manufacture, how the metals are to be used and the chemical composition specifying major alloying elements.

Various organisations in the United States have made attempts to classify metals. A few of the more well known are:

AA	The Aluminium association
ASTM	The American Society for Testing and Materials
ASM	The American Society for Metals
ASME	The American Society for Mechanical Engineers
CDA	The Copper Development Association
SAE	The Society of Automotive Engineers
AISI	The American Iron and Steel Institute
AWS	The American Welding Society
ANSI	The American National Standards Institute
ACI	The Steel Founders Society of America
AMS	Aerospace Materials Specification
MIL	Military Specification
WW	Federal Specification

A reference book that will aid persons in trying to understand the classification system of metals is the Unified Numbering System for Metals and Alloys. This is a cross index of chemically similar specifications. The Unified Numbering System (UNS) number assigned to a metal is not in itself a specification. The UNS number establishes no requirements for form, condition or properties. The number is a unified identification of metals and alloys for which limits have already been established.

The Unified Numbering System established 17 series of numbers for metals and alloys. A UNS number consists of a single letter (prefix) followed by five digits. In most cases the prefix letter is suggestive of the family of metals.

To show the efforts of different organisations to standardise the classification of an alloy, let us consider the term structural steel. The following are all specifications for an alloy that is somewhat similar: AISI 1020 (American Iron and Steel Institute), SAE 1020 (Society of Automotive Engineers), and ASTM A36-70a (American Society for Testing and Materials). A similar alloy is identified by the Unified Numbering System as 'UNS G10200'

The ASTM has written specifications for all materials. The specifications for steels from ASTM are similar to SAE and AISI. The SAE and AISI specifications are based primarily on chemical analysis. The ASTM A36-70a steel specification lists chemical analysis plus other restrictions such as:

1. Process of steel making: The steel would have to be made by one of three processes; open-hearth, basic oxygen, or electric arc.
2. Tensile properties: The tensile strength range will be from 58,000 to 80,000 pounds per square inch (psi). Yield point will be a minimum of 36,000 psi and ductility in terms of per cent elongation will have a minimum of 23 per cent for plates and bars or 21 per cent for shapes.

Each ASTM standard has a serial designation. A capital letter indicates the following general classification: A, ferrous metals; B, non-ferrous metals; C, ceramics and concrete; D, miscellaneous materials; E, miscellaneous subjects; F, end use materials; G, corrosion, deterioration; and degradation of materials; ES, emergency standards.

Look for this information when reading a specification designated by the American Society for Testing and Materials (Fig. 1.6).

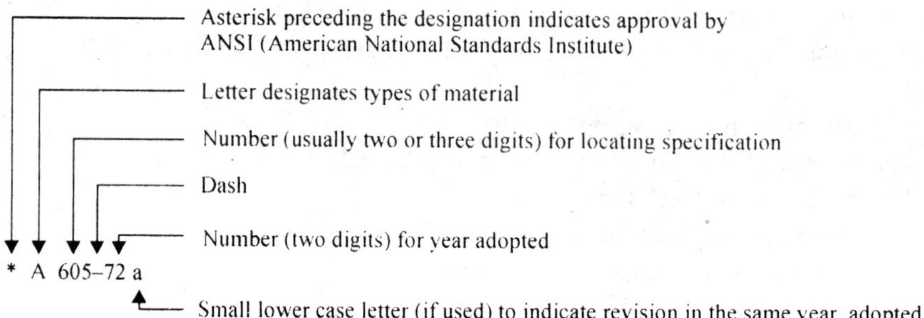

Fig. 1.6. A typical ASTM (American Society for Testing and Materials) number and how it may be analysed.

The metals discussed in this next section will be divided into two major groups. Those that are predominantly iron or have iron as the major alloying element will be called the ferrous groups. Alloys that have a small amount of iron, such as a fraction of one per cent, make up the non-ferrous groups. Typical metals in each groups are:

1. Ferrous alloys.
   (a) Carbon steel.
   (b) Alloy steel.
   (c) Tool steel.
   (d) Cast iron.
2. Non-ferrous alloys.
   (a) Aluminium alloys.
   (b) Magnesium alloys.
   (c) Copper alloys.
   (d) Nickel alloys.
   (e) Lead alloys.
   (f) Titanium alloys.

Carbon and Low Alloy Steel

Some of the carbon and low-alloy steels are classified by chemical compositions that were developed by the SAE and the AISI. Steels in this group are designated by a four digit number and certain letters. Typical examples of the prefix letters are:

A Basic open hearth alloy steel.
B Acid bessemer carbon steel.
C Basic open hearth carbon steel.
TS Tentative standard steel.
EX Numbers not yet adopted.

In the four digit series the first two numbers indicate the family of steels and the next two (sometimes three) give the approximately amount of carbon in hundredths of a per cent, as shown in Table 1.2.

Table 1.2. Typical four digit steel systems (SAE & AISI).

Number	*Type of steel*
10XX*	Plain carbon (plus minor amounts of impurities)
13XX	Low alloy manganese (1.75%)
31XX	Low alloy nickel (1.2%)—chromium (0.70%)
40XX	Low alloy molybdenum (0.25%)
41XX	Low alloy chromium (0.95%)—molybdenum (0.20%)
43XX	Low alloy nickel (1.80%)—chromium (0.50%)—molybdenum (0.25%)
46XX	Low alloy nickel (1.80%)—molybdenum (0.25%)
48XX	Low alloy nickel (3.50%)—molybdenum (0.25%)
50XX	Low alloy chromium (0.45%)
52XXX	Low alloy carbon (1.00 %)—chromium (1.00%)
61XX	Low alloy chromium (0.90%)—vanadium (0.12%) Min.
86XX	Low alloy nickel (0.55%)—chromium (0.50%)—molybdenum (0.25%)
87XX	Low alloy nickel (0.55%)—chromium (0.50%)—molybdenum (0.25%)
92XX	Low alloy manganese (1.00%)—silicon (2.00%)
93XX	Low alloy nickel (3.25%)—chromium (1.20)—molybdenum (0.12%)
94XX	Low alloy manganese (1.00%)—nickel (0.45%)—chromium (0.40%)—molybdenum (0.12%)
97XX	Low alloy nickel (0.55%)—chromium (0.17%)—molybdenum (0.20%)
98XX	Low alloy nickel (0.55%)—chromium (0.80%)—molybdenum (0.25%)

*XX—Denotes carbon content x 0.01

Other letters are sometimes used in classifying these steels which provide additional information helpful to the fabricator. Specific examples are:

Letter	Type of condition	Typical number
B	Boron steel	AISI 46B12
BV	Boron-vanadium steel	AISI 43BV14
H	Hardenability data	
J	Hardness parameter	
N	Nitriding steel	

Tool Steels

These alloys are specially made for the purpose of shaping other metals; for example, cutting, piercing, stamping, punching and bending. The system for classifying these alloys is based on:

1. Quenching medium (the material required to produce a fast cooling rate in the process of heat treating for the transformation of a desirable structure). One of the most desirable structures is called martensite (Fig. 1.7).
2. Composition or elements added to iron. This group may be divided into carbon tool steel, low alloy, and high alloy (where elements added to iron are more than 10 per cent of the total composition). Examples of these are shown in Table 1.3.
3. Service expectations of the steel. Examples of some of these services are high speed cutting, shock resisting and hot work.

4. Special purpose steels which do not fit into any of the above systems. Steels in this group may be required to resist corrosion, wear and abrasion.

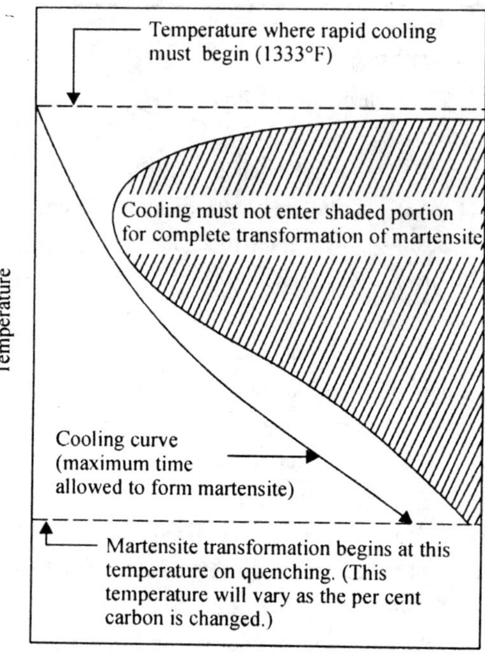

Fig. 1.7. Cooling requirements to obtain martensite from carbon tool steel.

Table 1.3. Showing typical tool steels and respective composition.

Group	*Composition*							*Applications*
	C	*Mn*	*Cr*	*V*	*W*	*Mo*	*Other*	
Carbon steel								
Water hardening	0.6–1.4	0.6						Chisels, punches, shear blades
Low alloy steel								
Shock resisting	0.5	0.8				0.5	Si-1	Pneumatic tools
Cold work			5.0	1.0		1.4	Ni-1.5	Blanking dies, thread rolling dies, piercing dies
High alloy steel								
Hot work	0.3		12.0		12.0			Hot forging dies, die casting, extruding
High speed	0.75		4.0	2.0	14.0		Co-5	Drill bits, cutting tools, hack saw blades

Cast Iron

The term cast iron is a very broad term since there are so many types of cast iron. Referring to a metal as cast iron signifies nothing. It is like thinking of a specific metal and referring to it as steel. Cast irons are composed of many different alloys and offer a wide range of mechanical properties. Some people think of cast iron as a brittle material that cannot be bent or shaped by force. This is only partly true, as shown in Table 1.4. While most cast irons are cast to finished shape as an economical necessity, it does not mean they cannot be re-shaped. Some nodular and malleable cast irons have ductility values similar to structural steel.

Table 1.4. Typical cast iron alloys and estimated mechanical properties.

Type of cast iron	*Tensile strength (psi)*	*Compression strength (psi)*	*Torsional strength (psi)*	*Modulus of elasticity* ($\times 10^6$)	*Elongation per cent in 2 inch*	*BHN**
Gray iron	22000 62500	83000 1,65,000	26000 88500	14–17	1.3	150 302
† Nodulur iron	55000 1,50,000	1,50,000	80000	18	2–40	130 320
Malleable	50000 1,20,000	1,50,000	90000	24	2–20	110 269

* BHN is Brinnel hardness number.

† Nodular iron is also called ductile iron.

Originally, cast irons were classified by the microshapes that appeared as a result of solidification (liquid to solid). These shapes, which make up the structure, could be affected by:

1. Carbon content.
2. Elements in addition to iron and carbon.
3. Cooling rate from liquid to solid during the casting process.
4. Heat treatment after casting.

Of all the elements that make up the composition of cast iron, carbon is the single most important. Carbon may exist as pure carbon, called graphite, or it may be dissolved with other elements. When carbon is dissolved in iron and slowly cooled it becomes iron carbide. If the carbon reaches a concentration of about 6.67 per cent, the structure is called cementite. Cast iron from a chemical analysis is mostly iron. Steel is almost all iron; however, the two may have completely different properties because of the shape and distribution of the particles that make up the structures. It is these particles, called microstructures, that determine how the different types of cast iron will behave when exposed to force, heat and a corrosive environment.

Gray Cast Iron

Gray iron castings are probably the most widely used of any of the cast iron group. Typical applications and advantages are shown in Table 1.5.

Table 1.5. Applications and related advantages in the use of gray iron castings.

Applications	*Advantages*
Machine frames and load supports	High compressive strength Good vibration dampener
Valve housings	Inexpensive to cast Good machinability

Contd...

Applications	*Advantages*
Small levers and valve handles	Lowest cost per part Corrosion resistant Strong
Automotive engine blocks	Competitive price for complex shapes Absorbs noise and vibration Retains tolerances High wear factor

A simple test for classifying gray iron castings has been devised by the American Society for Testing Materials (ASTM). Specifications for this test can be found in ASTM standards in section ASTM A48. One type of test bar is shown in Fig. 1.8. The load in pounds that is required to fracture the bar is divided by the area of the cross-section. This value is considered to be the strength of the bar in pound per square inch (psi). While loading, the specimen elongates. This provides information about the ductility of the specimen. One expression of ductility is called per cent elongation. It is found by dividing the change in length by the original length and multiplying by 100. Since gray iron castings are so brittle and it is difficult to obtain a true axial load, a special test has been devised by ASTM. The ASTM designation is A438-62 (Fig. 1.9).

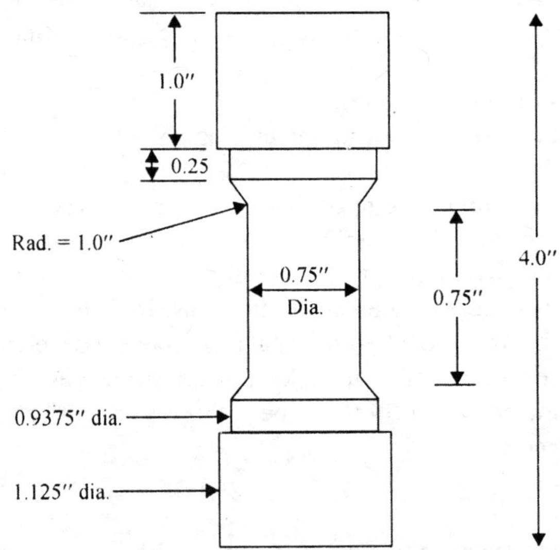

Fig. 1.8. Tensile test bar dimensions for gray iron castings.

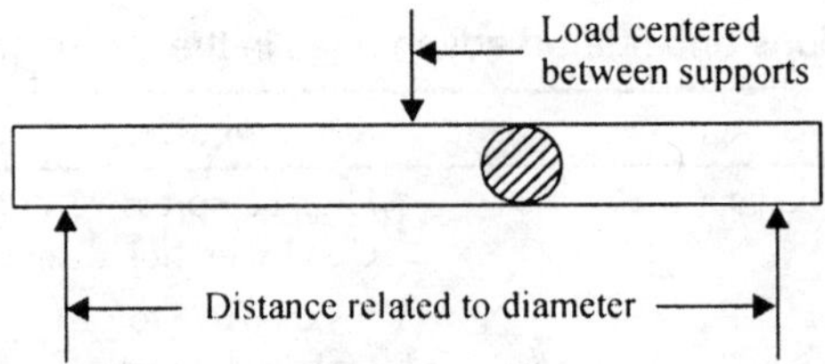

Fig. 1.9. A transverse test of gray cast iron. Tensile strength is computed or found from a table in the ASTM standards.

Estimates of typical loads that represent approximate tensile strength values are shown in Table 1.6.

Table 1.6. Showing test loads and respective tensile strength values.

	0.88-inch dia. 12 inch length	*1.2-inch dia. 18 inch length*	*2.0-inch dia. 24 inch length*	*Approximately tensile strength (× 1000 psi)*
Breaking	900	1800	6000	20
Loads	1400	2600	9100	40
	1925	3400	12500	60

Many indications are given that cast iron can have wide variations in properties. It is the small areas that are difficult to resolve with a microscope that determine the mechanical properties of most metals. These tiny shapes are called microstructures and may conform to definite shapes when exposed to stress and heat. In cast iron there are two major parts of the microstructure. One is pure carbon, called graphite and the other is the continuous area surrounding the graphite. The continuous area is called the matrix. It is interesting to note that the matrix can be changed every time the metal is heated to a specific temperature (below the melt). The graphite particles, once shaped, cannot be changed unless the casting is melted to permit the graphite to decompose. The ASTM and AFS have developed a set of standards to classify different types of cast iron. These standards reveal the microstructure of cast iron at 100 magnifications (100X). Typical shapes are shown in Fig. 1.10.

(a) (b) (c) (d) (e)

Fig. 1.10. Typical shapes of the microstructure of cast iron. (a) Flake graphite with a ferrite matrix; (b) flake graphite with a pealite matrix; (c) flake graphite with a martensite matrix; (d) temper carbon with ferrite matrix; and (e) graphite nodules in a pearlite matrix.

A true reproduction of the microstructure of metals is difficult to obtain. Before quality photographs can be made, the technician must carefully cut the sample without disturbing the adjacent crystals. Then the surface must be ground (protected by a coolant) with finer and finer abrasives to obtain a flat nearly scratch-free surface. After the final stage of grinding, the surface is polished with an abrasive that has finer particles than face powder. At this stage of the preparation the graphite structures will reveal their true shape with a microscope. If the matrix is to be studied, then the surface is usually etched with a chemical that will attack the surface to reveal minute detail. Gray cast iron usually has a matrix of ferrite, pearlite, or martensite. However, the graphite shapes may differ widely, as shown in Fig. 1.11.

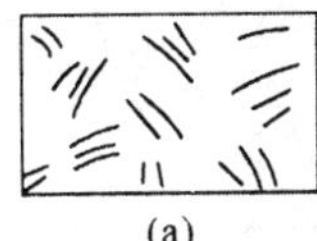

(a)

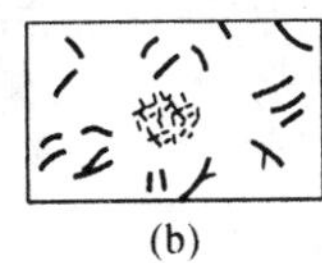

(b)

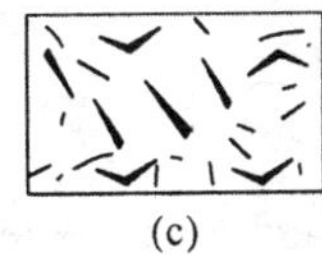

(c)

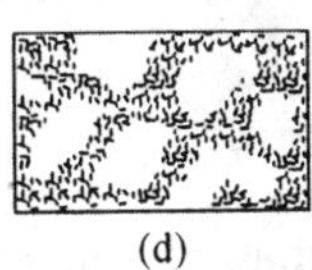

(d)

Fig. 1.11. Variations of graphite in gray cast iron. (a) Uniform flakes; (b) rosette grouping; (c) superimposed flake sizes; and (d) interdendritic segregation.

One of the most important alloying elements in cast iron, in addition to iron and carbon, is silicon. Silicon has a controlling influence upon the formation of graphite.

As an aid in demonstrating the influence of silicon in gray cast iron, the following formula can be used:

$$CE = TC + (\tfrac{1}{3} \times \%Si)$$

where,

CE = Carbon equivalent

TC = Total carbon

Si = Silicon

Observe how a cast iron low in carbon may have the structure of a much higher alloy with the addition of silicon (at 2.1 per cent).

$$\begin{aligned} CE &= TC + (\tfrac{1}{3} \times \%Si) \\ &= 2.75 + \tfrac{1}{3}(2.10) \\ &= 3.45 \text{ per cent} \end{aligned}$$

As a general practice the higher the carbon content the less the silicon requirement. Fig. 1.12 shows the ratio of silicon and carbon found in typical gray cast iron compositions.

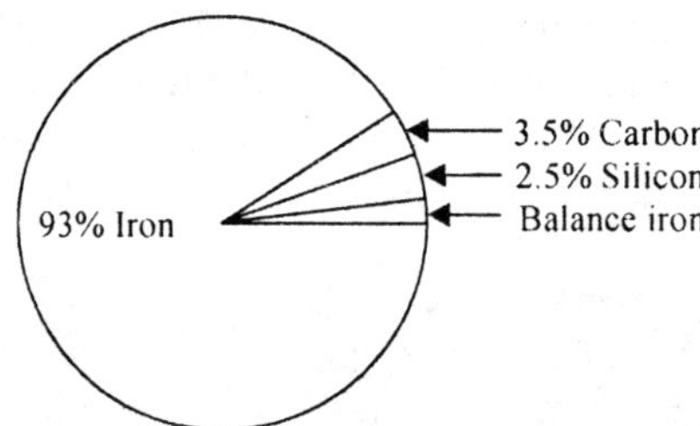

Fig. 1.12. The ration of carbon, silicon and iron content required to produce a gray iron casting of average tensile strength.

Mechanical Properties

It is interesting to investigate some of the mechanical properties of gray cast iron and compare them with those of structural steel. The low-strength gray irons have less than half the rigidity of steel when the stress is below the elastic range. This means that a gray iron section will stretch more or bend more than an identical bar of structural steel with the same load. Note that the term rigidity is a ratio of load versus deformation. In metallurgical terms, rigidity is called modulus of elasticity. The modulus of elasticity is found by dividing the load in the elastic range in pounds per square inch by strain (deformation) in inches per inch. The formula is written as:

$$\text{Modulus of elasticity (E)} = \frac{\text{Stress}}{\text{Strain}}$$

where,

Stress = Load psi

Strain = Deformation in inches/inch of gage length

Example 1. Find the modulus of elasticity of a tensile specimen made of gray iron that deforms 0.0003 inch/inch of length with a load of 4000 psi.

$$E = \frac{\text{Stress}}{\text{Strain}} = \frac{4000}{0.0003} = 13.3 \times 10^6 \text{ psi}$$

An informative experiment can be used to demonstrate the behaviour of gray cast iron in contrast to steel. It consists of a base made of ½ inch × 6 inch angle iron about 8 inch long. Drill a $\frac{5}{32}$ inch diameter hole to fit sample rods. Cut the rods about 8.0 inch long ± 0.001 inch tolerance. Cut a 1 inch diameter bar about 1.5 inch long for a counterweight. Insert the $\frac{5}{32}$ inch diameter rods, one at a time, into the angle iron base. With no load, adjust the dial indicator to zero, then attach the counterweight and read the deflection. The weight must not cause the rod to deflect beyond the elastic limit, otherwise the test is not valid. This test is shown in Fig. 1.13.

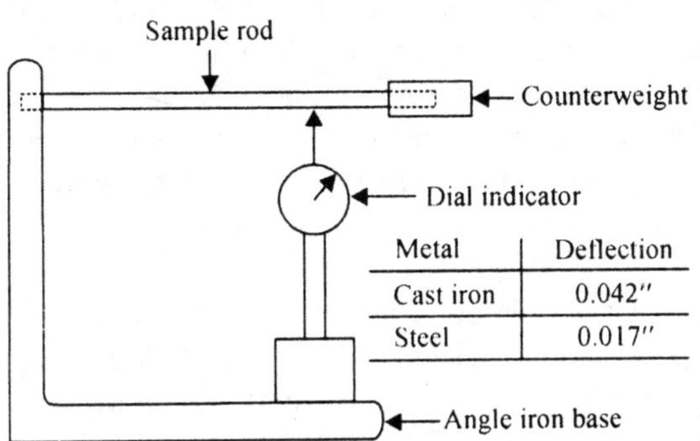

Fig. 1.13. Fixture to demonstrate the stiffness or modulus of elasticity of gray cast iron as compared to steel. The greater the deflection, the lower the modulus of elasticity.

With the deflection of each rod known (in inches), the modulus of elasticity for the gray cast iron may be determined by the formula:

Modulus of elasticity = E (of gray cast iron)

$$E = \frac{\text{Deflection of steel}}{\text{Deflection of cast iron}} \times \text{Modulus of elasticity of steel}$$

$$= \frac{0.017}{0.042} \times 29 \times 10^6$$

$$E = 11.74 \times 10^6 \text{ psi}$$

There are some interesting aspects of this test. The major factor that seems difficult for some people to comprehend is that the deflection of steel will be the same whether it is cold worked, quench hardness, or annealed. This is explained by the theory that the modulus of elasticity is dependent upon the type of metal rather than total strength. This test can be used to obtain the modulus of elasticity of any other metal by obtaining the deflection for the metal in question and substituting the value in the place of that for cast iron in the formula.

The cooling rate when gray cast irons are solidifying from a liquid is important because it is during this period that the type of graphite is formed. The amount, shape and size of graphite is determined primarily by the additional alloys present and the cooling rate. The composition and cooling rate may cause most of the carbon to form graphite. When this is the case, the matrix will be mostly ferrite. If more carbon is available to form iron carbide, then the matrix will be largely composed of pearlite.

Further cooling after solidification is also important. This is especially true when the temperature of the gray cast iron is in the red colour range (1300 ± 50°F). The cooling rate through this range will have an additional effect on the appearance and properties in the matrix.

White Cast Iron

The name white cast iron was probably given to this metal because of the appearance of the fractured surface. If both gray iron and white iron sections were broken, the white iron casting would be much lighter in colour than the gray iron section. White cast irons are characterised by high hardness and brittleness. These mechanical properties can be traced to the type of microstructure obtained during solidification. The low carbon content (from 2.0 to 2.5 per cent) when solidifying transforms carbon to iron carbide (called cementite). Most of the tonnage produced as white iron castings are later heat treated. After a special type of heat treatment, the white iron structure changes into a new form of cast iron with completely different mechanical properties. What was white iron now becomes malleable cast iron. There are a few job requirements that depend upon the hardness of white iron and its particular mechanical properties. White iron castings perform services such as rolls for crushing grain, rolls for crushing rock, rolls for making sheet metal and rolls for grinding cement.

Malleable Cast Iron

Malleable structures are not cast in this conditions. As mentioned earlier, the casting was originally white cast iron. Malleable castings may have about 2.0 per cent carbon and less than 1.0 per cent silicon. After being cast as white iron, the sections are heat treated at a temperature of 1675°F for 48 to 60 hours to decompose the iron carbide, as shown in Fig. 1.14.

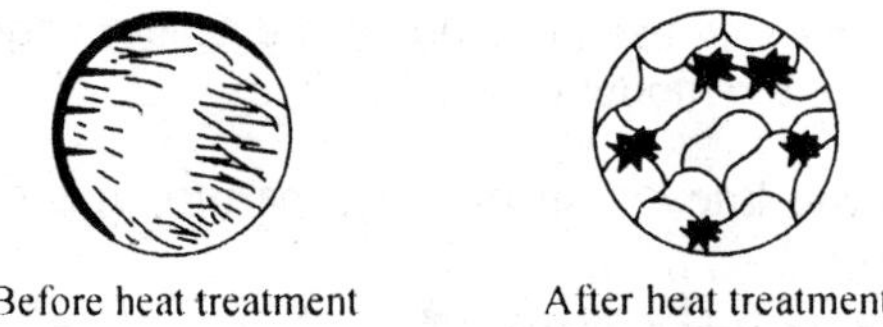

Fig. 1.14. Metallic phases that have been altered by many hours in a furnace.

After the iron carbide is decomposed, another long sequence of heat treatment begins. When the time requirement is completed in the 1675°F furnace, the parts are cooled from 1675°F to 1400°F in about 2 hours. Cooling from 1400°F to 1300°F may require 20 hours. Then the parts are taken from the furnace to cool in the air as shown in Fig. 1.15.

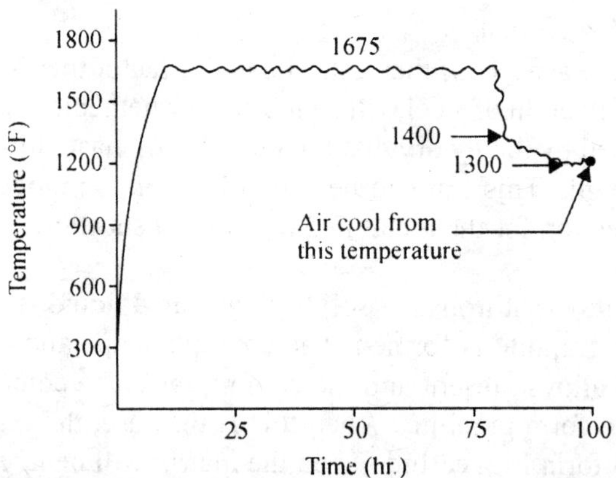

Fig. 1.15. The sequence required to transform white cast iron into malleable cast iron. Steps include: Slow heat to 1675°F, hold for 70 hrs., cool rapidly to 1400°F and cool 5°F to 15°F per hr., to 1300°F. Now air cool to room temperature and process is complete.

The slow cooling transforms the original arrangement of carbon in iron to grains of ferrite and irregular graphite shapes called temper carbon. Note that the term ferrite does not refer to the element pure iron. In most steel making and iron casting where carbon is present, ferrite will dissolve a small amount of carbon at any temperature.

Ductile Cast Iron

Ductile iron is the most recent of the cast irons. It has the advantages of low melting temperature and low cost and provides mechanical advantages similar to steel. The carbon content is about the same as gray cast iron, but the structure after solidification is quite different. Ductile cast iron (also called nodular iron and spherulitic graphite cast iron) is composed of small spherical shapes of graphite and sometimes a matrix similar to structural steel. Remember, the term matrix refers to the continuous phase in a microstructure. The matrix may consist of one or more of the such phases as ferrite, pearlite, or martensite, as shown in Fig. 1.16.

Fig. 1.16. Typical matrix microstructures of ductile iron. Ferrite etches as light grains, pearlite etches as a layer-like structure of ferrite (white) with iron carbide (black) and martensite etches as a gray needle-like structure.

The main reason for the success of ductile cast iron is that is tends to behave like the matrix structure. This is quite different from gray cast iron since the graphite shape and distribution dictates its mechanical properties. This means that ductile cast iron has the advantages of the cast irons during the casting process but can simulate the steels during service. On a chemical basis, ductile cast iron is about the same as gray cast iron. Ductile iron becomes different when a substance called an inoculant is added to the ladle to form nuclei for crystallisation. After the inoculant is added, the graphite collects in small spherical-shaped sections and solidifies, creating a low carbon matrix much like steel.

After the ductile cast iron part is cast, the following advantages can be expected:

1. Good machinability.
2. Resistance to wear.
3. High strength.
4. Good balance of toughness and ductility.
5. Can be worked to make a change in shape.
6. Can be easily welded.
7. Can be heat treated to change mechanical properties.

More information about ductile iron is provided in Table 1.7.

Table 1.7. Applications and selected mechanical properties of ductile cast iron.

Type	*Tensile strength (ksi)*	*Yield strength (ksi)*	*Per cent elongation*	*Hardness Brinell Std. Ball*	*Applications*
60-40-18	*80	55	18	207	Ingot moulds, pipe, valves, connecting rods

Contd...

Type	*Tensile strength (ksi)*	*Yield strength (ksi)*	*Per cent elongation*	*Hardness Brinell Std. Ball*	*Applications*
80-55-0	100	75	6	269	Gears, cams, pistons, dies, crankshafts
120-90-02	175	150	2	388	Pinions, gears, cams, dies, rollers, stearing gears, pump liners

* These numbers identify minimum values of tensile strength and yield strength in ksi (thousand pounds per square inch) and per cent of elongation. For example 60-40-18 indicates a cast iron with a tensile strength of 60000 lb/inch^2, yield strength of 40000 lb/inch^2 and 18% minimum elongation.

Alloy Cast Iron

Many types of special cast iron have been developed to compete on specific jobs. Examples of these improved features are:

1. Resistance to corrosion.
2. Resistance to heat.
3. Controlled thermal expansion.
4. High electrical resistivity.
5. Fixed magnetic properties.

Typical alloying elements that are added to iron and carbon to improve resistance to corrosion are silicon, nickel, chromium and copper. An important use of corrosion resistant castings is in water disposal systems in chemical plants. High nickel irons are used in pump housings, pipes and pipe fittings. These parts will show little deterioration with such liquids as sulphuric acid, hydrochloric acid, organic acids and caustics. Heat resistant cast irons can function at temperatures that cause structural steel to become soft and develop a thick layer of iron oxide. When cooled, this layer of iron oxide may fall off. This leaves the part materially reduced in size, as shown in Fig. 1.17.

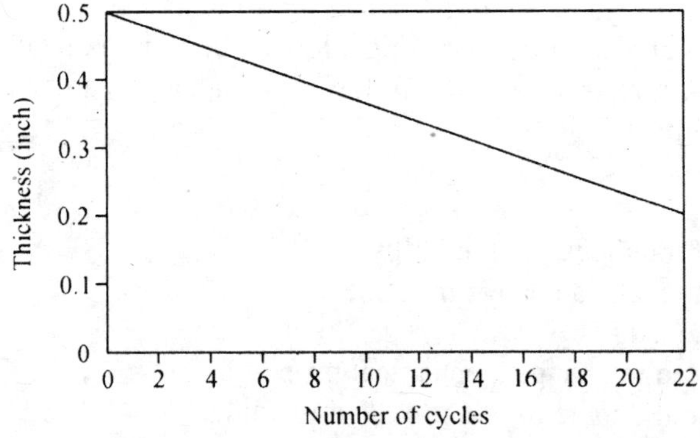

Fig. 1.17. Thickness loss from repeated heating and cooling cycles of carbon steel that is heated to 1600°F then cooled in air to room temperature.

To improve the service life and produce more desirable mechanical characteristics at high temperature, the following alloying elements are added:

Alloying elements	*Mechanical properties*
Silicon, chromium	Resistance to oxidising or scaling

Molybdenum, copper	High temperature strength
Nickel, vanadium	High temperature strength
Chromium	High temperature strength

Control of thermal expansion is necessary in some types of instrumentation. This control may be necessary over a wide range of temperatures. A nickel-iron alloy called 'invar' has been developed for the purpose of resisting the process of thermal expansion. Invar has only one-fifth the expansion of structural steel. Austenitic cast irons containing a high nickel content have high electrical resistivity. Heating elements are made from iron-nickel alloys. Magnetic clutch units are made by alloying iron and aluminium. These are just a few of the many alloys created to perform the many special tasks required in industrial applications.

Aluminium Alloys

Aerospace technology has made aluminium alloys some of the most important of the lightweight metals. Today aluminium products are common to almost everyone. Yet it is interesting to look into the last century when aluminium was just being developed. Commercial production of aluminium was not practical until the electrolysis of the molten bath was discovered. Today, mass production of aluminium can produce high purity (99.99 per cent) products or alloys with exacting controls. There are many applications of aluminium and aluminium alloys, such as:

1. Aircraft and spacecraft components.
2. Deoxidising steel.
3. Kitchenware.
4. Electrical transmission lines.
5. Making paint.
7. Structural assemblies.

Aluminium alloys can be produced with a wide range of mechanical properties. A few typical properties are shown in Fig. 1.18 through Fig. 1.21.

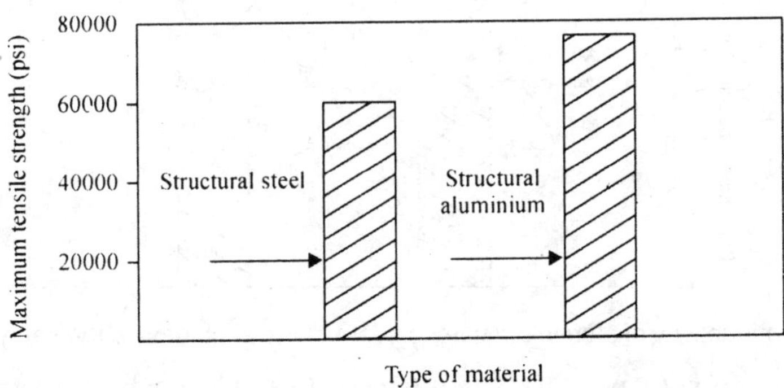

Fig. 1.18. A comparison of the strength of structural steel and structural aluminium, two metals used in construction.

It is a well-known principle in the field of metallurgy that ordinary ferrous metals exhibit a characteristics loss of toughness at some low temperature. The temperature at which normally ductile metals fail with a brittle fracture is called the transition temperature. The transition temperature may be near the freezing point of water or lower.

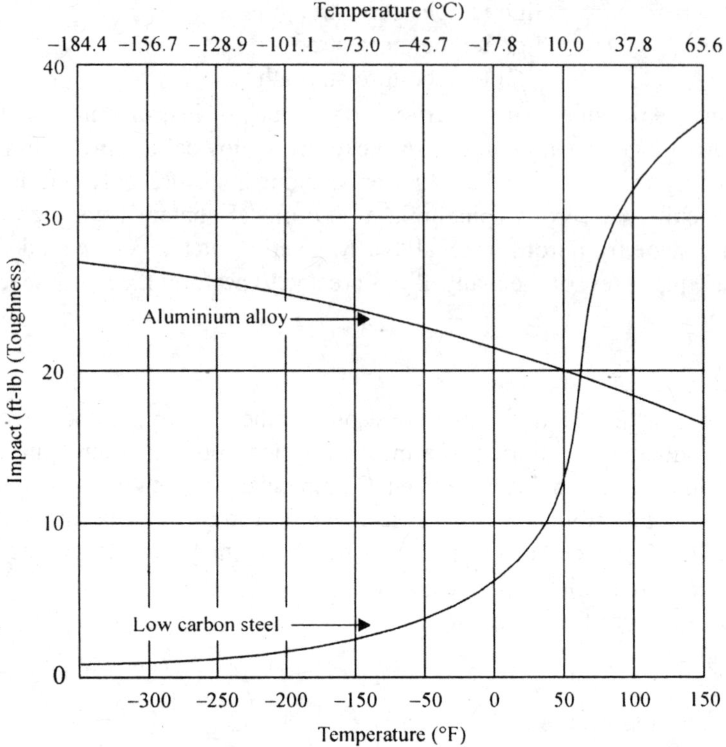

Fig. 1.19. Toughness of structural steel and aluminium alloys as related to impact test results. Note that the aluminium gets tougher as the temperature drops, while just the opposite is true for structural steel.

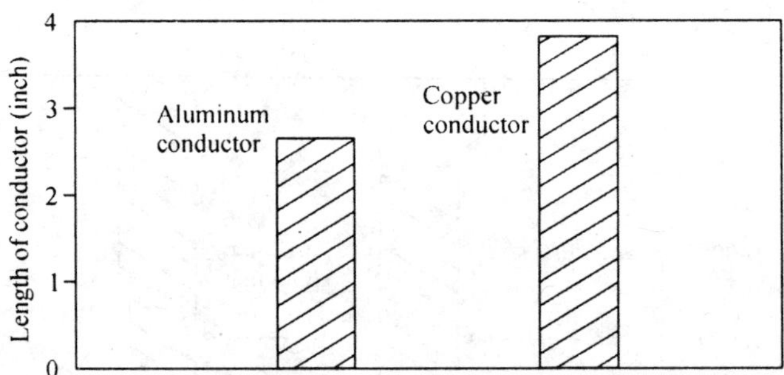

Fig. 1.20. Electrical conductivity of aluminium versus copper wire each having the same amount of resistance.

Aluminium alloys do not have a transition temperature. The colder the temperature the tougher they become.

A copper conductor has about 56 per cent of the resistance of the aluminium conductor. Consequently, it is possible to use a copper conductor of smaller diameter or longer length than would be possible with aluminium. Electrical resistance is one factor that establishes load-carrying capacity. Above ground transmission lines might favour the aluminium conductor because it has approximately 30 per cent of the weight of the copper conductor. Less weight means fewer power poles.

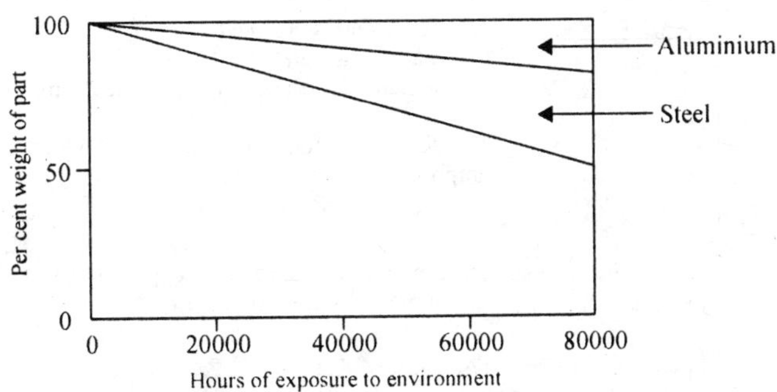

Fig. 1.21. Resistance to weather of aluminium versus structural steel.

Aluminium alloys are generally divided into two groups. One group is made by casting the required shapes. The other is referred to as wrought (forged, extruded, or rolled) material. The cast grades are designated by a two or three-digit system. Typical casting alloys are:

1. Sand-casting alloys 122, 214, 220, 355 and 356.
2. Permanent-mould-casting alloys 122, A132 and B195.
3. Die-casting alloys 13, 43, 218, 360, A360 and 380.

Wrought grades are classified by four-digit numbers such as:

Number	*Major alloys*
1xxx*	Aluminium
2xxx	Copper
3xxx	Manganese
4xxx	Silicon
5xxx	Magnesium
6xxx	Magnesium and silicon
7xxx	Zinc
8xxx	Other than above elements

* The x is substituted for a space that a number will occupy to specify a family of alloys.

The four-digit system may also use a series of letters and numbers to specify temper conditions. For example, using the aluminium-copper alloy (2024) note the significance of the letter and number after the dash.

Alloy	*Letter-number combination*	*Explanation (Temper designation)*
2024	—0	Annealed to full soft condition
2024	—F	As fabricated
2024	—W	Solution heat treated, unstable temperature
2024	—T4	Same as W only aged at room temperature
2024	—T6	Same as W only artificially aged*
2024	—T9	Same at T6 only followed by cold work

*Artificial ageing is done at elevated temperatures (180 to 220°F minute).

Magnesium Alloys

Magnesium is considered the lightest structural metal in production. Magnesium is about 36 per cent lighter than other light metals such as aluminium. There are almost unlimited quantities of magnesium in ocean water. Magnesium alloys are classified by a system of letters and digits, as shown in Fig. 1.22. The temper of wrought alloys is the same as that of aluminium alloys.

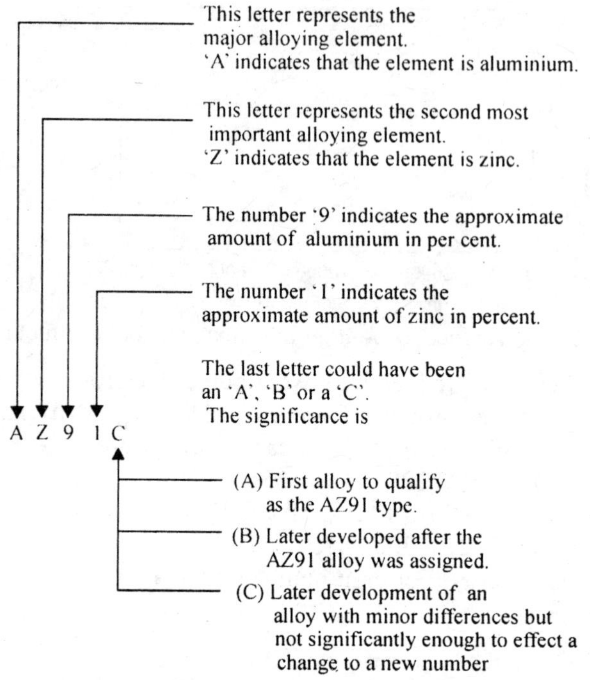

Fig. 1.22. A typical example of the magnesium numbering system.

A few typical cast and wrought alloys are listed in Table 1.8.

Table 1.8. Approximate composition and properties of a few magnesium alloys.

	Composition				*Estimated mechanical properties*			
Alloy temperature	*Al*	*Mn*	*Zn*	*Zr*	*Yield strength (ksi)*	*Tensile strength (ksi)*	*Per cent elongation*	*Impact energy (ft.-lb)*
AZ 81A-T4	7.6	0.13	0.7	–	14	40	12	13
AZ 916-T4	8.7	0.20	0.7	–	14	40	11	5
AZ 31B-H24	3.0	0.45	1.0	–	32	42	15	–
AZ 31C-F	3.0	0.45	1.0	–	19	35	14	–
HK 31A-H24	–	–	–	0.7	29	37	8	–

Copper Alloys

It is generally accepted that copper was first used by man several thousand years BC, probably because it was found on the earth's surface. Copper is successfully alloyed with many metals. Even though it is popular as an alloy, it has many job functions as a high-purity metal. Two of the most prominent uses of copper as a pure metal are to conduct electrical current and to transmit heat energy. Examine the lead cables on an arc welding power supply. They will usually be copper. Scrape the paint from an automobile radiator and analyse the metal. In most cases it will prove to be copper.

Pure copper is soft and ductile, but it may be work hardened. Copper can also be alloyed to improve mechanical properties. Certain alloys of copper can be heat treated. There are three standard grades of commercial copper as seen in Table 1.9.

Table 1.9. Typical classification of copper alloys.

Type	*ASTM designation*	*Typical uses*
Electrolytic tough pitch copper	ETP	Brazing rod, wire, forgings
Oxygen free, silver bearing	OFS	Plate, sheet, rods
Deoxidised phosphorised, low residual phosphorus	DLP	Tubes, general use

The copper development association has classified certain copper alloys. Some of these are given in Table 1.10.

Table 1.10. Selected copper alloys classified by the copper development association (CDA).

		Nominal composition (in %)						
CDA no.	*Name of alloy*	*Cu*	O_2	*Zn*	*Pb*	*Fe*	*Mn*	*Other*
101	Oxygen-free electrolytic copper	99.96						
110	Electrolytic tough pitch copper	99.9	0.04					
172	Beryllium copper	98.1						1.9 Be
187	Leaded copper	99.0			1.0			
260	Cartridge brass	70.0		30				
280	Muntz metal	60.0		40				
675	Manganese bronze	58.5		39		1.4	0.1	1 Sn
713	Copper nickel	75.0						25 Ni
752	Nickel silver	65.0		17				18 Ni
953	Aluminium bronze	89.0				1.0		10 Al
956	Silicon-aluminium bronze	91.0						7 Al, 2 Si
298	Copper-zinc brazing alloy	50.0		50				

Nickel Alloys

Nickel is a reasonably common metal and is comparable to iron in tensile strength, but its yield strength is about one-fourth that of iron. Low yield strength indicates high ductility. Pure nickel is used by the chemical industry and in electroplating processes. Nickel alloys are used primarily to resist chemical attack and to increase the temperature at which alloys can function. Typical uses of some nickel alloys are shown in Table 1.11.

Table 1.11. Applications and approximate composition of some nickel alloys.

Name of alloy	*Ni*	*C*	*Mn*	*Fe*	*S*	*Si*	*Cu*	*Cr*	*Other*	*Applications*
Monel	66	0.12	0.9	1.3	0.005	0.15	31			Corrosion resisting, strong, tough
Inconel	76	0.04	0.2	7.2	0.007	0.2	0.1	15		High temperature service

Contd...

Name of alloy	*Ni*	*C*	*Mn*	*Fe*	*S*	*Si*	*Cu*	*Cr*	*Other*	*Applications*
Ni-chrome	80	0.05	0.1	0.5		0.2		20		Heating elements
Chromel	90							10		Thermocouple wire
Alumel	94.5	2.5				1.0			Al	Thermocouple wire
Hastelloy A	57			20					Mo	Resist corrosion at high temperature
Hastelloy X	47		18					22	Mo	Service ceiling @ Fe 2000°F
									C	C
									W	
Invar	36			63						Very low expansion

Lead Alloys

Lead is another of the metals that appeared with the dawn of civilisation. Even today lead pipes are still intact that were installed several thousand years ago. Lead is widely known by its properties of high density, softness, ductility, low melting point, high corrosion resistance and lubricating qualities. The two elements that are most often alloyed with lead are antimony and tin. Table 1.12 shows typical applications of lead alloys.

Table 1.12. Typical information about lead alloys.

Name of alloy	*Composition per cent*	*Tensile strength (psi)*	*Yield strength (psi)*	*Per cent elongation*	*BHN*	*Application*
Pure lead	99.9	2400	1200	29	1.0	Liners for chemical tanks, alloy
Solder	80 Pb-20 Sn	5800	3600	16	11	Joining
Solder	50 Pb-50 Sn	6100	4800	60	14	General purpose solder
Soft solder	37 Pb-63 Sn	7500		32	14	Electrical solder
Babbitt	83 Pb-15Sb-1 As-1 Sn	10,350		2	20	Bearing liners

Titanium Alloys

Before World War II titanium was too expensive to be of widespread commercial use. In view of the good mechanical properties that titanium had to offer, the United States Bureau of Mines became involved in research with the production of titanium and its alloys. Later, a company called Albany Titanium Inc., made headlines in a newspaper. Company officers claimed they would produce a chemically pure version of the rare metal at a cost low enough to compete with other metals such as stainless steel.

Titanium has one of the advantages of steel in that it can transform to a new phase while in the solid state. This means that it can be heat treated to improve its mechanical properties. Titanium is known for its strength-to-weight advantage in temperatures as high a 800°F. It is ranked high in its ability to be immune to fatigue and corrosion. The stress-corrosion properties of titanium are as good in salt water as in normal atmosphere. Titanium is immune to chlorides, hydroxides and acids. Welding of titanium is satisfactory with adequate inert shielding. There are about 20 commercial alloys of titanium. For a matter of convenience in understanding this alloy, it will be classified into the following three groups.

Alpha alloys

Single-phase alloys with a hexagonal close-packed crystal structure. These alloys cannot be heat treated but may be work hardened. The fabricator must realise that the work-hardened structure will have attendent directional properties. For example, the work-hardened structure will have much less strength when placed in tension at 90° to the direction of original work hardening.

Beta alloys

These alloys have body-centered cubic crystal structure at room temperature. These alloys can be given high strength qualities by heat treatment. It is not uncommon to find these alloys with strengths of 215,000 psi and ductility of 5 per cent elongation after heat treatment. Heat treating is accomplished in two sequences. The first phase is called solution treatment. Heating to about 1450°F for 10 minute and cooling dissolve the alloys. The second part of the heat treating sequence is called precipitation treatment. During this time sub-microscopic particles grow or precipitate in size to become large enough to block slip planes. Temperature and time increase the size of the precipitate. If the precipitate becomes too large, the material loses strength. A typical for this sequence is to age for 48 hours at 900°F to cause precipitation (Fig. 1.23).

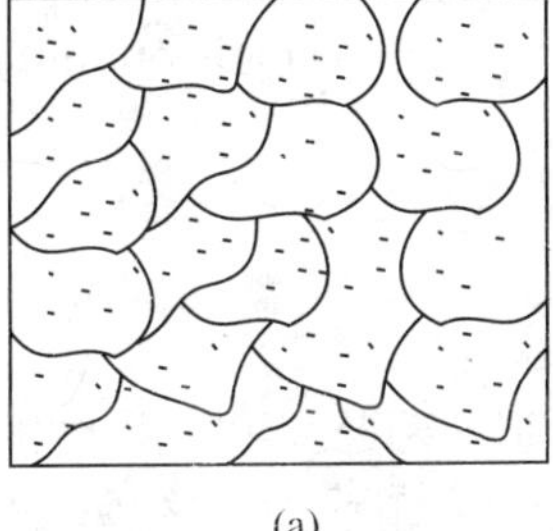

(a)

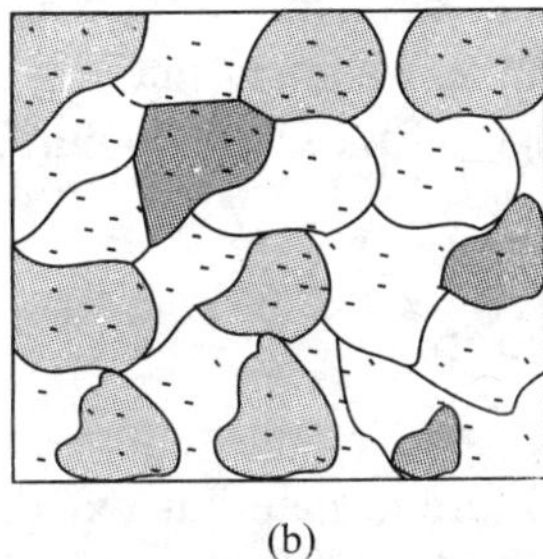

(b)

Fig. 1.23. A typical titanium alloy. (a) The result of solution heat treatment with equiaxed beta grains. (b) the same material except aged for 48 hrs. at 900°F. Structure is dark particles of precipitated alpha in beta grains.

Alpha-Beta alloys

Heat treatment for alpha-beta alloys is similar to that for beta alloys. These alloys can be heat treated by quenching from high temperature to prohibit transformation. After the quench the alloy is aged at a lower temperature to complete the heat treatment. A super alpha alloy for use in rotating engine components, such as the vanes on a turbine, are always heat treated. Heat treatment consists of 1 hr. at 1825°F and water quench, then heat to 1100°F for 8 hours, followed by air cooling. Typical properties from this heat treatment are shown in Table 1.13.

Table 1.13. Mechanical properties of heat treated titanium.

Exposure conditions				*Room temperature properties*			
Temp. (°F)	*Stress (ksi)*	*Time (hr.)*	*Creep (inch/inch)*	*Yield strength (ksi)*	*Tensile strength (ksi)*	*Elongation (%)*	*Reduction of area (%)*
900	42	100	0.0049	151.0	162.0	15.7	42.5
1000	10	1000	0.0071	161.0	171.0	12.6	34.9
1050	20	100	0.0168	158.5	167.2	16.3	39.2

Chapter 2

Nature of Metals and Alloys

INTRODUCTION

The fundamental engineering properties of materials are the direct result of the structure of the particular material. Moreover, such properties as strength and ductility are often sensitive to minute variations of structure, some of which are macroscopic, others microscopic and still others on the atomic scale. To control the properties of materials and intelligently use them to their optimum, the engineer must first have a working knowledge of material structure.

STRUCTURE OF MATERIALS

Basic Structure of Materials

As all materials are composed of the same basic components–protons, neutrons and electrons—it is amazing that so many different materials exist with such widely varying properties. This variation is explained by the many possible combinations of these units in a macroscopic assembly. The sub-atomic components, listed above, combine in different arrangements to form the various elemental atoms, each having a nucleus of protons and neutrons surrounded by the proper number of electrons to maintain charge neutrality. Atoms then combine in distinctive arrangements to form molecules or crystals. These units can then be assembled in differing amounts and configurations to form a microscopic-scale structure, or microstructure and ultimately an engineering component. The engineer, therefore, has at his or her disposal a wide variety of metals and non-metals that possess an almost unlimited range of properties. Because the properties of materials depend upon all levels of structure, as shown schematically in Fig. 2.1 it is important for the engineer to understand the entire structure spectrum, from atomic to macroscopic.

Atomic Structure

Experiments have revealed that atoms consist of a relatively dense nucleus composed of positively charged protons and neutral particles of nearly identical mass, known as neutrons. Surrounding the nucleus are the negatively charged electrons, which have only $1/1839$ the mass of a neutron and appear in numbers equal to the protons to maintain a net charge balance. Distinct particle groupings produce the known elements, ranging from the relatively simple hydrogen atom to unstable transuranium atoms over 250 times as heavy. Except for density and specific heat, however, the weight of atoms has relatively little influence on engineering properties.

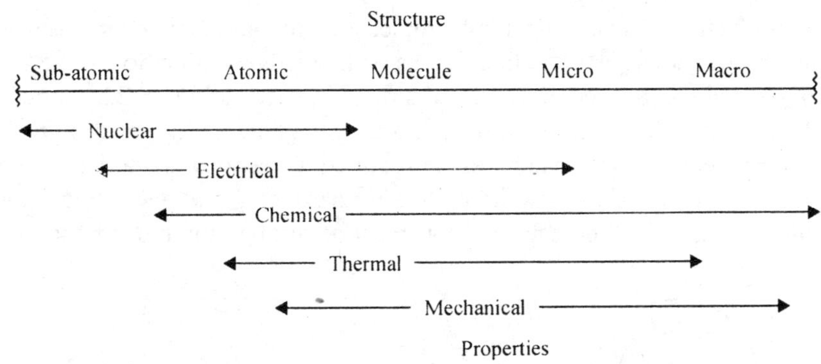

Fig. 2.1. General relationship of the structural level to various engineering properties.

The light electrons that surround the nucleus play a far more significant role in determining properties. Again, experiments reveal that the electrons are arranged in a characteristic structure consisting of shells and sub-shells, each possessing a distinctive energy. Upon absorbing a small amount of energy, an electron can jump from a low-energy shell near the nucleus to a higher-energy shell farther out. The reverse jump can occur with the release of a distinct amount, or quantum, of energy.

Each of the various shells and sub-shells can contain only a limited number of electrons. The first shell, nearest the nucleus, can contain only 2. The second shell can contain 8; the third, 32. Each shell and sub-shell is most stable when it is completely filled. For atoms containing electrons in the third shell and beyond, however, relative stability is associated with 8 electrons in the outermost layer.

If, in its outer shell, a normal atom has slightly less than the number of electrons required for stability (for example, seven in its third shell), it will readily accept an electron from another source. It will then have one electron more than the number of protons and becomes a negatively charged atom, or negative ion. Extra electrons may cause the formation of ions having negative charges of 1, 2, 3 and more. If an atom has a slight excess of electrons beyond the number required for stability (such as sodium, with one electron in the third shell), it will readily give up the excess electron and become a positive ion. The remaining electrons become more strongly bound, so that the removal of electrons is progressively more difficult. The number of electrons surrounding the nucleus of a neutral atom is called the atomic number. More important, however, are those electrons in the outermost shell (or sub-shell) known as valence electrons. These are influential in determining chemical properties, electrical conductivity, some mechanical properties, the nature of interatomic bonding, atom size and optical characteristics. Elements with similar electron configurations in their outer shells will tend to have similar properties.

Atomic Bonds

Atoms are rarely found as free and independent units; they are usually linked or bonded to other atoms in some manner as a result of interatomic forces. The electronic structure of the atoms influences the nature of the bond, which may be classed as primary (strong) or secondary (weak).

The simplest type of primary bond is the ionic bond. Electrons break free from atoms with excesses in their valence shell, producing positive ions and unite with atoms having an incomplete outer shell to form negative ions. The positive and negative ions have a natural attraction for each other, producing a strong bonding force. Fig. 2.2 illustrates the process for a bond between sodium and chlorine. In the ionic type of bonding, however, the atoms do not unite in simple pairs. All positively charged atoms attract all negatively charged atoms. Thus, for example, sodium ions surround themselves with negative chlorine ions. and chlorine ions surround themselves with positive sodium ions. The attraction is equal in all directions and results in a three-dimensional structure, such as in Fig. 2.3, rather than the simple link of a single bond. For stability in the structure, total charge neutrality must be maintained, so that equal numbers of positive and negative charges are required. General characteristics of materials joined by ionic bonds include moderate to high strength, high hardness, brittleness, high melting point and electrical insulating properties (all charge transport must be through ion movement).

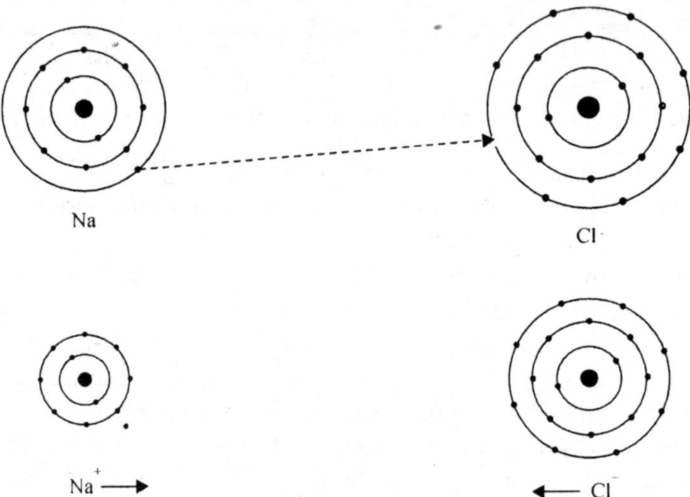

Fig. 2.2. Ionisation of sodium and chlorine, producing stable outer shells by electron transfer.

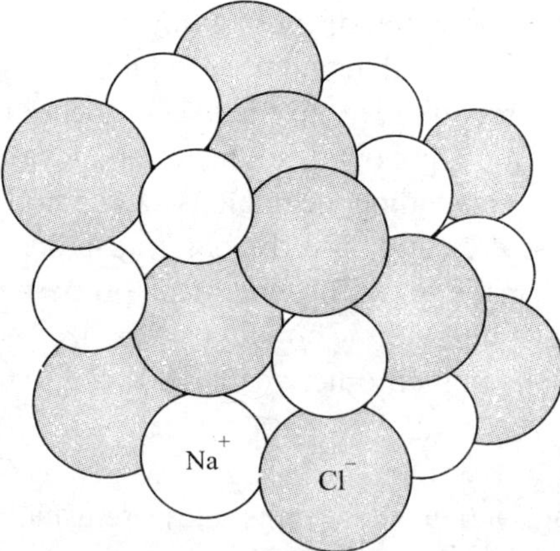

Fig. 2.3. Three-dimensional structure of the sodium chloride crystal. Note how the various ions are surrounded by ions of the opposite charge.

A second type of primary bond is the covalent type. Here the atoms being linked find it impossible to produce completed shells by electron transfer but achieve the same goal by electron sharing. Adjacent atoms share outer-shell electrons so that each achieves a stable electron structure. The shared negative electrons locate between the positive nuclei to form the bonding link. Fig. 2.4 illustrates this type of bond for chlorine, where two atoms, each containing seven valence electrons, share a pair to form a stable molecule. Stable molecules can also form from the sharing of more than one electron from each atom, as in the case of nitrogen in Fig. 2.5a. Atoms need not be identical (as in HF in Fig. 2.5b), the sharing need not be equal and an atom may share electrons with more than one other atom. For elements such as carbon (four valence electrons), one atom may share valence electrons with each of four neighbouring carbon atoms. The resulting structure becomes a network of bonded atoms (Fig. 2.5c) instead of a finite, well-defined molecule.

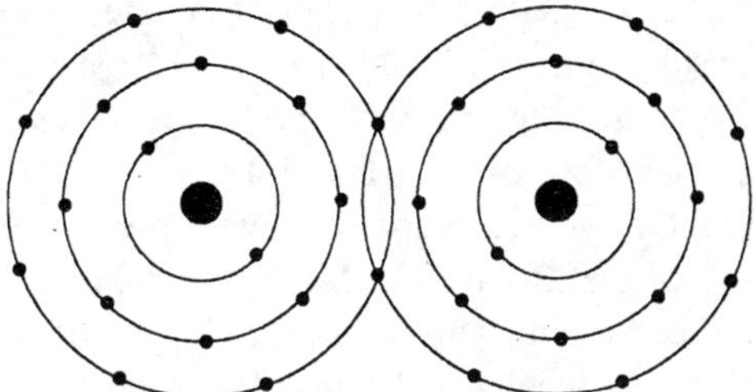

Fig. 2.4. Formation of a chlorine molecule by a covalent bond.

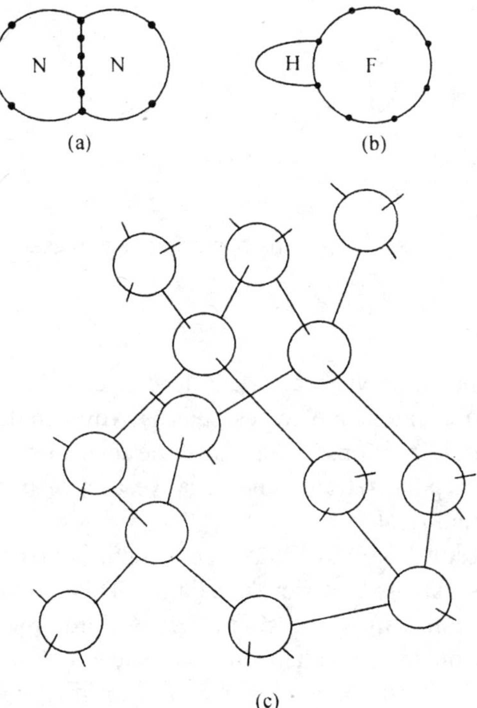

Fig. 2.5. Examples of covalent bonding in N_2, HF and diamond.

Like the ionic bond, the covalent bond tends to produce materials with high strength and high melting points. Atom movement within the material (deformation) requires the breaking of distinct bonds, thereby making the material characteristically brittle. Electrical conductivity depends upon bond strength, ranging from conductive tin (weak covalent bond) through semi-conductive silicon and germanium to insulating diamond. Engineering materials possessing ionic and covalent bonds tend to be ceramic (refractories or abrasives) or polymeric in nature.

A third type of primary bond can result when a complete outer shell cannot be formed by either electron transfer or electron sharing and is known as the metallic bond. If there are only a few valence electrons (one, two, or three) in each of a grouping of atoms, these electrons can be removed easily while the remainder are held firmly to the nucleus. The result is a structure of positive ions (nucleus and non-valence electrons) surrounded by a wandering assortment of universally shared valence electrons (electron cloud or gas), as in Fig. 2.6. These highly mobile 'free' electrons account for the high electrical and thermal conductivity as well as the opaque optical properties (free electrons can absorb light radiation energies) observed in metals. Moreover, they provide the 'cement' necessary to produce the positive–negative–positive attractions necessary for bonding. Bond strength and therefore material strength, varies over a wide range. More significant, however, is the observation that the positive ions can move within the structure without the breaking of distinct bonds. Materials bonded by metallic bonds can therefore be deformed by atom movement mechanisms and produce a deformed material every bit as strong as the original. This is the basis of metal plasticity, ductility and many of the shaping processes used in metal fabrication.

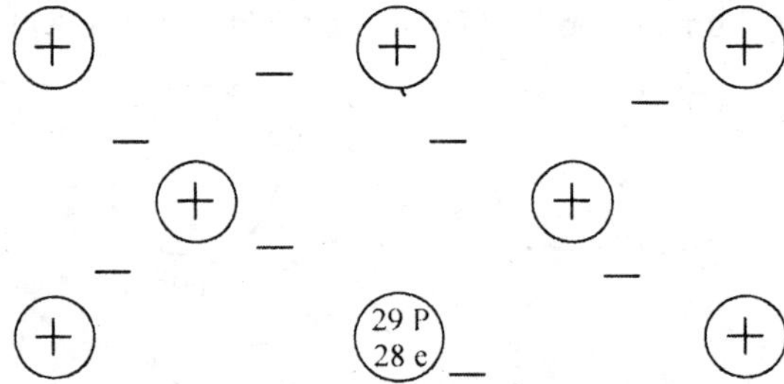

Fig. 2.6. Schematic of the metallic bond showing the positive ions and associated electron 'cloud' for the case of copper.

Secondary Bonds

Weak or secondary bonds, known as van der Waals forces, can link molecules that possess a non-symmetrical distribution of charge. Some molecules, such as hydrogen fluoride and water, can be viewed as electric dipoles, in that certain portions of the molecule tend to be more positive or negative than others (an effect referred to as polarisation). The negative part of one molecule tends to attract the positive part of another to form a weak bond.

Another weak bond can result from momentary polarisation caused by random movements of the electrons and the resulting momentary electrical unbalance. This random and momentary polarisation leading to attractive forces is called the dispersion effect. A third type of weak bond is the hydrogen bridge, where a small hydrogen nucleus, simultaneously attracted to the negative electrons of two different atoms, forms a three-atom link. Such bonds play a significant role in biological systems but, like all secondary bonds, are rarely of engineering significance.

Interatomic Distances

Since the space occupied by the electron shells is very large compared to the size of the actual electrons, most of the total volume of an atom is vacant space and the 'size of an atom' becomes somewhat undefined. The interatomic or bonding forces tend to pull the atoms together; at the same time, there are repelling forces between the positive nuclei. Some equilibrium distance exists where the forces of attraction and repulsion are equal.

Thus, atoms can be assigned a distinct size, the equilibrium distance between the centres of two neighbouring atoms being considered the sum of the atomic radii. The atomic radius is not a constant, however. Added thermal energy causes the atoms to vibrate about their equilibrium positions, transmitting the energy through the material (thermal conductivity). The repulsive force as atoms approach is much greater than the attractive force when they are separated. Thus, the vibrations, or oscillations, are non-symmetrical and the atoms are at larger than normal separations for-a-greater fraction of time. The result is a phenomenon that we observe macroscopically as thermal expansion. Removal of electrons from the outer shell will decrease the radius and the addition of electrons will increase it. Consequently, a negative atomic ion is larger than its base atom and a positive ion is smaller. Atomic radius also changes with the number of adjacent or nearest-neighbor atoms. With more neighbours, there is less attraction to any single neighbour atom and the interatomic distance is increased. Thus, iron has a slightly different atomic size in its two different crystal forms.

Atom Arrangements in Materials

At the next level of material structure, we find that the arrangement of atoms in a material has a significant effect on its properties. Depending on the manner of atomic grouping, materials are classified as having molecular structures, crystal structures, or amorphous structures.

Molecular structures have a distinct number of atoms that are held together by primary bonds, but they have only relatively weak bonds with other similar groups of atoms. Typical examples of molecules include O_2, H_2O and C_2H_4 (ethylene). Each molecule is free to act more-or-less independently, so that these materials exhibit relatively low melting and boiling points. Molecular materials tend to be weak, as the molecules can easily move past one another. Upon changes of state from solid to liquid or gas, the molecules remain intact as distinct entities.

Crystal structures are assumed by solid metals and most minerals. Here atoms are arranged in a regular geometric array known as a space lattice. These lattices are describable through a unit building block that is essentially repeated throughout space in a periodic manner. Such blocks are known as unit cells. In amorphous structures, such as glass, the atoms have a certain degree of local order but, when viewed as an aggregate, have a more disorganised atom arrangement than the crystalline solids.

Crystal Structure of Metals

From a manufacturing viewpoint, metals are the most important class of materials. Most often, they are the materials being processed and are used in the machines performing the processing. Consequently, in order to perform manufacturing operations intelligently, it is essential to have a basic knowledge of the fundamental nature of metals and their behaviour when subjected to mechanical or thermal treatment.

Crystal Structure of Metals

Any substance can be in one of the three states of aggregation: solid, liquid, or gaseous. Under the action of gravity, solids retain their shape, while liquids spread over and take the shape of the vessel which they fill. This definition is, however, insufficient to characterise the state of a substance.

For instance, solid glass softens on heating and gradually passes into the liquid state. The reverse transformation also occurs gradually, i.e., liquid glass becomes thicker with decreasing temperature and finally thickens into the solid state. There is no definite temperature ('point') at which glass passes from liquid to solid state and abruptly changes its properties. It is therefore appropriate to regard 'solid' glass as a strongly thickened liquid.

Consequently, the change from solid to liquid state or vice versa (as also from gaseous to liquid state) occurs at a definite temperature and causes an abruptly change in the properties of a substance.

Difference between the gaseous, liquid and solid states

In gases, there is no ordered position of particles (atoms and molecules); these particles move chaotically, collide and repulse and the gas tends to spread over the whole volume available.

In solids, there is a regular order of the position of atoms, the forces of mutual attraction and repulsion are balanced out and a solid body retains its shape.

In liquids, their particles (atoms and molecules) retain only what is called the short-range order, i.e., only a small number of atoms are positioned regularly in the space, but not the atoms in the whole volume, as is the case with solids. The short-range order is unstable; it can appear and disappear under the action of thermal fluctuations.

Thus, the liquid state is an intermediate one, as it were, between the solid and gaseous states. Under appropriate conditions, some substances can pass over directly from the solid to gaseous state without melting; the phenomenon is called sublimation (Fig. 2.7).

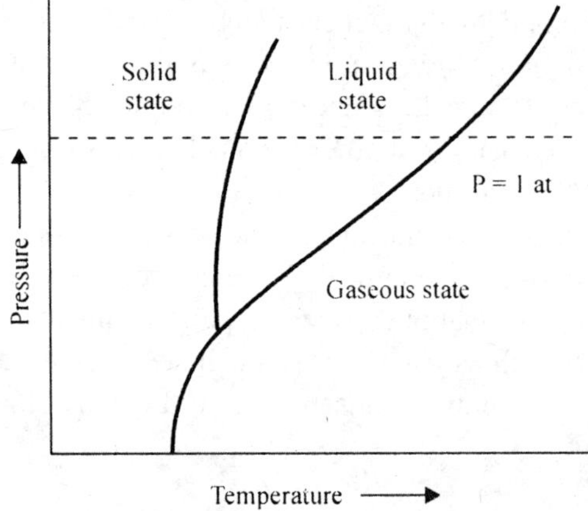

Fig. 2.7. Regions of solid, liquid and gaseous, depending on temperature and pressure.

An ordered, regular position of particles (atoms, molecules) in space is also characteristic of the crystalline state of a substance.

The crystalline structure of a substance may be imagined as a space lattice with atoms positioned at the sites (Fig. 2.8).

In the metals, the sites of the crystal lattice are occupied by positive ions, rather than by atoms, with free electrons moving in between; it is usually said, however, that the sites of the crystal lattice of metals are occupied by atoms.

In the next section we shall consider the arrangements of atoms (ions) that are typical of metals, i.e., the types of crystal lattice found in metals.

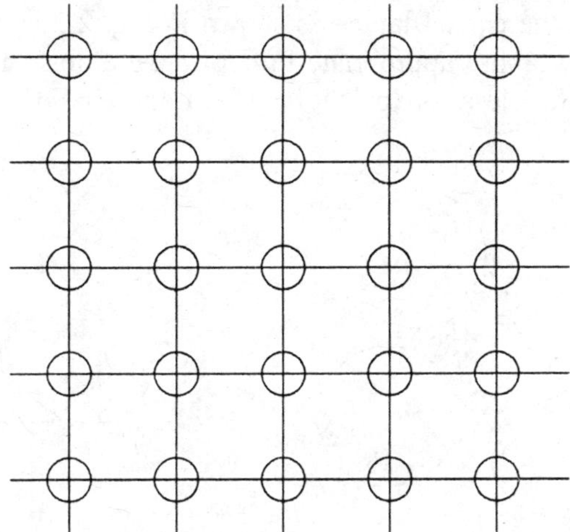

Fig. 2.8. An elementary crystal cell (simple cubic).

Crystal Lattices of Metals

As has been mentioned, the crystalline state is characterised primarily by an ordered regular arrangement of atoms in space. This implies that each atom in a crystal has the same number of nearest atoms–neighbours, spaced the same distance apart from the atom considered.

The tendency of atoms (ions) in metal to occupy positions as close as possible to one another, i.e., to arrange themselves as dense as possible results in that there are only a few possible combinations of mutual arrangement of atoms in crystals of metals.

There exist a number of schemes and methods for description of mutual arrangement, or packing, of atoms in crystals. The mutual arrangement of atoms in one of the crystal planes is shown in Fig. 2.7. The imaginary lines drawn through the centres of atoms form a lattice with atoms (positive ions) located in the lattice sites; this is what is called a crystal plane. Multiple reproduction of parallel crystal planes forms a space crystal lattice whose sites are locations of atoms (ions). The distances between the centres of neighbouring atoms are measured in angstroms (1Å = 1.10^{-8} cm) or in X-units (1 kX-unit = 1.00202 Å).

Mutual arrangement of atoms in space and interatomic distances are determined by X-ray structural analysis. The positions of atoms in a crystal can be very conveniently depicted in the form of space arrangements which are called elementary crystal cells. An elementary crystal cell is understood as the least combination of atoms which, upon multiple reproduction in space, gives a space crystal lattice.

The simplest type of crystal cell is the cubic lattice. In a simple cubic lattice, atoms are arranged (packed) not quite densely.

Atoms tend to occupy positions as close as possible to one another and thus form other types of crystal lattice: body-centred cubic (bcc, Fig. 2.9a), face-centred cubic (f.c.c., Fig. 2.9b) and close-packed hexagonal (cph, Fig. 2.9c).

The circles which depict atoms can be located in the centre and at the corners of a cube (body-centred cube), in the centres of cube faces and at the corners (face-centred cube), or in the form of a hexahedron with another hexahedron being half-inserted into it, with the three atoms of the upper plane of the latter hexahedron being inside the hexahedral prism (hexagonal lattice).

The method of depicting the crystal lattice as shown in Fig. 2.9 is only conditional (as any other possible method). It might be more appropriate to depict the atoms of crystal lattices as contacting spheres, but this method is often less convenient than the traditional one.

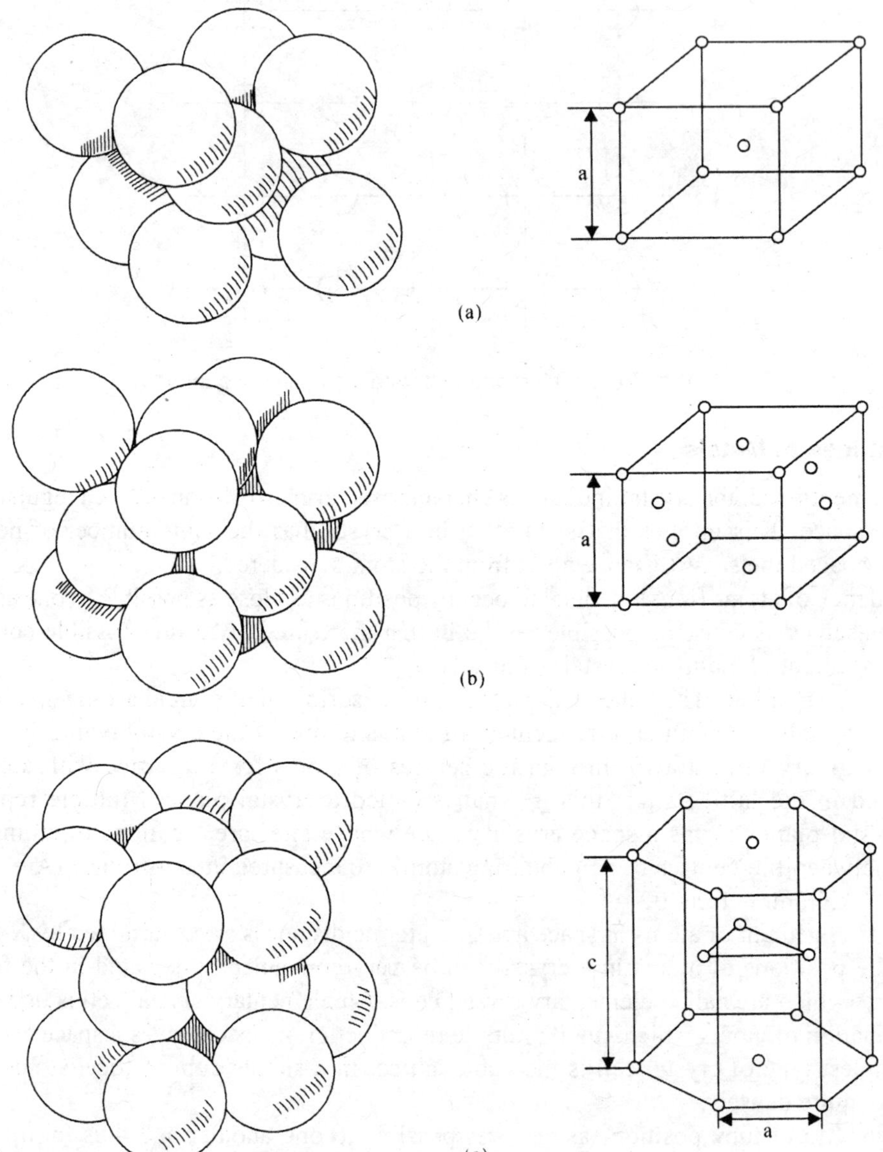

Fig. 2.9. Elementary crystal cells; (a) body-centred cubic; (b) face-centred cubic; and (c) close-packed hexagonal.

The dimensions of crystal lattices are characterised by the lattice parameters, or periods. A cubic lattice is determined by a single parameter—the length of cube edge a Fig. 2.9a, b. The lattice parameters have magnitudes of an order of the size of atoms and are measured in angstroms.

For instance, the lattice parameter of chromium, which has a body-centred cubic lattice, is 2.878 Å and that of aluminium, which has the structure of face-centred cube, is 4.041Å.

The lattice parameter is an important characteristic. Modern methods of X-ray analysis enable us to measure the lattice parameters to an accuracy of the fourth or even fifth decimal sign, i.e., one-ten-thousandth or one-hundred-thousandth of angstrom.

Considering the arrangement of crystal lattices (Fig. 2.9) and assuming that atoms are elastic spheres which touch one another, it may be concluded that there exist simple geometrical relationships between the lattice parameter a and the atomic diameter d.

For a body-centred cube

$$d = a\frac{\sqrt{3}}{2}$$

for a face-centred cube

$$d = a\frac{\sqrt{2}}{3}$$

Assuming that atoms are spheres, it may be calculated that atoms occupy 68 per cent of the volume in a body-centred cubic lattice and 74 per cent, in face-centred cubic and close-packed hexagonal lattices, i.e., atoms in the latter case are packed more closely.

The hexagonal lattice is typical of many metals Fig. 2.9c.

If the layers of atoms touch one another, i.e., the three atoms inside the lattice (Fig. 2.9c) touch the atoms on the upper and lower planes, the structure is called the close-packed hexagonal lattice.

The dimensions of a close-packed hexagonal lattice are characterised by a constant ratio $c/a = 1.633$. With other values of the c/a ratio the hexagonal lattice is not closely packed.

The face-centred cubic lattice and hexagonal lattice represent the most dense method of packing of spheres of the same diameter.

Some metals exhibit the tetragonal lattice (Fig. 2.10), which is characterised by that the edge c is not equal to edge a. The ratio of these parameters characterises what is called the tetragonality of a lattice. With $c/a = 1$, we have a cubic lattice. A tetragonal lattice (like a cubic lattice) may be simple, body-centred or face-centred, depending on the space arrangement of atoms.

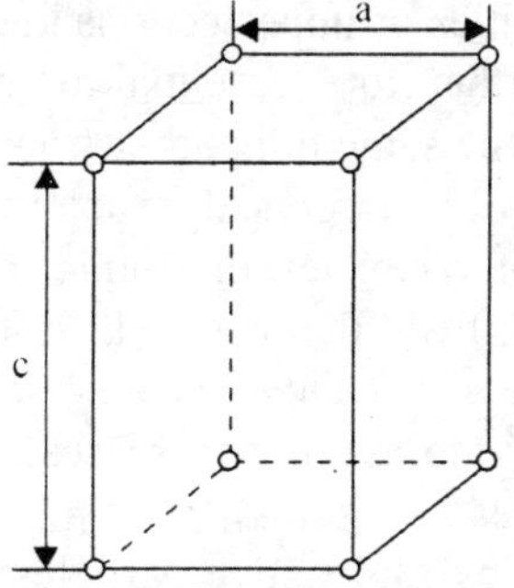

Fig. 2.10. A tetragonal lattice.

The properties of a particular metal or alloy depend appreciably on the number of 'touching' atoms in the crystal structure. This is determined by the number of nearest equidistant atoms for any atom.

The number-of-atoms which are at the nearest and equal distance from a given atom is called the co-ordination number. For instance, an atom in a simple cubic lattice has six nearest equidistant neighbours, i.e., the co-ordination number of the lattice is equal to six.

Real Structure of Metallic Crystals

Crystals of metals are usually small in size, so that a metallic article consists of a very large number of crystals. Such a structure is called polycrystalline. Individual crystals in a polycrystalline aggregate cannot have a regular shape. Such crystals of irregular shape are called grains, or crystallites.

Individual grains in a metal have different orientation of their crystal lattices (Fig. 2.11). In the general case, crystal lattices of grains are oriented at random, with any orientation in space being equally probable.

Fig. 2.11. A diagram to illustrate different orientation of crystal lattices in a pollycrystalline body.

Things, however, may be different in some particular cases. Plastic deformation of cold metal (rolling, drawing, etc.,) results in that the metal grains are mainly oriented in a particular direction (a texture is formed). The degree of primary orientation may be different, with random orientation being at one extreme and the same orientation of all the crystals, at the other.

If a metal is cooled very slowly during solidification, it can solidify into a single crystal, or monocrystal. There are special methods to grow single crystals. Single crystals of large size (a few hundred grams) are made for scientific studies and are also used in some branches of engineering (semi-conductors).

Research has shown that the internal crystalline structure of metal grains is far from being regular.

The nature and degree of irregularities or imperfections in the crystal structure determine to a large extent the properties of a metal. It is, therefore, essential to consider the types of imperfections in the crystal structure of metals or, what is the same thing, the structure of real crystals.

One of the types of imperfections in the crystal structure are non-occupied points in the crystal lattice, in other words, the presence of vacancies, or atomic 'holes' (Fig. 2.12a). This 'point' defect of the lattice is of great importance for diffusion processes in metals.

At room temperature, the number of vacancies is very small compared with the total number of atoms (roughly one vacancy per 10^{18} atoms), but increases appreciably with temperature, especially near the melting point (one vacancy per 10^4 atoms).

Some external effects (irradiation by radioactive particles, surface friction, etc.,) can increase the number of vacancies, i.e., loosen the metal. In certain extreme cases the number of vacancies may be as large as 1 to 3 per 100 atoms.

Another important defect in the crystal structure is dislocation. Imagine that an extra incomplete plane of atoms, which is called the extraplane, has for some or other reason appeared in a crystal lattice

(Fig. 2.13). The edge 3-3 of that plane forms a linear defect (imperfection) of the lattice, which is called the edge dislocation. An edge dislocation may extend to many thousands of lattice parameters and its Burgers vector (see further in this section) is perpendicular to the extraplane.

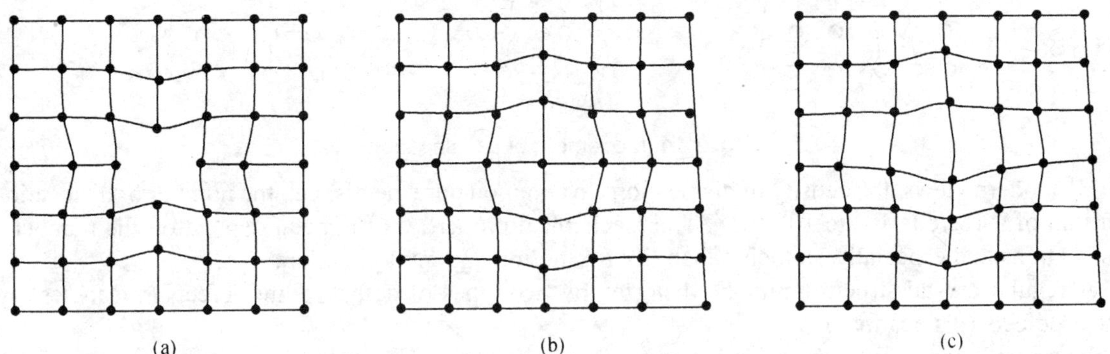

Fig. 2.12. A diagram of point defects; (a) vacancy; (b) substituted atom; and (c) interstitial atom.

Another type of dislocation is screw dislocation whose Burgers vector is parallel to the extraplane. In real metals, dislocations are mostly of a mixed type: edge dislocations in some portions and screw dislocations in others. A zone of elastic distortion of the lattice forms around a dislocation. The distance from the centre of a defect to the undistorted lattice is taken equal to the width of a dislocation, which is not large (a few atomic distances).

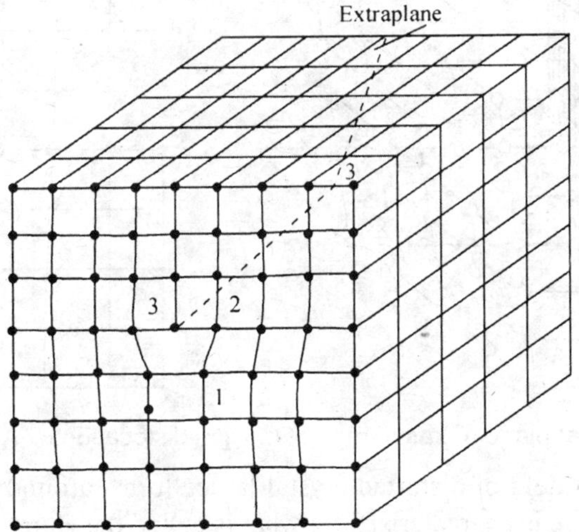

Fig. 2.13. Dislocation in a crystal lattice.

Because of the distortion of the lattice near a dislocation (Fig. 2.14a), the latter can easily move from its neutral position, the neighbouring plane passes to an intermediate position (Fig. 2.14b) and transforms into an extraplane (Fig. 2.14c) and forms a dislocation along the edge atoms. It is clear that a dislocation can move (more strictly, be passed on as the baton in a relay-race) along a certain plane (slip plane) which is normal to the extraplane.

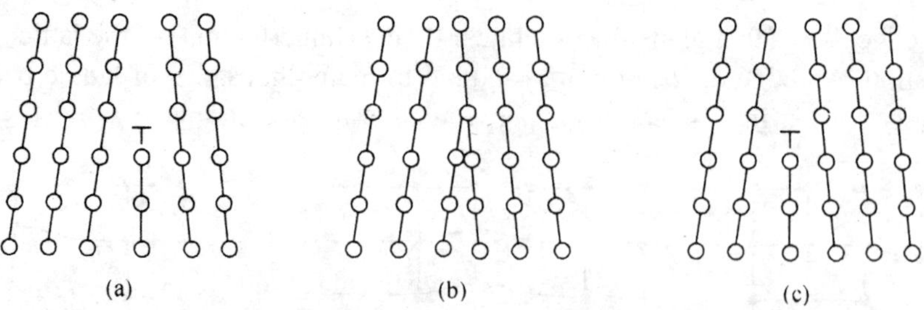

Fig. 2.14. Movement of a dislocation.

By modern views, the density of dislocations in common pure metals, i.e., the number of dislocations per unit of surface is 10^8 to 10^9 cm^{-2}. The mechanical properties of a metal depend on the number of dislocations and especially on their capability for motion and multiplication.

A regular crystal structure may be distorted by two types of defect: point defects (vacancies) and linear defects (dislocations).

Vacancies move continuously in a lattice when a neighbouring atom passes to a 'hole' and leaves empty its former location. An increase in temperature, i.e., in the thermal mobility of atoms, increases the number of such acts and adds to the number of vacancies.

Unlike vacancies, linear defects cannot move spontaneously and chaotically. A slight stress, however, is sufficient for a dislocation to start motion and form a plane and in the cross-section, a slip line C Fig. 2.15.

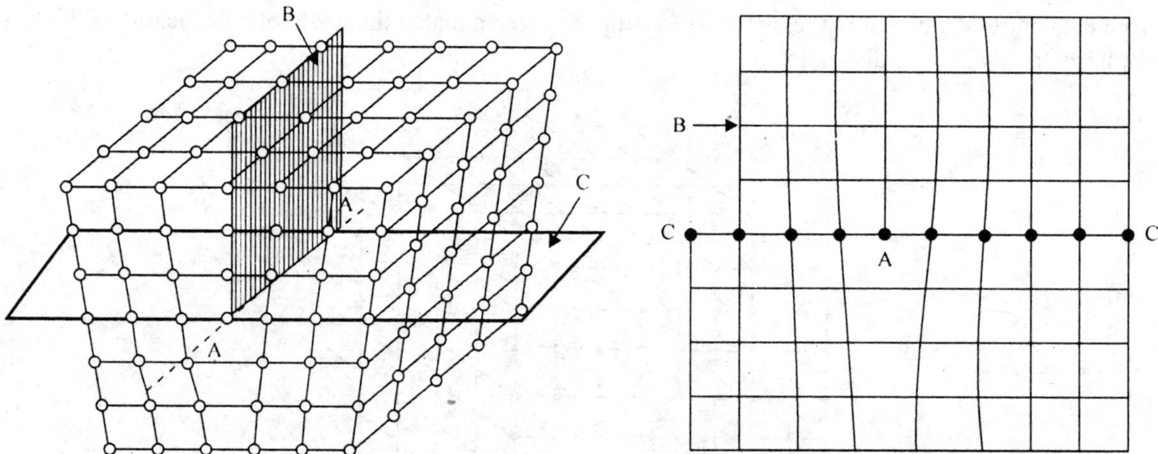

Fig. 2.15. A shear plane (C) as a trace of moving dislocation (A-A); B—extraplane.

As has been indicated, a field of distorted crystal lattice forms around a dislocation. The energy of distortion of the crystal lattice is characterised by what is called the Burgers vector.

If a contour ABCD is drawn around a dislocation ⊥ (Fig. 2.16), the portion BC will consist of six sections and the portion AD, of five. Their difference is BC — AD = b, where b is the magnitude of the Burgers vector. If a contour is drawn around a few dislocations (the distortion zones of the crystal lattice which overlap or merge), the total Burgers vector will be the sum of the Burgers vectors of the individual dislocations. The magnitude of the Burgers vector characterises the mobility of dislocations.

Apart from the dislocation structure (which may be quite diverse of great importance is the total characteristic of the number of dislocations, which is called the dislocation density. This is understood as the total length of dislocations (in centimetres) per cm^3 of metal volume, i.e., the dimension of dislocation density is ρ = cm/cm^3= cm^{-2}. The dislocation density of metals usually varies within 10^8–10^{13} cm^{-2} (a million kilometres per cubic centimetre).

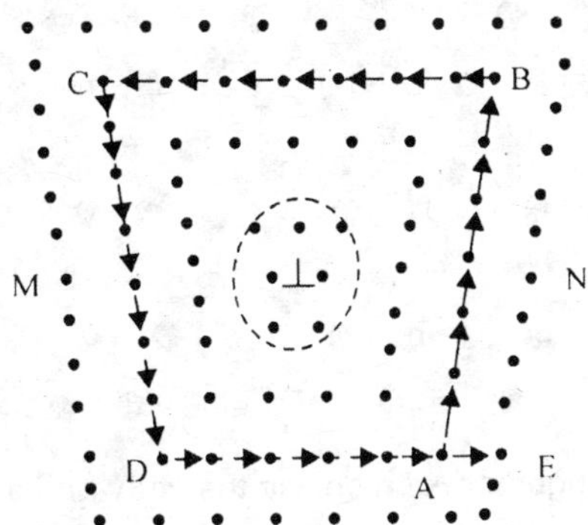

Fig. 2.16. A diagram to determine the Burgers vector for a linear dislocation.

A metallic grain is far from being a monolithic crystal composed of parallel atomic layers. Actually, it consists, as it were, of a mosaic of individual blocks whose size is $1 \cdot 10^{-5}$–$1 \cdot 10^{-3}$ cm (1000–1,00,000 Å) and whose crystal planes are turned through a small angle of a few minutes relative to one another. This structure of metal grains is called mosaic structure and the blocks that constitute it are called mosaic blocks. Various processes of metal treatment can change the sizes of mosaic blocks and their orientation thus changing the properties of the metal. Blocks can often be combined into larger aggregates which are called fragments. Each fragment contains a large number of blocks. Fragments are also misoriented relative to one another through angles of a few degrees.

Thus, grains in a metal are misoriented relative to one another through angles of a few tens of degrees. Grains can be composed of fragments misoriented only through a few degrees. Finally, fragments can consist of blocks misoriented only through small angles of a few minutes. This three-stage structure is not invariably found in all cases. In some particular cases metal grains may consist of fragments having no block structure, or only of blocks. A thermal process which causes division of grains into fragments is called the fragmentation, or polygonisation.

Mosaic blocks are usually very small and can be seen only in the electron microscope. The size of blocks and their misorientation angles are measured by methods of X-ray structural analysis.

If fragments are large, they can be easily seen in the optical microscope.

A real metallic crystal contains atomic-crystalline (vacancies and dislocations) and structural (blocks and fragments) imperfections. The former, i.e., vacancies and dislocations, are distributed non-uniformly and concentrate at the boundaries of grains, fragments and blocks.

As has been mentioned earlier, neighbouring grains are strongly misoriented (the misorientation angle being greater than 5°) and between them there is a boundary layer of a strongly distorted structure and a high concentration of atomic-crystalline defects (Fig. 2.17).

Misorientation of blocks is small (less than 1°). Blocks are bonded with one another and retain a regular, though distorted, crystal structure by means of dislocations.

The density of dislocations in that portion of the crystal is higher at a higher angle of misorientation between blocks. The properties of crystals of a metal are closely associated with many factors of its internal structure: the content (density) of vacancies and dislocations, their location (dislocation structure), the size and misorientation of the block structure (fine structure).

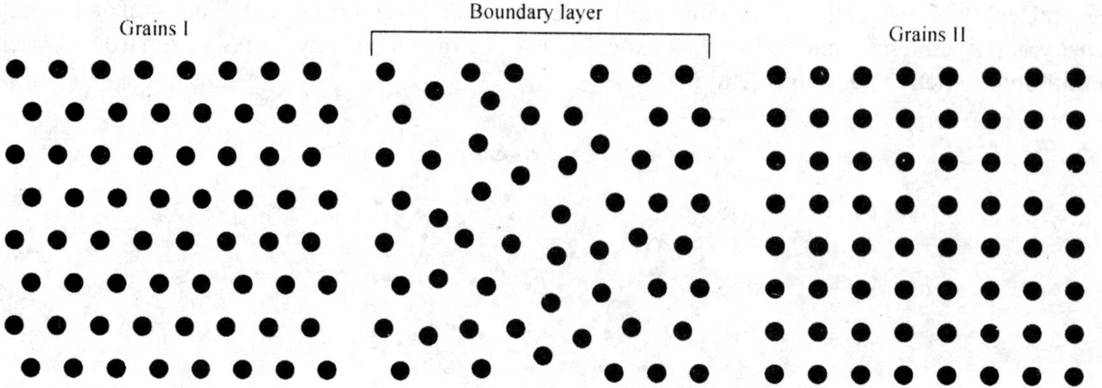

Fig. 2.17. A model of arrangement of atoms in the body and at the boundary of a grain.

Modern science possesses numerous experiemental means for analysis of fine and dislocation structures of metallic crystals. It should be also noted that the atoms on a branched so-called internal surface (at boundaries of grains, fragments and blocks) possess an elevated energy. Because of this, many of the processes, that will be discussed later, occur predominantly at the boundaries of grains (fragments or blocks).

Anisotropy of Properties of Crystals

Considering various planes, for instance, in a b.c.c. lattice, it may be easily concluded that they are filled with atoms to a different density. For instance, in the hatched square of a b.c.c. lattice (Fig. 2.18a) the centres of atoms are at the corners. Since each of these atoms simultaneously belongs to four squares, each of the squares of area a^2 shares one atom.

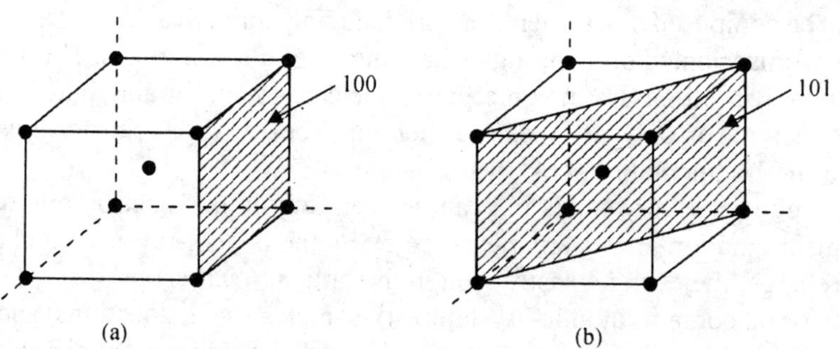

Fig. 2.18. Characteristic planes in a body-centred cubic lacttice.

In the hatched rectangle of the same lattice (Fig. 2.18b) there are two atoms per area of $a^2\sqrt{2}$, i.e., each of these atom shares an area of $a^2\sqrt{2}/2$, which is smaller than a^2, therefore, there are more atoms in that plane than in the cube side.

The properties of an individual crystal (single crystal) in a given direction differ from its properties in another direction and naturally, depend on how many atoms are met in that direction. The difference in the properties of crystals, depending on the direction of test, is called anisotropy. All crystals are anisotropic. Anisotropy is an inherent property of any crystal and is typical of crystalline structure.

A real metal is composed of a plurality of crystals. Individual crystals are only fractions of a millimetre in size, so that many thousands of crystals may be present in 1 cm^3 of metal volume. Since these crystals are randomly oriented, there is roughly the same quantity of differently oriented crystals in any direction. As a result, the properties of this polycrystalline body are the same in all directions, though the properties of every crystal, it is composed of, depend on the crystal orientation. This phenomenon is called quasi-isotropy.

TYPES OF DEFECTS IN CRYSTALS

Different types of defects are found in most of crystals. Crystals have mainly three types of defects:

Line Defects

Line defects involve the edge of extra plane of atoms. They are of two types: (i) edge dislocation; and (ii) screw dislocation.

Point Defects

The point defects are caused by missing atoms, extra atoms or displaced atoms. The point defects are classified as:
(i) Vacancies; (ii) substitutional impurity; (iii) interstitial defects; (iv) Frenkel defects; (v) Schottky defects; and (vi) phonon.

Vacancies

Vacancy is the simplest point defect, which involves a missing atom. Such defects are due to imperfect packing. They also arise due to thermal vibration of atoms at high temperature because at high temperature, an atom can jump out and go to lowest energy position, (Fig. 2.19). The vacancies may vary from one in 10^{18} atoms at room temperature to one in 10^4 atoms near melting point. Due to some external effects the vacancies may increase and in extreme cases may be 1 to 3 per cent.

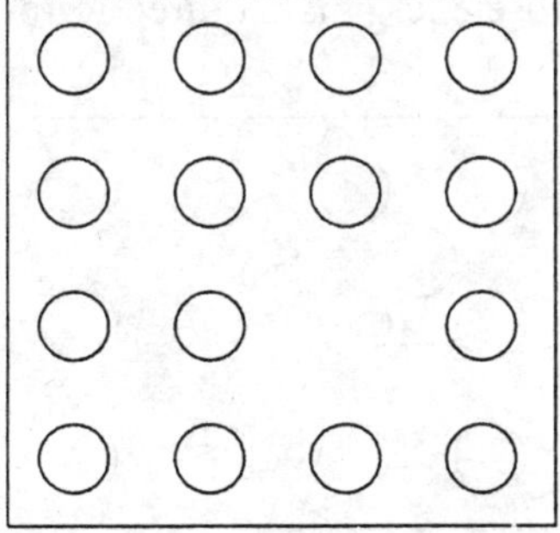

Fig. 2.19. Vacancy defect.

Substitutional defects

Sometimes a foreign atom replaces the parent atom of the lattice and thus occupies the position of parent atom (Fig. 2.20).

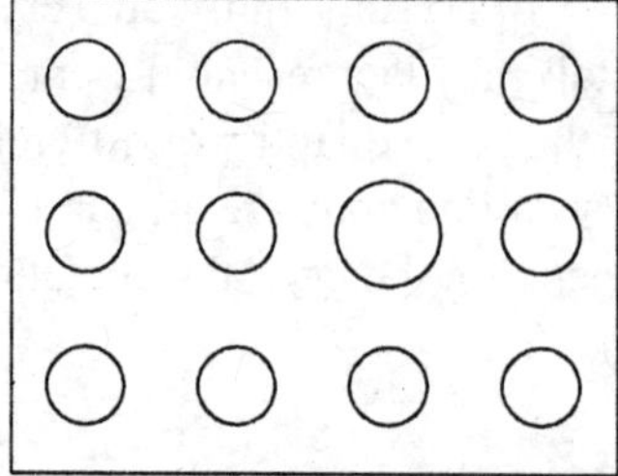

Fig. 2.20. Substitutional defects.

Interstitial defects

It is an addition of extra atom, of smaller diameter, within the crystal structure particularly where atomic packing factor is low. This results in atomic distortion. The foreign atom comes and occupies the space as shown in Fig. 2.21. This atom may form an alloy or may be present there simply as impurity.

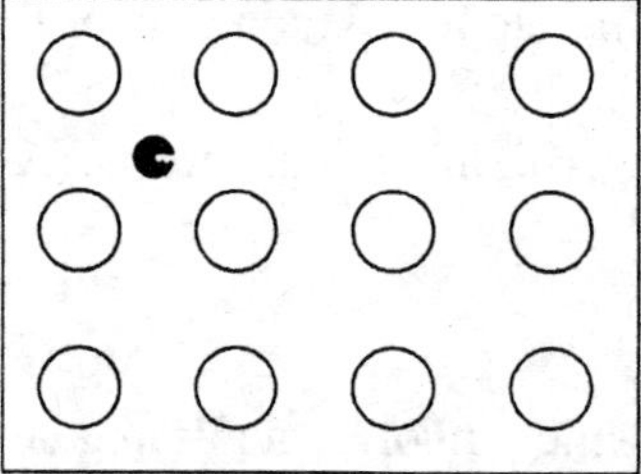

Fig. 2.21. Interstitial defects.

Frenkel defects

In this type of defect, an ion leaves its own place and occupies an interstitial site as shown in Fig. 2.22. These defects are less in number because energy is required to force an ion into new position.

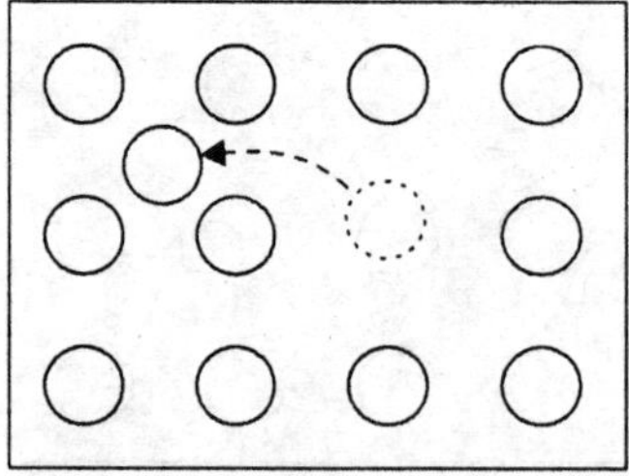

Fig. 2.22. Frenkel defects.

Schottky defects

These defects are similar to vacancies. One pair of ions of opposite charge is vacant (missing). Thus, this type of defect maintains a charge balance (Fig. 2.23).

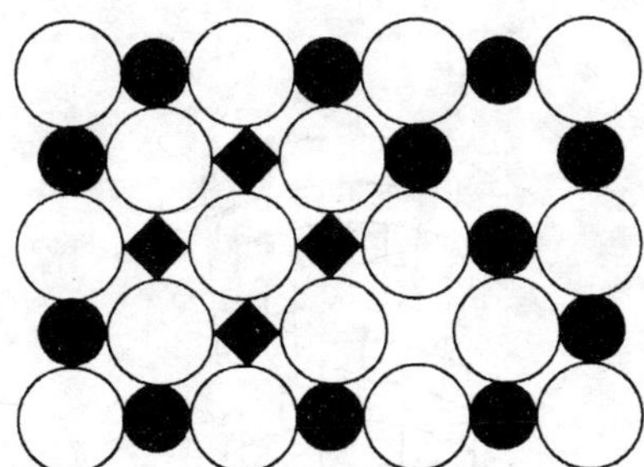

Fig. 2.23. Schottky defects.

Phonon

When temperature is raised, thermal vibrations are caused, which results in the defect of a symmetry and deviation in shape of atoms. This defects has much effect on the electric and magnetic properties.

Surface Defects

The imperfections between adjacent crystals of external surface of crystals are called surface defects. They are of two types: (i) grain boundaries; and (ii) stacking faults.

Grain boundary

Grain boundaries are surface defects at which crystals i.e., grains of different orientations in polycrystalline material get separated (Fig. 2.24a).

Tilt boundary

The crystal boundaries corresponding to misoreintation of less than a degree are described as tilt boundaries (Fig. 2.24b).

Twin boundary

In twin boundaries the atomic arrangement on one side of the boundary is mirror image of the arrangement of the atom on the other side (Fig. 2.24c).

Stacking faults

The planes of a crystal atoms are stacked in sequence as explained below.

Consider a (fcc) structure. Atoms are arranged in many repeated layers, as shown in Fig. 2.24.

1. Consider a layer of atoms as shown in black spheres, called A-layer.
2. The second layer of atoms exists above this bottom layer as (dotted) spheres, called B-layer. Here, all the atoms rest in pockets between the atoms of first layer and touches three atoms in the bottom layer.
3. In the third layer, the atoms are placed above the pocket of first layer that were not filled by second layer, called C-layer. These atoms are shown as (o).

The 4th, 5th and 6th layers are repeatation of 1st, 2nd and 3rd layers respectively. This sequence of planes is called A B C; A B C; A B C.

In hcp crystals the first two layers of planes are similar as explained above (as in fcc arrangement). But the atoms in the third layer are placed in the same way as in first layer i.e., the 3rd and 4th layer are just the repeatation of 1st and 2nd layer. This sequence is called A B, A B, A B.

Fig. 2.24. Crystal planes stacking. Black circles represent bottom layer of atoms. Dotted circles represent second layer nested on the first. Third Layer may be entered at closed dot position or open dot position.

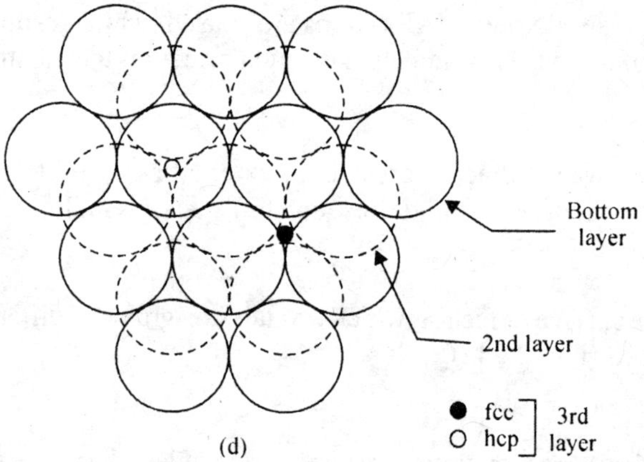

If the sequence is disturbed or atoms are stacked out of sequence, the stacking faults are developed. These stacking faults influence the behaviour of metals and alloys.

In Fig. 2.25(a), the sequence is B C A, B C A, B.

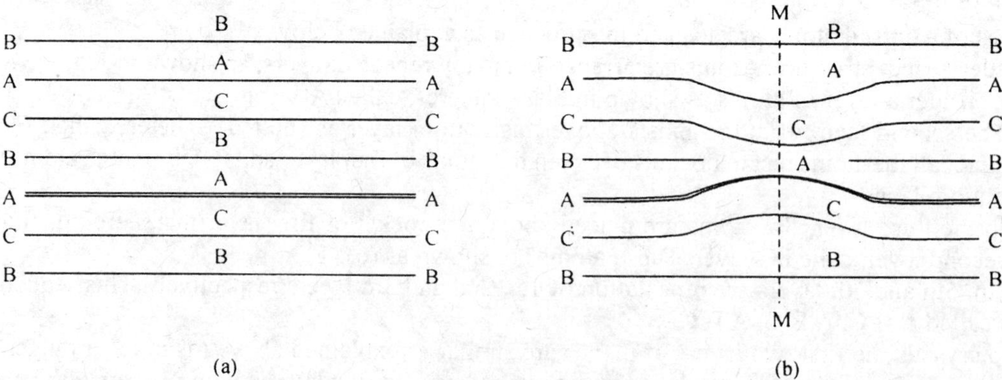

Fig. 2.25. Stacking Faults (a) Sequence of Planes not distrubed (b) Sequence is disturbed.

In Fig. 2.25 (b), the sequence is disturbed, some atoms are missing in one of the B-layer. The sequence of structure along vertical line MM becomes BCA, CAB, which was previously BCA, BCA, B—(when the sequence along MM was not disturbed i.e., it was fcc structure). Now the crystal will behave as hcp crystal along layers CA, CA (along line MM in the middle) while it will behave as fcc crystal along the sides. Similarly, if a layer of atoms comes in between A and B layers, as shown in Fig. 2.26 and if these incoming atoms arrange themselves as C-layer, the sequence gets distrubed and a sequence fault is developed. Now the layers in the centre along MM line are in the sequence as ABC, ACB, CA. The previous sequence (before the entry of C-layer) was ABC, ABC, ——

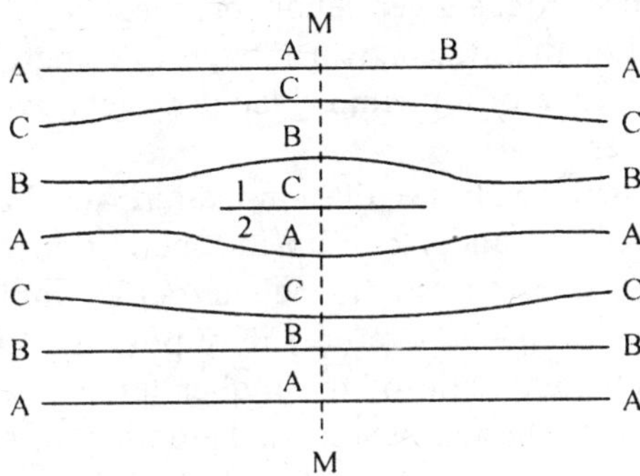

Fig. 2.26. Stacking fault.

This is just like missing two layers, i.e., A and B layer at 1 and 2 positions. This is called double stacking fault. All types of defects influence the characteristics of materials such as mechanical, electrical and chemical properties.

EQUILIBRIUM DIAGRAMS

As our study of engineering materials becomes more focused on specific metals and alloys, it is increasingly important that the natural characteristics and properties of the material be known for various environmental conditions. What is the basic structure of the material? Is the material uniform throughout, or is it a mixture of two or more distinct components? If there are multiple components, how much of each is present? Is there a component that may impart undesired properties or characteristics? What will happen if temperature is increased or decreased, pressure is changed, or chemistry is varied? The answers to these and other important questions can be obtained through the use of equilibrium phase diagrams.

Phases

Before we move to a discussion of these diagrams, it is imperative that a working definition of the term phase first be developed. As a starting definition, a phase is simply a form of material possessing a single characteristic structure and associated characteristic properties. Uniformity of chemistry, structure and properties is assumed throughout the phase. More rigorously, a phase is any physically distinct, chemically homogeneous and mechanically separable portion of a substance. In lay terms, a phase has a unique structure, uniform composition and well-defined boundaries or interfaces.

A phase can be continuous (like the air in a room) or discontinuous (like grains of salt in a shaker). A phase can be solid, liquid, or gas. In addition, a phase can be a pure substance or a solution, provided that the structure and composition are uniform throughout. Alcohol and water mix in all proportions and will therefore form a single phase when combined. Oil and water tend to form isolated regions with distinct boundaries and must be regarded as two distinct phases.

Phase Diagrams

Phase diagrams, also called equilibrium diagrams or constitution diagrams, are a very important tool in the study of alloys. They define the regions of stability of the phases that can occur in an alloy system under the condition of, constant pressure (atmospheric). The coordinates of these diagrams are temperature (ordinate) and composition (abscissa). Notice that the expression 'alloy system' is used to mean all the possible alloys that can be formed from a given set of components. This use of the word system differs from the thermodynamic definition of a system, which refers to a single isolated body of matter. An alloy of one composition is representative of a thermodynamic system, while an alloy system signifies all compositions considered together.

The interrelationships between the phases, the temperature and the composition in an alloy system are shown by phase diagrams only under equilibrium conditions. These diagrams do not apply directly to metals not at equilibrium. A metal quenched (cooled rapidly) from a higher temperature to a lower one (for example, room temperature) may possess phases or metastable phase and compositions that are more characteristic of the higher temperature than they are of the lower temperature. In time, as a result of thermally activated atomic motion, the quenched specimen may approach its equilibrium low-temperature state. If and when this occurs, the phase relationships in the specimen will conform to the equilibrium diagram. In other words, the phase diagram at any given temperature gives us the proper picture only if sufficient time is allowed for the metal to come to equilibrium.

Isomorphous Alloy Systems

The simplest of these systems is the isomorphous. In which only a single type of crystal structure is observed for all ratios of the components. A typical isomorphous system is represented in the phase diagram of Fig. 2.27. The binary alloy series in question is copper-nickel. These elements combine, as in all alloy systems of this type, to form only a single liquid phase and a single solid phase. Since the gaseous phase is generally not considered, there are only two phases involved in the entire diagram. Thus, the area in the figure above the line marked 'liquidus' corresponds to the region of stability of the liquid phase and the area below the solidus line represents the stable region for the solid phase. Between the liquidus and solidus lines is a two-phase area where both phases can co-exist.

The significance of several arbitrary points located on Fig. 2.27 will now be considered. A set of co-ordinatces—a temperature and a composition–is associaied with each point. By dropping a vertical line from point x until it touches the axis of abscissae, we find the composition 20 per cent Cu. Similarly, a horizontal line through the same point meets the ordinate axis at 500°C. (Note: Phase diagram temperatures are normally expressed in °C, rather than in K). Point x signifies an alloy of 20 per cent Cu and 80 per cent Ni at a temperature of 500°C. The fact that the given point lies in the region below the solidus line tells us that the equilibrium state of this alloy is the solid phase.

The structure implied is, therefore, one of solid-solution crystals and each crystal will have the same homogeneous composition (20 per cent Cu and 80 per cent Ni). Under the microscope, such a structure will be identical to a pure metal in appearance. In other respects, however, a metal of the given composition will have different properties. It should be stronger and have a higher resistivity than either pure metal and it will also have a different surface sheen or colour.

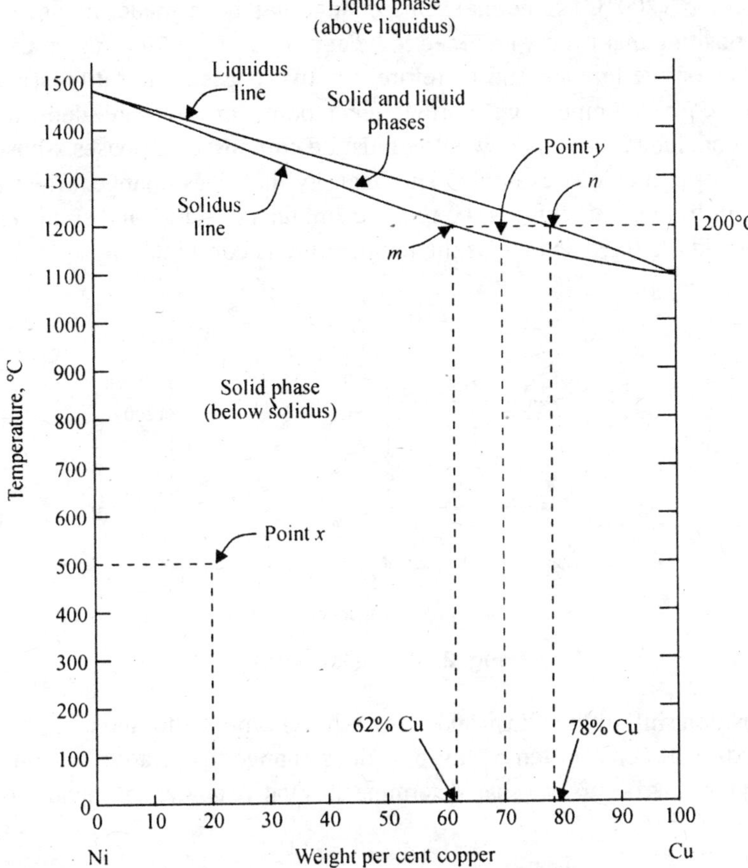

Fig. 2.27. The copper nickel phase diagram.

Let us turn our attention to the point *y*, which falls inside the two-phase region, the area bounded by the liquidus and solidus lines. The indicated temperature in this case is 1200° C, while the composition is 70 per cent Cu and 30 per cent Ni. This 70 to 30 ratio of the components represents the average composition of the alloy as a whole. It should be remembered that we are now dealing with a mixture of phases (liquid plus solid) and that neither possesses the average composition.

To determine the compositions of the liquid and the solid in the given phase mixture, it is only necessary to extend a horizontal line through point *y* until it intersects the liquidus and solidus lines. These intersections give the desired compositions. In Fig. 2.27, the intersections are at points *m* and *n* respectively. A vertical line dropped from point *m* to the abscissa axis gives 62 per cent Cu, which is the composition of the solid phase. In the same manner, a vertical line dropped from point *n* shows that the composition of the liquid is 78 per cent Cu.

Lever Rule

A very iroportant relationship, which applies to any two-phase region of a two-component or binary phase diagram, is the so-called lever rule. With regard to the lever rule, consider Fig. 2.28 which is an enlarged portion of the copper-nickel diagram of Fig. 2.27. Line *mn* is the same in the new figure as in the old and lies on the 1200°C isothermal. One change has been made, however and that is in the average alloy composition that is now assumed to be at points z or at 73 per cent Cu. Because this new composition still lies on the line *mn* and therefore in a two-phase region it must also be a mixture of liquid and solid at the given temperature. Further, since points m and n are identical in both Figs. 2.27 and 2.28 it can be concludcd that this new alloy must be composed of phases whose compositions are the same as those in the previous example. The quantity that does change, when the composition is shifted along line *mn* from point *y* to *z*, is the relative amounts of liquid and solid. The compositions of the solid and the liquid are fixed as long as the temperature is constant.

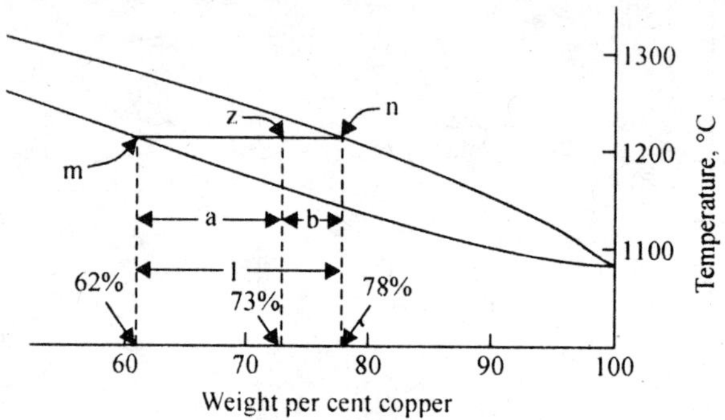

Fig. 2.28. The lever rule.

While a composition shift of the alloy (taken as a whole) cannot alter the compositions of the phases in a two-phase mixture at constant temperature. It does change the relative amounts of the phases. In order to understand how this occurs, we shall determine the relative amount of liquid and solid in an alloy of a given composition.

Eutectic Systems

The copper-silver phase diagram, (Fig. 2.29), can be taken as representative of eutectic systems. In systems of this type, there is always a specific alloy, known as the eutectic composition, that freezes at a lower temperature than all other compositions. Under conditions approaching equilibrium (slow-cooling), it freezes at a single temperature like a pure metal. In other respects, the solidification reaction of this composition is quite different from that of a pure metal since it freezes to form a mixture of two different solid phases. Hence, at the eutectic temperature, two solids form simultaneously from a single liquid phase. A transformation, where one phase is converted into two other phases, requires that three phases be in equilibrium.

Assuming constant pressure, three phases can only be in equilibrium at an invariant point, that is, at a constant composition (in this case the eutectic composition) and at a constant temperature (eutectic temperature). The eutectic temperature and composition determine a point on the phase diagram called the eutectic point, which occurs in the copper-silver system at 28.1 per cent Cu and 779.4°C.

In an eutectic system, the components do not have to crystalize in the same structure nor do they necessarily have to be chemically similar. If the two atomic forms are quite different chemically, however, then intermediate crystal structures are liable to form in the alloy system. Eutectics can still be observed in such systems, but they will not form between terminal phases, as shown in Fig. 2.29.

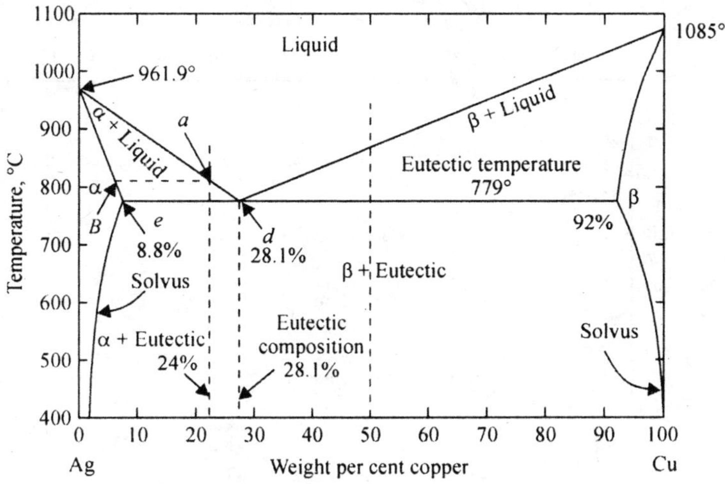

Fig. 2.29. Copper-silver phase diagram.

Microstructures of eutectic systems

Any composition of an isomorphous system at equilibrium and in the solid state consists of a single homogeneous group of solid-solution crystals. When viewed under the microscope, such a structure does not differ essentially from that of a pure metal. It is usually difficult, therefore, to tell much about the composition of these single-phased alloys from a study of their microstructure alone. On the other hand, the appearance under the microscope of an alloy from the two-phase field (solid) of an eutectic system is very characteristic of its composition.

Peritectic Transformation

The eutectic reaction, in which a liquid transforms into two solid phases, is just one of the possible three-phase reactions that can cccur in binary systems. Another that appears frequently involves a reaction between a liquid and a solid that forms a new and different solid phase. This three-phase transformation occurs at a peritectic point.

Fig. 2.30 shows the constitution diagram of the iron-nickel system. A peritectic point appears at the upper left-hand corner. An enlarged view of this region is given in Fig. 2.31. Before studying the peritectic reaction, some of the basic features of the alloy system in question should be considered. Iron and nickel have apparent atomic diameters that are almost identical (Fe, 0.2476 nm and Ni, 0.2486 nm). Since both iron and nickel belong to group VIII of the periodic table, these two elements are chemically similar. Both crystallise in the face-centred cubic system; nickel is face-centred cubic at all temperatures but iron only in the range 912°C to 1394°C. Conditions are thus ideal for the formation of a simple isomorphous system except for the fact that the stable crystalline form of iron is body-centred cubic at temperatures above 1394°C and below 912°C. It is not surprising, therefore, that iron-nickel alloys are face-centred cubic except for two small body-centred cubic fields at the upper and lower left-hand corners of the phase diagram. Fig. 2.30 also shows that a superlattice transformation, based on the composition $FeNi_3$, occurs in this system. The boundaries delineating this face-centred cubic order-disorder transformation appear at the lower right-hand side of the phase diagram.

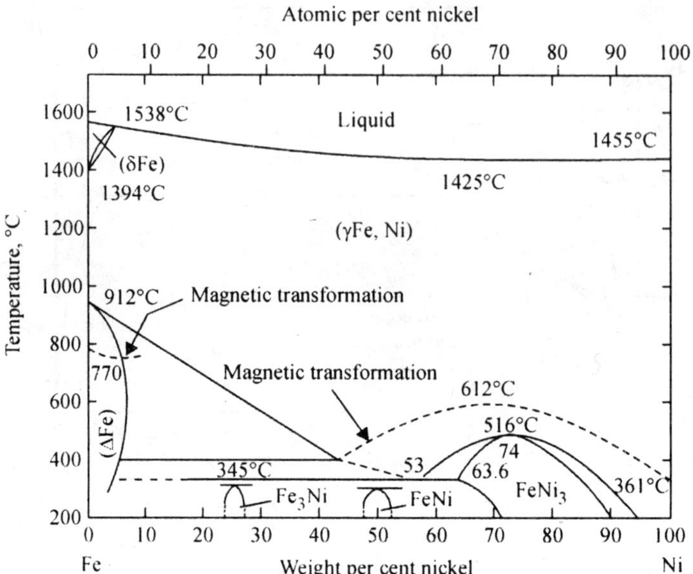

Fig. 2.30. Iron-nickel phase diagram.

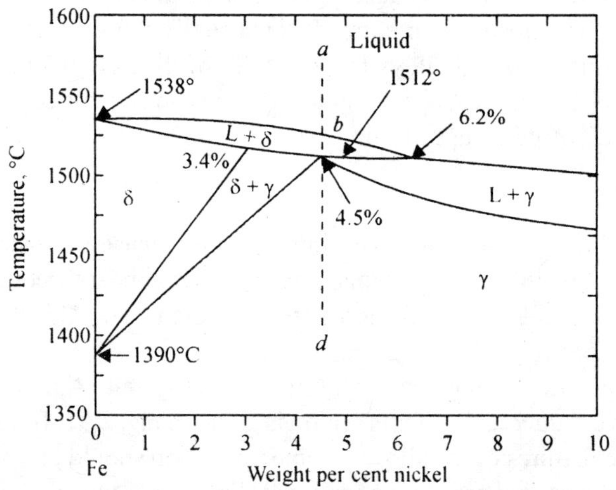

Fig. 2.31. The peritectic region of the iron-nickel phase diagram.

The addition of nickel to iron increases the stability of the face-centred cubic phase. As a consequence, the temperature range in which this crystalline phase is preferred expands with increasing nickel content and the boundaries separating the body-centred cubic fields from the face-centred field slope upward and downward respectively (with increasing nickel content).

Monotectices

Monotectics represent another form of three-phase transformation in which a liquid phase transforms into a solid phase and a liquid phase of different composition. Monotectic transformations are associated with miscibility gaps in the liquid state. A reaction of this type occurs in the copper-lead system at 955°C and 37.4 per cent Pb, as can be seen in Fig. 2.32. Notice the similarity between this monotectic and the eutectic of the copper-silver system (Fig. 2.29). The liquid miscibility gap lies just to the right of the monotectic point. Note that the copper-lead phase diagram also possesses an eutectic point at 327.5° C and 99.9 per cent Pb (0.1 per cent Cu). Because it lies so close to the composition of pure lead, it is not possible to show it on the scale of the present diagram. Finally, this system is representative of one composed of elements that do not mix in the solid state; at room temperature the solubility of copper in lead is less than 0.007 per cent, while the solubility of lead in copper is of the order of 0.002 to 0.005 per cent Pb. The terminal solid solutions are, accordingly, nearly pure elements.

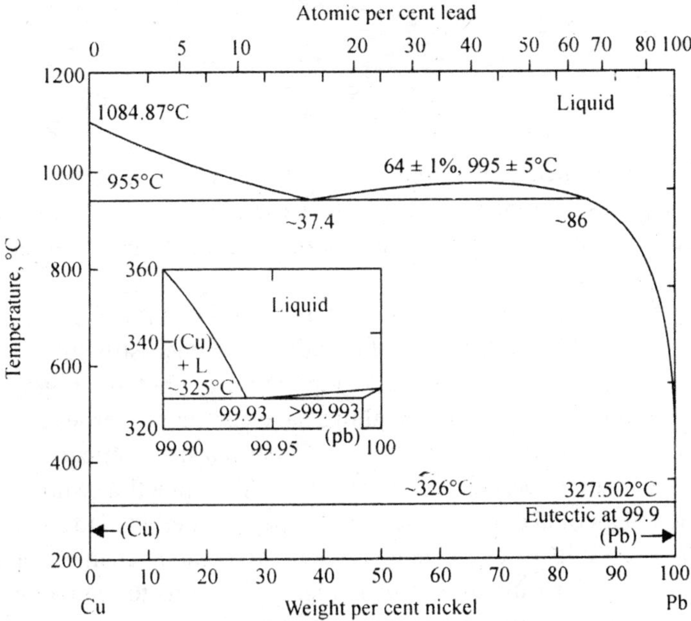

Fig. 2.32. Copper-lead phase diagram.

Other Three-Phase Reactions

The three basic types of three-phase reactions (eutectics, peritectics and monotectics) that we have studied so far involve transformations between the liquid and solid phases. They are, therefore, associated with the freezing or melting processes in alloys. Several other important three-phase reactions involve only changes between solid phases. The most important of these occurs at eutectoid and peritectoid points. At an eutectoid point, a solid phase decomposes into two other solid phases when it is cooled. The reverse is true at a peritectoid point where, on cooling, two solid phases combine to form a single solid phase. The similarity between the eutectoid and peritectoid transformations on the one hand and the eutectic and peritectic transformations on the other and is quite obvious. A very important eutectoid reaction occurs in the iron-carbon system.

Chapter 3

Metallography

INTRODUCTION

In the physical examination of metals, the observation of specimens in polished and etched condition under the microscope occupies an important place. The structure observed by means of a microscope is referred to as microstructure includes structures developed in order to be observed by the naked eye or under a low power microscope.

The word 'metallography' can be thought of as two words—metal and graph. Now a definition of metallography is rather simple. Metallography is the study of the microstructure of metals and alloys by making use of both the microscope and the metallograph to make photomicrographs.

A 'metallographer' has reason to believe that alloys have certain properties as a result of combining specific elements and thermal-mechanical exposure. Thermal-mechanical exposure is simply the application of stress which causes permanent deformation in conjunction with controlled heating and cooling. Any combination of thermal or mechanical cycles may be combined to create thermal-mechanical exposure. The net result is as follows: as a change is encountered in mechanical properties there will be a change or difference in the microstructure. A metallographer learns to read a photomicrograph like a traveller reads a road map. For example, a metal may have certain chemical proportions, yet the chemical analysis cannot determine hardness, strength or ductility. An alloy is hard, strong, ductile, tough or whatever because of its microstructure. A metallurgist can read a photomicrograph and make conclusions as to whether the part was correctly cast, properly forged, welded according to specifications or properly maintained in service.

This chapter will include metallographic laboratory practices and principles underlying metallographic structures. Common microstructural features that are part of the analysis of alloys will also be included.

OBJECTIVES OF A METALLOGRAPHER

Objective of a metallographer are:

To prepare a sample that may be studied at an appropriate magnification and make a photograph which is representative of what the structure really is.

This means that the true microstructure will be represented. This also means that the surface will have to be optically flat and free from man-made deletions or additions.

The entire sequence of making a photomicrograph requires:

1. Sectioning—removing a representative sample small enough to be conveniently handled in the laboratory.
2. Mounting—moulding a neutral material (plastic) around all but one surface.
3. Grinding—levelling the surface and reducing the thickness of the disturbed metal layer that was caused during sectioning.
4. Polishing—removing deep scratches and reducing the thickness of the disturbed metal layer caused during grinding.
5. Etching—chemically removing a few hundred atom layers and preferentially attacking the surface to produce a relief effect and a staining of the phases.
6. Photomicrography—producing a permanent record on photographic paper of the microstructural details.
7. Analysis—identification of phases present and microstructural characteristics.

Sectioning

The sectioning step in the preparation of a photomicrograph requires judgement and experience to remove a sample that is representative of the whole section. To be truly representative, more than one sample may have to be taken. For example, if a technician is to evaluate the effects of rolling an ingot structure, the following samples would have to be removed: surface sample, edge sample, internal sample perpendicular to the surface and internal sample parallel to the surface. Cast metals have different types of microstructure in specific areas. Much of the differences in cast structures is attributed to progressive solidification and cooling rate.

The most acceptable machines for removing sample specimens are called liquid cooled abrasive cut-off instruments. The slow-speed, diamond cutting wheel operating with only a few ounces of force produces less disturbed metal than any others. The most popular, from an economical point of view, is the high-speed abrasive wheel manually operated with liquid coolant.

If the use of a cut-off machine is not practical, then other means of sectioning will have to be used. Keep in mind that the sectioning sequence is to leave the workpiece free from man-caused damage. Some sections may be removed by cutting a notch, then fracturing the specimen free from the main piece by light hammer blows. If the material is soft enough, sawing may be recommended. Usually with sawing, a coolant is used to keep the disturbed surface metal to a minimum.

Mounting

The process of mounting is used primarily to facilitate preparation and analysis. The question of how much mounting material to use to produce the ideal height can be a problem. Sometimes it is an inconvenience to work with varying height mounts. Some laboratories use a measured amount or a height gauge. Neither of these account for the difference in volume of the metal specimen. A formula has been developed for use with phenolic resin and an inch and a quarter diameter moulds. The optimum height was selected as a matter of convenience and economy at 0.660 inch. By experiment it was found that 18 gram of phenolic resin made the correct height. The volume of the mount (without the metal specimen) is easy to calculate. It is pi $\times$ D^2 $\times$ 0.7854 $\times$ height. There is also a simple method to make the conversion of volume of metal to volume of mount – 1 ml. = 1 cm^3. Now all that is needed is to get the percentage that is phenolic resin and multiply that by the total weight of the original mount that was 100 per cent phenolic resin.

The formula looks like this:

$$W = [(V_o - V_m) \div V_o] \times 18$$

where W = weight of phenolic resin

V_o = volume of original mount (without metal specimen)

V_m = volume of metal specimen (measured by displacement of water in a graduate)

Also, during the process of mounting, edge preservation and optical flatness may be improved by simple techniques as shown in Fig. 3.1.

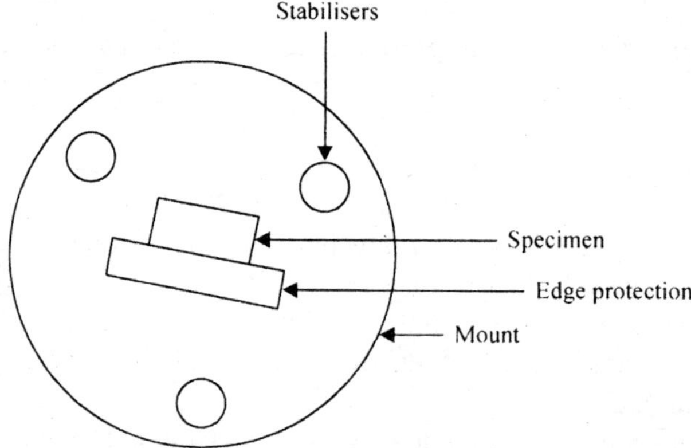

Fig. 3.1. Plan view of a metallographic mount showing edge protection and stabilisers to help keep section flat.

The back side of a plastic mount provides a convenient location for permanent identification. Marking is easily done with an inexpensive hard-tip vibrator. It is standard practice to mark the mount with the following data: type of metal, job number, date and person preparing the sample.

The type of material selected for the mount should be compatible to the entire sequence. The mount must be able to endure repeated attack by acids and elevated temperatures, remain non-conductive in electroyte and resist wear. Typical materials used to mount metallographic specimens are listed in Table 3.1.

Table 3.1. Metallographic mounting materials.

Moulding material	*Curing temp.*	*Minimum curing time (min.)*	*Ejection temperature*	*Minimum curing pressure (psi)*	*Transparent*	*Resistant to organic solvents*
Phenolic resin (bakelite)	135+°C 275+°F	7	As cured	2500	No	Yes
Thermoplastic resin (lucite)	140+°C 284+°F	Not required	RT	2500	Yes	No
EpoxyAir cure @ RT	480	RT	None	Semi-	Yes	
AcrylicAir cure @ RT	90	RT	None	Semi-	No	

Compression mounting techniques

The usual practice for compression mounting techniques is to place the mould cylinder on a table, push the lower piston through to the bottom, locate the specimen on the lower piston, add moulding material to the designated height, place the top piston in the cylinder and press firmly. The mould assembly is now ready to be pressurised and heated as shown in Fig. 3.2.

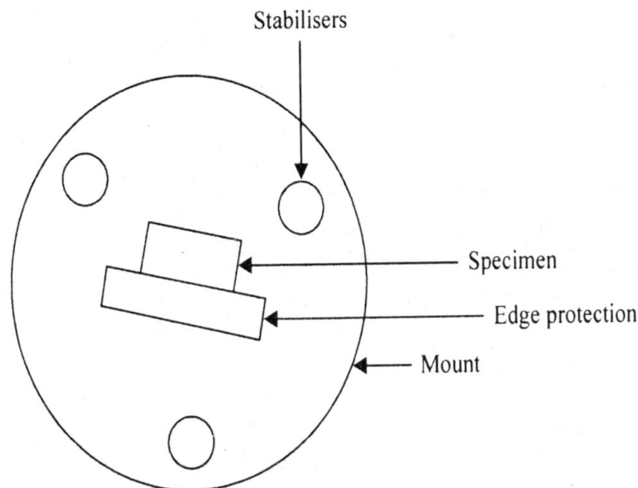

Fig. 3.2. Sketch of a metallographic moulding assembly. Note the thermometer well which is drilled to permit the temperature measurement in close vicinity to the plastic particles.

The mounting press may be a simple hydraulic ram and a furnace assembly. There are also many more complex types of mounting presses operating in the semi-automatic mode and fully automatic mode for mass production.

Phenolic resins are the most widely used mounting materials. They are available in three colours-black, red and green. They are a thermosetting compound and cannot be made liquid by heat and pressure after they once are treated. It is recommended that they be cooled by a heat sink for about 1 minute before releasing the pressure for ejection. Note: Pressure must be at a minimum of 2500 psi before being heated.

Rough and Fine Grinding

The grinding process must be done with care. At this stage of preparation the metallographer is attempting to remove the disturbed metal layers that occurred during normal sectioning. No mechanical process of cutting has been developed that will leave the surface free of disturbed metal. Since grinding is a mechanical process, it will cause plastic deformation on the surface. To this extent grinding will consist of starting with coarse abrasive grit and progressing to the finer grades for finishing. The finer the grit size the lesser the depth of disturbed metal. The average force holding the specimen against the abrasive grit is about 3 to 5 1b. Normal manual forces do not control the depth of deformed metal. Keep in mind that surface damage must be removed or reduced to a minimum in the final grinding stages. This is required because normal polishing and etching techniques remove very little surface metal. If surface damage occurs in grinding, the man-made artifacts will probably remain to be viewed in the photomicrograph.

The usual procedure in grinding is to select a grit size that is appropriate to the surface roughness of the part. After each abrasive grit sequence the specimen is washed thoroughly and inspected. Examples of grit size for the first grinding sequence are shown in Table 3.2.

Table 3.2. Suggested Grit sizes for beginning grinding operations.

Grid size	*Type of power driving abrasive*	*Type of lubricant*	*Type of surface*
40	Motor driven	Water	Fractured by impact
80	Motor driven	Water	Hack saw cut
150	Manual	Water	Abrasive cut-off
240	Manual	Water	Diamond cut-off

A customary procedure that will produce high quality surface finishes during the grinding stage might be the following (assuming an abrasive cut-off finish):

1. Use 240 grit, 5 1b of force and 8 to 10 strokes along the length of the abrasive grit for metals of average hardness. For soft metals use less force and fewer strokes.
2. Wash thoroughly, revolve specimen 45° to 90° from Step 1 and continue on the same abrasive for 12 to 16 strokes with a force related to the hardness of the metal.
3. Wash thoroughly, revolve the specimen 45° to 90° and continue for 18 to 20 strokes on the 320 grit abrasive. The force against the abrasive is again related to the hardness of the metal.
4. Revolve the specimen and continue until all previous abrasive marks have disappeared.
5. Wash thoroughly, revolve the specimen, move to 400 grit abrasive and continue grinding until the 320-grit scratches have been removed. The best method to determine if the 320 grit scratches have disappeared is to wash and dry the specimen and inspect it with a microscope at 100 magnifications.
6. Repeat Step 5 but go to 600 grit abrasive paper. The last few strokes should have less force against the abrasive paper. If extremely soft metals are being finished, it is recommended that new abrasive paper not be installed for finishing strokes.

All of these steps are to be done with a liberal supply of water to flush away loose surface particles. The technician will have to determine when it is economically feasible to change the abrasive grit to keep the best balance of quality versus cost.

One of the major questions in the field of metallography—for each step in the preparation of the specimen—is, how much time does it take to produce acceptable quality work? Of course this will differ between individuals and especially so with different metals. Each time a specimen is prepared, the operator should be looking for key points that will serve as end-of-sequence indicators. In grinding, for example, a typical end-of-sequence indicator is the disappearance of first-step scratches with a uniform pattern of second-step scratches. Manual grinding with silicon carbide abrasive grits are somewhat standardised by starting with a 240 grit and finishing with a 600 grit. Realise that a new abrasive grit has to be broken in (in use for several strokes) before it will produce an optimum surface finish.

For each step of any sequence job the technician who will excel will ask himself how he can best attain the objective of his work. In grinding the objective is to reduce the depth of disturbed metal. The depth can never be reduced to zero, but it can be reduced to the point where the microstructure does not noticeably reveal any artifacts. An artifact is the name of a structure that was caused by preparation steps and is not representative of the metallic structure.

While working to attain metallographic quality, the technician will have to develop a feel for certain things. For example, he must determine how much force is ideal. This will change from manual to power driven abrasives and from one abrasive material to another. Too little force on the abrasive causes minimum metal removal, which amounts to time wasting. Extremely heavy force may cause non-uniform patterns, in soft metals it may embed abrasive particles from a new paper.

Probably one of the most serious mistakes that is made while grinding is that of producing a rounded or faceted surface. This is easy to do if the specimen is allowed to tip as it is moved on the abrasive grit The technician should frequently inspect the surface for evidence of this, as shown in Fig. 3.3.

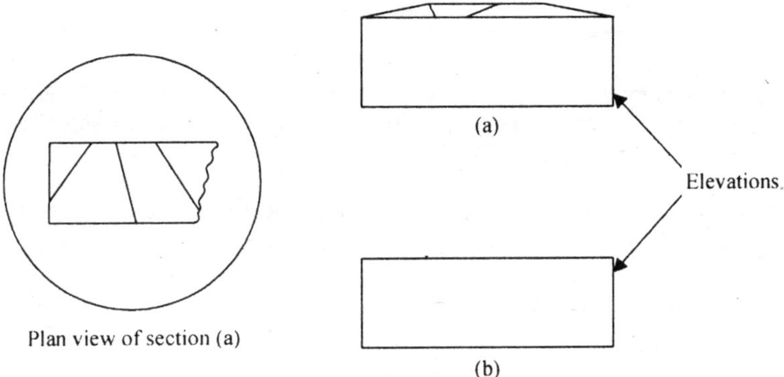

Fig. 3.3. Surfaces attained during grinding. (a) Specimen has been allowed to tip, which produces a series of surfaces (faceted surface); and (b) surface was held flat during grinding.

There are several types of grinding wheels, consisting of silicon carbide (SiC), aluminium oxide (Al_2O_3), emery (Al_2O_3-Fe_3O_4) and diamond particles, which are the hardest of all. Another very hard abrasive (nearly ten on the mho's scale) is boron carbide. On the mho's scale, diamond rates as ten. Diamond abrasive is becoming more desirable as the wages of technicians continue to rise. The diamond grades are quite indispensable in the polishing of quench-hardened steels and alloys that fall in the 60 R_c hardness and above. If diamond abrasives were less vulnerable to contamination, they would be more in demand since they reduce polishing time. Diamond abrasives are not recommended in the finishing stages for softer metals.

Hand grinding

Hand grinding is usually considered as the procedure that will yield the highest quality. Belt, disc and surface grinding with power equipment will reduce the time spent in the initial stages. Surface grinding attachments also may be used to create automatic operations.

Hand Polishing

Polishing starts the next step in the preparation of metallographic specimen. Technicians should keep in mind that polishing may be divided into two stages. The first is rough polishing, which is used to remove the deep scratches from grinding. The second is final polishing, which removes scratches from rough polishing. The objective during this sequence is to remove the disturbed metal and leave the surface optically flat. One of the most important factors to remember during this stage is that contamination destroys quality. This must be stressed at all times in the metallographic laboratory.

Standard polishing equipment consists of a horizontal rotating wheel covered with a specific type of cloth. The rotating wheel is wet with distilled water and loaded with an abrasive grit. The abrasive size is regulated by the type of metal and whether the polishing stage is beginning or finishing. The better quality finish is produced by holding the specimen flat against the wheel and rotating it slowly in a direction opposite to the rotation of the wheel. Rotation of the specimen like this during polishing will help to avoid directional polishing or comet tails, as shown in Fig 3.4.

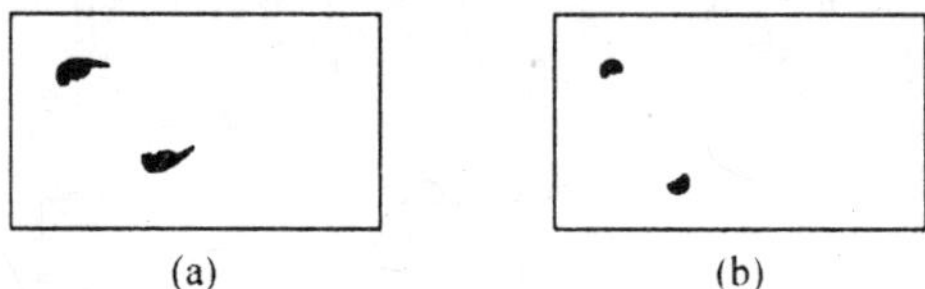

Fig. 3.4. Metallographic polishing causing surface metal removal from one side of heterogenous structures. (a) Comet tail type profile results from directional polishing; and (b) the same surface after polishing by rotating the specimen during the sequence.

Some metallographers prefer to hold the specimen at the three o'clock position on the polishing wheel while turning the specimen a few degrees as it polishes.

Control factors during polishing stage

General control factors during the polishing stage are:

1. Amount of abrasive applied.
2. Regulation of amount of water on the cloth.
3. Surface speed or revolutions per minute of the wheel.
4. Force applied to the specimen to make contact with the polishing wheel.

Amount of abrasive

The general tendency during polishing is to add too much abrasive. The highest quality surface finish will be produced after fresh abrasive has had time to break up into consistent smaller-sized particles. When the original abrasives are graded, they are not absolutely the same size. With continued use the particles strike the metal specimen and are fragmented to smaller sizes. This may occur over a short period of time, for example, 60 to 90 seconds when polishing the harder metals. Of course the amount of force used on the specimen surface regulates fragmentation. If polishing wheels are overloaded with abrasive and left standing, oxidation may harden the residue and scratch the next specimen polished. If this occurs, the wheel should be flooded with demineralised or distilled water while turning slowly. After several seconds of applying water to soften the build-up, the wheel may be scraped with a plastic or metal disk. Finish the operation by adding a generous supply of water and turning the wheel at maximum speed for a few seconds.

Regulation of water

Regulation of water is important if quality surfaces are to be developed. It is simple to determine how much water to add. The correct amount is indicated by an inspection of the specimen surface. Hold the specimen on the wheel firmly for a few seconds then lift it and observe the evaporation time. The moisture should disappear in about 10 to 15 seconds. Too much moisture will cause pitting around inclusions and soft phases. If the cloth is too dry the surface metal will be burnished or cold worked. Burhishing creates a beautiful jewellry-type luster and at the same time an artificial surface. This might please a jeweler but not a metallographer.

Polishing wheel speed

The polishing wheel speed should be selected by considering the hardness of the specimen. The softer the specimen the slower the wheel speed. The standard 8 inch diameter wheel may turn up to 1100 rpm. Some are variable speed and others are two speed (1100 and 550 rpm), (Table 3.3).

Table 3.3. Wheel speed for metallographic polishing.

Metal	*Revolutions per minute*	
	Rough polishing	*Final polishing*
Soft steel	550	550 and slower
Hard steel	1100	550
Copper alloys	550	550 and slower
Aluminium alloys	550	550 and slower

Specimen-to-wheel force

Specimen-to-wheel force is another factor that will help contribute to a quality surface. It should be recognised that there is no one single force that will be correct. Sometimes the technician uses excessive force to fragment the newly applied abrasives and then a reduced force. Other times a light force at the final polishing sequence may permit some of the particles to roll between the cloth and the wheel. This could produce a maze of scratches and general confusion as to how they appeared on the surface. Some metallographers use a heavy force (approximately 5 lb) at the beginning of the polishing cycle and then gradually reduce the load to about 1 lb at the finish of the cycle. Force on the specimen has to be learned by experience and balanced by size of the mount, surface area of the specimen, hardness of the specimen, amount of moisture on the wheel and type of abrasive.

Polishing cloths

A careful selection of polishing cloths may help to reduce the cost of polishing while improving the quality of the job. Basically the polishing cloths should be selected because of their longevity or service life and their ability to load and carry abrasive and remain reasonably free of contamination that may cause scratches on the surface of the specimen. There are many different types of cloths, but a few are easy to select. Generally the rough-polishing sequence can use a cloth with a short nap. This could be a canvas or nylon surface. The reason for the short nap is to permit greater specimen surface contact to increase cutting action. Whatever type cloth is selected, it should be stretched tight over the wheel. To facilitate the process of changing polishing cloths, some have pressure-sensitive adhesive backing. In any event the bare wheel surface has to be cleaned during every change of cloth.

Intermediate stages of polishing could be accomplished with a felt or billiard type of cloth. These cloths are also used for final polishing with higher hardness specimen (R_c35 and harder). (Table 3.4).

Final polishing is generally accomplished by a low-napped cloth and fine alumina. The harder metals such as steel (R_c20 and harder) can be finished with 1.0 micron (μ) alumina.

The softer non-ferrous metals such as aluminium, copper and lead alloys need to be finished with 0.05 μ alumina. The technician will need to develop a polishing sequence regulated by:

1. Amount of magnification.
2. Type of metal.
3. Time factor.
4. Cost factor.

5. Susceptibility to cloth contamination.

Many questions arise regarding the type of abrasive that is best. There is no short answer for this. (Table 3.5 for additional information).

Table 3.4. A listing of typical polishing cloths.

Type of cloth	*Type of operation recommended*
Canvas	For general rough polishing
Red felt	For rough polishing on steel
Billiard cloth	For rough polishing on steel
Nylon (napless cloth)	Use with diamond abrasives for maintaining flatness and preventing relief
Synthetic rayon (fibres are bonded to cotton backing)	For final polishing with diamond and aluminium oxide abrasives
Wood broadcloth (short pile)	For final polishing with aluminium oxide
Synthetic velvet	For final polishing of soft metals

Table 3.5. Types of abrasives for metallographic polishing.

Type	*Stage of polishing*	*Metal to be polished*	*Cost*	*Freedom from contamination*	*Maximum recommended polishing time each stage*
Aluminium oxide, Al_2O_3	All	All softer metals	Low	Yes	2 minutes
Magnesium oxide, MgO	Final	Aluminium and magnesium	Low	No	2 minutes
Chromic oxide, Cr_2O_3	All	Steel and cast iron	Medium	No	3 minutes
Iron oxide, Fe_2O_3	Rough and med	Steel and cast iron	Low	Yes	3 minutes
Diamond	Rough Med Final	Boron carbide Sintered tungsten Carbide-steel, cast iron	High	No	2 minutes

Remember, the metallographer should analyse each polishing sequence to look for signs that will indicate completion. Continuation of work after the particular sequence is finished is a waste of time and money. The final polishing step should be considered completed when a 100X magnification reveals a highly reflective surface and gray lines. A deep scratch will leave a wide black line.

Manual polishing operations have the potential to produce the greatest level of quality attainable. There are certain jobs that should be done in an automatic or semi-automatic mode. This type of polishing is desirable when a large quantity of specimens make it appropriate. To meet this demand, other polishing procedures were developed. The vibropolishing method will produce a quality finish on several specimens at the same time. This process uses a cloth-covered wheel that vibrates in a horizontal fashion beneath several specimens. An extra weight presses the specimens down on the vibrating surface. The abrasive is usually a 0.05 μ alumina with enough water to make a thin slurry. Polishing time may require several minutes to hours.

The electrolytic method may be as simple as a beaker containing an electrolyte, with direct current connected to a metal cathode (negative) and to the specimen (positive). Anodic dissolution results in metal removal from the specimen surface. Electropolishing does not disturb remaining surface atoms. This makes electropolishing ideally suited for some specimens, especially the soft metals or those that are easily scratched. The disadvantages are preferential attack in multiphase alloys and chemical attack on non-metallic inclusions, plus the possible hazards involved.

Actually the application of chemicals is a process of polishing or removing surface atoms without disturbing surface structure. Sometimes the only possibility of removing scratches is to mechanically polish and then chemically polish in several repeated treatments.

Etching

The objective of etching is to accent the structural characteristics of a specimen. Examination of most specimens in the polished condition reveals a mirror-like surface. Etching by swabbing or immersion will stain and preferentially etch away the surface to reveal the presence of phases unnoticed before. Etching by swabbing is usually done at room temperature while some etchants for immersion may be heated to as high as 180° F.

Before being etched a specimen should be inspected for surface scratches. If the polishing operation was satisfactory, only minor scratches will be seen. Then the surface should be searched for structural integrity, which will include porosity, inclusions and cracks. On some non-ferrous alloys grain boundaries will show more prominently before etching (Table 3.6).

Table 3.6. Metallographic etchants.

Metal	*Objective in etching*	*Chemicals required*	*Application*	*Time*	*Temperature*
Aluminium alloys	General	0.5% HF & H_2O	Swab	15 sec.	RT
Aluminium alloys	Macro	10 g NaOH, 90 ml H_2O	Immerse	5 sec.	160° F
Copper alloys	Accent coring	NH_4OH, H_2O_2, H_2O (equal parts) Fresh H_2O_2	Swab	5–30 sec.	RT
Copper alloys	Darken beta brass	5 g, 2.5 ml HCl, 95 ml Alcohol	Swab	3 sec. Repeated to required results	RT
Lead alloys	Details in photography	Acetic acid 3 parts, HNO_3 4 parts, H_2O 16 parts	Immerse	4–30 min.	40-42° C
Lead alloys	Details in lead-antimony	Acetic acid 3 parts, H_2O_2 (30%) 1 part	Swab	5–15 sec.	RT

Contd...

Metal	*Objective in etching*	*Chemicals required*	*Application*	*Time*	*Temperature*
Tin alloys	Details in Babbitt	10 g. $FeCl_3$, 2 ml. HCl, 95 ml H_2O	Immerse	½–5 min.	RT
Zinc	Macro details alloys	50% HCl and H_2O accent grain	Swab	Few sec.	RT
Steel	Accent Heat	2–5% HNO_3 and treating	Swab Ethyl alcohol	5–60 sec.	RT

It is good practice to examine all metallographic specimens in the final stages of polishing before etching. Inspection at this time may reveal some structural characteristics that will show up better than after etching. For example, in a heat-treated steel tiny surface cracks will be more prominent when contrasted on a mirror-like background.

The cracks will be seen as dark irregular lines. Also some types of inclusions may be revealed in greater contrast, like free graphite in cast iron. Some non-ferrous alloys have distinct grain boundaries before etching, especially if the view can be made with polarised light.

Another technique that may be used, is the etch-polish procedure. This requires several sequences of light etching and polishing, one after another. The etching times should be evaluated to determine the optimum time to produce desired details at the required magnification. Repeated etching and polishing is continued until at least two etch-polish sequences show no increase in quality. The etch-polish procedure tends to produce less disturbed metal than most regular polishing operations.

Photomicrography

The basic tool in photomicrography is the metallurgical microscope. Metallurgical microscopes may range from a system of simple low-power objective and eyepiece lenses with an incandescent light reflecting on the specimen to a very complex array. Low magnification (10X to 25X) can sometimes give more structural information. This range can be obtained by removing the eyepiece lens and focusing the image on a ground glass. Each type has its own advantages and limitations. Fig. 3.5 provides more information. More complex systems require the use of many components before the image is formed. The most important part of a microscope is the objective lens. Some photography is done using only the objective lens to magnify the structure. When ordinary light is passed through a curved lens, the rays are broken up into specific colours. These colours bend or refract an amount equal to their respective wavelengths (Fig. 3.6).

Magnification

Light passing through a particular curved lens causes the image to become larger than the real object. The selection of proper magnification is an important factor in the critical analysis of a microstructure. Total magnification is determined by power of the objective lens, power of the eyepiece, distance between the objective and the eye lens (called tube length) and distance from eyepiece to focusing screen or film. To calculate the magnification factor for a photomicrograph, the following formula may be used:

$$M_{\dagger} = \frac{Dm}{250}$$

where,

M_t = magnification total

D = projection distance mm

m = magnification power of objective and eyepiece at a projection distance of 250 mm

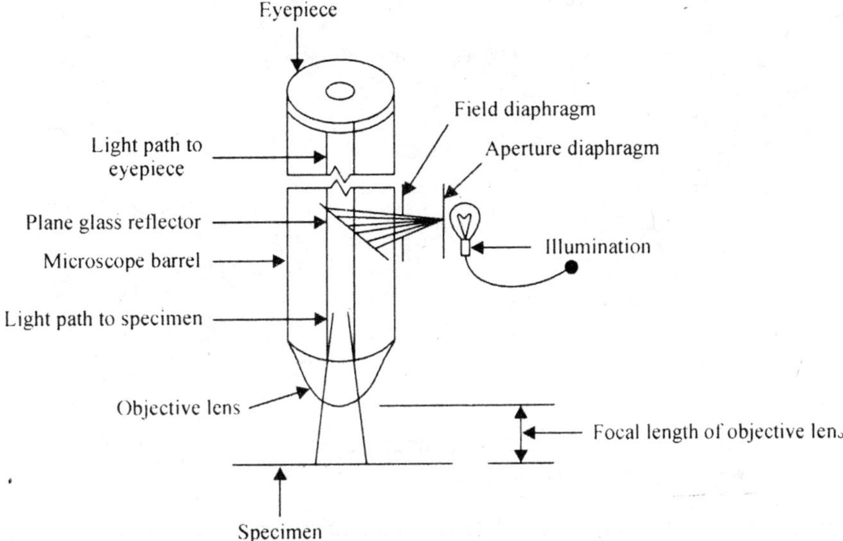

Fig. 3.5. Typical lens system for low-power analysis. Usual low-power magnifications range for 25X to 100X.

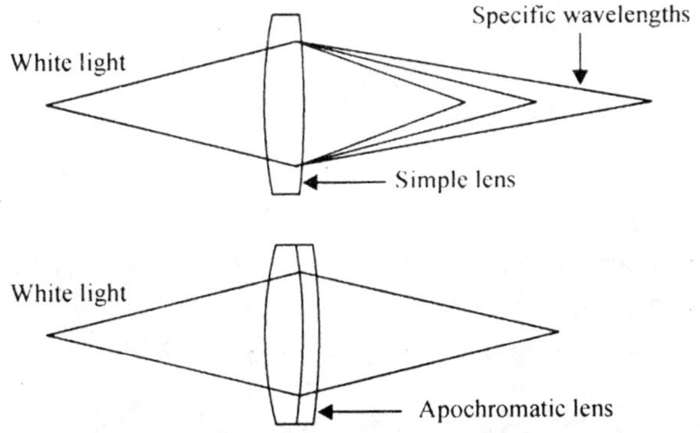

Fig. 3.6. Schematic diagram illustrating the phenomenon of longitudinal chromatic aberration.

The major reason for using high magnification is to resolve fine details in the microstructure. Resolving power is directly governed by the amount of light entering the objective. The higher the magnifying power of a lens, the closer it is brought to the surface of the specimen. Fig. 3.7 shows the lens to specimen distance with different magnifications. The light-gathering power of a lens may be expressed by the formula:

$$NA = n \times \sin u$$

where,

NA = numerical aperture

n = index of refraction

u = angle of light aperture ÷ 2

In using the angle shown in Fig. 3.7, the numerical aperture of the lens would be (in air, n = 1.00)

$$1.00 \times \sin 25° = 1.00 \times 0.423 = 0.42$$

If cedar oil were applied to the lens so that air would be excluded, the numerical aperture would be (in cedar oil, n = 1.50)

$$1.50 \times \sin 25° = 1.50 \times 0.423 = 0.63$$

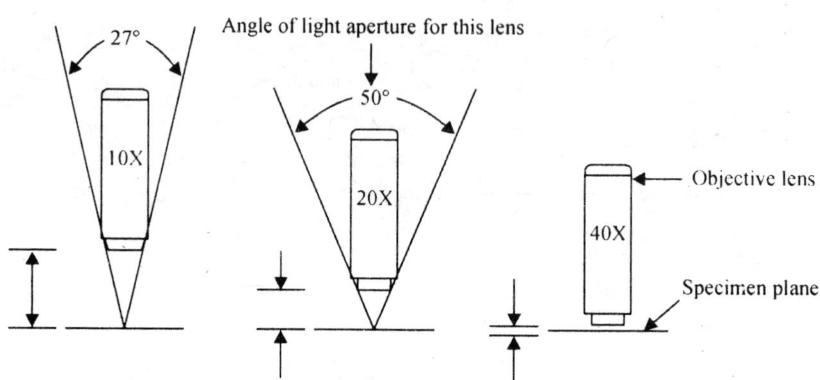

Fig. 3.7. Lens to specimen distance with different magnifications. Also notice how the angle of light aperture decreases with the lower magnification lenses.

The ability of a lens to produce sharp details is dependent upon two factors: numerical aperture and wavelength of the light illuminating the material. Thus the ability of a lens to produce fine detail (true power) can be expressed by the formula:

$$\text{Lens power} = \frac{\lambda}{\text{NA} \times 2}$$

where

λ = wavelength of light

NA = numerical aperture

Highest quality photomicrography with the least amount of opportunity for error is accomplished by using minimum acceptable magnification. If higher than normal magnification or photomicrographs is required, it can be obtained, but plan on spending considerable time to get consistent quality. Important considerations are:

1. Eyepiece corrected for field curvature and colour rendition. Maximum recommended 10X.
2. Objective lens with oil immersion. Minimum suggested numerical aperture of 1.25.
3. Short wavelength of light for illumination source. Example, green filter for incandescent light.

Quality photomicrography requires knowledge of how to maintain the optical system. The following ideas might be helpful. First the optical room should be kept as clean as possible. If contamination does appear on the lens, attempt to remove it by an air syringe or a brush-syringe combination. If more thorough cleaning is required after the air syringe treatment, dampen a lens tissue with lens cleaner and rub lightly. Finish with a dry lens tissue. If a lens has been used with oil immersion, it should be cleaned when through or as soon as possible. First wipe excess oil with the lens tissue and follow with a dampened lens tissue. Immediately wipe dry while the lens still has some vapour from the damp tissue. Use no liquids other than lens cleaner on any optics.

Analysis of Photomicrographs

Interpreting photomicrographs is similar to studying fossils as a means of reconstructing historical happenings. Micrographs reflect past events such as segregation during cooling, locating areas where stress has been applied, exposure to temperature, time at temperature, rate of temperature change and exposure to a controlled atmosphere.

During the preparation stages, these evaluations may be helpful in indicating progress of the sequence. A metallographer should learn to look for signs that indicate progress. Time is not a good measure of when a certain stage of work has been completed. Every time a specimen is prepared, the technician should study the surface condition to determine how much progress is being made. For example, during sectioning the metal may show slightly darkened lines on the surface of the cut, as seen in Fig. 3.8. This is an indication (on steels) that the part has been heated to a temperature that will result in rapid oxidation.

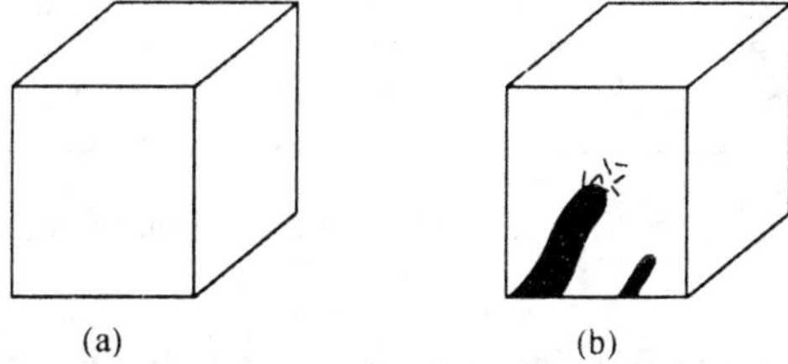

Fig. 3.8. Surfaces of a steel specimen that has been heated during cutting. (a) This is what a quality metallographic cut should look like (no evidence of heated or disturbed metal); and (b) this cut reveals oxidation reaction from overheating.

High temperature during the cut-off operation may be caused by feeding too fast, feeding too slow, wrong cutting wheel or lack of coolant. An extremely fast feed rate causes friction build-up faster than the coolant can dissipate the heat. If the cut-off wheel dwells on the surface with little or no pressure, it will become dull. Further attempts to cut will start heat build-up in the metal surfaces. If the operator suspects a dull wheel, then a sharpening stone could be used as shown in Fig. 3.9.

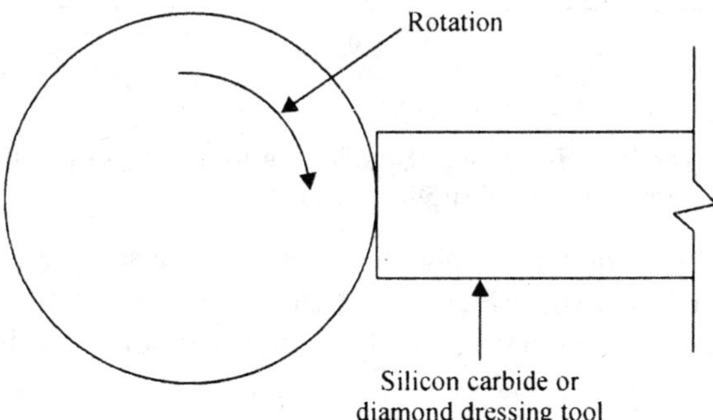

Fig. 3.9. A dull cutting wheel is being sharpened by removing the outer surface of the wheel with a dressing tool. Cutting wheels will develop a glaze unless properly matched to the type and condition of metal.

Proper feed rate may be determined after a few cuts on different materials. The technician should tune his ears to the correct sound on each specific type of metal. In regards to coolant, a generous supply should be used since it is recirculated. The more coolant supplied, the less opportunity there is for heating the metal during cut-off.

Changing the metals structure by heat transformation or cold working is a common problem during sectioning. If the change is inconsistent with the quality requirement, then a slow speed diamond cutting wheel may be the next alternative, as shown in Fig. 3.10.

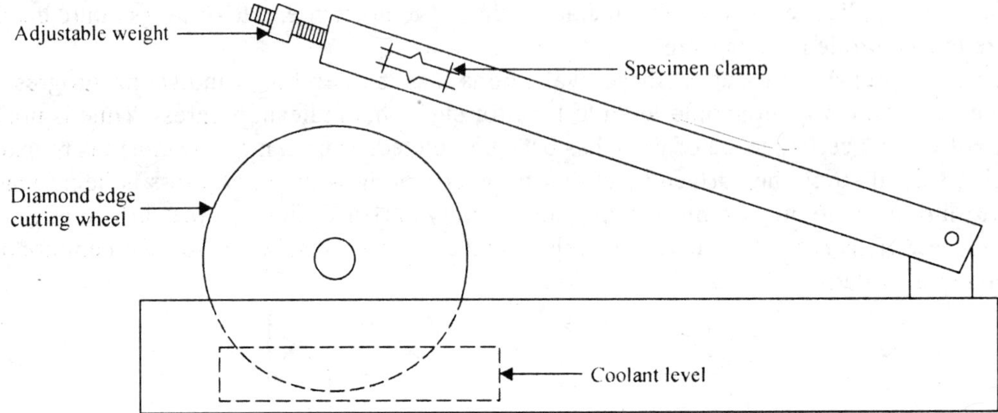

Fig. 3.10. A slow-speed cut-off machine. This type of machine requires only ounces of force on the diamond cutting wheel.

After the sectioning sequence is finished, the technician should analyse the surface finish. Deep scratches should be removed with a power abrasive belt or disc with coolant. Abrasive patterns during hand grinding should be studied to determine when the sequence is finished. For example, when going from 240 grit to 320 grit abrasive, the scratch pattern should be parallel and of consistent darkness, as seen in Fig. 3.11.

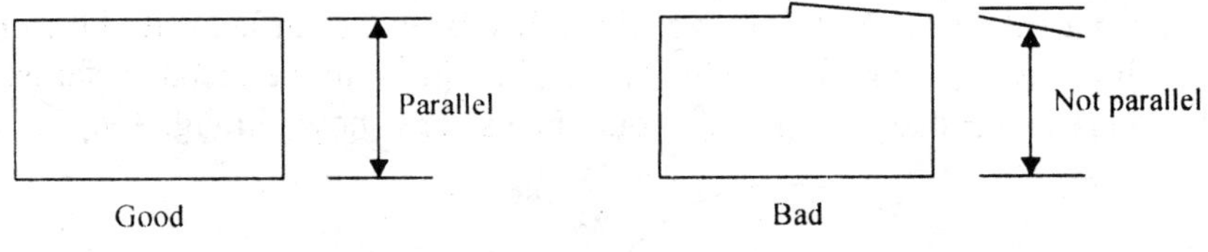

Fig. 3.11. A good and bad example of grinding. Specimens that are not parallel (top and bottom surfaces) create an inconvenience in focusing on the upright microscope stage.

Since the major objective in grinding is to obtain an optically flat surface, frequent inspections should be made. Also it is important that the two surfaces of the mount remain parallel. If the metal appears to have more than one plane surface, the condition should be corrected before polishing as shown in Fig. 3.12.

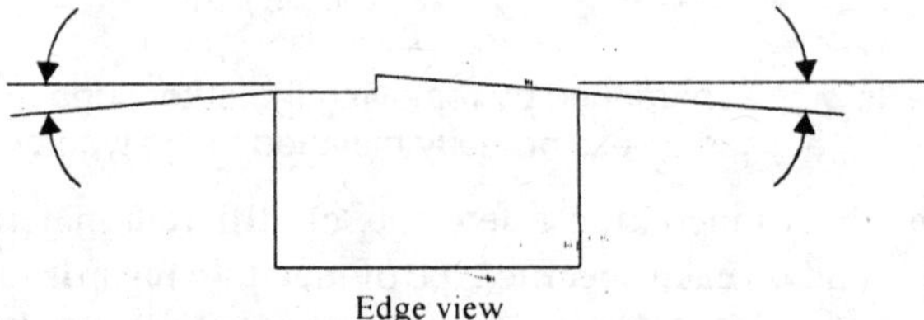

Fig. 3.12. Multisurfaces (faceted) that occured when specimen was being prepared during grinding stages.

If a specimen mounted in plastic has a severely faceted surface, it may be advisable to fracture the plastic, remove the metal, level it on a power driven wheel and re-mount it in plastic. In the inspection stage, after grinding and polishing, the specimen should possess a lustre except for inclusions. Some typical inclusions are gas cavities, non-metallic formations and graphite Fig. 3.13. To the inexperienced observer a surface such as gray cast iron appears to have tiny pits or surface roughness. Many people have spent extra hours of polishing time trying to remove this natural characteristic. Again a word of caution—for each stage of work, look for indicators that tell when the operation is complete.

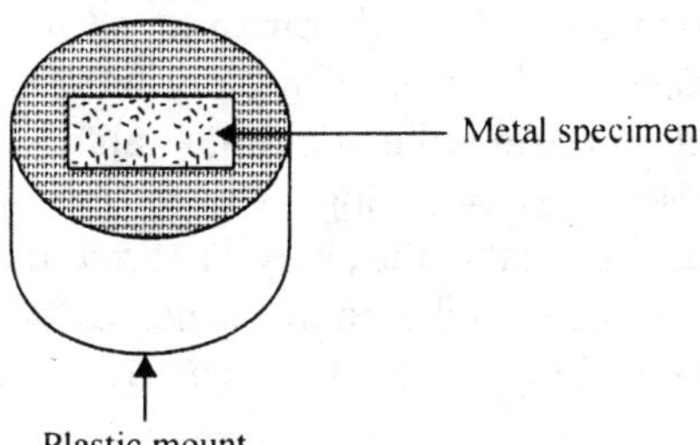

Fig. 3.13. A mirror-like surface appears on a gray cast iron surface after polishing except for the graphite flakes. Graphite is brittle so it may become dislodged. Also, it will not polish to a lustre.

Segregation

During normal cooling when a metal changes from a liquid to a solid, the alloying elements tend to separate. This is called segregation. In the process of solidification atoms from the liquid collect in a very orderly arrangement around a nucleus or core. The first atoms to collect around the core or nucleus of a crystal make up the constituents that have the highest melting point. This is not related to the melting temperature of the individual elements that make up the alloy. For example, iron has a melting temperature of about 2800°F, carbon has a melting temperature of about 6700°F and the greatest amount of carbon that can be dissolved in a solid solution is about 2 per cent which has a melting temperature of about 2066°F. It is not difficult to visualise the first constituent to solidify with an iron-carbon alloy, as shown in Fig. 3.14.

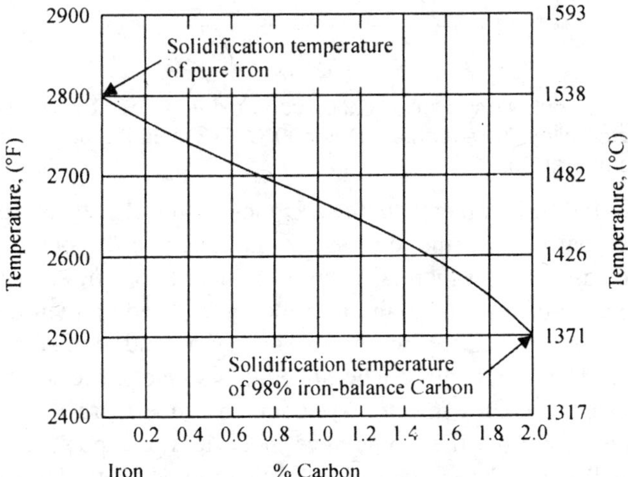

Fig. 3.14. Diagram to show the result of alloying a high melting temperture element (carbon melts at 6700°F) with iron that melts at 2800°F).

Later during the process of solidification, other combinations of elements are attracted to the solid front to complete the crystal formation. The process of selective solidification of multiple elements can be seen on the photomicrograph. The composition of a single crystal usually changes from the central section to the grain boundary. Changes in composition with a magnitude of 0.01 can be detected by viewing a photomicrograph that is representative of this condition. Look for the grain boundary to appear as a different shade of darkness than the central region.

As the temperature gradients of the constituents that make up a single crystal become farther apart, the final solid consists of a cored structure. The higher melting alloy is surrounded by the lower melting alloy. The lower melting alloy makes up the shell of the crystal. The faster the rate of cooling during solidification, the greater the effects. Consider the aluminium alloy (90 per cent aluminium plus Si, Cu and other elements) which is representative of coring.

Knowledge of phase diagrams is essential in the analysis of microstructures. From the phase diagram it is possible to associate names of phases and amount or percentage of specific phases (by volume). Consider the alloy aluminium (90 per cent) silicon (balance), as shown in Fig. 3.15.

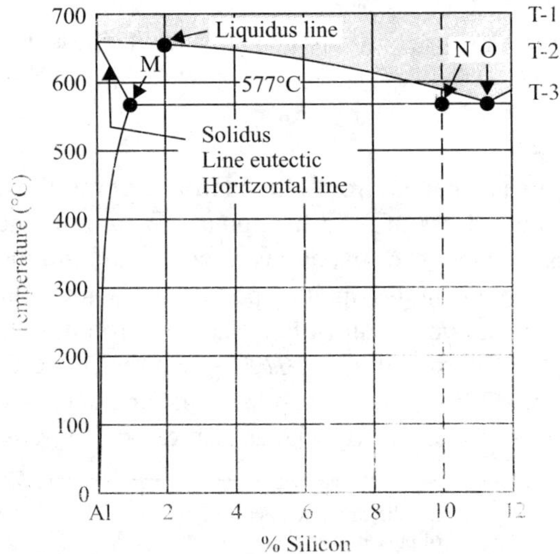

Fig. 3.15. Segregation of elements during solidification can be determined by a phase diagram. Dotted line at 10.0% silicon is the selected alloy. Temperture T-1 is 700°C, T-2 is 600°C and T-3 is 580°C. Line T-3 is 3°C above the solidification temperature.

At 700°C the alloy would be completely melted. Upon cooling slowly, the alloy would approach the intersection where the 10 per cent Si line touches the liquidus line. If cooling below the liquidus line continued, atoms would lose enough thermal energy to change to a new phase called a solid. In the liquid phase atoms have been given enough thermal energy to speed up their movement and establish a greater distance between atoms. At the temperature T-l any combination of aluminium and silicon shown in the graph will be liquid. The liquid will be a homogeneous single phase constituent. Moving down the dashed line, which represents the alloy selected, we come to temperature T-2. The liquid is still a single-phase, homogeneous constituent but the atoms have much less energy. At some temperature between T-2 and T-3 the metal will begin the process of solidification. With the process of solidification will come a release of energy and a decrease in volume. Now there exists a two-phase structure, part liquid and part solid and the amount of each can be determined if a temperature is specified.

A good example of how the phase diagram could help a metallographer analyse the microstructure of an aluminium-10 per cent silicon alloy is to estimate the per cent that will solidify before the eutectic temperature is reached. With a quick glance at the Al-Si phase diagram, it is possible to see that most of the solidified alloy will consist of eutectic composition. The proper procedure is to estimate the distance from the eutectic composition to the composition of the selected alloy. Investigate the following examples to see how this works:

1. If the selected alloy were 11.6 per cent silicon, the solidified phase would be 100 per cent eutectic composition.
2. If the selected alloy were about 5.8 per cent silicon, the solidified phase would be about 50 per cent eutectic composition.
3. If the selected alloy were between 5.8 and 11.6 per cent silicon, the solidified phase would be between 50 and 100 per cent eutectic composition.

These are just round estimates. Let us write a procedure to do this in a more technical manner. Using the 10 per cent Si-balance aluminium alloy and at the temperature of T-3 (58°C), let us find the per cent of the whole structure that is solid. To determine the per cent solid that exists at any temperature is simple. Compare the part of the horizontal line that represents the solid to the whole line. This sounds complicated but it is really easy if the approach can be defined by a formula. For example, find the amount (by area) of solid that is shown in Fig. 3.15 at T-3. To solve this problem, use the following procedure:

1. Locate the composition and draw a vertical line through the diagram (vertical dashed line at 10 per cent silicon).
2. Locate the temperature and draw a horizontal line through the phase field.
3. Label these intersections.
   (a) Point M where the solidus line touches the horizontal line.
   (b) Point N where the horizontal line touches the composition line (composition line is dashed line).
   (c) Point O where the horizontal line touches the liquidus line.

Note: Solid portion is touching liquidus line; liquid portion is touching solidus line.

4. Formula: Per cent solid $= \dfrac{\text{Length solid} \times 100}{\text{Length whole}}$

5. Computation:

$$\text{Per cent Solid} = \frac{\text{Length solid} \times 100}{\text{Length whole}}$$

$$= \frac{O - N \times 100}{O - M}$$

$$= \frac{12.5 - 10 \times 100}{12.5 - 1.6} = \frac{2.5 \times 100}{10.9} = 23 \text{ per cent}$$

Key: To find values of M, N and O, drop a perpendicular to the base line and read per cent Si. M = 1.6, N = 10.0, and O = 12.5

Composition of the crystal can be determined by constructing a horizontal line at a specific temperature. To solve this problem use the following procedure:

1. Construct the horizontal line through the liquid + solid region.
2. Read the composition of solid by a perpendicular dropped from the intersection of the horizontal line and the solidus line.
3. Composition of the liquid last to solidify is determined by dropping a perpendicular from the intersection of the horizontal line and the liquidus line.

For example, find the composition of the nucleus or core of the first crystal to solidify (Fig. 3.16). Compare this to the composition of the grain boundary region of the last crystal to solidify in Fig. 3.16.

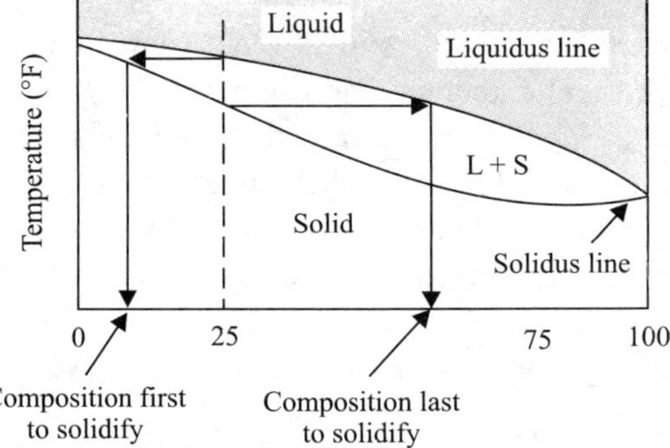

Fig. 3.16. Graphical construction to determine the composition of specific regions of a solidified metal.

Cast Structures

The microstructure of cast metals is determined by the composition and cooling conditions. Probably the most basic structure is that of a solidified pure metal or element. Crystalline geometry or grain shape and size of the solidified pure metal is controlled by the number of nuclei, growth rate and other factors. Further thermal exposure after solidification may also have an effect on the structure. The average casting may have a medium coarse grain because of the slow cooling rate. These grains will be composed of a series of dendritic formations. The term dendrite refers to a particular growth direction with lateral formations. The term means tree like in that the original growth is thought of as the trunk of a tree and the lateral formations are the branches. A schematic illustration of dendrites is shown in Fig. 3.17.

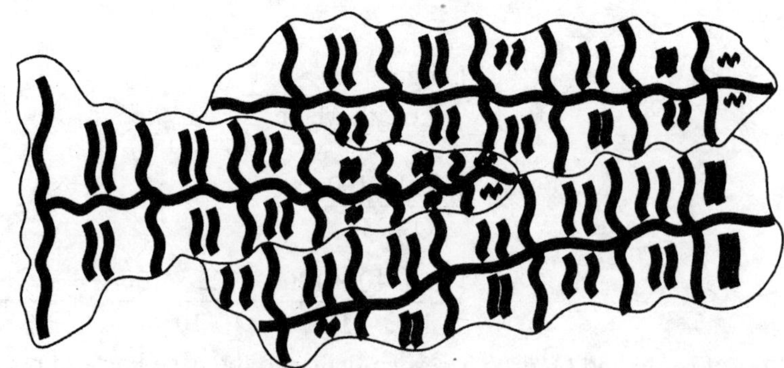

Fig. 3.17. A schematic illustration of a group of grains with dendritic formations.

The original dendritic formation may be affected by very rapid cooling rates. A dendritic formation may form and be re-melted by the sudden release of energy that is exchanged when liquid metals become solid. This energy release during solidification was discussed earlier and is called latent heat of fusion.

In a cast structure where the mould wall is cold, the grains or crystals tend to grow in directional pattern causing the dendrites to become long and narrow.

It is interesting to note that in polishing a pure metal that is supposed to be homogeneous, the abrasive action immediately brings out a relief pattern. The relief pattern is created because some crystals are softer (because of crystallographic orientation) and polish faster.

During etching the reagent will attack certain crystals to a greater extent. The areas that have more metal dissolved by the reagent become rougher and get darker. This phenomenon is attributed to the particular orientation of the dendrites or crystallographic axis. Also, certain patterns that appear within the crystals could be attributed to crystallographic formations.

As alloys solidify, the mechanism is different than that of pure metals. Alloys almost invariably segregate during solidification. This means that the chemistry of a cast ingot will vary. In the process of segregation high-melting mixtures or constituents will solidify first. This changes the percentage of alloys in the remaining liquid.

To demonstrate the process of segregation during solidification, let us select a medium carbon steel of about 0.4 per cent carbon. In the first stages of solidification, a speck of solid will appear. (Refer to the phase diagram for iron-carbon alloys to determine the composition). Use the procedure given in Fig. 3.16 and apply it to the phase diagram for iron-carbon. Then check with Fig. 3.18 to see if the values agree.

Transformation of Solidified Alloys

In tracing the cooling process of the 0.4 per cent carbon steel, start at a temperature immediately below final solidification range. The alloy is shown on the iron-carbon phase diagram as a single phase (austenite). Austenite is described as a solid solution of decomposed carbon in ferrite. The term iron is acceptable only as a pure element. Ferrite can dissolve a small amount of carbon and still retain its identity as ferrite.

A phase diagram is constructed on the basis of equilibrium conditions. This means that temperature represents phases after an infinitely long period of time. The 0.4 per cent carbon alloy would eventually become homogeneous just below the solidification temperature. Let us decide that by the time it cools through the austenitic range it reaches a zone called upper critical temperature and has become a single phase austenite.

Now, with further cooling, the austenite becomes unstable. The first transformation product is ferrite, so the beginning of the first crystal is rich in iron. As more ferrite transforms from austenite, the remaining austenite attains a higher ratio of carbon. When the concentration of carbon reaches 0.8 per cent and the temperature drops to 1333°F, the remaining austenite will transform at one temperature. Carbon in the form of carbide precipitates, leaving the surrounding zone enriched in iron called ferrite. When austenite contains 0.8 per cent carbon, the transformed structure will be (with slow cooling) a layer-like formation called pearlite.

A crystal that has completed this transformation is shown in Fig. 3.19. It is interesting to know that austenite grains on cooling do not ordinarily change size. On the other hand, cooling rate does affect the spacing of the pearlite layers. Very slow temperature change in the metal, for example cooling of austenite in a furnace to room temperature, will produce pearlite with wide spacing. Cooling in air may cause the spacing of the pearlite to become so close that at low magnification the structure appears to be without the ferrite-iron carbide layers.

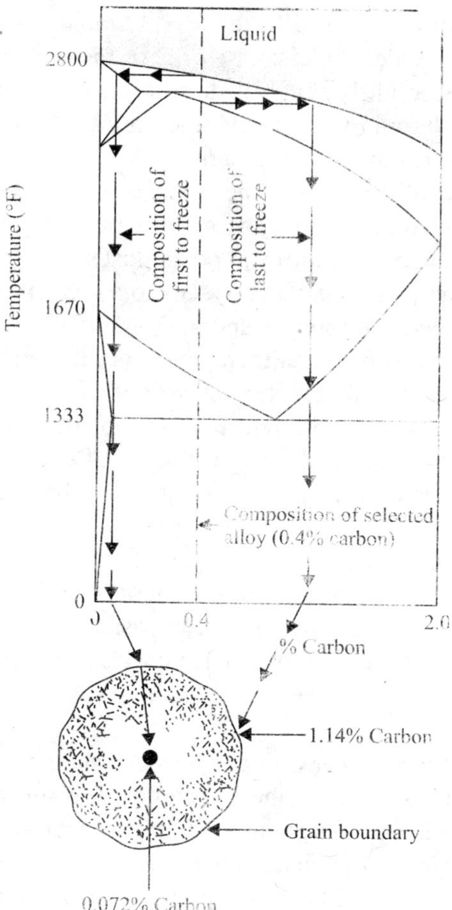

Fig. 3.18. Sketch of a crystal taken from a 0.4% carbon steel. Compare the composition of the first particles to solidify with those at the grain boundary.

Hot Working of Alloys

The hot working process requires a high enough stress to cause a metal to deform plastically while at high temperature. The temperature must be less than melting but above the re-crystallisation range. Hot working has been found to increase the quality of many of the cast products. Examples of improved quality are obtained by

1. Making castings more homogenous.
2. Reducing dendritic segregation.
3. Improving grain geometry.
4. Re-orienting inclusions to improve mechanical properties in the direction of rolling.

Cold Working of Alloys

Some metal products are finished by cold working to provide certain combinations of properties. There is a considerable change between the structure of a cold-worked metal and a hot-worked metal.

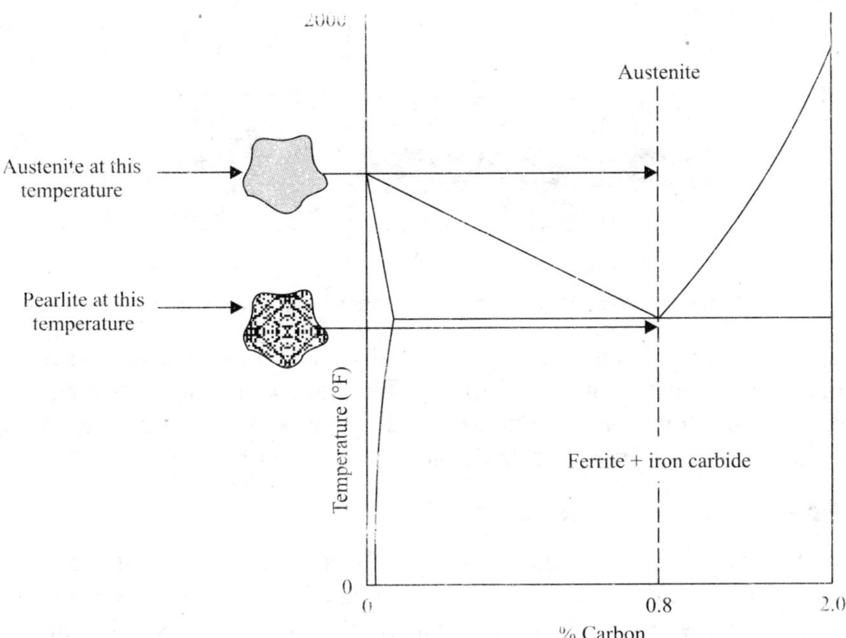

Fig. 3.19. A sketch of an iron-carbide crystal transforming to pearlite from slowly cooled austenite.

Annealing Processes

Annealing of metals is sometimes the only way to retain the required ductility so the product can be fabricated. Full annealing softens, improves ductility and on some alloys refines the grain size. Some annealing temperatures only require a mild form of stress relief. The metallographer can visually read the process of annealing by evaluating the microstructure.

During high temperature annealing of some steels, the surface chemistry may be changed. This can take place by a process of diffusion while the metal is solid, especially at higher temperatures.

Diffusion may take place at any temperature. The process of diffusion is the movement of atoms within a solid structure. The higher the temperature that surrounds a metal the faster the rate of diffusion. The environment adjacent to the surface can affect the microstructure. The best method in understanding this is to do it. Obtain a 2.0 inch long piece of ¼ inch square mild steel. This is low-carbon steel. Pack it in a steel box filled with bones or wood and cover it with a tight lid. Heat it in a furnace or with a torch to about 2000°F for 15 minutes and air cool. Cut the piece so a cross-section may be studied with a microscope

Continue with the carburised ¼ inch square section by reversing the atmosphere to that of an oxidising nature. Heat with an oxyacetylene torch adjusted to a highly oxidising flame to near the melting temperature. Heat for 5 minutes, air cool and prepare as before (Fig. 3.20).

Quench Hardening Process

The process of quench hardening imparts many mechanical properties to steel. In the as-quenched state the steel may be so brittle that the metal is almost doomed to failure. The metallographer should recognise quenched structures and those of varying degrees of stress relief called tempering.

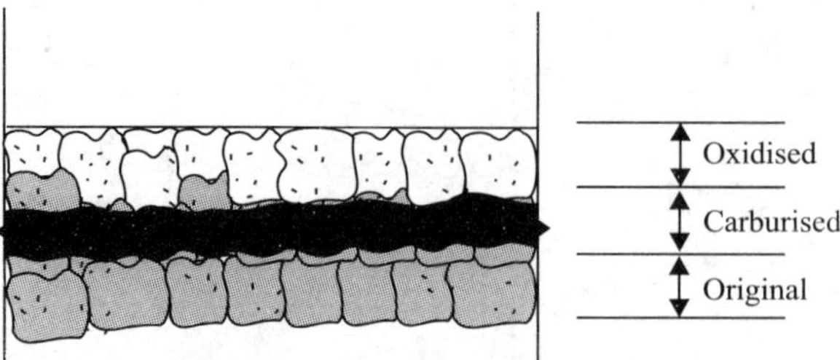

Fig. 3.20. Oxidised surface, carburised layer and original base metal structure as affected by environment.

If the cooling of austenite is rapid to below the transformation of pearlite, such as quenching in molten lead, a new phase called bainite may form. The minimum temperature for bainite transformation is above where martensite forms. Bainite is classified by two types according to the temperature at which it forms. These two types are called upper bainite and lower bainite.

Heat-Treated Non-Ferrous Microstructures

Non-ferrous metals are heat treated differently than steels. The first sequence consists of solution treatment, which is used to permit the formation of a solid solution. A solid solution is a homogeneous single crystalline phase where alloying elements are decomposed. This process is quite sluggish and usually takes about 2 hours or more. Then the metal is quenched to retain the saturated solid solution at room temperature. The second sequence consists of ageing the metal to precipitate fine particles from the solid solution. The first precipitates are sub-microscopic, but with higher temperatures and longer time the precipitate particles increase in size.

Metallurgical Microscope

Metallic specimens, in contrast to transparent specimens used in microscopic work on plant and animal life, have to be examined by reflected light. Consequently, the sub-stage condenser and mirror udereath the table are not required. A diagrammatic arrangement of a metallurgical microscope is shown in Fig. 3.21. The instrument consists of a three-limbed foot of sufficient size and shape to support the body. The limb, pivoted in order to obtain suitable inclinations of the body, carries at the upper portion a coarse adjustment, consisting of a rack and pinion device and a fine adjustment to facilitate final focusing of the object. These adjustments operate on the body tube which contains a draw-tube by means of which the distance between the eyepiece and the objective can be varied.

The draw-tube carrying the eyepiece (23.3 mm diameter) is engraved in millimetres which indicate the distance from the back of the objective to the end of the tube. The body tube carries a suitable vertical illuminator, on to the end of which the objective screws (36 threads per inch). In the more expensive microscopes a revolving nose-piece carrying three-objectives enables a quick change of objective.

The two most important optical parts of the microscope are the objective for resolving the structure of the metal and the ocular or eyepiece for enlarging the image formed by the objective.

The eyepiece is so named because it is nearest the eye and for bench work the Huygenian form is most common. It is made in a variety of powers, such as × 5, × 8, × 10, × 12.5, marked on the top, these designations meaning that the lenses magnify the image formed by the objective by these amounts. Other types known as compensating and projection eyepieces are employed in photo-micrography.

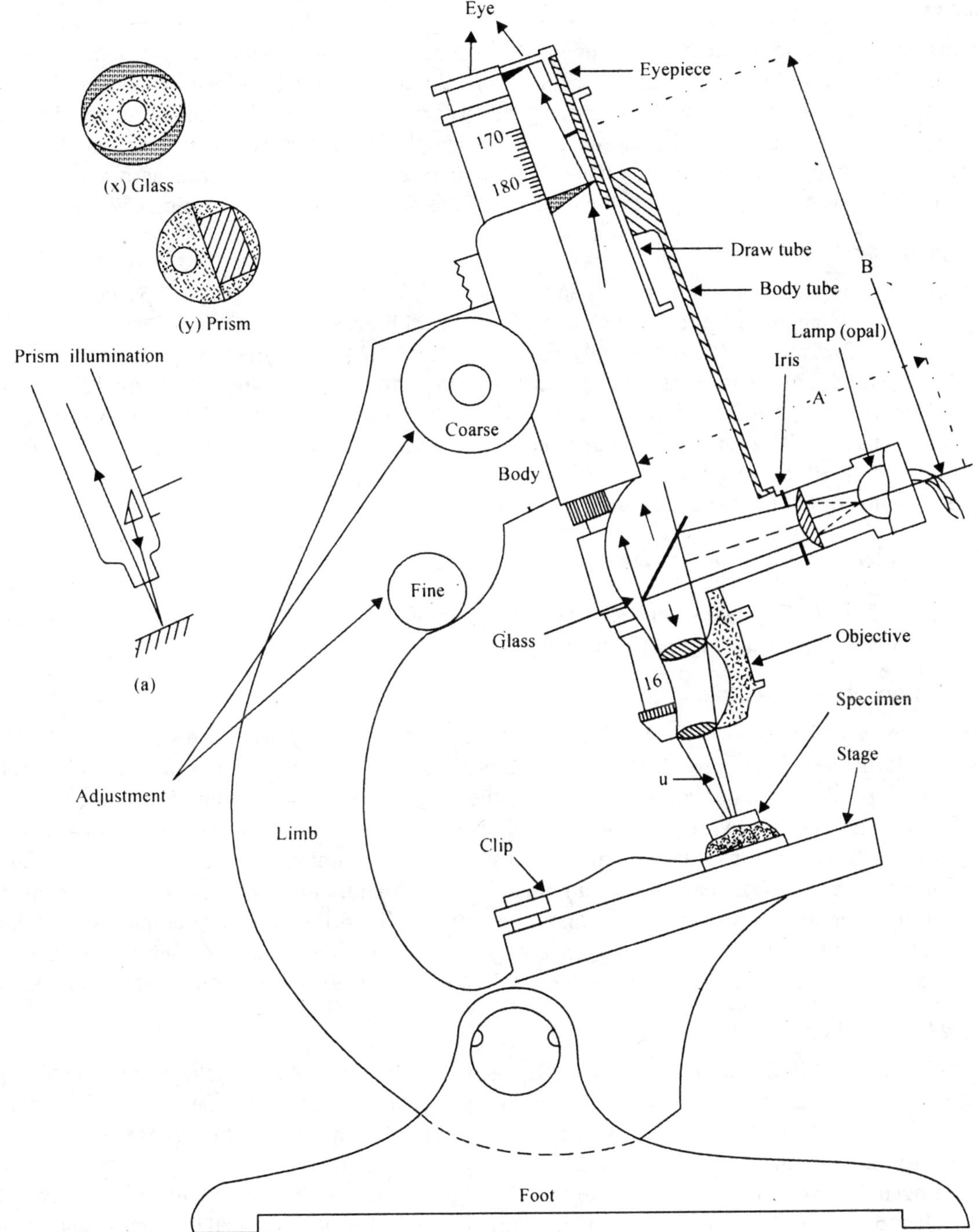

Fig. 3.21. Details of a metallurgical microscope.

Objective

The combination of lenses nearest the specimen is called the 'objective' and their quality determines the final result. High-class lenses are corrected for chromatic and spherical aberrations and produce an image free from colour (apochromatic) when used with a correct tube length. The magnification depends upon the focal length, such that the shorter the focal length the higher the magnification. The focal length of an objective is usually marked on the side and, in rarer cases, the numerical aperture (NA). The focal length, however, does not indicate the distance of the objective from the specimen.

Magnifying power

This is always expressed in linear diameters and offers no difficulty when the image is projected on to a screen, since the enlarged image of an accurately divided millimetre scale can be measured. In visual examination we find more difficulty, mainly due to the fact that the apparent image seen by the eye cannot be measured. This image is assumed to exist at 10 inch from the eye, which for normal sight is the least distance for distinct vision. If an object 1 mm diameter is placed at 10 inch from the eye it looks a certain size and a magnification of ×15 means that the microscope makes it appear as if it were an object 15 mm diameter at 10 inch distance.

A rough value of the magnification of a bench microscope is given by:

$$\text{Magnification} = \frac{D}{F} \times \text{power of eyepiece}$$

where, D = distance from back of objective to eyepiece
F = focal length of objective.

Resolving power of objective

Resolving power is the property by which an objective shows distinctly separated two small adjacent bands in the structure of an object. This is usually expressed as the number of lines per inch that can be separated. It depends on the numerical aperture and the wavelength of the light employed, e.g., a larger number of lines are resolved with blue than with yellow light. Resolution is particularly important when dealing with certain microconstituents of metals consisting of fine laminations, for with poor resolution these appear as one uniform area; whereas an objective with high numerical aperture (NA) reveals the duplex nature of the structure. Naturally the magnification must be sufficient to render the resolved lines visible (that is useful magnification) but a higher magnification obtained by extending the tube length or by a high-power eyepiece, does not increase the resolution and is really empty or useless magnification.

Vertical illumination

The simplest form of illuminator consists of a thin glass slip held in an adjustable mount situated just behind the objective as shown in Fig. 3.21. The incident light strikes the plate at 45° and is partially reflected down onto the specimen. The rays are returned by reflection, passing through the objective and glass plate to form the final image. In certain cases a right-angled prism is substituted for the glass plate, being arranged to cover only half the area of the tube Fig. 3.21a. The prism transmits much more light than the glass plate, but resolution is not so great. From Fig. 3.21 it will be noticed that an iris diaphragm is fitted to reduce glare, its opening should be the smallest at which it will work without encroaching on the field of view (i.e., just fills the back lens of the objective). For best results, also, critical illumination should be employed such that distances A and B are equal.

The photography of the specimen entails a few modifications in the arrangement of the equipment. A high-power illuminant such as an arc light is necessary together with a camera placed behind an eyepiece of the projection type. In most cases the microscope is arranged in a horizontal position and for metallography the 'inverted' type with the stage on top is favoured.

The procedure consists in focusing the image on a ground glass screen which is then replaced by a photographic plate contained in a holder. Exposure is made while using a definite colour screen (green, yellow), with subsequent development and printing.

For photomicrography the type of plate commonly used is that known as the Ortho process plate. These slow plates give a high contrast, but are sensitive only to yellow and blue light, consequently they are usually developed in red light. They consist essentially of a piece of flat glass coated with gelatine which has been impregnated with silver salts (chloride or bromide). These salts are affected by light in such a manner that, after exposure, the salt is rapidly reduced to metallic silver by the action of certain organic reagents, known as developers. The unexposed salts are only slowly reduced by the developer and these portions of the negative become transparent.

A typical developer consists of the following:

Metol	1	gram
Hydroquinone	4	gram organic reducers
Sodium sulphite (crystalline)	35	gram preservative
Sodium carbonate (crystalline)	35	gram alkali to enable reducers to work
Potassium bromide	0.8	gram re-strains action on unexposed areas
Water	500	cc

Dissolve in the above order and use diluted with equal quantity of water for plates, undiluted for papers.

After development, the unreduced silver salts are removed in order to render the plate permanent. The usual solvent or 'fixer' is sodium thiosulphate (hypo), 400 gram per litre of water. To prevent oxidation of the hypo 25 gram of potassium metabisulphite is added, i.e., 'acid hypo.' After fixing the plate is washed for 10 minutes and dried.

The treatment of gas-light printing papers is similar to that given to plates.

Phase Contrast Microscope

In the phase contrast microscope the specimen is illuminated by a hollow cone of light and the objective incorporates a glass phase plate having an annulus which advances by quarter of a wavelength the directly reflected light passing through it relative to the scattered light transmitted by the rest of the plate, so that high areas in the specimen appear light. In this way very small differences in level of the order of 50– 250 $\times 10^{-10}$ metres can be detected and this may also be helpful in interpreting electron micrographs.

Electron Microscope

The optical microscope reaches its limit of performance at about × 2000 diameter owing to the detail size being comparable with the wavelength of light. The use of ultra-violet light, with its shorter wavelength, extends the magnification limit somewhat, but the use of electrons with a wavelength some 40000 times smaller still vastly extends the useful range to a present commercial limit of × 75000. The 'optical' systems in the electron and optical microscope are analogous except that in the former the focusing is done with energised coils, Fig 3.22. The main limitation from the metallurgical aspect is the fact that the electron beam must pass through the object and this necessitates the use of a thin plastic or carbon replica of the etched surface. The anodic film on aluminium can be separated from the parent metal and be used to show fine detail. Inclusions can be fixed in a replica by dissolving the matrix after the replica has been fixed on the surface. With such an extraction replica electron diffraction photographs can be taken of particular particles to enable their identity to be deduced, in addition to the normal electron micrograph. Recent development involves the use of thin foils, formed by anodic dissolution of the sample, for direct observation of internal structure and imperfections and is complementary to replica technique which is limited to surface topography.

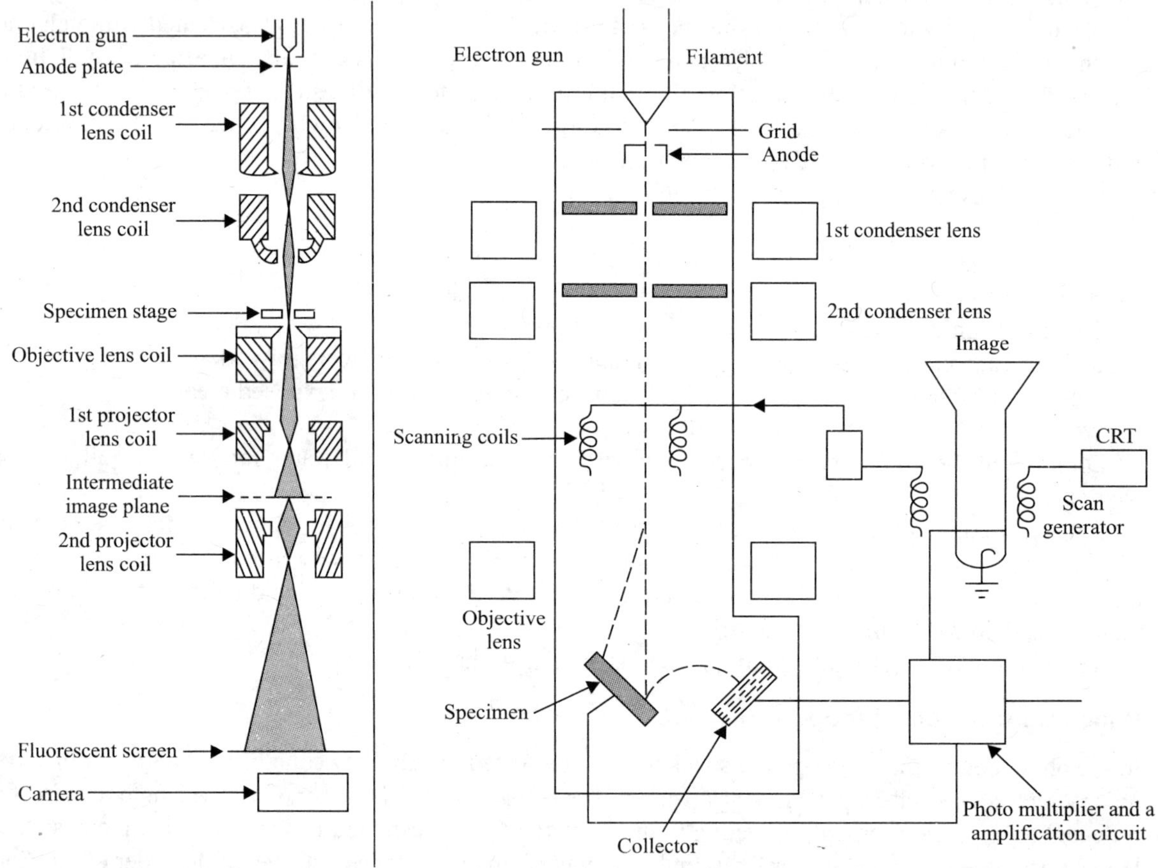

Fig. 3.22. Diagram of electron microscope.

An important extension of electron microscopy is the scanning microscope which is basically the electron-optic equivalent of the reflecting-light microscope. The most common use of the SEM is concerned with the examination of surfaces and magnifications of × 80,000 are possible, maintaining a remarkable depth of focus. A finely focused beam of electrons is moved by the scan coils in a square-raster over the specimen surface. Secondary electrons generated in the surface layers of the specimen are collected in a scintillator/detector system and the detected signal is displayed on the observation cathode-ray tube which is monitored in step with the scan-signal as shown in Fig. 3.22.

Chapter 4

Iron Carbon and Alloy System

INTRODUCTION

The study of the interrelation of phases in an alloy system at different temperatures and for different alloy compositions is of importance in understanding the characteristics of alloys. It would indeed be a difficult and cumbersome task to tabulate the interrelation and composition of co-existing phases at all temperatures in an alloy system. Therefore, this information is given in the form of a diagram. In most cases, these relations are given for equilibrium conditions. The diagram shows the composition of the phases that are in equilibrium at all temperatures and is referred to as an equilibrium or phase diagram. Some prefer to employ the name constitution diagram, but it is probably well to apply this latter term to diagrams that show the phase relations under special circumstances of quasi equilibrium.

An equilibrium diagram may be defined as a plot of the composition of phases as a function of temperature in any alloy system under equilibrium conditions.

A study of the iron-carbon system is valuable not only because it helps explain the properties of steels but also because it provides a means of understanding solid-state reactions in general.

The iron-carbon diagram shown in Fig. 4.1 is not a complete diagram in that it is plotted only for concentrations (in weight per cent) of less than 12 per cent carbon. It does, however, cover that part of the diagram of primary interest in the study of steels and cast irons. It is currently accepted, with minor changes, as the standard iron-carbon diagram. This version of the diagram is complicated because it shows the equilibrium relationships not only between iron and carbon (graphite) but also between iron and two iron-carbides; cementite (Fe_3C) and the Hagg carbide ($Fe_{2.2}C$). Included in this diagram are three nearly horizontal lines in the upper left-hand part of the diagram. These lines represent the boundaries between the liquid and gaseous phases at three pressures; 1.0, 0.1 and 0.01 atmospheres. They thus indicate the dependence of the boiling point temperature on composition at each of the indicated pressures. The two carbides are metastable. In other words, carbon or graphite is more stable than the carbides. However, the decomposition of the carbides does not occur easily and, from a practical point of view, one may consider that, in plain carbon steels, the breakdown of the carbides to form iron and graphite does not occur. Furthermore, when a steel is slowly cooled from the liquid to the solid state, cementite is normally the easiest to nucleate so that when carbon is precipitated from solid solutions of alpha (body-centred cubic) or gamma (face-centered cubic), the resulting precipitate is almost always cementite, if the temperature is above 350°C. It should be noted that once cementite has formed, it is very stable. As shown on the phase diagram, the Hägg carbide is actually the more stable carbide below about 350°C. This constituent, however, has been identified in purified iron specimens carburised below 350°C. Hägg carbides have been observed to transform to cementite when steels containing them were heated above 500°C. It is still not clear whether cementite transforms to the Hägg carbide in a reasonable period of time on cooling. In any event, it will be assumed that in the study of steels one may reasonably use the simplified phase diagram shown in Fig. 4.2.

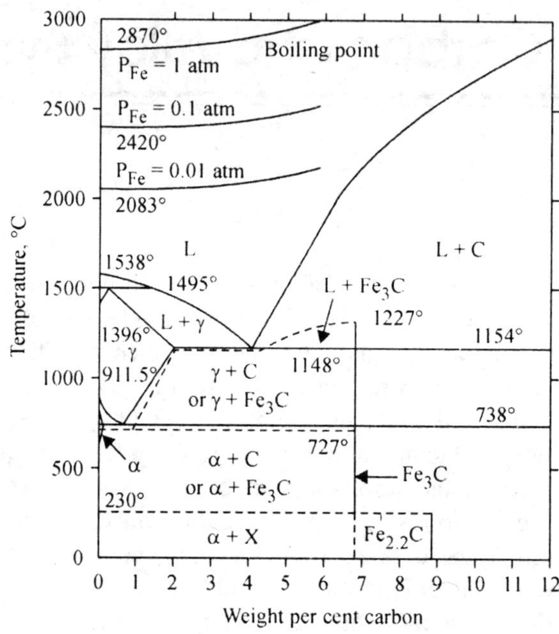

Fig. 4.1. The iron-carbon phase diagram.

The phase diagram in Fig. 4.2 is also not a complete diagram in that it is only plotted for concentrations (in weight per cent) less than 6.67 per cent carbon, the composition of Fe_3C, or cementite. Cementite is an intermetallic compound with negligible solubility limits and the iron-carbon diagram can be divided at its composition into two independent parts. That part of the diagram containing carbon concentrations higher than 6.67 per cent has little commercial significance and is usually ignored. The iron-carbon diagram in Fig. 4.2 is characterised by three invariant points: a peritectic point at 0.17 per cent C and 1495°C, an eutectic point at 4.32 per cent C and 1154°C and an eutectoid point at 0.77 per cent C and 727°C. The peritectic transformation occurs at very high temperatures and in steels of very low-carbon concentration (the upper left-hand corner of the phase diagram).

The peritectic transformation has only secondary effects on the structures of steels at room temperature. All compositions that, as they freeze, pass through the peritectic transformation region, enter the single-phase face-centred cubic field. Except for coring effects (segregation) caused by the complexities of the peritectic transformation, these alloys are equivalent to higher carbon compositions (above 0.53 per cent C) that freeze directly to the gamma or face-centered cubic phase. With the assumption that sufficient time is allowed as the alloys cool to permit diffusion to bring about a homogeneous solid solution, we may disregard the peritectic transformation in considerations of phase transformations that occur at lower temperatures.

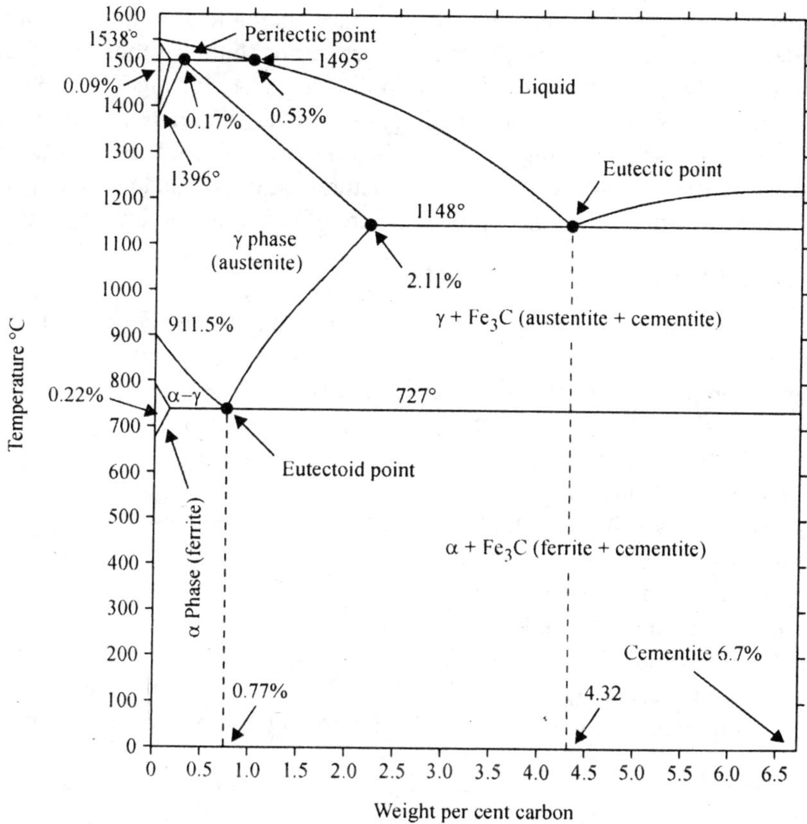

Fig. 4.2. The Fe-Fe_3C metastable phase diagram.

The face-centred cubic solid solution, or gamma phase, is given the identifying name austenite. A study of the phase diagram shows that all compositions containing less than 2.11 per cent C pass through the austenite region on cooling from the liquid state to room temperature. Alloys in this interval are arbitrarily classed as steels. Actually, most carbon steels contain less than 1 per cent carbon, with the greatest tonnages produced in the range 0.2 to 0.3 per cent C (structural steels used in buildings, bridges, ships, etc.). Only in very rare instances is steel used with more than 1 per cent C (razor blades, cutlery, etc.) and then the composition never rises more than a few tenths of one per cent above one per cent. Compositions above 2 per cent are classified as cast irons. However, it is necessary to notice that cast irons of commerce are not simple alloys of iron and carbon, for they contain relatively large quantities of other elements, the most prevalent of which is silicon. In general, cast irons are better considered as ternary alloys of iron, carbon and silicon. They also differ in another important respect from steels. The presence of silicon promotes the formation of graphite in them. As a result, cast irons may contain carbon in the form of both graphite and cementite. This fact signifies an important difference between cast irons and steels, for the latter contain only combined carbon in the form Fe_3C.

The phase transformations associated with the eutectic point at 4.32 per cent C are interesting and useful in a detailed study of the structures in cast irons.

A detailed analysis of the phase transformations at the peritectic and eutectic points of the iron-carbon system are not essential in a study of steels. We shall focus our attention, therefore, on the phase transformations that occur in the eutectoid region of the phase diagram. This study is not simple, because we have to consider not only equilibrium phase transformations, but also those that occur under non-equilibrium conditions. With the assumption of slow-cooling, the phase changes in steels can be predicted with considerable accuracy using the equilibrium diagram. On the other hand, when the transformations do not occur under equilibrium conditions, such as when the metal is cooled rapidly, new and different metastable phases and constit ents form. Since these structures are important to the theory of the hardening of steels, the kinetics and principles of their formation are of considerable importance.

CLASSIFICATION OF EQUILIBRIUM DIAGRAMS

Equilibrium diagrams may be classified according to the relation of the components in the liquid and solid states as follows:

1. Components completely soluble in the liquid state.
   (a) Completely soluble in the solid state.
   (b) Partly soluble in the solid state.
   (c) Insoluble in the solid state.
2. Components partially soluble in the liquid state.
   (a) Completely soluble in the solid state.
   (b) Partly soluble in the solid state.
3. Components completely insoluble in the liquid state.
   (a) Completely insoluble in the solid state.

The first type of equilibrium diagram in which the components are completely soluble in the liquid state is most common in engineering alloys and will be given the most consideration in the following discussion.

Cooling Curves

One method of determining the temperatures at which phase changes occur in a system consists of following the temperature as a function of time as different alloys in the system are very slowly cooled. The data obtained in this manner form a cooling curve for each of the alloys. This method is particularly useful in studying the changes that occur during the solidification of alloys and, in some instances, may be used to advantage in determining transformations subsequent to solidification. However, in the latter case, the method may not be sufficiently sensitive; hence, other procedures may be employed.

If the equilibrium diagram of the antimony-bismuth system is to be determined, a series of alloys containing different amounts of antimony and bismuth will be made. Each sample is heated until molten and uniform in composition and then it is allowed to cool very slowly. A thermocouple may be employed to determine the temperature as cooling proceeds.

Consider first a sample of pure antimony cooling slowly from the molten state. The cooling curve for this alloy is shown at (a) in Fig. 4.3. A hold occur in the cooling curve beginning at x and ending at y. Examination of the metal in the crucible will indicate that solidification begins at the time corresponding to point *x* and is completed at the time corresponding to point *y*, i.e., solidification occurs at the constant temperature T_{sb} (1167°F). The temperature remains constant during the process of solidification because of the liberation of heat of fusion.

Consider an alloy of antimony and bismuth containing 25 per cent bismuth and 75 per cent antimony. This homogeneous molten alloy is allowed to cool very slowly and observations of temperature and time are made. The cooling curve for this alloy is shown at (b) in Fig. 4.3. Cooling continues in a uniform manner until the temperature reaches the value of 1095°F at which a break in the curve is encountered. At this temperature, the rate of cooling decreases. Examination will show that solidification begins at the break, or point *x* and continues until the temperature corresponding to point y, 800°F, is reached. At point *y*, the cooling rate changes again. This point y is designated as a recovery. Solidification occurs in this case over a range of temperature. The structure of the solid material at temperatures below 800°F consists of homogeneous grains i.e., all grains, are exactly of the same composition. It is a one-phase alloy, namely, a solid solution of antimony and bismuth.

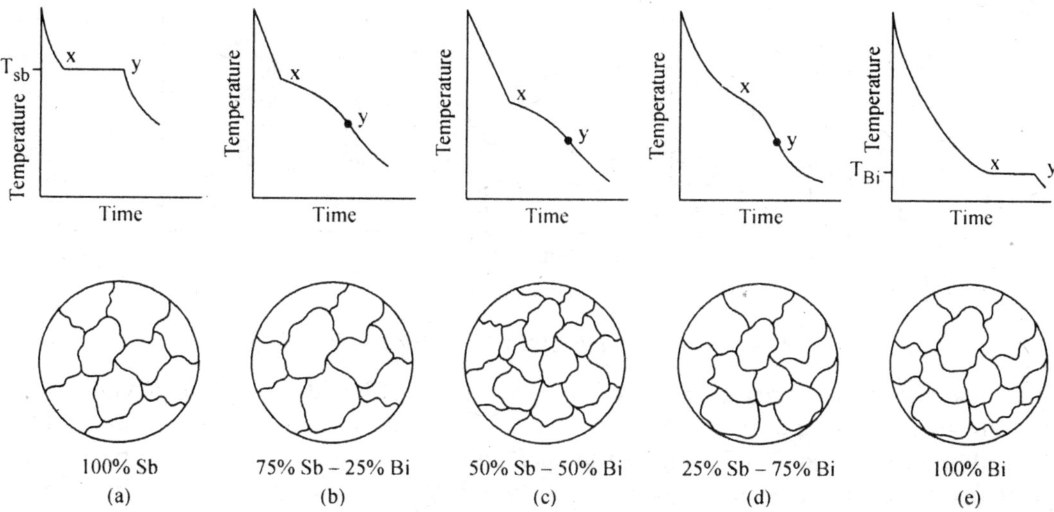

Fig. 4.3. Cooling curves for antimony-bismuth system.

Consider an alloy containing 50 per cent antimony and 50 per cent bismuth. These constituents are melted in a crucible and allowed to cool slowly. The cooling curve of this alloy is shown at (c) in Fig. 4.3. The break, x (980°F), and the recovery, y (640°F), occur at lower temperatures than for the previous alloy. Examination of the structure of the solidified alloy will reveal that it consists of only one phase, namely, a solid solution of bismuth and antimony.

An alloy containing 75 per cent bismuth and 25 per cent antimony will produce a cooling curve as shown at (d) in Fig. 4.3. This curve has characteristics similar to those found in the preceding two cases. The break, x (800°F) and the recovery, y (560°F), occur at lower temperatures than in the preceding cases. The structure of the solid alloy will consist of homogenous grains of solid solution.

The cooling curve of pure bismuth is shown at (e) in Fig. 4.3. In this case, a hold occurs in the cooling curve at a temperature of 520°F. Solidification begins at point x and is completed at point y. One may conclude from this discussion that pure metals solidify at constant temperature and their cooling curves exhibit only a hold, whereas the cooling curves of solid solutions exhibit a break and a recovery.

The results obtained from the cooling curves may now be plotted in a diagram of temperature as a function of composition called the equilibrium or phase diagram, Fig. 4.4. The temperature at which solidification begins, as indicated by all points designated x in the cooling curves, will form the liquidus curve. The temperatures at which solidification is completed, namely, all points *y* of the cooling curves, form the solidus curve of the equilibrium or phase diagram.

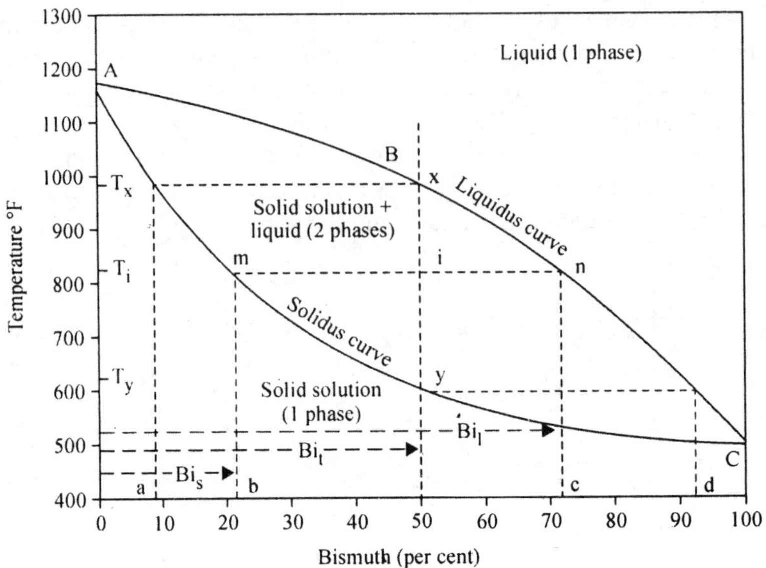

Fig. 4.4. Equilibrium diagram of antimony and bismuth.

The liquidus curve is, in reality, a plot of the composition of liquid that will be in equilibrium with solid at any given temperature. Similarly, the solidus curve indicates the composition of solid that will be in equilibrium with liquid at any given temperature. This is in accord with the phase rule. At a temperature of Ti in the equilibrium diagram of the antimony-bismuth system (Fig. 4.4), the composition of the liquid and the composition of the solid solution are determined by the intersection of this temperature horizontal with the liquidus and solidus curves, respectively. Any alloy that exists within the area enclosed by the liquidus and solidus curves will have one degree of freedom and will be considered as univariant.

COMPONENTS COMPLETELY SOLUBLE IN LIQUID AND SOLID STATES

A system that illustrates on equilibrium diagram in which there is complete solubility in the liquid and solid states is that of the antimony-bismuth system shown in Fig. 4.4. Consider an alloy containing 50 per cent antimony and 50 per cent bismuth. As cooling of this alloy occurs from a temperature greater than 980°F, solidification will begin when the temperature reaches the liquidus curve at point x. The composition of the first solid to form may be determined from the equilibrium diagram by passing a horizontal line from the point x through the two-phase region to the left until the first solid line, namely, the solidus curve, of the equilibrium diagram is reached. The intersection of the horizontal line corresponds to a solid solution containing 91 per cent antimony and 9 per cent bismuth. When the temperature has reached a value corresponding to point i on the equilibrium diagram, the composition of the solid that will be in equilibrium with the liquid can be found by extending a horizontal line at this temperature from the vertical line representing the composition of the alloy to the left until it intersects the solidus curve at m. The composition of the liquid at this same temperature can be determined by extending the same temperature horizontal to the right until it intersects the liquidus curve at n. The intersection points m and n correspond to the compositions b and c of the solid and liquid, respectively.

If equilibrium prevails, as it must in interpreting an equilibrium diagram, all of the solid solution must be of composition b and all of the liquid must be of composition c. This is different from the compositions that prevailed at higher temperatures. This means that there has been an adjustment in the composition of the solid and of the liquid during cooling from temperature T_x to temperature T_i. This adjustment in composition takes place by the process of diffusion. If cooling is too rapid, the required changes of composition cannot occur and the compositions of the phases will not correspond to those indicated by the equilibrium diagram. The rate of cooling must be sufficiently slow to permit complete diffusion and re-adjustment of the concentration.

As cooling continues from the temperature T_i to temperature T_y, solidification continues. Upon reaching temperature T_y, the composition of the liquid corresponds to the point d on the concentration axis and the composition of the solid solution corresponds to that of the alloy being studied; hence, solidification is completed at this temperature, as shown by the application of the lever law. In this system and, in general, in systems that possess this type of phase diagram, all alloys that exist below the solidus curve will consist of homogeneous solid solutions. Examples of binary alloys that form this type of diagram include nickel-copper, gold-silver, chromium-molybdenum and tungsten-molybdenum.

COMPONENTS COMPLETELY SOLUBLE IN LIQUID STATE AND INSOLUBLE IN SOLID STATE

A system in which the components are completely soluble in the liquid state and insoluble in the solid state is illustrated by bismuth and cadmium.

The cooling curve for pure bismuth is shown at (a) in Fig. 4.5. Solidification begins at the temperature corresponding to x and is completed at the same temperature at point *y*. An alloy containing 20 per cent cadmium and 80 per cent bismuth will have the cooling curve shown at (b) in Fig. 4.5. This curve exhibits a break at point x at a temperature lower than the temperature at which solidification of pure bismuth occurred. The cooling curve then continues at a lower rate of cooling until point y is reached, where there is a hold. The hold continues for a certain length of time until point z is reached and solidification is completed. The structure of this solid alloy will consist of large grains of pure bismuth formed during the cooling from x to y and, in addition, an intimate mechanical mixture of crystals of bismuth and cadmium which formed at the constant temperature corresponding to the time interval from y to z. This latter combination is a eutectic.

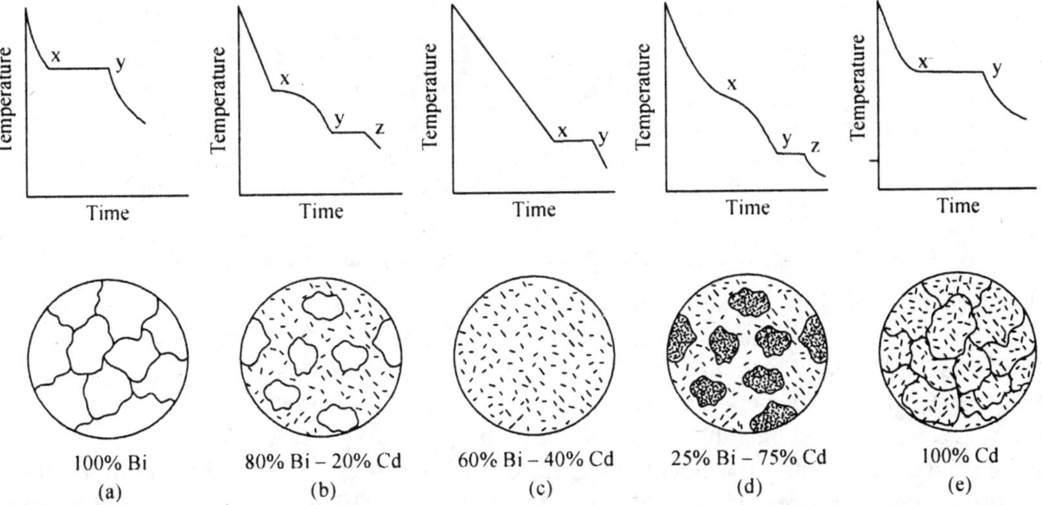

Fig. 4.5. Cooling curves for bismuth-cadmium system.

The cooling curve for an alloy containing 40 per cent cadmium and 60 per cent bismuth is shown at (c) in Fig. 4.5. This cooling curve is unique in that it has only a hold, occupying the time interval, x-y. This alloy in the solid state is composed entirely of an intimate mechanical mixture of grains of cadmium and grains of bismuth, forming the eutectic. This is the eutectic alloy of this system.

An alloy containing 75 per cent cadmium and 25 per cent bismuth will have a cooling curve as shown at (d) in Fig. 4.5. Solidification begins at the break, x. A eutectic begins to form at point y and solidification is completed at point z. The solid alloy will consist of large grains of cadmium and the eutectic.

The cooling curve of pure cadmium will exhibit only a hold shown at (e) in Fig. 4.5, solidification beginning at point x and being completed at point y. The temperatures at which solidification begins and ends for all of these alloys are plotted in the temperature-composition diagram to form the equilibrium diagram shown in Fig. 4.6. Had other alloys in this system been taken with less than 20 per cent cadmium or more than 75 per cent cadmium, the cooling curves for these alloys would also have had a horizontal line or a hold corresponding to the eutectic temperature. This indicates that no solubility exists between bismuth and cadmium in the solid state. It is important to indicate that there is probably no case in which there is not at least some slight solubility of one element in another. However, this solubility may be so slight that it cannot be readily detected. For practical consideration, these alloys are considered as being insoluble in the solid state.

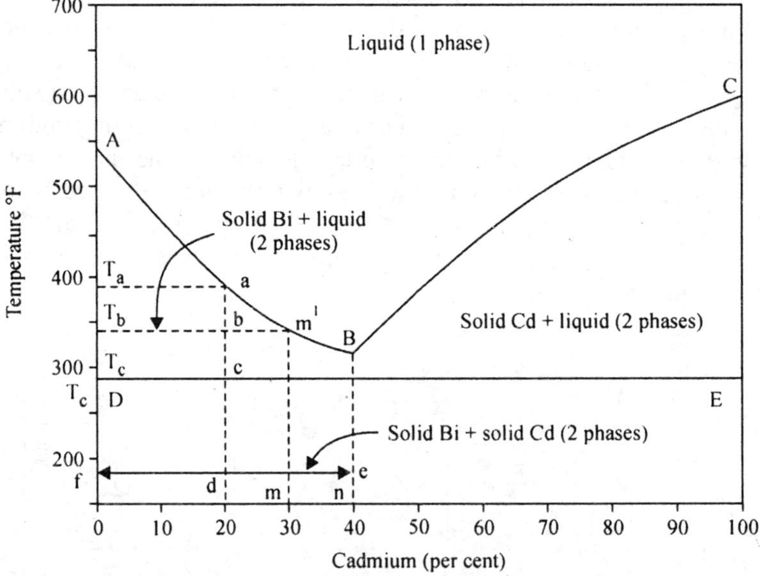

Fig. 4.6. Equilibrium diagram of bismuth and cadmium.

The phase regions of the bismuth-cadmium system are as follows:

Above curve ABC—homogeneous liquid solution—one phase.

Area ABD — solid bismuth + liquid — two phases.

Area CBE — solid cadmium + liquid — two phases.

Below DBE — solid bismuth + solid cadmium — two phases.

In the solid state, one may consider the regions of different structural constituents which should not be confused with the phase regions. All alloys below the line DB will consist of solid bismuth plus eutectic. All alloys below the line BE will consist of solid cadmium plus eutectic.

When an alloy containing 20 per cent cadmium and 80 per cent bismuth cools from above temperature T_a in Fig. 4.6, crystals of pure bismuth begin to form when temperature T_a is reached. Bismuth continues to solidify as the temperature changes from a to b. At temperature T_b, crystals of pure bismuth will be in equilibrium with a liquid of composition m indicated by the intersection of this horizontal temperature line with the liquidus curve, AB, at point m′. Relative amounts of solid and liquid at this temperature may be determined by means of the lever-arm principle.

As cooling continues, more crystals of bismuth form. Just as the temperature reaches T_c, the alloy will be composed of large crystals of bismuth in equilibrium with a liquid of composition n. The cooling curve for an alloy of composition n (40 per cent cadmium and 60 per cent bismuth), c in Fig. 4.6, shows that solidification occurs at only one temperature with the formation of eutectic. Therefore, the liquid existing at temperature T_c will solidify at this temperature to form an intimate mechanical mixture of crystals of bismuth and crystals of cadmium which makes up the eutectic. The solidification of alloys to the right of point n may be analyzed in a similar manner and will be shown to consist of large grains of cadmium existing together with the eutectic.

The relative amounts of structural constituents can be found by applying the lever-arm principle; for example, the proportion of structural constituents in the alloy containing 20 per cent cadmium and 80 per cent bismuth at a temperature corresponding to d in Fig. 4.6 is as follows:

Per cent bismuth = $\overline{de}/\overline{fe} \times 100$

Per cent eutectic = $\overline{fd}/\overline{fe} \times 100$

The uniqueness of the eutectic with respect to temperature and composition is a further illustration of the application of the phase rule. At the eutectic temperature of the bismuth-cadmium system (Fig. 4.6), three phases are in equilibrium, namely, liquid containing 40 per cent cadmiuim and 60 per cent bismuth, solid pure cadmium and solid pure bismuth. In this state, there are no degrees of freedom and the system is invariant. When one of the phases is eliminated, as by removing heat to complete solidification of the liquid, the temperature decrease. Two phases can coexist over a range of temperature.

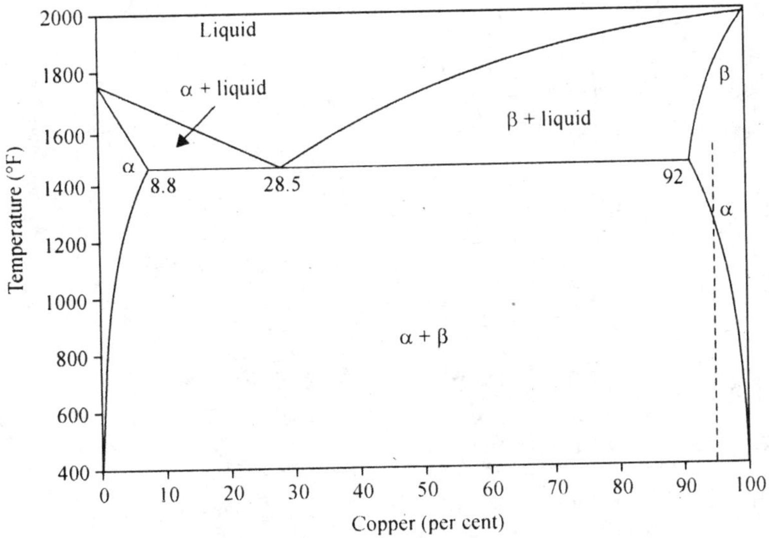

Fig. 4.7. Equilibrium diagram of silver and copper.

COMPONENTS COMPLETELY SOLUBLE IN LIQUID STATE, PARTIALLY SOLUBLE IN SOLID STATE

The most common and, therefore, the most important type of system is that in which there is partial solubility in the solid state. One example of a system of this type is silver-copper, Fig. 4.7. The changing solubility in the solid state may be somewhat confusing to the beginner. Therefore, in order to simplify the discussion, a hypothetical system of elements A and B will be considered. The cooling curves for the pure components and a series of alloys in the system A-B are shown in Fig. 4.8. The cooling curves shown at (b), (c), (g) and (h) in Fig. 4.8 are of the type observed in Fig. 4.3, in which solid solutions were formed. It is, therefore, to be expected in this system A-B that solid solutions will form in these particular alloys. The cooling curves shown at (d) and (f) are similar to those found in the cooling curves of Fig. 4.5 in which a eutectic was formed. The cooling curve shown at (e) in Fig. 4.8 is characteristic of eutectic formation. It is important to note that the point *y* in (c) and (g) of Fig. 4.8 occurs at the same temperature as the holds in (d), (e) and (f).

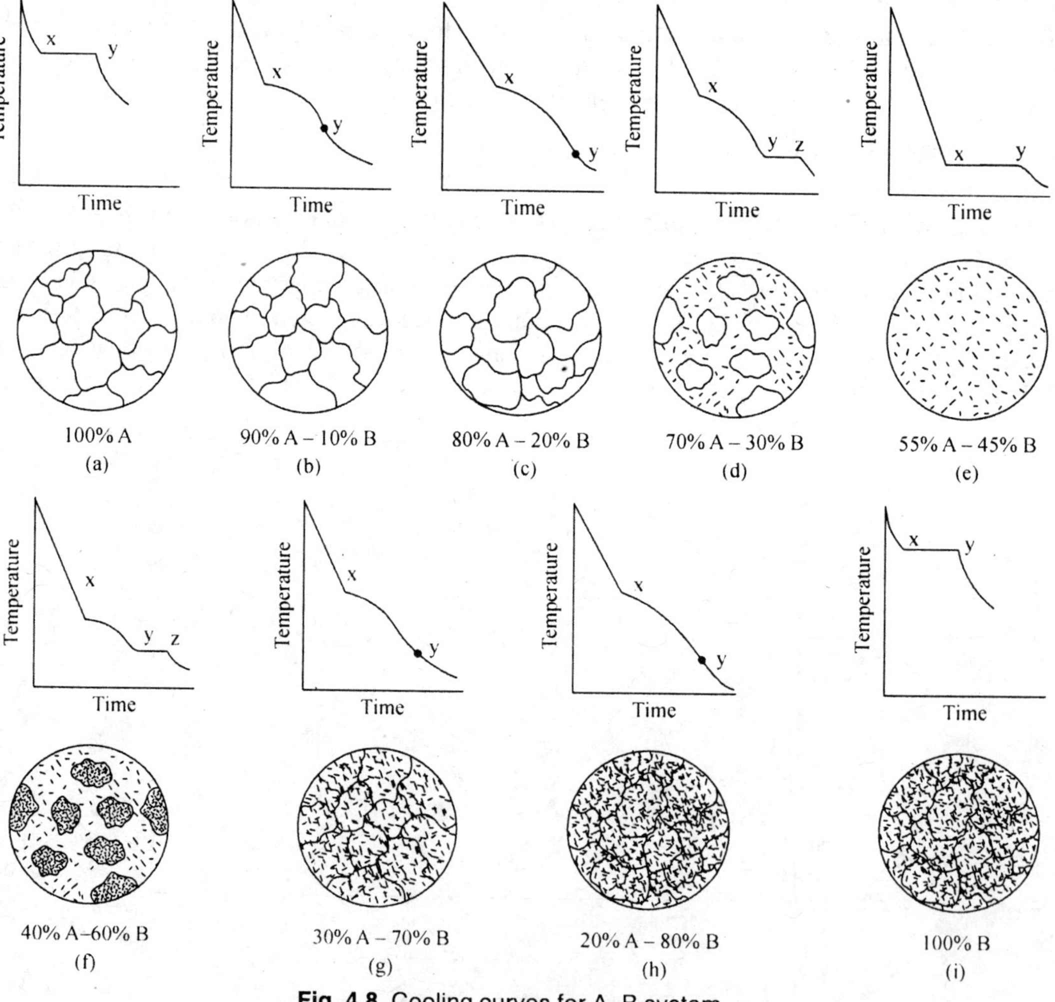

Fig. 4.8. Cooling curves for A–B system.

The temperatures at which solidification begins (all points x) and the temperatures at which solidification is completed (all points y) are plotted on the temperature-composition diagram to form the equilibrium or phase diagram as shown in Fig. 4.9. The phase regions of this system are as follows:

Above DKG — homogeneous liquid solution—one phase.
Region DEK — alpha solid solutions + liquid—two phases.
Region GKF — beta solid solution + liquid—two phases.
Below DE — alpha solid solution—one phase.
Below GF — beta solid solution—one phase.
Below EKF — alpha solid solution + beta solid solution—two phases.

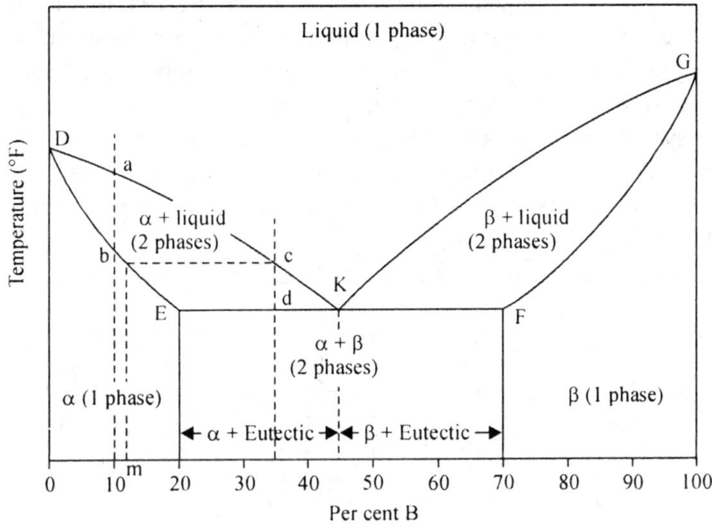

Fig. 4.9. Equilibrium diagram of A and B.

The solidification of any alloy containing 10 per cent B and 90 per cent A wlll begin at temperature a and will be completed at temperature b. In this case, the alloy will consist entirely of alpha solid solution. The solidification of an alloy containing 35 per cent B and 65 per cent A will begin at temperature c with the formation of alpha solid solution of composition m. The alpha solid solution will continue to form until the temperature d is reached. During the cooling from c to d, the composition of alpha solid solution will change from m and become richer in B. At temperature d, the alloy will be composed of alpha solid solution containing 20 per cent B and 80 per cent A in equilibrium with liquid of eutectic composition containing 45 per cent B and 55 per cent A. This liquid will solidify at constant temperature to form the eutectic consisting of grains of alpha solid solution and grains of beta solid solution. The regions in which the different structural constituents exist in the solid state are indicated in Fig. 4.9.

As indicated previously, most diagrams of this type are not as simple as that illustrated in Fig. 4.9. In most instances, there is a change in solubility in the solid state as indicated in the diagram for the silver-copper system, Fig. 4.7. In this system, the solubility of copper in silver and silver in copper changes as the temperature decreases below that of the eutectic temperature. An alloy containing 95 per cent copper and 5 per cent silver will be composed of beta solid solution at a temperature of 1400°F. Upon reaching the temperature indicated by point a on cooling, the solubility of silver in copper reaches its saturation value; in cooling below this, the solubility becomes less and, therefore, alpha solid solution begins to precipitate from the beta solid solution. The phases present in the final alloy will be alpha and beta solid solutions. The curves bounding the alpha + beta region are called the *solvus* curves.

Other alloy systems that form equilibrium diagrams of this type include copper-tin, copper-zinc, copper-beryllium, aluminium-copper and magnesium-aluminium.

SYSTEMS FORMING COMPOUNDS

It is common to find in certain systems the existence of compounds that may be represented by formulae of the type A_xB_y. Systems of this type include silver-lithium, silver-magnesium, aluminium-calcium, aluminium-cobalt, aluminium-copper, aluminiumiron, aluminium-lithium, aluminium-magnesium, aluminium-manganese, aluminium-nickel, arsenic-copper, arsenic-iron, gold-zinc, boro-niron, beryllium-copper, beryllium-nickel, bismuth-magnesium, iron-carbon, carbon-tungsten, calcium-magnesium, calcium-lead, columbium-iron, columbium-nickel, cadmium-copper, cadmium-antimony, copper-magnesium, copper-phosphorus, copper-antimony, copper-silicon, iron-phosphorus, iron-silicon, magnesium-silicon. A system in which a compound is formed is shown in Fig. 4.10. In this case, it is observed that a compound, Mg_2Si, is formed. This diagram can be interpreted by considering the two systems, Mg-Mg_2Si and Mg_2Si-Si. A system in which a compound is formed and some solubility exists in the solid state is shown in Fig. 4.11.

In some instances, the compound dissociates before melting occurs.

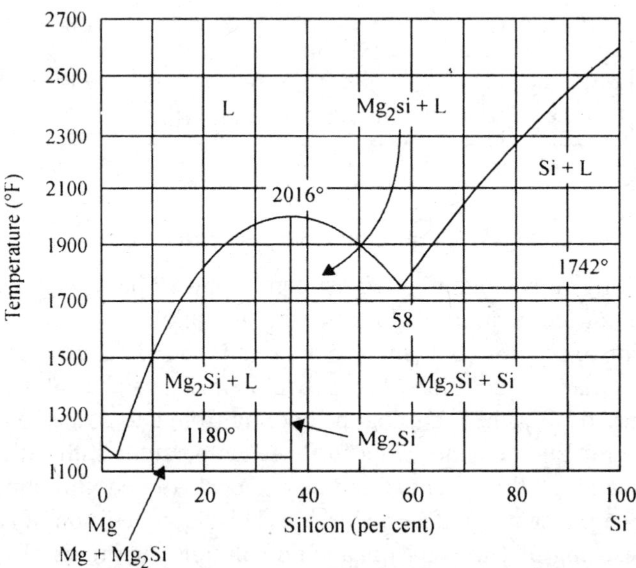

Fig. 4.10. Equilibrium diagram of magnesium and silicon.

PERITECTIC REACTION

A peritectic reaction is essentially an inversion of a eutectic reaction. In the latter, there is a transformation from one phase to two phases on cooling. In a peritectic reaction, there is transformation from two phases to one phase on cooling. An example of a system in which a compound decomposes before melting is shown in Fig. 4.12. In this case, an alloy containing 34.64 per cent bismuth will transform on heating at a temperature of 700°F into gold and liquid. In the cooling of such an alloy, pure gold will crystallise from, the liquid first; then at a temperature of 700° F, a reaction will occur to form the compound Au_2Bi. This would take place at constant temperature, forming a hold in the cooling curve. In this instance, cooling continues as soon as the gold and liquid react completely to form the compound Au_2Bi. Such reactions are usually slow. The compound Au_2Bi forms around the surface of each crystal of gold. The formation of a layer of Au_2Bi around every gold crystal prevents direct contact between the liquid and the gold crystals. Continued reaction will be very slow because it will depend upon diffusion of the atoms of the liquid through the compound to the gold crystals. This reaction is called a peritectic reaction because of this action, the Greek 'peri' meaning 'around.'

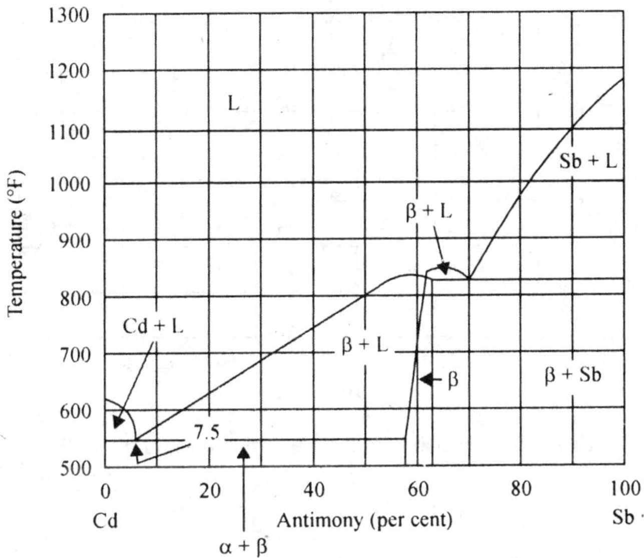

Fig. 4.11. Equilibrium diagram of cadmium and antimony.

A peritectic reaction may also occur when there is reaction between liquid and solid during the process of cooling to form a solid solution in the place of a chemical compound. This case may be illustrated by the higher temperature region of the iron-rich portion of the iron-carbon equilibrium diagram shown in Fig. 4.13. In this case, if an alloy containing 0.18 per cent carbon slowly cools from the liquid region, delta solid solution is formed. When a temperature of 2715°F is reached, there is a reaction between the liquid and the delta solid solution to form gamma solid solution. The relative proportions of phases present before and after a peritectic reaction may be determined by the application of the lever arm principle.

EUTECTOID TRANSFORMATIONS

In some systems, a transformation may occur within the solid state. A solid solution may transform to an intimate mechanical mixture of two or more phases. In a binary system, a mixture of two phases forms from the single solid solution. This will occur at a unique temperature and the proportions of the two phases will be unique for the system under consideration. A structure of this type is called a eutectoid, an intimate mechanical mixture of two or more phases having a definite temperature of transformation in the solid state and a definite composition. An example of this is shown in a portion of the copper-aluminium system in Fig. 4.14. In this case, during the cooling of an alloy containing 11.8 per cent aluminum, the remainder copper, the beta solid solution transforms at a temperature of 1050°F to an intimate mechanical mixture of alpha solid solution and gamma solid solution.

This forms a eutectoid. A more common example of a eutectoid of considerable commercial importance is found in the iron-iron carbide system.

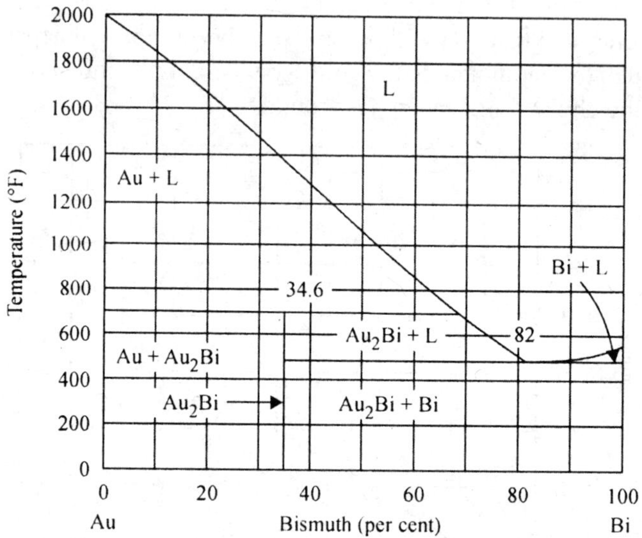

Fig. 4.12. Equilibrium diagram of gold and bismuth.

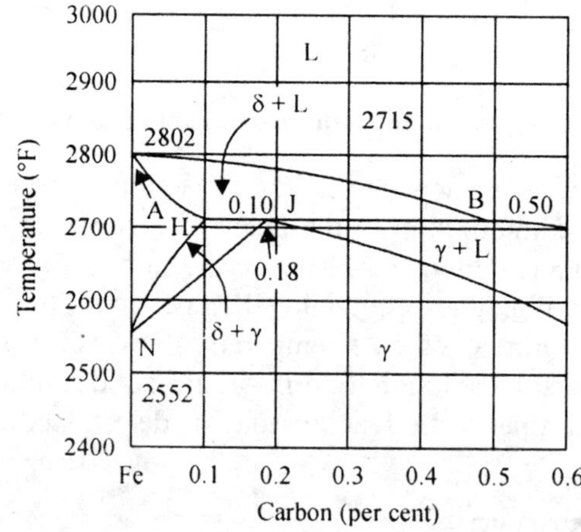

Fig. 4.13. Iron-rich portion of ir on-iron carbide equilibrium diagram.

Systems may also possess a transformation that is similar to a peritectic reaction, but the reaction occurs within the solid state. These are called peritectoid reactions. An example of such a reaction is shown in the nickel-zinc system, Fig. 4.15. If an alloy containing 48 per cent zinc, the remainder nickel, slowly cools from a temperature above 1500°F, the beta prime phase forms from the alpha and beta phases. Since all of this reaction occurs in the solid state, it is called a peritectoid.

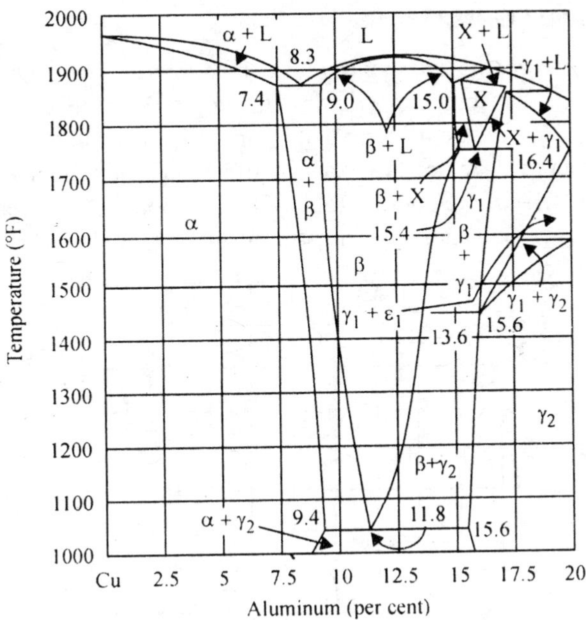

Fig. 4.14. Copper-rich portion of copper-aluminium equilibrium diagram.

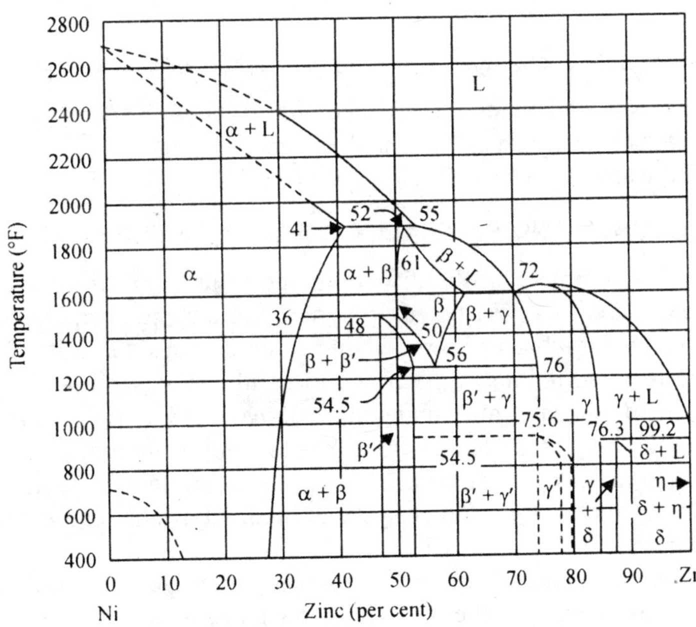

Fig. 4.15. Equilibrium diagram of nickel and zinc.

Combinations of these transformations are usually found in the more complex diagrams. An example of one in which eutectics, partial solubility, eutectoids, peritectic and peritectoid transformations occur is in the copper-tin system.

ORDER/DISORDER TRANSFORMATIONS

Another type of transformation occurs within the solid state in some systems. Solid solutions are composed of atoms of different elements distributed in a random fashion on a space lattice. Sometimes the atoms in a solid solution within a specific range of composition may, at a certain temperature, assume a definite arrangement to form an ordered structure. An example of a system in which an ordered structure occurs on cooling is shown in Fig. 4.16 of the copper-gold system. Ordered structures occur at 25 and 50 atom per cent (50.9 and 75.5 wt. per cent) gold, at temperatures of 745 and 795°F, respectively.

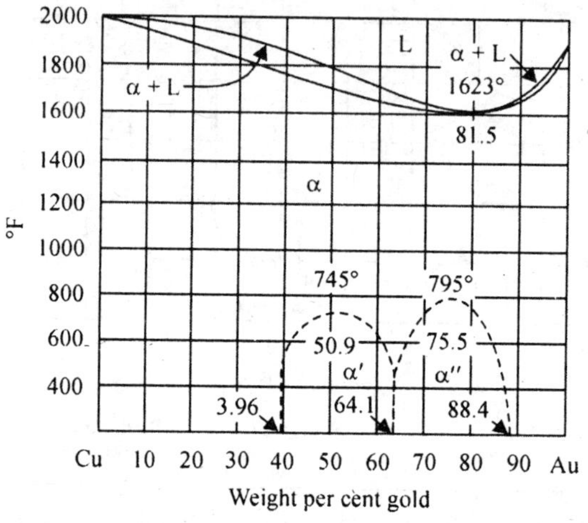

Fig. 4.16. Copper-gold system (order-disorder transformation).

In each case, when the phase boundary is reached on cooling, the ordering occurs. However, there will be a two-phase region on each side of the α′ and α″ areas. The interpretation of this feature of phase diagrams does not require any new principles.

COMPONENTS INSOLUBLE AND PARTIALLY SOLUBLE IN THE LIQUID STATE

Some systems show either complete insolubility or limited solubility of the components in the liquid state. Many of these cases are not of very great importance to the engineer. However, there are a few cases which have some engineering value and they should be considered briefly. An example of a system in which there is no solubility in the liquid or solid states is shown in the system, copper-molybdenum, Fig. 4.17. In this case, at a temperature above 4748°F, two liquids exist: one, copper and the other, molybdenum. When the temperature falls to 4748°F, molybdenum crystallizes, leaving liquid copper. The copper, in turn, solidifies at 1981°F. Such a system gives rise to a layer structure, the denser metal being at the bottom. Other systems of this type include copper-tungsten, silver-tungsten and silver-iron.

A system in which there is partial solubility in the liquid state is typified, by the zinc-bismuth system shown in Fig. 4.18. At temperatures above the dome in this diagram, a homogeneous liquid exists. As the temperature falls and reaches the boundary of the region of two melts, two liquid phases exist: one, rich in zinc, the other, rich in bismuth. Subsequent cooling will be in accordance with the rules previously set forth. Other systems of this type include copper-lead and zinc-lead, aluminium-lead.

In the system iron-copper, there are two distinct liquids at high temperatures, but, as the temperature decreases, a homogeneous liquid region is formed in the equilibrium diagram.

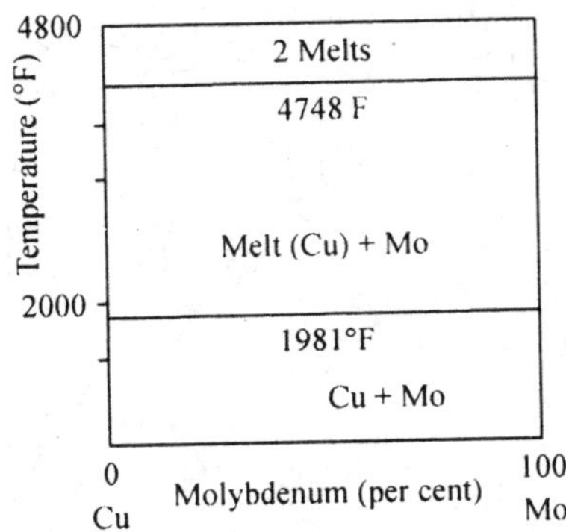

Fig. 4.17. Equilibrium diagram of copper and molybdenum.

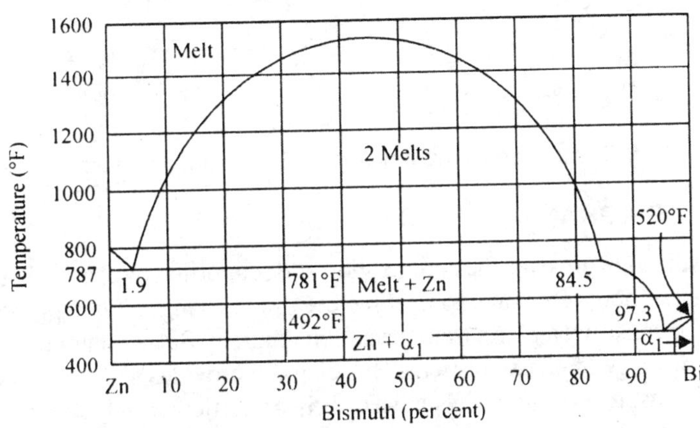

Fig. 4.18. Equilibrium diagram of zinc and bismuth.

It has been possible to make homogeneous alloys of those metals which are not soluble in the liquid state by the methods of powder metallurgy. These materials are reduced to powders, compacted and heated to temperatures below the melting range. This treatment produces a homogeneous alloy which may have properties of engineering importance.

TERNARY PHASE DIAGRAMS

The simple binary systems are seldom utilised in engineering applications. The majority of alloys contain more than two components. Representing the phase relations that exist over a range of temperatures in systems containing more than two components presents a real problem. In alloys containing three components (ternary systems), the phase diagram consists of a three dimensional figure, the base of which is an equilateral-composition triangle and the temperature is plotted vertically. In the place of curves to indicate phase boundaries, surfaces are encountered. These diagrams will have liquidus surfaces, solidus surfaces, etc. An example of a ternary diagram of very simple character is illustrated in Fig. 4.19 for the Cd-Pb-Bi system. The representation of phases in equilibrilum for systems containing more than three components usually requires sections in which the proportions of certain of the components 9are held constant. The details of these more complex systems are beyond the scope of this text.

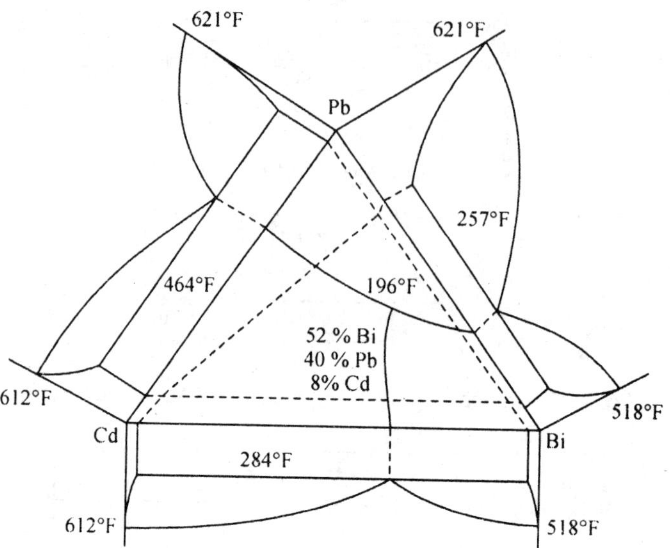

Fig. 4.19. Ternary equilibrium diagram of bismuth, cadmium and lead.

TRANSFORMATION DIAGRAMS

While the equilibrium diagram presents the composition and proportion of phases that will be in equilibrium at any temperature, it does not present a graphical picture of the changes in the proportions of phases. It is, therefore, convenient to construct a phase transformation diagram. Such a diagram is a plot of temperature as a function of the percentage of phases present. This diagram provides a means by which the formation of phases and structural constituents can be visualised. It is of particular value in studying alloys in which complex transformations occur, particularly within the solid state. The transformation diagram consists of a rectangular figure in which the temperature is represented on the ordinate and the percentage of phases or structural constituents is represented on the abscissa. If any phase or structural constituent occupies 100 per cent of the structure, it is represented by the full width of the transformation diagram. As an example, consider an alloy of 10 per cent B and 90 per cent A in the equilibrium diagram of Fig. 4.9. At temperature *a*, solidification of the alloy begins with the formation of the solid solution alpha. This gives point *a* in the transformation diagram shown in Fig. 4.20. As cooling continues, the amount of alpha solid solution increase until temperature *b* is reached. At this point, the entire structure is composed of grains of alpha solid solution, as indicated at *b* in the transformation diagram.

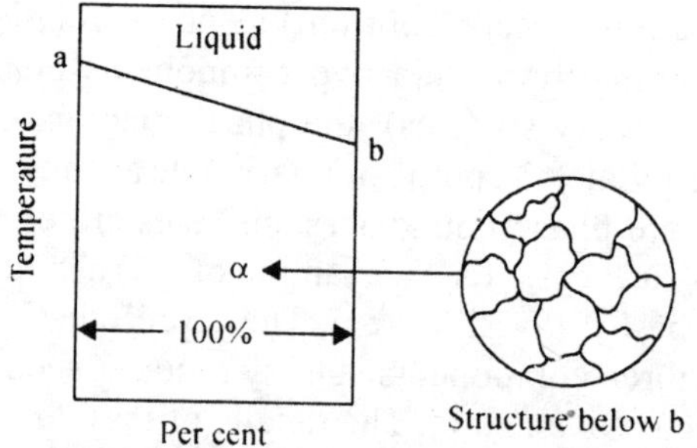

Fig. 4.20. Phase transformation diagram for 90 per cent A–10 per cent B alloy.

If one considers an alloy contaning 35 per cent B and 65 per cent A, solidification begins at temperature c with the formation of grains of alpha solid solution. The beginning of solidification is indicated at *c* in the transformation diagram of Fig. 4.21. As cooling continues, solidification of the alpha solid solution proceeds until temperature *d* is reached. Upon reaching this temperature, the structure consists of the following proportion of phases:

$$\frac{45-35}{45-20} \times 100 = \frac{10}{25} \times 100 = 40 \text{ per cent alpha}$$

$$\frac{35-20}{45-20} \times 100 = \frac{15}{25} \times 100 = 60 \text{ per cent liquid}$$

Therefore, at *d* in Fig. 4.21, 40 per cent of the full width is occupied by the alpha solid solution (proeutectic α) and 60 per cent by liquid. The liquid at temperature d is of eutectic composition and transforms to the eutectic, composed of alpha and beta solid soluions. A vertical line is drawn in the transformation diagram to sepaate the eutectic from the alpha solid solution that formed in cooling c to d. A horizontal line is drawn in the transformation diagram at temperature d. This horizontal line must be divided in accordance with the proportion of alpha and beta in the eutectic.
This is as follows:

$$\text{alpha} = \frac{70-45}{70-20} \times 100 = \frac{25}{50} \times 100 = 50 \text{ per cent}$$

$$\text{beta} = \frac{45-20}{70-20} \times 100 = \frac{25}{50} \times 100 = 50 \text{ per cent}$$

A vertical line is drawn at the point delineating the two phases in the eutectic. The percentage of the entire structure occupied by the eutectic alpha, i.e., the alpha solid solution forming a part of the eutectic, may be computed by taking 50 per cent of the amount of liquid of eutectic compositions existing at temperature *d*, or 60 × 0.50 = 30 per cent.

The result of these computation is presented in Fig. 4.21. The principles given here may be utilized in the analysis of equilibrium diagrams of a more complex character.

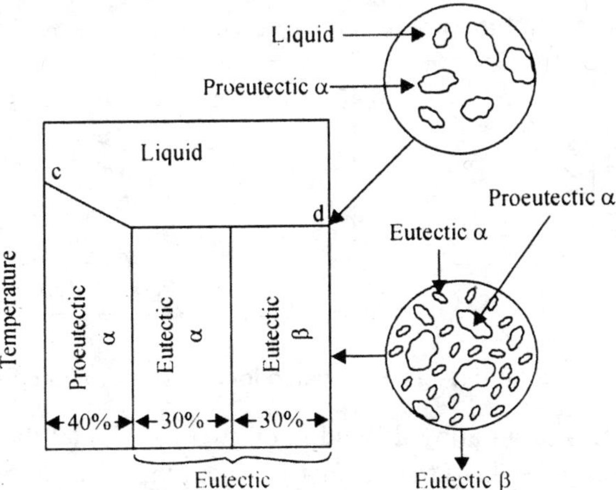

Fig. 4.21. Phase transformation diagram for 65 per cent A–35 per cent B alloy.

DENDRITIC STRUCTURE IN ALLOYS

The dendritic structure cannot be detected readily because of the high purity of the material. In alloys in which two phases crystallise from the liquid state, the dendritic structure is usually very easily detected. If a section of a cast two-phase alloy is machined to smooth surface and etched with a suitable reagent, a tree-like pattern may be revealed. This condition is, in reality, a segregation and may be called dendritic segregation. If one looks at this structure under the microscope, it is observed that the phase that solidifie first forms the main stem and branches of the dendrite, while the other phase fills the interstices. The interstices of the dendrite also contain impurities which are not readily soluble in the phase that first solidified. The dendritic structure in alloys can be destroyed most easily by hot working and, in some cases, by heat treatment. Dendritic segregation in a cast alloy may be a source of weakness. Attempts are made to eliminate this structure as far as possible by forging or by heat treatment.

CORING

In discussing equilibrium diagrams, it was assumed that cooling took place under equilibrium conditions. If the rate of diffusion of the components of a particular alloy is very low and the rate of cooling too rapid, there will not be sufficient time for equilibrium to be established. Under this condition the composition of the phases will not be as represented by the equilibrium diagram. Non-equilibrium cooling may lead to a variation in this composition of a solid solution which crystallises from the melt. In this case, the interior of the crystal will be richer in the principal constituent than the other layers, producing a condition known as coring.

PROPERTIES OF ALLOYS

The equilibrium diagrams give some indication of the relative properties of alloys in a given system. It is difficult and in most cases impossible, to predict the absolute values of the properties of a series of alloys. For a system in which there is complete solubility in the solid state, as in Fig. 4.4, the hardness, density, electrical conductivity and corrosion resistance may be expected to vary as shown in Fig. 4.22. The relation of these properties to composition for a system in which there is complete insolubility in the solid state is illustrated in Fig. 4.22. In a system in which there is partial solubility, the relation of composition to properties is a combination of the two preceding cases as shown in Fig. 4.23.

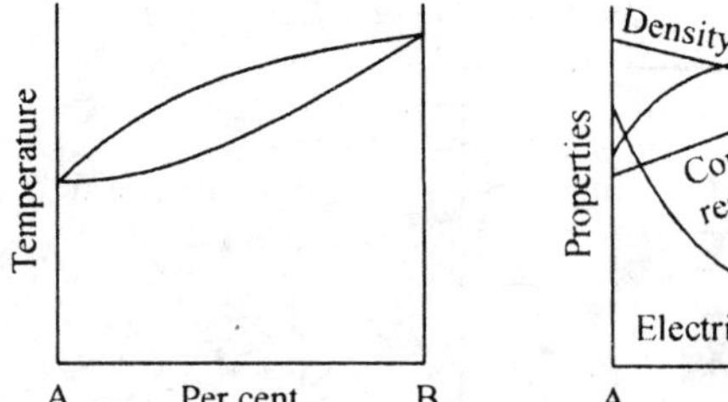

Fig. 4.22. Relation of properties to composition for system with complete solid solubility.

The mechanical properties of an alloy depend upon two factors: (i) the properties of the phase or phases of which it is composed; and (ii) the manner in which the several phases are associated to form the structure. The second factor is possibly the more important. In two-phase alloys in which one phase surrounds the grains of the other, the alloy as a whole usually has the properties of the continuous phase.

Since the strength or hardness of any crystalline substance depends upon the resistance to slip on certain crystallographic planes, it is obvious that any mechanism or factor that tends to prevent slip within a crystal will bring about an increase of strength or hardness.

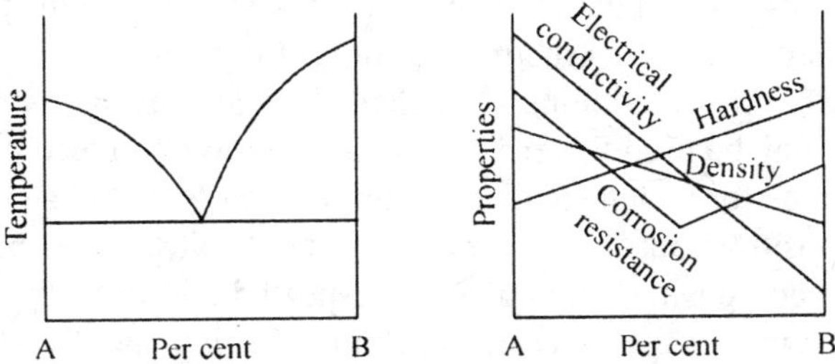

Fig. 4.23. Relation of properties to composition for system with solid insolubility.

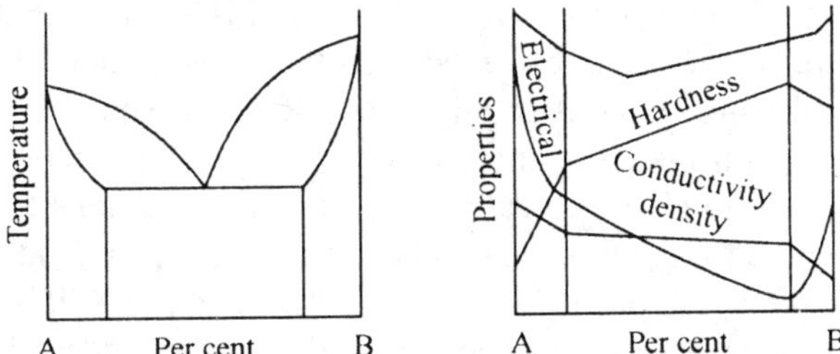

Fig. 4.24. Relation of properties to composition for system with partial solid solubility.

Metals and alloys of engineering importance consist of an aggregate of grains of varying orientation. The interference offered by the grain boundaries precludes the possibility of one continuous slip throughout the whole mass of the metal or alloy. Those metals or alloys that have a complex space lattice may be expected to have higher strength and greater hardness than those with a simpler lattice. An illustration of this fact may be found in the intermetallic compounds which have complex space lattices and are harder than the pure metals of which they are composed.

In a given metal, resistance to slip may be increased by: (i) an alteration of the space lattice through the formation of a compound or a solid solution; (ii) a distortion of the lattice structure by plastic deformation (strain hardening); and (iii) a precipitation of another phase within the lattice structure (precipitation hardening).

THERMAL ANALYSIS— HEATING AND COOLING CURVES

If the equilibrium diagram is to be used, there must be some method for correlating the composition of a piece of metal of uncertain but suspected composition with the diagram. Much can be done by microscopical examination but the most useful means may be by thermal analysis. The use of this depends on the fact that if a substance is being heated or cooled through a phase change then, at the temperature of the change, there will be a difference in the absorption or dispersion rate of thermal energy because the thermal characteristics of the phases will differ. If a marked amount of latent heat has to be absorbed or given up, then a marked arrest point will appear on the temperature time curve for the metal. On the other hand, if there is no noticeable latent heat change involved, then a change of gradient may be the only sign of the phase change and considerable care may be required if it is to be observed. Under normal atmospheric pressure, these temperature inflections will always take place at the exact temperature of the phase change, provided that sufficient time is allowed for the change to take place and that the rate of heating or cooling is not so rapid that super-heating or undercooling is caused.

Thermal analysis offers the only reliable means for plotting the liquidus temperature of an alloy of known composition and, hence, for constructing an equilibrium diagram. It is neither the only method nor the best method for verifying or examining solid phase change effects. Having obtained an equilibrium diagram, thermal analysis may be used to verify the composition of a piece of alloy rnetal known to be composed of metal of the type for which the diagram applies. Note that this is not a means for analysing an alloy of unknown composition, but simply for confirming the absence or presence and quantity of an element in an alloy whose composition is known or suspected within fairly close limits. The limits of composition, within which thermal analysis can be attempted, will depend on how much is known of the nature and complication of the particular alloy system.

Thermal analysis is usually performed on a small sample of the metal which, if the metal is not to be melted, has a hole drilled into its interior (or cast in, if the metal is being melted) to carry a thermocouple. The thermocouple is carefully packed into the hole so that any change of temperature of the metal immediately influences the thermocouple. A small furnace of a type which can be heated to the necessary temperature at a reasonably controlled rate and then can be cooled at a rate sufficiently low to prevent excessive undercoolirig is used to give the heating and cooling conditions for the specimen.

A heating and cooling curve for the furnace will be of the general type indicated in Fig. 4.25a but this will be modified within the metal sample and might be as shown in Fig. 4.25b, which shows two types of 'arrest' points and their characteristics during heating and cooling. Under ideal conditions the level of their arrests would be the same on both curves but it is difficult to keep the heating rate down to a sufficiently slow and yet uniform rate. So the heating curve may Be Less reliable, possibly giving slightly elevated readings. Cooling is more uniformly controllable since it is dependent on the insulation of the furnace in relation to the uniformity of atmospheric temperature and so may give more reliable results. A curve of the type shown in Fig. 4.25b may be difficult to interpret, if the changes are slight and closely spaced and so a special method of recording is used. This is the so-called 'inverse-rate' curve which is plotted as the times taken for temperature changes of a constant amount of increment and decrement. Such curves for the furnace have a smooth uniform shape but sharp peaks appear at the change points for the specimen, as shown in Fig. 4.25c. If cooling is a little rapid, the phase change temperature may be depressed below the true temperature by an appreciable amount; an effect known as 'undercooling'. If in such circumstances the exothermal reaction is strong, then a phenomenon known as "recalescence" may occur. In the latter case the temperature falls below the change temperature before transformation begins and then tends to rise again towards the true value as transformation proceeds (Fig. 4.25c).

For the most accurate results another system of thermal analysis recording is sometimes employed, although it entails more complicated work in setting up. This is the 'differential' method using a neutral test piece, which shows no phase changes in the relevant range, as a reference body alongside the test specimen in the furnace. In this case any temperature differences which may appear between the two bodies as a result of phase changes in the specimen are noted and then plotted against the temperature of the reference body.

PROEUTECTOID TRANSFORMATIONS OF AUSTENITE

The microstructures obtained when a steel is slowly cooled from the austenitic region depend on the original carbon concentration of the steel. If this concentration is less than 0.77 per cent, the eutectoid composition, then the microstructure will contain two primary constituents: proeutectoid ferrite and pearlite. If the carbon concentration equals that of the eutectoid, the structure will consist only of pearlite and if it is higher than 0.77 per cent it will contain proeutectoid cementite and pearlite.

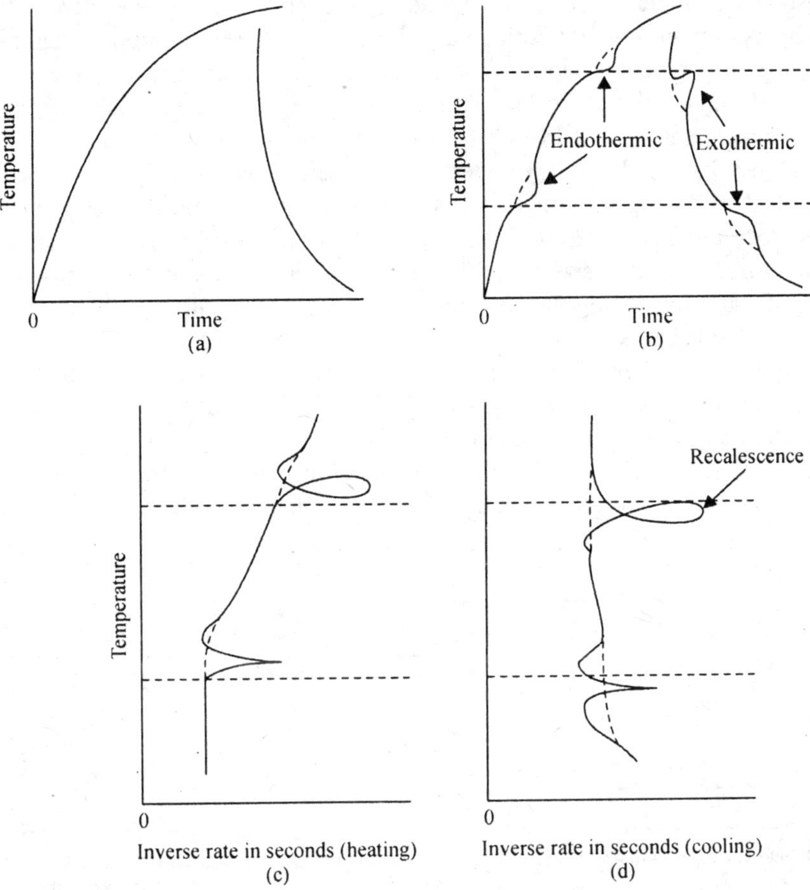

Fig. 4.25. Heating and cooling curves: (a) furnace heating and cooling curves; (b) heating and cooling curves of an alloy in the furnace. Phase changes cause steps according to the nature and magnitude of the energy change involved; and (c) inverse rate curves.

Both the proeutectoid ferrite and the proeutectoid cementite nucleate preferentially and heterogeneously on the austenite grain boundaries. There are two basic reasons for this: the boundaries contain energetically favourable sites for nucleation and they are regions where the diffusion rates are higher. When a ferrite nucleus forms at an austenitic grain boundary, it normally develops a different orientation relationship with respect to the austenite grains on the two sides of the boundary.

TRANSFORMATION OF AUSTENITE TO PEARLITE

When the carbon concentration of an iron-carbon alloy is 0.77, it transforms on slow cooling to the eutectoid structure, pearlite. In practice, steels are almost always cooled so that the temperature falls continuously. Thus, annealed steels are slowly cooled by leaving them in a furnace after the power to the furnace has been shut off, or a normalised steel is cooled by removing the red hot metal from the furnace and allowing it to cool in air. In continuous cooling processes, the nature of the reaction changes with the decreasing temperature. This change is particularly marked when the rate of cooling is appreciable).

Fig. 4.26 illustrates a simple experimental method of studying isothermal austenitic transformations. As may be seen by examining Fig. 4.27 (an enlarged section of the iron-carbon diagram containing the eutectoid point), a steel of eutectoid composition is austenitic at temperatures above 727°C. Thus, if a small furnace containing a crucible filled with a molten salt mixture is held at 730°C, eutectoid-composition steel specimens placed in this bath can be maintained in the austenite phase and at a temperature just above that at which they undergo the eutectoid reaction. This furnace is shown on the left in Fig. 4.26. Just to the right of this unit is a similar furnace that also contains a salt bath but maintains it at a temperature below 727°C. In an experiment of this type, it is convenient to use specimens in the shape of flat discs about the size of a dime. A short length of temperature-resistant wire forms a convenient handle for use in moving the specimens from furnace to furnace. Metal pieces of this size, when placed in liquid salt baths, very quickly attain the temperature of the bath. Consequently, if a specimen originally in the left-hand furnace is quickly removed and inserted into the furnace to its right, we can assume that its temperature changes instantly from just above the eutectoid temperature to just below it. Since austenite is no longer stable below 727°C, it is now able to decompose into other phases. If this solid-state reaction is carried out at temperatures not too far below the eutectoid temperature, the reaction occurs by nucleation and growth and is, therefore, time dependent.

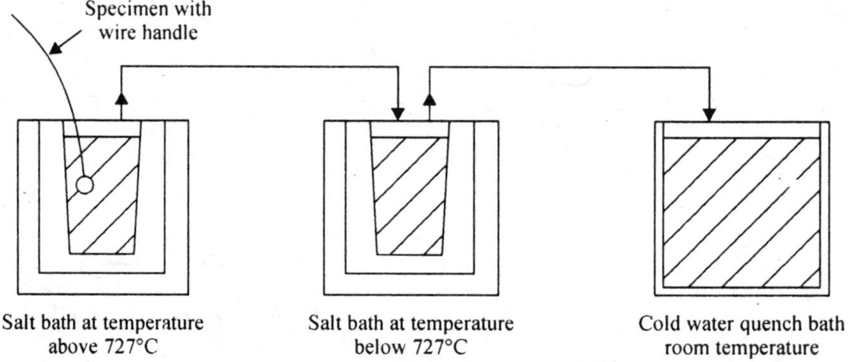

Fig. 4.26. Simple experimental arrangement for determining the kinetics of isothermal austenitic transformations.

The isothermal decomposition of the austenite is usually followed with the aid of a number of specimens (generally about ten), all of which are quenched simultaneously from the upper temperature to the lower temperature. Individual specimens are then removed from the second bath at increasing time intervals (usually measured on a logarithmic scale) and quickly cooled to room temperature by quenching them in a bath of cold water. This latter fast-cooling effectively stops the isothermal reaction and any austenite still untransformed, at the instant of the quench, undergoes a martensitic (athermal) transformation as the specimen approaches room temperature. It suffices to state here that it is of the same basic type as the iron-nickel transformation. Fortunately, steel martensite has a different appearance under the microscope than the high-temperature reaction products of austenite. After a suitable metallographic polish and etch, the specimens, prepared as outlined above, may be examined metallographically and the relative amount of the isothermal-transformation product determined in each case.

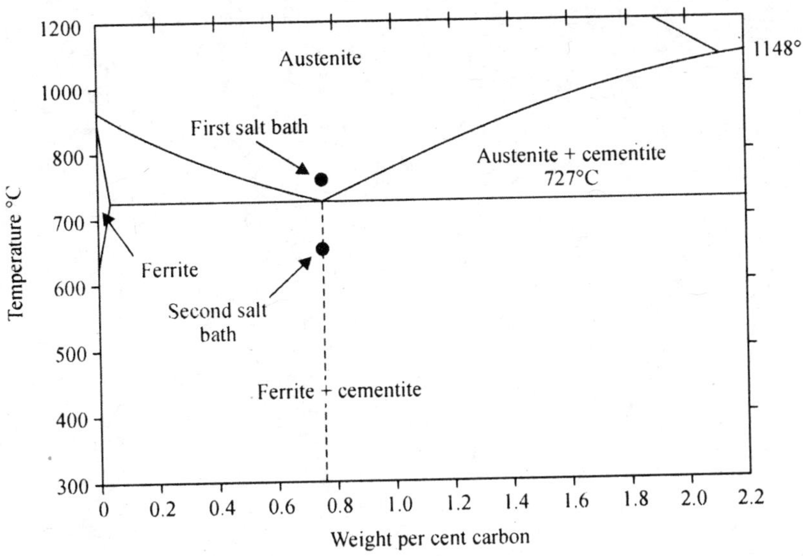

Fig. 4.27. The eutectoid section of the iron-carbon diagram.

If austenite is allowed to transform isothermally at temperatures just below 727°C, the reaction product is the same as that predicted by the iron-carbon diagram for a very slow, continuous cooling process. As may be seen by examining Fig. 4.27, the stable phases below the eutectoid temperature are ferrite and cementite and the eutectoid structure is a mixture of these phases. This constituent, called pearlite, consists of alternate plates of Fe_3C and ferrite, with ferrite being the continuous phase.

SECTION II

Heat Treatment Processes

Chapter 5

Basic Concepts of Heat Treatment of Steel

INTRODUCTION

Heat treatment is an important operation in the manufacturing process of many machine parts and tools. Only by heat treatment it is possible to impart the high mechanical properties to steel required for the normal operation of modern machinery and tools. Heat treatment is usually classified into four typical groups.

The first group comprises first-order or re-crystallisation annealing which is employed to relieve internal stresses, reduce the hardness and to increase the ductility of strain-hardened metal. It is also known as stress-relieving. At first, upon an increase in the heating temperature, the elastic distortions of the crystal lattice are eliminated (recovery). At higher temperature, new grains form and begin to grow (re-crystallisation). Due to recovery and re-crystallisation, the metal is softened and regains its high ductility.

It is important to point out that the heating temperature for first-order annealing is not associated with phase transformation temperatures and only depends upon the recovery or re-crystallisation temperatures peculiar to each metal. The second group, second-order or full annealing, which involves phase re-crystallisation, consists in heating alloys above the temperatures required for phase transformation. This is followed by slow cooling. Full annealing substantially changes the physical and mechanical properties and may refine a coarse-grained structure.

In the third group, hardening, alloys are heated above the phase transformation temperature and are then rapidly cooled (quenched) to retain at room temperature, either a phase stable at high temperatures or various stages of decomposition or transformation of this phase. Thus, hardening causes the formation of unstable (non-equilibrium) structures.

The fourth group, tempering, involves the re-heating of a hardened alloy (steel) to a temperature below that required for phase transformation so as to bring it nearer to an equilibrium state.

ANNEALING

In general, the term annealing refers to any heating and cooling operation that is usually applied to induce softening. More specifically, however, annealing may be divided into two operations, namely, full annealing and process annealing. In full annealing, the steel is heated usually to about 100°F (40°C) above the upper critical temperature and held for the desired length of time, followed by very slow cooling as in the furnace. The purpose of full annealing is three-fold: to soften the steel and improve ductility; to relieve internal stresses caused by previous treatment; and to refine the grain.

In some cases, the properties obtained by normalising (air cooling) from just the critical temperature satisfy the design requirements. In this event, the normalising treatment is employed as a mild quench .

A stress-relief anneal may be employed to remove residual stresses in large steel castings and welding structure. Parts are heated to temperature below the A_1 (1000–1200°F or 550–650°C), are held for a period of time and are then slowly cooled. Times and temperature vary with the condition of component

In process annealing, the steel is heated to a temperature below or close to the lower critical temperature, followed by any desired rate of cooling. Its principal purposes are to soften the steel partially and to obtain release of internal stresses. In this treatment, grain refinement by phase transformation is not accomplished as it is in full annealing. Process annealing is used extensively in the treatment of sheet and wire. For this purpose, the steel is heated to a temperature between 1020 and 1200°F (249 and 649°C).

The purpose of annealing is to obtain softeness, improve machinability, increase or restore ductility and toughness, relive internal stresses, reduce or eliminate structural inhomogeneity, refine grain size and to prepare steel for subsequent heat treatment. The term annealing, when used in connection with the treatment of steel, will mean full annealing. When annealing is applied for the purpose of refining grain, the steel is heated to a temperature 50 to 100 F (10 to 40°C) above the upper critical point. Even though the heating temperature may be considerably above the upper critical point, the treatment is still called 'annealing' if the steel is slowly cooled from this temperature.

In special cases, annealing may be accomplished by isothermal transformation at temperatures above the pearlite nose of the isothermal transformation diagram. This process is called isothermal or cyclic annealing and decreases the total time required for an annealing operation, provided adequate facilities are available.

NORMALISING

In general, the term normalising refers to the heating of steel to approximately 100°F (40°C) above the upper critical temperature, followed by cooling in still air. The purpose of normalising may be to refine grain structure prior to hardening the steel, to harden the steel slightly, or to reduce segregation in castings or forgings. It is common practice to normalise a steel at a temperature of about 100°F (40°C) above the upper critical point for the purpose of refining the grain.

A hypoeutectoid steel consisting of a structure of ferrite and coarse pearlite may be made easier to machine if the ferrite and cementite are more finely distributed. A very soft steel has a tendency to tear in machining; therefore, the slight increase in hardness obtained by normalising leads to a more brittle chip and thus improves machinability.

When a steel casting solidifies and cools in the mould, a certain amount of segregation occurs and the austenite grains are rather coarse. This structure is not conducive to the best properties. The situation, however, may be altered by heat treatment. If heated to just above the critical range, grain refinement will be very slow and adjustment of composition will be extremely slow. At higher temperatures, the reduction of segregation will be much more rapid because the rate of diffusion is higher. Therefore, the steel is normalised at a temperature well above the upper critical point. At this temperature, although the grain size may by larger than it was just above the critical temperature, some refinement will have taken place. The rate of cooling from this higher temperature will be somewhat immaterial since the steel will be given a further treatment to improve the grain size later.

HARDENING BY QUENCHING

By controlling the manner in which the transformation of austenite occurs, it is possible to increase greatly the strength and hardness of steel. Maximum hardness may be attained by cooling the material at a rate equal to or greater than, the critical cooling rate. The increase in hardness attainable with different plain carbon steels is shown in Fig. 5.1. These curves show that the maximum per cent increase of hardness by quenching is obtained in steel containing between 0.35 and 0.70 per cent carbon. Above about 0.55 per cent carbon, there is a small increase in the hardness value obtainable by quenching. This may be explained on the basis of the amount of eutectoid that is present. The existence of proeutectoid ferrite (ferrite formed between Ar_3 and Ar_1), which does not contribute to the hardness of the steel, would naturally account for the lower hardness obtained in the lower-carbon steels; whereas the existence of proeutectoid cementite in hypereutectoid steels generally will not contribute greatly to increased hardness.

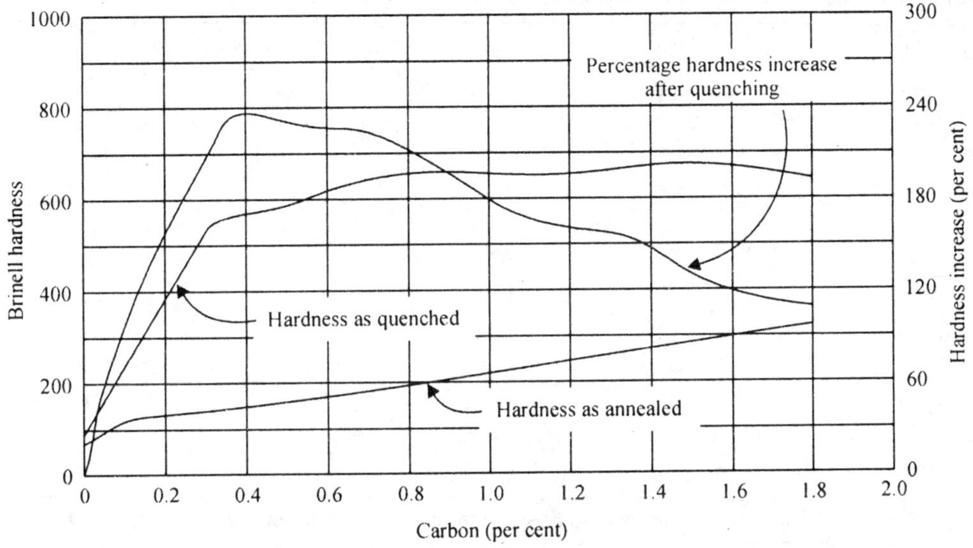

Fig. 5.1. Relation between Brinell hardness and carbon content.

The term hardening as used in the heat treatment of steel refers to that process of cooling by which the steel is made hard. The steel is quenched in some medium, whereby heat is removed from the steel at the desired rate. It is not always possible or convenient to employ a quenching medium that will produce that rate of cooling which will give the steel a required hardness. In hardening, it is usually customary to quench the steel into certain liquids which are readily available.

The cooling rate for hardening is controlled by the selection of the proper quenching medium. Quenching media include water, dilute sodium hydroxide solution, sodium chloride brine, oil-water emulsion and air. Quenching characteristics of liquid coolants are controlled by the following factors:

1. Temperature of the medium.
2. The specific heat: the amount of heat necessary to raise the temperature one degree per unit weight.
3. The heat of vapourisation: the amount of heat necessary to vapourise unit weight.
4. Thermal conductivity of the medium.
5. Viscosity.
6. Agitation: the rate of movement of piece or flow of coolant.

The relative cooling rates that may be obtained with some quenching media are shown in Table 5.1. As can be seen in the table, there is a wide range of cooling rates available. Where extremely rapid cooling is required, water spray may be used. It is general practice to quench machinery steel in oil. Water quenching of steel containing more than 0.35 per cent carbon may be dangerous because of the liability of cracking and warping. When austenite transforms, there is an increase in volume. If the transformation occurs at a very low temperature, as in the formation of martensite, this increase, in volume will set up very high stresses in the steel which may be of sufficient magnitude to cause the development of microscopic cracks or visible rupture.

Certain steels may not develop maximum potential hardness even though they are cooled at a rate equal to, or greater than the critical cooling rate. This situation may be attributable to the presence of retained austenite in the quenched steel.

Table 5.1. Relative cooling rates for some quenching media.*

Quenching medium	*Cooling rate relative to that of water at 65°F†* (1328–1022°F)	*(At 392°F)*
Aqueous solution 10% NaOH	2.06	1.36
Aqueous solution 10% H_2SO_4	1.22	1.49
Water at 32°F	1.06	1.02
Water at 65°F	1.00	1.00
Mercury	0.78	0.62
Water at 77°F	0.72	1.11
Oil (rapeseed)	0.30	0.055
Water at 122°F	0.17	0.95
Water at 212°F	0.044	0.71
Liquid air	0.039	0.033
Air	0.028	0.007

* Determined at centre of a 4 mm diameter Nichrome ball through the temperature ranges indicated during quenching from 1580°F.

†Cooling rates for water at 65°F are 3260°F per second through range 1328 to 1022°F and 810°F per second at 392°F.

INTERRUPTED QUENCH

Large sections of certain steels in which high surface hardness is desired require water quenching. However, the high stresses induced by such a rapid quench may be objectionable. This objection may be overcome in some cases by time quenching. The procedure is to quench the steel into water for a pre-determined length of time after which it is transferred to an oil-quenching bath for completion of the transformation. The results of the application of this treatment to an SAE 52100 steel are discussed below.

Bars 2 ½ inch in diameter and 6 inch long were heated to 1550°F (843°C). Each bar was quenched into water for a specified length of time and then transferred to the oil-quenching tank. The results of this treatment are shown in Fig. 5.2.

By the usual quenching procedure where the cooling rate is greater than the critical, transformation to martensite occurs at relatively low temperatures. In a heavy section, this transformation will not take place simultaneously throughout the section. This may lead to serious internal stresses and possibly cracking of the piece. This difficulty, however, may be eliminated or the danger reduced by a procedure called martempering. In this treatment the steel is cooled rapidly to a temperature just above the M_8 and held until the temperature becomes uniform across the section. Then the piece is allowed to cool in air. In this way, the transformation tends to occur more uniformly throughout the section. The steps in this treatment are illustrated in Fig. 5.3.

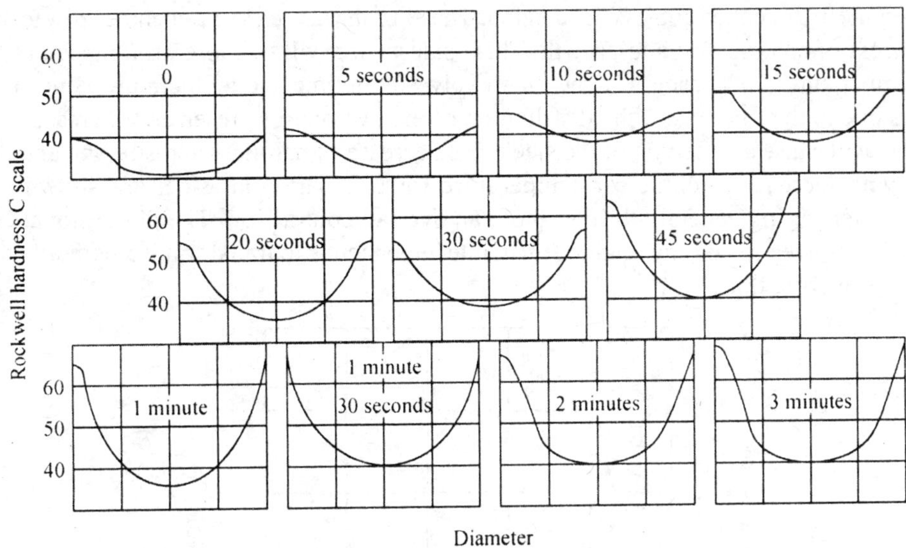

Fig. 5.2. Hardness distribution obtained in time-quenching bars of SAE 52100 steel. Quenched in water for times indicated and then into oil.

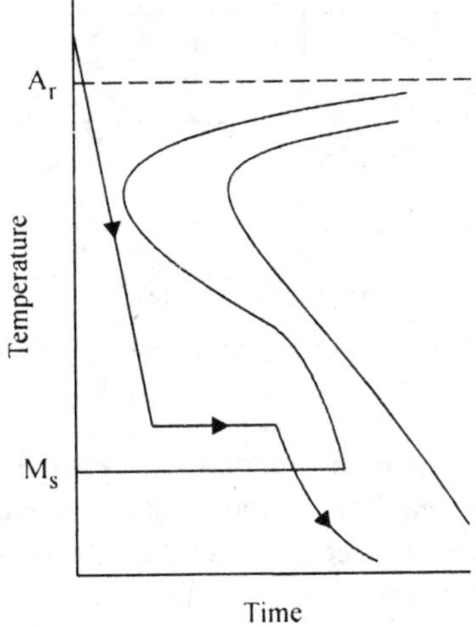

Fig. 5.3. Cooling conditions for martempering in relation to continuous cooling transformation curves.

TEMPERING

Tempering is the process of heating a hardened steel to any temperature below the lower critical temperature and then cooling at a desired rate. The object of tempering is to reduce the hardness and to relieve the internal stresses of a quenched steel in order to obtain greater ductility than is associated with the high hardness of the quenched steel. If retained austenite is present in the hardened structure, tempering may serve to cause isothermal conversion to bainite. If a fully hardened steel, that is, one in which the structure consists entirely of martensite, is heated to successively higher temperatures above room temperature, sub-microscopic particles of cementite will be expelled from the body centred tetragonal lattice and the lattice will become body centred cubic. With continued heating, the particles of cementite grow to microscopic size by a process of coalescence and growth. This coalescence will be more pronounced the higher the tempering temperature. This process is one that involves both time and temperature. Since the hardness of the steel is due to the fineness of the distribution of the two phases present, an increase in the size of these particles will cause a softening of the steel. The decrease of hardness and strength and the increase in ductility with increasing tempering temperature for one particular steel are shown in Fig. 5.4. Dimensionally, tempering generally brings about an over-all contraction. The dimensional change may be influenced by conversion of any retained austenite which tends to transform isothermally to bainite as a function of tempering time.

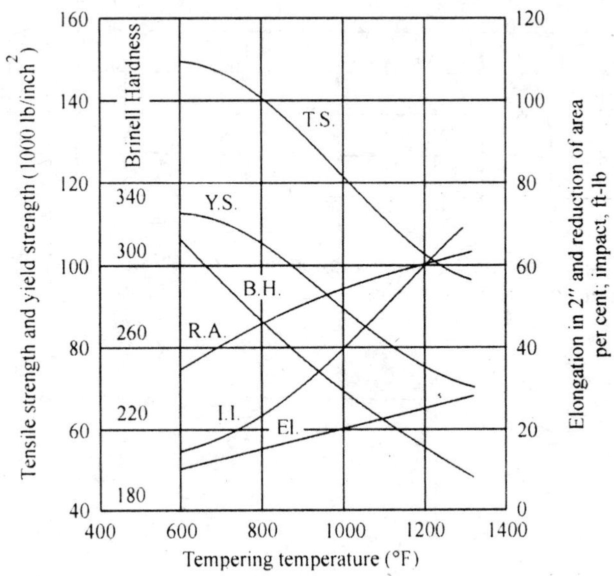

Fig. 5.4. Properties of SAE 1045 steel quenched in water from 1475–1525 F and tempered at temperatures indicated.

If the steel is tempered at a temperature just below the Ac_1 point, the cementite will collect into small spheroids surrounded by ferrite. This treatment is known as *spheroidising*. It is employed as a toughening treatment and in high-carbon steels machinability may be improved. A comparison of the mechanical properties of annealed and spheroidised steel is given below:

Property	*Pearlitic (1 per cent carbon)*	*Spheroidised (1 per cent carbon)*
Yield point (lb/inch^2)	85000	40000
Tensile strength (lb/inch^2)	1,50,000	78000
Reduction of area (per cent)	12–15	57
Brinell hardness	300	156

The most common method of treating a steel to obtain particular mechanical properties is to quench the steel in the proper medium from just above the Ac_3 temperature, followed by tempering at the proper temperature. The plain carbon steels containing less than about 0.35 per cent carbon are usually quenched in water; those containing 0.35 to 0.55 per cent carbon, in oil or water; those containing in excess of

0.55 per cent carbon, in oil. The choice may be influenced by other factors, such as section size and shape. A noted exception is with high-carbon, low-manganese steels which may also be water-quenched.

Since the process of tempering involves the diffusion coalescence and growth of cementite particles throughout the structure, time will be required for the softening to occur at any given temperature. This effect is illustrated in Fig. 5.5 which is a plot of hardness vs. time for four tempering temperatures. The time required at the tempering temperature is usually about 20 to 30 minutes. However, at the lower temperatures, a somewhat longer time may be necessary.

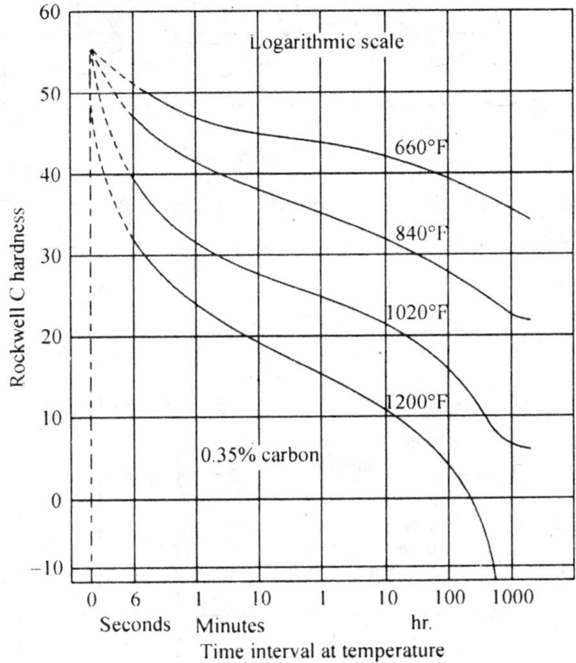

Fig. 5.5. Hardness vs. time for tempering a 0.35 per cent carbon steel at four temperatures.

Fig. 5.6 shows the variation of the toughness of a steel with tempering temperature as indicated by the Izod impact values for different tempering temperatures. The toughness gradually increases with increase of tempering temperature, but between 450 and 650°F (232 and 343°C), the Izod value decreases very noticeably. Above this temperature range, the toughness again increases to a very high value. This range is referred to as the brittle-tempering range and is particularly noticeable in certain alloy steels. Such brittleness is probably due to the formation of martensite or bainite from retained austenite.

The toughness of some steels when tempered to about 1150°F (621°C) and very slowly cooled is inferior to the toughness of the same steel when quenched from the tempering temperature. This phenomenon, called temper brittleness, is illustrated in Fig 5.7. The chromium-nickel steels and others of relatively high carbon content show this characteristic. The cause of temper brittleness is not clear, but it is probably attributable to the precipitation of some phase from ferrite. Slow cooling from the temperature range of 1000 to 1200°F (538 to 649°C) permits this precipitation, while quenching retains this phase in solution. Temper brittleness can be removed by re-heating the steel to the tempering temperature followed by quenching. Molybdenum tends to decrease the effect of temper brittleness.

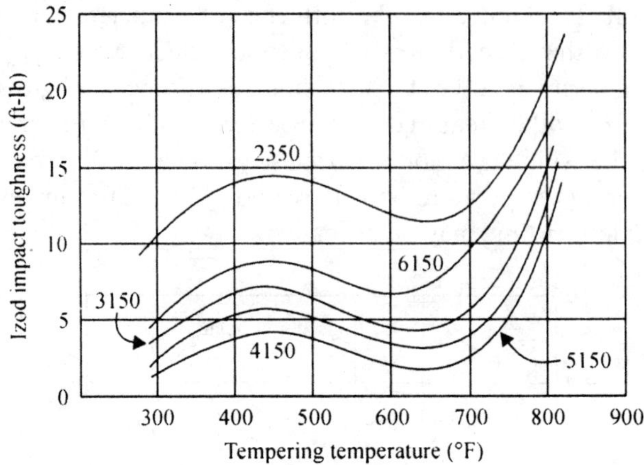

Fig. 5.6. Impact toughness vs. tempering temperature for five alloy steels fully hardened before tempering.

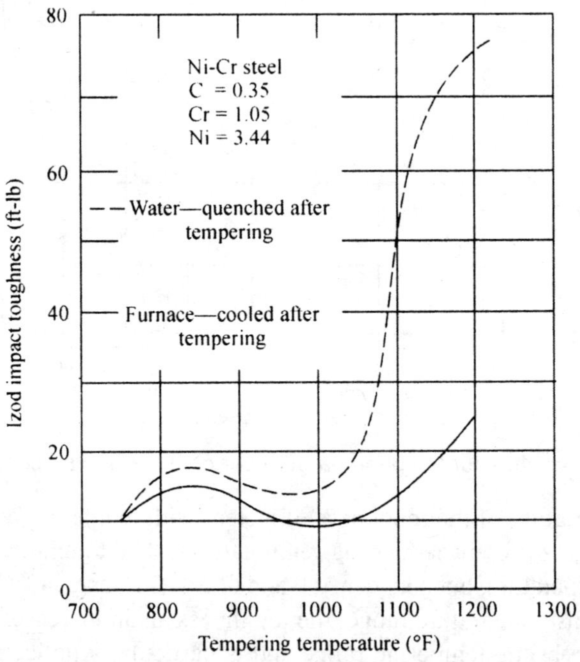

Fig. 5.7. Impact toughness vs. tempering temperature for a temper-brittle steel.

AUSTEMPERING

The quench and temper method of hardening steel to a desired hardness value has the hardness value has the inherent disadvantage of the possibility of producing cracks. In order to obtain the desired hardness, it may be necessary to quench the steel in oil since air cooling would render the steel softer than the desired hardness. After quenching in oil, the steel may be too hard and it must be tempered. If one could obtain a quenching medium that would cool the steel at the desired rate, there might be less tendency to crack. However, this procedure is not practical in most instances, so another method may be followed for certain steels. By referring to the isothermal transformation diagram, it will be noted that the desired properties can be secured by transformation at some specific temperature. The steel must be cooled at a rate that will avoid passing through the nose of the continuous cooling transformation curve with resultant formation of fine pearlite. Cooling must be stopped and the temperature held at that value for the proper time to produce the desired structure. In this way, none of the austenite transforms to martensite; hence, the formation of microscopic cracks is prevented. This process, called austempering, is illustrated in Fig. 5.8. Only those steels for which the initial cooling rate can be fast enough in the desired section to prevent the formation of fine pearlite can be treated in the manner.

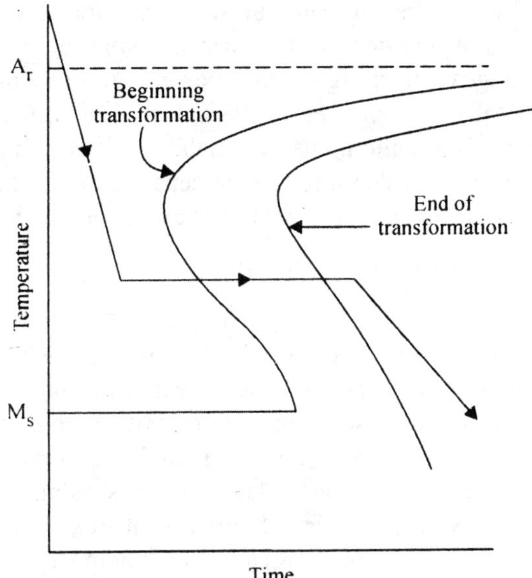

Fig. 5.8. Cooling conditions for austempering in relation to isothermal transformation curves.

Improvement of the dynamic properties by the use of austempering cannot be attained in all steels. Its application will depend in a large degree on the character of the isothermal transformation curve, upon the size of the pieces treated and upon the final hardness desired.

HARDENABILITY

The maximum hardness obtainable after quenching steels of different carbon content of the piece was transformed to the fully hardened state. With small sections, such hardnesses will be obtained; but with larger pieces, the centre of a bar may not become as hard, because transformation will occur at a higher temperature. The effect of increasing the size of bar on the uniformity of hardening can be shown by quenching a series of bars of different diameters into the same quenching medium from the same temperature. Each bar is sectioned and the variation of hardness is measured from surface to surface along a diameter. Even in the one-half-inch diameter bar, the centre of the piece is somewhat softer than the outside surface. The one-inch diameter piece shows almost the same surface hardness, but the centre hardness is very much less. The larger diameter pieces have a much lower hardness at the surface and in the interior. The effect of size on the hardening of steel is related to the term hardenability. Hardenability may be defined as susceptibility to hardening by quenching. A material that has higher hardenability is said to harden more uniformly throughout than one that has lower hardenability.

In times of critical material shortages, an understanding of hardenability has made possible the direct substitution of an available steel having a hardenability curve closely approaching that of the steel previously found satisfactory for a given part.

Hardenability is directly related to the isothermal transformation diagrams. A steel with a low critical cooling rate will harden deeper on quenching than a steel with a high critical cooling rate. Even though the centre of a large bar cools slower than the surface, the rate may be equal to or greater than the critical rate; hence, full hardening will result. Any condition that shifts the isothermal transformation curve to the right will improve hardenability. A steel can be made to harden deeper by coarsening the austenitic grain, since such treatment will shift the beginning of transformation to longer times. In doing this, one must remember that the other characteristics typical of a steel composed of coarse grains will be obtained which may not be to the advantage of the design. It may be better to select another steel, some alloy steel, which even though composed of fine grains may harden deeper by virtue of the inherent effect of the alloying element on the rate of transformation. The hardenability of a steel may be obtained by determining the diameter of the bar that will just harden through the centre. This is accomplished by heating round bars of different diameter to the austenitising temperature and quenching into oil or water.

CASEHARDENING AND SURFACE TREATMENT

The parts of some equipment must have a very hard, wear-resistant surface and also possess great toughness. Such a combination is usually not possible in a piece of heat-treated steel. If it is treated to give maximum surface hardness, it is too brittle. If it is treated for maximum toughness, it will not be hard enough. Consequently, several processes, known as casehardening, have been developed by which this combination can be attained commercially. These include carburising, cyaniding, carbonitriding, nitriding and localised surface hardening. The surface hardness obtained by the first three processes (carburising, carbonitriding and cyaniding) depends upon heat treatment after the composition of the case has been altered. The fourth process (nitriding) alters the composition of the case in such a way that the compounds formed are inherently hard. The last process of casehardening (surface hardening) depends entirely upon heat-treating the surface of a hardenable steel.

Carburising

Carburising is a process by which the carbon content of the surface of a steel is increased. Since the object of casehardening is to obtain a hard, wear-resistant surface with a tough interior, the first consideration is the selection of a low-carbon steel which will be tough. The relation between carbon content, strength and hardenability has already been discussed. The low-carbon steels are tougher than the medium or high-carbon steels, but their hardenability is quite low. Therefore, any means by which the carbon content of a low-carbon (0.15 per cent carbon) steel can be raised to eutectoid proportions will produce the desired results, provided the steel is properly heat-treated. The use of alloy steels may prove to be beneficial by directly improving toughness, hardenability, or wear-resistance.

There are three general methods of carburising, depending upon the form of the carburising medium: first, solid or pack carburising, employing solid carburising material; second, gas carburising, employing suitable hydrocarbon gases; and third, liquid carburising, employing fused baths of carburising salts.

The choice of the method that is to be used in carburising depends mostly upon the characteristics of the case which are desired, the equipment available and the quantity of parts which is to be handled.

Cold treatment is often required in connection with the hardening of carburised parts. This is particularly true of those alloy steels that tend toward austenite retention such as AISI-SAE 3310, 4320, 4620, 4820 and 9310. The greatest improvement in hardness takes place after the first cold treatment and is most pronounced near the surface where the carbon content is a maximum.

Grain Size

The austenitic grain size also has considerable influence on the carburising characteristics of the steel and the response to hardening.

Some steels when carburised and heat-treated do not harden uniformly. Casehardening by carburising is used extensively for all sorts of machine parts, including cams, piston pins, gears, pump shafts, etc. In some applications, only certain portions of the part require a hard surface. This can be done either by slowly cooling the part after carburising, machining away the surface of those portions that are not to be hardened and then heat-treating; or by copper plating those portions that are not to be carburised.

Cyaniding

Closely related to carburising is the process known as cyaniding in which an extremely thin case (usually less than 0.010 inch) is produced. This process is similar to liquid carburising except that the bath used is generally fused sodium cyanide. The steel is immersed in a molten bath of sodium cyanide at a temperature between 1475 and 1600°F (800 and 870°C) for a period of between 30 minutes and 3 hours, depending upon the depth of case required. The depth of penetration obtained by cyaniding is shown in Fig. 5.9.

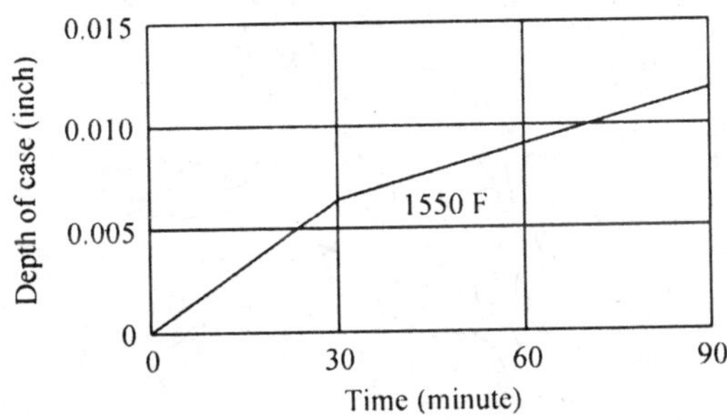

Fig. 5.9. Depth of penetration vs. time in cyaniding.

Sodium cyanide decomposes in two possible ways on heating in contact with air:

1. $2NaCN + 2O_2 \longrightarrow Na_2CO_3 + CO + 2N;\ 2CO \rightleftharpoons CO_2 + C$
2. $2NaCN + O_2 \longrightarrow 2NaCNO$

also

$$NaCN + CO_2 \longrightarrow NaCNO + CO;$$
$$3NaCNO \longrightarrow NaCN + Na_2CO_3 + C + 2N$$

Sodium cyanide is oxidised to form sodium cyanate which further dissociates to liberate carbon and nitrogen. The last two elements, in turn, are absorbed by the iron.

In order to obtain hardness after cyaniding, it is necessary to quench directly into oil or water from the cyaniding bath. In cyaniding, both carbon and nitrogen are absorbed at the surface of the steel. Nitrogen imparts inherent hardness, whereas the increased carbon content makes the surface of the steel respond to a quenching treatment. Probably the greatest use of cyaniding is for parts that are to be subjected to relatively light loads and that require improvement in the surface-water resistance. The containers for the molten cyanide are usually pressed steel and the area above the container should be well ventilated to allow adequate removal of fumes.

Carbonitriding

Carbonitriding is a modification of the gas carburising process with the addition of anhydrous ammonia gas to the furnace atmosphere to cause both carbon and nitrogen to be absorbed by the surface of the steel at the carbonitriding temperature. The process is also known as dry cyaniding, gas cyaniding, nitrocarburising and ni-carbing. Although a wide variety of gas mixtures are possible for this process, depending on the work load and furnace construction, a common atmosphere consists of 15 per cent anhydrous ammonia gas with 5 per cent natural gas in a carrier atmosphere of endothermic gas having a furnace dew point of 20 to 25°F (–6 to – 4°C). Carbonitriding is conducted at lower temperatures than gas carburising. The temperatures range from 1200 to 1625°F (650 to 885°C) with 1550°F (845°C) being most common. The absorption of nitrogen is greater at the lower temperatures, whereas the process closely approaches gas carburising at the higher temperatures with a minimum transfer of nitrogen. Fig. 5.10 shows the relationship between temperature, time and case depth. In view of the relatively low temperatures used in carbonitriding compared with those used in gas carburising, it is seldom economical to specify a case depth greater than 0.020 inch.

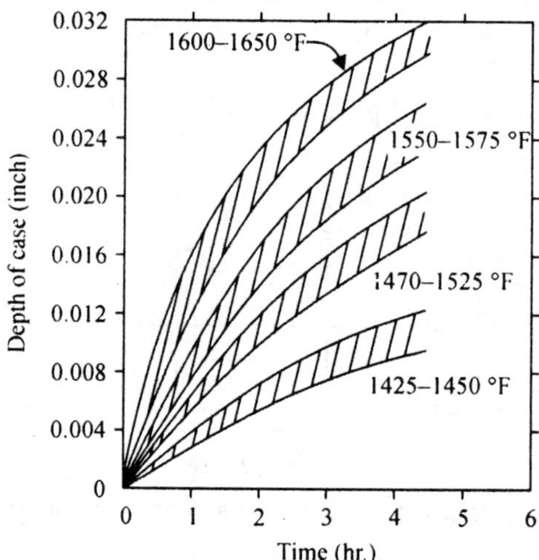

Fig. 5.10. Depth of penetration vs. time and temperature in carbonitriding.

Nitrogen increases the case hardenability of steel, increases the solubility of carbon in austenite and lowers the lower critical temperature. It is possible to harden small carbonitrided parts by cooling from a temperature as low as 1200°F (650°C) by transferring the part to an atmosphere cooling chamber without the necessity of oil quenching, thus minimising distortion. The increased hardenability made possible by the alloying effect of nitrogen permits the oil quenching of carbonitrided plain carbon steels that would otherwise require drastic water quenching to develop effective hardening. A comparison of the hardenability of carbonitrided and carburised plain carbon steel is shown in Fig. 5.11. In view of the high carbon and nitrogen content of carbonitrided cases, it is often necessary to employ a cold treatment at –100°F (–75°C) to convert retained austenite and to increase the surface hardness.

Nitriding

Nitriding is a casehardening process by which the surface hardness of certain alloy steels may be increased by heating in contact with a nitrogeneous medium, without the necessity of quenching. The process involves the formation of hard, wear-resistant nitrogen compounds on the surface of the steel by absorption of nascent nitrogen. Prior to nitriding, it is essential that the required core properties be developed by heat treatment of the steel in its rough machined state. In such heat treatment the tempering temperature must be at least 100°F (40°C) above the intended nitriding temperature. Furthermore, the structure must consist of fully tempered martensite free of ferrite. If considerable machining is required to bring the parts to dimension after the core has been heat treated, the part must be stress-relieved at a temperature above the temperature at which the nitriding is to be done. The presence of residual stresses leads to non-uniform nitriding and an unsatisfactory case. The surface of the steel must be free of decarburisation or there will be excessive growth and the case will be very brittle and susceptible to cracking and spalling. If the surface of the steel is contaminated with such substances as oil, the absorption of nitrogen will be impeded. Nitrogen absorption is restricted by highly buffed or burnished surfaces.

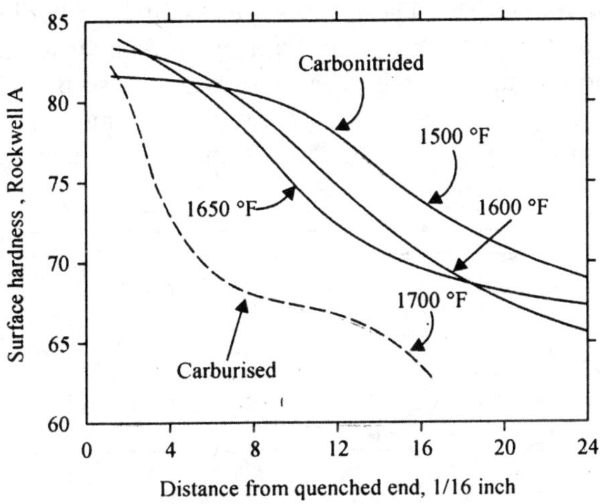

Fig. 5.11. End-quench hardenability of SAE 1020 steel carbonitrided vs. carburised.

The gas nitriding process consists of placing properly racked and supported steel parts in an alloy retort or specially constructed furnace heated to a temperature between 930 and 1050°F (500 and 565°C), while anhydrous ammonia gas is circulated around the parts to be intruded. Retorts or boxes for nitriding are usually constructed of Inconel or high-nickel alloys and generally incorporate a fan to assure adequate circulation of ammonia in intimate contact with the work. Anhydrous ammonia gas partially dissociates or cracks on the surface of the workpiece at the nitriding temperature in accordance with the following reaction:

$$NH_3 \rightleftharpoons N + 3H$$

$$2N \rightleftharpoons N_2$$

$$2H \rightleftharpoons H_2$$

The nascent nitrogen combines with the elements in the steel to form nitrides. Both nascent nitrogen and hydrogen are rapidly converted to their molecular forms. The process is controlled by metering the flow of ammonia gas and checking the exhaust gas for percentage of dissociation or cracking. Decreasing the flow of ammonia gas increases the percentage of dissociation. In the single-stage nitriding process, the dissociation is generally maintained at 30 per cent and the temperature is controlled at 975°F (525°C). In long nitriding cycles there is a tendency to build up a high nitrogen content on the surface, known as a 'white layer.' This compound layer may extend to a depth of 0.002 inch and is very brittle, thus requiring removal by finish grinding or lapping if chipping or spalling is to be avoided.

A two-stage nitriding process has been developed, known as the 'Floe process', which makes it possible to minimise the white layer to a depth of from about 0.0002 to 0.0005 inch. The first stage is generally accomplished at a temperature of 975°F (525°C) for a period of from 5 to 10 hours. After this time, the temperature is raised in the second stage to 1025°F (550°C) and the dissociation is increased to 80 per cent for the remainder of the cycle. Dissociated ammonia from an external cracking unit must be admitted to produce the high per cent of dissociation in the second stage. The second stage serves as a diffusion cycle permiting the high nitrogen surface layer to diffuse inward in a manner similar to the diffusion in gas-carburised cases. Although gas nitriding is the most widely used production process of nitriding, salt baths have been developed that are useful for small lots of parts or for shallow cases. These baths are essentially alkali cyanides, which partially oxidise to form cyanates and then liberate active nitrogen. Anhydrous ammonia is also bubbled through salt baths in a modification of the process in which case nitriding probably takes place directly from the active nitrogen which is derived from the ammonia. The variation of hardness produced by nitriding is shown in Fig. 5.12 for various time cycles. During the process of nitriding, the dimensions of a part will be found to increase somewhat.

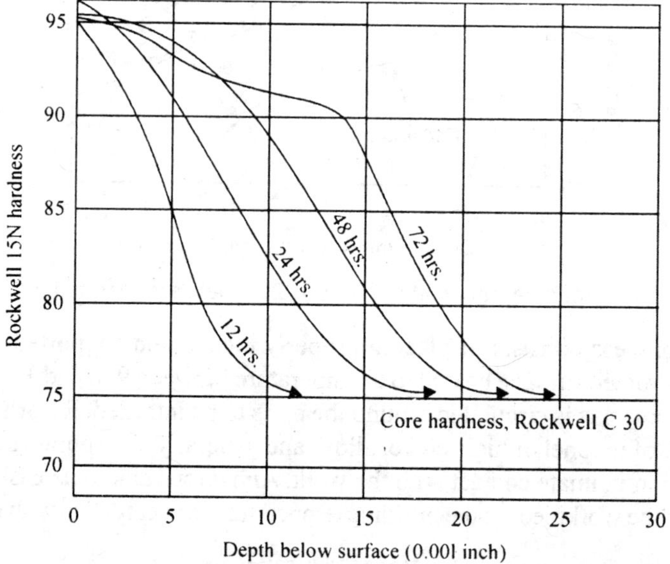

Fig. 5.12. Hardness vs. depth below surface of nitrided nitralloy 135 modified for various time cycles at 975°F and 30 per cent dissociation.

Therefore, allowance must be made for this growth in the original dimensions. The growth is of the order of 0.002 inch in diameter and 0.0005 inch/inch in length for a solid round bar. Once the growth pattern has been established for a given part and the nitriding cycle, the growth will remain constant for subsequent lots, provided all processing variables are constant. Parts may be placed in service in the 'as-nitrided' condition or may be finished to close dimensions by grinding, lapping, or honing. Since the nitriding temperature is relatively low, there is less chance for warpage to occur than in other hardening processes. However, distortion may be serious if the part has not been properly stress-relieved before nitriding. This can be very serious if the surface area of one side of a part is greater than the other and internal stresses are present at the time of nitriding. This condition will cause undue warping.

Areas on which nitriding is not desired may be plated with copper, bronze, or nickel. Proprietary paints containing tin oxide and a vehicle may also be applied. The tin oxide is reduced in the nitriding atmosphere to metallic tin which effectively prevents nitriding in selected areas.

Steels for Nitriding

The addition of such elements as aluminium, chromium, vanadium, titanium, tungsten, molybdenum and manganese to steel greatly improves the nitriding characteristics. Most of the steels that are commonly used for nitriding contain combinations of aluminium, chromium, molybdenum, vanadium and in some instances, nickel. The carbon content is somewhat higher than is employed for carburising grades to provide support for the hard, brittle case. Steels in the SAE 4000 series respond well to nitriding, but do not develop the high surface hardness typical of special steels containing aluminium.

Steels of special composition which are used principally for the purpose of nitriding are referred to as nitralloys. N 135, N 135 modified and EZ steels have excellent properties and, therefore, are used not only for nitriding, but also for machine parts where high strength is required. The presence of nickel in nitriding steels tends to reduce the hardness of the case somewhat, but improves toughness and strength. Nickel also tends to impart greater hardness and strength to the core by the mechanism of precipitation hardening taking place during the nitriding cycle to produce a core hardness of Rockwell C 40–44. For improved machinability, the sulphur or selenium-bearing steels may be employed. Hardenable stainless steels may be nitrided. The austenitic stainless steels nitride with difficulty to shallow case depths and low hardness, because of the presence of the impervious oxide film on the steel. Corrosion resistance of the stainless steels is greatly reduced by nitriding. In addition to high hardness and wear resistance, nitrided steels are more corrosion-resistant and have a higher endurance limit than most of the plain carbon, carburised steels. Such parts as gauges, bushings, forming dies, bearing parts, cylinder liners, dies, air craft crankshafts, wrist pins, valve slides and sleeves, gears, thrust washers, etc., are surface-hardened by nitriding. The hardness and wear resistance of valve parts that are operated at elevated temperatures can be maintained by using nitrided steel The cost of the nitriding process is a limitation to its use. When extreme hardness is required which cannot be obtained by carburising, then nitriding becomes feasible. In general, the time required to complete the process is much longer than with carburising, requiring from 50 to 90 hours, to produce a maximum case depth of from 0.020 to 0.030 inch. Furthermore, nitriding requires close control by regulating ammonia flow and dissociation.

Surface Treatment of Steel

Coatings may be applied to the surface of iron and steel to improve corrosion, heat, or wear resistance; to re-build worn or undersized parts; or merely to serve as an ornamental finish. Metallic coatings may be classified as; (i) hot-dipped; (ii) electroplated; (iii) impregnated; (iv) sprayed; (v) faced; or (vi) clad. The principal non-metallic coatings include: (i) oxide; (ii) phosphate; and (iii) vitreous enamel.

Heat Treatment of Cast Irons

The operations of heat treatment that can be employed to modify the properties or structures of cast irons are stress relieving, annealing, quenching, tempering and surface hardening.

The relief of casting stresses is accomplished by heating within the range 800–1000°F (430–540°C), holding one hour per inch of section and then following by furnace cooling. As a general rule, this treatment will not change the structure or promote graphitisation. The temperature should, however, be reduced for gray cast irons containing appreciable amounts of graphitisation agents and increased when strong carbide stabilising elements are present. The operation promotes dimensional stability in castings prior to machining or service.

Annealing operations for cast iron may be classed as: (i) full annealing; (ii) graphitising; and (iii) malleableising, depending on the purpose. In the broad sense, the term annealing implies softening. Machinability of iron castings may be improved by annealing processes which cause graphitisation of hard massive proeutectoid cementite or cause decomposition of the eutectoid cementite. The former is accomplished by full annealing, which consists of heating to above the upper critical temperature in the range 1500–1800°F (815–985°C). The exact time and temperature are governed by the presence of carbide stabilising elements. After the proeutectoid cementite is broken up by graphitisation, castings are furnace-cooled to allow formation of coarse pearlite. Decomposition of pearlite in gray cast iron by a process known as graphitising consists of annealing below the lower critical temperature in the range 1200–1500°F (650–815°C) until the combined carbon is reduced by decomposition of the eutectoid cementite to form ferrite and graphite. The rate of graphitisation will be great in high-silicon irons. Malleableising is a specialised annealing operation in the manufacture of malleable cast iron.

Quenching and tempering increase the hardness and abrasion resistance of gray cast iron. However, other mechanical properties are not, in general, improved over the as-cast product. Tensile, transverse and impact strength, together with deflection, are all lowered by quenching. Tempering permits recovery to values corresponding to the as-cast condition. Alloy cast iron with fine graphite flakes dispersed in a pearlitic matrix offers the best response to hardening, giving hardness values as high as 600 Brinell. Castings that are to be quenched should be designed with all precautions kept in mind as to drastic section changes. An austenitising range of 1450–1600°F (790–870°C) is suitable for most gray cast iron grades, with the lower temperatures giving highest strength properties in a given composition. The usual quenching medium is oil, although many alloy grades will form a martensitic structure by air hardening. The effects of tempering at various temperatures are shown in Fig. 5.13. Surface hardening of regular grades of gray cast iron by means of flame or induction hardening is common for certain applications. Alloy irons containing aluminium as an essential alloying element may also be nitrided.

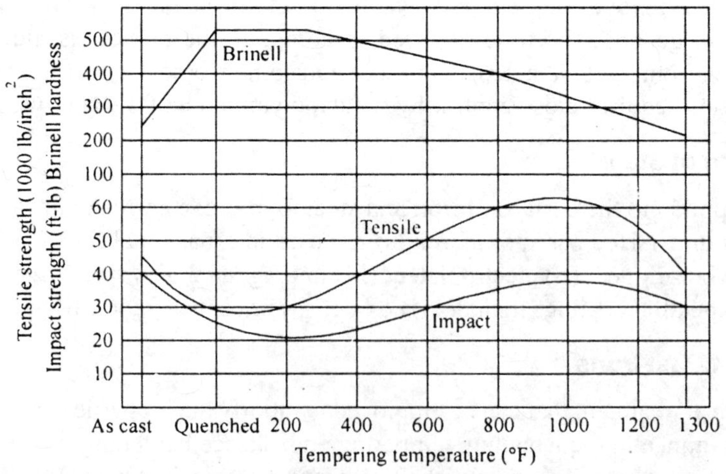

Fig. 5.13. Properties of quenched and tempered gray cast iron.

Chapter 6

Theory of Heat Treatment of Steel

INTRODUCTION

The theory of heat treatment of steel is understood as the analysis of the processes of structure formation (on transformations) and the particularities of structural state of alloys (in non-equilibrium).

This chapter deals with the theory of heat treatment of steel, based on the general theory of phase transformations in undercooled systems.

FORMATION OF AUSTENITE

The transformation of pearlite into austenite can only take place on a very slow heating, as follows from the Fe-C constitutional diagram. Under common heating conditions, the transformation is retarded and results in overheating, i.e., occurs at temperatures slightly higher than those indicated in the Fe-C diagram.

When overheated above the critical point, pearlite transforms into austenite, the rate of transformation being dependent on the degree of overheating. Fig. 6.1 shows the time of transformation at various temperatures (depending on the degree of overheating). The form of the curves shows that the transformation takes place faster (in a shorter time) at a higher temperature and occurs at a higher temperature on a quicker heating. For instance, on quick heating and holding at 780°C, the pearlite-to-austenite transformation is completed in 2 minutes and on holding at 740°C, in 8 minutes.

The diagram in Fig. 6.1 is plotted in time-temperature co-ordinates and, therefore, heating curves can be drawn on it. The superposition of heating or cooling curves on the diagram of transformations at constant temperature cannot give accurate quantitative relationships, but may be useful for qualitative analysis of the processes. The method will be widely used in the further discussion.

A pensil in the figure corresponds to the heating of steel at a definite rate, say v_2. It intersects the lines of the beginning and end of transformation in points a″ and b″. Therefore, with continuous heating at a rate of v_2, we shall fix the transformation occurring in the temperature interval from point a″ to b″. With slower heating, say, at a rate of v_1, the curves will be intersected at lower tmperatures (points a′ and b′) and the transformation will also take place at lower temperatures.

Irrespective of the heating rate, the transformation occurs immediately after the passage through the equilibrium critical point A_1, the temperature interval between a′ and b′ or between a″ and b″ on heating with a rate of v_1 or v_2 determines not the physical beginning and end of the transformation, but the temperature interval within which the main mass of pearlite changes to austenite. Points a and b are practically determined as the points at which the content of the new (or respectively of the old) phase is 5 (or 1) per cent.

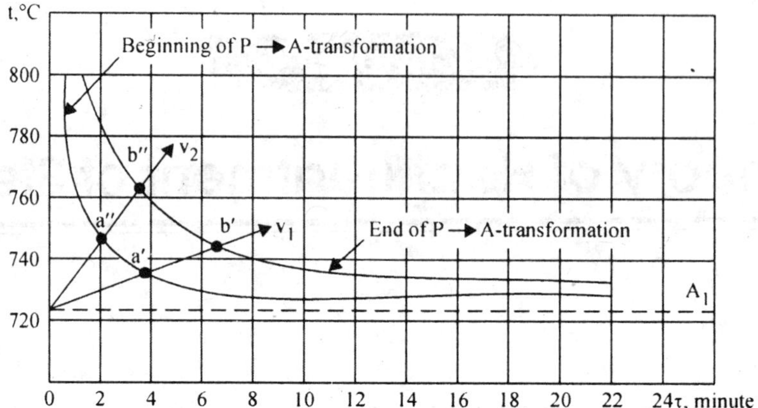

Fig. 6.1. Transformation of pearlite (P) into austenite (A) at constant tmperature. Steel with 0.86 per cent C.

The curves of the beginning and end of transformation asymptotically approach the horizontal line A_1. With heating at an infinitely slow rate, the curves of the beginning and end of the transformation will merge and intersect the horizontal line A_1 in infinity (after an infinitely long time) and the pearlite-to-austenite transformation will occur at a single 'point', i.e., at a constant temperature. This evidently would be the case of equilibrium transformation in the Fe-C constitutional diagram. Unlike equilibrium transformations, real transformations occur at tmperatures above A_1 and within a temperature interval, rather than at a fixed constant temperature, this temperature interval lying higher at a higher heating rate.

The end of the transformation is characterised by the formation of austenite and disappearance of pearlite (ferrite + cementite). This austenite is, however, inhomogeneous even in the volume of a single grain. In places earlier occupied by lamellae (or grains) of pearlitic cementite the content of carbon is greater than in places of ferritic lamellae. This is why the austenite just formed is inhomogeneous.

In order to obtain homogeneous austenite, it is essential in heating not only to pass through the point of the end of pearlite-to-austenite transformation, but also to overheat the steel above that point and to allow a holding time to complete the diffusion processes in austenitic grains.

The rate of homogenisation of austenite appreciably depends on the original structure of the steel, in particular on the dispersity and particle shape of cementite. The transformations described occur more quickly when cementite particles are fine and therefore, have a large total surface area.

COARSENING OF AUSTENITIC GRAIN

At the beginning of pearlite-to-austenite transformation the first grains of austenite form at the boundaries between the ferrite and cementite—the two structural constituents of pearlite. Since these boundaries are very developed, (the area of ferrite-cementite boundaries in 1 cm^3 of pearlite is a few square metres), the transformation starts from formation of a multitude of fine grains. Therefore, at the end of the transformation the austenite will be composed of a great multitude of fine grains whose size characterises what is called the original austenitic grain size. Further heating (or holding) upon the transformation will cause coarsening of austenitic grains. The process of grain coarsening is spontaneous, since the total surface area of grains diminishes (the surface energy decreases) and a high temperature can only accelerate the rate of this process.

In that connection, two types of steel are distinguished: inherent fine-grained and inherent coarse-grained, the former being less liable to grain coarsening than the latter.

Fig. 6.2 shows how the size of grain varies in these two types of steel which differ from each other in the kinetics of grain coarsening.

The transition through the critical point A_1 is associated with an abrupt refinement of grain. Austenitic grains in fine-grained steel do not grow on subsequent heating up to 950–1000°C; above that temperature, the factors that prevent grain coarsening are eliminated and the grains rapidly grow in size. In coarse-grained steel, the factors that might prevent grain coarsening are inactive and the process of grain coarsening begins immediately after the transit through the critical point.

As may be seen from Fig 6.2 the size of grain in inherent fine-grained steel is smaller at temperatures slightly above the critical point (Ac_1), but may be larger at higher temperatures. Therefore, the size of grain in a given piece of metal is not indicative of the inherent grain size.

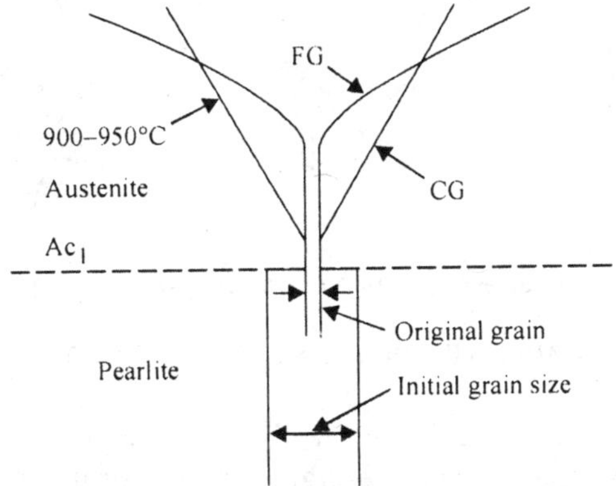

Fig. 6.2. Grain coarsening in inherently coarse-grained (CG) and inherently fine-grained (FG) steel.

The size of grains formed in a steel by heat treatment is called the actual grain size. Thus, a distinction should be made between:

1. Original grain, i.e., the size of austenitic grains immediately after the pearlite-to-austenite transformation.
2. Inherent (natural) grain, i.e., the liability of austenite to grain coarsening.
3. Actual grain, i.e., the size of austenitic grains under the given particular conditions.

The size of pearlitic grains at the same temperature of the austenite-to-pearlite transformation depends on that of the austenitic grains from which they have formed (Fig. 6.3). Austenitic grains grow only during heating (but are not refined in subsequent cooling), because of which the highest temperature a steel is heated to in the austenitic state and the inherent grain size of that steel determine the final grain size.

The properties of a steel are affected only by the actual grain size and not by the inherent grain size. If two steels of the same grade (one inherently coarse-grained and the other, fine-grained) have the same actual grain size upon heat treatment at different temperatures, their properties will also be the same; if otherwise, many properties of the two steels will also be different.

Coarsening of austenitic grain in steels has almost no effect on the statistic characteristics of mechanical properties (hardness, rupture resistances, yield limit, elongation), but can appreciably reduce the impact toughness, especially at a high hardness (low-temperature tempering). This is due to the fact that grain coarsening raises the ductile-to-brittle transition temperature.

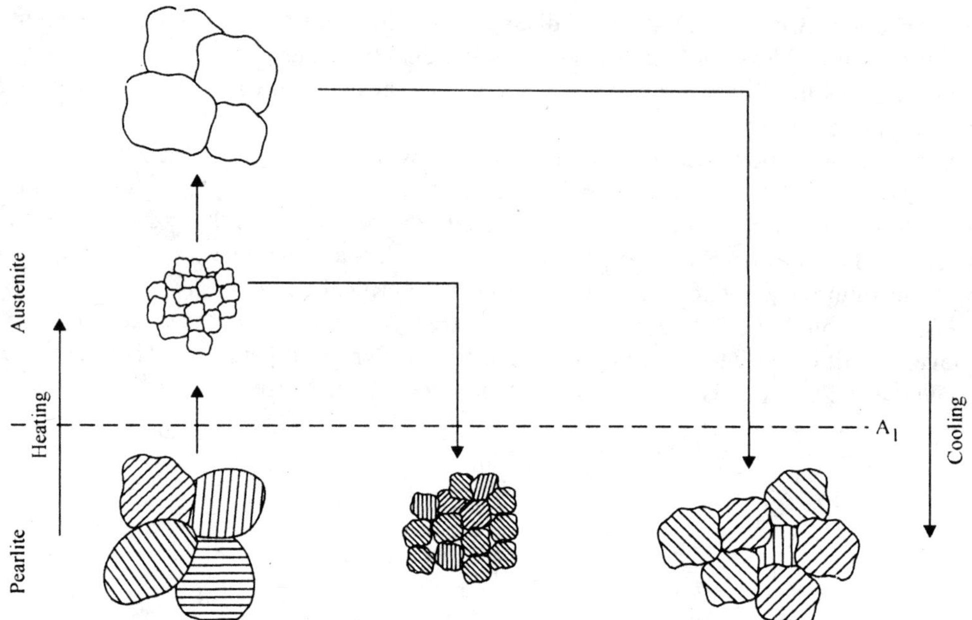

Fig. 6.3. Variations in grain size of pearlite depending on heating in the austenitic range.

Wherease the actual grain size affects the properties of steel, the inherent grain size is decisive for the processes of hot working.

Inherently fine-grained steel is insensitive to overheating, i.e., intensive grain coarsening begins at appreciably higher temperatures than in a coarse-grained steel. For that reason, the temperature interval of hardening for the former is substantially wider than for the latter. Inherently fine-grained steel can be rolled (or forged) at higher temperatures and the process (rolling or forging) can be finished at a higher temperature without danger of forming a coarse-grained structure. As a rule, all grades of killed steel are made inherently fine-grained and all rimming steels, inherently coarse-grained.

DECOMPOSITION OF AUSTENITE

The austenite-to-pearlite transfromation is essentially the decomposition of austenite (solid solution of carbon in γ-iron) into almost pure α-iron and cementite:

$$Fe_{\gamma}\,(C) \longrightarrow Fe_{\alpha} + Fe_3C$$

At the equilibrium temperature A_1, the transformation is impossible, since the free energy of the original austenite is equal to that of the final product, pearlite.

The transformation can only start at a certain undercooling when the free energy of the ferrite-carbide mixture (pearlite) is lower than that of austenite (Fig. 6.4). The lower the transformation temperature, the higher the degree of undercooling and the greater the difference in free energies and the transformation proceeds at higher rate. In the pearlitic transformation, the new phases sharply differ in their composition from the initial phase; they are ferrite which is almost free of carbon, and cementite which contains 6.67 per cent C. For that reason the austenite-to-pearlite transformation is accompanied with the diffusion, re-distribution of carbon. The rate of diffusion sharply diminishes with decreasing temperature, therefore, the transformation should be retarded at a greater undercooling.

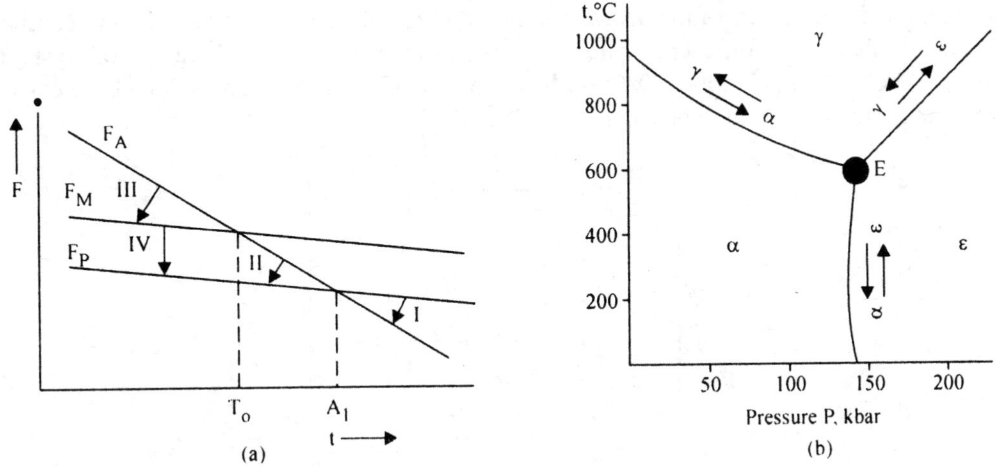

Fig. 6.4. (a) Changes in free energy of austenite (F_A), martensite (F_M), and pearlite (F_P) with temperature variations; and (b) the regions of α, γ and ε-iron depending on temperature and pressure.

Thus, we have come to an important conclusion that undercooling (lowering the transformation temperature) may have two opposite effects on the rate of transformation.

On the one hand, a lower temperature (greater undercooling) gives a greater difference in free energies of austenite and pearlite ($\Delta F = F_A - F_P$), thus accelerating the transformation; on the other hand, it diminishes the rate of carbon diffusion, D, and thus slows down the transformation (Fig. 6.5). The combined effect is that the rate of transformation first increases as undercooling is increased to a certain maximum and then decreases with further undercooling (solid line in Fig. 6.5).

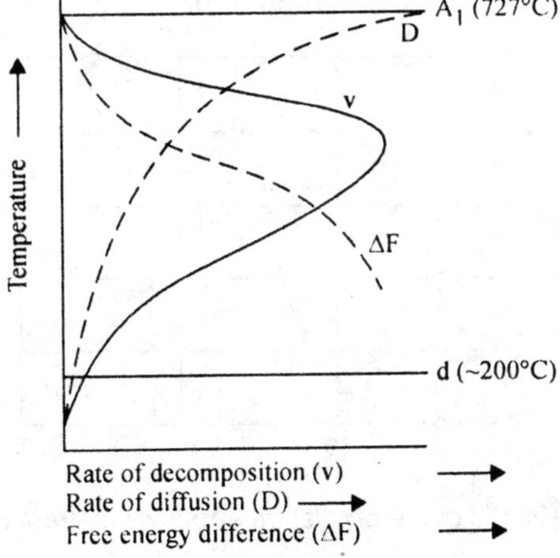

Fig. 6.5. Rate of decomposition of austenite (solid line) depending on temperature degree of undercooling.

At 727°C (A_1) and below 200°C (d), the rate of transformation is zero, since at 727°C the free energy difference is zero and below 200°C the rate of carbon diffusion is zero (more strictly, too low for the transformation to proceed). It has been investigated that the formation of pearlite is the process of nucleation of pearlite and growth of pearlitic crystals.

Therefore, the different rate cf the pearlitic transformation at various degrees of undercooling is due to the fact that undercooling differently affects the rate of nucleation, N, and the rate of crystal growth, G (Fig. 6.6). At point A_1 and below 200°C, both parameters of crystallisation, N and G, are equal to zero and have a maximum at an undercooling of 150–200°C.

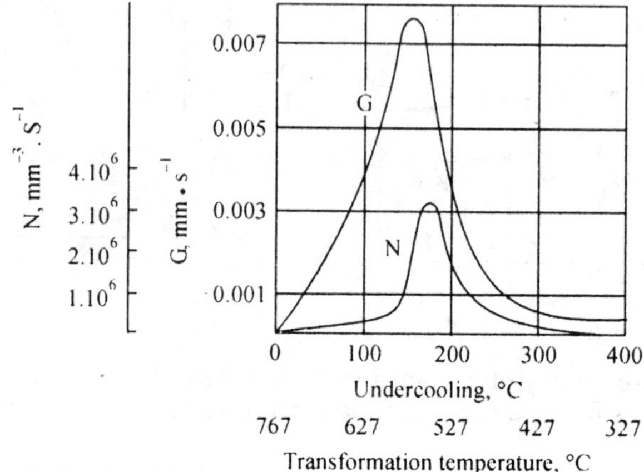

Fig. 6.6. Rate of nucleation (N) and crystal growth (G) of pearlite depending on temperature (degree of undercooling).

It follows from the foregoing that as soon as the conditions are favourable, i.e., austenite is undercooled below A_1, therefore, $F_P \ll F_A$ and the diffusion of carbon is not zero, centres of crystallisation appear, which give rise to crystals. This process occurs with time and can be represented in the form of so-called kinetic curve of transformation, which shows the quantity of pearlite that has formed during the time elapsed for the beginning of the transformation (Fig. 6.7).

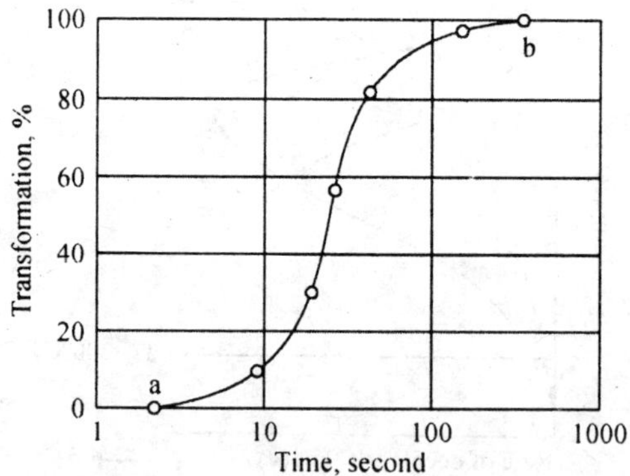

Fig. 6.7. Kinetic curve of austenite-to-pearlite transformation.

The initial stage is characterised by a very low rate of transformation; this is what is called the incubation period. Point a of the curve marks the moment when the beginning of the transformation is detected (usually this corresponds to the formation of 1 per cent pearlite). The rate of transformation increases with the progress in the transformation. Its maximum approximately corresponds to the moment when roughly 50 per cent of austenite has transformed into pearlite. The rate of transformation then diminishes and the transformation stops at point b.

The rate of transformation depends on undercooling . At low and high degrees of undercooling the transformation proceeds slowly, since N and G are low (Fig. 6.6), in the former case, owing to a low difference in free energies and in the latter, due to a low diffusion mobility of atoms. At the maximum rate of transformation the kinetic curves have sharp peaks, and the transformation is finished in a short time interval. Fig. 6.7 shows a series of kinetic curves similar to that in Fig. 6.8 but related to different temperatures (different degrees of undercooling).

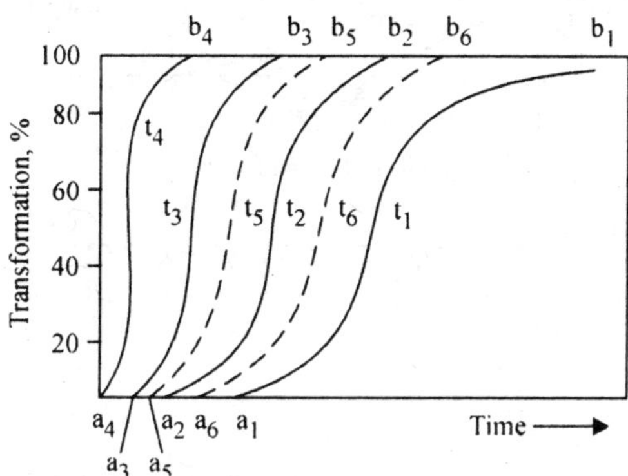

Fig. 6.8. Kinetic curves of austenite-to-pearlite transformation at various temperatures ($t_{10} > t_2 > t_3 \ldots$).

At a high temperature, t_1 (slight undercooling), the transformation proceeds slowly and the time of the incubation period (from the origin of coordinates to point a) and the time of the transformation proper (from the origin of coordinates to point b) are long. At a lower temperature of the transformation, i.e., a deeper undercooling, the rate of transformation is greater, and the time of the incubation period and of the transformation is shorter. The rate of transformation is at its maximum at temperature t_4 and again decreases at lower temperatures.

Having determined the time of the beginning of austenite-to-pearlite transformation (incubation period) and the time of the end of transformation at various degrees of undercooling, we can construct diagram like that shown in Fig. 6.9. The left-hand curve (a curve) determines the time of the beginning of the transformation i.e., the time during which austenite still exists in the undercooled state, and the section 'arom the axis of ordinates' to the a curve is the measure of its finstability. This section is the shortest at a temperature of 500–600°C (t_4), i.e., the transformation begins in a shortest time at that temperature.

The right-hand curve (b curve) shows the time needed to complete the transformation at a given degree of undercooling. As is seen, this time is the shortest at the same temperature t_4 (500–600°C).

Note that the abscissa of the diagram is logarithmic (log τ). This is done for more convenience, since the rates of formation of pearlite appreciably differ (thousands of seconds near the critical point A_1 and only one or two seconds at the bend of the curve).

The horizontal line M in the diagram determines the temperature of the diffusionless martensitic transformation. The martensitic transformation occurs by a different mechanism and will be discussed in the next section. The diagram in Fig. 6.9 gives the time of austenite-to-pearlite transformation for a given degree of undercooling, i.e., it is assumed that the transformation takes place at a constant temperature. Therefore, it is essentially the diagram of isothermal austenitic transformation. Diagrams of this type are usually called TTT-diagrams (time-temperature-transformation), or C-curves, owing to the specific shape of the curves.

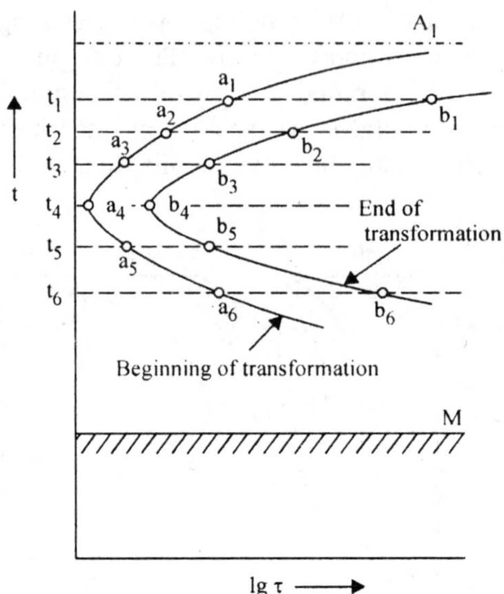

Fig. 6.9. TTT-diagram of austenite-to-pearlite transformation obtained from the kinetic curves in Fig. 6.8.

The structure and properties of the products of austenite decomposition depend on the temperature at which the transformation has taken place.

At high temperatures, i.e., low degrees of undercooling, a coarse-grained mixture of ferrite and cementite is formed which is easily discernible in the microscope. This structure is called pearlite.

At lower temperatures and therefore, greater degrees of undercooling, more disperse and harder products are formed. The pearlitic structure of this finer type is called sorbite.

At still lower temperatures (near the bend of the C-curve), the transformation products are even more disperse, so that the lamellar structure of the ferrite-cementite mixture is only discernible in the electron microscope. This structure is called troostite.

Thus, pearlite, sorbite and troostite are the structures of the same nature (ferrite + cementite) but a different dispersity of ferrite and cementite.

Pearlitic structures may be of two types: either granular (in which cementite is present in the form of grains) or lamellar (with cementite platelets). Homogeneous austenite always transforms into lamellar pearlite. Therefore, heating to a high temperature sets up favourable conditions for the formation of a more homogeneous structure and thus promotes the appearance of lamellar structures. Inhomogeneous austenite produces granular pearlite at all degrees of undercooling, therefore, heating to a low temperature (below Ac_3 for hypereutectoid steels) results in the formation of granular pearlite on cooling. The formation of granular cementite is probably promoted by the presence of undissolved particles in austenite, which serve as additional crystallisation nuclei.

However, initial heating to a temperature up to 900°C gives lamellar pearlite, with more disperse structures being obtained at lower temperatures.

The structure of the same steel at the same temperature of transformation but upon a low-temperature heating (780°C) is granular pearlite. Cementite grains that are formed in the structure are finer at a lower temperature of transformation.

Consequently, the size of cementite particles depends on the temperature of austenitic transformation and their form depends on the temperature of heating (the temperature of austenitisation).

The transformations at temperatures above and below the bend of the C-curve differ in the kinetics and the shape of decomposition products.

Above the bend of the C-curve, i.e., at small degrees of undercooling, the transformation begins from a few centres and pearlitic crystals freely grow to interference. Below the bend of the C-curve there forms an acicular microstructures, since acicular platelets cannot grow freely and the transformation mainly occurs through the formation of new crystals.

The acicular structure that forms below the bend of the C-curve is called bainite. The transformation of austenite into bainite is in some aspects similar to the pearlitic and martensitic transformations.

As has been given earlier, the rate of isothermal transformation of austenite is determined by the rates of nucleation and crystal growth, N and G.

The rate of crystal growth is quite sensitive to various changes in the steel structure and can be strongly affected by the metallurgical nature of the steel, the degree of deoxidation, the presence of undissolved particles, the homogeneity of austenite, and the size of austenitic grains. The presence of undissolved particles increases the value of G, since these particles serve as additional nuclei.

The value of G diminishes with increasing size of austenitic grains. Crystallisation nuclei appear mainly on grain boundaries, therefore, the conditions for nucleation are worse in coarse-grained steel where the extent of grain boundaries is smaller.

The factors mentioned have, however, no effect on the nucleation rate, N. The value of N depends only on steel compostion and for a steel of a given compostion, is a natural characteristic depending only on the degree of undercooling.

We have discussed the austenitic-to-pearlite transformation in steels whose composition is close to eutectoid. If the content of carbon in a steel differs from the eutectoid value, the pearlitic transformation will be preceded with the precipitation of ferrite or cementite (as follows from the iron-carbon constitutional diagram).

In hypoeutectoid steels, the transformation of austenite begins with the formation of ferrite and the saturation of the remaining γ-solution with carbon, and in hypereutectoid steels, with the precipitation of cementite and depletion of the austenite of carbon. Under equilibrium conditions, the decomposition of austenite into ferrite and cementite (pearlitic transformation) begins when the content of carbon in the austenite, remained upon precipitation of excess ferrite and cementite, corresponds to point S (0.8 per cent).

Consider the transformations under conditions of undercooling. Line SE in the iron-carbon diagram is the boundary of limiting saturation of austenite with cementite; this means that cementite can precipitate from austenite to the right of line SE and evidently, to the right of the extension of that line downwards into the region of undercooled austenite. Similarly, the extention of line GS is the boundary of saturation of undercooled austenite with ferrite.

If this is so, the simultaneous precipitation of ferrite and cementite from austenite is only possible when the austenite is supersaturated with both components, i.e., to the left of line GG′ and to the right. of line EE′ (in the shaded triangle of Fig. 6.10). This also means that the eutectoid will have a strictly definite composition only in crystallisation without undercooling (at point S), and the composition of eutectoid (pearlite) under real conditions in characterised by a concentration interval between the lines E′S and SG′.

The eutectoid which forms from undercooled austenite and has a concentration differing from the eutectoid value is called quasi-eutectoid.

As follows from the diagram in Fig. 6.10, the quasi-eutectoid in hypereutectoid steels contains more than 0.8 per cent carbon and that in hypoeutectoid steels, less than 0.8 per cent, the deviation from this value being greater at a lower temperature of transformation. Therefore, the lower the temperature of transformation, the less the excess ferrite (or cementite) precipitates before the pearlitic transformation begins. At temperatures near the bend of the C-curve and at lower temperatures, decompostion of austenite begins without precipitation of excess phases.

These processes are plotted in the diagram of isothermal decomposition of austenite.

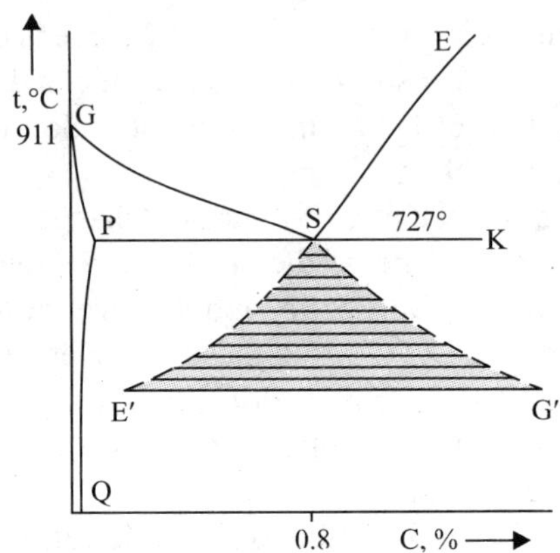

Fig. 6.10. Composition of eutectoid depending on the degree of undercooling below A_1.

The complete form of the diagram of isothermal austenitic decomposition for a hypoeutectoid steel is shown in Fig. 6.11. The diagram shows the curve of the beginning of ferrite precipitation, the curves of the beginning and finish of the pearlitic transformation, and horizontal lines A_3, A_1, M and M_f. Such a diagram can provide the principal information on isothermal transfromation in the given steel for any degree of undercooling. For instance, on undercooling to 650°C, the transformation begins in a definite time with the precipitation of ferrite from the solution. Ferrite precipitates for a certain time, after which the decomposition of austenite into pearlite begins and the process ends on the curve describing the finish of the transformation. If austenite is quickly cooled to 550°C, the transformation will start just with the formation of pearlite. The transformation at 550°C occurs much more rapidly than at 650°C.

If we take a hypereutectoid steel instead of hypoeutectoid, the decompostion of austenite at small degrees of undercooling will be preceded with precipitation of cementite.

The correlation between the nature of isothermal decomposition of austenite, the content of carbon and the temperature can be expressed in a generalised diagram, such as that shown in Fig. 6.12.

In that diagram, the decomposition of austenite takes place in a temperature interval between the horizontal lines A_1 and d. The temperatures corresponding to lines e and d (points of a particular carbon content) and their physical sense is also explained. In modern interpretation, the rate of diffusion of iron and alloying elements above point e is sufficient for realising the appropriate phase transformations, and above point d, only the rate of carbon diffusion is sufficient for the purpose. Therefore, only diffusionless (martensitic) transformations are possible below point d, while the transformations between points e and d occur with the diffusion of carbon but without displacement of metallic atoms. The transformations between the horizontal lines A_1 and e produce lamellar structures and those between e and d, acicular structures. The inclined line M determines the temperature of the beginning of diffusionless martensitic transformation. The diffusion pearlitic decomposition occurs without precipitation of either ferrite or cementite in region I; to the left of line SE′, the process is preceded by the precipitation of ferrite, and to the right of line SG′ by the precipitation of cementite.

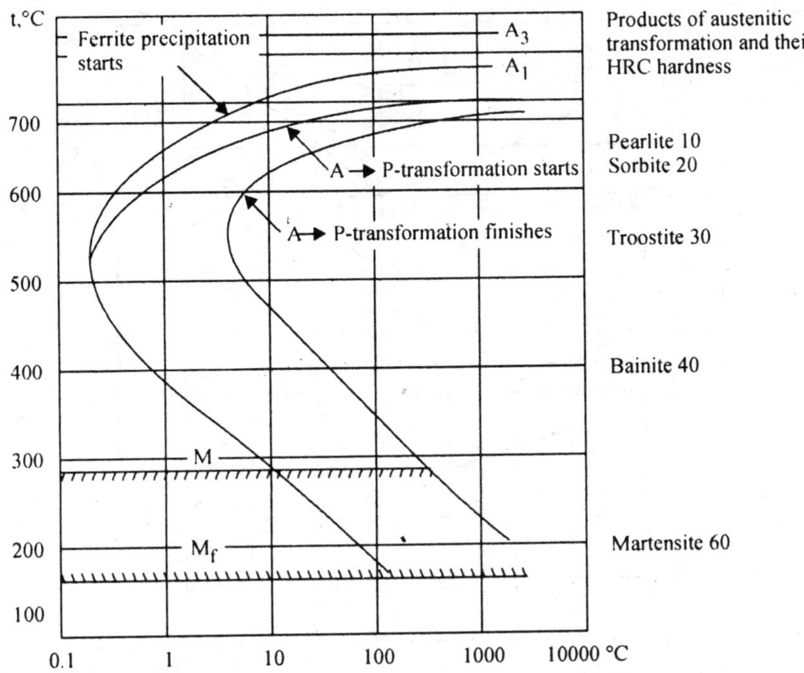

Fig. 6.11. Diagram of isothermal decomposition of austenite in hypoeutectoid steel (0.4 per cent C).

Consequently, the following regions of austenitic transformation are possible, depending on the content of carbon and the degree of undercooling:

I — Austenite-to-pearlite transformation.

II — Precipitation of ferrite followed by austenite-to-pearlite transformation.

III — Precipitation of cementite followed by austenite-to-pearlite transformation,

IV — Austenite-to-bainite transformation.

V — Austenite-to-martensite transformation and decomposition of retained austenite into bainite.

VI — Austenite-to-martensite transformation.

VII — Undercooled austenite remains non-transformed.

Having discussed the transformation of austenite at a constant temperature and various degrees of undercooling, we can now analyse the process of austenitic transformation in continuous cooling, when a steel heated to the austenitic state is cooled at different rates.

The diagrams of isothermal decomposition of austenite are constructed in temperature-time coordinates, as are the cooling curves. Let the cooling curves be superimposed on the isothermal decomposition diagram (Fig. 6.13).

Line v_1, which characterises slow cooling, intersects the diagram lines at points a_1 and b_1. The transformation will occur at these temperatures (determined by points a_1 and b_1) and will form coarse-lamellar pearlite of a low hardness.

With a quicker cooling (curves v_2 and v_3), the diagram lines are intersected at lower temperatures (points a_2, b_2 or a_3, b_3) and the transformation products will be more disperse.

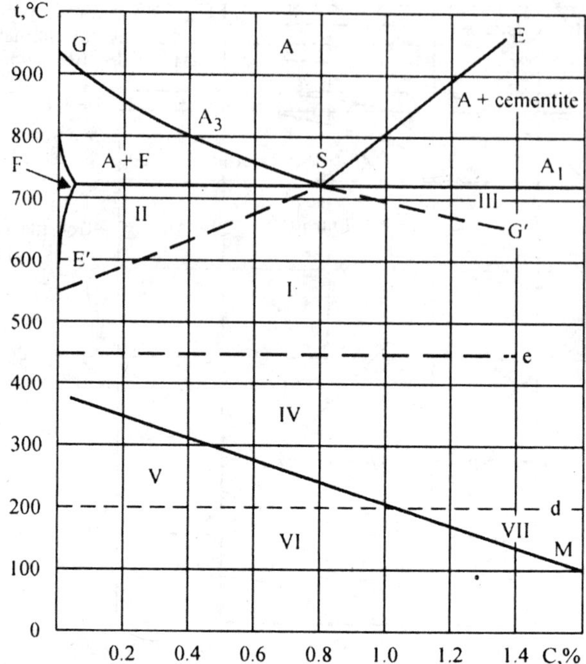

Fig. 6.12. Generalised diagram of transformation of undercooled austenite in carbon steel.

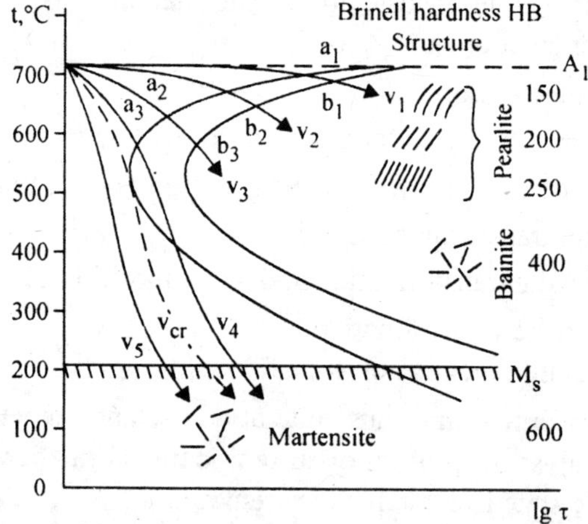

Fig. 6.13. Superposition of cooling curves on C-curve of austenite decomposition; the figure also schematically shows the structures formed and their hardness.

As follows from this construction, at a greater rate of cooling, the transformation occurs at a lower temperature and forms more disperse and harder products. At a high rate of cooling (v_5), the transformation has no time to proceed in the upper temperature range, the austenite will be undercooled to a low temperature and will transform into martensite, such a cooling will result in hardening.

Therefore, to harden steel, it should be cooled at a high rate so that austenite has no time to decompose the upper temperature range. As seen from Fig. 6.13 at cooling rates higher than v_{cr} (v_{cr} is the cooling curve that is tangent to the bend of the C-curve) the martensitic transformation takes place and at lower cooling rates, austenite decomposes in the upper temperature range.

The lowest cooling rate needed to undercool austenite up to martensitic transformation is called the critical rate of hardening. If a steel is to be hardened, it should be cooled at a rate not less than the critical. The critical rate v_{cr} is lower for steels whose curve of the beginning of transformation passes farther to the right. In other words, with a lower rate of austenite-to-pearlite transformation, it is easier to undercool the austenite to the temperature of martensitic transformation and the critical rate of hardening will be lower. If cooling is done at a rate slightly below the critical, the austenite will undergo only a partial transformation in the upper temperature range and the structure will consist of the products of transformation in the upper temperature range (troostite) and martensite.

More accurate estimations of the transformations at a continuously varying temperature are made by using the so-called thermokinetic, or anisothermal, diagrams of austenitic transformations, which characterise the transformation at various cooling rates.

Though isothermal transformation diagrams can give us much knowledge on the nature of transformations, the transformations only seldom can be achieved under isothermal conditions in practice.

With a very quick transforamtion, the process of decomposition of austenite can and usually does occur as the specified temperature is attained. This is why the isothermal transformation diagrams are quite inaccurate for holding times less than 10 seconds.

When large sections are heat treated, it is impossible to meet another important condition essential for the constructuion of a diagram, i.e., quick cooling to the given temperature. Isothermal austenite transformation diagrams are of great theoretical significance but are superseded by anisothermal diagrams in practical cases of selecting heat treatment conditions.

The cooling conditions have been studied quite thoroughly. It is possible in laboratories to imitate the conditions of cooling large massive articles (Fig. 6.14) and in various quenchant (Fig. 6.15) on small specimens. During cooling of specimens, the points of the beginning and end of transformations are determined by dilatometric (by changes in the dimensions of specimens) or magnetic (austenite is non-magnetic and its products are magnetic) method. The experimentally found cooling curves and the points of transformation are then plotted in a temperature-time diagram and the sections of like transformations are combined into regions (Fig. 6.16).

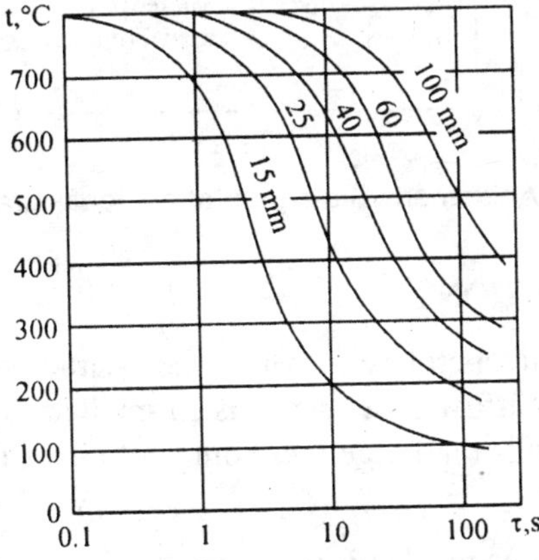

Fig. 6.14. Cooling curves for the centre of cylindrical specimens quenched in water. Numbers at curves are the diameters of specimens.

Up to this point, we have considered only schematic diagrams of austenitic transformation. For exhaustive information on the transformation in a particular grade of steel, however, use is made of both types of the diagram and of some additional data: grade and composition of the steel, heating temperature, austenitic grain size, properties (at least hardness) of decomposition products, and proportion of the constituents. All these data are combined as, for instance, is done in Fig. 6.17, which shows the diagrams of isothermal and anisothermal decomposition of austenite in steel Grade 45X.

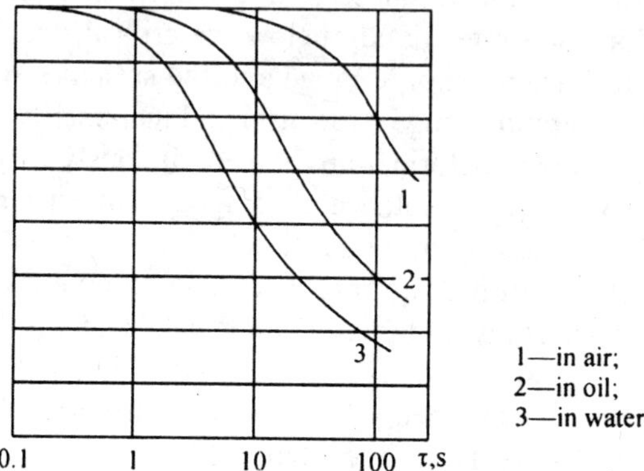

Fig. 6.15. Cooling curves for the centre of cylindrical specimens quenched in various media.

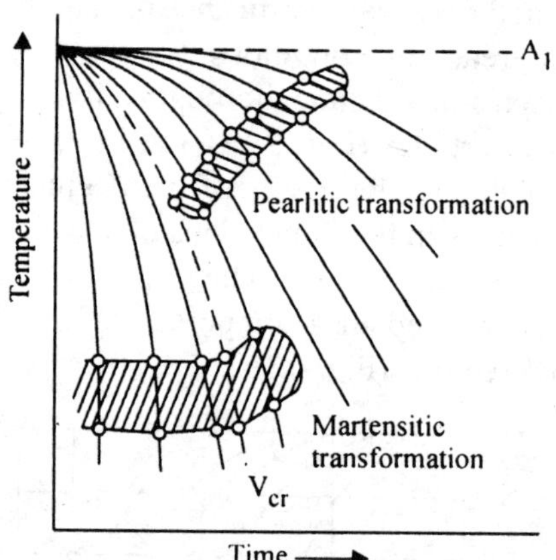

Fig. 6.16. Anisothermal diagram of austenitic transformation.

MARTENSITIC TRANSFORMATION

If austenite is undercooled to a temperature at which the γ-lattice is unstable, despite the presence of dissolved carbon, but the rate of diffusion of carbon is so small (owing to the low temperature) that can be neglected, then the iron lattice is rearranged without evolution of carbon:

$$Fe_{\gamma}(C) \longrightarrow Fe_{\alpha}(C)$$

This transformation differs in its mechanism and in the nature of transformation products from the eutectoid decomposition of austenite that has been discussed earlier.

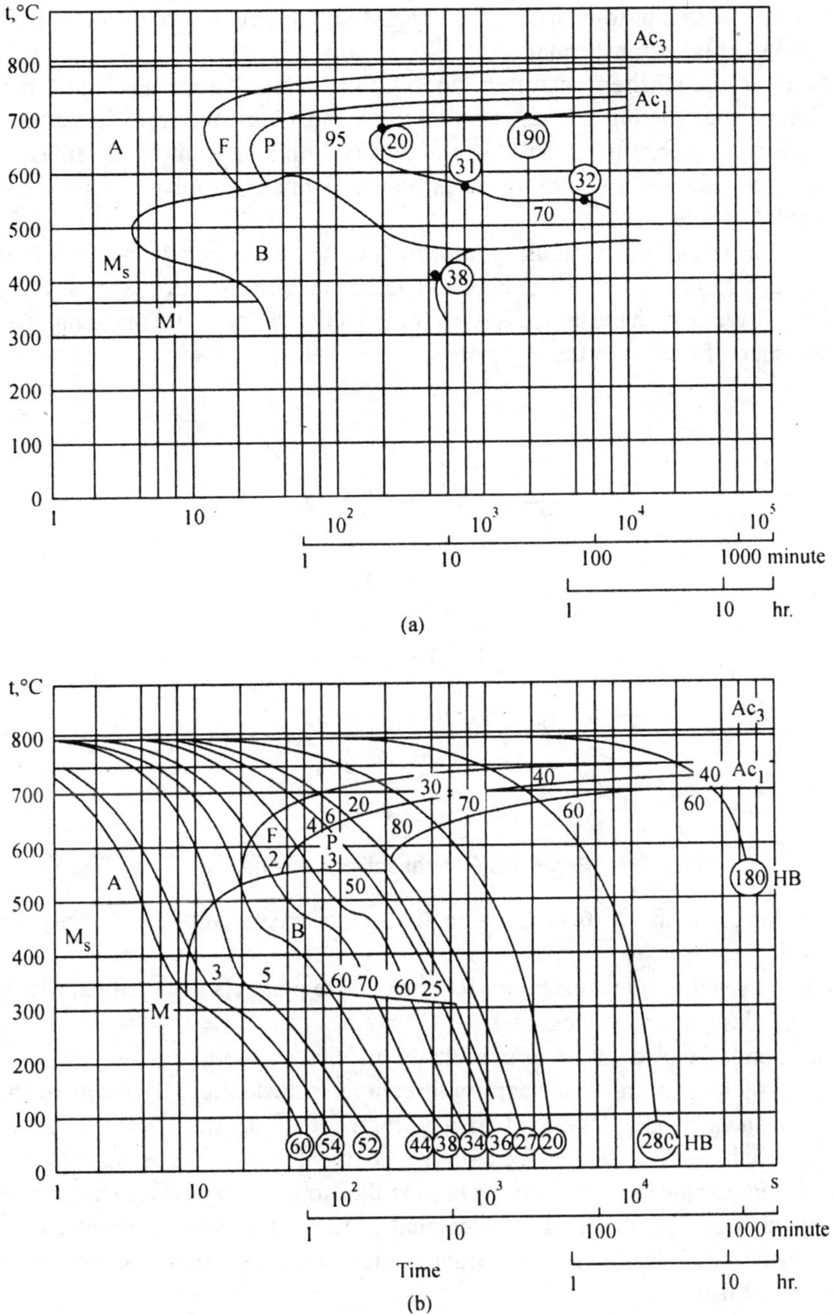

Fig. 6.17. Diagrams of (a) isothermal and (b) anisothermal (thermokinetic) transformation of austenite in steel 45X. Steel composition: 0.44 per cent C, 0.22 per cent Si, 0.80 per cent Mn, and 1.04 per cent Cr. Numbers at curves are percentages of a structural component upon transformation; circled numbers are HRC values of decomposition products.

As we recall, the austenite-to-pearlite transformation appreciably relies, on diffusion and may be regarded as a diffusion-type transformation.

In the austenite-to-martensite transformation, only the crystal lattice undergoes rearrangement, but the concentration of the reacting phases remains the same. Thus, the transformation is diffusionless. Martensite in steel is a supersaturated solid solution of carbon in α-iron of the same concentration as in the original austenite. Since the solubility of carbon in α-phase is only 0.01 per cent martensite is a supersaturated solid solution.

Since the transformation is diffusionless, carbon does not precipitate from the solution and only iron atoms are rearranged from f.c.c. to b.c.c. lattice. The crystal structure of martensite is characterised by tetragonality, i.e., the c/a ratio in the lattice is greater than unity Fig. (6.18). This tetragonality of martensite is due to the presence of carbon in the solution.

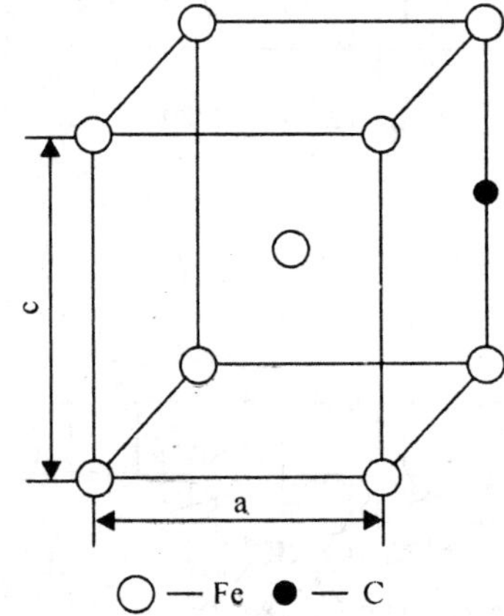

Fig. 6.18. Crystal cell of martensite.

As seen from the graph in Fig. 6.19 tetragonality of martensitic lattice is directly proportional to the content of carbon.

Martensite has a specific microstructure. Its crystals are platelets oriented parallel to one another or tilted at definite angles (60 or 120 degrees).

As has been found, the orientation of martensite platelets is due to the fact that martensite can form only on certain crystal planes and along certain directions in austenite. This oriented transformation can be regarded as a shear or displacement of a volume of metal in a definite plane, which occurs simultaneously with the $\gamma \longrightarrow \alpha$-transformation. The transformation is associated with an appreciable displacement of metal atoms in space, though neither the atoms change places nor the distances between them appreciably change. This oriented displacement of atoms during transformation produces a relieved surface in microsections. Thus, martensitic transformation differs from other phase transformations in the following typical features:

1. It is diffusionless, i.e., is associated only with lattice rearrangement and the composition of the final phase (martensite) is the same as that of the original phase (austenite).
2. It is oriented, i.e., the new phase (martensite) is regularly oriented relative to the old phase (austenite); the shearing effect of the transformation produces relieved surface. Transformations characterised by these two particularities are referred to the class of martensitic transformations.

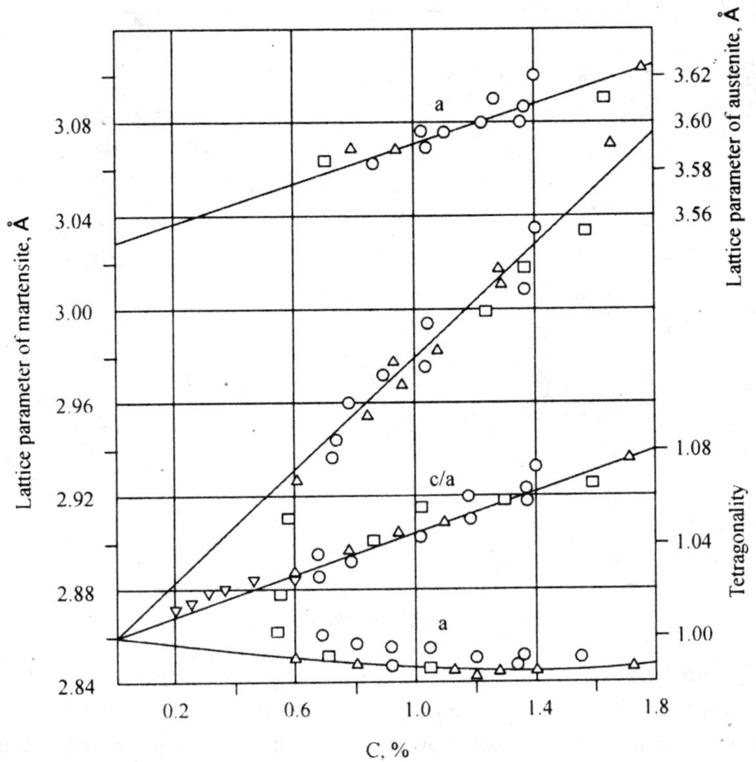

Fig. 6.19. Effect of carbon on lattice parameters of martensite and austenite and c/a ratio of martensitic lattice.

Martensitic transformations have been discovered in many metals and alloys and in this section we shall discuss only the martensitic transformation in steels.

The martensitic transformation in steel has the two typical features mentioned above and also some other specific features not found in other alloys. In steels, this transformation is irreversible, that is, can only proceed in the direction $Fe_{\gamma}(C) \longrightarrow Fe_{\alpha}(C)$, but cannot occur in the reverse direction by the same diffusionless mechanism. Further, martensitic crystals form in steel, irrespective of temperature, in an extremely short time interval (of an order of 10^{-7} s, i.e., practically instantaneously).

It has been established experimentally that the transformation consists in practically instantaneous formation of a portion of martensitic platelets, rather than of a single platelet (each platelet forms in a time interval of approximately 1.10^{-7} second and the whole portion consisting of hundreds or thousands of crystals, in 1.10^{-3} second), after which the transformation comes to a stop. With further cooling, the transformation is renewed by the formation of new portions and so on. The process can be represented graphically in the form of a kinetic curve of transformation which is called martensitic curve. Such a curve is schematically shown in Fig. 6.20 where the ordinate gives the quantity of martensite that has formed (measured by a magnetic or some other method) and the abscissa is temperature.

As seen from the curve, the transformation in cooling starts at point M_s. This point determines the temperature of the beginning of austenite-to-martensite transformation in a given steel.

The quantity of martensite increases as the temperature is lowered. The end of the transformation is determined by point M_f. At this temperature, a certain quantity of retained austenite remains in the structure. Further cooling below M_f causes no transformation nor changes the quantity of retained austenite.

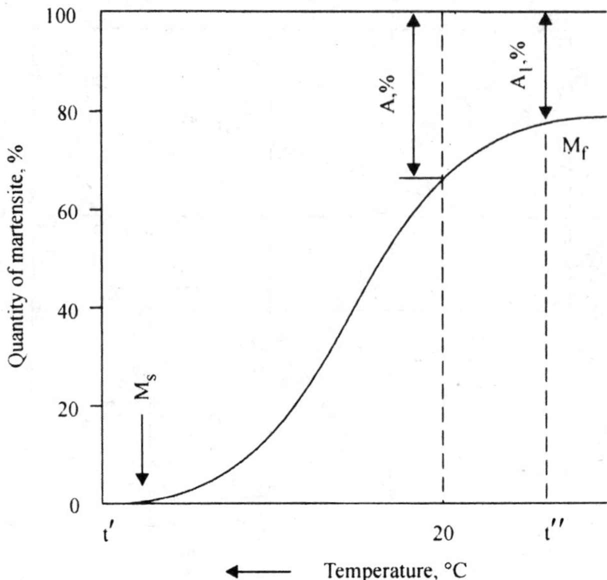

Fig. 6.20. Martensitic curve.

The position of the martensitic points (M_s and M_f) appreciably depends on steel composition. The temperature range of martensitic transformation lowers with increasing content of carbon.

The general effect of some or other factor on the temperatures of martensitic transformation can be represented in the form of so-called martensitic diagrams. In particular, the effect of carbon is shown in Fig. 6.21.

Carbon appreciably lowers the martensitic points. At a carbon content above 0.5 per cent, the temperature range of martensitic transformation extends to sub-zero temperatures, i.e., such steels cannot be made fully martensitic by continuous cooling to room temperature.

Some alloying elements also can lower the martensitic point so that certain grades of alloy steel containing a sufficiently high percentage of carbon and alloying elements have their M_s point blow 0°C, and common hardening produces a purely austenitic structure in them. It then follows that the temperature of martensitic transformation mainly depends on steel composition (composition of austenite).

Let us now discuss some other factors which may affect the martensitic transformation.

In the first place, it should be examined how the temperature of martensitic transformation is affected by cooling rate. Experiments with continuous cooling of austenite at various rates from v_{cr} (which is roughly 150°C/s for carbon steel) to approximately 10000°C/s failed to lower the temperature of the beginning of martensitic transformation.

It can be concluded therefore that the temperature of martensitic transformation is independent of cooling rate.

Though the cooling rate has no effect on the position of the martensitic point, it can definitely affect the course of the martensitic transformation. At temperatures slightly lower than M_s, a slower cooling gives a greater degree of transformation. This is due to the ability of austenite to isothermally transform into martensite at temperatures slightly less than M_s.

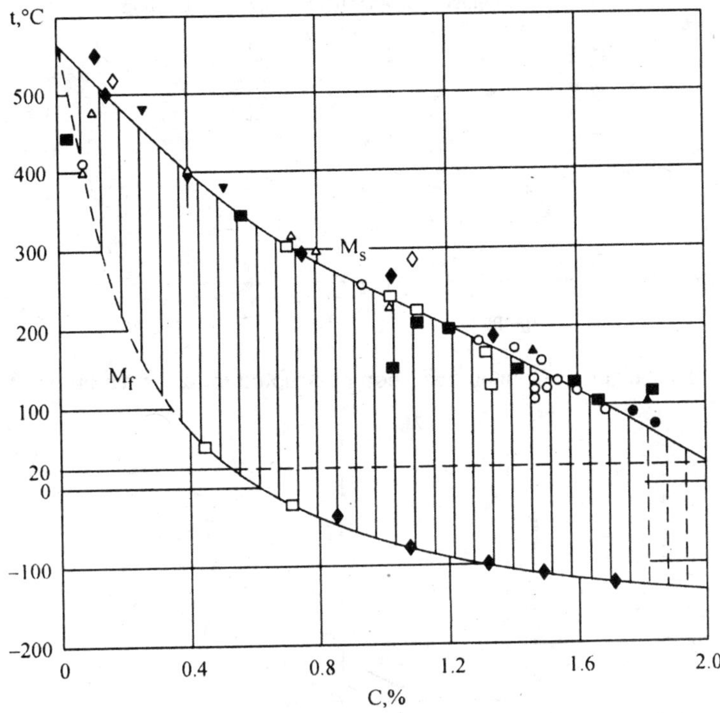

Fig. 6.21. Martensitic diagram (effect of carbon on the martensitic transformation points).

If the cooling is interrupted and an isothermal holding is allowed within the martensitic range Fig. 6.22, a certain, though small, amount of martensite forms during the holding time. This makes it possible to distinguish between athermal martensite which forms in continuous cooling and isothermal martensite which forms at a constant temperature. The latter differs from athermal martensite both in the appearance (microstructure) and the properties (the last circumstance has not been thoroughly studied). In common steels, the formation of isothermal martensite quickly comes to a stop, i.e., only a small quantity of isothermal martensite is formed, so that martensite in real steels is mostly of the athermal type. A holding time in the martensitic range (as also at temperatures 100–200°C above M_s) produces the effect of stabilisation of austenite which essentially consists in that the austenite-to-martensite transformation upon the holding time is renewed not immediately but only after a certain undercooling (Fig. 6.23) and the final structure contains more austenite upon final cooling, i.e., less martensite is formed.

The effect of stabilisation can be explained by the relaxation of the stresses which are essential for realisation of the martensitic transformation. This is why externally applied stresses cause martensitic transformation but if they are eliminated (for instance, by disintegrating a piece of steel into fine (monocrystalline particles), the martensitic transformation will not take place.

The structure of martensite in hardened steel is typically acicular. The austenite which can co-exist with martensite at room temperature in the structure is called retained austenite. Since the steel structure contains little austenite, the whole field is occupied by martensitic needles. With an appreciable quantity of retained austenite (20–30 per cent or more), it is revealed metallographically as bright fields between martensitic needles.

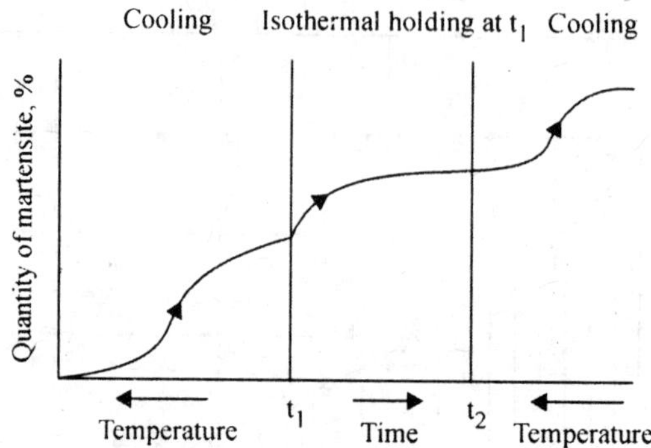

Fig. 6.22. Martensitic curve. Cooling has been interrupted for a certain time at temperature t_1.

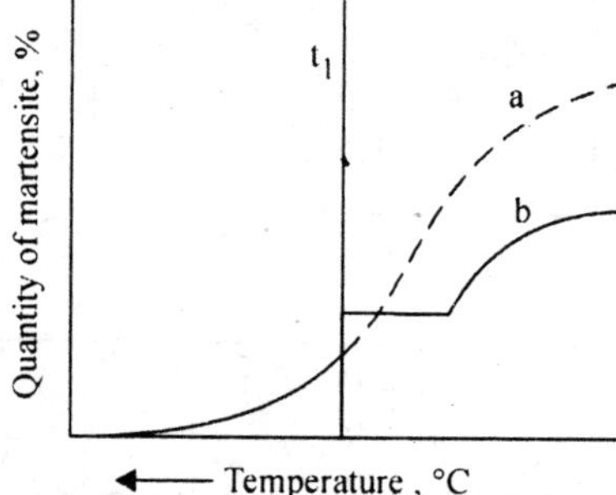

Fig. 6.23. Martensitic curve; a—continuous cooling; and b—quenching interrupted in the martensitic range (at t_1).

The quantity of retained austenite which can be fixed in a steel by hardening depends on the position of the martensitic point, being greater at a lower position of that point. This is why carbon, which lowers the martensitic point, increases the content of retained austenite (Fig. 6.24).

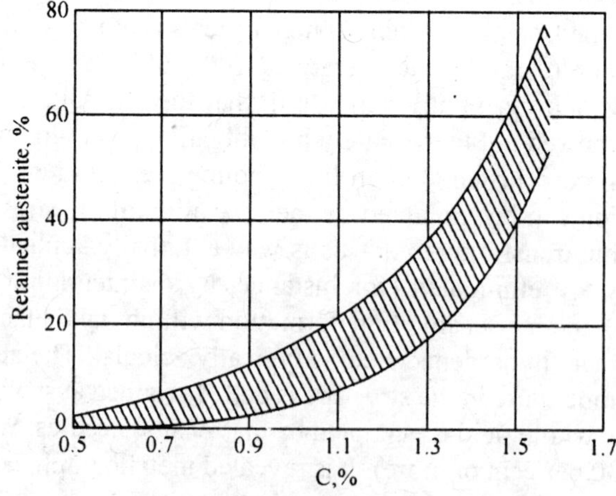

Fig. 6.24. Effect of carbon on the quantity of retained austenite in hardened steel.

The effects of various factors on the kinetics of the two types of martensitic transformation are summarised in Table 6.1.

Table 6.1. Effects of various factors on kinetic characteristics of athermal and isothermal martensitic transformations.

External factors	*Affected kinetic characteristic*	*In athermal martensitic transformation*	*In isothermal martensitic transformation*
Increasing cooling rate	Position of M_s point	Not affected or is raised	Lowered
Decreasing temperature	Rate of growth of martensitic crystal	Not affected or is accelerated	Decelerated
Stresses	Rate (degree) of transformation	Accelerated	Slowed down
Grain coarsening	Position of M_s point	Not affected	Lowered

BAINITIC TRANSFORMATION

The bainitic transformation of undercooled austenite occurs in a temperature range between the pearlitic and the martensitic range and for that reason is often called an intermediate transformation. It has features of both the pearlitic and the martensitic transformation (therefore it should not be related to the principal types of transformation).

The decisive feature of the bainitic transformation is that it occurs in the temperature interval where there is practically no diffusion (self-diffusion) of iron, but the diffusion of carbon is quite intensive, i.e., the interval of bainitic transformation is above Chernov's d point but below e point (Fig. 6.12). Bainitic transformation occurs by a more complicated mechanism than pearlitic or martensitic transformation. If austenite is undercooled to a suitable temperature (below point e), its carbon is re-distributed by diffusion so that some portions of austenite are enriched in carbon and the other are depleted. This inhomogeneity of the concentration causes the appearance of stresses and since the martensitic point of the carbon-depleted portions is above the temperature of the isothermal holding, a plastic deformation will induce the $\gamma \longrightarrow \alpha$-transformation by the martensitic reaction. This $\gamma \longrightarrow \alpha$-change by the martensitic type is the specific feature of the bainitic transformation, which is confirmed by the fact that bainitic transformation produces a relived surface on polished microsections.

Thus, though the $\gamma \longrightarrow \alpha$-change proper in the bainitic transformation takes place by the diffusionless mechanism, it is prepared by diffusion processes occurring in the austenite; these diffusion processes determine the rate of the bainitic transformation.

The portions of austenite which have been enriched in carbon undergo no changes and remain austenitic upon cooling from the temperature of isothermal holding to room temperature (or they may partially undergo the martensitic transformation).

Carbide particles which are found in the structure of steel upon bainitic transformation precipitate after the $\gamma \longrightarrow \alpha$-change, which fact indicates that the re-distribution of carbon does not result in full depletion of carbon in individual austenitic portions.

Whereas the described mechanism is valid for the whole temperature range of bainitic transformation, a change of temperature within that range may cause appreciable quantitative differences. At high temperatures (near point e), the effect of concentrations re-distribution is more appreciable than at lower temperatures (near point M_S). This is why a distinction is made between upper and lower bainite which differ from one another both in microstructure.

TRANSFORMATIONS ON TEMPERING

The transformations occurring during heating of hardened steel have been established quite reliably by studying the microstructure, crystal lattice and the physical and mechanical properties of astempered steel and of the variation of these properties in the course of tempering.

The original structure of hardened steel prior to tempering consists of tetragonal martensite and austenite. Since martensite is the structure of the highest specific volume and austenite, the structure of the least specific volume, the transformations on tempering should be associated with volume changes, i.e., the volume of metal will decrease upon martensitic transformation (contraction of specimen) and increase upon austenitic transformation (expansion of specimen). The transformations in tempering can be fixed by means of a dilatometric curve (Fig. 6.25). If an annealed specimen is heated in a dilatometer and there are no transformations in the metal, the instrument will plot a horizontal line. Curve 1 of a hardened steel specimen deviates from horizontal line 2; which is indicative of certain transformations. As is seen in the dilatometric curve of Fig. 6.25 no transformations occur to a temperature of about 80°C. With further heating to 200°C, the steel specimen contracts in length. This is an indication of what is called the first transformation on tempering. As has been detected by X-ray analysis, the c parameter of the martensitic lattice gradually diminishes within this temperature interval and the c/a ratio tends to unity.

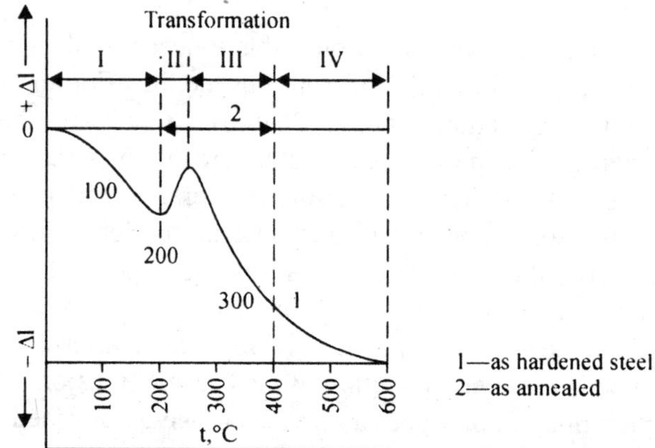

Fig. 6.25. Dilatometric curve of tempering of carbon steel (1.2 per cent C).

The martensite that forms in this low-temperature tempering, in which the c/a ratio is not equal but very close to unity is called temper martensite. Therefore, the first tempering transformation is the transformation of tetragonal martensite into almost cubic temper martensite.

As has been indicated earlier, the sole cause for the tetragonality of martensite is dissolution of carbon; thus, a decrease in tetragonality can be explained by precipitation of carbon from the solution.

The high-carbon phase that precipitates from the solution is extremely fine (a few atomic radii thick) carbide platelets which are coherently bonded with the solid-solution matrix. As has been established by radiographic and magnetothermal analysis, the metastable carbide that forms on low-temperature tempering differs from cementite. It has a hexagonal lattice and a formula close to Fe_2C and is described as ε-carbide in the specialist literature. At higher temperatures of tempering (300–400°C), an $\varepsilon \longrightarrow Fe_3C$ carbide transformation takes place (sometimes via an intermediate stage of χ-carbide, i.e., by the reaction $\varepsilon \longrightarrow \chi \longrightarrow Fe_3C$).

Thus, the first transformation on tempering produces temper martensite which is a heterogeneous mixture of supersaturated α-solution (of an inhomogeneous concentration) and non-isolated carbide particles. Heating to a temperature above 200°C will cause a different transformation resulting in expansion of the steel. This is what is called the second transformation on tempering which is confined to a temperature range from 200 to 300°C. In this interval, the retained austenite changes to a heterogeneous mixture of supersaturated α-solution and carbide. In other words, retained austenite transforms into temper martensite. This is diffusion transformation and in its nature resembles the bainitic transformation of primary austenite.

At the end of the second transformation, i.e., at 300°C, α-solid solution still contains around 0.15–0.20 per cent C. The contraction that occurs on further heating (Fig. 6.25) is indicative of the complete precipitation of carbon from the solution and of the relaxation of the internal stresses that have appeared owing to the prior transformations which involved volume changes. At the same time, carbides separate from the matrix and form cementite (Fe_3C). All these changes characterise what is called the third transformation on tempering.

In other words, the third transformation on tempering involves definite changes which result in stress relaxation and transformations of carbides. The third transformation is finished at 400°C, with the steel structure consisting of ferrite and cementite. A further rise in temperature can cause coagulation of ferrite and cementite particles, which can be readily seen in microstructures at large magnifications.

The transformations occurring on tempering are graphically illustrated in Fig. 6.26 where I is the curve of precipitation of carbon from the solution, which occurs mainly at low temperatures (in high-carbon steel), but extends onto a wide temperature range; II is the curve of decomposition of retained austenite, which occurs within a limited temperature interval depending on steel composition; III is the curve of stress relaxation which develops most intensively at temperatures within 300–400°C; and IV is the curve of coagulation of carbide particles which begins at low temperatures of tempering (150°C), but proceeds most intensively at temperatures above 400°C.

The indicated temperature ranges relate to tempering of carbon steel with slow heating. With fast heating, the temperatures of these transformations are shifted upwards along the temperature scale.

All the transformations just discussed occur within martensitic platelets, because of which the acicular structure is retained up to quite appreciable temperatures of tempering.

Tempering at temperatures above 400°C forms a mixture of ferrite and cementite; this is why the structures formed on tempering have the same names as those obtained through direct decomposition of austenite into ferrite and cementite.

A steel tempered at 350–500°C has the structure of troostite and that tempered at 500–600°C, the structure of sorbite. These structures differ in the hardness and dispersity of cementite particles.

It should also be noted that a ferrite-cementite mixture formed by decomposition of austenite substantially differs from a mixture obtained from martensite. The former (hardening troostite or hardening sorbite) contains lamellar cementite and the latter (if formed at temperatures above 400°C) has the granular form of cementite. This difference in the form of cementite determines the difference in the properties of these two mixtures.

EFFECT OF HEAT TREATMENT ON STEEL PROPERTIES

Heat treatment can appreciably change the properties of a steel, its effect on the mechanical properties being most important.

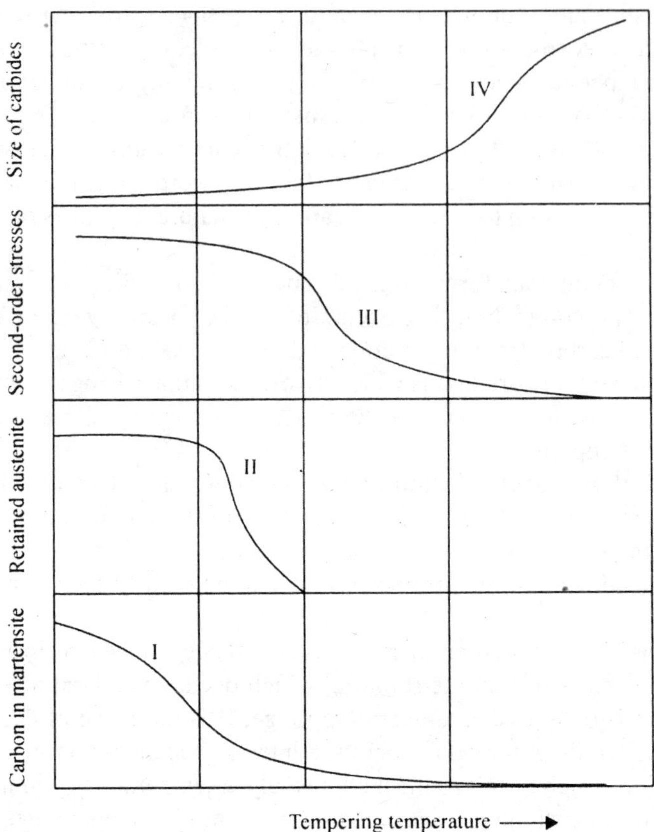

Fig. 6.26. Curves to illustrate the processes occurring in tempering.

A steel as annealed, normalised or tempered (t_{tem} > 400°C) consists of lamellar ferrite and carbide (cementite) inclusions. Ferrite has a low strength and high ductility, whereas comentite has a high hardness (around 800 HB) and zero values of elongation and reduction. The fact that iron-base alloys with more than 0.01 per cent carbon have an elevated strength and reduced plasticity should evidently be attributed to the strengthening effect of carbide inclusions. Since plastic strain (in some or other method of stressing) is possible only in ferrite, the strengthening effect of hard carbide inclusions can be assumed to act as follows. At a low concentration of cementite inclusion (Fig. 6.27a) plastic deformation can develop relatively unobstructedly and the properties of the material are characterised by a low hardness.

With a larger number of cementite particles, for instance, if cementite particles are refined by heat treatment (Fig. 6.27b) they will distort the crystal lattice in adjacent grains and impede the movement of dislocations, thus producing a strengthening effect in the steel. On the contrary, if cementite particles are coarsened (Fig. 6.27c), certain volumes of ferrite will be freed for dislocation movement and the steel will be more capable of plastic deforming. The quantity of carbide particles of a constant size depends on the carbon content in steel (directly proportional to the carbon content in carbon steels). In carbon steel with 0.35 per cent C, cementite occupies approximately 5 per cent of the mass (or volume); steel with 0.7 per cent C contains 10 per cent of cementite, etc. This is why the values of strength in steels increase and the values of ductility diminish with increasing carbon content.

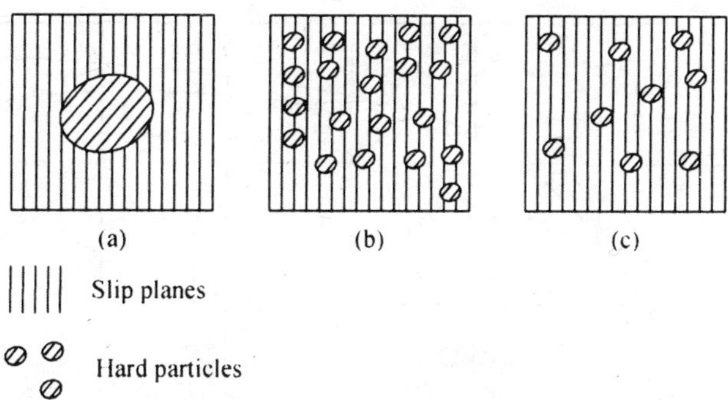

Fig. 6.27. Schemes to illustrate the effect of hard particles on strength and ductility of steel.

For the same carbon content, the number of carbide particles, and therefore, the area of phase boundaries increase with refinement of carbides. The refining effect is obtained by heat treatment. For instance, the size of particles of the strengthening phase (cementite) in as-normalised steel may be depicted as in Fig. 6.27a and their state upon hardening and tempering at various temperatures, as in Fig. 6.27c; the refinement of the carbide component upon this treatment explains why hardened and tempered steel has a higher strength than in normalised state.

A higher temperature of tempering produces coarser cementite particles and thus lowers the strength. The same effect is observed on lowering the cooling rate (in hardening) and on raising the temperature of isothermal decomposition.

Thus, we have explained the variations of hardness in annealed (normalised) and tempered steel which has the structure of ferrite-cementite mixture of various dispersity. It is, however, impossible to explain the high hardness of martensite in this manner. The point is that elementary crystal cells in martensite are distorted, shear displacements in its structure are almost impossible, and martensite can be hardly deformed plastically.

At a higher carbon content in steel, the tetragonal lattice of martensite is more distorted and martensite has a higher hardness. The martensite in a steel with 0.1 per cent C has a hardness of approximately 30 HRC. With 0.7 per cent C, the hardness of martensite attains the maximum value (64 HRC) and does not substantially increase with further increase in the carbon content (curve 2 in Fig. 6.29). However, this curve does not always characterise the hardness of hardened steel since the latter usually contains a certain quantity of retained austenite. If heating for hardening is done above point Ac_3 and all the carbon passes into the solid solution, the hardness of hardened steel will diminish with an increase of the carbon content above 0.8 per cent, since the quantity of retained austenite increases quite rapidly (curve 1 in Fig. 6.28 see also Fig. 6.24).

At heating for hardening to a conventional temperature, i.e., to Ac_1 + 30°C, the same quantity of carbon in all hypereutectoid steels passes into the solution (in accordance with line SE in the Fe–C diagram), therefore, the same quantity of retained austenite is fixed, and all hypereutectoid steels acquire the same hardness upon this heat treatment (curve 2 in Fig. 6.28).

The ductility (toughness) of steel is of no less importance than hardness. The ductility and toughness are usually lower at higher values of hardness. For the same hardness value, however, the values of ductility and toughness may appreciably vary depending on the structure and the size of martensitic platelets.

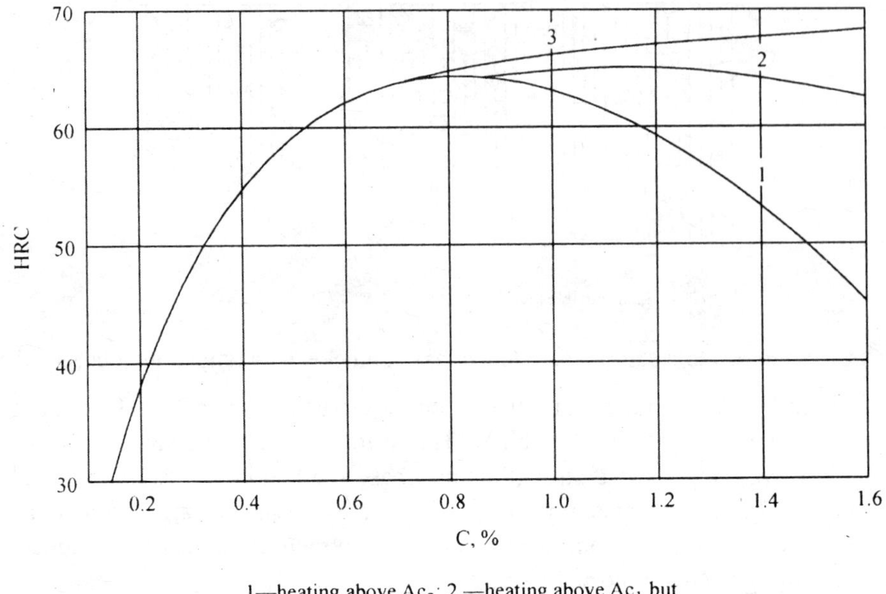

1—heating above Ac_3; 2.—heating above Ac_1 but below Ac_3 (770°C); 3—microhardness of martensite

Fig. 6.28. Effect of carbon and hardening temperature on hardness of steel.

The mechanism of diffusionless oriented martensitic transformation that has been discussed and is responsible for the structure of martensite being substantially dependent on the original austenitic structure. Both the displacements on plastic deformation and martensitic platelets develop within austenitic grains from one edge to another. This means that longer martensitic platelets can form in larger austenitic grains. Large martensitic needles have formed in a Large grain of austenite and finer needles, in smaller grains. The plastic properties and especially the toughness of martensite and of the products of its decomposition (up to the tempering temperatures at which the microstructure remains acicular) are appreciably impaired with structure coarsening (whereas the hardness is almost independent of the size of martensitic needles). This dependence of the properties of heat-treated steel on the size of martensitic platelets is of great importance.

Thus, in order to obtain the best combination of mechanical properties in hardened steel, it is essential to form a fine-acicular martensitic structure, which is possible only if the original austenitic structure is also fine-grained.

As has been noted, a lower temperature of isothermal decomposition of austenite increases the dispersity of ferrite-cementite particles and thus increases the hardness. This is why pearlite, i.e., the product of transformation of austenite at 650–700°C, has a lower hardness than sorbite which forms on decomposition of austenite at 600–650°C, and so on. Approximate values of the hardness of various structures obtained on decomposition of austenite may be found in Fig. 6.13.

Tempering is the final operation of heat treatment, which forms the final desired properties in steel parts. For that reason, the properties of tempered steels should be discussed in more detail. The variations of hardness of steels with different carbon contents depending on the temperature of tempering, are shown in Fig. 6.29.

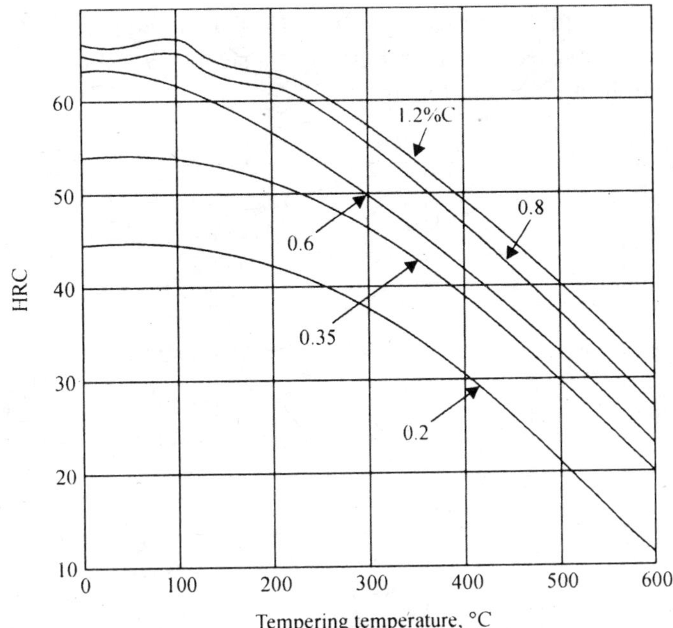

Fig. 6.29. Effect of tempering temperature on hardness of carbon steels with various content of carbon.

Hardness changes on tempering owing to certain structural changes. Heating at a temperature up to 100°C gives only a slight (by 1–2 HRC numbers) increase in hardness (and only in high-carbon steels) owing to the change from tetragonal martensite to temper martensite. With a further increase of the tempering temperature, the hardness diminishes due to the coarsening of carbide particles and the removal of carbon from α-solid solution. The linear relationship between hardness and temperature is disturbed in the temperature range of 200–250°C, i.e., where retained austenite begins to transform. At these temperatures, the decline in hardness slows down and can even increase somewhat in high-carbon steels where retained austenite transforms into a harder temper martensite. The general trend is, however, that hardness diminishes with increasing temperature of tempering, as also do the other strength characteristics (σ_t, $\sigma_{0.2}$), whereas the ductility indices (δ, ψ)increase (Fig. 6.30). The variations of these properties with tempering temperature are, however, far from being monotonous. The marks on the left hand scale in Fig. 6.30 correspond to the properties of as-hardened (untempered) steel and those on the right-hand scale, to those of annealed steel. Tempering at 300°C increases the ultimate strength and the elastic limit. These characteristics are lower in as-hardened steel (owing to the stressed state of the metal) and in that tempered at a lower temperature.

The ductility characteristics (δ, ψ) increase with increasing temperature of tempering. The highest ductility (ψ) is obtained on tempering at 600–650°C, whereas tempering at higher temperatures gives no increase in ψ. Note also that the whole complex of mechanical properties of a steel tempered at 600–650°C is higher than in as-annealed state.

The better mechanical properties of a steel upon hardening and high-temperature tempering compared to the same steel in as-annealed or normalised state (higher values of $\sigma_{0.2}$, ψ, and a_n for the same ultimate strength) can be explained by the different structure of the temper sorbite (pearlite) and the sorbite formed on hardening (as has been noted, the former is granular and the latter, lamellar). A double heat treatment procedure, consisting in hardening followed with high-temperature tempering, appreciably improves the whole complex of mechanical properties of a steel. It is called improving heat treatment and is the principal type of heat treatment for structural steels.

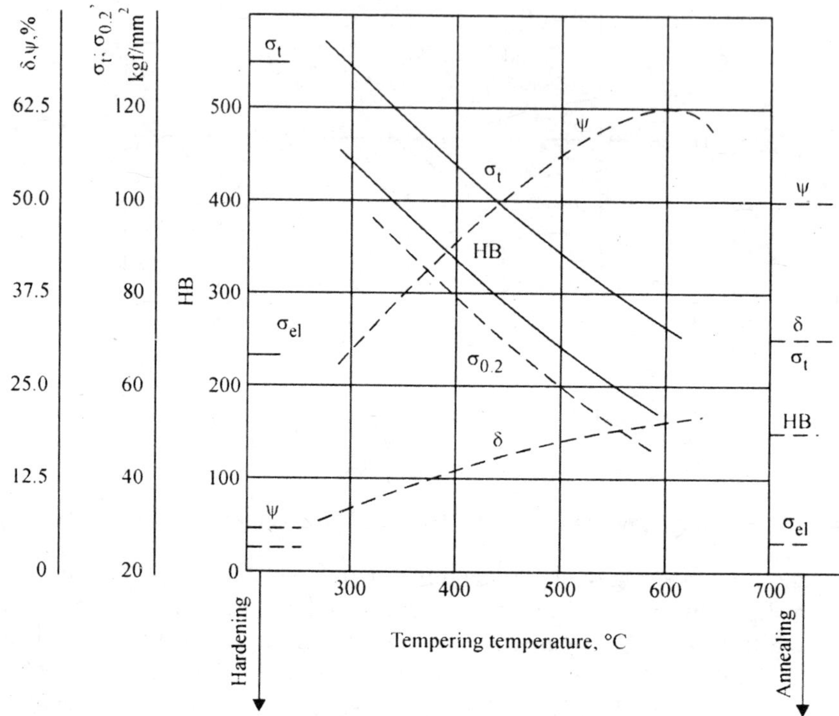

Fig. 6.30. Effect of tempering temperature on mechanical properties of steel grade 40.

The effect of tempering temperature on the impact toughness of steels is as follows. A hardened carbon steel retains a low value of impact toughness (as measured in bending impact tests) up to a temperature of 400°C; at higher temperatures of tempering, the impact toughness intensively increases and attains its maximum at 600°C. In some grades of alloy steel, tempering at a temperature of roughly 300°C can lower the initial value of toughness, which again increases on tempering at temperatures above 450–500°C.

We have considered the effect of tempering temperature on the properties of steels. The tempering temperature is the most essential factor affecting the properties of tempered steel. Tempering is associated with diffusion process, because of which a sufficient holding time can be favourable for the transformations that occur at given temperature.

A long tempering can be made shorter but carried out at a higher temperature. If the temperature and time of tempering are balanced so that the tempered steel has the same hardness, the other mechanical properties of the steel will be almost the same.

In contrast to some alloy steels, the mechanical properties of carbon steels (and many other types) do not depend on the rate of cooling from the temperature of tempering. The properties of steel on tempering depend only on the temperature and time of tempering.

THERMOMECHANICAL TREATMENT

As has been noted, thermomechanical treatment (thermoplastic working) is a combination of two methods of strengthening plastic deformation and phase transformation. As applied to steels, thermomechanical treatment (TMT) consists in strain hardening of austenite followed by the transformation of austenite into another structure.

The most common processes of thermomechanical treatment are as follows Fig. 6.31. High-temperature thermomechanical treatment (HTMT), in which austenite is deformed at temperatures above the re-crystallisation point (t_r in Fig. 6.31), usually above the critical points and then cooled quickly.

Low-temperature thermomechanical treatment (LHMT) in which austenite is deformed at a temperature below t_r, i.e., in the unstable state of undercooling below the critical points and then cooled to cause phase transformation.

Preliminary thermomechanicl treatment (PTMT) in which the steel is strain hardened prior to being heat treated and then heated up quickly to the austenitic state and cooled, but so as to retain to some or other extent the strain-hardening effect.

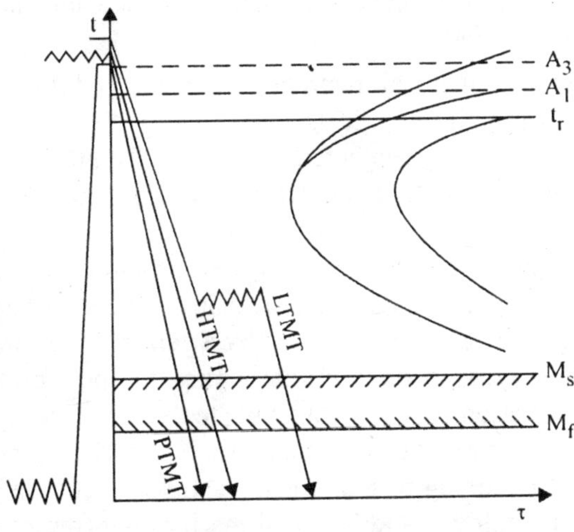

Fig. 6.31. Classification diagram of thermomechanical treatment.

There also are many other varieties of thermomechanical treatment differing in the conditions of heating or cooling, type of deformation and other minor details.

In all types of thermomechanical treatment the changes that are caused in austenite by plastic deformation are transferred (either partially or completely) to martensite.

Thus, a steel is strengthened due to the martensitic reaction and also due to the structural defects produced in deformed austenite and inherited by martensite.

A typical procedure of thermomechanical treatment is low-temperature thermomechanical treatment. Austenite is undercooled to 500–600°C and deformed to cause the strain-hardening effect. A heavier deformation, and therefore, a greater strain-hardening effect, produce a higher strength of martensite (Fig. 6.32). Very heavy deformation (80–90 per cent) are employed to obtain the maximum strength. Since deformation is made at a low temperature (substantially lower than the common temperatures for hot working), the steel retains a high ductility (more strictly, a high resistance to deformation), so that the necessity of a heavy deforming is associated with certain technological difficulties.

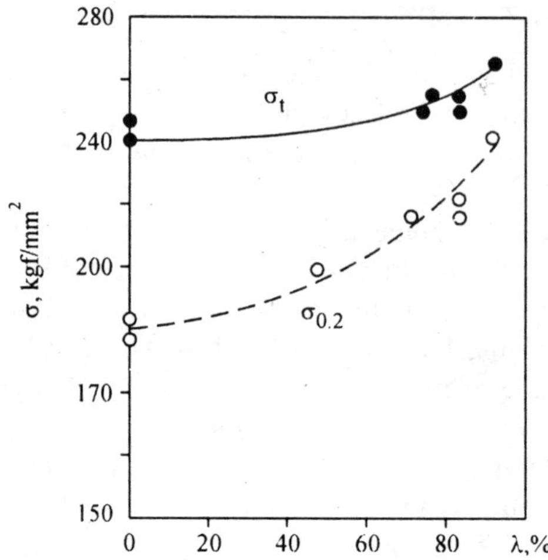

Fig. 6.32. Effect of deformation (λ) in low-temperature thermomechanical treatment on strength of steel (0.3% C, 2.2% Cr, 1.5% Ni, 0.3% V, and 0.3% Mo).

In all procedures of thermomechhanical treatment, the metal is deformed, transforms to martenisite and then tempered to a given strength (ductility).

In certain cases, the deformed steel is subjected to isothermal annealing to bainite or even to pearlitic transformation. The latter method, called 'controlled rolling', has found wide application in the manufacture of high-strength low-alloy steels.

In all procedures of thermomechanical treatment, deformation precedes transformation. There is, however, a class of heat treatment procedures based on an inverse order of operations, i.e., transformation followed by strain hardening. They can be combined by a general name mechanico-thermal treatment.

If a high strength (martensite) is obtained by hardening, the possibilities for strengthening of this structural state by plastic deformation are not great owing to the low ductility of martensite. Nevertheless, a deformation of 3–5 per cent in martensite can produce an additional strengthening effect of 10–20 per cent. This heat treatment process (hardening to martensite + slight plastic deformation + low-temperature tempering) is also sometimes employed and is called marforming (the name implies that martensite is deformed, in contrast to ausforming where deformation is applied to austenite).

A procedure in which the steel is transformed to fine-lamellar pearlite (troostite) and then plastically deformed is called patenting. In patenting, heavy deformations are essential for obtaining high mechanical properties. It should be noted that patenting of high-carbon steels (1 per cent C) with a heavy deformation (above 95 per cent) can produce the highest value of real ultimate strength obtainable in industrial products: 450 kgf/mm^2 or almost one-third of the theoretical ultimate strength. Such a strength, however, is obtained in thin wire only.

Chapter 7

Stress-Relief Heat Treating of Steel

INTRODUCTION

Stress-relief heat treating is used to relieve stresses that remain locked in a structure as a consequence of a manufacturing sequence. This definition separates stress relief heat treating from postweld heat treating in that the goal of postweld heat treating is to provide, in addition to the relief of residual stresses, some preferred metallurgical structure or properties. For example, most ferritic weldments are given postweld heat treated to improve fracture toughness of heat-affected zones. Moreover, austenitic and non-ferrous alloys are frequently postweld heat treated to improve resistance to environmental damage.

STRESS-RELIEF HEAT TREATING

Stress-relief heat treating is the uniform heating of a structure or portion thereof to a suitable temperature below the transformation range (Ac_1 for ferritic steels), holding at this temperature for a pre-determined period of time, followed by uniform cooling. Care must be taken to ensure uniform cooling, particularly when a component is composed of variable section sizes. If the rate of cooling is not constant, new residual stresses can result that are equal to or greater than those for which the heat treating was intended to relieve. Stress-relief heat treating can reduce distortion and high stresses from welding that affect service performance. The presence of residual stresses can lead to stress corrosion cracking near welds and in regions of component that has been cold strained during processing.

RESIDUAL STRESSES AND ITS SOURCES

Residual stresses in a ferritic steel cause significant reduction in resistance to brittle fracture. In a material that is not prone to brittle fracture, such as an austenitic stainless steel, residual stresses can be sufficient to provide the stress necessary to promote stress corrosion cracking even in environments that appear to the benign.

There are many sources of residual stress; they can occur during processing of the material from ingot to final product form. Residual stresses can be generated during rolling, casting or forging; during forming operations such as shearing, bending and machining; and during fabrication, in particular, welding. Residual stresses are present whenever a component is stressed beyond its elastic limit and plastic flow occurs. Bending a bar during fabrication at a temperature where recovery cannot occur (cold forming, for example), will result in one surface location containing residual tensile stresses, whereas a location

180° away will contain residual compressive stresses. Quenching of thick sections results in high residual compressive stresses on the surface of the material. These high compressive stresses are balanced by residual tensile stresses in the internal areas of the section.

Grinding is another source of residual stresses; these can be compressive or tensile in nature, depending on the grinding operation. Although these stresses tend to be shallow in depth, they can cause warping of thin parts.

Causes of Residual Stresses

The cause of residual stresses that has received the most attention in the open literature is welding. The residual stresses associated with the steep thermal gradient of welding can occur on a macroscale over relatively long distances (reaction stresses) or can be highly localised (microscale) (Fig. 7.1). Welding usually results in localised residual stresses that approach levels equal to the yield strength of the material at room temperature.

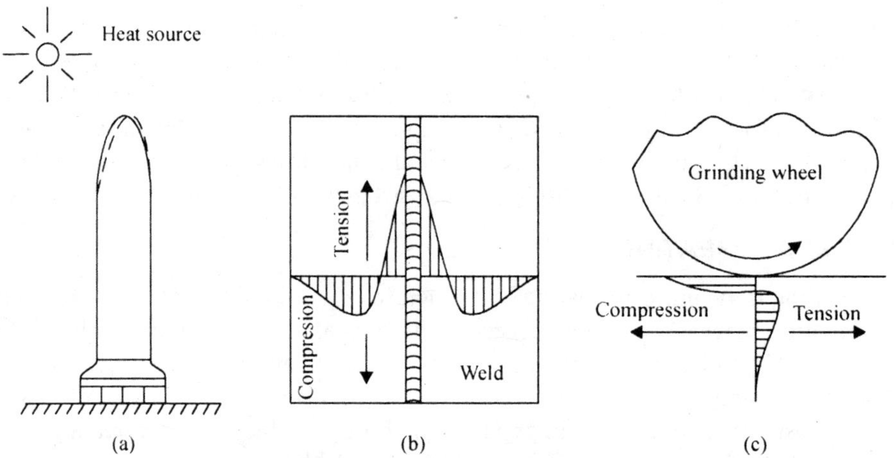

(a) Thermal distortion in a structure due to heating by solar radiation; (b) Residual stresses due to welding; and (c) Residual stresses due to grinding.

Fig. 7.1. Examples of the causes of residual stresses.

Factors Influence Relief of Residual Stresses

A number of factors influence the relief of residual stresses, including level of stress, permissible (or practicable) time for their relief, temperature, and metallurgical stability.

The relief of residual stresses is a time-temperature related phenomenon (Fig. 7.2), parametrically correlated by the Larson-Miller equation:

$$\text{Thermal effect} = T(\log t + 20)\,(10^{-3})$$

where T is temperature (Rankin) and t is hours. It is evident in Fig. 7.2 that similar relief of residual stresses can be achieved by holding a component for longer periods of time at a lower temperature. For example, holding a piece at 595°C (1100°F) for 6 hours, provide the same relief of residual stress as heating at 650°C (1200°F) for 1 hour.

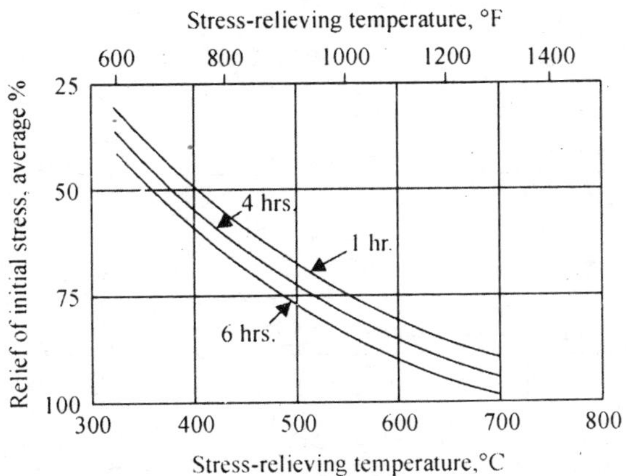

Fig. 7.2. Illustration of the relationship between time and temperature in the relief of residual stresses in steel.

Relief of residual stresses represents typical stress-relaxation behaviour, in which the material undergoes microscopic (sometimes even macroscopic) creep at the stress-relief temperature. Creep-resistant materials, such as the chromium-bearing low-alloy steels and the chromium-rich high-alloy steels, normally require higher stress-relief heat treating temperatures than conventional low-alloy steels. Typical stress-relief temperatures for low-alloy ferritic steels are between 595 and 675°C (1100 and 1250°F). For high-alloy steels, these temperatures may range from 900 to 1065°C (1650 to 1950°F).

For high-alloy steels, such as the austenitic stainless steels, stress relieving is sometimes done at temperatures as low as 400°C (750°F). However, at these temperatures, only modest decreases in residual stress are achieved. Residual stresses can be significantly reduced by stress-relief heat treating those austenitic materials in the temperature range from 480 to 925°C (900 to 1700°F). At the higher end of this range, nearly 85 per cent of the residual stresses may be relieved. Stress relief heat treating in this range, however, may result in sensitising susceptible material.

This metallurgical effect can lead to stress-corrosion cracking in service. Frequently, solution-annealing temperatures at around 1065°C (1950°F) are used to achieve a reduction of residual stresses to acceptably low values.

Some copper alloys may fail by stress-corrosion cracking due to the presence of residual stresses. These stresses are usually relieved by mechanical or thermal stress-relief treatments. Stress-relief heat treating tends to be favoured because it is more controllable, less costly, and also provides a degree of dimensional stability. Stress-relief heat treating of copper alloys is usually carried out at relatively low temperatures, in the range from 200 to 400°C (390 to 750°F).

Resistance of a material to the reduction of its residual stresses by thermal treatment can be estimated with a knowledge of the influence of temperature on its yield strength. Fig. 7.3 provides a summary of the yield strength to temperature relationship for three generic classes of steels. The room temperature yield strength of these materials provides an excellent estimate of the level of localised residual stress that can be present in a structure.

To relieve the residual stress requires that the component be heated to a temperature where its yield strength approaches a value that corresponds to an acceptable level of residual stress. Holding at this temperature can, through the reduction of strain due to creep, further reduce the residual stress. Uniform cooling after residual stress heat treating is mandatory if these levels of residual stress are to be maintained.

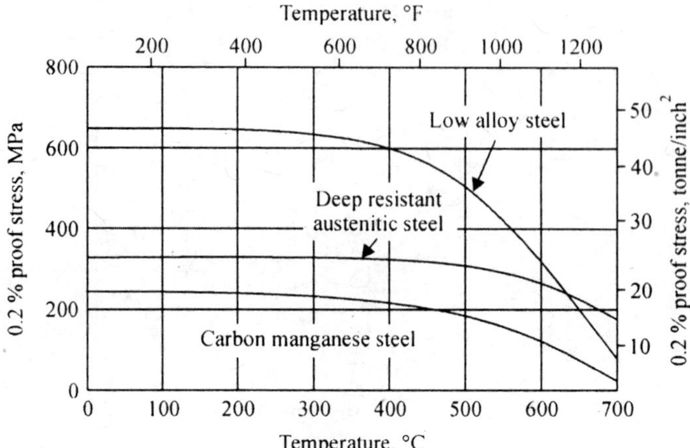

Fig. 7.3. Variation of the yield strength with temperature for three generic classes of steel.

Chapter 8

Annealing of Steel

INTRODUCTION

Annealing is a generic term denoting a treatment that consists of heating to and holding at a suitable temperature followed by cooling at an appropriate rate, primarily for softening of metallic materials. It is also applied to produce desired changes in other properties or in microstructure. Steels may be annealed to facilitate cold working or machining, to improve mechanical or electrical properties, or to promote dimensional stability. The choice of an annealing treatment that will provide an adequate combination of such properties at minimum expense often involves a compromise. Terms used to denote specific types of annealing applied to steels are descriptive of the method used, the equipment used or the condition of the material after treatment.

Full annealing: This process consists of heating hypoeutectoid steels, 30–50°C above the upper critical point i.e., A_3 line and holding it for sufficient amount of time at this temperature. This ensures that the metal is heated thoroughly and phase transformation has taken place throughout the whole volume. The material is then cooled slowly in the furnace. During the process of heating, fine grained austenite is obtained and thus, fine grained structure is produced on cooling. This imparts much improvement in mechanical properties, high ductility and high toughness.

Heating rate is usually 100°C/hr. and holding time is one hour per tonne of metal. The cooling rate is usually 10–100°C per hr., for alloy steel and 200°C/hr., for carbon steels. Once austenite is decomposed into pearlite, the further cooling rate can be increased or the material can be cooled in air.

Process annealing or partial annealing: In process annealing steel is heated to lower critical temperature i.e., slightly above Ac_1, line and then cooled slowly. For hypoeutectoid steels, only process annealing is applied. It is also applied for hypoeutectoid steels where hardness is reduced and machinability is improved. The steel is partially re-crystalised. Only pearlite is transformed to austenite. Ferrite iron is partially deformed into austenite. Process annealing is applied when hot working has been done correctly and there is no coarse grain structure.

For process annealing hypereutectoid steel is heated to 10–30°C above Ac_1. Practically complete re-crystalisation takes place. Granular pearlite is obtained instead to lamellar.

Heating and holding periods are followed by slow cooling. Steel containing granular pearlite (after process annealing) has lower hardness and tensile strength. It has got higher per cent elongation and reduction in area. Machinability is high i.e., cutting speed is high with good surface finish.

Isothermal Annealing: Steel is heated as it is heated in case of full annealing, then it is rapidly cooled to a temperature 100–500°C below point A_1. The steel is held at this temperature for long time so that γ-iron is completely decomposed. Then it is cooled in air. Isothermal annealing has few advantages, such as: (i) annealing time is reduced; (ii) more homogeneous structure is obtained because during isothermal heating temperature is maintained throughout the cross-section; and (iii) machinability is improved.

Diffusion annealing (homogenisation): Steel is heated to high temperature 1000–1200°C. At this temperature composition homogenise throughout the section of steel. It is held for 8–20 hours, at this temperature. The total heating, holding and cooling time ranges from 50 to 120 hours. This type of annealing produces coarse grain structure. But, by subsequent plastic working, fine grain structure can be produced.

After diffusion annealing, the castings undergo full annealing or normalising to improve their properties of to refine grain structure. Ingots and large castings are homogenised (Diffusion annealed).

Re-crystalisation annealing: Cold worked steel is heated to a temperature above re-crystalisation temperature, it is held at this temperature for some time and then cooled. This reduces strain hardening between cold working operations. It may also be applied as final heat treatment.

In this process, there is re-crystalisation of ferrite and coalescences (come together and form one) as well as spheroidisation of pearlite particles. There is increase in ductility and plastic deformation becomes easy, in this type of annealing. *Stress relief annealing (low temperature annealing):* The steel parts are heated in the range of 350–700°C, held at this temperature for several hours and then cooled slowly. Residual stresses are relieved which were developed in previous processing, due to non-uniform cooling or non-uniform plastic deformation. The residual stresses developed in welding process or in any other manufacturing process, are also relieved.

BASIC CONCEPTS

Critical Temperatures

The critical temperatures that must be considered in discussing annealing of steel are those that define the onset and completion of the transformation to or from austenite. For a given steel, the critical temperatures depend on whether the steel is being heated or cooled. Critical temperatures for the start and completion of the transformation to austenite during heating are denoted, respectively, by Ac_1 and Ac_3 for hypoeutectoid steels and by Ac_1 and Ac_{cm} for hypereutectoid steels.

These temperatures are higher than the corresponding critical temperatures for the start and completion of the transformation from austenite during cooling, which are denoted, respectively, by Ar_3 and Ar_1 for hypoeutectoid steels and by Ar_{cm} and Ar_1 for hypereutectoid steels. These critical temperatures converge to the equilibrium values Ae_1, Ae_3 and Ae_{cm} as the rates of heating or cooling become infinitely slow. Fig. 8.1 illustrates the positions of the Ae_1, Ae_3 and Ae_{cm} lines on the equilibrium phase diagram for plain carbon steels. The presence of other alloying elements will also have marked effects on these critical temperatures.

Table 8.1 provides critical temperatures for selected steels, measured at heating and cooling rates of 28°C/hr. (50°F/hr.). The equilibrium critical temperatures generally lie about midway between those for heating and cooling at equal rates. Because annealing may involve various ranges of heating and cooling rates in combination with isothermal treatments, the less specific terms A_1, A_3 and A_{cm} are used here in discussing the basic concepts.

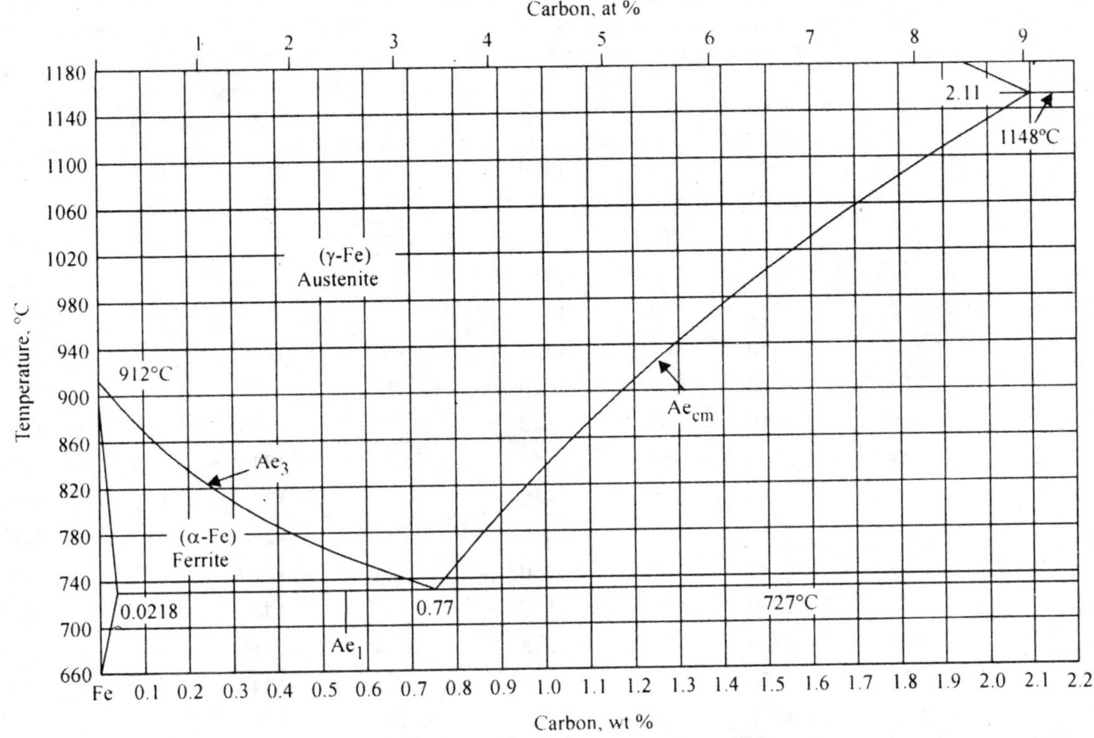

The equilibrium transformation temperatures Ae_1, Ae_3 and Ae_{cm} are labeled on the diagram.

Fig. 8.1. Fe-Fe_3C phase diagram, showing the temperature range of interest for annealing plain carbon steels.

Table 8.1. Approximate critical temperatures for selected carbon and low-alloy steels.

	Critical temperatures on heating at 28°C/hr. (50°F/hr.)				*Critical temperatures on cooling at 28°C/hr. (50°F/hr.)*			
	Ac_1		*Ac_3*		*Ar_3*		*Ar_1*	
Steel	*°C*	*°F*	*°C*	*°F*	*°C*	*°F*	*°C*	*°F*
1010	724	1335	877	1610	849	1560	682	1260
1020	724	1335	846	1555	816	1500	682	1260
1030	727	1340	813	1495	788	1450	677	1250
1040	727	1340	793	1460	757	1395	671	1240
1050	727	1340	768	1415	741	1365	682	1260
1060	727	1340	746	1375	727	1340	685	1265
1070	727	1340	732	1350	710	1310	691	1275
1080	729	1345	735	1355	699	1290	693	1280
1340	716	1320	777	1430	721	1330	621	1150
3140	735	1355	766	1410	721	1330	660	1220
4027	727	1340	807	1485	760	1400	671	1240

(Contd...)

	Critical temperatures on heating at 28°C/hr. (50°F/hr.)				Critical temperatures on cooling at 28°C/hr. (50°F/hr.)			
	Ac_1		Ac_3		Ar_3		Ar_1	
Steel	°C	°F	°C	°F	°C	°F	°C	°F
4042	727	1340	793	1460	732	1350	654	1210
4130	757	1395	810	1490	754	1390	693	1280
4140	732	1350	804	1480	743	1370	679	1255
4150	743	1370	766	1410	729	1345	671	1240
4340	724	1335	774	1425	710	1310	654	1210
4615	727	1340	810	1490	760	1400	649	1200
5046	716	1320	771	1420	732	1350	682	1260
5120	766	1410	838	1540	799	1470	699	1290
5140	738	1360	788	1450	727	1340	693	1280
5160	710	1310	766	1410	716	1320	677	1250
52100	727	1340	768	1415	716	1320	688	1270
6150	749	1380	788	1450	743	1370	693	1280
8115	721	1330	838	1540	788	1450	671	1240
8620	732	1350	829	1525	768	1415	660	1220
8640	732	1350	779	1435	727	1340	666	1230
9260	743	1370	816	1500	749	1380	713	1315

Annealing Cycles

In practice, specific thermal cycles of an almost infinite variety are used to achieve the various goals of annealing. These cycles fall into several broad categories that can be classified according to the temperature to which the steel is heated and the method of cooling used. The maximum temperature may be below the lower critical temperature, A_1 (sub-critical annealing); above A_1 but below the upper critical temperature, A_3 in hypoeutectoid steels or A_{cm} in hypereutectoid steels (intercritical annealing); or above A_3 (full annealing).

Because some austenite is present at temperatures above A_1, cooling practice through transformation is a crucial factor in achieving desired microstructures and properties. Accordingly, steels heated above A_1 are subjected either to slow continuous cooling or to isothermal treatment at some temperature below A_1 at which transformation to the desired microstructure can occur in a reasonable amount of time. These generalised treatments are illustrated schematically in Fig. 8.2. Under certain conditions, two or more such cycles may be combined or used in succession to achieve the desired results. The success of any annealing operation depends on proper choice and control of the thermal cycle, based on the metallurgical principles discussed in the following sections.

Heating Below A_1

Sub-critical annealing does not involve formation of austenite. The prior condition of the steel is modified by such thermally activated processes as recovery, recrystallisation, grain growth and agglomeration of carbides. The prior history of the steel is, therefore, an important factor.

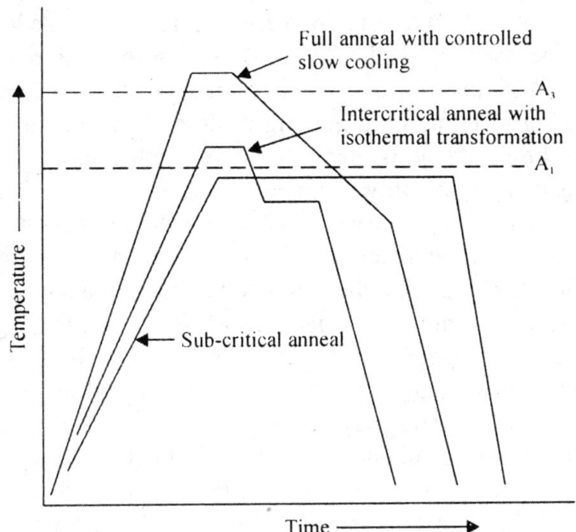

Fig. 8.2. Schematic representation of some basic annealing schedules for a hypoeutectoid steel.

In as-rolled or forged hypoeutectoid steels containing ferrite and pearlite, sub-critical annealing can adjust the hardnesses of both constituents, but excessively long times at temperature may be required for substantial softening. The sub-critical treatment is most effective when applied to hardened or cold worked steels, which re-crystallise readily to form new ferrite grains. The rate of softening increases rapidly as the annealing temperature approaches A_1. Cooling practice from the sub-critical annealing temperature has very little effect on the established microstructure and resultant properties.

Heating Above A_1

Austenite begins to form when the temperature of the steel exceeds A_1. In hypoeutectoid steels, the equilibrium structure in the intercritical range between A_1 and A_3 consists of ferrite and austenite and above A_3 the structure becomes completely austenitic. However, the equilibrium mixture of ferrite and austenite is not achieved instantaneously. Undissolved carbides may persist, especially if the austenitising time is short or the temperature is near A_1, causing the austenite to be inhomogeneous. In hypereutectoid steels, carbide and austenite co-exist in the intercritical range between A_1 and A_{cm} and the homogeneity of the austenite depends on time and temperature. The degree of homogeneity in the structure at the austenitising temperature is an important consideration in the development of annealed structures and properties. The more homogeneous structures developed at higher austenitising temperatures tend to promote lamellar carbide structures on cooling, whereas lower austenitising temperatures in the intercritical range result in less homogeneous austenite, which promotes formation of spheroidal carbides.

Decomposition of Austenite

Austenite formed when steel is heated above the A_1 temperature transforms back to ferrite and carbide when the steel is cooled below A_1. The rate of austenite decomposition and the tendency of the carbide structure to be either lamellar or spheroidal depend largely on the temperature of transformation. If the austenite transforms just below A_1, it will decompose slowly. The product then may contain relatively coarse spheroidal carbides or coarse lamellar pearlite, depending on the composition of the steel and the austenitising temperature. This product tends to be very soft. However, the low rate of transformation at temperatures just below A_1 necessitates long holding times in isothermal treatments, or very low cooling rates in continuous cooling, if maximum softness is desired. Isothermal treatments are more efficient than slow continuous cooling in terms of achieving desired structures and softness in the minimum amount of time. Sometimes, however, the available equipment or the mass of the steel part being annealed may make slow continuous cooling the only feasible alternative.

As the transformation temperature decreases, austenite generally decomposes more rapidly and the transformation product is harder, more lamellar and less coarse than the product formed just below A_1. At still lower transformation temperatures, the product becomes a much harder mixture of ferrite and carbide and the time necessary of complete isothermal transformation may again increase. Temperature time plots showing the progress of austenite transformation under isothermal (IT) or continuous cooling transformation (CCT) conditions for many steels have been widely published and illustrate the principles just discussed. These IT or CCT diagrams may be helpful in design of annealing treatments for specific grades of steel, but their usefulness is limited because most published diagrams represent transformation from a fully austenitsed, relatively homogeneous condition, which is not always desirable or obtainable in annealing. However, transformation diagrams (either IT or CCT) that represent cooling from a specific austenitising treatment may be developed by use of the techniques. Such diagrams provide the information necessary for design of effective annealing.

Cooling After Transformation

After the austenite has been completely transformed, little else of metallurgical consequence can occur during cooling to room temperature. Extremely slow cooling may cause some agglomeration of carbides and, consequently, some slight further softening of the steel, but in this regard such slow cooling is less effective than high-temperature transformation. Therefore, there is no metallurgical reason for slow cooling after transformation has been completed and the steel may be cooled from the transformation temperature as rapidly as feasible in order to minimise the total time required for the operation.

If transformation by slow continuous cooling has been used, the temperature at which controlled cooling may be stopped depends on the transformation characteristics of the steel. However, the mass of the steel or the need to avoid oxidation are practical considerations that may require retarded cooling to be continued below the temperature at which austenite transformation ceases.

Effect of Prior Structure

The finer and more evenly distributed the carbides in the prior structure, the faster the rate at which austenite formed above A_1 will approach complete homogeneity. The prior structure, therefore, can affect the response to annealing. When spheroidal carbides are desired in the annealed structure, pre-heating at temperature just below A_1 sometimes is used to agglomerate the prior carbides in order to increase their resistance to solution in the austenite on subsequent heating. The presence of undissolved carbides or concentration gradients in the austenite promotes formation of a spheroidal, rather than lamellar, structure when the austenite is transformed. Pre-heating to enhance spheroidisation is applicable mainly to hypoeutectoid steels but also is useful for some hypereutectoid low-alloy steels.

Austenitising Time and Dead-Soft Steel

Hypereutectoid steels can be made extremely soft by holding for long periods of time at the austenitising temperature. Although the time at the austenitising temperature may have only a small effect on actual hardness (such as a change from 241 to 229 HB), its effect on machinability or cold forming properties may be appreciable.

Long-term austenitising is effective in hypereutectoid steels because it produces agglomeration of residual carbides in the austenite. Coarser carbides promote a softer final product. In lower carbon steels, carbides are unstable at temperatures above A_1 and tend to dissolve in the austenite, although the dissolution may be slow.

Steels that have approximately eutectoid carbon contents generally form a lamellar transformation product is austenitised for very long periods of time. Long-term holding at a temperature just above the A_1 temperature may be as effective in dissolving carbides and dissipating carbon-concentration gradients as is short-term holding at a higher temperature.

GUIDELINES FOR ANNEALING

The metallurgical principles discussed above have been incorporated by *Payson* into the following seven rules, which may be used as guidelines for development of successful and efficient annealing schedules:

Rule 1: The more homogeneous the structure of the as-austenitised steel, the more completely lamellar will be the structure of the annealed steel. Conversely, the more heterogeneous the structure of the as-austenitised steel, the more nearly spheroidal will be the annealed carbide structure.

Rule 2: The softest condition in the steel is usually developed by austenitising at a temperature less than 56°C (100°F) above A_1 and transforming at a temperature (usually) less than 56°C (100°F) below A_1.

Rule 3: Because very long times may be required for complete transformation at temperatures less than 56°C (100°F) below A_1, allow most of the transformation to take place at the higher temperature, where a soft product is formed and finish the transformation at a lower temperature, where the time required for completion of transformation is short.

Rule 4: After the steel has been austenitised, cool to the transformation temperature as rapidly as feasible in order to minimise the total duration of the annealing operation.

Rule 5: After the steel has been completely transformed, at a temperature that produces the desired microstructure and hardness, cool to room temperature as rapidly as feasible, to decrease further the total time of annealing.

Rule 6: To ensure a minimum of lamellar pearlite in the structures of annealed 0.70 to 0.90 per cent carbon (C) tool steels and other low-alloy medium-carbon steels, pre-heat for several hours at a temperature about 28°C (50°F) below the lower critical temperature (A_1) before austenitising and transforming as usual.

Rule 7: To obtain minimum hardness in annealed hypereutectoid alloy tool steels, heat at the austenitising temperature for a long time (about 10 to 15 hours), then transform as usual.

These rules are applied most effectively when the critical temperatures and transformation characteristics of the steel have been established and when transformation by isothermal treatment is feasible.

ANNEALING TEMPERATURES

For many annealing applications, it is sufficient simply to specify that the steel be cooled in the furnace from a designated annealing (austenitising) temperature. Temperatures and associated Brinell hardnesses for simple annealing of carbon steels are given in Table 8.2 and similar data for alloy steels are presented in Table 8.3.

Table 8.2. Recommended temperatures and cooling cycles for full annealing of small carbon steel forgings.

(Data are for forgings up to 75 mm (3 inch) in section thickness. Time at temperature usually is a minimum of 1 hour, for sections up to 25 mm (1 inch) thick; ½ hr. is added for each additional 25 mm (1 inch) of thickness).

	Annealing temperature		*Cooling cycle (a)*				*Hardness range,*
			°C		*°F*		
Steel	*°C*	*°F*	*From*	*To*	*From*	*To*	*HB*
1018	855–900	1575–1650	855	705	1575	1300	111–149
1020	855–900	1575–1650	855	700	1575	1290	111–149
1022	855–900	1575–1650	855	700	1575	1290	111–149
1025	855–900	1575–1650	855	700	1575	1290	111–187
1030	845–885	1550–1625	845	650	1550	1200	126–197
1035	845–885	1550–1625	845	650	1550	1200	137–207
1040	790–870	1450–1600	790	650	1450	1200	137–207
1045	790–870	1450–1600	790	650	1450	1200	156–217
1050	790–870	1450–1600	790	650	1450	1200	156–217
1060	790–845	1450–1550	790	650	1450	1200	156–217
1070	790–845	1450–1550	790	650	1450	1200	167–229
1080	790–845	1450–1550	790	650	1450	1200	167–229
1090	790–830	1450–1525	790	650	1450	1200	167–229
1095	790–830	1450–1525	790	655	1450	1215	167–229

(a) Furnace cooling at 28°C/hr. (50°F).

Heating cycles that employ austenitising temperatures in the upper ends of the ranges given in Table 8.3 should result in pearlitic structures. Predominantly spheroidised structures should be obtained when lower temperatures are used.

Table 8.3. Recommended annealing temperatures for alloy steels (furnace cooling).

	Annealing temperature		*Hardness (max.),*
Steel	*°C*	*°F*	*HB*
1330	845–900	1550–1650	179
1335	845–900	1550–1650	187
1340	845–900	1550–1650	192
1345	845–900	1550–1650	–
3140	815–870	1500–1600	187
4037	815–855	1500–1575	183
4042	815–855	1500–1575	192
4047	790–845	1450–1550	201
4063	790–845	1450–1550	223
4130	790–845	1450–1550	174
4135	790–845	1450–1550	–
4137	790–845	1450–1550	192
4140	790–845	1450–1550	197

(Contd...)

Steel	*Annealing temperature* °C	°F	*Hardness (max.), HB*
4145	790–845	1450–1550	207
4147	790–845	1450–1550	–
4150	790–845	1450–1550	212
4161	790–845	1450–1550	–
4337	790–845	1450–1550	–
4340	790–845	1450–1550	223
50B40	815–870	1500–1600	187
50B44	815–870	1500–1600	197
5046	815–870	1500–1600	192
50B46	815–870	1500–1600	192
50B50	815–870	1500–1600	201
50B60	815–870	1500–1600	217
5130	790–845	1450–1550	170
5132	790–845	1450–1550	170
5135	815–870	1500–1600	174
5140	815–870	1500–1600	187
5145	815–870	1500–1600	197
5147	815–870	1500–1600	197
5150	815–870	1500–1600	201
5155	815–870	1500–1600	217
5160	815–870	1500–1600	223
51B60	815–870	1500–1600	223
50100	730–790	1350–1450	197
51100	730–790	1350–1450	197
52100	730–790	1350–1450	207
6150	845–900	1550–1650	201
81B45	845–900	1550–1650	192
8627	815–870	1500–1600	174
8630	790–845	1450–1550	179
8637	815–870	1500–1600	192
8640	815–870	1500–1600	197
8642	815–870	1500–1600	201
8645	815–870	1500–1600	207
86B45	815–870	1500–1600	207
8650	815–870	1500–1600	212
8655	815–870	1500–1600	223
8660	815–870	1500–1600	229
8740	815–870	1500–1600	202
8742	815–870	1500–1600	–

(Contd...)

Steel	Annealing temperature °C	Annealing temperature °F	Hardness (max.), HB
9260	815–870	1500–1600	229
94B30	790–845	1450–1550	174
94B40	790–845	1450–1550	192
9840	790–845	1450–1550	207

When an alloy steel is annealed to obtain a specific microstructure, greater precision is required in specifying temperatures and cooling conditions for annealing. Table 8.4 presents, for a variety of standard alloy steels, typical schedules for such annealing operations. In isothermal annealing to produce a pearlitic structure, particularly in forgings, an austenitising temperature as much as 70°C (125°F) higher than that indicated in Table 8.4 may be selected in order to decrease the austenitising time.

For most steels, as indicated in Table 8.4, annealing may be accomplished by heating to the austenitising temperature and then either: (i) cooling in the furnace at a controlled rate; or (ii) cooling rapidly to and holding at, a lower temperature for isothermal transformation. Both procedures result in virtually the same hardness; however, considerably less time is required for isothermal transformation.

Uniformity of Temperature

One potential contribution to the failure of an annealing operation is a lack of knowledge of the temperature distribution with the load of steel in the furnace. Furnaces large enough to anneal 18 to 20 tonnes of steel at a time are not uncommon. The larger the furnace, the more difficult it is to establish and maintain uniform temperature conditions throughout the load and the more difficult it is to change the temperature of the steel during either heating or cooling. Furnace thermocouples indicate the temperature of the space above, below or beside the load, but this temperature may differ by 28°C (50°F) or more from the temperature of the steel itself (Fig. 8.3), especially when the steel is in a pipe or box or when bar or strip is packed in a dense charge in a quiescent atmosphere. When these conditions exist, the distribution of temperature throughout the load during heating and cooling should be established by placing thermocouples among the bars, forgings or coils. Regulation of the furnace during the annealing operation should be based on the temperatures indicated by these thermocouples, which are in actual contact with the steel, rather than on the temperatures indicated by the furnace thermocouples.

SPHEROIDISING

Steels may be spheroidised—that is, heated and cooled to produce a structure of globular carbides in a ferritic matrix—by the following methods:

1. Prolonged holding at a temperature just below Ae_1*.
2. Heating and cooling alternately between temperatures that are just above Ac_1 and just below Ar_1.
3. Heating to a temperature above Ac_1 and then either cooling very slowly in the furnace or holding at a temperature just below Ar_1.
4. Cooling at a suitable rate from the minimum temperature at which all carbide is dissolved, to prevent reformation of a carbide network and then re-heating in accordance with method 1 or 2 above (applicable to hypereutectoid steel containing a carbide network).

*It is difficult to remain consistent with designations for critical temperatures. When heating with prolonged holding is being discussed, the critical temperatures of interest should be the equilibrium temperatures Ae_1 and Ae_3. Terminology becomes more arbitrary in discussions of heating and cooling at unspecified rates and for unspecified holding times.

Table 8.4. Recommended temperatures and time cycles for annealing of alloy steels.

Steel	Austenitising temperature °C	Austenitising temperature °F	Conventional cooling(a): Temperature °C From	Temperature °C To	Temperature °F From	Temperature °F To	Cooling rate °C/hr.	Cooling rate °F/hr.	Time, hr.	Isothermal method(b): Cool to °C	Cool to °F	Hold, hr.	Hardness (approx.), HB
To obtain a predominantly pearlitic structure(c)													
1340	830	1525	735	610	1350	1130	11	20	11	620	1150	4.5	183
2340	800	1475	655	555	1210	1030	8.3	15	12	595	1100	6	201
2345	800	1475	655	550	1210	1020	9.1	15	12.7	595	1100	6	201
3120 (d)	885	1625	–	–	–	–	–	–	–	650	1200	4	179
3140	830	1525	735	650	1350	1200	11	20	7.5	660	1225	6	187
3150	830	1525	705	645	1300	1190	11	20	5.5	660	1225	6	201
3310(e)	870	1600	–	–	–	–	–	–	–	595	1100	14	187
4042	830	1525	745	640	1370	1180	11	20	9.5	660	1225	4.5	197
4047	830	1525	735	630	1350	1170	11	20	9	660	1225	5	207
4062	830	1525	695	630	1280	1170	8.3	15	7.3	660	1225	6	223
4130	855	1575	765	665	1410	1230	20	35	5	675	1250	4	174
4140	845	1550	755	665	1390	1230	14	25	6.4	675	1250	5	197
4150	830	1525	745	670	1370	1240	8.4	15	8.6	675	1250	6	212
4320(d)	885	1625	–	–	–	–	–	–	–	660	1225	6	197
4340	830	1525	705	565	1300	1050	8.3	15	16.5	650	1200	8	223
4620(d)	885	1625	–	–	–	–	–	–	–	650	1200	6	187
4640	830	1525	715	600	1320	1110	7.6	15	15	620	1150	8	197
4820(d)	–	–	–	–	–	–	–	–	–	605	1125	4	192
5045	830	1525	755	665	1390	1230	11	20	8	660	1225	4.5	192
5120(d)	885	1625	–	–	–	–	–	–	–	690	1275	4	179
5132	845	1550	755	670	1390	1240	11	20	7.5	675	1250	6	183
5140	830	1525	740	670	1360	1240	11	20	6	675	1250	6	187
5150	830	1525	705	650	1300	1200	11	20	5	675	1250	6	201
52100(f)	–	–	–	–	–	–	–	–	–	–	–	–	–
6150	830	1525	760	675	1400	1250	8.4	15	10	675	1250	6	201

(Contd...)

Steel	Austenitising temperature °C	°F	Conventional cooling(a) Temperature °C From	°C To	°F From	°F To	Cooling rate °C/hr.	°F/hr.	Time, hr.	Isothermal method(b) Cool to °C	°F	Hold, hr.	Hardness (approx.), HB
8620(d)	885	1625	–	–	–	–	–	–	–	660	1225	4	187
8630	845	1550	735	640	1350	1180	11	20	8.5	660	1225	6	192
8640	830	1525	725	640	1340	1180	11	20	8	660	1225	6	197
8650	830	1525	710	650	1310	1200	8.4	15	7.2	650	1200	8	212
8660	830	1525	700	655	1290	1210	8.4	15	8	650	1200	8	229
8720(d)	885	1625	–	–	–	–	–	–	–	660	1225	4	187
8740	830	1525	725	645	1340	1190	11	20	7.5	660	1225	7	201
8750	830	1525	720	630	1330	1170	8.4	15	10.7	660	1225	7	217
9260	860	1575	760	705	1400	1300	8.4	15	6.7	660	1225	6	229
9310(e)	870	1600	–	–	–	–	–	–	–	595	1100	14	187
9840	830	1525	695	640	1280	1180	8.4	15	6.6	650	1200	6	207
9850	830	1525	700	645	1290	1190	8.4	15	6.7	650	1200	8	223
To obtain a predominantly ferritic and spheroidised carbide structure													
1320(d)	805	1480	–	–	–	–	–	–	–	650	1200	8	170
1340	750	1380	735	610	1350	1130	5.5	10	22	640	1180	8	174
2340	715	1320	655	555	1210	1030	5.5	10	18	605	1125	10	192
2345	715	1320	655	550	1210	1020	5.5	10	19	605	1125	10	192
3120(d)	790	1450	–	–	–	–	–	–	–	650	1200	8	163
3140	745	1370	735	650	1350	1200	5.5	10	15	660	1225	10	174
3150	750	1380	705	645	1300	1190	5.5	10	11	660	1225	10	187
9840	745	1370	695	640	1280	1180	5.5	10	11	650	1200	10	192
9850	745	1370	700	645	1290	1190	5.5	10	11	650	1200	12	207

(a) The steel is cooled in the furnace at the indicated rate through the temperatue range shown. (b) The steel is cooled rapidly to the temperature indicated and is held at that temperature for the time specified. (c) In isothermal annealing to obtain pearlitic structure, steels may be austenitised at temperatures up to 70°C (125°F) higher than temperatures listed. (d) Seldom annealed. Structures of better machinability are developed by normalising or by transforming isothermally after rolling or forging. (e) Annealing is impractical by the conventional process of continuous slow cooling. The lower transformation temperature is markedly depressed and excessively long cooling cycles are required to obtain transformation to pearlite. (f) Predominantly pearlitic structures are seldom desired in this steel.

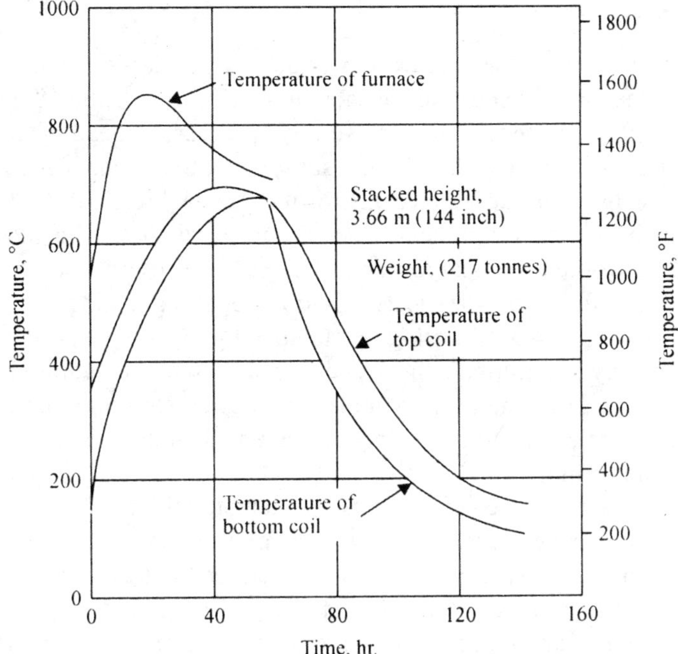

Fig. 8.3. Heating and cooling cycle for process annealing of a load of coiled low-carbon steel sheet.

The rates of spheroidising provided by these methods depend somewhat on prior microstructure, being greatest for quenched structures in which the carbide phase is fine and dispersed. Prior cold work also increases the rate of the spheroidising reaction in a subcritical spheroidising treatment.

For full spheroidising, austenitising temperatures either slightly above the Ac_1 temperature or about midway between Ac_1 and Ac_3 are used. If a temperature slightly above Ac_1 is to be used, good loading characteristics and accurate temperature controls are required for proper results; otherwise, it is conceivable that Ac_1 may not be reached and thus that austenitisation may not occur. Because time and temperature affect austenitisation and thereby influence the number of undissolved carbides from which nucleation and coalescence of the spheroids occur, close control of temperature is necessary. For example, if it is determined that spheroidisation of a given steel will require an austenitising temperature of 750°C (1385°F), a deviation of 11°C (20°F) may easily result in an incompletely spheroidised structure.

The spheroidised structure is desirable when minimum hardness, maximum ductility or (for high-carbon steels) maximum machinability is important. Low-carbon steels are seldom spheroidised for machining, because in the spheroidised condition they are excessively soft and 'gummy', cutting with long, tough chips. When low-carbon steels are spheroidised, it is generally to permit severe deformation. For example, when 1020 steel tubing is being produced by cold drawing in two or three passes, a spheroidised structure will be obtained if the material is annealed for ½ to 1 hour, at 690°C (1275°F) after each pass. The final product will have a hardness of about 163 HB. Tubing in this condition will be able to withstand sever deformation during subsequent cold forming.

As with many other types of heat treatment, hardness after spheroidising depends on carbon and alloy contents. Increasing the carbon or alloy content, or both, results in an increase in the as-spheroidised hardness, which generally ranges from 163 to 212 HB (Table 8.4).

PROCESS ANNEALING

As the hardness of steel increases during cold working, ductility decreases and additional cold reduction becomes so difficult that the material must be annealed to restore its ductility. Such annealing between processing steps is referred to as 'in-process' or simply 'process' annealing. It may consist of any appropriate treatment. In most instances, however, a sub-critical treatment is adequate and least costly and the term 'process annealing' without further qualification usually refers to an in-process subscritical anneal. It is often necessary to specify process annealing for parts that are cold formed by stamping, heading or extrusion. Hot worked high-carbon and alloy steels also are process annealed to prevent them from cracking and to soften them for shearing, turning or straightening.

Process annealing usually consists of heating to a temperature below Ae_1soaking for an appropriate time and then cooling, usually in air. In most instances, heating to a temperature between 11 and 22°C (20 and 40°F) below Ae_1 produces the best combination of microstructure, hardness and mechanical properties. Temperature controls are necessary only to prevent heating the material above Ae_1 and thus defeating the purpose of annealing.

When process annealing is performed merely to soften a material for such operations as cold sawing and cold shearing, temperatures well below Ae_1 normally are used and close controls are unnecessary.

In the wire industry, process annealing is used as an intermediate treatment between the drawing of wire to a size slightly larger than the desired finished size and the drawing of a light reduction to the finished size. Wire thus made is known as 'annealed in process' wire. Process annealing is used also in the production of wire sufficiently soft for severe upsetting and to permit drawing the smaller sizes of low-carbon and medium-carbon steel wire that cannot be drawn to the desired small size directly from the hot rolled rod. Process annealing is more satisfactory than spheroidise annealing for a material that, because of its composition or size (or both), cannot be drawn to finished size because it either lacks ductility or does not meet physical requirements. Also, material that is cold sheared during processing is process annealed to raise the ductility of the sheared surface to a level suitable for further processing.

Examples

A change to process annealed material was required, to overcome fabrication problems in two production operations involving 1040 and 1045 steels. Spheroidised 1040 steel was too soft for cold shearing; and when spheroidised 1045 steel was subsequently cold headed into stub shafts, excessive machining problems were encountered because of burrs and poor finish.

However, a spheroidised material may be necessary because of the sequence or severity of forming operations, or because of the mechanical properties desired in the end product. Thus, to eliminate cracks developed in process annealed 1035 steel during cold working for producing large-head carriage bolts, the maximum ductility provided by spheroidising was required. In another operation, involving 1038 steel, spheroidising was necessary so that the hardness of the end product was in the specified range of 78 to 88 HRB.

ANNEALED STRUCTURES FOR MACHINING

Different combinations of microstructure and hardness, considered together, are significant in terms of machinability. For instance, Fig. 8.4 shows that a partially spheroidised 5160 steel shaft was machined (by turning) with much less tool war and better surface finish than the same steel in the annealed condition with a pearlitic microstructure and a higher hardness.

Based on many observations, optimum microstructures for machining steels of various carbon contents are usually as follows:

Carbon, %	*Optimum microstructure*
0.06 to 0.20	As rolled (most economical)
0.20 to 0.30	Under 75-mm (3-inch) diam, normalised; 75–mm dia and over, as, rolled
0.30 to 0.40	Annealed, to produce coarse pearlite, minimum ferrite
0.40 to 0.60	Coarse lamellar pearlite to coarse spheroidite
0.60 to 1.00	100 % spheroidite, coarse to fine

The type of machining operation is also a factor. For example, certain gears were made from 5160 steel tubing by the dual operation of machining in automatic screw machines and broaching of cross slots. The screw-machine operations were easiest with thoroughly spheroidised material, but a pearlite structure was more suitable for broaching. A semi-spherodised structure prove to be a satisfactory compromise.

Semi-spheroidised structures can be achieved by austenitising at lower temperatures and sometimes at higher cooling rates, than those used for achieving pearlitic structures. The semi-spherodised structure of the 5160 steel tubing mentioned above was obtained by heating to 790°C (1450°F) and cooling at 28°C/hr. (50°F/hr.) to 650°C (1200°F). For this steel, austenitising at temperature of about 775°C (1425°F) results in more spheroidisation and less pearlite.

Medium-carbon steels are much more difficult to spheroidise fully than high-carbon steels such as 1095 and 52100. In the absence of excess carbides to nucleate and promote the spheroidising reaction, it is more difficult to achieve complete freedom from pearlite in practical heat treating cycles.

At lower carbon levels, structures consisting of coarse, blocky pearlite in a ferrite matrix often are found to be the most machinable. In some alloy steels, this type of structure can best be achieved by heating to temperatures well above Ac_3 to establish a coarse austenite grain size, then holding below Ar_1 to allow coarse, lamellar pearlite to form. This process sometimes is referred to as 'cycle annealing' or 'lamellar annealing'. For example, forged 4620 steel gears were heated rapidly in a five-zone furnace to 980°C (1800°F), cooled to 625 to 640°C (1160 to 1180°F) in a water-cooled zone and held at that temperature for 120 to 150 minutes. The resulting structure—coarse, lamellar pearlite in a ferrite matrix—had a hardness of 140 to 146 HB.

TYPES OF FURNACES

Furnaces for annealing are of two basic types: batch furnaces and continuous furnaces. Within either of these two types, furnaces can be further classified according to configuration, type of fuel used, method of heat application and means by which the load is moved through, or supported in, the furnace. Other factors that must be considered in furnace selection are cost, type of annealing cycle, required atmosphere and physical nature of parts to be annealed. In many cases, however, the annealing cycle used is dictated by the available equipment. Batch-type furnaces are necessary for large parts such as heavy forgings and often are preferred for small lots of a given part or grade of steel and for the more complex alloy grades requiring long cycles. Specific types of batch furnaces include car-bottom, box, bell and pit furnaces. These furnaces may be controlled manually or may be equipped with programmed controllers that permit automatic-operation. Continuous furnaces such as roller-hearth, rotary-hearth and pusher types are ideal for isothermal annealing of large quantities of parts of the same grade of steel. These furnaces can be designed with various individual zones, allowing the work to be consecutively brought to temperature, held at temperature and cooled at the desired rate.

FURNACE ATMOSPHERES

Electric furnaces used with air atmospheres and gas furnaces used with atmospheres consisting of the products of combustion, cannot be regulated for complete elimination of oxidation of the steel being treated. Only atmospheres independent of the fuel are generally considered satisfactory for clean or bright annealing. Excessive oxidation during annealing usually is prevented by the use of controlled atmospheres in conjunction with a suitable furnace that is designed to exclude air and combustion gases from the heating chamber. The gases and gas mixtures used for controlled atmospheres depend on the metal being treated, the treatment temperature and the surface requirements of the parts being annealed. The need to eliminate decarburisation as well as oxidation is often a significant factor in the selection of annealing atmospheres.

The gas most widely used as a protective atmosphere for annealing is exothermic gas. This gas is inexpensive, the raw materials for making it are readily available and the results obtained with it are generally excellent. Hydrocarbon gases such as natural gas, propane, butane and coke-oven gas are commonly burned in an exothermic-gas producer, creating a self-supporting, heat-producing combustion reaction. A commonly used exothermic gas mixture contains 15 per cent H_2, 10 per cent CO, 5 per cent CO_2, 1 per cent CH_4 and 69 per cent N_2. This gas is used for bright annealing of cold rolled low-carbon steel strip. It will decarburise medium-carbon and high-carbon steels, however, because of the carbon dioxide and water vapour it contains.

Exothermic gas sometimes is refrigerated to reduce its moisture content, particularly in geographic areas where the temperature of the water used for cooling is high. When decarburisation of workpiece surfaces must be prevented, water vapour and carbon dioxide must be completely removed from the gas. Purified exothermic gas, with its carbon dioxide and water vapour removed, has many applications in heat treatment of steel without decarburisation

Purified rich exothermic gas, formed by partial combustion, is used for short cycle annealing and process annealing of medium and high-carbon steels of the straight-carbon and alloy types. For long-cycle batch annealing, however, this gas is unsuitable, because its high carbon monoxide content results in soot deposits on the work and be cause of the possibility of surface etching as a result of the relatively long time for which the work is in the critical low-temperature range where gas reactions can occur. In short-cycle annealing, these effects are minimal and the high-CO gas is then desirable because of its high carbon potential. The fairly lean purified gas formed by more complete combustion is used for long- cycle annealing of medium and high-carbon steels of the straight-carbon and alloy types and for batch and continuous annealing of low-carbon steel strip for tin plating.

Other atmospheres commonly used in annealing include endothermic-base, dissociated ammonia, vacuum and nitrogen-base atmospheres, which consist of nitrogen plus additions such as hydrogen, methane, methanol and carbon monoxide.

ANNEALING OF SHEET AND STRIP

In terms of total tonnage of material processed, annealing of sheet and strip during production of steel-mill products represents the major use of annealing. Because such annealing is done to prepare the material for further processing (such as additional cold rolling or fabrication into parts) and because the temperatures employed are usually below the A_1 temperature, the more specific terms 'sub-critical annealing' and 'process annealing' are approperiate, although common practice is to use the term 'annealing' without qualification.

In annealing of sheet and strip, two techniques predominate: the batch process and the continuous process. In the batch process (also called 'box annealing'), coils or cut lengths of sheet are placed on an annealing base and covered with containers that are sealed to hold the appropriate atmosphere. A furnace is then placed over the covered steel. A protective atmosphere is introduced within the inner covers to protect the steel from oxidation and is circulated through the coils by use of fans and convector plates. Heating is provided by the outer furnace and may be done either through use of radiant tubes or by direct firing. The charge is heated to the required temperature and held for a period of time that will result in the desired properties. The outer furnace is then removed and the coils are allowed to cool under the inner covers. When the temperature has been reduced to the point where oxidation of the steel will not occur, the inner covers are removed and the steel is forwarded for further processing.

In the continuous process, steel coils are uncoiled and drawn through a furnace where they are subjected to the annealing cycle under a protective atmosphere. After the sheet or strip has been cooled and removed from the furnace, further in-line processing (such as hot dip galvanising) may be done, or the steel may be cut into sheets. In general, however, the steel is re-coiled and then forwarded as in the batch process.

In addition to the obvious differences in equipment, the batch process and the continuous process differ considerably in several other ways. Batch annealing may require up to a week because of the large mass of material being treated, whereas continuous annealing is accomplished in about five minutes. Differences are also evident in the temperatures employed, with the batch process generally being conducted at lower temperatures. Because in batch annealing it is difficult to ensure that the temperature is uniform throughout the charge (which may consist of several hundred tonnes of steel), the continuous process offers the potential of more uniform properties. The short annealing times of the continuous process, however, frequently result in hardness levels slightly higher than those of similar material annealed by the batch process.

Cold Rolled Plain Carbon Sheet and Strip

The usual method of manufacturing cold rolled sheet and strip is to produce a hot rolled coil, pickle it to remove scale (oxide) and cold roll it to the desired final gage. Cold rolling may reduce the thickness of the hot rolled material in excess of 90 per cent, which increases the hardness and strength of the steel but severely decreases its ductility. If any large amount of subsequent cold working is to be done, the ductility of the steel must be restored.

Annealing of cold rolled steel normally is designed to produce a re-crystallised ferrite microstructure from the highly elongated, stressed grains resulting from cold work. During heating of the steel and in the first segment of the holding portion of the cycle, the first metallurgical process to occur is recovery. During this process, internal strains are relieved (although little change in the microstructure is evident), ductility is moderately increased and strength is slightly decreased.

As annealing continues, the process of re-crystallisation occurs and new, more equiaxed ferrite grains are formed from the elongated grains. During re-crystallisation, strength decreases rapidly, with a corresponding increase in ductility. Further time at temperature causes some of the newly formed grains to grow at the expense of other grains; this is termed grain growth and results in modest decreases in strength and small (but often significant) increases in ductility. Most plain carbon steels are given an annealing treatment that promotes full re-crystallisation, but care must be taken to avoid excessive grain growth, which can lead to surface defects (such as 'orange peel') in formed parts.

The rates at which the metallurgical processes noted above proceed are functions of both the chemical composition and the prior history of the steel being annealed. For example, small amounts of elements such as aluminium, titanium, niobium, vanadium and molybdenum can decrease the rate at which the steel will re-crystallise, making the annealing response sluggish and therefore necessitating either higher temperatures or longer annealing times to produce the same properties. Although the presence of these alloying elements is generally the result of deliberate additions intended to modify the properties of the sheet (as in the case of aluminium, titanium, niobium and vanadium), some elements may be present as residual elements (molybdenum, for example) in quantities great enough to modify the response to annealing. Conversely, larger amounts of cold work (greater cold reductions) will accelerate the annealing response. Therefore, it is not possible to specify a single annealing cycle that will produce a particular set of mechanical properties in all steels; the chemical composition and the amount of cold work also must be taken into account.

Cold rolled plain carbon steels are produced to a number of different quality descriptions. Commercial quality (CQ) steel is the most widely produced and is suitable for moderate forming. Drawing quality (DQ) steel is produced to tighter mechanical-property restrictions for use in more severely formed parts. Drawing quality special killed (DQSK) steel is produced to be suitable for the most severe forming applications. Structural quality (SQ) steel is produced to specified mechanical properties other than those of the above three grades. Typical annealing cycles for all possible combinations of composition, cold reduction and grade cannot be listed here. However, typical batch-annealing temperatures range from 620 to 690°C (1150 to 1270°F) for the coldest point in the charge. Cycle times vary with the grade desired and the size of the charge, but total times (from the beginning of heating to removal of the steel from the furnace) can be as long as one week.

Continuous-annealing cycles are of shorter duration and are conducted at higher temperatures than batch-annealing cycles. In some applications, the annealing temperature may exceed A_1. Typical cycles are 40 seconds at 700°C (1290°F) for cold rolled commercial quality sheet and 60 seconds at 800°C (1470°F) for drawing quality special killed sheet. Most continuous annealing of cold rolled sheet includes an overageing treatment designed to precipitate carbon and nitrogen from solution in the ferrite and to reduce the likelihood of strain ageing. Overageing for 3 to 5 minute at 300 to 450°C (570 to 840°F) accomplishes the desired precipitation of carbon and nitrogen.

Batch annealing and continuous annealing differ slightly in the properties they produce. Typical average properties of batch-annealed and continuous annealed commercial quality plain carbon steel are as follows:

Annealing process	*Yield strength*		*Elongation, %*
	MPa	*ksi*	
Batch	210	30.4	43.0
Continuous	228	33.0	41.7

High-strength Cold Rolled Sheet and Strip

High-strength cold rolled sheet and strip are growing in importance due to their high load-bearing capacities. Strength of sheet and strip can be increased through modifications of chemical composition and/or selection of different annealing cycles, but these methods result in decreased ductility. Plain carbon steels, produced by conventional techniques, may be batch annealed or continuous annealed under conditions that result only in recovery or partial re-crystallisation. Typical batch-annealing cycles of this type employ soak temperatures of 425 to 480°C (800 to 900°F) and various soak times. High-strength low-alloy (HSLA) steels containing alloying elements such as niobium, vanadium and titanium also may be produced as cold rolled grades. The additional alloying produces a stronger hot rolled steel, which is strengthened even more by cold rolling. Cold rolled HSLA steels may be recovery annealed to produce higher-strength grades or re-crystallisation annealed to produce lower-strength grades. Successful production of cold rolled HSLA steel requires selection of the appropriate combination of steel composition and hot rolled strength, amount of cold reduction and type of annealing cycle. Table 8.5 presents typical properties after recovery or re-crystallisation annealing, as appropriate, for a family of cold rolled sheet products employing titanium as the principal strengthening element.

Table 8.5. Typical properties of titanium-strengthened cold rolled steel.

Minimum yield strength		*Yield strength*		*Tensile strength*		*Elongation*
MPa	*ksi*	*MPa*	*Ksi*	*MPa*	*ksi*	*%*
275	40	325	47	415	60	30
345	50	380	55	490	71	26
415	60	455	66	545	79	23
485	70	525	76	605	88	19
550	80	615	89	670	97	17
690	100	745	108	800	116	12
825	120	885	128	905	131	10
965	140	1040	151	1050	152	5

Hot Dip Galvanised Products

Hot dip galvanised products are produced on lines that process either pre-annealed (batch annealed) or full hard coils. Lines for processing full hard coils incorporate an in-line annealing capability so that annealing and hot dip galvanising can be accomplished in a single pass through the line. This in-line annealing, like continuous annealing of uncoated steel, generally results in slightly higher strength and slightly lower ductility than batch annealing. Maximum strip temperatures are below the A_1 temperature for commercial quality steel, but temperature in excess of 845°C (1550°F) are required for DQSK grades. Galvanising of pre-annealed steel results in properties similar to those of ungalvanised material.

The atmosphere in a continuous galvanising line, in addition to protecting the sheet from oxidation, must remove any oxides present on the strip to promote metallurgical bonding between the steel and the zinc.

Tin Mill Products

Tin mill products are distinguished from their cold rolled sheet mill counterparts chiefly by the fact that they are produced in lighter gages (0.13 to 0.38 mm, or 0.005 to 0.015 inch) and by the fact that some of them are coated with tin or chromium and chromium oxide for corrosion resistance. The sequence used for processing single reduced tin mill produts is similar to that for cold rolled sheet—that is, pickling, cold reducing, annealing and temper rolling of hot rolled coils. Double-reduced products are cold rolled an additional 30 to 40 per cent follwing annealing (this step replaces temper rolling). Whereas much of the tonnage produced in tin mills is batch annealed, a considerable amount is continous annealed (facilities for continuous annealing currently are more prevalent in tin mills than in sheet mills).

Because tin mill products tranditionally have been produced at facilities separate from sheet mills and because applications for these products are different from those for cold rolled sheet, tin mill products have been assigned separate designations for indicating the mechanical properties developed during annealing. A list of these temper designations is given in Table 8.6.

Table 8.6. Temper designations for steel tin mill products.

Designation	*Hardness aim, HR30T*
Batch (box) annealed products	
T–1	52 max
T–2	50 to 56
T–3	54 to 60
T–4	58 to 64
T–5	62 to 68
T–6	67 to 73
Continuously annealed products	
T–4 CA	58 to 64
T–5 CA (TU)	62 to 68
T–6 CA	67 to 73
Double-reduced products	
DR–8	73
DR–9	76
DR–9M	77
DR–10	80

Open-Coil Annealing

Open-coil annealing is done in batch furnaces, involves loose rewinding of a cold reduced coil to provide open spaces between successive laps. This allows the controlled atmosphere gases to be drawn between the laps, providing faster and more uniform heating and cooling than are obtained with tightly wound coils. In addition, by control of the hydrogen content and dew point of the atmosphere, decarburising conditions can be established. The carbon content of the steel can thereby be reduced to low levels for such materials as enameling steel and electrical steel. Loose rewinding of coils for open-coil annealing is done on a turntable having a vertical mandrel. As the coil is wound, a twisted wire spacer is inserted between the laps. This spacer remains in the coil during annealing and is removed after the coil has been removed from the furnace. The coil is then tightly rewound and is ready for temper rolling.

ANNEALING OF STEEL FORGINGS

Annealing of forgings is most often performed to facilitate some subsequent operation—usually machining or cold forming. The type of annealing required is determined by the kind and amount of machining or cold forming to be done as well as the type of material involved. For some processes it is essential that the microstructure be spheroidal, while for others spheroidal structures may not be necessary or even desirable.

Annealing of Forgings for Machinability

In many cases, a structure suitable for machining can be developed in low-carbon steel forgings by transferring the forgings directly from the forging operation to a furnace heated to a proper transformation temperature, holding them at this temperature for a time sufficiently long to permit all the austenite to transform, then cooling in air. In this process, the effective austenitising temperature is the finishing temperature of forging, not the initial forging temperature. This process is capable of producing reasonably uniform structures in forgings of uniform sections. However, in forgings shaped such that some portions are cooler than others, this difference in finishing temperature will cause the structures to be dissimilar. This process generally will not produce a spheroidal structure except in high-alloy steels containing large amounts of carbide-forming elements. If a lamellar structure is suitable for subsequent operations, however, this process can minimise energy usage and lower costs by reducing processing and handling time

In many instances where the product or subsequent process requires a more consistent hardness, forgings can be sub-critical annealed by heating to a temperature between 11 and 22°C (20 and 40°F) below Ae_1, holding sufficiently long (determined by degree of softening required) and then cooling in air (or equivalent). Care should be taken to maintain the temperature below Ae_1 to prevent formation of austenite, which would require a much lower cooling rate.

In forgings produced from higher-carbon steels with or without significant amounts of alloying elements, a spheroidal structure generally is preferable for high-speed machining operations. Direct transfer of high-carbon steel forgings to a furnace for transformation sometimes can be used as the preliminary step of an annealing cycle and as a means of preventing the possibility of cracking in deep-hardening steel parts, but seldom will produce satisfactory properties alone. Most annealing of high-carbon steel forgings is done either in a batch furnace or in a continuous tray pusher furnace. Typical schedules for spheroidising 52100 steel in a batch furnace are as follows:

1. Austenitise by holding at least 2 hours, at 790°C (1450°F), furnace cool at 17°C/hr. (30°F/hr.) to 595°C (1100°F) then air cool.
2. Austenitise by holding at least 2 hours, at 790°C (1450°F), cool as rapidly as practical to 750°C (1380°F), cool at 6°C/hr. to 675°C (1250°F), then air cool.
3. Austenitise by holding at least 2 hours, at 790°C (1450°F), cool as rapidly as practical to 690°C (1275°F), transform isothermally by holding at this temperature for 16 hours, then air cool.

In all instances, the load should be distributed to promote uniform heating and cooling. Use of circulating fans in the furnace chamber will greatly aid in producing a product that is uniform in both hardness and microstructure.

A typical continuous furnace for annealing steel forgings might consist of five or six zones. An example of a specific spheroidise annealing treatment in such a furnace is given in the next section.

Annealing of Forgings for Cold Forming and Re-forming

If a steel forging or blank requires further cold forming, it may be necessary to soften it in order to enhance its plastic-flow characteristics. In general, this type of annealing is done only to the extent that the forming operation requires—that is to satisfy dimensional, mechanical and tool-life requirements as well as to prevent cracking and splitting. Much intermediate annealing is done successfully, but cold forming processes are best performed on parts with totally spheroidised microstructures—especially for parts made of high-carbon steels

In one plant, both 5160 and 5260 and 52100 steels have been successfully spheroidised with a common cycle in a six-zone tray pusher furnace. In this cycle, the temperatures in the six zones are 750, 750, 705, 695, 695 and 680°C (1380, 1380, 1300, 1280, 1280 and 1260°F). Time in each zone is 150 minutes. This process yields 5160 steel forgings with hardnesses of 170 to 190 HB and 52100 steel parts with hardnesses of 175 to 195 HB, both suitable for cold or warm re-strike operations.

Low-carbon steels generally can be cold formed successfully after being heated to temperatures near A_1 and then being cooled through 675°C (1250°F) at a controlled rate. In one plant, 5120 steel annealed 1 to 2 hours, at 745°C (1375°F) and slow cooled has been cold formed successfully. Large quantities of 1008, 1513, 1524, 8620 and 8720 steels are being cold formed after annealing cycles consisting of 1 to 6 hours, at 720°C (1325°F) followed by slow cooling. The severity of the forming operation, as well as the grade of steel and history of the part, determines the extent of annealing required. Batch furnaces, continuous tray pusher furnaces and continuous belt furnaces are being used successfully to perform these types of annealing operations on low-carbon steels.

Any part that contains significant stresses resulting from cold forming or re-strike operations should be reviewed for some type of stress-relief process. Stress relieving usually is done by means of time-temperature cycles that result in slight reduction in hardness. These cycles often consist of 1 hour, at 425 to 675°C (800 to 1250°F).

Annealing to Obtain Pearlitic Microstructures

Forgings—especially plain and alloy high-carbon steel forgings—sometimes are isothermally annealed to produce a pearlitic microstructure that is preferred for a subsequent process. In steels that are to be induction hardened, for example, the carbide distribution of a fine pearlitic structure offers excellent preparation for optimum control in selective hardening while producing a reasonably machinable core structure. Isothermal annealing to obtain fine pearlite can be performed in batch or continuous furnaces; however, temperature control and uniformity are more critical thar in conventional slow cooling cycles, because a particular microstructure and a particular hardness level usually are desired. In one plant, a continuous belt-type furnace is used for isothermal annealing of 1070 steel forgings. The forgings are uniformly heated for 30 minutes at 845°C (1550°F), cooled to 675°F (1250°F) and held for 20 minutes, then rapidly cooled. The microstructure produced is essentially fine lamellar pearlite with a hardness of 219 to 228 HB. The hardness and the structure can be modified by adjusting the transformation temperature.

Hardness After Annealing

Figs 8.4 and 8.5 presents data on distribution of hardness after annealing. These data are based on production annealing of forgings, bars, tubes and rings made of 1045, 4140, 4340, 8640H and 52100 steels. The details of each annealing cycle used are given also.

ANNEALING OF BAR, ROD AND WIRE

Significant tonnages of bar, rod and wire are subjected to thermal treatments that decrease hardness and prepare the material for subsequent cold working and/or machining. For low-carbon steels (up to 0.20 per cent C), short-time sub-critical annealing often is sufficient for preparing the material for further cold working. Steels with higher carbon and alloy contents require spheroidising, to impart maximum ductility.

Most carbon and alloy steel coiled products can be successfully spheroidised. In batch annealing, it is helpful to use higher-than-normal temperatures (such as 650°C, or 1200°F) during initial heating for purging, because the higher initial temperature promotes a lower temperature gradient in the charge during subsequent heating into the temperature range between A_1 and A_3. Use of a higher purge temperature also promotes agglomeration of the carbides in the steel, which makes them more resistant to solution in the austenite when the charge temperature is finally elevated. These undissolved carbides will be conducive to the formation of a spheroidal rather than a lamellar structure when transformation is complete.

Parts	Number of tests
1045, forgings	50
4140, tube and bar	25
4140, forgings	25
4340, forgings	25
8640H, bar	50
52100, tube	400

Average

Hardness, HB: 100 150 200 250 300

1045 Steel forgings were heated at 790°C (1450°F) in a batch-type muffle furnace, furnace cooled at 11°C/hr. (20°F/hr.) to 650°C (1200°F), then air cooled. Specified maximum hardness was 207 HB. Hardness was measured on polished flash line; data cover a four-year period

4140 Seamless tubes and bars were annealed in a continuous car-bottom furnace to produce a predominantly lamellar structure. Bar diameter ranged from 47.6 to 203 mm(1.875 to 8.000 inch.); tube wall thickness ranged from 16.0 to 35 mm (0.629 to 1.379 inch). The steel was austenitised at 885°C (1625°F), furnace cooled to 760°C (1400°F), furnace cooled at 11°C/hr. (20°F/hr) to 635°F), and air cooled

4140 Steel forgings for automotive transmissions were annealed in a batch furnace. Forgings were held 5 hrs., at 675°C (1250°F). Hardness was measured on polished flash line. Specified hardness range was 170 to 241 HB.

4340 Steel forgings for aircraft piston engines were annealed in a batch furnace. Forgings were held 8 hr., at 650°C (1200°F). Hardness was measured on polished flash line. Specified hardness range was 170 to 241 HB.

8640H Hot rolled steel bars, 17.5 mm (11/16 inch) in diameter, cold heading quality, in coils, were spheroidise annealed to produce minimum hardness. Data represent as received material.

52100 Steel seamless tubes were austenitised at 790°C (1450°F), rapidly furnace cooled to 750°C (1380°F), cooled at 6°C/hr. (10°F/hr.) to 695°C (1280°F), and air cooled.

Fig. 8.4. Variation in Brinell hardness of annealed plain carbon and low-alloy steels.

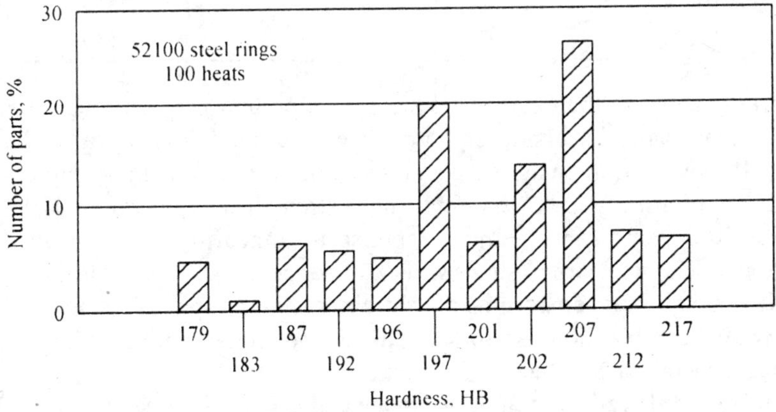

Rings were heated for 3 hr, at 790°C (1450°F), rapidly cooled to 725°C (1340°F), cooled at 8°C/hr. (15°F/hr) to 695°C (1280°F), and air cooled. Hardness measurements were taken on rings located at extreme positions in furnace load. Treatment resulted in a spheroidised structure. Composition: 0.90 to 1.05C, 0.95 to 1.25 Mn, 0.50 to 0.70 Si, 0.90 to 1.15 Cr

Fig. 8.5. Distribution of hardness for annealed rings made of a modified 52100 steel.

A knowledge of the temperature distribution in the furnace and in the load can be a major factor in achieving a good, consistent response to spheroidisation. Temperature distribution and control are much more critical in batch and vacuum furnaces, which may handle loads of up to 27 tonnes (30 tonnes), than in continuous furnaces, in which loads of only 900 to 1800 kg (2000 to 4000 lb) may be transferred from zone to zone. Test thermocouples should be placed strategically at the top, middle and bottom (inside and outside) of the charge during development of cycles. In spheroidising, to minimise formation of pearlite on cooling, it is important to ensure that no part of the charge be allowed to approach A_3. Conversely, if temperatures only slightly above A_1 are used and temperature controls are inaccurate because of poor placement of thermocouples, it is probable that the A_1 temperature will not be attained and that no austenitisation will occur.

Table 8.7 gives typical mechanical properties that can be obtained in hypoeutectoid plain carbon steels by spheroidising in accordance with rules 2 and 3. Recommended temperatures and times for lamellar and spheroidise annealing of hypoeutectoid alloy steels are presented in Table 8.4.

Table 8.7. Typical mechanical properties of spheroidised plain carbon steels.

	Tensile strength			
	Hot rolled		*Spheroidised*	
Steel	*MPa*	*ksi*	*MPa*	*ksi*
1010	365	53	295	43
1018	450	65	365	53
1022	470	68	385	56
1030	585	85	415	60
1038	600	87	485	70
1045	675	98	515	75
1060	860	125	550	80
1065	910	132	600	87
1524	510	74	450	65
1541	710	103	540	78

Prior cold working increase the degree of spheroidisation and provides even greater ductility. For example, 4037 steel in the as-rolled condition normally can be spheroidised to a tensile strength of about 515 MPa (75 ksi). If, however, the material is drawn 20 per cant and then spheroidised (referred to as 'spheroidise annealed in-process'), the resulting tensile strength will be around 470 MPa (68 ksi).

Although prior cold work can enhance response to annealing, caution must be observed in spheroidising cold worked plain carbon steels with 0.20 per cent C or less. Unless a reduction of at least 20 per cent is applied, severe grain coarsening may be observed after spheroidising. Such grain coarsening is the result of a critical combination of strain and annealing temperature peculiar to the steel and may severely impair subsequent performance.

In the wire industry, a wide variety of 'in-process' annealing operations have been evolved for rendering coiled material suitable for further processing that may require formability, drawability, machinability or a combination of these characteristics. One large wire mill reports current use of 42 separate and distinct annealing cycles, the majority of which represent compromises between practical considerations and optimum properties. For example, annealing temperatures below those that might yield optimum softness sometimes must be used in order to preclude scaling of wire coils, which often can occur even in controlled-atmosphere furnaces. Even slight scaling may cause the coil wraps to stick together, which can impede coil pay-off in subsequent operations.

'Patenting' is a special form of annealing that is peculiar to the rod and wire industry. In this process, which usually is applied to medium and higher-carbon grades of steel, rod or wire products are uncoiled and the strands are delivered to an austenitising station. The strands are then cooled rapidly from above A_3 in a molten medium—usually lead at about 540°C (1000°F)—for a period of time sufficient to allow complete transformation to a fine pearlitic structure. Both salt baths and fluidised beds have also been used for this purpose. This treatment increases substantially the amount of subsequent wiredrawing reduction that the product can withstand and permits production of high-strength wire. Successive drawing and patenting steps may be employed, if necessary, in order to obtain the desired size and strength level.

Austenitising for patenting can be accomplished in oil, gas, or electric furnaces; in high-temperature lead or salt baths; or by induction or direct resistance heating. As an alternative to quenching in molten lead, continuous air cooling often is employed. Such 'air patenting' is less expensive than 'lead patenting' but results in coarser pearlite and often more proeutectoid ferrite, a microstructure that is less desirable from the standpoint of drawing high-strength wire.

ANNEALING OF PLATE

Plate products are occasionally annealed to facilitate forming or machining operations. Annealing of plate is usually done at sub-critical temperatures and long annealing times generally are avoided. Maintaining adequate flatness can be a significant problem in annealing of large plates.

Annealing of Tubular Products

Tubular products known as 'mechanical tubing' are used in a variety of applications that can involve machining or forming. For these products, which are made from various grades of steel, annealing is a common treatment. In most annealing cycles, sub-critical temperatures and short annealing times are used to reduce hardness to the desired level. High-carbon grades, such as type 52100 for bearings, generally are spheroidised to facilitate machining. Tubular products manufactured in pipe mills are rarely annealed. These products normally are used in the as-rolled, the normalised of the quenched and tempered condition.

AUSTENITISING TEMPERATURES FOR HARDENING CARBON AND LOW-ALLOY STEELS

Temperatures recommended for austenitising carbon and low-alloy steels prior to hardening are given in Table 8.8 (for direct-hardening grades) and Table 8.9 (for carburised steels). Table 8.9 is applicable to carburised steels that have been cooled slowly from the carburising temperature and are to be furnace hardened in a subsequent operation.

For most applications, the rate of heating to the austenitising temperature is less important than other factors in the hardening process, such as maximum temperature attained throughout the section, temperature uniformity, time at temperature and rate of cooling. The thermal conductivity of the steel, the nature of the furnace atmosphere (scaling or non-scaling), thickness of section, method of loading (spaced or stacked) and the degree of circulation of the furnace atmosphere all influence the rate of heating of the steel part to the requir ed temperature selected from Tables 8.8 and 8.9.

Table. 8.8. Austenitising temperatures for direct-hardening carbon and alloy steels (SAE).

Steel	Temperature °C	°F	Steel	Temperature °C	°F	Steel	Temperature °C	°F
Carbon steels			1146	800–845	1475–1550	50B50	800–845	1475–1550
1025	855–900	1575–1650	1151	800–845	1475–1550	50B60	800–845	1475–1550
1030	845–870	1550–1600	1536	815–845	1500–1550	5130	830–855	1525–1575
1035	830–855	1525–1575	1541	815–845	1500–1550	5132	830–855	1525–1575
1037	830–855	1525–1575	1548	815–845	1500–1550	5135	815–845	1500–1550
1038(a)	830–855	1525–1575	1552	815–845	1500–1550	5140	815–845	1500–1550
1039(a)	830–855	1525–1575	1566	855–885	1575–1625	5145	815–845	1500–1550
1040(a)	830–855	1525–1575	Alloy steels			5147	800–845	1475–1550
1042	800–845	1475–1550	1330	830–855	1525–1575	5150	800–845	1475–1550
1043(a)	800–845	1475–1550	1335	815–845	1500–1550	5155	800–845	1475–1550
1045(a)	800–845	1475–1550	1340	815–845	1500–1550	5160	800–845	1475–1550
1046(a)	800–845	1475–1550	1345	815–845	1500–1550	51B60	800–845	1475–1550
1050(a)	800–845	1475–1550	3140	815–845	1500–1550	50100	775–800(c)	1425–1475(c)
1055	800–845	1475–1550	4037	830–855	1525–1575	51100	775–800(c)	1425–1475(c)
1060	800–845	1475–1550	4042	830–855	1525–1575	52100	775–800(c)	1425–1475(c)
1065	800–845	1475–1550	4047	815–855	1500–1575	6150	845–885	1550–1625
1070	800–845	1475–1550	4063	800–845	1475–1550	81B45	815–855	1500–1575
1074	800–845	1475–1550	4130	815–870	1500–1600	8630	830–870	1525–1600
1078	790–815	1450–1500	4135	845–870	1550–1600	8637	830–855	1525–1575
1080	790–815	1450–1500	4137	845–870	1550–1600	8640	830–855	1525–1575
1084	790–815	1450–1500	4140	845–870	1550–1600	8642	815–855	1500–1575
1085	790–815	1450–1500	4142	845–870	1550–1600	8645	815–855	1500–1575
1086	790–815	1450–1500	4145	815–845	1500–1550	86B45	815–855	1500–1575
1090	790–815	1450–1500	4147	815–845	1500–1550	8650	815–855	1500–1575
1095	790–815(a)	1450–1500(b)	4150	815–845	1500–1500	8655	800–845	1475–1550
Free-cutting carbon steels			4161	815–845	1500–1550	8660	800–845	1475–1550
1137	830–855	1525–1575	4337	815–845	1500–1550	8740	830–855	1525–1575
1138	815–845	1500–1550	4340	815–845	1500–1550	8742	830–855	1525–1575

(Contd...)

Steel	Temperature °C	°F	Steel	Temperature °C	°F	Steel	Temperature °C	°F
1140	815–845	1500–1550	50B40	815–845	1500–1550	9254	815–900	1500–1650
1141	800–845	1475–1550	50B44	815–845	1500–1550	9255	815–900	1500–1650
1144	800–845	1475–1550	5046	815–845	1500–1550	9260	815–900	1500–1650
1145	800–845	1475–1550	50B46	815–845	1500–1550	94B30	845–885	1550–1625
						94B40	845–885	1550–1625
						9840	830–855	1525–1575

(a) Commonly used on parts where induction hardening is employed. All steels from SAE 1030 up may have induction hardening applications. (b) This temperature range may be employed for 1095 steel that is to be quenched in water, brine or oil. For oil quenching, 1095 steel may alternatively be austenitised in the range 815 to 870°C (1500 to 1600°F). (c) This range is recommended for steel that is to be water quenched. For oil quenching, steel should be austenitised in the range 815 to 870°C (1500 to 1600°F).

Table. 8.9. Re-heating (austenitising) temperatures for hardening of carburised (a) carbon and alloy steels (SAE).

Steel	Temperature °C	°F	Steel	Temperature °C	°F	Steel	Temperature °C	°F
Carbon steels			1527	760–790	1600–1650	4626	815–845	1500–1550
						4718	815–845	1500–1550
1010	760–790	1400–1450	Free-cuttingcarbon steels			4720	815–845	1500–1550
1012	760–790	1400–1450				4815	800–830	1475–1525
1015	760–790	1400–1450	1109	760–790	1400–1450	4817	800–830	1475–1525
1016	760–790	1400–1450	1115	760–790	1400–1450	4820	800–830	1475–1525
1017	760–790	1400–1450	1117	760–790	1400–1450	8115	845–870	1550–1600
1018	760–790	1400–1450	1118	760–790	1400–1450	8615	845–870	1550–1600
1019	760–790	1400–1450				8617	845–870	1550–1600
1020	760–790	1400–1450						
1022	760–790	1400–1450	Alloy steels			8620	845–870	1550–1600
1513	760–790	1600–1650	3310	790–830	1450–1525	8622	845–870	1550–1600
1518	760–790	1600–1650	4320	830–845	1525–1550	8625	845–870	1550–1600
1522	760–790	1600–1650	4615	815–845	1500–1550	8627	845–870	1550–1600
1524	760–790	1600–1650	4617	815–845	1500–1550	8720	845–870	1550–1600
1525	760–790	1600–1650	4620	815–845	1500–1550	8822	845–870	1550–1600
1526	760–790	1600–1650	4621	815–845	1500–1550	9310	845–870	1550–1600

(a) Carburising is commonly carried out at 900 to 925°C (1650 to 1700°F); slow cooled and re-heated to given austenising temperature.

Scaling, decarburisation and other undesirable surface reactions related to furnace heating are dependent on furnace atmosphere. If these reactions are to be minimised or avoided, furnace heating must be performed in a protective atmosphere. A description of furnace atmospheres suitable for hardening is given in the section of this volume entitled 'Furnace Atmospheres and Carbon Control'. That section also discusses atmosphere generation and control. Information on austenitising in molten salt baths, especially when austenitising is to be followed by austempering in salt, is given in the article entitled 'Austempering of Steel'. See also the article on liquid carburising.

Other subjects related to the hardening of carbon and low-alloy steels, and dealt with in separate articles, are as follows: quenching, induction hardening flame hardening, austempering of steel and martempering of steel. A discussion of quenching from the carburising temperature is given in the article on gas carburising.

Chapter 9

Normalising of Steel

INTRODUCTION

Normalising of steel is a heat treating process that is often considered from both thermal and microstructural standpoints. In the thermal sense, normalising is an austenitising heating cycle followed by cooling in still or slightly agitated air. Typically, the work is heated to a temperature about 55°C (100°F) above the upper critical line of the iron-iron carbide phase diagram, as shown in Fig. 9.1; that is, above Ac_3 for hypoeutectoid steels and above A_{cm} for hypereutectoid steels. To be properly classed as a normalising treatment, the heating portion of the process must produce an austenitic phase (face-centred cubic crystal structure) prior to cooling. Typical normalising temperatures for many standard steels are given in Table 9.1.

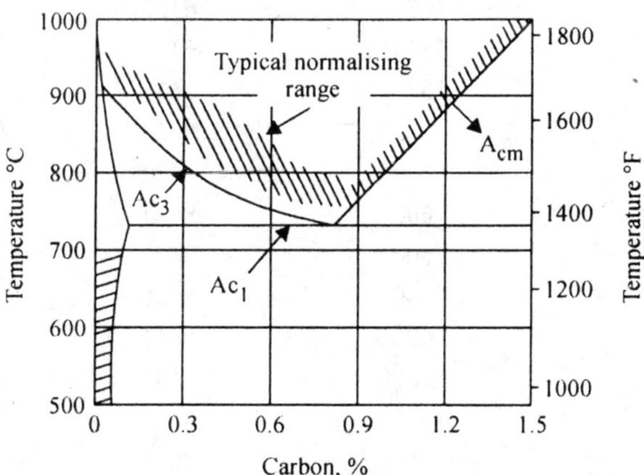

Fig. 9.1. Partial iron-Iron carbide phase diagram, showing typical normalising range for plain carbon steels.

Normalising is also frequently thought of in terms of microstructure. The areas of the microstructure that contain about 0.8 per cent carbon are pearlitic (lamellae of ferrite and iron carbide). The areas that are low in carbon are ferritic (body-centered cubic atomic structure). In hypereutectoid steels, proeutectoid iron carbide (iron carbide in excess of that within the pearlite structure) can be present in the microstructure. Air-hardening steels are excluded from the class of normalised steels because they do not exhibit the 'normal' pearlitic microstructure that characterises normalised steels.

Table 9.1. Typical normalising temperatures for standard carbon and alloy steels.

Based on production experience, normalising temperature may vary from as much as 27°C (50°F) below to as much as 55°C (100°F) above indicated temperature. The steel should be cooled in still air from indicated temperature.

	Temperature			*Temperature*	
Grade	*°C*	*°F*	*Grade*	*°C*	*°F*
Plain carbon steels			Standard alloy steels (continued)		
1015	915	1675	4817	925	1700
1020	915	1675	4820	925	1700
1022	915	1675	5046	870	1600
1025	900	1650	5120	925	1700
1030	900	1650	5130	900	1650
1035	885	1625	5132	900	1650
1040	860	1575	5135	870	1600
1045	860	1575	5140	870	1600
1050	860	1575	5145	870	1600
1060	830	1525	5147	870	1600
1080	830	1525	5150	870	1600
1090	830	1525	5155	870	1600
1095	845	1550	5160	870	1600
1117	900	1650	6118	925	1700
1137	885	1625	6120	925	1700
1141	860	1575	6150	900	1650
1144	860	1575	8617	925	1700
Standard alloy steels			8620	925	1700
1330	900	1650	8622	925	1700
1335	870	1600	8625	900	1650
1340	870	1600	8627	900	1650
3135	870	1600	8630	900	1650
3140	870	1600	8637	870	1600
3310	925	1700	8640	870	1600
4027	900	1650	8642	870	1600
4028	900	1650	8645	870	1600
4032	900	1650	8650	870	1600
4037	870	1600	8655	870	1600
4042	870	1600	8660	870	1600
4047	870	1600	8720	925	1700
4063	870	1600	8740	925	1700
4118	925	1700	8742	870	1600
4130	900	1650	8822	925	1700

(Contd...)

Grade	Temperature °C	°F	Grade	Temperature °C	°F
4135	870	1650	9255	900	1650
4137	870	1600	9260	900	1650
4140	870	1600	9262	900	1650
4142	870	1600	9310	925	1700
4145	870	1600	9840	870	1600
4147	870	1600	9850	870	1600
4150	870	1600	50B40	870	1600
4320	925	1600	50B44	870	1600
4337	870	1600	50B46	870	1600
4340	870	1600	50B50	870	1600
4520	925	1600	60B60	870	1600
4620	925	1700	81B45	870	1600
4621	925	1700	86B45	870	1600
4718	925	1700	94B15	925	1700
4720	925	1700	94B17	925	1700
4815	925	1700	94B30	900	1650
			94B40	900	1650

A broad range of ferrous products can be normalised. All of the standard low-carbon, medium-carbon, and high-carbon wrought steels can be normalised, as well as many castings. Austenitic steels, stainless steels and maraging steels either cannot be normalised, or usually are not normalised.

PURPOSE OF NORMALISING

The purpose of normalising varies considerably. Normalisation may increase or decrease the strength and hardness of a given steel in a given product form depending on the thermal and mechanical history of the product. Actually, the functions of normalising may overlap with or be confused with those of annealing, hardening and stress relieving. Improved machinability, grain-structure refinement, homogenisation, and modification of residual stresses are among the reasons for which normalising is done. Homogenisation of castings by normalising may be done in order to break up or refine the dendritic structure and facilitate a more even response to subsequent hardening. Similarly, for wrought products, normalisation can oblitreate banded grain structure due to hot rolling, as well as large grain size or mixed large and small grain size due to forging practice. The details of normalising treatments applied to three typical production parts are given in Table 9.2, which also lists the reasons for normalising and gives some of the mechanical properties obtained in the normalised and tempered condition. Comparisons of typical hot rolled or annealed mechanical properties versus typical normalised properties are presented in Table 9.3. Depending on the mechanical properties required, normalising may be substituted for conventional hardening when the size or shape of the part is such that liquid quenching might result in cracking, distortion or excessive dimensional changes. Thus, parts that are of complex shape or that incorporate sharp changes in section may be normalised and tempered, provided that the properties obtained are acceptable.

INFLUENCE OF HEATING AND COOLING DURING NORMALISING

The rate of heating generally is not critical for normalising; on an atomic scale, it is immaterial. In parts having great variations in section size, however, thermal stress can cause distortion.

Time at temperature is critical only in that it must be sufficient to cause homogenisation. Sufficient time must be allowed for solution of carbides, if present and/or for movement of alloy atoms to obtain a desired final structure. Generally, time sufficient for complete austenitisation is all that is required. One hour at temperature, after the furnace recovers, per inch of part thickness, is considered to be very liberal. Parts often can be austenitised adequately in much less time (with a saving in energy).

The rate of cooling significantly influences both the amount of pearlite and the size and spacing of the pearlite lamellae. At higher cooling rates, more pearlite forms, and the lamellae are finer and more closely spaced. Both the increased amount of pearlite and the greater fineness of the pearlite result in higher strength and higher hardness. Conversely, lower cooling rates result in softer parts. The effect of mass on hardness (via its effect on cooling rate) is illustrated by the data in Table 9.4. In any part having both thick and thin sections, the potential exists for variations in cooling rate, and thus for variations in strength and hardness as well as an increase in the probability of distortion or even cracking. Cooling rate sometimes is enhanced purposely with fans, to increase strength and hardness of parts or to decrease the time required, following the furnace operation, for sufficient cooling of parts to permit convenient handling.

Table 9.2. Typical applications of normalising and tempering.

Part	*Steel*	*Heat treatment*	*Properties after treatment*	*Reason for normalising*
Cast 50-mm (2-inch) valve body, 19 to 25 mm (3/4 to 1 inch) in section thickness	Ni-Cr-Mo	Full annealed at 955°C (1750°F), normalised at 870°C (1600°F), tempered at 665°C (1225°F)	Tensile strength, 620 MPa (90 ksi); 0.2% yield strength, 415 MPa (60 ksi); 60 ksi); 60 ksi); in 50 mm or 2 inch, 20%; reduction in area, 40%	To meet mechanical-property requirements
Forged flange	4137	Normalised at 870°C (1600°F), tempered at 570°C (1060°F)	Hardness, 200 to 225 HB	To refine grain size and obtain required hardness
Valve-bonnet forging	4140	Normalised at 870°C (1600°F) and tempered	Hardness, 220 to 240 HB	To obtain uniform structure, improved machinability and required hardness

After parts have cooled evenly to 'black heat' below Ar_1 (the parts are no longer red, as when they were removed from the furnace), they may be water or oil quenched to decrease the total cooling time. In heavy sections, cooling of the centre material to 'black heat' may require considerable time. Thermal shock, residual thermally induced stress, and resultant distortions are factors to be considered. The microstructure remains essentially unaffected by the increased cooling rate, provided that the entire mass is below the lower critical temperature, Ar_1, although changes involving precipitates may occur.

Table 9.3. Properties of selected carbon and alloy steels in the hot rolled, normalised and annealed conditions.

AISI grade (a)	Condition or treatment	Tensile strength MPa	Tensile strength ksi	Yield strength MPa	Yield strength ksi	Elongation (b) %	Reduction in area, %	Hardness, HB	Izod impact strength J	Izod impact strength ft-lb
1015	As rolled	420	61	315	46	39.0	61	126	111	82
	Normalised at 925°C (1700°F)	425	62	325	47	37.0	70	121	116	85
	Annealed at 870°C (1600°F)	385	56	285	41	37.0	70	111	115	85
1020	As rolled	450	65	330	48	36.0	59	143	87	64
	Normalised at 870°C (1600°F)	440	64	345	50	35.8	68	131	118	87
	Annealed at 870°C (1600°F)	395	57	295	43	36.5	66	111	123	91
1022	As rolled	505	73	360	52	35.0	67	149	81	60
	Normalised at 925°C (1700°F)	485	70	360	52	34.0	68	143	117	87
	Annealed at 870°C (1600°F)	450	65	315	46	35.0	64	137	121	89
1030	As rolled	550	80	345	50	32.0	57	179	75	55
	Normalised at 925°C (1700°F)	520	76	345	50	32.0	61	149	94	69
	Annealed at 845°C (1550°F)	465	67	340	50	31.2	58	126	69	51
1040	As rolled	620	90	415	60	25.0	50	201	49	36
	Normalised at 900°C (1650°F)	590	86	375	54	28.0	55	170	65	48
	Annealed at 790°C (1450°F)	520	75	355	51	30.2	57	149	44	33
1050	As rolled	725	105	415	60	20.0	40	229	31	23
	Normalised at 900°C (1650°F)	750	109	425	62	20.0	39	217	27	20
	Annealed at 790°C (1450°F)	635	92	365	53	23.7	40	187	17	13
1060	As rolled	815	118	485	70	17.0	34	241	18	13
	Normalised at 900°C (1650°F)	775	113	420	61	18.0	37	229	13	10
	Annealed at 790°C (1450°F)	625	91	370	54	22.5	38	179	11	8
1080	As rolled	965	140	585	85	12.0	17	293	7	5
	Normalised at 900°C (1650°F)	1010	147	525	76	11.0	21	293	7	5
	Annealed at 790°C (1450°F)	615	89	375	55	24.7	45	174	6	5
1095	As rolled	965	140	570	83	9.0	18	293	4	3
	Normalised at 900°C (1650°F)	1015	147	500	73	9.5	14	293	5	4
	Annealed at 790°C (1450°F)	655	95	380	55	13.0	21	192	3	2

(Contd...)

AISI grade (a)	*Condition or treatment*	*Tensile strength*		*Yield strength*		*Elongation (b)*	*Reduction in area,*	*Hardness,*	*Izod impact strength*	
		MPa	*ksi*	*MPa*	*ksi*	*%*	*%*	*HB*	*J*	*ft-lb*
1117	As rolled	485	71	305	44	33.0	63	143	81	60
	Normalised at 900°C (1650°F)	465	68	305	44	33.5	54	137	85	63
	Annealed at 860°C (1575°F)	430	62	280	41	32.8	58	121	94	69
1118	As rolled	520	76	315	46	32.0	70	149	109	80
	Normalised at 925°C (1700°F)	480	69	320	46	33.5	66	143	104	76
	Annealed at 790°C (1450°F)	450	65	285	41	34.5	67	131	106	79
1137	As rolled	625	91	380	55	28.0	61	192	83	61
	Normalised at 900°C (1650°F)	670	97	395	58	22.5	49	197	64	47
	Annealed at 790°C (1450°F)	585	85	345	50	26.8	54	174	50	37
1141	As rolled	675	98	360	52	22.0	38	192	11	8
	Normalised at 900°C (1650°F)	705	103	405	59	22.7	56	201	53	39
	Annealed at 815°C (1500°F)	600	87	355	51	25.5	49	163	34	25
1144	As rolled	705	102	420	61	21.0	41	212	53	39
	Normalised at 900°C (1650°F)	665	97	400	58	21.0	40	197	43	32
	Annealed at 790°C (1450°F)	585	85	345	50	24.8	41	167	65	48
1340	Normalised at 870°C (1600°F)	835	121	560	81	22.0	63	248	93	68
	Annealed at 800°C (1475°F)	705	102	435	63	25.5	57	207	71	52
3140	Normalised at 870°C (1600°F)	890	129	600	87	19.7	57	262	54	40
	Annealed at 815°C (1500°F)	690	100	425	61	24.5	51	197	46	34
4130	Normalised at 870°C (1600°F)	670	97	435	63	25.5	60	197	86	64
	Annealed at 865°C (1585°F)	560	81	360	52	28.2	56	156	62	46
4140	Normalised at 870°C (1600°F)	1020	148	655	95	17.7	47	302	23	17
	Annealed at 815°C (1500°F)	655	95	415	61	25.7	57	197	55	40
4150	Normalised at 870°C (1600°F)	1155	168	735	107	11.7	31	321	12	9
	Annealed at 815°C (1500°F)	730	106	380	55	20.2	40	197	25	18
4320	Normalised at 895°C (1640°F)	795	115	465	67	20.8	51	235	73	54
	Annealed at 850°C (1560°F)	580	84	425	62	29.0	58	163	110	81

(Contd...)

AISI grade (a)	Condition or treatment	Tensile strength MPa	Tensile strength ksi	Yield strength MPa	Yield strength ksi	Elongation (b) %	Reduction in area, %	Hardness, HB	Izod impact strength J	Izod impact strength ft-lb
4340	Normalised at 870°C (1600°F)	1280	186	860	125	12.2	36	363	16	12
	Annealed at 810°C (1490°F)	745	108	470	69	22.0	50	217	51	38
4620	Normalised at 900°C (1650°F)	575	83	365	53	29.0	67	174	133	98
	Annealed at 860°C (1575°F)	510	74	370	54	31.3	60	149	94	69
4820	Normalised at 860°C (1580°F)	755	110	485	70	24.0	59	229	110	81
	Annealed at 815°C (1500°F)	680	99	465	67	22.3	59	197	93	69
5140	Normalised at 870°C (1600°F)	795	115	470	69	22.7	59	229	38	28
	Annealed at 830°C (1525°F)	570	83	295	43	28.6	57	167	41	30
5150	Normalised at 870°C (1600°F)	870	126	530	77	20.7	59	255	32	23
	Annealed at 825°C (1520°F)	675	98	355	52	22.0	44	197	25	19
5160	Normalised at 860°C (1575°F)	955	139	530	77	17.5	45	269	11	8
	Annealed at 810°C 1495°F)	725	105	275	40	17.2	31	197	10	7
6150	Normalised at 870°C (1600°F)	940	136	615	89	21.8	61	269	36	26
	Annealed at 815°C (1500°F)	665	97	410	60	23.0	48	197	27	20
8620	Normalised at 910°C (1675°F)	635	92	355	52	26.3	60	183	100	74
	Annealed at 870°C (1600°F)	535	78	385	56	31.3	62	149	112	83
8630	Normalised at 870°C (1600°F)	650	94	430	62	23.5	54	187	95	70
	Annealed at 845°C (1550°F)	565	82	370	54	29.0	59	156	95	70
8650	Normalised at 870°C (1600°F)	1025	149	690	100	14.0	45	302	14	10
	Annealed at 795°C (1465°F)	715	104	385	56	22.5	46	212	29	22
8740	Normalised at 870°C (1600°F)	930	135	605	88	16.0	48	269	18	13
	Anealed at 815°C (1500°F)	695	101	415	60	22.2	46	201	40	30
9255	Normalised at 900°C (1650°F)	935	135	580	84	19.7	43	269	14	10
	Annealed at 845°C (1550°F)	775	112	485	71	21.7	41	229	9	7
9310	Normalised at 890°C (1630°F)	905	132	570	83	18.8	58	269	119	88
	Annealed at 845°C (1550°F)	820	119	440	64	17.3	42	241	79	58

(a) All grades are fine grained except for those in the 1100 series, which are coarse grained. (b) In 50 mm or 2 inch.

Table 9.4. Effect of mass on hardness of normalised carbon and alloy steels.

Grade	Normalising temperature		Hardness, HB, for bar with diameter, mm (inch), of:			
	°C	°F	13(½)	25(1)	50(2)	100(4)
Carbon steels, carburising grades						
1015	925	1700	126	121	116	116
1020	925	1700	131	131	126	121
1022	925	1700	143	143	137	131
1117	900	1650	143	137	137	126
1118	925	1700	156	143	137	131
Carbon steels, direct-hardening grades						
1030	925	1700	156	149	137	137
1040	900	1650	183	170	167	167
1050	900	1650	223	217	212	201
1060	900	1650	229	229	223	223
1080	900	1650	293	293	285	269
1095	900	1650	302	293	269	255
1137	900	1650	201	197	197	192
1141	900	1650	207	201	201	201
1144	900	1650	201	197	192	192
Alloy steels, carburising grades						
3310	890	1630	269	262	262	248
4118	910	1670	170	156	143	137
4320	895	1640	248	235	212	201
4419	955	1750	149	143	143	143
4620	900	1650	192	174	167	163
4820	860	1580	235	229	223	212
8620	915	1675	197	183	179	163
9310	890	1630	285	269	262	255
Alloy steels, direct-hardening grades						
1340	870	1600	269	248	235	235
3140	870	1600	302	262	248	241
4027	905	1660	179	179	163	156
4063	870	1600	285	285	285	277
4130	870	1600	217	197	167	163
4140	870	1600	302	302	285	241
4150	870	1600	375	321	311	293
4340	870	1600	388	363	341	321
5140	870	1600	235	229	223	217
5150	870	1600	262	255	248	241

(Contd...)

Grade	Normalising temperature		Hardness, HB, for bar with diameter, mm (inch), of:			
	°C	°F	13(½)	25(1)	50(2)	100(4)
5160	860	1575	285	269	262	255
6150	870	1600	285	269	262	255
8630	870	1600	201	187	187	187
8650	870	1600	363	302	293	285
8740	870	1600	269	269	262	255
9255	900	1650	277	269	269	269

Carbon Steels

Table 9.1 lists typical normalising temperatures for some standard grades of carbon steel. These temperatures can be interpolated to obtain values for carbon contents not listed. Steels containing 0.20 per cent C or less usually receive no treatment subsequent to normalising. However, medium-carbon or high-carbon steels are often tempered after normalising, cold working or machining. Whether or not tempering is desirable depends primarily on carbon content and section size. Table 9.3 presents typical mechanical properties of selected carbon and alloy steels in the hot rolled, normalised and annealed conditions. A low-carbon or medium-carbon steel of large section size subjected to the same treatment.

Alloy Steels

For alloy steel forgings, rolled products and castings, normalising is commonly used as a conditioning treatment before final heat treatment. Normalising also refines the structures of forgings, rolled products and castings that have cooled non-uniformly from high temperatures. Table 9.1 lists typical normalising temperatures for some standard alloy steels.

Alloy carburising steels, such as 3310 and 4320, usually are normalised at temperatures higher than the carburising temperature, to minimise distortion in carburising and to improve machining characteristics. Carburising steels of the 3300 series sometimes are double normalised with the expectation of minimising distortion; these steels are tempered at about 650°C (1200°F) for up to 15 hours, to reduce hardness to below 223 HB for machinability. Carburising steels of the 4300 and 4600 series usually can be normalised to a hardness not exceeding 2C7 HB, and therefore, need not be tempered for machinability.

Hypereutectoid alloy steels, such as 52100, are normalised for partial or complete elimination of carbide networks, thus producing a structure that is more susceptible to 100 per cent spheroidisation in the subsequent spheroidise annealing treatment. The spheroidised structure provides improved machinability and a more uniform response to hardening.

FORGINGS

When forgings are normalised before carburising or before hardening and tempering, the upper range of normalising temperatures is used. However, when normalising is the final heat treatment, use is made of the lower range of temperatures.

Furnaces

Either batch-type or continuous furnaces may be used for normalising steel forgings. In a continuous furnace, forgings to be normalised are usually placed in shallow pans and a pusher mechanism at the loading end of the furnace transports the pans through the furnace. Furnace burners, located on both sides of the furnace, fire below the hearth, and combustion products rise along the walls of the work-zone muffle and exhaust into the roof of the furnace. No atmosphere control is used. Combustion products enter the work zone through ports lining both sides of the entire hearth. A typical furnace is 9 m (30 ft) long and has 18 gas burners (or 9 oil burners) on each side. For purposes of temperature control, such a furnace is divided into three 3 m (10 ft) zones, each having a vertical thermocouple extending into it through the roof of the furnace.

Processing

Small forgings are usually normalised as received from the forge shop. They are placed or piled loosely on the pans to a maximum depth of 75 mm (3 inch). A typical furnace has five pans in each of the three furnace zones. Heating is adjusted so that the work reaches normalising temperature in the last zone. After passing through the last zone, the pans are discharged onto a cooling conveyor. The work, while still in the pans, is cooled in still air to below 480°C (900°F); it is then discharged into tote boxes, in which it cools to room temperature. Total furnace time is approximately 3½ hours, but during this period the work is held at the normalising temperature for only 1 hour.

Normalising of large open-die forgings usually is performed in batch-type furnaces pyrometrically controlled to narrow temperature ranges. Forgings are held at the normalising temperature long enough to allow complete austenitising and carbide solution to occur (usually one hour per inch of section thickness), and then are cooled in still air.

Axle-Shaft Forging

In forging an axle shaft made of fine-grain 1049 steel, only one end of the forging bar was heated to upset the wheel-flange section. When the part was examined in cross-section from the flanged end to the cold end, the following metallurgical conditions were revealed. The hot worked flanged area of the axle exhibited a fine-grain structure as a result of the hot working at the forging temperature (approximately 1100°C or 2000°F). However, a section adjacent to the flange, which also had been heated to the forging temperature but which had not been hot worked, exhibited a coarse-grain structure. Nearer the cool end of the shaft, a zone that reached a temperature of about 700°C (1300°F) exhibited a spheroidised structure. The cold end of the shaft retained its initial fine grain size throughout the forging operation.

In subsequent operations, this shaft was to be mechanically straightened, machined, and induction hardened. Because of the mixed grain structure, these operations posed several problems. The coarse-grain area adjacent to the flange was extremely weak in the transverse direction, and there was a possibility that fracture would occur if this section were subjected to a severe straightening operation. The spheroidised area would not respond adequately to induction hardening, because the solution rate of this type of carbide formation was too sluggish for the relatively rapid rate of induction heating. Furthermore, the mixed metallurgical structure would present difficulties in machining. Consequently, normalising was required in order to produce a uniformly fine-grain structure throughout the axle shaft prior to straightening, machining and induction hardening.

Low-Carbon Steel Forgings

In contrast to the medium-carbon axle shaft discussed in the preceding paragraphs, forgings made of carbon steels containing 0.25 per cent C or less are seldom normalised. Only severe quenching from above the austenitising temperature will have any significant effect on their structure or hardness.

Structural Stability

Normalising and tempering is also a preferred treatment for promoting the structural stability of low-alloy heat-resistant alloys, such as AMS 6304 (0.45 C, 1 Cr, 0.5 Mo, 0.3 V), at temperatures up to 540°C (1000°F). Wheels and spacer rings used in the 'cold' ends of aircraft gas turbine engine compressors are typical of parts subjected to such treatment to promote structural stability.

Effects on Mechanical Properties

Differences in mechanical properties obtained by normalising and tempering and by quenching and tempering result from differences in the rate of cooling from the austenitising temperature, and hence are functions of the hardenability of the steel and the section size of the part. For air cooling, just as for liquid quenching, a larger section size requires a higher alloy content if a given hardness is to be maintained. Attainment of the same hardness in a normalised part as in a liquid-quenched part also requires a higher alloy content, to compensate for the smaller hardening response elicited by air cooling compared with that brought about by liquid quenching.

Hardness is not the only property that is affected by differences in cooling rate. Other mechanical properties may be expected to differ from the normalised and tempered condition to the quenched and tempered condition, even when surface hardness is about the same for both treatments. For example, although tensile strength will be about the same, yield strength, elongation, and reduction in area will be lower in the normalised and tempered condition than in the quenched and tempered condition.

Multiple Normalising Treatments

Multiple normalising treatments are employed: (i) to obtain complete solution of all lower-temperature constituents in austenite by the use of high initial normalising temperatures (for example, 925°C or 1700°F); and (ii) to refine final pearlite grain size by the use of a second normalising treatment at a temperature closer to the Ac_3 temperature (for example, 815°C or 1500°F) without destroying the beneficial effects of the initial normalising treatment.

Locomotive-axle forgings made of carbon steel to AAR Specification M-126, class F (ASTM A236, class F), containing 0.45 to 0.59 per cent C and 0.60 to 0.90 per cent Mn, are double normalised to obtain a uniformly fine grain structure along with other exacting mechanical-property requirements. Forgings made of a low-carbon steel (0.18 per cent C) with 1 per cent Mn intended for low-temperature service are double normalised to meet sub-zero impact requirements.

BAR AND TUBULAR PRODUCTS

Frequently, the finishing stages of hot-mill operations employed in making steel bar and tube produce properties that closely approximate those obtained by normalising. When this occurs, normalising is unnecessary and may even be inadvisable. Nevertheless, the reasons for normalising bar and tube products are generally the same as those applicable to other forms of steel.

There is an additional reason, however, for normalising bar and tube products. When these products are cold finished by a sequence of cold reductions with high sub-critical anneals between passes, some spheroidisation occurs. In such instances, the product is sometimes normalised before the last cold reduction pass. Normalising eliminates the spheroidisation that earlier passes and anneals may have generated and restores the pearlitic structure beneficial to machinability in low-carbon and medium-carbon grades of carbon or alloy steel.

Tubes are easier to normalise than bars of equivalent diameter, because the lighter section thickness of tubes permits more rapid heating and cooling. These advantages help minimise decarburisation and promote more nearly uniform microstructures in tube products.

Furnaces

Continuous furnaces of the roller-hearth type are widely used for normalising tube and bar products, especially in long lengths. Batch-type furnaces or other types of continuous furnaces are satisfactory if they provide some means for rapid discharge and separation of the load to permit free circulation of air around each tube as it cools. Continuous furnaces should have at least two zones, one for heating and one for soaking. Cooling facilities should be ample, so that uniform cooling can proceed until complete transformation has occurred. If tubes are packed or bundled during cooling from a high temperature, the purpose of normalising is defeated and a semi-annealed or a tempered product results.

Generally, protective atmospheres are not used in roller-hearth continuous furnaces for normalising bar or tube products. The scale that forms during normalising is removed by acid pickling or abrasive blast cleaning.

Alloy Steels

Although the principles involved in normalising alloy steel bar, tube and pipe products are the same as those for normalising carbon steels, application of these principles is sometimes more complex for alloy steels. Some alloy grades require more care in heating, to prevent cracking from thermal shock. They also require longer soaking times because of lower austenitising and solution rates. For many alloy steels, rates of cooling in air to room temperature must be carefully controlled. Certain alloy steels are forced-air cooled from the normalising temperature in order to develop specific mechanical properties. This is a normalising treatment only in the microstructural sense discussed near the beginnings of this chapter.

CASTINGS

In industrial practice, steel castings may be normalised in car-bottom, box, pit or continuous furnaces; the same heat treating principles apply to each type of furnace.

Loading

Furnaces are loaded with castings in such a manner that each casting will receive an adequate and uniform heat supply. This may be accomplished by stacking castings in regular order or by interspersing large and small castings so that load concentration in any one area is not excessive. At normalising temperatures, the tensile strength of steel is greatly reduced, and heavy unequal sections may become distorted unless bracing and support are provided. Accordingly, small and large castings may be arranged so that they support each other.

Loading Temperature

When castings are charged, the temperature of the furnace should be such that the thermal shock will not cause metal failure. For the higher-alloy grades of steel castings, such as C5, C12 and WC9, a safe furnace temperature for charging is 315 to 425°C (600 to 800°F). For lower-alloy grades, furnace temperatures may be as high as 650°C (1200°F). For cast carbon steels and low-alloy steels with low carbon contents (low hardenability), castings may be charged into a furnace operating at the normalising temperature.

Heating

After the furnace has been charged, the temperature is increased at a rate of approximately 220°C/hr. (400°F/hr.) until the normalising temperature is reached. Depending on steel composition and casting configuration, a reduction in the rate of heating to approximately 28 to 55°C/hr. (50 to 100°F/hr.) may be necessary to avoid cracking. Extremely large castings should be heated more slowly to prevent development of extreme temperature gradients.

Soaking

After the normalising temperature has been reached, castings are 'soaked' at this temperature for a period of time that will ensure complete austenitisation and carbide solution. The duration of the soaking period may be pre-determined by microscopic examination of specimens held for various times at the normalising temperature.

Cooling

After the soaking period, the castings are unloaded and allowed to cool in still air. Use of fans, air blasts or other means of accelerating the cooling process should be avoided.

SHEET AND STRIP

Hot rolled steel sheet and strip (about 0.10 per cent C) are normalised primarily to refine grain size, minimise directional properties and develop desirable mechanical properties. Uniformly fine equiaxed ferrite grains normally are obtained in hot rolled sheet and strip by finishing the final hot rolling operation above the upper transformation temperature. However, if part of the hot rolling operation is performed on steel that has transformed partially to ferrite, the deformed ferrite grains usually will re-crystallise and form abnormally coarse-grain patches during the self-anneal induced by coiling or piling at temperatures of 650 to 730°C (1200 to 1350°F). Also, relatively thin hot rolled material, if it is inadvertently finished well below the upper transformation temperature and coiled or piled while it is too cold to self-anneal, may possess directional properties. These conditions are unsuitable for some types of severe press drawing applications and may be corrected by normalising.

Normalising also may be used to develop high strength in alloy steel sheet and strip if the products are sufficiently high in carbon and alloy contents to enable them to transform to fine pearlite or martensite when cooled in air from the normalising temperature. In general, the hardened material is tempered to attain an optimum combination of strength and ductility. Typical mechanical properties of normalised 4130, modified 4335 and modified 4340 steel sheet are given in Table 9.5.

Table 9.5. Typical mechanical properties of normalised alloy steel sheet.

	Thickness		*Tensile strength*		*Yield strength(a)*		*Elongation(b),*	*Hardness,*
Grade	*mm*	*inch*	*MPa*	*ksi*	*MPa*	*ksi*	*%*	*HRC*
4130	4.9	0.193	835	121	585	85	14	25
4335(c)	4.6	0.180	1725	250	1240	180	8	48
4340(c)	2.0	0.080	1860	270	1345	195	7	50

(a) At 0.2% offset. (b) In 50 mm or 2 inch. (c) Modified: 0.40% Mo, 0.20% V.

Processing

The normalising operation consists of passing the sheet or strip through an open, continuous furnace in which the material is heated to a temperature approximately 55 to 85°C (100 to 150°F) above its upper transformation temperature, 845 or 900°C (1550 to 1650°F), thus obtaining complete solution of the original structure with the formation of austenite, and then air cooling the material to room temperature.

Furnace Equipment

Normalising furnaces are designed to heat and cool sheets singly or two in a pile. They are built in the form of long, low chambers, and usually comprise three sections: a pre-heating zone (12 to 20 per cent of the total length); a heating, or soaking zone (about 40 per cent of the total length); and a cooling zone, which occupies the remaining 40 to 50 per cent of the length.

Heating Arrangements

Normalising furnaces usually are heated with gas or oil and do not employ protective atmospheres. Therefore, sheets are scaled during heat treatment. Burners are arranged along each side of the heating zone; they usually are above the conveyor, but occasionally are both above and below it. The furnace roof, which is higher in the pre-heating and soaking zones than in the cooling zone, is usually built in sections. In most furnaces, both the pre-heating zone and the cooling zone are heated by the hot gases from the heating zone. However, both of these zones may be equipped with burners for more accurate temperature control. Air is excluded by regulating the draft to maintain a slight pressure within all zones.

Conveyor-Type Furnaces

In modern furnaces of the conveyor type (the only type suitable for treating short lengths), sheets are carried through each of the three zones on rotating disks made of heat-resistant alloys. These disks have polished surfaces, which prevent them from scratching the sheets, and are staggered to ensure uniform heating. The disks are mounted on water-cooled shafts, which are driven by variable-speed motors through chains and sprockets or shafts and gears. These furnaces may be up to 2.5 m (100 inch) wide and from 27 to 61 m (90 to 200 ft) long. Fuel consumption is 2.3 to 5.2 million kJ per tonne (2.0 to 4.5 million Btu per tonne) of steel treated, and production rates vary from 2.7 to 10.9 tonnes (3 to 12 tonnes) per hour.

Normalising in a three-zone conveyor type furnace equipped with pyrometric controls is a relatively simple operation. If scratching of sheets is to be avoided, the sheets are brought to the charging table and hand laid, one or more at a time, on a rider or conveyor sheet. Heavy sheets are normalised singly, but lighter sheets may be stacked two in a pile. To control heating and retard scaling, single sheets may be laid on a rider sheet and covered with a cover sheet. Sheets are carried by disk-rollers into the pre-heating zone, where they absorb heat rapidly because of the large temperature differential between the sheets and the interior of the furnace and because of the large surface-to-volume ratio. As the sheets become heated and the temperature differential is reduced, the rate of heat absorption slackens. After traveling 4½ to 6 m (15 to 20 ft), the sheets enter the soaking zone at a temperature several degrees below the normalising temperature. Heating is completed in the soaking zone, which is maintained at a constant temperature, and sheets are held at the required temperature for a time sufficient to convert the microstructure to austenite before they are passed into the cooling zone. The sheets emerge from the cooling zone at a temperature that can be varied between 150 and 540°C (300 and 1000°F), and are conveyed for a short distance on the runout table, where, after being cooled rapidly in air, they are carefully removed from the rider sheet. The trip through such a furnace is carried out at a uniform speed of 0.03 to 0.10 m/s (5 to 20 ft/minute) and requires 5 to 20 minute to complete.

Catenary Furnaces

The catenary, or free-loop, type of furnace is designed for continuous normalising of cold reduced steel unwound from coils; it does not have rolls or any other type of conveyor for supporting the material passing through the heating zone. The heating zones of catenary furnaces range in length from 6 to 15 m (20 to 50 ft). The pre-heating and cooling zones usually are shorter than those in conveyor-type furnaces, and for some kinds of work may be omitted entirely. At their exit ends, catenary furnances may incorporate pickling or other descaling equipment for removing surface oxides formed on the steel during normalising.

Chapter 10

Hardening of Steel

INTRODUCTION

Hardening is a heat treating process in which steel is heated to a temperature above the critical point held at this temperature and then quenched (rapidly cooled in water), oil or molten salt baths.

The present chapter will deal mainly with the practical aspects of hardening. The structure of steel as a result of hardening may be regarded as a starting structure which is capable of modification by subsequent tempering. The result of the hardening depends on several factors, each of which may be of decisive importance. Hence it is of the utmost importance that each of the various steps of the hardening process is given the same care and consideration.

HEATING MEDIA

During the heating-up stage it is necessary to provide against any unintentional carburisation or decarburisation. Should this happen the superficial hardness measured after hardening will be misleading, which may result in the choice of an incorrect tempering temperature. Variations in the carbon content of the surface layers may cause the steel to crack during hardening. If the surface oxidation is too severe it is possible that the tolerances set for the tool might not be kept after descaling and grinding. To protect the surface of the tool the necessary precautions to be observed depends on the type of furnace used.

Salt-Bath Furnaces

These offer good protection against variations in the superficial carbon content, owing to the short heating time required and the neutral character of the bath. The bath must of course be maintained in good condition. As a rule, manufacturers of heat-treatment salts do not specify the composition of the salts. On the other hand they give detailed information on their properties and uses. For purposes of information, however, some typical compositions and working temperatures for neutral salt baths are given below:

45 per cent NaCl + 55 per cent KCl	675–900°C
20 per cent NaCl + 80 per cent $BaCl_2$	675–1060°C
100 per cent $BaCl_2$	1025–1325°C

During use the bath takes up iron from the material being heat treated. This iron is oxidised since the bath is in contact with the atmosphere. The iron oxide thus formed has decarburising action on the steel charge and hence the bath must be regenerated. Follow the recommendations given by the salt manufacturers. Barium chloride baths with certain additions are used as heating media for high-speed steel and certain other high-alloy steels. Such baths may be regenerated by means of a few pieces of silica brick, which combines with the iron dissolved in the bath. The sludge thus formed must be removed regularly. By means of a graphite rod immersed into the salt bath it is easy to check whether iron is present. Any iron oxide in the bath will be reduced by the graphitic carbon and small bright beads of metallic iron will form on the rod. The carburising or decarburising action of a salt bath may be tested by means of steel foil which is dipped into the bath for a few minutes and then quenched. If the foil is soft after quenching, the bath is decarburising; if more brittle than it normally should be the bath is carburising. Foil of varying carbon content is available.

Gas or Oil-fired Muffle Furnaces

These allow the amounts of air of combustion to be controlled. Hence the atmosphere inside the muffle can be adjusted so that for the treatment temperature and grade of steel concerned the superficial carbon content of the charge will not be appreciably affected.

Electrically-Heated Muffle Furnaces

These, if not operated with controlled atmospheres, should be used with annealing boxes in which the steel charge is packed with some protective material, which should be as neutral as possible. Unfortunately there is no general-purpose substance so versatile that is suitable as a packing material for all grades of steel, since these materials are only neutral towards steels of a definite carbon content and at a definite temperature. One purpose of the packing material is to prevent the ingress of air to the steel. For this purpose cast-iron chips are very suitable since the air that is sucked into the box on account of temperature variations preferentially oxidises the chips. Practical tests and scientific investigations have shown that cast-iron chips are practically inert towards plain-carbon steel and low-alloy steel, containing carbon between 0.6 and 1 per cent. They are also inert towards high-alloy chrome steels of the type D 2 and D 6, provided the customary hardening temperatures are employed. With regard to hot-work tool steel containing carbon around 0.3 per cent cast-iron chips have a slightly carburising action. At about 1000°C there is already a danger that cast-iron chips may begin to sinter and they should not be used at temperatures above 1050°C. If the tool is wrapped in a thin layer of newsprint before being packed in cast-iron chips the surface is also protected against mechanical damage.

Wood-charcoal or coke fines (but not carburising compounds) act as decarburisers on plain-carbon and low-alloy steels containing medium to high carbon, but act as carburisers on high-alloy chrome steels and hot-work steels. However, it is the temperature and not the composition of the steels that is the factor that determines whether carburisation or decarburisation will take place.

Under certain conditions borax may be used as a protective agent against decarburisation. The borax powder is sprinkled over the surface to be protected and on melting, it forms a covering film. Before hardening, this film should be brushed off, otherwise it retards the rate of cooling of the steel. Borax is often used to protect the engraved surface of coining dies when they are being hardened. On water-hardening steels the borax film spalls off when the steel is quenched and leaves a bright clean surface.

There are also some commercially available protective pastes on the market. These pastes are applied to the protected surface prior to heat treating. The protective efficiency of these pastes is best assessed by testing them on a piece of steel of the grade for which they are recommended. Heat-resisting steel foil has also been put to use as a protective measure against decarburisation. The foil is supplied as sheet or bags in which the steel parts are wrapped.

Controlled Atmospheres

These specially produced protective gases have shown themselves, both technically and economically, to be highly suitable when used for the heat treatment of costly tools and component parts. This is particularly so for large-scale production hardening. The gas is made in a separate plant, whence it is led into the hardening furnace which must be gas-tight to prevent any leakage. Protective gas for hardening is made according to the following three main principles:

1. Exothermic gas is produced by the exothermic combustion (i.e., without the addition of heat) of gas and air; the combustion not going to completion. As the gas contains large volumes of CO_2 and H_2O the greater part of these must be removed by chemical absorption and drying, respectively. If both H_2O and CO_2 are removed, the gas contains CO, H_2 and N_2. This is called monogas.
2. Endothermic gas is produced by endothermic combustion, i.e., energy must be supplied. The gas, which is usually propane or propylene, is mixed in carefully balanced proportions with air and passed through a gas generator in which a so-called carrier gas is produced. Table 10.1 gives a selection of carbon potentials that have been used with various steels with good results.
3. Inert gas is protective gas that, as regards its carbon, oxygen and nitrogen contents, remains unreactive to the steel. From a chemical point of view it is only the inert gases, such as argon, that truly satisfy this criterion. The most commonly used unreactive gas consists mainly of nitrogen. On a large scale it is produced by the exothermic combustion of crude gas and air and then cleaned according to various methods.

A fairly simple method applicable to air-hardening steels is to place the tools in a gastight box fitted with an inlet and an outlet tube. Nitrogen gas from a nitrogen cylinder is led through the box during the whole treatment cycle. When the box is taken out of the furnace the rate of cooling may be increased by placing the box in a current of air from a fan. A large number of such hardening treatments have been applied to tools made from steel H13.

Table 10.1. Selection of carbon potential for various steels.

Grade of steel AISI, BS	*SS*	*Carbon content %*	*Carbon potential %*
W1	1880	1.0	0.80
–	2092	1.0	0.80
O1	2140	0.90	0.75
–	2550	0.55	0.45
A2	2260	1.0	0.70
D2	2310	1.50	0.70
D6	2312	2.00	0.80
H13	2242	0.40	0.30

This method is a simplified form of 'hardening in a protective atmosphere' and the necessary equipment can be installed without large capital expenditure.

Analysis and Control of the Furnace Atmosphere

The following four methods are employed for this purpose. They measure and control the carbon potential either directly or indirectly.

Direct: By measuring the specific resistance of a thin iron wire.
Indirect: By measuring the dew point.
Indirect: By means of an infrared analyser.
Indirect: By means of an oxygen probe.

Vacuum Hardening

This has become increasingly popular during the last two decades. The charge is placed in a cold furnace which is then evacuated by means of vacuum pumps. For most cases an operating pressure of 10^{-2} torr is acceptable. Should a higher vacuum be required, diffusion pumps are necessary which will increase the cost of the furnaces.

After the charge has been heated and austenitised, the furnace and contents are cooled to room temperature, usually in a nitrogen atmosphere but in special cases argon is used. New furnace types have been developed in which the charge can be quenched straight into oil while under vacuum. For this purpose special oils have been developed. These oils can be heated to about 200°C, which enables an austempering treatment to be carried out.

The greatest benefit resulting from vacuum hardening is that the surfaces of the parts comprising the charge are not altered by the hardening treatment, i.e., the parts are completely free from surface decarburisation or oxidation.

RATE OF HEATING

If the rate of heating to the hardening temperature is very fast the steel may crack or warp, since sections having different dimensions heat up at different speeds. These differences may be diminished by pre-heating the tool to a temperature usually lying close below the transformation temperature A_{c1} of the steel. When heating for hardening in a muffle furnace, pre-heating is not so necessary as when a salt bath is used. Even in this latter case pre-heating may be omitted if the parts to be treated are small and symmetrical. However, if a salt bath is used, a pre-heat to about 100°C is always advisable in order to remove any moisture adhering to the steel.

During the initial heating-up stage the surface of the steel is at a higher temperature than the centre. The closer the temperature of the steel approaches the present temperature, the smaller is this temperature difference, which is shown in Figs. 10.1 and 10.2. In actual practice it can therefore be assumed that when the surface has reached the temperature of the furnace, the steel is heated right through.

This assumption runs contrary to the generally widespread belief that the surface of the steel reaches the temperature of the furnace considerably faster than the centre. This belief is based on the observation that the corners of steel objects become warm more quickly than other sections.

This diagrams in Figs. 10.1 and 10.2 were obtained by inserting thermocouples at the centre of test specimens, thereby ensuring that the heating effect from the end surfaces would not influence the readings. That the 'corner effect' really does exist has been shown in a heating trial carried out on a large die block. One thermocouple was located at a corner, 50 mm below the surface; another at the middle of the die, also 50 mm below the surface and a third thermocouple right at the centre. As is seen from Fig. 10.3 the corner heated up much more quickly than did the other parts of the block. As regards the other surfaces it is seen that even in this large section the temperature as registered in the middle followed closely that at the surface, once the latter had reached the temperature of the furnace.

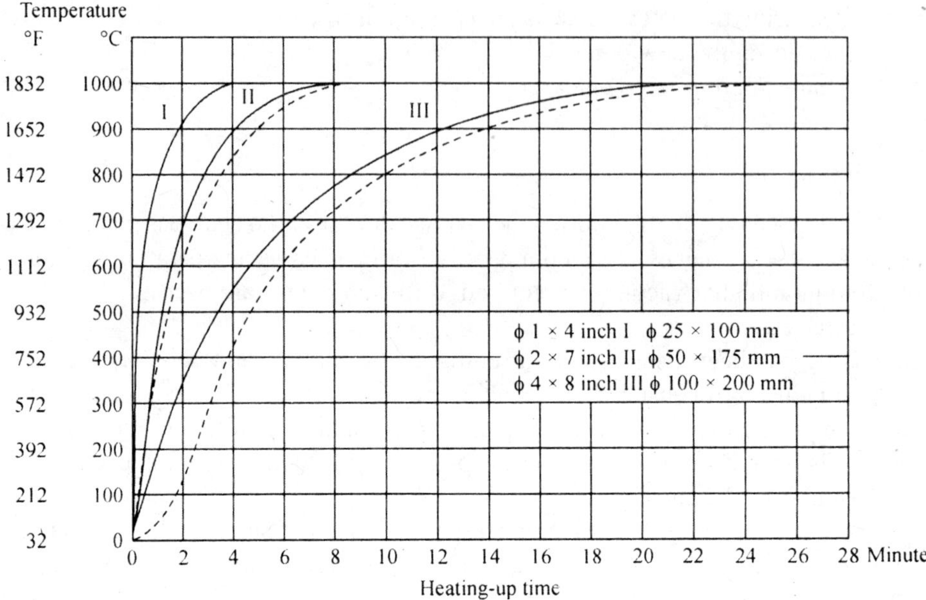

Fig. 10.1. Time-temperature curves for steel bars heated in a salt bath at 1000°C. Full line: temperature at surface. Dashed line: temperature at centre.

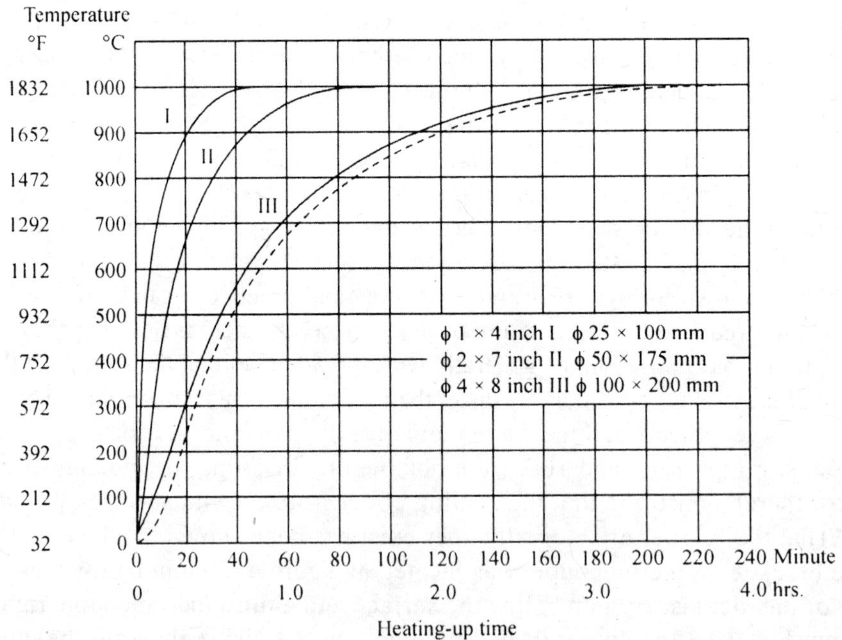

Fig. 10.2. Time-temperature curves for steel bars packed in cast-iron chips, heated in a muffle furnace at 1000°C. Diameter of boxes: 50, 75 and 140 mm. Full line: temperature at surface. Dashed line: temperature at centre.

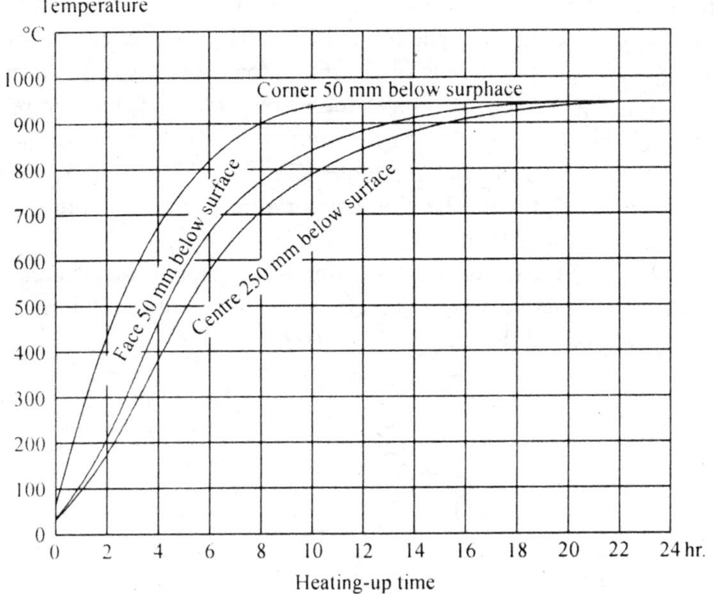

Fig. 10.3. Time-temperature curves for different positions in a die block, 2300 × 950 × 500 mm, on being heated in an oil-fired muffle furnace.

The heating-up time may be calculated from standard expressions. Based on trials and with the help of these expressions the heating-up time necessary in an oil-fired muffle furnace has been calculated for various steel dimensions and plotted in a diagram as a function of the cross-sectional area (Fig. 10.4). This diagram is valid for plain-carbon and low-alloy steels. A pre-supposition is that the length of the part is at least twice the size of the diameter or diagonal. The dimension 950 × 500 mm (Fig. 10.3) is equivalent to a round bar 700 mm in diameter. According to Fig. 10.4 the heating-up time for a round bar of this diameter is 20 hrs. which is in quite good agreement with the 22 hours indicated in Fig. 10.3.

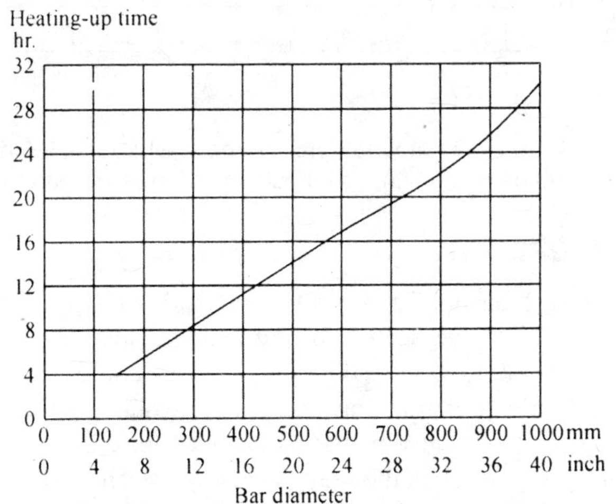

Fig. 10.4. Time for heating to hardening temperature, 850°C, for bars of varying diameters up to 1000 mm. Length 2 × diameter. Starting temperature 20°C. Oil-fired muffle furnace, calculated values.

The time of heating-up depends not only on the dimensions of the steel but but also on the heating capacity of the furnace and the degree of packing. Hence it is obviously necessary to gain experience with specific furnaces and to compile tables or diagrams for bars of various dimensions. An example illustrating this is shown in Fig. 10.5 which traces the heating curve of the centre of a 75 mm diameter bar made from steel D 2. Pre-heating was carried out in an 80 kW convection furnace measuring 800 mm diameter by 2000 mm. The final heating took place in a 160 kW furnace of the same dimensions. Note the rather long time to reach the pre-heating temperature (1.5 hour) compared with the time taken to reach the hardening temperature (about 0.75 hour).

This is due to the increase in the rate of heat transfer as the temperature increases. Another five trials were made with the same steel but of different dimensions. The results containing the heating times from 550°C to 1000°C are reproduced in the curve in Fig. 10.6.

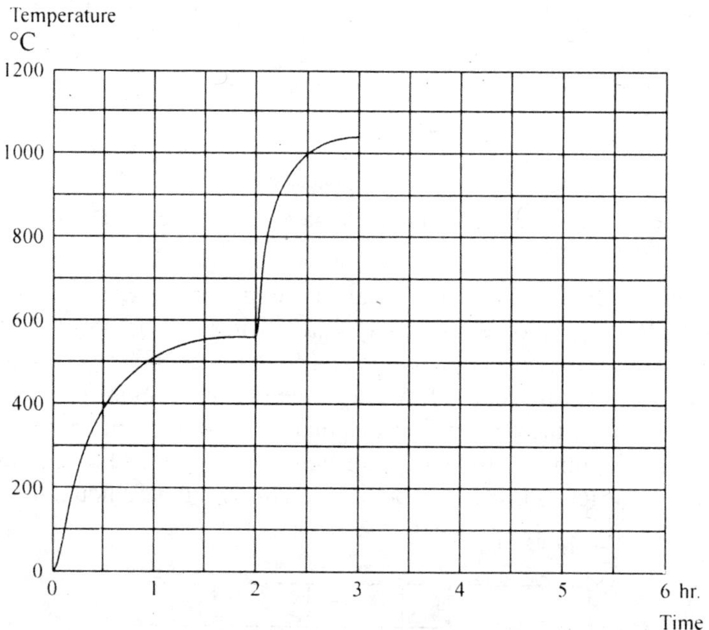

Fig. 10.5. Heating curve for a 75 mm diameter bar, pre-heated in a furnace 800 mm diameter × 200 mm, rated at 80 kW. Final heating in a furnace of the same dimensions, rated at 160 kW.

HARDENING TEMPERATURE

For each grade of steel, on the basis of a series of practical trials, a range of temperature has been established to which the steel is to be heated for hardening. This temperature range, also called the quenching range, is chosen so as to give maximum hardness and at the same time maintain a fine-grained structure in the steel. A lower hardening temperature would give rise to the formation of pearlite and bainite and the hardness at the surface would be inadequate.

As the temperature increases, so does the grain size and also the amount of retained austenite. The specimens were tempered at 200°C in order to bring out more clearly the retained austenite. At 920°C and 970°C the retained austenite may be discerned as light angular areas.

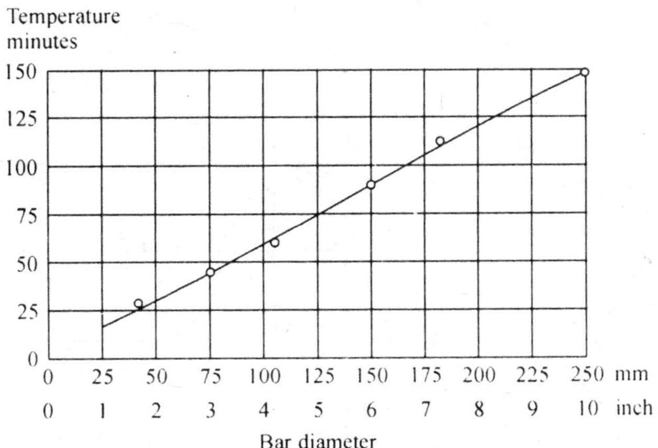

Fig. 10.6. Heating curve for round bars of varying diameters from 550 to 1000°C, in a furnace 800 mm diameter × 2000 mm, rated at 160 kW.

Based on actual trials the quenching range for this steel has been fixed at 850–880°C.

As has been previously mentioned, by raising the hardening temperature the hardenability of steel can be increased. This increase is due to the greater amount of carbide going into solution and to increased grain size. When high-speed steel is being hardened a temperature of some tens of degrees below the melting range is used. Only a slight increase above the ordinary temperature will cause incipient fusion round the carbides. This molten phase will 'permeate' through the grain boundaries and hence reduce the toughness. A further increase in temperature may lead to pronounced melting. The steel is then said to be burnt, the danger of melting increases with the degree of segregation. The larger the dimension the greater the segregation. For this reason a lower hardening temperature should be chosen for high-speed steel tools made from a heavy sections. Even for dimensions corresponding to 100 mm diameter a hardening temperature near the lower limit of the customary temperature range should be chosen.

HOLDING TIME AT TEMPERATURE

When the steel has reached the hardening temperature it is austenitic provided that the temperature has been correctly chosen. The time of holding at the hardening temperature depends on the desired degree of carbide dissolution and acceptable grain size. Since the amount of carbide is different for different types of steel the time of holding is also dependent on the grade of steel.

However, the holding time is not only dependent on the hardening temperature but also on the rate of heating. With very slow heating, hypo-eutectoid steels are completely austenitic immediately above A_{c3} and hence no holding time should be necessary. With more rapid heating some holding time should be required to ensure temperature equalisation and carbide dissolution. Alternatively, a higher temperature might be used.

The influence of the rate of heating and the holding time on carbide dissolution, grain growth and hardness after hardening was carried out both isothermally and continuously. In the first case the steel was heated at the rate of 130°C/s to the temperature in question, held there for certain pre-determined times and then cooled rapidly. The continuous heating was carried out at various constant rates, ranging from 0.05 to 2400°C/s. Fig. 10.7 shows such as diagram. With a continuous heating rate of, for example, 0.22°C/s homogeneous austenite was obtained when 860°C was reached after about 4000 seconds (i.e., about 1 hr.). If the heating rate had been 1°C/s the temperature and time required would have been

880°C and 14 minutes, respectively. With a rate of 10°C/s, 920°C and 1.5 minutes would have been required, etc.

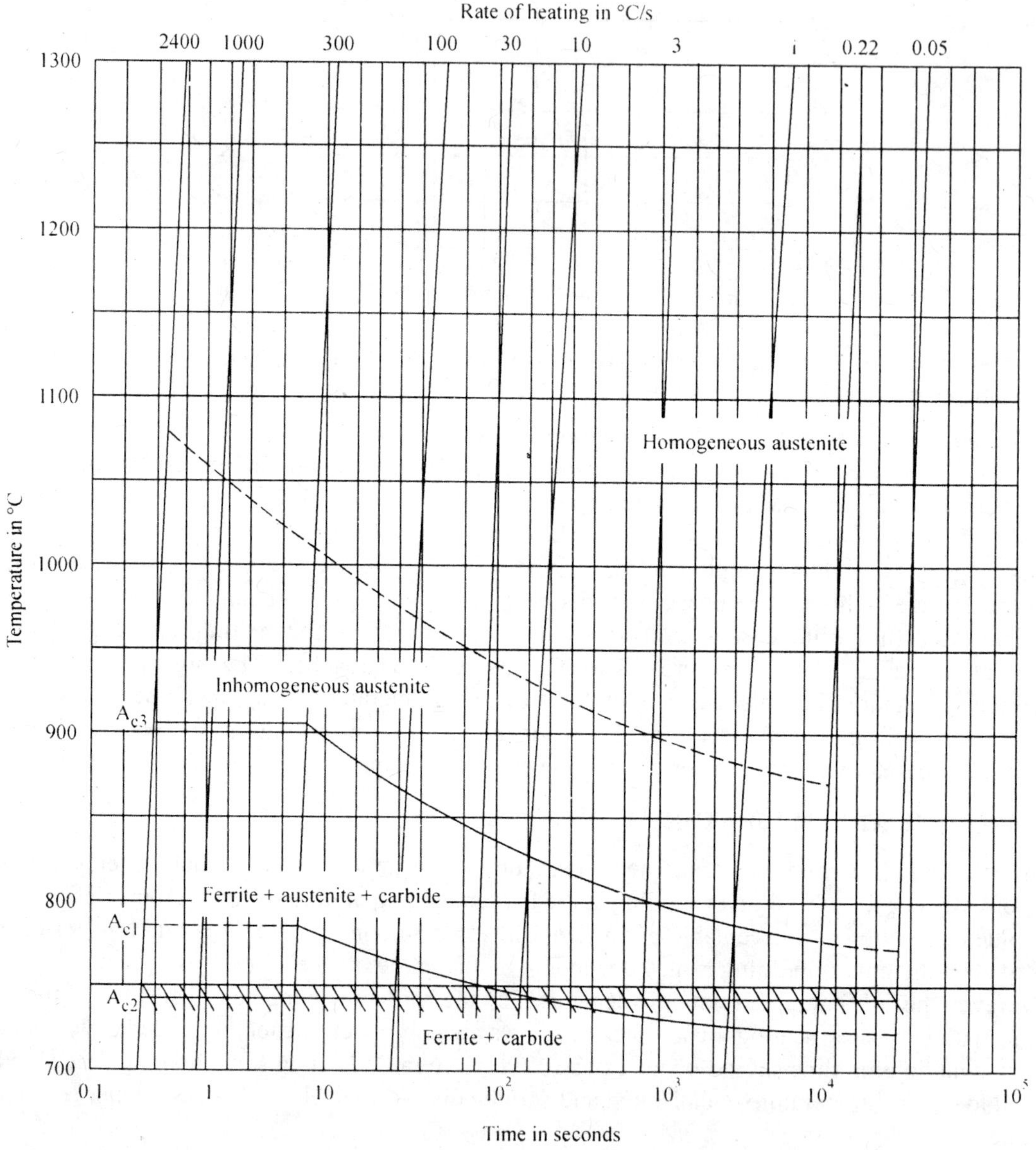

Fig. 10.7. Time-temperature-austenitising diagram (continuous heating) for steel Cf 53)0.51% C, 0.38% Si, 0.64% Mn, 0.022% Al, 0.005% N). Initially fully annealed for 5 hours. at 700°C, furnace cooled.

However, the important feature with the austenitising treatment is whether maximum hardness is attained on quenching to room temperature. This information may be gleaned from Fig. 10.8 which shows that the maximum hardness, 770 HV, would be obtained with a heating rate of 1°C/s up to 875°C (heating-up time about 14 minutes). This indicates that in this instance a homogeneous austenitic structure would be required to give maximum hardness. If the heating rate were 10°C/s as temperature of slightly more than 900°C would be required for maximum hardness. Heating-up time in this instance would be 5 minutes. Thus the dissolution of carbide in this steel is so rapid that no holding time should be required for the tests reported above.

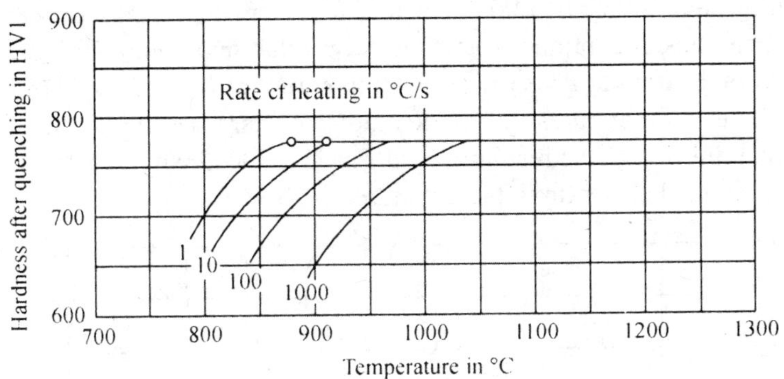

Fig. 10.8. Hardness after quenching as a function of the rate of heating and austenitising temperature for fully annealed steel, type Cf 53.

The amount of dissolution is also dependent on the initial structure of the steel. According to Fig. 10.8 the structure of the steel investigated consisted of spheroidised cementite. If, on the other hand, a start were made on a hardened and tempered structure such as that shown in Fig. 10.9, a heating rate of 1000°C/s i.e., shorter than one second's heating-up time, could be used and on quenching the steel to room temperature a hardness of 770 HV would still be attained. Most low-alloy hypoeutectoid steels require only similarly short time for carbide dissolution, as do straight carbon steels containing about the same carbon content.

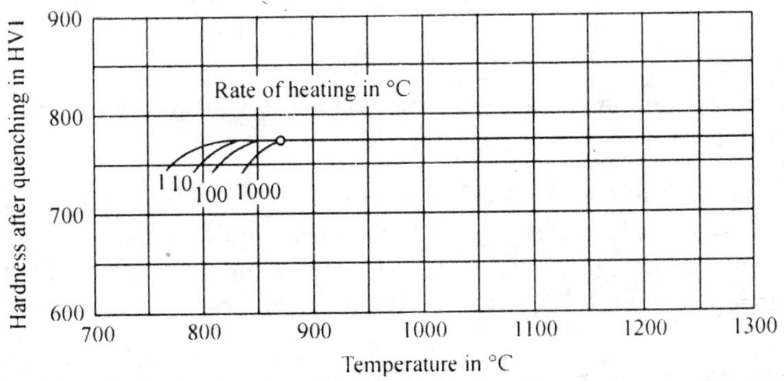

Fig. 10.9. Hardness after quenching as a function of the rate of heating and austenitising temperature for hardened and tempered steel, Cf 53.

Hypoeutectoid steels, containing pearlite and excess ferrite are hardened by heating to a temperature slightly (30–50°C) above point A_{c3}. This will enable homogenous austenite to be obtained which will be transformed into martensite. Upon heating within the interval between the critical points (A_{c1}– A_{c3}), ferrite is partially retained with the austenite. The austenite is transformed into martensite after subsequent quenching but the ferrite remains unchanged. The result will be incomplete hardening since the segregates of the free ferrite in the structure of the hardened steel reduce its hardness.

The presence of ferrite, in addition to the brittle 'framework' of martensite crystals, does not allow noticeable toughness to be obtained.

The situation is somewhat more complicated for eutectoid and hyper-eutectoid steels. Consider, for example the ball-bearing steel 100 Cr 6 (1.00 per cent C, 0.22 per cent Si, 0.24 per cent Mn, 1.52 per cent Cr). Fig. 10.10 which applies to continuous heating, shows that maximum hardness, i.e., 900 HV after quenching, is attained on heating to a very narrow temperature range and furthermore that the range is displaced towards higher temperatures as the heating rate increases. If the steel is quenched from temperature higher than this range the resulting hardness is reduced which is due to the presence of an increasing amount of retained austenite in the structure.

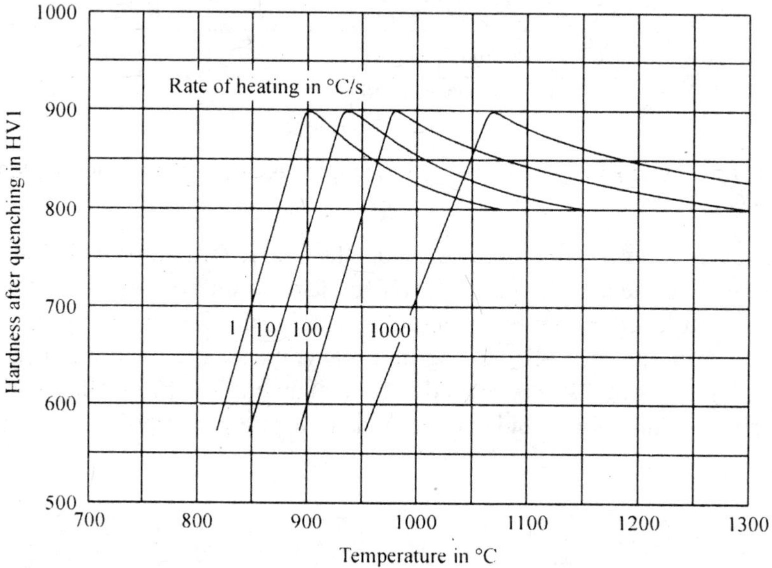

Fig. 10.10. Hardness after quenching as a function of the rate of heating and austenitising temperature for steel 100 Cr 6: initially fully annealed.

This state of affairs is more clearly illustrated in Fig. 10.11 which reproduces details from the isothermal diagram. If the steel is heated rapidly to 950°C a holding time of 2–4 seconds is sufficient for the attainment of maximum hardness on quenching. At a temperature of 900°C the holding time would be 30–60 seconds and at 850°C, between 800 and 1200 seconds—13 and 20 minutes—must be allowed.

Should a temperature of 800°C be used, a holding time of several hours must be reckoned with in order that the steel may attain maximum hardness on quenching. Hence it is rather obvious that in practice a hardening temperature of around 850°C should be used for this steel. However, this inference is not immediately obvious on referring to the heating diagram in Figs. 10.12 and 10.13 from which much higher temperatures ought to be chosen in order to obtain either homogeneous or inhomogeneous austenite. For this steel, however, such temperatures would result in too much retained austenite after quenching.

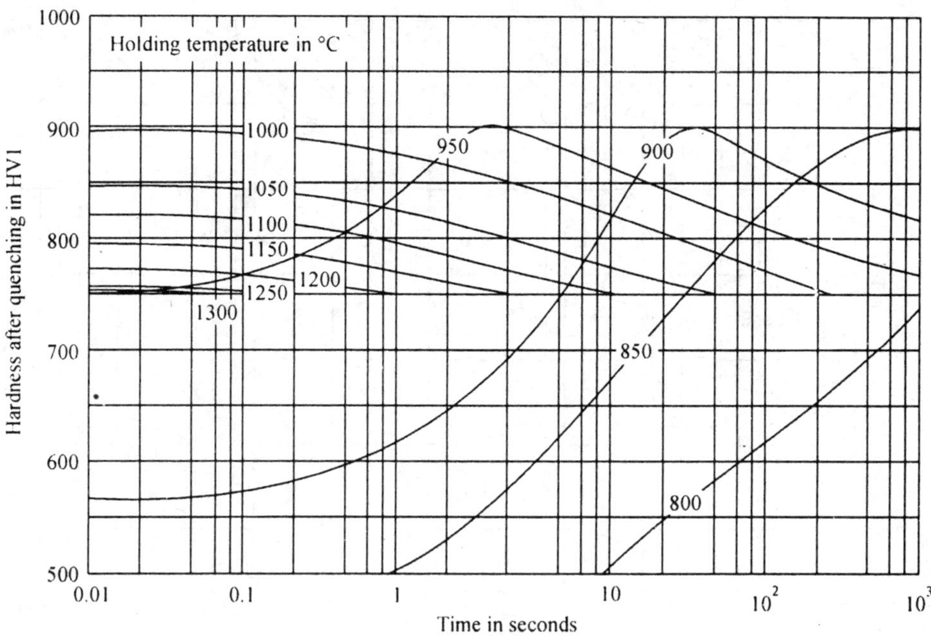

Fig. 10.11. Hardness after quenching as a function of holding time and austenitising temperature after a heating rate of 130°C/s for steel 100 Cr 6; initially fully annealed.

Similar characteristics are displaced by high-chromium steels, i.e., the hardenable stainless steels as well as tool steels of type D 2. Fig. 10.14 illustrates a practical case in point from which it is seen that 1020–1040°C should be a suitable hardening temperature range for D 2 and a holding time of about 20 minutes. For the example submitted, the heating-up time was about 10 minutes.

The diagram in Fig. 10.15 applies to the hot-work steel X 38 CrMoV 51 which is approximately equivalent to H 13. From this diagram it is seen that maximum hardness on quenching, 750 HV, is attained after holding for 10^3 seconds (17 minutes) at 1000°C. If the temperature is raised to 1100°C, 60 seconds, holding time is sufficient. At 1300°C less than one second is required and the holding time may be increased to 17 minutes without causing any variation in the hardness. In order to see what might occur in such a case the diagram in Fig. 10.16 should be studied. In it the grain size is set off against the temperature and holding time. Earlier it has been pointed out that the grain size should preferably not be larger than 7 ASTM and consequently a holding time of 17 minutes at 1000°C is quite on the safe side in this respect.

The tests as reported in *Atlas zur Warmebehandlung der Stahle* were carried out with tubular test pieces, D_0/D_i = 8/7 mm and were produced from bar stock 10–20 mm diameter. In such dimensions the carbides in the steel are very small and evenly distributed, which is apparent from the accompanying micrographs. Especially for the high-alloy steels the times for dissolution will thus be considerably shorter than for the heavy dimensions since the carbide particle size increases as the dimensions of the steel stock increases. On the other hand the heating-up times for heavy sections are generally much longer than for smaller dimensions. However, it should be brought to general attention that the maximum hardness, as given in the diagrams, may be up to 100 HV higher than what is attainable in practice.

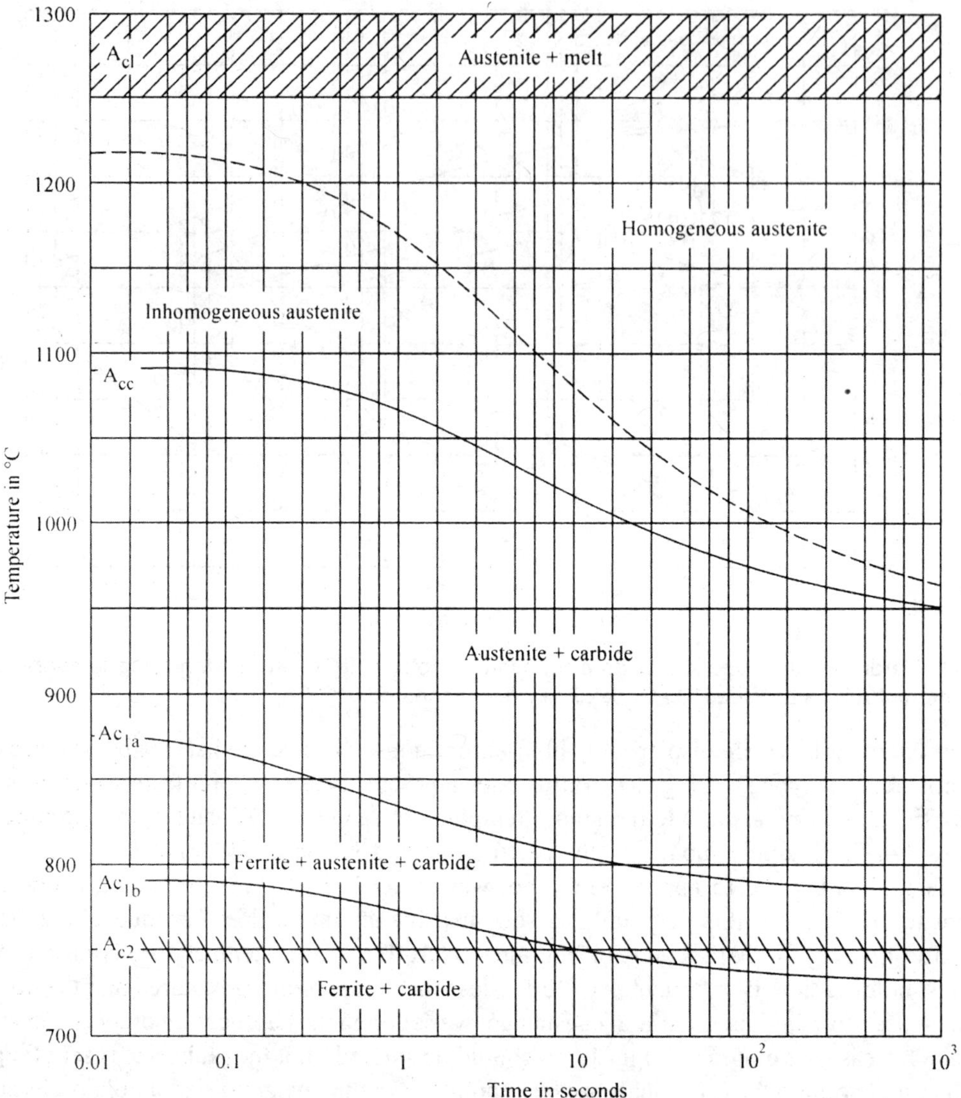

Fig. 10.12. Time-temperature-austenitising diagram (isothermal) for steel 100 Cr 6. Rate of heating to holding temperature: 130°C/s.

Plain-carbon and low-alloy structural steels which contain easily dissolved carbides require only a few minutes' holding time after they have reached the hardening temperature. In order to make certain that there has been sufficient carbide dissolution, a holding time of 5–15 minutes is quite sufficient.

For medium-alloy structural steels a holding time of 15–25 minutes is recommended, irrespective of the dimension.

For flame and induction hardening which enable the setting of accurately determined heating times, considerably higher hardening temperatures are used than for conventional hardening. Besides, the time of holding can be reduced to a few minutes and in some instances even to a second or so.

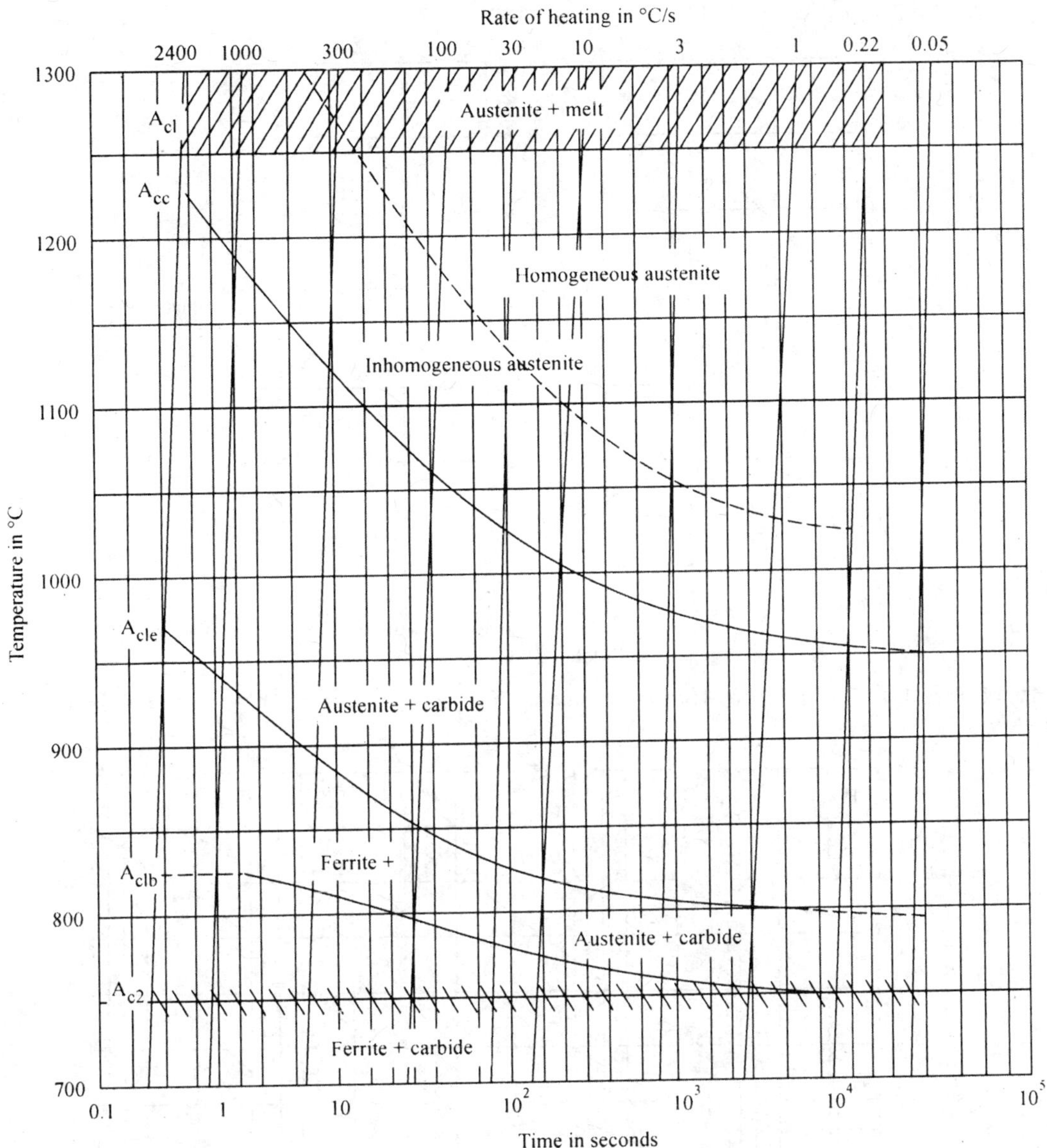

Fig. 10.13. Time-temperature-austenitising diagram (continuous heating).

For low-alloy tool steels a definite holding time is necessary to ensure that the required hardness will be attained. A suggested figure for the holding time is 0.5 minutes per millimetre of section thickness; the minimum and maximum time, however, should be 10 and 30 minutes respectively. Why the holding time should increase with section thickness is due to the fact that the size of the carbide grains and hence their reluctance to go into solution increases as the section dimension increases. Besides, most low-alloy steels generally have insufficient hardenability which, however, may be improved, particularly in heavy sections, by promoting the increased dissolution of carbides. This may be brought about by increasing the time of holding at temperature.

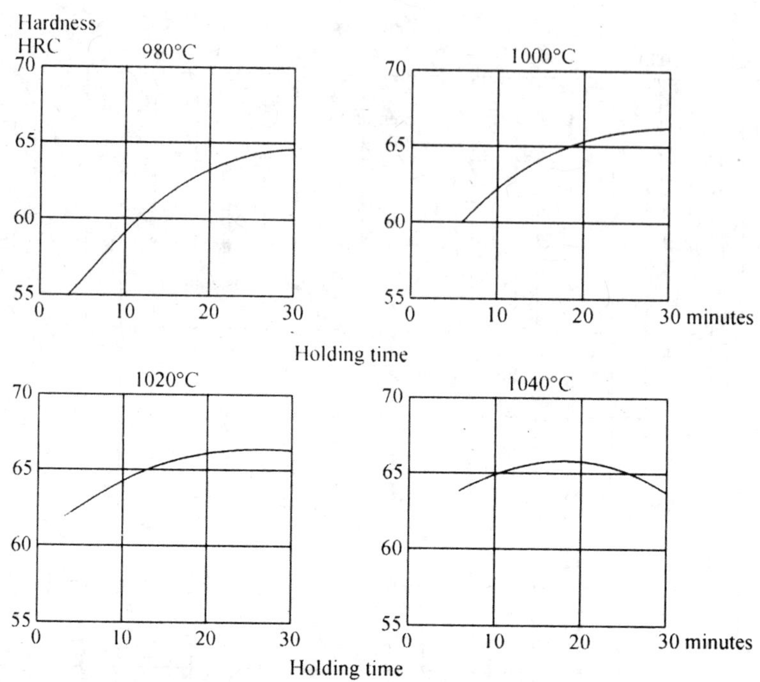

Fig. 10.14. Steel D 2. Dependence of hardness on hardening temperature and holding time. Test-piece 30 mm diameter.

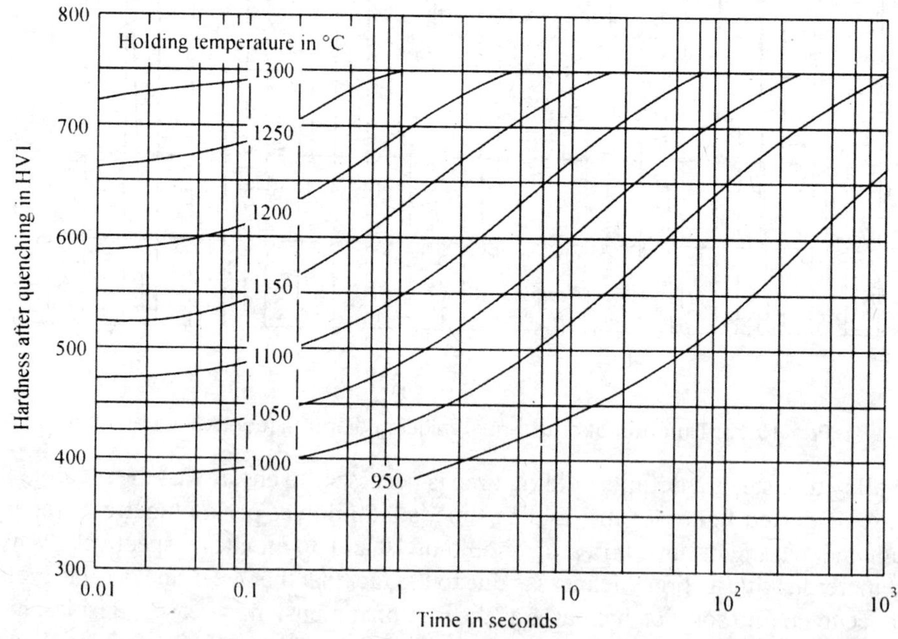

Fig. 10.15. Hardness after quenching as a function of the austenitising temperature and holding time for steel X 38 CrMoV 51 (H 13) after heating.

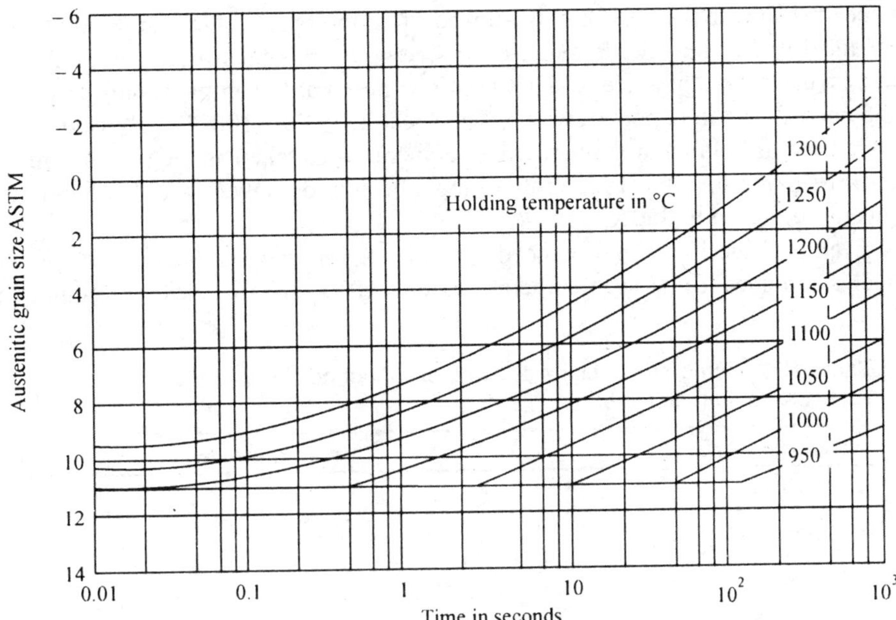

Fig. 10.16. Austenitic grain size as a function of the holding time and austenitising temperature for steel X 38 CrMoV 51 (H 13) after heating at a rate of 130°C/s; initially fully annealed.

For low-alloy tool steels a definite holding time is necessary to ensure that the required hardness will be attained. A suggested figure for the holding time is 0.5 minutes per millimetre of section thickness; the minimum and maximum time, however, should be 10 and 30 minutes respectively. Why the holding time should increase with section thickness is due to the fact that the size of the carbide grains and hence their reluctance to go into solution increases as the section dimension increases. Besides, most low-alloy steels generally have insufficient hardenability which, however, may be improved, particularly in heavy sections, by promoting the increased dissolution of carbides. This may be brought about by increasing the time of holding at temperature.

High-alloy chrome steels require the longest time of all the tool steels. To a very great extent, however, the holding time for those steels is dependent on the hardening temperatures (Fig. 10.14), which applies to steel D 2. Each hardening temperature requires a definite holding time in order that maximum hardness should be obtained on hardening. Too short a time results in a lower hardness, caused by an insufficient amount of carbide going into solution; too long a time also results in a lower hardness due to the presence of retained austenite. A suggested figure for the holding time is 0.5 minute per millimetre of section thickness, with minimum and maximum times of 10 minute and 1 hour respectively. Hot-work steels contains carbides that do not go into solution until about 1000°C. At these high temperatures grain growth is fairly rapid and for this reason the time of holding must be limited. If the heating is carried out in a salt bath, in which it is easy to see when the steel has reached the desired temperature, 15–30 minutes is a normal holding time which is by an large independent of the dimensions of the part being treated. If the tool is packed ind a box it is more difficult for the heat treater to judge when it is at the right temperature. Under such circumstances the best temperature to use is the one at the lower limit of the hardening range since a longer holding time may then be used, provided it does not exceed 1 hour.

The high-speed steels are the most highly alloyed of all tool steel. Their hardening temperatures lie between 1200 and 1300°C. In order to avoid excessive grain growth, which may easily occur at such high temperatures, the current practice is to use holding times a of few minutes only. In the case of high-speed steel, the normal heat treatment of which is pre-heating to 800–1000°C followed by immersion in the salt bath, the time of holding at the hardening temperature is taken as the time of immersion in the salt bath. The times given in Table 10.2 apply to tools heated to conventional hardening temperatures after having been pre-heated to 850°C.

The times given in Table 10.2 can be exceeded by 50 per cent without adverse consequences. When high-speed steel is hardened from lower temperatures, e.g., 1000–1100°C, the times of immersion should be trebled.

Table 10.2. Hardening of high-speed steel. Influence of section thickness on immersion time.

Section thickness	*Immersion time*
mm	*minutes*
Less than 10	3
10–20	4
20–30	5
30–40	6
40–50	7

Holding Time vs Heating-Up Time Over A_{c3}

For the sake of simplicity the times given in Table 10.3 refer to holding times, after pre-heating, in a salt bath—short heating-up time—and in a muffle furnace—very long heating-up time, specially if pack heating. The heating-up time necessary, in the case of a muffle furnace, to raise the temperature of the steel through the final ten or so degrees increases sharply with increasing section thickness. Carbide dissolution, however, begins some 50–100°C below the ordinary hardening temperature and this fact should be taken into consideration when using long heating times.

For the large die block referred to in Fig. 10.3 the heating-up time between 825 and 850°C is about 4 hour. Hence, when heating in a muffle furnace it should be possible, on economical grounds, to reduce the time of holding to half of that previously reported; always remembering to keep to the stipulated minimum number of minutes.

Table 10.3. Heating up times between 950 and 1000°C from Figs. 10.2 and 10.3.

Diameter of section *mm*	*Salt bath* *minutes*	*Muffle furnace* *minutes*
25	1	15
50	4	30
100	8	60

When heating-up a piece of steel in a muffle furnace the heat treater might well make use of the very rough and ready approximation of the holding time for all steels, except high-speed steels, viz: 20 minutes. This time is quite independent of the section thickness and the hardening temperature should be the one in the middle of the recommended temperture range. Therefore it is important for reasons of economy to study the characteristics of the furnaces at the disposal of the heat treater and as suggested previously, to draw up recommended times for the heating-up of different dimensions and batch weights.

METHODS OF COOLING

The cooling methods used for hardening depend on the grade of steel, the shape of the part and the properties to be imparted to the steel. The three following methods are described below: direct quenching to martensite, martempering and austempering (Fig. 10.17).

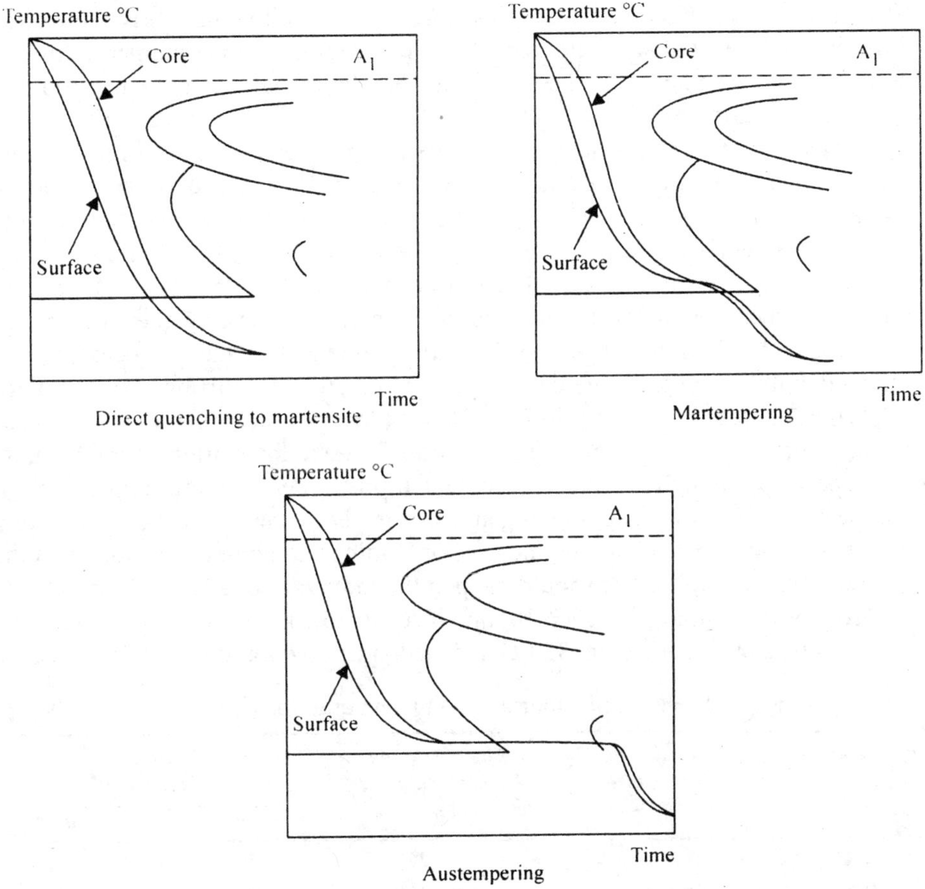

Fig. 10.17. Diagrams showing principles of different methods of cooling for hardening.

Direct Quenching to Martensite

Direct quenching was doubtless the original hardening method and it is still the commonest one. According to this method the steel is cooled or quenched straight from the hardening temperature to room temperature or to the slightly higher temperature of the cooling medium being used. This medium may be water, oil or air.

Martempering

The principle underlying martempering is most clearly understood by referring to a TTT diagram (Fig. 10.17). In this hardening treatment the cooling takes place in two stages. First the tool is quenched in a molten salt or metal bath kept at a temperture somewhat above that at which the austenitic structure of the steel starts to transform to martensite (i.e., the M_s-temperature). The steel is held there until temperature equalisation is complete. Next, the tool is allowed to cool freely in air, which results in the formation of martensite. Due to this method the temperature difference between the surface and the centre during the martensite formation is much smaller than it is in conventional hardening. Consequently the austenite transforms to martensite practically simultaneously throughout the various parts of the tool even if the latter has widely different section thicknesses. As a result there is less residual stress and a minimum of distortion. The final structure and properties obtained by martempering the steel are, in principle, the same as those obtained from conventional hardening but the amount of retained austenite is generally larger after martempering.

In order to carry out martempering successfully, the composition of the steel must be such that the steel obtains a martensitic structure—if this is what is required—when it has been quenched in the martempering bath and then air cooled. Therefore this heat-treatment process is confined mainly to the oil-hardening and air-hardening grades. Table 10.4 lists the temperatures of quenching and martempering as applicable to some popular tool steels. Included in the table are also the largest dimensions suitable for martempering, i.e., such as will develop full superficial hardness on being quenched in a clean, well-stirred martempering bath. 300 mm is set as the maximum diameter. In some instances larger dimensions can be martempered but in that case large-capacity baths and effective stirring equipment are required.

The heating to the temperature indicated in Table 10.4 is carried out in the same way as for conventional hardening. When sufficiently austenitised, the tool is transferred to the martempering bath, the capacity of which should be large enough to take up the heat from the tool without appreciably raising the temperature of the bath. The tool is kept moving and submerged in the bath long enough for the core to reach the temperature of the bath. However, the time of holding should not be so long that the austenite starts to transform to bainite, for if this should happen the tool will not attain full hardness. A suitable holding time is usually 1–3 minutes per each 10 mm of section thickness. The shorter time is applicable to low hardening temperatures, say around 850°C and the longer time for higher temperatures, ca 1000°C.

Table 10.4. Martempering, austenitising and martempering temperatures.

Designation of steel AISI		*Austenitising temperature*	*Martempering temperature*	*Maximum diameter*	
BS	*SS*	*°C*	*°C*	*mm*	*inch*
O1	2140	810–840	225–250	50	2
A2	2260	950–980	225–250	150	6
D2	2310	1000–1020	225–500	200	8
D6	2312	960–1000	225–500	200	8
–	2550	820–840	225–300	125	5
S1	2710	880–920	350–400	25	1
H13	2242	1000–1050	300–500	300	12
H10A		1999–1050	300–500	100	4
M2	2722	1220–1240	450–550	100	4

Since the steel is still in the austenitic state when it is removed from the bath it can be subjected to any necessary straightening operation during the subsequent air cooling. This capacity is particularly valuable when hardening long tools such as reamers, mandrels, etc. The straightening is commenced as soon as the tool has been taken out of the bath but generally not above about 350°C and should be completed before the temperture has fallen to about 150°C. After the tool has been removed from the bath and straightened, if necessary, it is allowed to cool in air. It is during this cooling that the hard martensite is formed. Then, as in conventional hardening; the tool is tempered when the temperature has fallen to 75°C or 50°C. The curve in Fig. 10.18 shows the profile of the complete heat treatment programme for a roll made from steel D 2, diameter 103 mm, length 1000 mm. The thermocouple that indicated the temperture was located at the centre of the roll. After being pre-heated to 550°C the roll was heated in a protective atmosphere of nitrogen gas in a pit furnace rated at 160 kW. In this instance the hardening temperture was high, viz. 1030°C and the time of holding at this rather high temperture was 30 minutes. The roll was quenched in a salt bath at 220°C, held there for 45 minutes and then allowed to cool freely in air. The total heat-treatment time was some 8 hrs. To this should be added the time required for tempering plus the time taken for straightening, if the latter operation should have been deemed necessary.

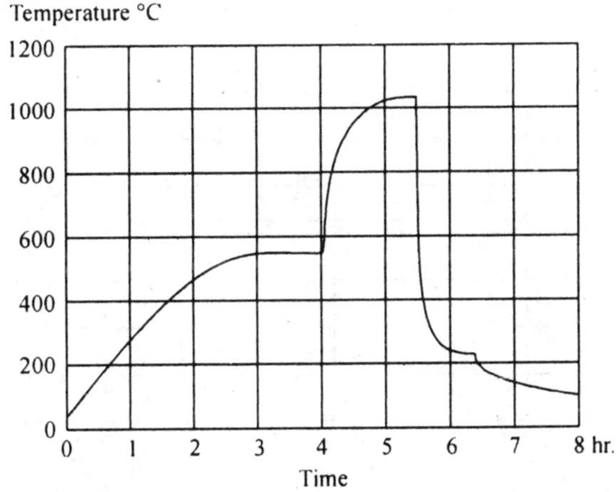

Fig. 10.18. Heating and cooling programme for hardening of a roll, diameter 103 mm, made from steel D2.

Austempering

By applying a heat treatment termed austempering it is possible, in certain cases, to obtain greater toughness than that obtained after conventional hardening and tempering to the same hardness. There are no excessively large or harmful stresses induced in the tools and, consequently, distortion is small. As in the case of martempering, the parts to be hardened are quenched in a salt bath, the temperature of which is kept above that of the martensite formation. The holding time in the bath, however, is so long that the austenite is practically transformed to bainite (Fig. 10.17). The resulting hardness depends on the temperature of the salt bath; a high temperture gives a low hardness. Diagrams of isothermal transformations (TTT diagrams) often indicate the hardness obtained at different temperatures after the formation of 98–99 per cent bainite. Conventional punching tools made, for example from steel O1 are seldom able to utilise fully the increased toughness resulting from austempering. Very thin-walled tools, such as hollow punches austempering to 58–60 HRC, have given remarkably good service. This also applies to the type of punching tools used for making tin cans. When worn, the tool is renewed by upsetting and then ground to size. Fig. 10.19 exemplifies the austempering temperature-time programme for steel O1 .

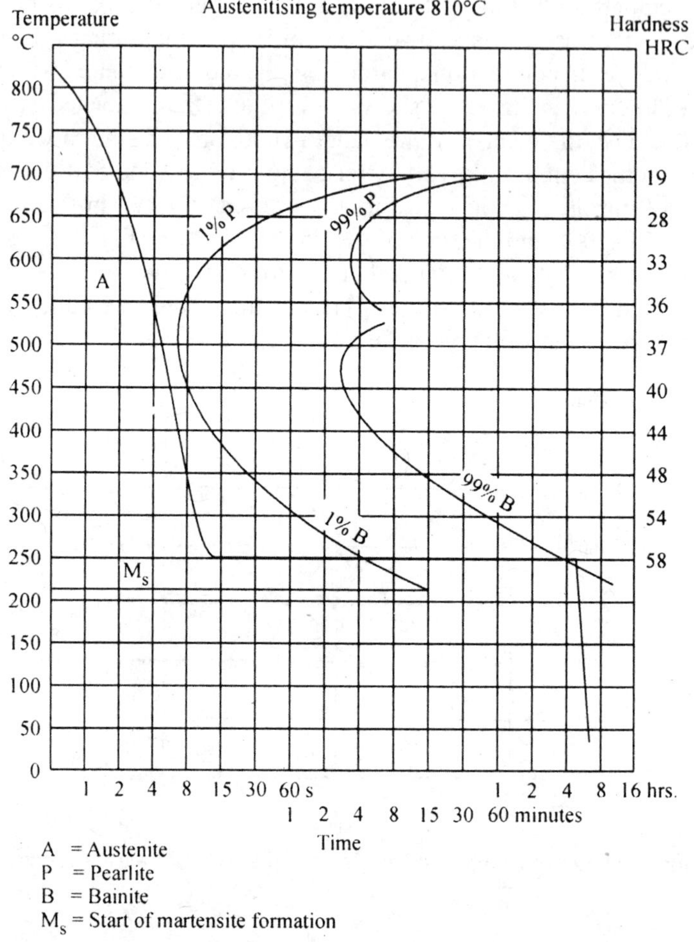

Fig. 10.19. Austempering of steel O1.

All steels and dimensions that are capable of being martempered as described in the preceding section can also be austempering. In addition, it is common practice to austemper certain low-alloy structural steels, e.g., spring steel, up to about 10 mm section and plain-carbon steel up to a maximum diameter of 5 mm. The length of time necessary for complete transformation restricts the application of the austempering treatment of high-alloy steels, since the time needed will often be uneconomically long. Cooling from the austempering temperature takes place freely in air. No tempering is required.

Austempering usually improves the toughness, particularly in that hardness range in which conventionally hardened and tempered steel are susceptible to a reduction in their impact strength. In ordinary low-alloy steels this reduction generally occurs when they are tempered around 300°C.

Owing to the rather high treatment costs, austempering is used mainly for expensive tools for which there are stringent requirements of dimensional accuracy and where the risk of cracking must be kept to a minimum. The austempering of machine components and structural parts is resorted to only when carefully conducted tests have shown that the generally unavoidable distortion during austempering is so small that the cost of this process is justified.

Patenting

The actual patenting treatment may be regarded as a variant of austempering. It is applied mainly to unalloyed and low-alloy steel wire with a carbon content of 0.6–1.1 per cent. The treatment takes place at about 500°C in a molten lead bath through which the wire passes continuously. At this temperture very fine lamellar pearlite is formed. In spite of its high hardness this structure has shown itself to be well suited for wire drawing since it can accommodate large reductions of area without fracture.

QUENCHING MEDIA

When hardening tools and machine components, the heat treater generally aims at obtaining a martensitic structure, at least in the surface layers of the steel. Hence the rate of cooling must be controlled so that the formation of ferrite, pearlite or bainite is avoided.

Since the dimensions of the part have generally been decided beforehand, the depth of hardness must be determined by a judicious choice of steel and cooling medium. The choice of the latter for hardening is often just as important as the choice of the former. With regard to the steel, a knowledge of its hardenability is a good aid when choosing a suitable, grade. For the cooling medium, however, there are not available the same good criteria concerning the cooling capacity of the medium. Hence the simple rule as usually applied: unalloyed steel is quenched in water; alloy steels in oil and high-alloy steels cooled in air.

When it becomes desirable to introduce less sophisticated steel grades and more automation of the heat treating process, specialised knowledge is required both of the hardenability of steel and the cooling capacity of cooling media. A successful final result is also due in large measure to the mutual interplay between these two factors.

Three stages of quenching

During the quenching of steel in liquid media the whole operation may be split up into three stages, viz.,

1. The vapour-blanket stage.
2. The boiling stage.
3. The convection stage.

Fig. 10.20 shows a typical cooling curve for an oil and at the same time shows what happens at the surface of a steel that is being quenched from the hardening temperature.

During the vapour-blanket stage a thin vapour film forms which is heat insulating and thus prevents heat abstraction. Therefore the curve has a rather flat appearance in this range.

During the boiling stage there is a high rate of heat abstraction which in part is due to the energy consumed by the vapour formation and in part to the vigorous agitation. During the convection stage the heat abstraction takes place more slowly and the curve is flat once again. In practically all literature on quenching media it is stressed that the shorter the vapour-blanket stage is kept the better will be the cooling effect. This is going to be examined in more detail in the following section.

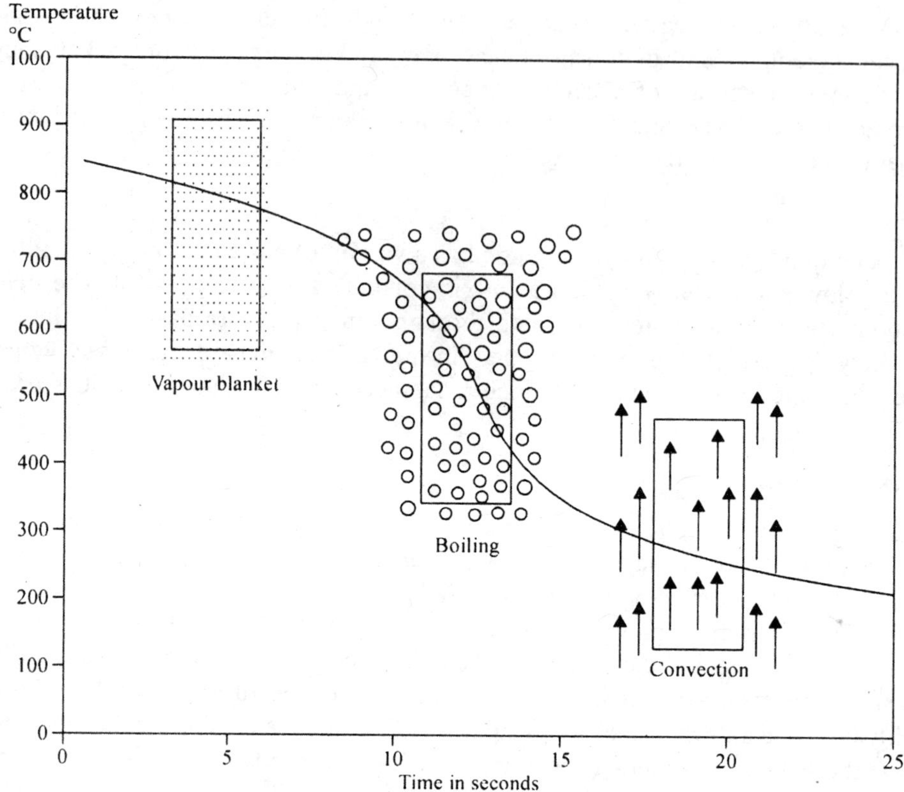

Fig. 10.20. Three stages of the cooling curve.

Methods for evaluating quenching media

There are a great number of different methods for evaluating quenching media. An example of the type of apparatus used for such tests is shown in Fig. 10.21.

Originally a silver ball, 20 mm in a diameter, was employed, but for several reasons a cylindrical body of either steel or silver is used now a days. A thermocouple is located at the centre of the testing body and is connected to a recording instrument. This instrument traces a record of the progress of the temperature change as the testing body (after being heated to the required temperture) is immersed in the cooling medium. Cooling curves obtained in this way are shown in Fig. 10.22. However, for a number of reasons they are rather difficult to interpret and put to practical use. The interpretation is facilitated if the rate of cooling during the whole course of the cooling process is registered, in which case the curves take the shape shown in Fig. 10.23. From looking at Fig. 10.22 it is clearly seen that brine (3 per cent common salt solution) produces the greatest cooling effect, also that the cooling capacity is reduced as the curves are displaced towards longer times. To a certain extent this observation is verified by Fig. 10.24 which shows the hardness profile from surface to centre of testing specimens, 50 mm in diameter, of a steel equivalent to 50 M 7. The specimens were hardened inthe various quenching media mentioned, except brine. Of the various quenching media tested, water gave the highest hardness; of the oils, quench oil A yielded better results than oil B, but the oil emulsion gave a higher hardness than oil A. However, this last point is not directly obvious from Figs. 10.22 or 10.23.

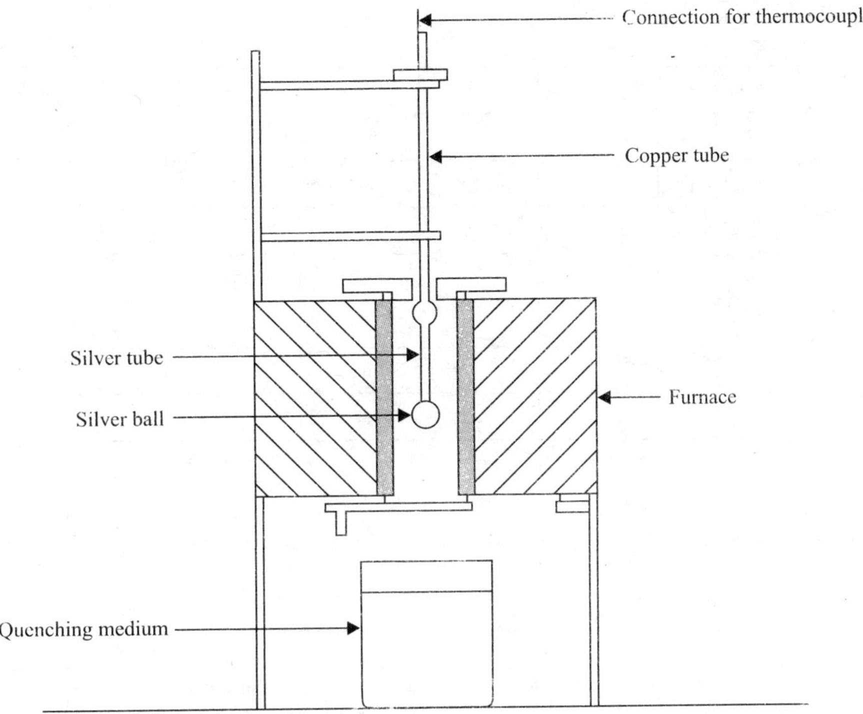

Fig. 10.21. Equipment for evaluating quenching media using a silver ball.

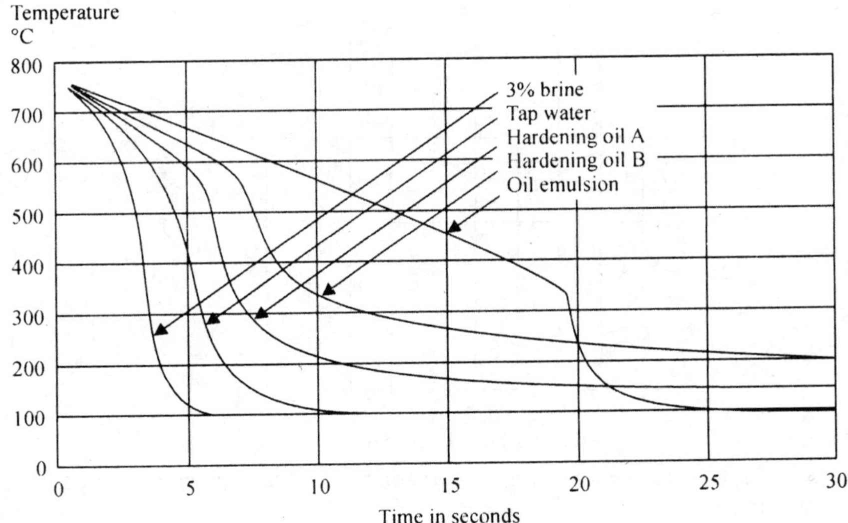

Fig. 10.22. Cooling curves for various quenching media, derived by means of a silver ball. Testing temperture 40°C.

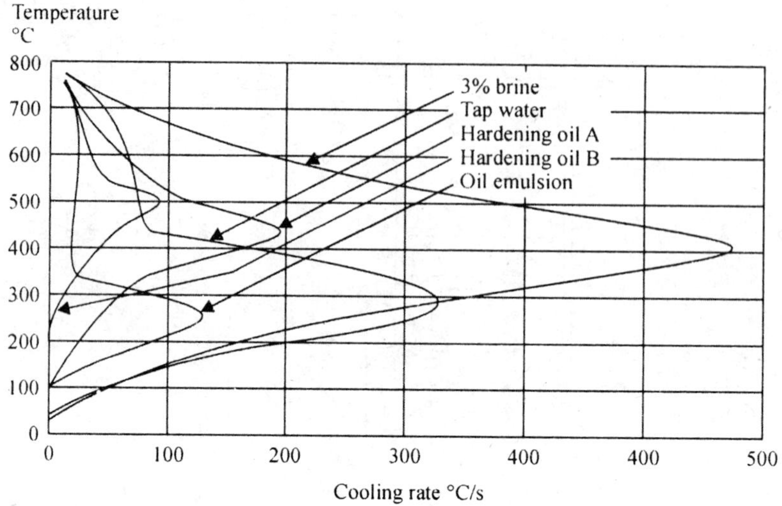

Fig. 10.23. Cooling rate of various quenching media tested by means of a silver ball.

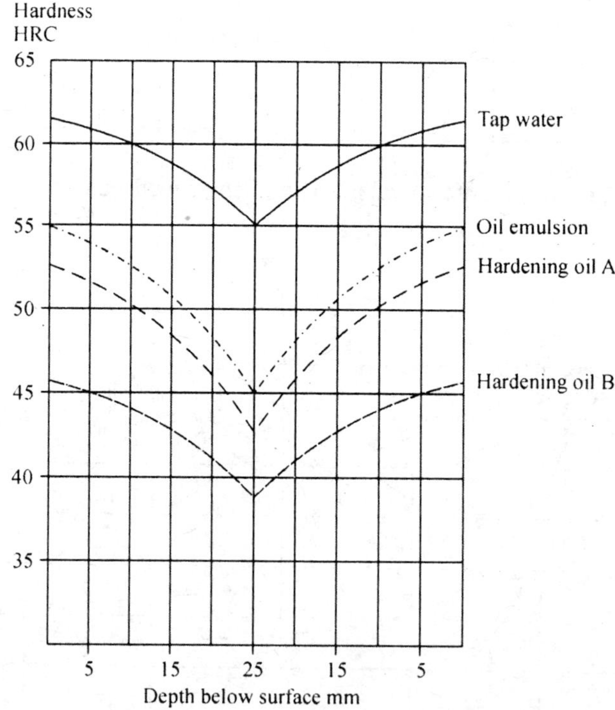

Fig. 10.24. Hardness profiles across test bars, 50 M diameter, in steel equivalent to 50 mm 7, after hardening in various quenching media.

Water

Water is probably the oldest cooling medium used for hardening and it has remained the major coolant throughout the ages. Pure water, however, is rather unsuitable as a cooling medium since its greatest cooling efficiency occurs at 300°C, i.e., the temperture at which martensite formation starts in many steels. By adding 5–10 per cent of common salt or soda to the water, its cooling capacity is increased very considerably and at the same time its greatest heat-extracting capacity occurs at 500°C. There is rapid fall in cooling capacity as the temperature of the water rises above 60°C. The overall best result is obtained when the water temperture is between 20°C and 40°C. The great drawback with water, is that the rate of cooling is high in the temperature range of martensite formation. This exposes the steel to the simultaneous influence of transformation stresses and thermal stresses, the combined effects of which will increase the risk of crack formation.

The danger of cracking occurring during water quenching may be reduced if the steel is removed from the water when it has cooled to same 200–400°C and is rapidly transferred to an oil bath. Incidentally, this is a very neat way of increasing the depth of hardening of oil-hardening, low-alloy steels.

Oil

Oil cooling is much slower than water cooling. Oil is not a well-defined cooling medium as regards cooling capacity. The simplest and cheapest hardening oil is a standard mineral oil with the lowest possible viscosity (spindle oil). By introducing certain additives to such oils it has been possible to produce oils having considerably increased cooling capacity. Furthermore, there are oils that may be used at elevated temperatures, e.g., at 200°C. The rate of cooling is generally greatest at about 600°C and is relatively slow inthe range of martensite formation. Since oil has a rather low capacity for heat extraction relative to water, its use as a coolant for medium to low-alloy steels is restricted to light sections. For example, on quenching steel O1 or En 19C (42 CrMo4) in still oil, it is not certain that full superficial hardness is obtained in dimensions around 100 mm. A reliable and tested way of increasing the cooling capacity of oil is by agitating vigorously the bath or the charge. This simple technique should not be underrated.

Mixtures of Water and Oil (Emulsions)

By emulsifying water and 'water-soluble' oil in various proportions it is possible to obtain cooling media of various cooling capacities. In this respect, however, such media might be inferior to the oil itself. An emulsion containing 90 per cent water and 10 per cent oil has a much higher cooling rate between 400 and 500°C. This may produce a satisfactory hardening result for a steel having its bainite nose within this temperature range.

If water is inadvertently added to ordinary quenching oil this may lead to cracking on hardening particularly if the steel is a deep-hardening one because then the martensite formation will start at the surface considerably in advance of the centre which, when it transforms, will increase the stresses in the surface. If the water and oil are not properly emulsified, the former having collected at the bottom of the tank instead, a rapid heating-up of the water followed by steam generation may give rise to an explosion.

The presence of water may be ascertained by dipping a narrow-bore tube, preferably of glass and temporarily closed at its upper end, into the tank at its lowest point. On opening the tube the operator lets the quenching fluid rise in the tube. The opening is then again closed, the tube withdrawn and its contents examined.

Polymer Quenching Media

By mixing water with about 10 per cent of water-soluble polymers (e.g., polyalkylene glycol or polyvinyl alcohol) a quenching medium is obtained that has a cooling capacity between that of water and oil.

Intensive research and development work in this field is being pursued at several centres. Besides other advantages arising from the use of this quenching medium, a reduced dependence on oil and an eliminated risk of fire are gained.

Air

Low-alloy steels in light sections and high-alloy steels may be successfully hardened by means of compressed air or still air. The advantages of using air are that distortion is negligible and that the steel can easily be straightened during the cooling process. So that it should cool evenly the part is rotated in a steady current of compressed air. One drawback here is that the surface may be oxidised during the cooling. Now a days it is not so usual for structural parts to be made in air-hardening steel on account of the high cost of the necessary alloying elements and also for the fact that such steels must inevitably be annealed after forging or rolling before subsequent machining. A typical composition of a conventional air-hardening structural steel is: 0.30 per cent C, 1.2 per cent Cr, 4 per cent Ni, 0.3 per cent Mo. In dimensions up to 100 mm diameter such a steel would be wholly martensitic after air hardening.

Salt Baths

The most popular salt baths used for cooling purposes consist of approximately equal parts of sodium nitrite and potassium nitrate. They are used in the temperture range 160–500°C. There are also salts for 500–600°C. A salt bath is the ideal quenching medium for a steel having a reasonably good hardenability and not too heavy a section. The cooling capacity down to about 500°C is high and then decreases as the temperture of the steel continues to fall. The lower the temperature of the bath and the greater the agitation, the better is its cooling capacity (Fig. 10.25). In Fig. 10.26 is illustrated the effect of agitation on the rate of cooling and on the hardness of a plain-carbon steel containing 0.5 per cent °C.

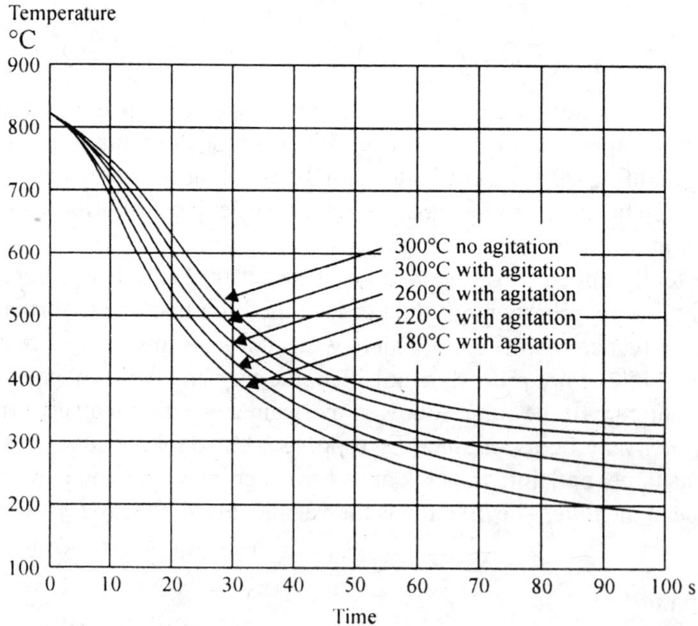

Fig. 10.25. Influence of temperature of quenching bath on rate of cooling when martempering cylindrical bars of stainless steel, 37 mm diameter × 100 mm.

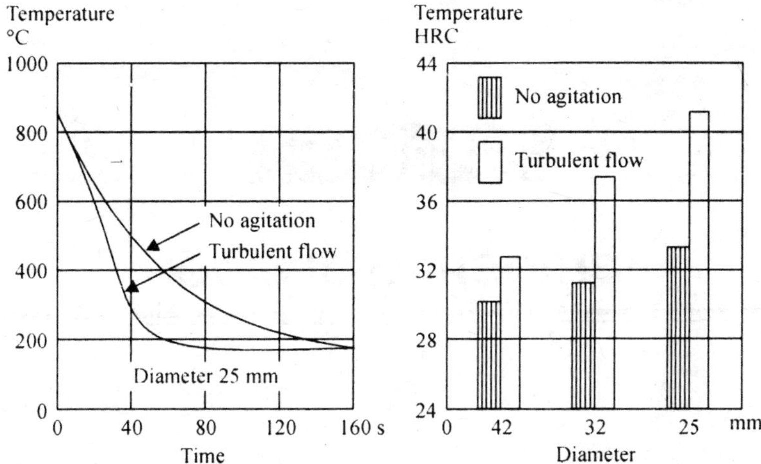

Fig. 10.26. Influence of turbulent flow on rate of cooling and on hardness after quenching in salt bath at 190°C. Specimen of 0.5 per cent plain carbon steel.

If the salt bath becomes contaminated its cooling efficiency is very much reduced and when the bath is stirred the foreign particles are kept in suspension in the molten salt. They are apt to attach themselves to the parts being treated and hence impede heat transfer. In such cases it is better to let the impurities sink to the bottom and instead move the parts smartly about in the clear supernatant molten salt. A recommended holding time in the salt bath is 1–3 minutes per centimetre of section thickness. The shorter time applies to low hardening temperatures and light sections; the longer time to high temperatures and heavy sections.

The cooling capacity may be increased by adding water to the salt bath. The amount of water must be kept within close limits and should amount to 0.3–0.5 per cent. The cooling capacity will then be approximately doubled. The water, as steam, leaves the surface of the bath continuously and should therefore be replenished, either all the time or as periodic additions.

Chapter 11

Tempering of Steel

INTRODUCTION

Tempering of steel is a process in which previously hardened or normalised steel is heated to a temperature below the transformation range and cooled at a suitable rate, primarily to increase ductility and toughness. Steels are tempered by re-heating after hardening to obtain specific values of mechanical properties and to relieve quenching stresses and ensure dimensional stability. Tempering usually follows quenching from above the critical temperature; however, tempering also is used to relieve the stresses and reduce the hardness developed during welding and to relieve stresses induced by forming and machining.

PRINCIPAL VARIABLES

Variables associated with tempering that affect the microstructure and the mechanical properties of a tempered steel include temperature; time at temperature; cooling rate from the tempering temperature; and composition of the steel, including carbon content, alloy content and residual elements. In a steel quenched to a microstructure consisting essentially of martensite, the iron lattice is strained by the carbon atoms, producing the high hardness of quenched steels. On heating, the carbon atoms diffuse and react in a series of distinct steps that eventually form Fe_3C or an alloy carbide in a ferrite matrix of gradually decreasing stress level. The properties of the tempered steel are determined primarily by the size, shape, composition and distribution of the carbides that form, with a relatively minor contribution from solid-solution hardening of the ferrite. These changes in microstructure usually decrease hardness, tensile strength and yield strength but increase ductility and toughness.

Under certain conditions, hardness may remain unaffected by tempering or may even be increased as a result of it. For example, tempering of a hardened steel at very low tempering temperatures may cause no change in hardness but may achieve a desired increase in yield strength. Also, those alloy steels that contain one or more of the carbide-forming elements (chromium, molybdenum, vanadium and tungsten) and capable of 'secondary hardening'—that is, they may become somewhat harder as a result of tempering.

The tempered hardness values for several quenched steels are presented in Table 11.1. Temperature and time are interdependent variables in the tempering process. Within limits, lowering temperature and increasing time usually can produce the same result as raising temperature and decreasing time. With few exceptions, tempering is done at temperatures between 175 and 700°C (350 and 1300°F) and for times from 30 minutes to 4 hours.

Table 11.1. Typical hardnesses of various carbon and alloy steels after tempering.

Data were obtained on 25 mm (1 inch) bars adequately quenched to develop full hardness.

Grade	Carbon content, %	*Hardness, HRC, after tempering for 2 hrs. at* °C: 205 °F: 400	260 500	315 600	370 700	425 800	480 900	540 1000	595 1100	650 1200	Heat treatment
Carbon steels, water hardening											
1030	0.30	50	45	43	39	31	28	25	22	95(a)	Normalised at 900°C (1650°F); water quenched from 830–845°C (1525–1550°F); average dew point, 16°C (60°F)
1040	0.40	51	48	46	42	37	30	27	22	94(a)	
1050	0.50	52	50	46	44	40	37	31	29	22	
1060	0.60	56	55	50	42	38	37	35	33	26	Normalised at 885°C (1625°F); water quenched from 800–815°C (1475–1500°F); average dew point, 7°C (45°F)
1080	0.80	57	55	50	43	41	40	39	38	32	
1095	0.95	58	57	52	47	43	42	41	40	33	
1137	0.40	44	42	40	37	33	30	27	21	91(a)	Normalised at 900°C (1650°F); water quenched from 830–855°C (1525–1575°F); average dew point, 13°C (55°F)
1141	0.40	49	46	43	41	38	34	28	23	94(a)	
1141	0.40	55	50	47	45	39	32	29	25	97(a)	
Alloy steels, water hardening											
1330	0.30	47	44	42	38	35	32	26	22	16	Normalised at 900°C (1650°F); water quenched from 800–815°C (1475–1575°F); average dew point, 16°C (60°F)
2330	0.30	47	44	42	38	35	32	26	22	16	
3130	0.30	47	44	42	38	35	32	26	22	16	
4130	0.30	47	45	43	42	38	34	32	26	22	Normalised at 885°C (1625°F); water quenched from 800–855°C (1475–1575°F); average dew point, 16°C (60°F)
5130	0.30	47	45	43	42	38	34	32	26	22	
8630	0.30	47	45	43	42	38	34	32	26	22	
Alloy steels, oil hardening											
1340	0.40	57	53	50	46	44	41	38	35	31	Normalised at 870°C (1600°F); oil quenched from 830–845°C (1525–1550°F); average dew point, 16°C (60°F)
3140	0.40	55	52	49	47	41	37	33	30	26	
4140	0.40	57	53	50	47	45	41	36	33	29	
4340	0.40	55	52	50	48	45	42	39	34	31	Normalised at 870°C (1600°F); oil quenched from 830–845°C
4640	0.40	52	51	50	47	42	40	37	31	27	

(Contd...)

Grade	Carbon content, %	Hardness, HRC, after tempering for 2 hrs. at °C: 205 / °F: 400	260 / 500	315 / 600	370 / 700	425 / 800	480 / 900	540 / 1000	595 / 1100	650 / 1200	Heat treatment
8740	0.40	57	53	50	47	44	41	38	35	22	(1525–1575°F); average dew point, 13°C (55°F)
4150	0.50	56	55	53	51	47	46	43	39	35	Normalised at 870°C (1600°F); oil quenched from 830–870°C (1525–1600°F); average dew point, 13°C (55°F)
5150	0.50	57	55	52	49	45	39	34	31	28	
6150	0.50	58	57	53	50	46	42	40	36	31	
8650	0.50	55	54	52	49	45	41	37	32	28	Normalised at 870°C (1600°F); oil quenched from 815–845°C (1500–1550°F); average dew point, 13°C (55°F)
8750	0.50	56	55	52	51	46	44	39	34	32	
9850	0.50	54	53	51	48	45	41	36	33	30	

(a) Hardness, HRB

Tempering Temperature

Several empirical relationships have been made between the tensile strength and hardness of tempered steels such that measurement of hardness is commonly used to evaluate the response of a steel to tempering. Fig. 11.1 shows the effect of tempering temperature on the hardness, tensile and yield strengths, elongation and reduction in area of a plain carbon steel (AISI 1050) held at temperature for 1 hr. Note that both room temperature hardness and strength decrease as the tempering temperature is increased. Ductility at ambient temperatures, as measured by either elongation or reduction in area, increases with tempering temperature.

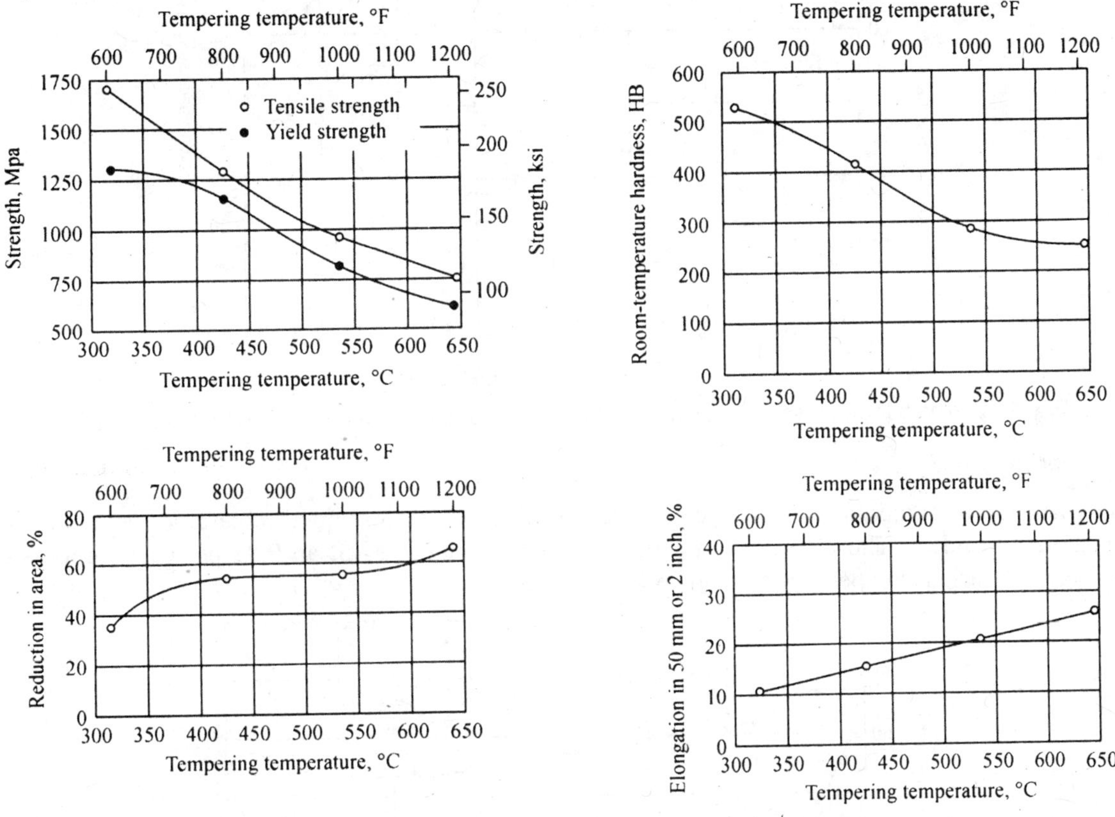

Properties summarised for one heat of 1050 steel that was forged to 38 mm (1.50 inch) in diameter, then water quenched and tempered at various temperatures. Composition of heat: 0.52 C, 0.93 Mn.

Fig. 11.1. Effect of tempering temperature on room-temperature mechanical properties of 1050 steel.

Whereas elongation and reduction in area increase continuously with tempering temperature, toughness, as measured by a notched-bar impact test, varies with tempering temperature for most steels as shown in Fig. 11.2. Tempering at temperatures from 260 to 320°C (500 to 610°F) decreases impact energy to a value below that obtained at about 150°C (300°F). Above 320°C, impact energy again increases with increasing tempering temperature. Both plain carbon and alloy steels respond to tempering in this manner. The phenomenon of impact-energy minima centered around 300°C (570°F) is called '500°F temper embrittlement' or 'blue brittleness'. This phenomenon will be discussed more thoroughly in a subsequent section on temper embrittlement.

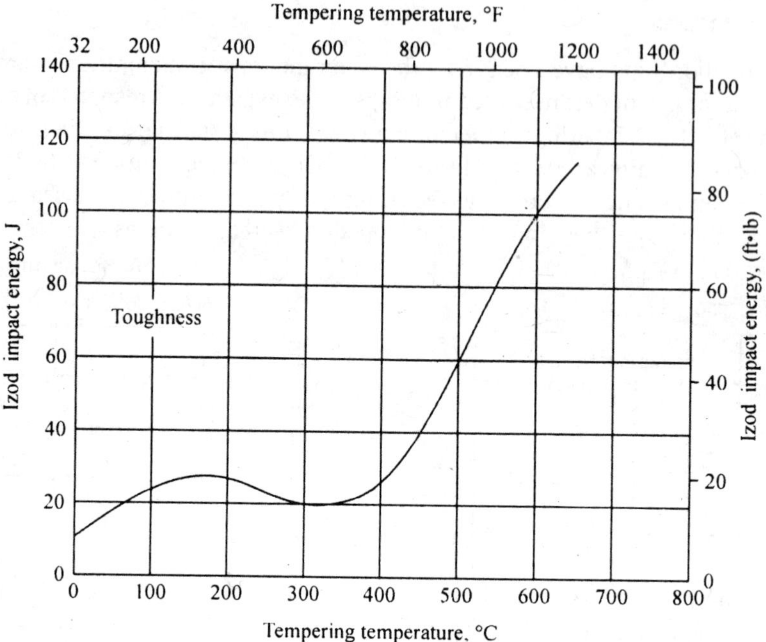

Fig. 11.2. Notch toughness as a function of tempering temperature for 4140 steel tempered 1 hr.

Tempering Time

The diffusion of carbon and alloying elements necessary for the formation of carbides is temperature and time dependent. The effect of tempering time on the hardness of an 0.82 per cent carbon steel tempered at various temperatures is shown in Fig. 11.3.

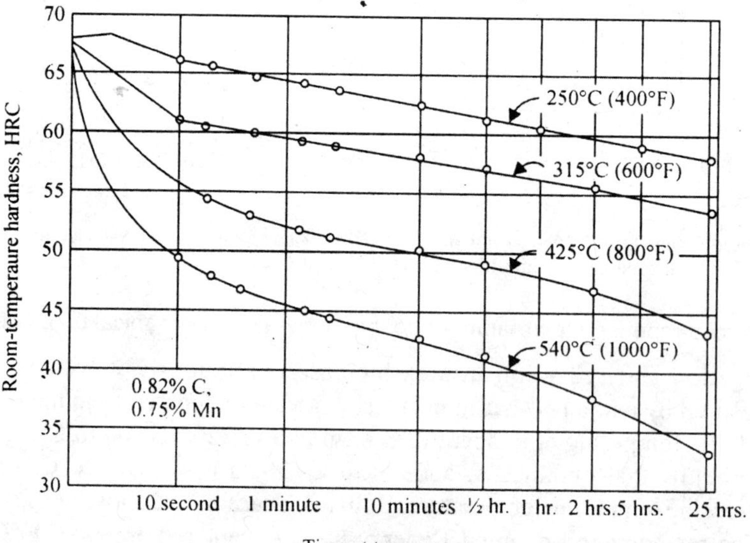

Fig. 11.3. Effect of time at four tempering temperatures on room-temperature hardness of quenched 0.82% carbon steel.

The changes in hardness are approximately linear over a large portion of the time range when the time is presented on a logarithmic scale. Rapid changes in room-temperature hardness occur at the start of tempering in times less than 10 seconds. Less rapid, but still large, changes in hardness occur in times from 1 to 10 minutes and smaller changes occur in times from 1 to 2 hours. For consistency and less dependency on variations in time, components generally are tempered for 1 to 2 hours. The levels of hardness produced by very short tempering cycles, such as in induction tempering, would be quite sensitive to both the temperature achieved and the time at temperature.

By use of an empirical tempering parameter, the approximate hardnesses of quenched and tempered low and medium-alloy steels can be predicted. The parameter is T (c + log t), where T is temperature in degrees Kelvin, *t* is time in seconds and *c* is a constant that depends on the carbon content of the steel. Reasonably good correlations are obtained except when significant amounts of carbide-forming elements or large amounts of retained austenite are present.

Cooling Rate

Another factor that can affect the properties of a steel is the cooling rate from the tempering temperature. Although tensile properties are not affected by cooling rate, toughness (as measured by notched-bar impact testing) can be decreased if the steel is cooled slowly through the temperature range from 375 to 575°C (705 to 1065°F), especially in steels that contain carbide-forming elements. Elongation and reduction in area may be affected also. This phenomenon is called 'temper embrittlement' and is discussed further in a subsequent section of this chapter.

CARBON CONTENT

The effect of carbon content on the properties of tempered steel is illustrated in Fig. 11.4, which presents hardness data for 14 carbon steels that were tempered at temperatures ranging from 205 to 705°C (400 to 1300°F) and for times from 10 minutes to 24 hrs. The principal effect of carbon content is on asquenched hardness. Fig. 11.5 illustrates the relationship between carbon content and the maximum hardness that can be obtained upon quenching. The relative difference in hardness as compared with as-quenched hardness is retained after tempering. Fig. 11.6 shows the combined effect of time, temperature and carbon content on the hardness of four carbon-molybdenum steels of different carbon contents. Fig. 11.7 shows the hardness of these steels after tempering for 1 hour, as a function of tempering temperature; the effect of carbon content is clearly evident.

ALLOY CONTENT

The main purpose of adding alloying elements to steel is to increase the steel's ease of hardenability—that is, its ability to form martensite on quenching from above its critical temperature. The general effect of alloying elements on tempering is a retardation of the rate of softening, especially at the higher tempering temperatures. Thus, to obtain a given hardness in a given period of time, alloy steels require higher tempering temperatures than do carbon steels. Alloying elements can be characterised as carbide-forming or non-carbide-forming. Elements such as nickel, silicon, aluminium and manganese, which have little or no tendency to occur in the carbide phase, remain essentially in solution in the ferrite and have only a minor effect on tempered hardness.

Hardening due to the presence of these elements occurs mainly through solid-solution hardening of the ferrite. The carbide-forming elements, chromium, molybdenum, tungsten, vanadium, tantalum, niobium and titanium, retard the softening process by formation of alloy carbides.

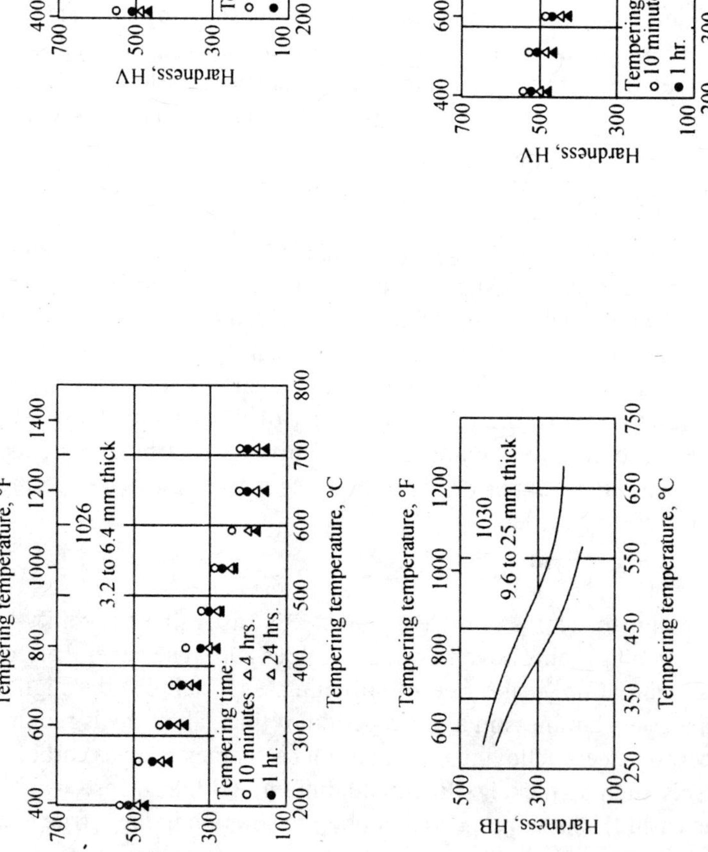

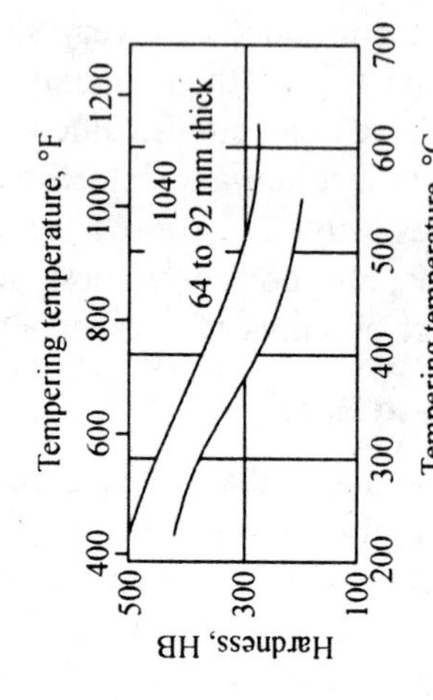

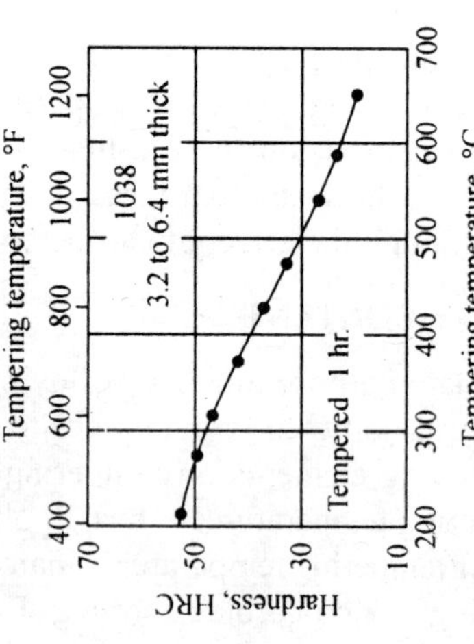

Fig. 11.4. Influence of tempering temperature on room-temperature hardness of quenched carbon steels.

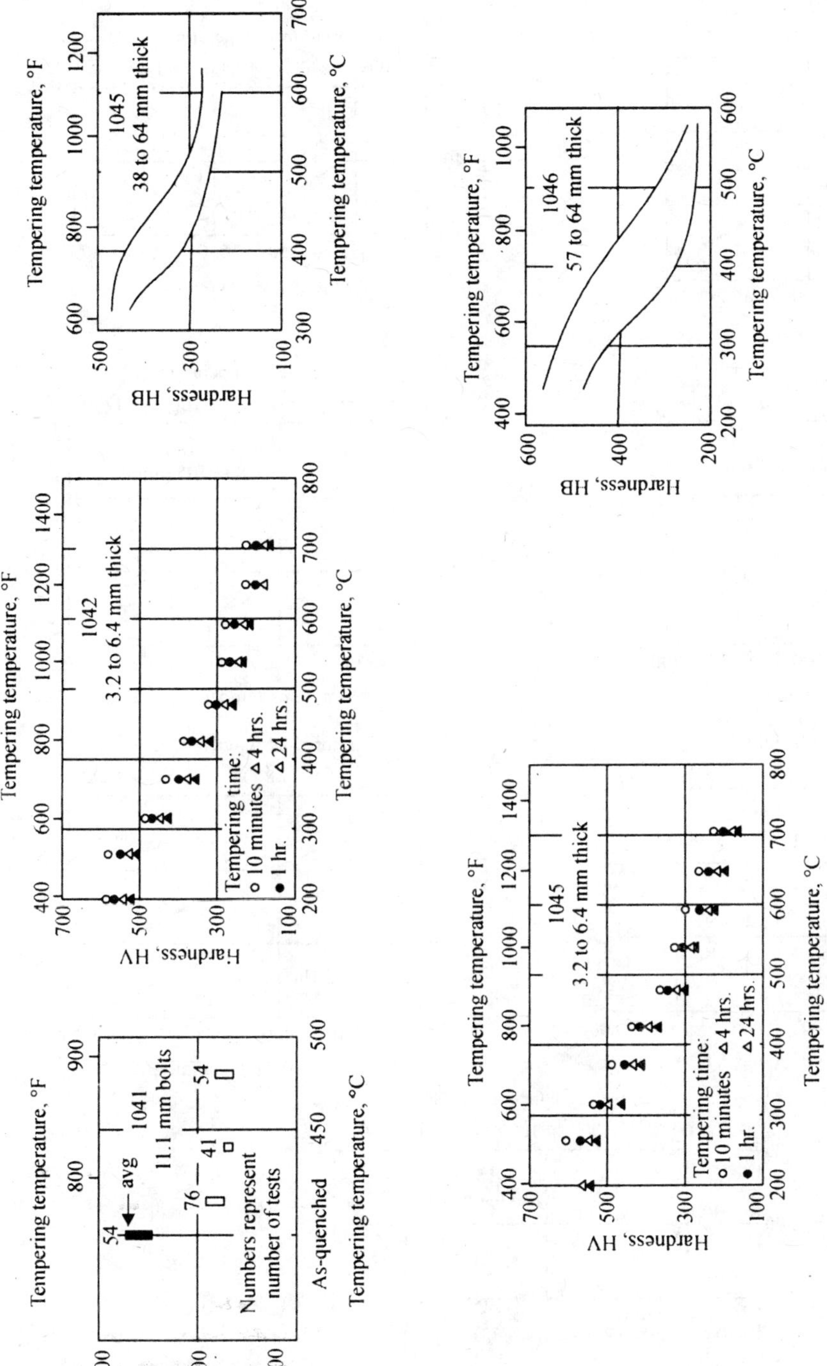

Fig. 11.4 (Continued). Influence of tempering temperature on room-temperature hardness of quenched carbon steels.

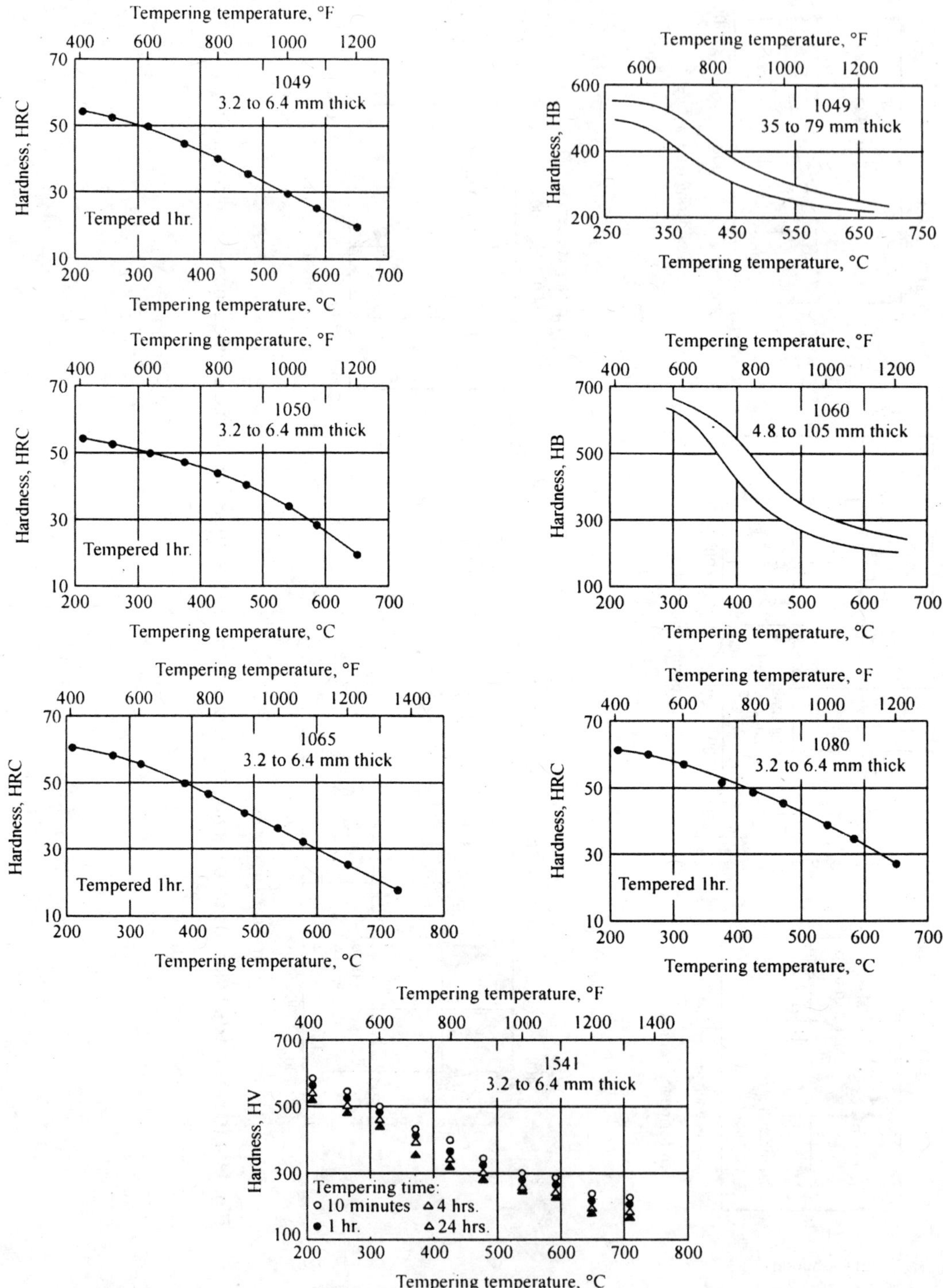

Fig. 11.4 (Continued). Influence of tempering temperature on room-temperature hardness of quenched carbon steels.

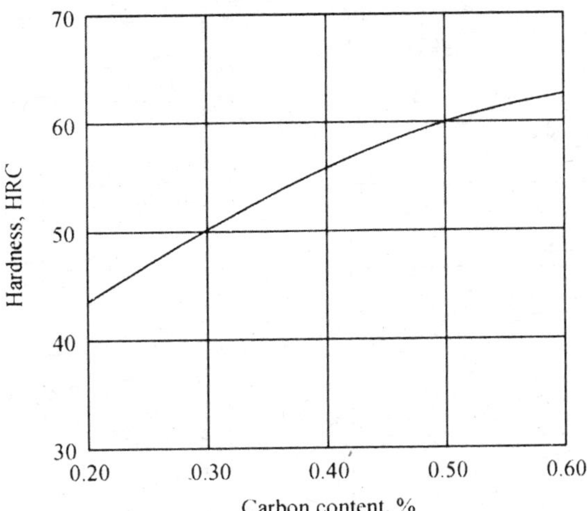

Fig. 11.5. Relationship between carbon content and room-temperature hardness for steels comprising 99.9% untempered martensite.

The effect of the carbide-forming elements is minimal at low tempering temperatures where Fe_3C forms; however, at higher temperatures, alloy carbides are formed and hardness decreases slowly with tempering temperature. Under certain conditions, such as with highly alloyed steels, hardness may actually increase. This latter effect is known as 'secondary hardening'.

The effect of molybdenum content on the tempering behaviour of a 0.35 per cent carbon steel is shown in Fig. 11.8. As the alloy content increases, the magnitude of the secondary-hardening effect increases. Synergistic effects of various combinations of alloying elements can occur: chromium tends to produce secondary hardening at a lower temperature than does molybdenum; and the combination of chromium and molybdenum produces a rather flat tempering curve, with the peak hardness occurring at a somewhat lower temperature than when only molybdenum is present. H11 steel is a widely used hot working die steel that contains nominally 0.35 per cent C, 5 per cent Cr, 1.5 per cent Mo and 0.4 per cent V. Fig. 11.9 shows the room-temperature hardness of H11 as a function of tempering temperature. A very flat tempering curve results because of the specific combination of the three carbide-forming elements.

Tools Steels and Stainless Steels

Extensive data on tempering of tool steels (including H11) and martensitic stainless steels are given in the chapter on heat treating of tool steels and heat treating of stainless steels and heat-resisting alloys.

Other Alloying Effects

Alloying elements produce a number of other effects besides providing ease of hardening and secondary hardening. The higher tempering temperatures used to temper alloy steels presumably permit greater relaxation of residual stresses and improve properties. Furthermore, the hardenability of alloy steels may permit a less drastic quench to be used so that as-quenched stresses are lower and cracking prior to tempering is minimised. The higher hardenability of alloy steels may also permit the use of a lower carbon content to achieve a given strength level but with improved ductility and toughness.

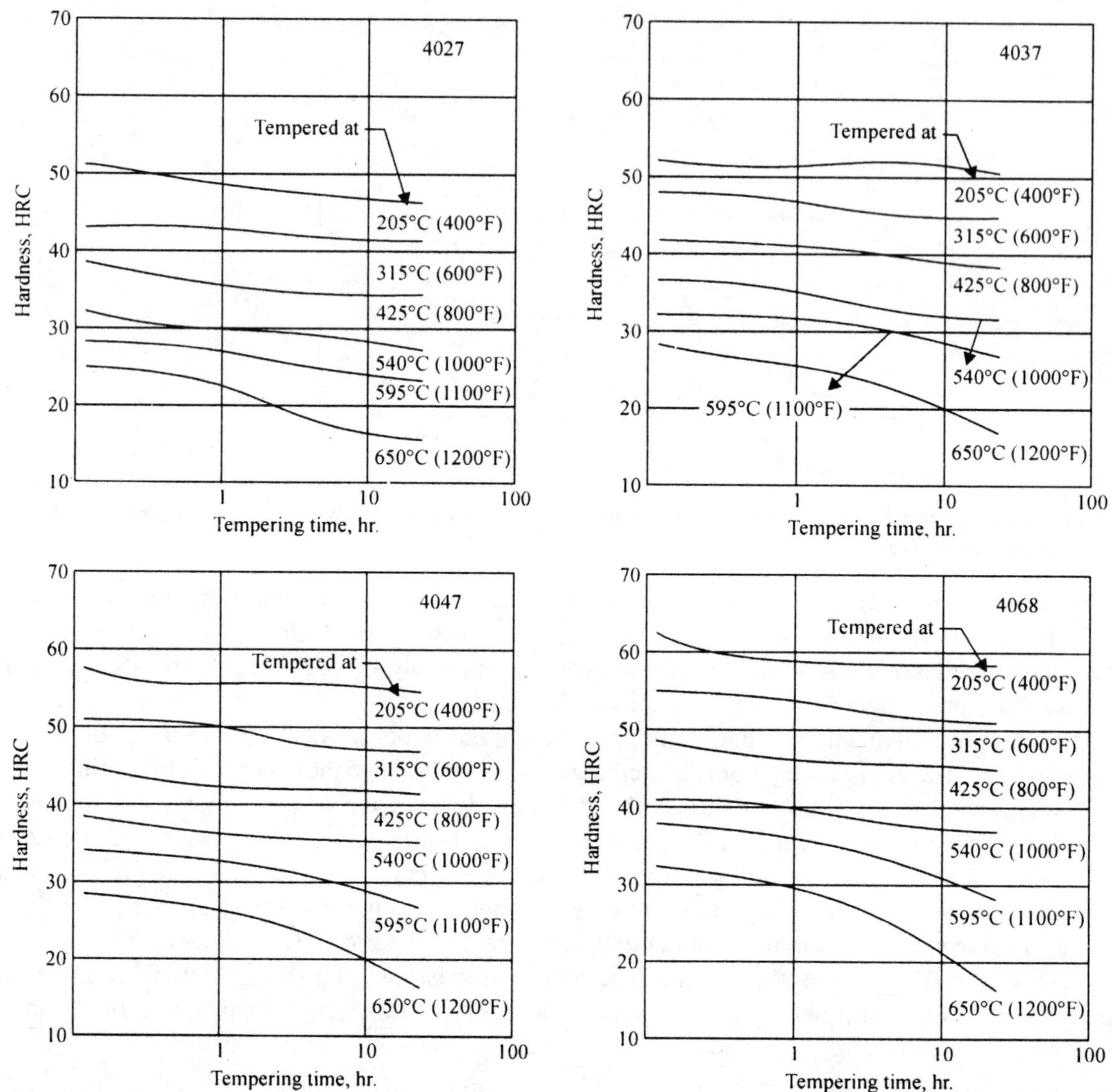

Fig. 11.6. Effect of tempering time at six temperatures on room-temperature hardness of C-Mo steels of different carbon contents but all with prior martensitic structures.

Residual Elements

Residual elements—those not intentionally added to a steel—can cause embrittlement. The elements that are known to cause embrittlement are tin, phosphorus, antimony and arsenic.

TEMPERING PROCEDURES

Tempering can be accomplished by soaking entire parts in the furnace for periods of time sufficient to bring the tempering mechanism to the desired point to completion or by selective heating of certain portions of the part to achieve toughness or plasticity in those areas.

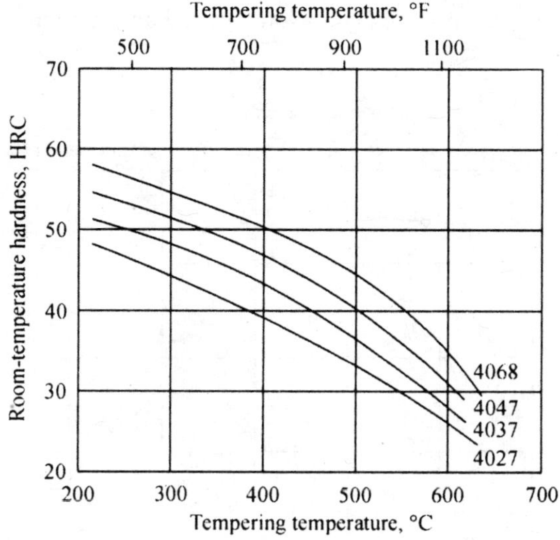

Tempering time: 1 hr. at temperature. 4068 is former standard classification.

Fig. 11.7. Effect of carbon content and tempering temperature on room-temperature hardness of four molybdenum steels.

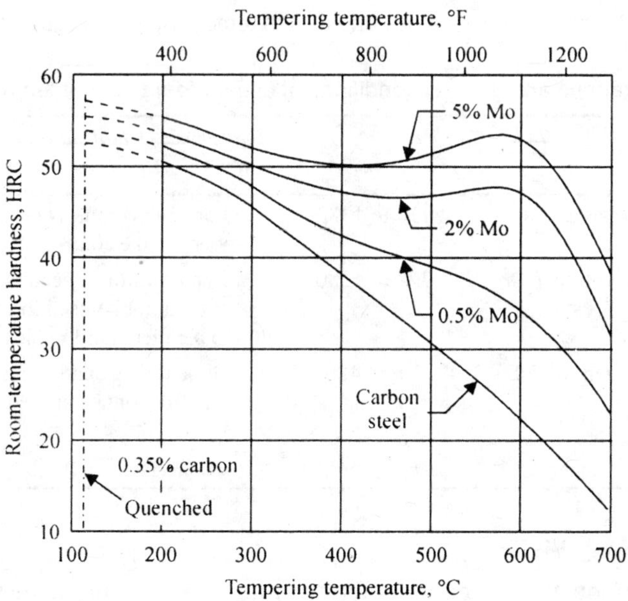

Fig. 11.8. Influence of molybdenum content on softening of quenched 0.35% carbon steels with increasing tempering temperature.

Bulk Processing

Bulk processing may be done in convection furnaces or in molten salt, hot oil or molten metal baths. Selection of the type of furnace depends primarily on number and size of parts and on desired temperature.

Table 11.2 gives temperature ranges, most likely reasons for use and fundamental problems of these four types of equipment.

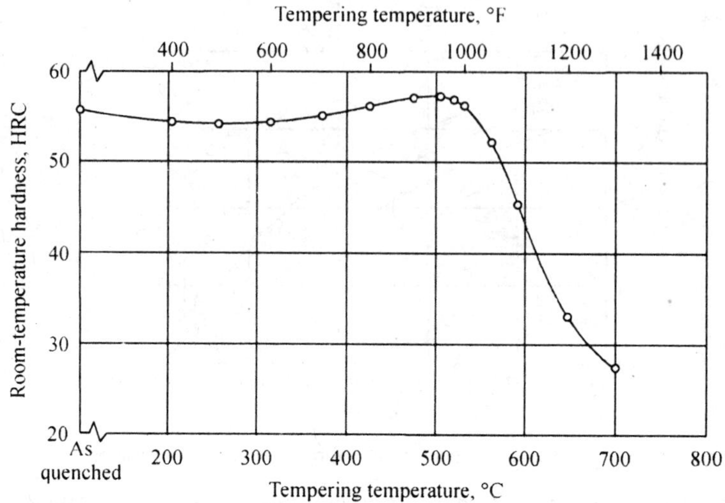

All specimens air cooled from 1010°C (1850°F) and double tempered 2 + 2 hr. At temperature

Fig. 11.9. Variation of room-temperature hardness with tempering temperature for H11 Mod steel.

Table 11.2. Temperature ranges and general conditions of use for four types of tempering equipment .

Type of equipment	*Temperature range*		*Service conditions*
	°C	*°F*	
Convection furnace	50 to 750	120 to 1380	For large volumes of nearly common parts; variable loads make control of temperature more difficult
Salt bath	160 to 750	320 to 1380	Rapid, uniform heating; low to medium volume; should not be used for parts whose configurations make them hard to clean
Oil bath	Up to 250	Up to 480	Good if long exposure is desired; special ventilation and fire control are required
Molten metal bath (high	Above 390	Above 735	Very rapid heating; special fixturing is required density)

Selective Tempering Techniques

Selective tempering techniques are used to soften specific areas of fully hardened parts or to temper areas that were selectively hardened previously.

The purpose of this treatment is to improve the machinability, the toughness or the resistance to quench cracking in the selected zone.

Induction and flame tempering are the most commonly utilised selective techniques because of their controllable local heating capabilities. Immersion of selected areas in molten salt or molten metal can be accomplished, but with somewhat less control.

Special Processes

Special processes occasionally are employed to achieve specific properties such as those derived from steam treating or use of protective atmospheres.

The tempering mechanism in certain steels is enhanced by cyclic heating and cooling. A particularly important procedure employs cycles between subzero temperatures and the tempering temperature to increase the transformation of retained austenite. The term used for this procedure, 'multiple tempering', also is applied to procedures that employ intermediate thermal cycles to soften parts for straightening prior to the actual tempering operation designed to achieve the desired degree of toughness and plasticity.

EQUIPMENT FOR TEMPERING

Steel usually is tempered in either an air (convection) furnace or a salt bath. Molten metal baths, oil baths and flame or induction heating units are also used.

Convection Furnaces

The most commonly employed tempering method utilises the recirculating or forced-air convection furnace and the equipment most commonly used in conjunction with convection furnaces includes continuous belt conveyor, roller rail or dog beam pusher systems. Batch equipment such as box or pit types are also used.

Forced recirculating air is the most common and efficient method of tempering because it lends itself to a wide selection of furnace designs to accommodate a variety of products and capacities. Moreover, the metallurgical results are very good in terms of price per unit weight of yield.

Convection furnaces generally are designed for tempering temperatures of 150 to 750°C (300 to 1380°F). For temperatures up to 550°C (1020°F), recirculated hot air is supplied to the product from a chamber separate from the work-holding area, to avoid uneven heating by radiation. For temperatures of 550 to 750°C, either forced convection or radiant heating is used depending on the metallurgical requirements of the product. To obtain closer control of metallurgical properties, the recirculated forced hot air is employed; but for greater efficiency, radiant heating is used, because transfer of radiant heat is greater as the temperature approaches 750°C (1380°F).

The most important phase of convection-furnace design is determination of the proper amount of forced-air supply. The objective of the blower is to furnish enough hot air to the complete work area so that it is efficiently used to heat the product in the shortest time thermophysically allowed. The type of product and material being processed will dictate the forced-air supply required, which is measured at the operating temperature. Consultation with fan manufacturers can help achieve maximum efficiency of heat transfer.

Heat for the furnace can be supplied by electricity, gas or oil. In most furnace designs, a dual heat-source capability can be built in, such as gas and electricity. This allows for more than one choice of utility when there is a shortage or a cost advantage, of one or the other.

Temperature control is accomplished by positioning a thermocouple at the hot-air side of the recirculating system close to the product. When this technique is used, there is minimum danger of overheating and loads of various sizes can be handled. This method also allows the duration of processing (holding time) to be varied by movement of the thermocouple location—but only within the limits of the furnace size (and/or conveyor speed for continuous-type furnaces). Temperatures generally are held within ±5°C (±9°F).

Efficient use of a continuous furnace cannot be attained when production quantities are small or when parts vary in size, shape and mechanical requirements; the batch furnace is better suited for work of this type. When a continuous furnace is used for such applications, production time is lost in raising or lowering furnace temperature. Sometimes, when the process variables must be changed, a dummy load must be placed in the furnace to accelerate a desired reduction of temperature, or production must be stopped until the temperature is stabilised.

Salt Bath Furnaces

Salt bath furnaces may be employed for tempering at 160°C (320°F) and above. Good heat transfer and natural convection in the bath promote uniformity of workpiece temperature.

All moisture must be removed from parts before they are immersed in the molten salt, because hot salt reacts violently with moisture. If dirty or oily parts are immersed in the bath, the salt will become contaminated and require more frequent rectification.

All parts tempered in salt must be cleaned soon after being removed from the bath, because the salt that clings to them is hygroscopic and may cause severe corrosion. Parts with small or blind holes from which salt is difficult to clean should not be tempered in salt.

Salt Bath Compositions

Salt bath compositions and operating temperature ranges presented in Table 11.3 pertain to baths in common use for tempering.

Class 1 and class 2 salts are reasonably stable and seldom require rectification. If chlorides are added by carry-over from a higher-temperature bath, they will cause an increase in the viscosity of the bath. Chlorides can be removed by filtering through fine screens or by cooling and settling out the insoluble chlorides as a sludge. Occasionally, carbonates become excessive. These can be removed by reaction with dilute nitric acid-base rectifiers. Upper temperature limits must not be exceeded, or salt will become highly oxidising, even toward alloy steels.

Class 3 salts seldom require rectification. However, their high melting points (about 560°C, or 1040°F) severely restrict their useful temperature range. Also, they are decarburising to steels at temperatures above about 700°C (1300°F).

Class 4 salts, which are all-chloride neutral salts, are quite stable. They seldom require rectification but are restricted to temperatures above 590°C (1100°F).

Class 4A salts are close relatives of class 4 salts but contain calcium chloride, which lowers the minimum working temperature to 550°C (1025°F). The upper limit for these salts is more restricted than that for class 4 salts.

All of the salts for these baths are commercially available.For further information, see the articles on martempering, on austempering and on heat treating of tool steels.

Safety precaution

Introduction of cyanide salts or other reducing agents into nitrite tempering baths will cause violent explosions.

Oil Bath Equipment

Oil bath equipment for tempering may be similar in design to that used for salt baths, or a steel tank over large hotplate-type burners will serve satisfactorily.

Table 11.3. Compositions and operating temperatures for salt baths used in tempering.

Class	*Composition of bath, %*								*Operating temperature*		*Operating temperature*	
	Sodium nitrite	*Sodium nitrite*	*Potassium nitrite*	*Sodium carbonate*	*Sodium chloride*	*Potassium chloride*	*Barium chloride*	*Calcium chloride*	*°C*	*°F*	*°C*	*°F*
1	37–50	0–10	50–60	–	–	–	–	–	165–595	325–1100	635	1175
2	–	45–57	45–57	–	–	–	–	–	290–595	550–1100	650	1200
3	–	–	–	45–55	–	45–55	–	–	620–925	1150–1700	938	1720
4	–	–	–	–	15–25	20–32	50–60	–	595–900	1100–1650	940	1725
4a	–	–	–	–	10–15	25–30	40–45	15–20	550–760	1025–1400	790	1450

Submerged electric heating elements may also be employed. Stirring is essential for temperature uniformity and satisfactory oil life. Simple, oven-type temperature controls may be employed, but localised over-heating must be avoided to prevent fire and rapid decomposition of the oil. A standard thermometer of the proper range may be used to check the temperature of the oil.

Low-temperature tempering in a hot oil bath is a simple and inexpensive method that is especially useful for holding work at temperature for long periods of time. The practical temperature limit is about 120°C (250°F) without special ventilation and fire-protection equipment and about 250°C (480°F) with such precautions, which may include extremely efficient ventilators or inert-gas blanketing systems. When a tempering temperature above 205°C (400°F) is required, a salt bath is usually preferable to an oil bath. Oils for tempering must resist oxidation and have a flash point well above the operating temperature. The most commonly used oils are high-flash-point paraffinic oils with antioxidant additives. Oils used for martempering are also satisfactory for tempering.

Molten Metal Baths

Molten metal baths for tempering have largely been replaced by salt baths. When employed, commercially pure lead, which melts at about 327°C (620°F), has proved to be the most generally suitable of all metals and alloys. For special applications, however, lead-base alloys having lower melting points are used.

Lead oxidises readily. Although lead itself will not adhere to clean steel, adherence of lead oxide to steel surfaces is a problem, especially at higher tempering temperatures. Within the range of temperatures usually employed, a film of molten salt will protect the surface of the lead bath and the work will be easily cleaned. Above 480°C (900°F), granulated carbonaceous material, such as charcoal, may be used as a protective cover. Because of its high thermal conductivity, lead is useful for rapid local heating and selective tempering. A typical application is the tempering of a ball joint. The part is carburised and quenched to a minimum case hardness of 59 HRC and a core hardness of 30 to 40 HRC. The thread and taper then are tempered in lead to produce a maximum case hardness of 40 HRC.

Because of the high specific gravity of lead, parts tempered in molten lead will float unless held down by fixtures. All parts and fixtures must be dry when immersed in the bath, to prevent formation of steam in and resultant violent expulsion of, the molten lead. Precautions also must be taken to protect personnel from industrial lead poisoning; hoods and ventilating equipment are required.

Temperature Control

For either gas or electric heat, properly adjusted potentiometers of the on-off type will control the tempering temperature within ±6°C (±10°F) at the thermocouple location. With proportioning controls, these instruments can maintain temperatures within ±1°C (±2°F) at the thermocouple location.

SELECTION OF TEMPERING EQUIPMENT

Selection of equipment for tempering is based principally on: (i) temperature requirements; and (ii) quantity and similarity of the work to be treated. Temperature requirements are dictated by prior heat treatment and by the properties to be developed by tempering.

PROCESS CONTROL

Variation in hardness after tempering is most frequently the result of differences in prior microstructure, as discussed previously. When prior microstructure is the same, control of temperature is the most important factor in control of the tempering process.

In general, control of tempering temperature to within ±3°C (±5°F) is adequate and is within the practical limits of most furnace and molten-bath equipment. Temperature variations are seldom permitted to exceed ±6°C (±10°F) unless mechanical-property requirements are correspondingly broad.

Variations in room-temperature hardness of two cast steels after production tempering are shown in Figs. 11.10, 11.11 and 11.12.

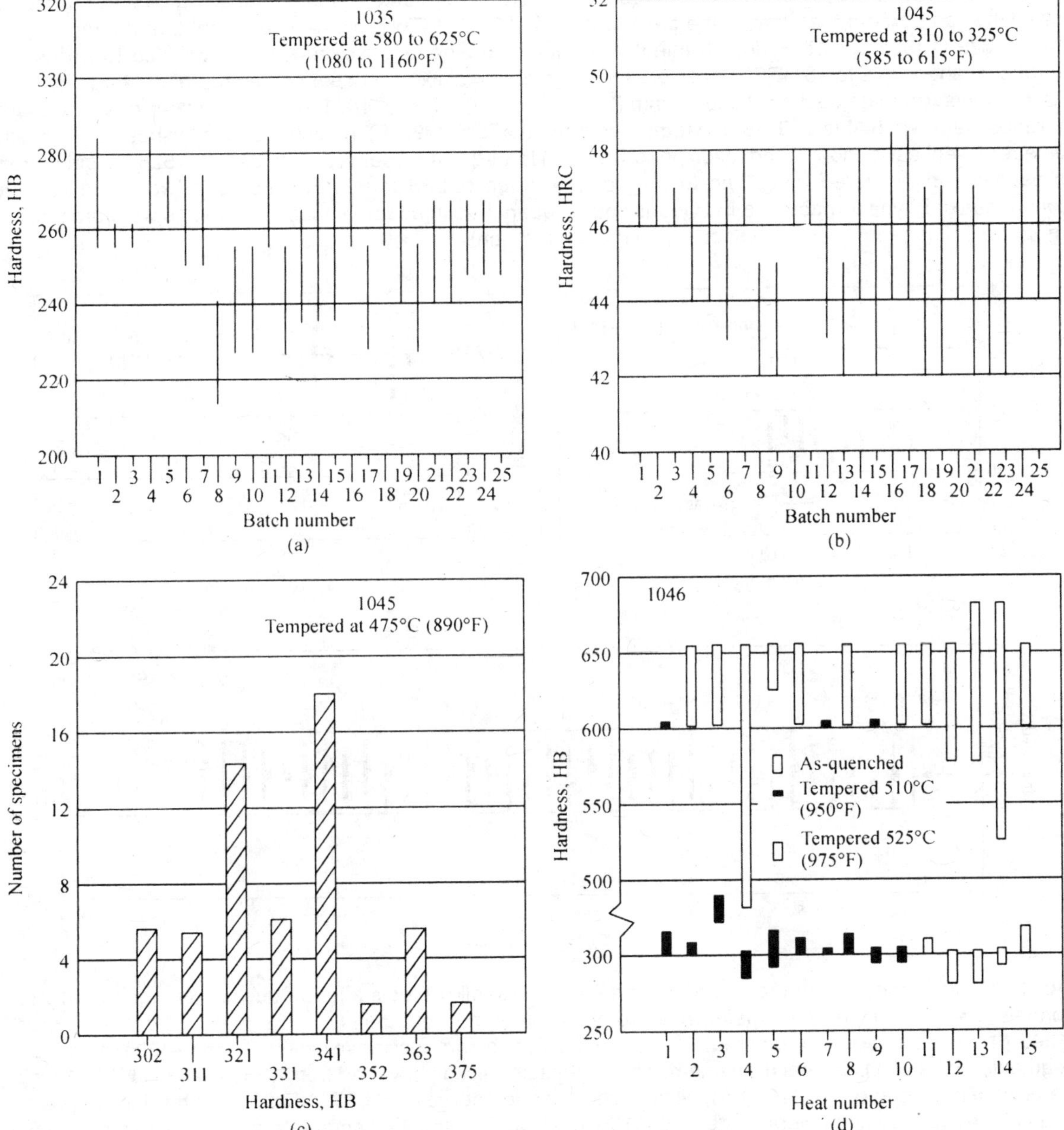

Fig. 11.10. Variations in room-temperature hardness of three carbon steels after production tempering. (a) Automotive steering-arm forgings made of fine-grain 1035 steel. **(Continued).**

Fig. 11.10. (Continued). Section thickness varied from 16 to 29 mm (5/8 to 1 1/8 inch). Forgings were austenitised at 825°C (1520°F) in oil-fired pusher conveyor furnace, held 45 minutes quenched in water at 21°C (70°F) and tempered 45 minutes at 580 to 625°C (1080 to 1160°F) in oil-fired link-belt furnace to required hardness range of 217 to 285 HB. Hardness was checked hourly with a 5% sample; readings were taken on polished flash line of 29 mm (1 1/8 inch) section. Survey of furnace revealed temperature variation at 605°C (1120°F) of +8, –4°C (+15, –7°F). Data represent forgings from four mill heats of steel and cover a six-week period; (b) woodworking cutting tools forged from 1045 steel. Section of cutting lip was hardened locally by gas burners that heated the steel to 815°C (1500°F). Tools were oil quenched and tempered at 305 to 325°C (585 to 615°F) for 10 minutes in electrically heated recirculating-air furnace to a desired hardness range of 42 to 48 HRC. Data were recorded during a six-month period and represent forgings from 12 mill heats; (c) plate section, 19 to 22 mm (3/4 to 7/8 inch) thick, of 1045 steel were water quenched to a hardness range of 534 to 653 HB and tempered 1 hr. at 475°C (890°F) in continuous roller-hearth furnaces. Data represent a two-month production period; and (d) forged 1046 steel heated to 830°C (1525°F), quenched in caustic and tempered from 1 hr. to a hardness range of 285 to 321 HB. Forgings were heated in a continuous belt-type furnace and individually dump quenched in agitated caustic. Forgings weighed 9 to 11 kg (20 to 24 lb) each; maximum section, 38 mm (1 1/2 inch).

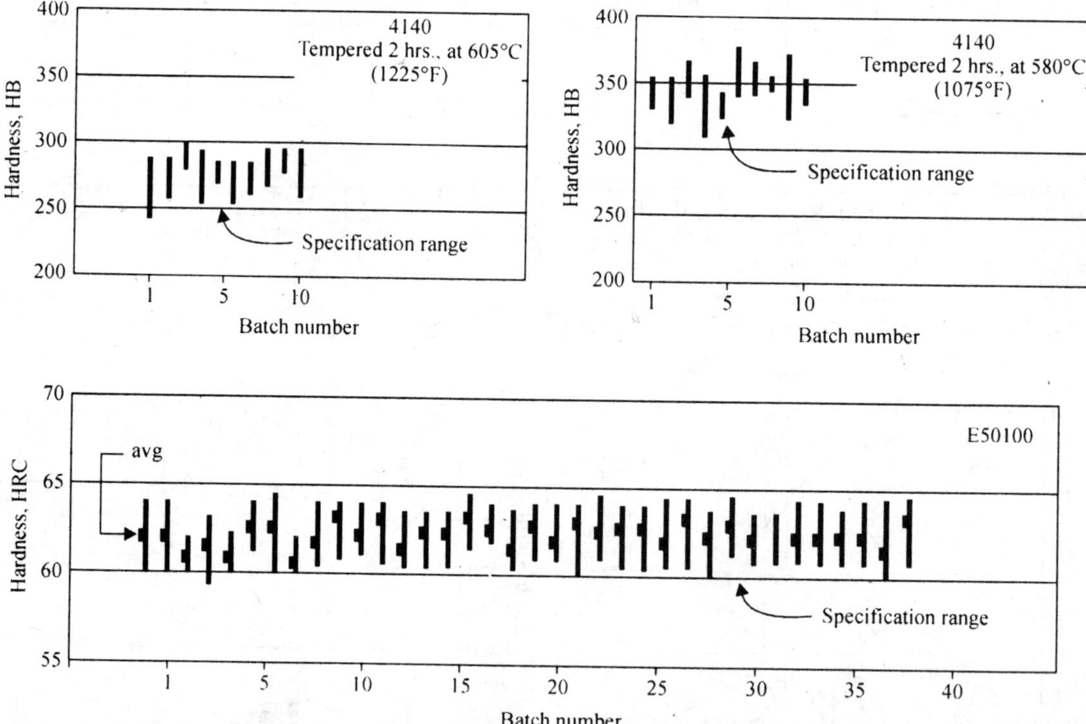

Fig. 11.11. Variations in room-temperature hardness of two alloy steels after production tempering. (a) valve bonnets, 75 mm (3 inch) in diameter, made of 4140 steel from one mill heat. Parts were heated at 870°C (1600°F), oil quenched and tempered at 605°C (1125°F) to a hardness specifiation of 255 to 302 HB; (b) valve segments made of 4140 steel from one mill heat. Section thickness varied from 13 to 25 mm (1/2 to 1 inch). Parts were heated at 870°C (1600°F), oil quenched and tempered at 580°C (1075°F) to a hardness of 321 to 363 HB; and (c) needles rollers made of E50100 steel wire varying in diameter from 2.2 to 2.6 mm (0.086 to 0.103 inch). Data represent 36 batches of parts (nine mill heats of steel) tempered to a hardness specification of 60 to 64 HRC. Each batch represents a minimum furnace load to 270 kg (600 lb).

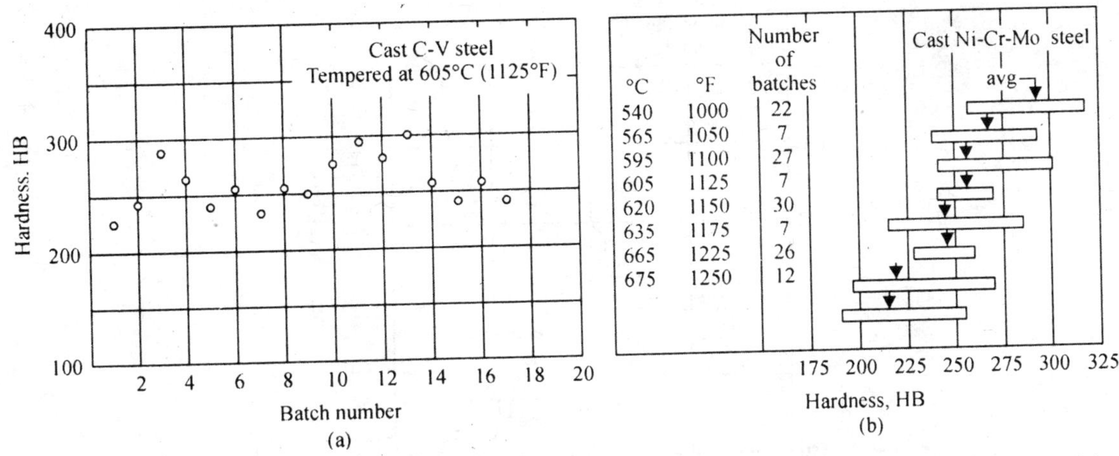

Fig. 11.12. Variations in room-temperature hardness of two cast steels after production tempering: (a) cast C-V steel containing 0.25 to 0.30 C, 0.70 to 0.80 Mn, 0.08 to 0.12 V and 0.40 to 0.60 Si, 25 mm (1 inch) thick, water quenched from 870°C (1600°F) and then tempered at 605°C (1125°F); and (b) cast Ni-Cr-Mo steel containing 0.28 to 0.33 C, 0.70 to 0.90 Mn, 0.40 to 0.70 Ni, 0.40 to 0.60 Cr, 0.15 to 0.25 Mo and 0.40 to 0.60 Si. Spread of hardness readings for various tempering temperatures.

TEMPERING OF SPECIAL MICROSTRUCTURES

Carburised Components

Tempering of carburised parts represents a special instance in that the combination of toughness, strength, hardness, residual stress and retained austenite all must be considered when selecting tempering time and temperature. Core properties cannot always be controlled by tempering when trying to achieve maximum case properties and a favourable compressive residual stress pattern may be retained only at the expense of over-all toughness. The effect of tempering on the hardness of carburised cases is shown in Fig. 11.13 and the influence of tempering on residual stress is illustrated in Fig. 11.14.

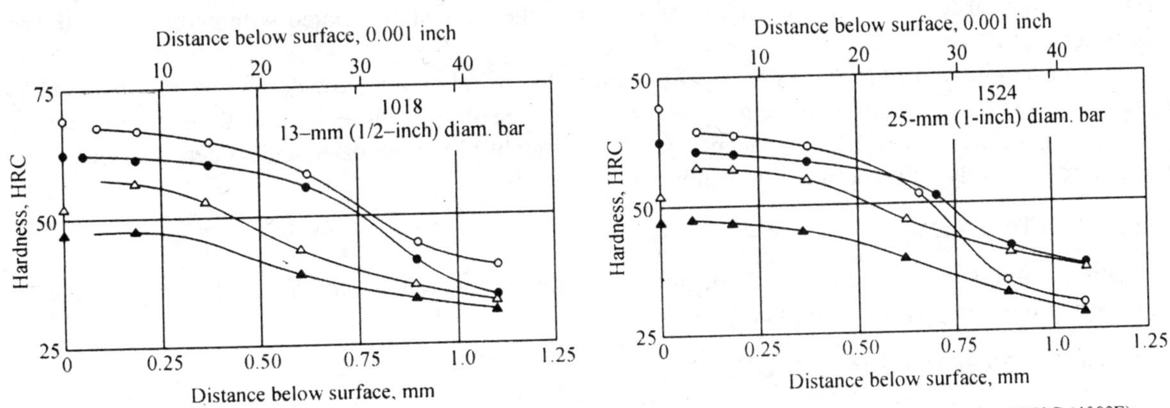

Carburised for 4.5 hrs., at 925°C (1700°F), oil quenched and tempered: ○ As-quenched; ● tempered at 205°C (400°F); △ tempered at 315°C (600°F); ▲ tempered at 425°C (800°F).

Fig. 11.13. Effect of tempering on room-temperature hardness of carburised cases.

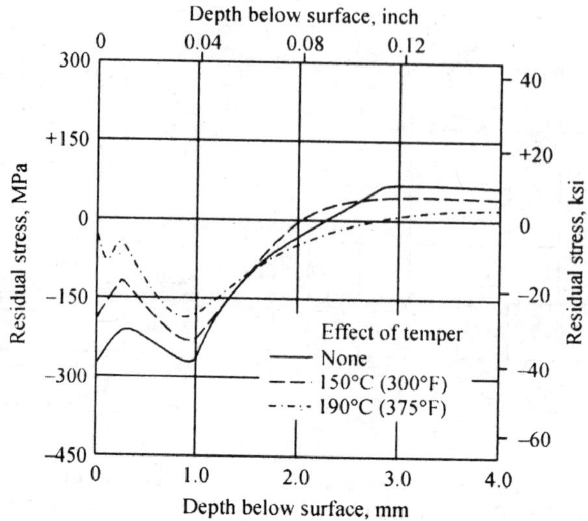

Bars of 8617 steel, 19 mm (0.75 inch) inch diameter, were carburised, direct oil quenched, and tempered for 1 hr., at the indicated temperature.

Fig. 11.14. Effect of tempering on residual stress in carburised steel.

The transformation of retained austenite and the resultant change in the relative volume of case and core are primarily responsible for the change in residual stress as a function of temperature. The effect of retained austenite on the performance of components is still a controversial issue. Reduction of retained austenite is apparently desirable for resistance to grinding abuse and to provide dimensional stability, but retained austenite appears to be beneficial for contact-fatigue durability.

Non-martensitic Structures

Tempering of microstructures other than martensite and retained austenite also represents special applications of tempering. Reactions of structures containing substantial amounts of lower bainite are relatively similar to that of martensite in terms of the phenomena associated with carbide growth and coalescence.

Upper bainite and fine pearlite formed by controlled or relatively slow cooling simply respond by carbide growth and eventual ferrite re-crystallisation. The softening associated with tempering in such instances is shown in Fig. 11.15. Impact properties of normalised and tempered and hardened and tempered structures at nearly equal hardness are shown in Fig. 11.16.

Induction Tempering

Extensive production experience has demonstrated the commercial success of induction heating for tempering in several types of applications.

SPECIAL TEMPERING PROCEDURES

Selective Tempering

Selective or localised tempering is applied to parts in which adjacent areas must have significantly different hardnesses. Chisels, punches, the upset ends of cold formed riverts and the threaded portions of carburised parts are typical examples. Localised tempering is also employed in pre-heating and post-heating of weld areas when a lowering of the hardness in the heat-affected zone is desired.

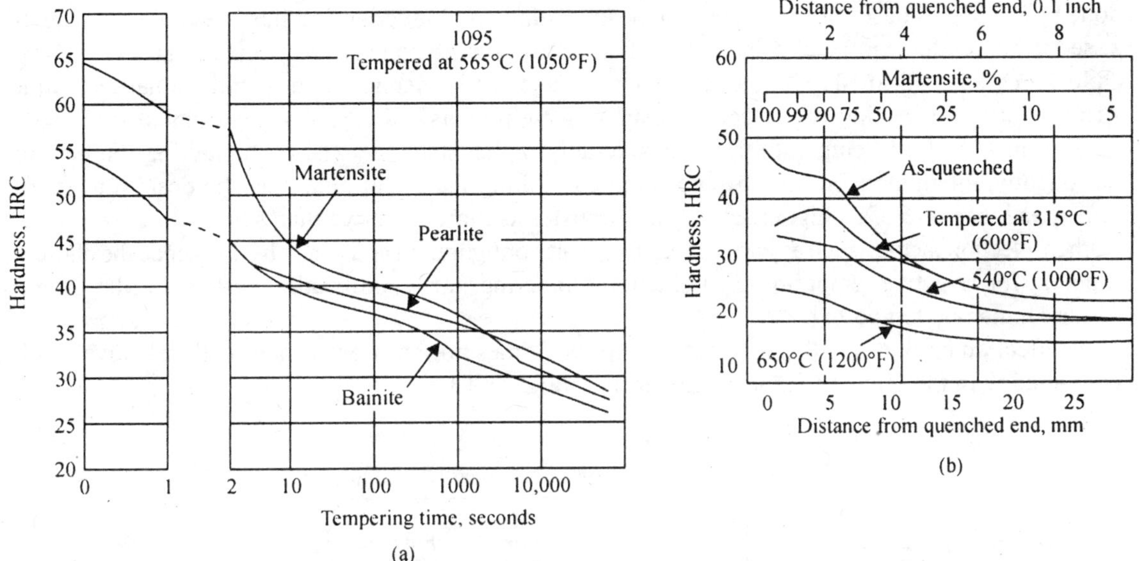

Fig. 11.15. Effect of prior microstructure on room-temperature hardness after tempering: (a) 1095 steel tempered at 565°C (1050°F) for various periods of time; and (b) room-temperature hardness before and after tempering and amount of martensite present before tempering, in 4320 steel end-quenched hardenability specimens tempered for 2 hrs.

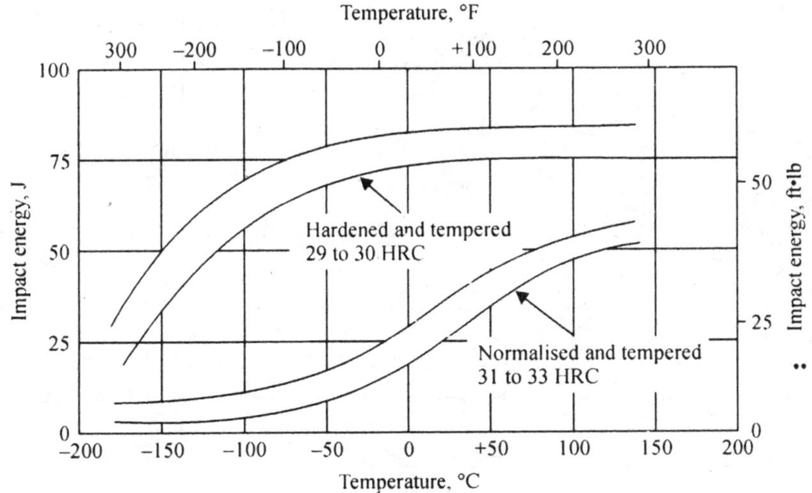

Variation in Charpy keyhole-notch impact energy with temperature for 4340 steel hardened and tempered to 29 to 30 HRC or normalised and tempered to 31 to 33 HRC

Fig. 11.16. Effect of microstructure on notch toughness.

Selective tempering entails heating a restricted area to the required tempering temperature without heating the remainder of the part to this temperature. Induction heating coils, special flame heads and immersion in lead or salt baths are employed to achieve this selective heating. Selective tempering also is done by use of defocused lasers or electron beam equipment. Induction and flame techniques generally are used in high-volume production applications and are easiest to control. Deeper penetration is achieved with low-frequency (3 to 10 kHz) induction heating and salt immersion than with the other techniques.

To obtain rapid heating for selective tempering by immersion in salt or lead baths, it is usually necessary that the bath temperature be considerably higher than the desired tempering temperature. Consequently, the immersion time becomes the controlling factors in obtaining the desired tempering results. Lead, because of its higher rate of heat transfer, is more effective than salt.

Other selection factors, such as ease of fixturing, part configuration and cost, also influence the choice of equipment for selective tempering. In induction tempering, the same heating system can be used for both hardening and tempering.

The effect on hardness of the very short tempering times normally associated with selective heating is compared with the effect of furnace tempering in Fig. 11.17.

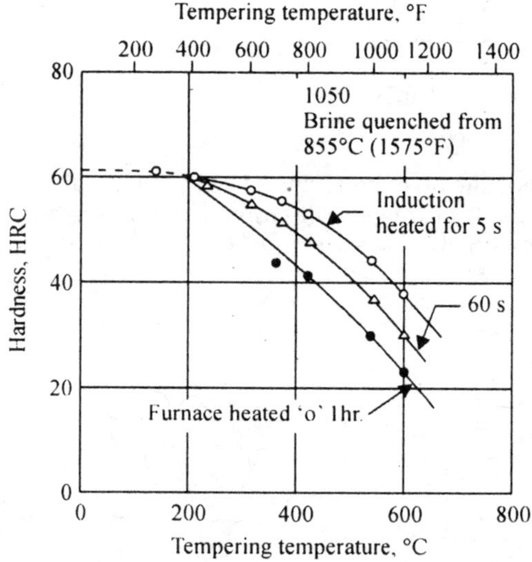

Fig. 11.17. Variations of room-temperature hardness with tempering temperature for furnace and induction heating.

Multiple Tempering

Multiple tempering is principally used: (i) to relieve the quenching and straightening stresses in irregularly shaped carbon and alloy steel parts and thereby lessen distortion; (ii) to eliminate retained austenite and improve dimensional stability in such parts as bearing components and gage blocks; and (iii) to improve yield and impact strengths without decreasing hardness.

USE OF FIXTURES

Many high-strength steel components having tensile strength greater than 1720 MPa (250 ksi) are finish machined before final heat treatment. To minimise distortion and satisfy stringent dimensional requirements, some of these components, such as cylinders, pressure vessels and thin parts, are held in fixtures during both hardening and tempering or during tempering only.

External rings, internal mandrels, jacks, screws, weights, wedges, dies and other mechanical devices are used to assist in dimensional correction.

CRACKING IN PROCESSING

Because of their carbon or alloy contents, some steels are likely to crack if they are permitted to cool all the way to room temperature during or immediately following the quenching operation. This likelihood is caused by generation of high tensile residual stresses during quenching, due to thermal gradients, abrupt changes in section thickness, decarburisation or other hardenability gradients. Accordingly, for carbon steels containing more than 0.4 per cent C and alloy steels containing more than 0.35 per cent C, it is recommended that the parts be transferred to tempering furnaces before they cool below 100 to 150°C (212 to 300°F).

Alternately, many heat treating operations use quenching oil for the tempering operation (martempering) or to avoid cooling below 125°C (255°F). Steels that are known to be sensitive to this type of cracking include 1060, 1090, 1340, 4063, 4150, 4340, 52100, 6150, 8650 and 9850.

Other carbon and alloy steels are generally less sensitive to this type of delayed quench cracking but may crack as a result of part configuration or surface defects. These steels include 1040, 1050, 1137, 1144, 4047, 4132, 4640, 8632, 8740 and 9840. Some steels, such as 1020, 1038, 1132, 4130, 5130 and 8630 are not sensitive.

Before being tempered, parts should be quenched to room temperature to ensure transformation of most of the austenite to martensite and to achieve maximum as-quenched hardness. Austenite retained in low-alloy steels will, upon heating for tempering, transform to an intermediate structure, reducing over-all hardness. However, in medium to high-alloy steels containing austenite-stabilising elements (nickel, for example), retained austenite may transform to martensite upon cooling from tempering and thus such steels may require additional tempering (double tempering) for relief of transformation stresses.

TEMPER EMBRITTLEMENT

When carbon or low alloys steels are slow cooled from tempering above 575°C (1065°F), or are tempered for extended times between 375 and 575°C (705 and 1065°F), a loss in toughness occurs that manifests itself in reduced notched-bar impact strength compared to that resulting from normal tempering cycles and relatively fast cooling rates (Fig. 11.19). The cause of temper embrittlement is believed to be precipitation of compounds containing trace elements such as tin, arsenic, antimony and phosphorus along with chromium and/or manganese.

The intergranular nature of the fracture suggests that the embrittlement occurs at the prior austenite grain boundaries. Although manganese and chromium cannot be restricted, reduction of the other elements and quenching from above 575°C are the most effective remedies for this type of embrittlement. It has also been found that small amounts of molybdenum retard embrittlement.

A second form of embrittlement, called 'blue brittleness', occurs when steels are tempered at about 300°C (570°F) after being hardened. This is illustrated in Fig. 11.18. Generally, for steels containing potent carbide formers such as chromium, tempering between 200 and 370°C (390 and 700°F) should be avoided.

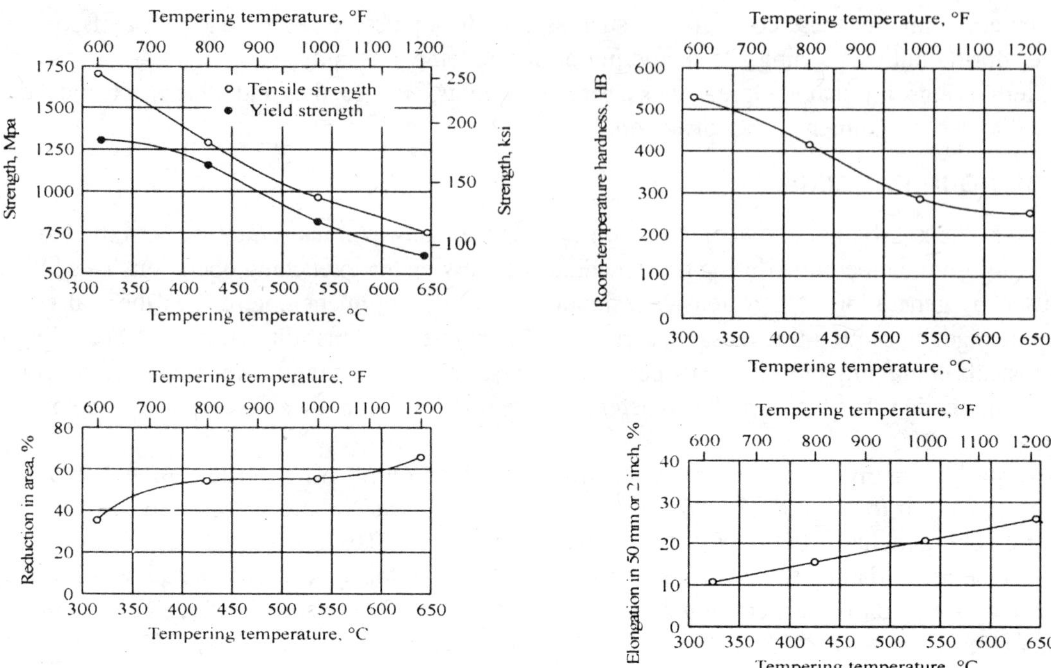

Properties summarised for one heat of 1050 steel that was forged to 38 mm (1.50 inch) in diameter, then water quenched and tempered at various temperatures. Composition of heat: 0.52 C, 0.93 Mn.

Fig. 11.18. Effect of temper embrittlement on notch toughness.

HYDROGEN EMBRITTLEMENT

Selection of tempering temperature and the resultant hardness or plasticity, must include consideration of the potential problem of hydrogen embrittlement should the part be exposed to hydrogen through electroplating, phosphating or other means, or if environmental conditions cause cathodic absorption of hydrogen during service.

Generally, the restricted notch ductility of steels with hardness above 40 HRC presents ideal conditions for development of stress concentrations in parts containing notches or defects that would, in the presence of relatively low hydrogen concentrations, lead to failure of parts at stresses far below the nominal tensile strength of the material. Thus, tempering should be carried out to achieve hardness below 40 HRC if the part will be subjected to relatively high stresses and probable exposure to hydrogen.

Chapter 12

Austempering of Steel

INTRODUCTION

Austempering is the isothermal transformation of a ferrous alloy at a temperature below that of pearlite formation and above that of martensite formation. Steel is austempered by being:

1. Heated to a temperature within the austenitising range, usually 790 to 870°C (1450 to 1600°F).
2. Quenched in a bath maintained at a constant temperature, usually in the range of 260 to 400°C (500 to 750°F).
3. Allowed to transform isothermally to bainite in this bath.
4. Cooled to room temperature, usually in still air.

The fundamental difference between austempering and conventional quenching and tempering is shown schematically in Fig. 12.1. Austempering of steel and hardenable grades of cast iron offers several potential advantages:

1. Increased ductility or notch toughness at a given hardness (Table 12.1).
2. Reduced distortion, which lessens subsequent machining time, stock removal and cost.
3. The shortest over-all time cycle to through harden within the hardness range of 35 to 55 HRC, with resulting savings in energy and capital investment.

Table 12.1. Mechanical properties of 1095 steel heat treated by three methods.

Specimen no.	*Heat treatment*	*Hardness, HRC*	*Impact strength*		*Elongation in 25 mm or 1 inch, %*
			J	*ft.lb*	
1	Water quenched and tempered	53.0	16	12	–
2	Water quenched and tempered	52.5	19	14	–
3	Martempered and tempered	53.0	38	28	–
4	Martempered and tempered	52.8	33	24	–
5	Austempered	52.0	61	45	11
6	Austempered	52.5	54	40	8

For true austempering, the metal must be: (i) cooled from the austenitising temperature to the temperature of the austempering bath fast enough so that no transformation of austenite occurs during cooling; and (ii) held at bath temperature long enough to ensure complete transformation of austenite to bainite. Modifications of these procedures, constituting departures from true austempering, are discussed in the section on modified austempering, later in this chapter.

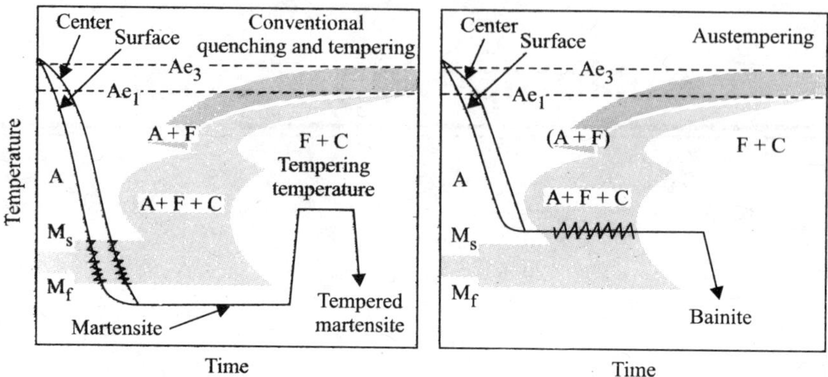

Fig. 12.1. Schematic comparison of time-temperature transformation cycles for conventional quenching and tempering and for austempering.

QUENCHING MEDIA FOR AUSTEMPERING

Molten salt is the quenching medium most commonly used in austempering, because: (i) it transfers heat rapidly; (ii) it virtually eliminates the problem of a vapour phase barrier during the initial stage of quenching; (iii) its viscosity is uniform over a wide range of temperature; (iv) its viscosity is low at austempering temperatures (near that of water at room temperature), thus minimising dragout losses; (v) it remains stable at operating temperatures and is completely soluble in water, thus facilitating subsequent cleaning operations; and (vi) the salt can be easily recovered from wash water so that there is no discharge to drain. Formulations and characteristics of two typical salt quenching baths are given in Table 12.2. The high-range salt is suitable for austempering only, whereas the wide-range salt may be used for austempering, martempering and modifications thereof.

Table 12.2. Compositions and characteristics of salts used for austempering.

	High range	*Wide range*
Sodium nitrate	45 to 55%	15 to 25%
Potassium nitrate	45 to 55%	45 to 55%
Sodium nitrite	–	25 to 35%
Melting point (approx.)	220°C (430°F)	150 to 165°C (300 to 330°F)
Working temperature range	260 to 595°C (500 to 1100°F)	175 to 540°C (345 to 1000°F)

Water Additions to Salt

The quenching severity of a nitrate-nitrite salt can be increased significantly by careful addition of water. Agitation of the salt is necessary to disperse the water uniformly.

Periodic water additions are needed to maintain required water content. Water can be added with complete safety as follows:

1. Water may be misted at a regulated rate into a vigorously agitated area of the molten bath.
2. In installations where the salt is pump-circulated, returning salt is cascaded into the quench zone. A controlled fine stream of water may be injected into the cascade of returning salt.
3. The austempering bath may be kept saturated with moisture by introducing steam directly into the bath. The steam line should be trapped and equipped with a discharge to avoid emptying condensate directly into the bath.

Water should never be added to a salt bath from a pail or dipper. Oil is seldom used for austempering, because of its chemical instability and resulting change in viscosity at austempering temperatures. Because of its persistent vapour phase, oil is a slower quenchant than salt at elevated temperatures and presents a fire hazard.

STEELS FOR AUSTEMPERING

The selection of steel for austempering must be based on transformation characteristics as indicated in time-temperature transformation (TTT) diagrams. Three important considerations are: (i) the location of the nose of the TTT curve and the time available for bypassing it; (ii) the time required for complete transformation of austenite to bainite at the austempering temperature; and (iii) the location of the M_s point.

As indicated in Fig. 12.2, 1080 carbon steel possesses transformation characteristics that provide it with limited suitability for austempering. Cooling from the austenitising temperature to the austempering bath must be accomplished in about 1 second to avoid the nose of the TTT curve, and thus prevent transformation to pearlite during cooling. Depending on the temperature, isothermal transformation in the bath is completed within a period ranging from a few minutes to about 1 hour. Because of the rapid cooling rate required, austempering of 1080 can be successfully applied only to thin sections of about 5 mm (0.2 inch) maximum.

5140 low-alloy steel is well suited to austempering, as indicated by the TTT curve for this steel shown in Fig. 12.2. Approximately 2 second is allowed in which to bypass the nose of the curve, and transformation to bainite is completed within 1 to 10 minute at 315 to 400°C (600 to 750°F). Parts made of 5140 steel or of other steels with similar transformation characteristics are adaptable to austempering in larger section sizes than are feasible for 1080 steel, because of the greater time allowed for bypassing the nose of the curve.

In addition to the steels previously indicated (1080 and 5140), steels adaptable to austempering include:

1. Plain carbon steels containing 0.50 to 1.00 per cent C and a minimum of 0.60 per cent Mn.
2. High-carbon steels containing more than 0.90 per cent C and, possibly, a little less than 0.60 per cent Mn.
3. Certain carbon steels (such as 1041) with a carbon content of less than 0.50 per cent but with manganese content in the range from 1.00 to 1.65 per cent.
4. Certain low-alloy steels (such as the series 5100 steels) containing more than 0.30 per cent C; the series 1300 to 4000 steels with carbon contents in excess of 0.40 per cent; and other steels, such as 4140, 6145 and 9440.

Some steels, although they have sufficient carbon or alloy content to be hardenable, are borderline or impractical for austempering because: (i) transformation at the nose of the TTT curve starts in less than 1 second, thus making it virtually impossible to quench other than thin sections in molten salt without forming some pearlite; and (ii) they require excessively long periods of time for transformation. A typical example of a steel belonging to the first category is 1034, whose transformation characteristics are shown in Fig. 12.2. Transformation characteristics for 9261 (also shown in Fig. 12.2) indicate no difficulty in quenching past the nose of the curve, but the time required for isothermal transformation to bainite (about 24 hr.) is excessive. Other alloy steels having similarly excessive transformation times include those in series 4300,4600 and 4800.

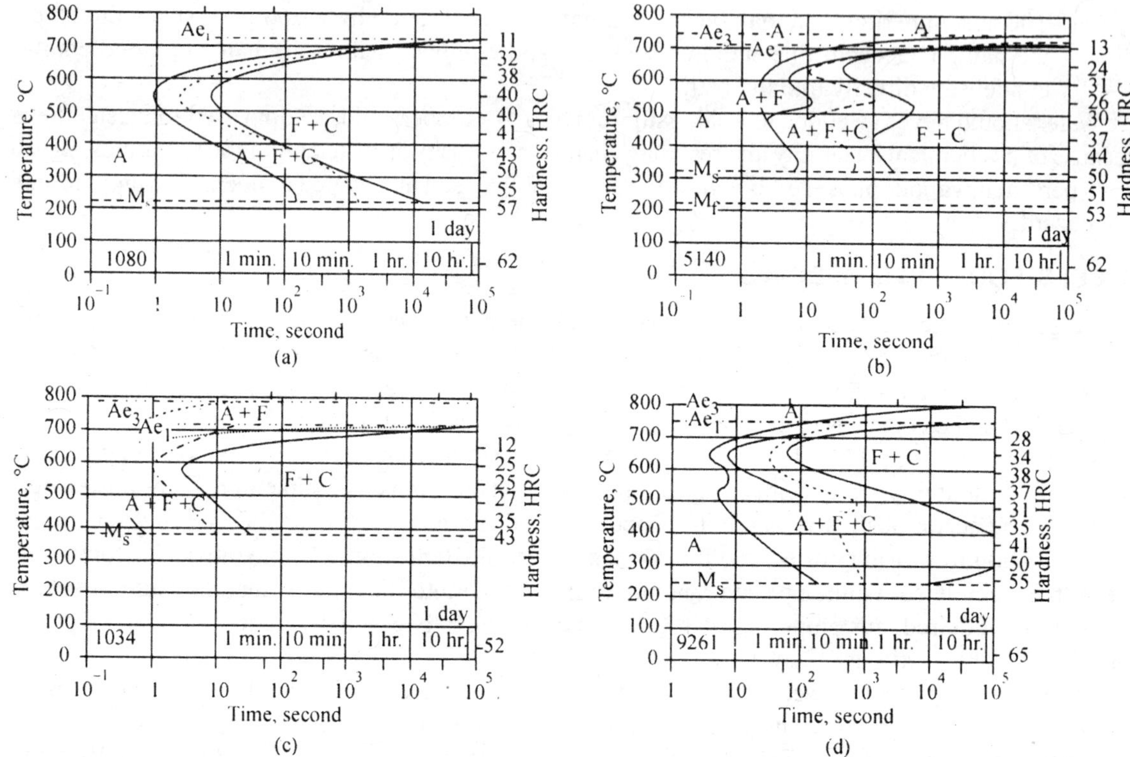

1080, limited suitability for austempering because pearlite reaction starts too soon near 540°C (1000°F); 5140, well suited to austempering; 1034, impossible to austemper because of extremely fast pearlite reaction time at 540 to 595°C (1000 to 1100°F); 9261, not suited to austempering because of slow reaction to bainite at 260 to 400°C (500 to 750°F).

Fig. 12.2. Transformation characteristics of 1080, 5140, 1034 and 9261 steels, in relation to their suitability for austempering.

Austenitising Temperature

As the austenitising temperature of a high-carbon steel increases, the M_s temperature decreases because of more complete solution of carbon. The direct effects of alloying elements on the M_s point are much less pronounced than the effect of carbon. However, carbide-forming elements (such as molybdenum and vanadium) can tie up the carbon as alloy carbides and prevent complete solution of carbon. The approximate M_s temperature, in degrees Centigrade, of a completely austenitised steel can be calculated by means of the following formula:

$$\begin{aligned} M_s = 538 &- (361 \times \text{per cent C}) \\ &- (39 \times \text{per cent Mn}) \\ &- (19 \times \text{per cent Ni}) \\ &- (39 \times \text{per cent Cr}) \end{aligned}$$

Expressed in degrees Fahrenheit, the formula is:

$$M_s = 1000 - (650 \times \text{per cent C}) - (70 \times \text{per cent Mn}) - (35 \times \text{per cent Ni}) - (70 \times \text{per cent Cr})$$

Austenitising temperature has a significant effect on the time at which transformation begins. As the austenitising is increased above normal (for a specific steel), the nose of the TTT curve shifts to the right because of grain coarsening. In Fig. 12.2 for example, approximately ¾ second is allowed for quenching 1080 steel in order to avoid the nose of the curve. However, this is based on an austenitising temperature of 790°C (1450°F); higher austenitising temperatures move the TTT curve to the right, allowing more time before transformation begins. Practical use is sometimes made of this phenomenon in order to process compositions or section sizes that would otherwise be borderline for austempering. However, the coarser grain sizes resulting from higher austenitising temperatures may be detrimental to some desired properties. Therefore, it is recommended that standard austenitising temperature be given preference for austempering. If experience with specific compositions and parts proves that advantages can be gained from the use of a higher temperature and that no harm will be incurred from grain coarsening, higher austenitising temperatures may be employed.

SECTION THICKNESS LIMITATIONS

Maximum thickness of section is important in determining whether or not a part can be successfully austempered. For example, a fully bainitic structure can be obtained almost as easily in a 3-mm (⅛ inch) thick part that is 0.6 m (24 inch) long by 50 mm (2 inch) wide as in a 3-mm (⅛ inch) thick part that is only 75 mm (3 inch) long by 25 mm (1 inch) wide, assuming that all other conditions are equal and that proper facilities for handling the longer part are available.

For 1080 steel, a section thickness of about 5 mm (0.2 inch) is the maximum that can be austempered to a fully bainitic structure. Carbon steels of lower carbon content will be restricted to a proportionately lesser thickness. Lower carbon steels containing boron, however, can be successfully austempered in heavier sections. In some alloy steels, section thicknesses up to about 25 mm (1 inch) can be austempered to fully bainitic structures.

Nevertheless, sections of carbon steel significantly thicker than 5 mm (0.2 inch) are regularly austempered in production when some pearlite is permissible in the microstructure. This is demonstrated in Table 12.3, which lists section sizes of austempered parts made of various steels.

Table 12.3. Hardness of various steels and section sizes of austempered parts.

	Section size		*Salt temperature*		*M_s temperature(a)*		
Steel	*mm*	*inch*	*°C*	*°F*	*°C*	*°F*	*Hardness, HRC*
1050	3(b)	0.125(b)	345	655	320	610	41 to 47
1065	5(c)	0.187(c)	(d)	(d)	275	525	53 to 56
1066	7(c)	0.281(c)	(d)	(d)	260	500	53 to 56
1084	6(c)	0.218(c)	(d)	(d)	200	395	55 to 58
1086	13(c)	0.516(c)	(d)	(d)	215	420	55 to 58
1090	5(c)	0.187(c)	(d)	(d)	–	–	57 to 60

(Contd...)

Steel	Section size mm	inch	Salt temperature °C	°F	M_s temperature(a) °C	°F	Hardness, HRC
1090 (e)	20(c)	0.820(c)	315(f)	600(f)	–	–	44.5 (avg)
1095	4(c)	0.148(c)	(d)	(d)	210(g)	410(g)	57 to 60
1350	16(c)	0.625(c)	(d)	(d)	235	450	53 to 56
4063	16(c)	0.625(c)	(d)	(d)	245	475	53 to 56
4150	13(c)	0.500(c)	(d)	(d)	285	545	52 max
4365	25(c)	1.000(c)	(d)	(d)	210	410	54 max
5140	3(b)	0.125(b)	345	655	330	630	43 to 48
5160 (e)	26(c)	1.035(c)	315(f)	600(f)	255	490	46.7 (avg)
8750	3(b)	0.125(b)	315	600	285	545	47 to 48
50100	8(c)	0.312(c)	(d)	(d)	–	–	57 to 60

(a) Calculated; (b) sheet thickness; (c) diameter of section; (d) salt temperature adjusted to give maximum hardness and 100 per cent bainite; (e) modified austempering; microstructure contained pearlite as well as bainite; (f) salt with water addition; and (g) experimental value.

The effects of section thickness on the hardness of austempered 1090 and 5160 steels are shown in Fig. 12.3. Hardness remains reasonably consistent to the center of a 17-mm (0.680-inch) diameter of 1090, but it becomes erratic when the diameter is increased to 21 mm (0.820 inch). A similar difference is evident for 5160 steel when the diameter is increased from 24.6 to 26 mm (0.967 to 1.035 inch).

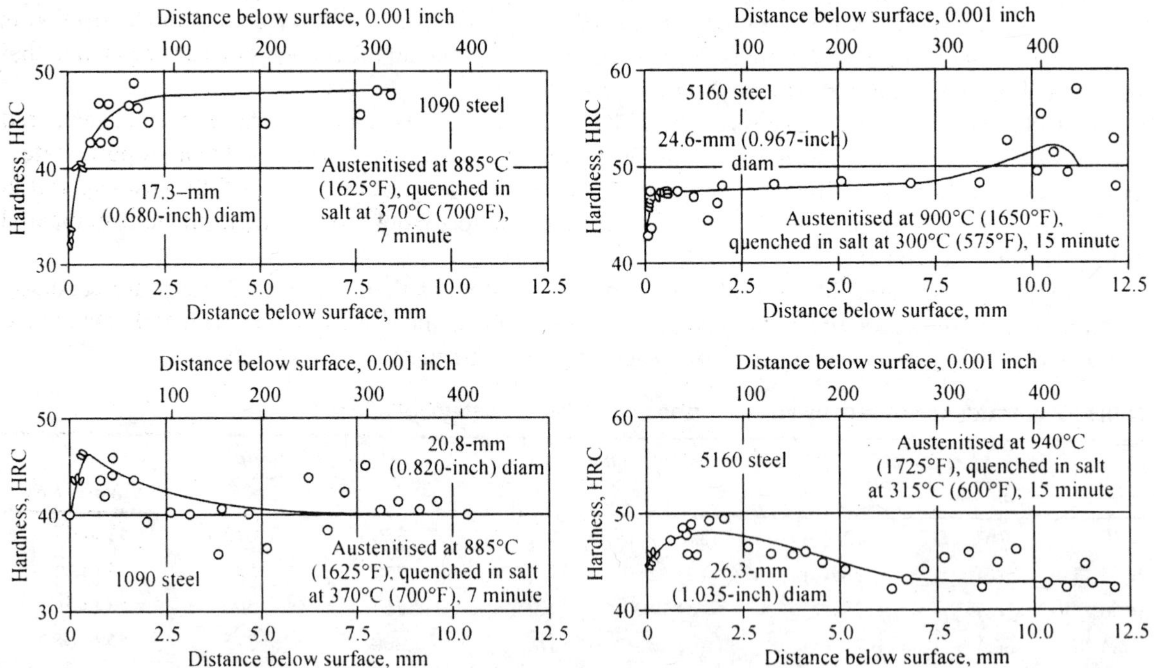

The 5160 steel was quenched in agitated salt containing some water. HRC values were converted from microhardness readings taken with a 100-gram load. Low values of surface hardness result from decarburisation. High hardness at center of 24.6-mm (0.967-inch) 5160 is due to segregation.

Fig. 12.3. Effect of section thickness on the hardness of austempered carbon and alloy steels.

The lower surface hardnesses indicated for 1090 and 5160 were the result of surface decarburisation. The high core hardness of the 24.6 mm (0.967 inch) diameter of 5160 is attributed to chemical segregation in the center portion of the bar.

APPLICATIONS

Austempering usually is substituted for conventional quenching and tempering for either or both of two reasons: (i) to obtain improved mechanical properties (particularly higher ductility or notch toughness at a given high hardness); and (ii) to decrease the likelihood of cracking and distortion. In some applications austempering is less expensive than conventional quenching and tempering. This is most likely when small parts are treated in an automated setup wherein conventional quenching and tempering comprise a three-step operation—that is, austenitising, quenching and tempering. Austempering requires only two processing steps, austenitising and isothermal transformation in an austempering bath.

The range of austempering applications generally encompasses parts fabricated from bars of small diameter or from sheet or strip of small cross section. Austempering is particularly applicable to thin-section carbon steel parts requiring exceptional toughness at a hardness near 50 HRC.

In austempered carbon steel parts, reduction in area is usually much higher than in conventionally quenched and tempered parts, as indicated in the following tabulation for 5-mm (0.180 inch) diam bars of 0.85 per cent C plain carbon steel:

Austempered mechanical properties	
Tensile strength	1780 MPa (258 ksi)
Yield strength	1450 MPa (210 ksi)
Reduction in area	45%
Hardness	50 HRC
Quenched and tempered mechanical properties	
Tensile strength	1795 MPa (260 ksi)
Yield strength	1550 MPa (225 ksi)
Reduction in area	28%
Hardness	50 HRC

The mechanical properties of sway bars made of 1090 steel and hardened by these two processes are listed in Table 12.4 (see also Table 12.1).

Table 12.4. Comparison of typical mechanical properties of austempered and of oil quenched and tempered sway bars of 1090 steel.

Property(a)	*Austempered at 400°C (750°F)(b)*	*Quenched and tempered(c)*
Tensile strength, MPa (ksi)	1415 (205)	1340 (200)
Yield strength, MPa (ksi)	1020 (148)	895 (130)
Elongation, %	11.5	6.0
Reduction of area, %	30	10.2
Hardness, HB	415	388
Fatigue cycles (d)	105000(e)	58600(f)

(a) Average value; (b) six tests; (c) two tests; (d) fatigue specimens 21 mm (0.812 inch) in diameter; (e) seven tests; range, 69050 to 137000; and (f) eight test; range, 43120 to 95220.

It is more important that austempered parts possess desired mechanical properties than that they have a 100 per cent bainitic structure. In Table 12.3 it is evident from the hardness values indicated that several of the austempered steels have mixed structures. Higher-than-normal hardness indicates that some martensite has formed, and below-normal hardness indicates the presence of some pearlite. The formation of pearlite is more common and results from a quenching speed too slow for complete avoidance of the nose of the TTT curve.

In industrial austempering practice, a sizable percentage of applications are successful with less than 100 per cent bainite. In fact, 85 per cent bainite has been found to be satisfactory for some applications. Obviously, austempering is often 'modified to some degree in commercial application, and whether or not the metallurgical properties obtained conform to those obtained in true austempering is at least partially ignored if the treated parts meet service requirements. However, variations in harden-ability from heat to heat may give rise to erratic results from variations in the amount of pearlite when borderline conditions are involved in modified austempering.

Table 12.5 presents processing data for a number of specific parts made of various plain carbon, alloy and carburising steels, these data are representative of austempering practice in at least a dozen different manufacturing plants.

Table 12.5. Typical production applications of austempering.

Parts listed in order of increasing section thickness.

		Maximum section thickness		*Parts per unit weight*		*Salt temperature*		*Immersion*	*Hardness,*
Part	*Steel*	*mm*	*inch*	*kg*	*lb*	*°C*	*°F*	*time, min*	*HRC*
Plain carbon steel parts									
Clevis	1050	0.76	0.030	770/kg	350/lb	360	680	15	42
Follower arm	1050	0.76	0.030	412/kg	187/lb	355	675	15	42
Spring	1080	0.79	0.031	220/kg	100/lb	330	625	15	48
Plate	1060	0.81	0.032	88/kg	40/lb	330	630	6	45 to 50
Cam lever	1065	1.02	0.040	62/kg	28/lb	370	700	15	42
Plate	1050	1.02	0.040	0.1 kg	¼ lb	360	675	15	42
Type bar	1065	1.02	0.040	141/kg	64/lb	370	700	15	42
Tabulator stop	1065	1.22	0.048	440/kg	200/lb	360	680	15	45
Lever	1050	1.27	0.050	–	–	345	650	15	45 to 50
Chain link	1050	1.52	0.060	573/kg	260/lb	345	650	15	45
Shoe-last link	1065	1.52	0.060	86/kg	39/lb	290	550	30	52
Shoe-toe cap	1070	1.52	0.060	18/kg	8/lb	315	600	60	50
Lawn mower blade	1065	3.18	0.125	0.3 kg	⅔ lb	315	600	15	50
Lever	1075	3.18	0.125	24/kg	11/lb	385	725	5	30 to 35
Fastener	1060	6.35	0.250	110/kg	50/lb	310	590	25	50
Stabiliser bar	1090	19.05	0.750	4.5 kg	10 lb	370	700	6 to 9	40 to 45
Boron steel bolt	10B20	6.35	0.250	100/kg	45/lb	420	790	5	38 to 43

(Contd...)

Part	Steel	Maximum section thickness mm	inch	Parts per unit weight kg	lb	Salt temperature °C	°F	Immersion time, min	Hardness, HRC
Alloy steel parts									
Socket wrench	6150	–	–	0.06 kg	⅛ lb	365	690	15	45
Chain link	Cr-Ni-V(a)	1.60	0.063	110/kg	50/lb	290	550	25	53
Pin	3140	1.60	0.063	5500/kg	2500/lb	325	620	45	48
Cylinder liner	4140	2.54	0.100	3 kg	7 lb	260	500	14	40
Anvil	8640	3.18	0.125	0.3 kg	3/4 lb	370	700	30	37
Shovel blade	4068	3.18	0.125	–	–	370	700	15	45
Pin	3140	6.35	0.250	100/kg	45/lb	370	700	45	40
Shaft	4140(b)	9.53	0.375	0.1 kg	¼ lb	385	725	15	35 to 40
Gear	6150	12.70	0.500	0.9 kg	2 lb	305	580	30	45
Carburised steel parts									
Lever	1010	3.96	0.156	6.8 kg	15 lb	385	725	5	30 to 35(c)
Shaft	1117	6.35	0.250	66/kg	30/lb	385	725	5	30 to 35(c)
Block	8620	11.13	0.438	132/kg	60/lb	290 to 315	550 to 600	30	50(c)

(a) Contains 0.65 to 0.75% C; and (b) leaded grade; and (c) case hardness.

EQUIPMENT AND PROCESSING PROCEDURES

Austenitising prior to austempering can be performed in molten salt baths, in one of several types of standard heat treating furnaces (including box, retort and shaker-hearth furnaces), or in special furnaces (such as direct-fired tunnel types). In the types of austenitising furnaces that employ a gaseous atmosphere, atmosphere control is extremely important, because the formation of scale on parts will inhibit the rapid heat exchange that is essential in a successful austempering operation.

Molten Salt Baths for Austenitising

Molten salt baths for austenitising are widely used for batch-type operation: they can also be mechanised for continuous operation. Heating in salt for batch-type operation is advantageous because the film of adhering salt prevents scaling during transfer from the austenitising operation to the austempering bath. Salt bath furnaces heated by either immersed or submerged electrodes are most common, although externally heated pots also are used.

Because most austempering is performed in molten salt, the salt bath used for austenitising must be compatible with the austempering salt. Chloride salts are recommended for austenitising: the composition and characteristics of a typical chloride salt bath for austenitising are as follows:

Sodium chloride	45 to 55 per cent
Potassium chloride	45 to 55 per cent
Melting temperature	650 to 675°C (1200 to 1250°F)
Working temperature range	705 to 900°C (1300 to 1650°F).

Austenitising salt baths should not contain cyanide, because it reacts violently with quenching salts. Carbonate compositions and mixtures containing barium salts are not recommended, because they are difficult to separate from quenching salts.

Shaker-Hearth Furnaces

Sheker- Hearth with atmosphere control are widely used for austenitising, especially where large quantities of small parts are being processed. In a typical automated installation, the workpieces enter the shaker from a feeder and, after being heated to the austenitising temperature, are dropped through a chute into the austempering bath, thus avoiding exposure in air and to possibility of scaling. The parts are transported through the austempering bath on variable-speed mesh or plate belt-type conveyors at a speed that allows complete transformation before they are automatically transferred to a constant-speed conveyorised washing and drying setup.

Feeders for shaker-hearth furnaces vary widely in design, depending on the size, shape and fragility of workpieces. In one type of equipment, parts are fed by gravity from the hopper onto a reciprocating hearth, on which they are transported through the shaker-hearth furnace and then into the austempering furnace. This mechanism is used for small round or flat parts that will not be distorted by the vibrating action. Other methods of feeding to the moving hearth include 'Walmil'-style ram volumetric feeders, weigh loaders, and special magnetic conveyor-and-bin combinations which disentangle parts that bundle.

Rotary Retort Furnaces

Rotary retort furnaces are widely used for products in the saw-chain, concrete, explosive drive and air driven nail industries and in similar applications where both hardness and ductility are prime factors in the end product and reasonably large quantities of similar small parts are being processed without the threat of damage by the tumbling action of the retort. One of the advantages of the rotary retort, which advances the work by means of the internal spiral, is the absolute control of the retention time within the heated chamber. This control ensures uniformity of metallurgical properties of the product. Excellent atmosphere control is also attained, because the work progresses through a sealed retort, is sealed at the quench chute by the quench media, and the process has a restricted part entry area. Consistent production rates and uniformity of results are obtained by controlling the loading rate with a weigh loader that is fed from a storage hopper by discharging the workpieces to a loading elevator that in turn transports the load through a chute into the partially sealed entry of the retort.

Belt Furnaces

Belt furnaces may use either a cast-link or mesh-type part-carrier belt. In either instance, the belt is totally contained within the heated chamber, with the work being fed into the chamber by means of either a shaker or vibrating hearth onto the moving surface of the belt.

The mesh-belt type of furnace is primarily used to process relatively thin sheet-metal stampings that would become distorted if loaded directly onto a heavy cast-link belt or would become tangled or distorted if processed through a rotary retort. Continuous automated parts feeding is accomplished by use of specialised magnetic conveyor feeders. These feeders strip the parts that tend to tangle from one edge of the pile in the feeding bin and carry them up to a vibratory trough. The trough in turn feeds the workpieces to the hearth that enters the furnace. The cast-link belt types are used to handle heavier parts that would not be subject to distortion upon loading onto the hot cast-link belt. Continuous feeding is accomplished with Walmil volumetric feeders, automated orienters, or simple bin and vibratory trough types, depending upon part configuration. Each feeding system delivers parts to the hearth, which enters the furnace.

Belt-type furnaces ensure controlled retention time for repeatability of metallurgical properties of the product. The integrity of the atmosphere is controlled by sealing the furnace. A flexible seal about the shaker or vibrator hearth feeding into the chamber restricts atmosphere exhaust and the opening for parts entry. The discharge chute into the molten salt also is sealed.

Conveyorised transport through the austempering salt bath and wash tanks as described for shaker hearth furnaces is also employed with both the rotary retort and belt furnace systems.

Direct-Fired Tunnel-Type Furnaces

Direct-fired tunnel-type furnaces are employed for heat treating large quantities of similar parts that must be hung from hooks or fixtures to minimise distortion. In a typical installation, the hooks or fixtures carrying the work travel in front of the operators and can be easily loaded. The work is conveyed through the tunnel-type heating zone, automatically removed after heating and dropped into the austempering bath, and then conveyed to washing and drying. During austenitising, provided by the controlled gas-to-air ratio of the burners. The entire operation is automatic except for the hand loading of parts onto hooks.

Austempering Furnaces

Simple salt baths, either fuel-fired or electrically heated, are suitable for low-production austempering. Circulation can be provided by a propeller-type mixer. Most of these furnaces have no provision for cooling and depend on heat losses to equal the heat input acquired from workpieces.

Sufficient production quantities warrant the use of furnaces specifically designed for austempering. These furnaces usually provide for automatic handling. The salt bath units are heated with immersion heaters, immersed electrodes, or gas-fired immersion tubes. To maintain a constant temperature, heat is dissipated by forced air that cools the pot sidewall or by immersed plate coils. Both heating and cooling are automatically controlled. A salt-separating system is required to separate chloride salts dragged over from the austenitising furnace. Chlorides, scale and other contaminants are removed by:

1. Gravity separation: providing a separate chamber for the suspensions to settle; settled solids are then removed by manual desludging or sludge pans.
2. Continous filtration: suspended contaminants are continuously passed through filter baskets; the baskets are periodically removed and dumped.

To provide uniform quenching, one or more propeller agitators are used to agitate the salt.

Racking and Handling

Racking and handling procedures for austempering are not necessarily different from those used for other heat treating operations. Racking or other means of keeping parts separated is often essential to the process. To minimise distortion, delicate parts must be handled with particular care.

For heating in tunnel-type or other special furnaces, a standard form of racking is usually employed. (Provision must also be made for lowering of parts into the austempering bath). The well-known Christmas tree-type of rack is used for parts that can be hung. Some parts are strung on wires that can be suspended into the austenitising and quenching media. Heavier parts, or parts that cannot be hung, may be placed in baskets provided that rapid and uniform quenching can be obtained. Parts must not be stacked together, because this will result in non-uniform quenching.

Racks, baskets and other fixtures may be made of carbon steel, but if they are expected to last for many cycles it is usually more economical to make them of a heat-resisting alloy such as 35 Ni 18 Cr or another nickel-base alloy.

Washing and Drying

Salts used in the austempering bath are water soluble, and no special equipment is needed for their removal. Normal practice is to use clean, hot water, with agitation, for dissolving and removing salt from austempered parts. Cold water may be used, but its dissolving action is much slower. Hot water may not suffice to dissolve all salt from parts with deep blind holes, small crevices, laps, hidden corners, or ledges. If it is not completely removed, salt may bleed from such crevices or joints long after heat treating. Quenching into hot water directly from the austempering salt bath is often utilised to blow out trapped salt. Large parts may be cleaned with high pressure steam. However, rapid cooling may cause distortion.

CONTROL OF PROCESS VARIABLES

The achievement of desired results from austempering processes depends on control of the bath temperature and the immersion time of the work being processed, and on agitation of the bath.

Temperature Control of Austempering Baths

Temperature control of austempering baths is important because:

1. Bath temperature determines the hardness and other properties obtained in the part being heat treated.
2. Unintended freezing of a salt bath can result in unwarranted expense.
3. At temperatures exceeding 455°C (850°F), nitrate-nitrite salt mixtures will cause pitting attack of steel pots and, more importantly, steel work-pieces.
4. Local temperatures above 595°C (1100°F) anywhere in a nitrate-nitrite salt bath can cause violent exothermic reactions in the salt that may result in fire or explosion even, after the source of furnace heat has been shut off. Equipment manufacturers should be consulted as to their recommendations regarding safe average bath temperatures.

The potential hazard of nitrate-nitrite salt mixtures overheating as the result of failure of a thermocouple or control instrument must be recognised, and appropriate safeguards must be provided. A second instrument and thermocouple are recommended. These may be set at 28°C (50°F) higher than the desired operating temperature and arranged to take over in the event of malfunction.

Usually, a variation in bath temperature of ±6°C (±10°F) is permissible. However, variations greater than ±6°C have been known to cause unacceptable variations in the hardness of austempered parts.

Time in the Bath

Time in the bath should be sufficient to allow complete transformation. Allowing parts to remain in the bath for longer than the required time will increase the cost of the operation, but it is not harmful to mechanical properties (Table 12.6). In one instance, parts that had fallen from racks were unintentionally held at 265 °C (510°F) for several months. Subsequent tests revealed that their properties were not harmed by the prolonged immersion.

Bath Agitation

Bath agitation can be a significant variable in austempering, because it affects quenching speed. Mechanical stirring, pumping and air agitation have all proved successful for specific operations. Pumping provides the greatest amount of salt movement, although it requires a more expensive installation. In many applications, the other methods have provided a sufficiently rapid quench.

Dimensional Control

Parts can usually be produced with less dimensional change by austempering than by conventional quenching and tempering. Austempering may be the best way to hold close tolerances without extensive straightening or machining after heat treatment.

The parts had been austenitised in a shaker-hearth furnace at 830°C (1525°F) and quenched in an agitated salt bath at 290°C (550°F). Examination disclosed excessive martensite and un-dissolved carbides in the microstructure. Over-spheroidisation prior to hardening was determined to be the cause of brittleness. This resulted in the formation of abnormally large carbides that were not dissolved during the 11 minute furnace cycle, thus lowering the carbon content of the matrix and raising the M_s temperature. This caused partial transformation to martensite instead of 100 per cent lower bainite. The problem was solved by prehardening and tempering prior to the regular heat treating cycle.

Table 12.6. Effect of austempering time on hardness of three steels.

Steel	*Austempering temperature* °C	°F	*Rockwell hardness(a)*						
			Austempering time, min						
			30	60	90	120	240	300	360
1095 (b)	230(c)	450(c)	91	90	90	90	90	90	90
	265(d)	510(d)	90	89	89	89	89	89	89
			Austempering time, min						
			1	2	5	10	20	40	80
8735 (e)	260(f)	500(f)	51	51	49	49	48	48	47
	315(f)	600(f)	49	45	46	46	46	46	46
	370(f)	700(f)	40	39	38	38	38	38	37
8750 (g)	260(f)	500(f)	58	56	53	51	52	52	51
	315(f)	600(f)	58	52	48	47	47	47	47
	370(f)	700(f)	54	42	39	39	39	38	39

(a) Rockwell 15-N hardness values for the 1095 steel; Rockwell C hardness values for the 8735 and 8750 steels; (b) steel contained 0.90% C; specimen thickness, 0.25 mm (0.010 inch). Each hardness value is an average of 12 specimens; range of test values did not exceed one point on Rockwell 15-N scale; (c) time for 100% transformation was 170 min; (d) time for 100% transformation was 85 min; (e) steel contained 0.37% C; specimen size, 16 by 32 by 2 min (0.622 by 1.250 by 0.087 inch); (f) time for complete transformation was 5 to 10 min; and (g) steel contained 0.49% C; specimen size, 25 by 25 by 3 mm (1 by 1 by 1/8 inch).

MODIFIED AUSTEMPERING

Modifications of austempering practice give rise to mixed structures of pearlite and bainite and quite common in industrial practice. The amounts of pearlite and bainite may vary widely in different modifications of processing.

Patenting, a treatment used in the wire industry, is a significant and useful form of modified austempering, in which austenitised wire or rod is continuously quenched into a bath maintained at 510 to 540°C (950 to 1000°F) and held in the bath for periods ranging from 10 second (for small wire) to 90 seconds (for rod). Patenting provides a combination of moderately high strength and high ductility.

As indicated in Fig. 12.4 by the line designated 'Modified practice', the process varies from true austempering in that the quenching rate, instead of being rapid enough to avoid the nose of the TTT curve, is sufficiently slow to intersect the nose, which results in the formation of fine pearlite.

Similar practice is usefully employed in applications involving plain carbon steels when a hardness between about 30 to 42 HRC is desirable or acceptable. The hardness of plain carbon steel quenched at a rate that intersects the nose of the TTT curve will vary with carbon content Fig. 12.5.

Modified practices can be applied to parts having sections thicker than are normally considered practicable for austempering. These practices, however, also are subject to certain metallurgical limitations, such as the following which apply to the modified austempering of 1080 steel:

1. The part to be quenched must be of sufficient mass or weight that cooling to the quench bath temperature cannot be accomplished in the time (about ¾ second) allowed for avoiding the nose of the TTT curve: otherwise, the part will undergo true austempering and may be harder than desired. The temperature of the quenching medium should not be raised above 370°C (700°F) in an effort to retard the cooling rate, or temper brittleness may result. For small parts, isothermal annealing at 565°C (1050°F) is preferred to austempering for obtaining hardness in the range of 30 to 42 HRC.
2. The part to be quenched is subject to a maximum, as well as a minimum, weight limit. If more than about 20 second is required for the center of the part to transform, or if there is arise in bath temperature, some upper bainite may form, resulting in a variable hardness and mixed structure.
3. The lower temperature limit of the quench bath depends on the weight of the part. For pieces weighing 1 to 2 kg (2 to 4 lb), the lower limit of bath temperature is about 330°C (625°F).

From the above it is evident that modified practices are limited by several critical factors and that some trial and error is necessary in developing an optimum cycle for parts of specific composition and section thickness.

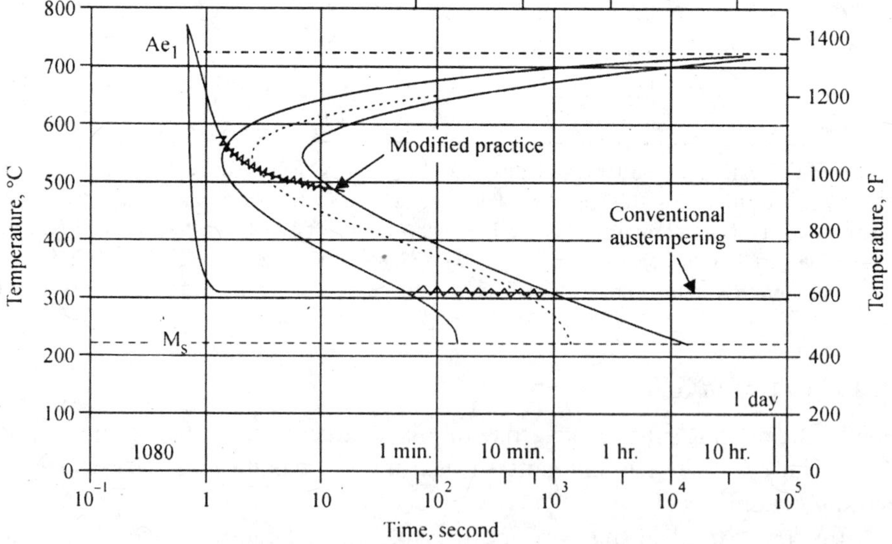

Fig. 12.4. Time-temperature transformation diagram for 1080 steel, showing difference between conventional and modified austempering.

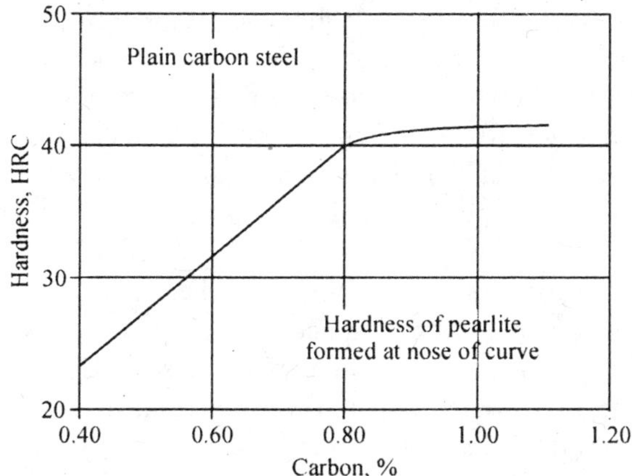

Fig. 12.5. Effect of carbon content in plain carbon steel on the hardness of fine pearlite formed when the quenching curve intersects the nose of the time-temperature diagram for isothermal transformation.

Another form of modified austempering also entails the use of a special technique but produces results that are similar to those obtained in true austempering. This modification is employed for parts that, because of their size and the type of steel of which they are made, are difficult to quench rapidly enough to avoid the nose of the TTT curve before transformation begins. In such applications, the cooling rate can be increased by first quenching the part into a bath that is slightly above the M_s temperature. For 1080 steel, this is about 260°C (500 °F), as shown in Fig. 12.4. The part is held at this temperature for only the brief time necessary to obtain temperature equalisation throughout the section and is then transferred to the austempering bath at a higher temperature and allowed to transform isothermally in the normal manner.

WASHING AND DRYING

Plain water, preferably warm or hot, is most suitable for cleaning austempering salts from parts. A single overflowing tank can be used. In automated systems, it is common practice to use two conveyorised tanks, the first heated by incoming parts and the second separately heated. Fresh water is added to the second tank, with the overflow going to the first tank. In some cases, the concentrated salt solution is pumped back to the salt tank with available salt-recovery systems to recover both the salt and the water. This recovery reduces salt and water consumption and alleviates the disposal problem. The parts should be dried by warm forced air or treated with a rust preventive immediately after washing.

SAFETY PRECAUTIONS

In operating an austempering installation, the same precautions should be taken that are appropriate for operation of any other hot liquid bath. These include the wearing of gloves and face masks. Several other special precautions are unique to operation of nitrate nitrite baths:

1. Local temperatures above 595°C (1100°F) anywhere in a nitrate-nitrite salt bath can cause violent exothermic reactions in the salt that may result in fire or explosion even after the source of furnace heat has been shut off. (Overheating might be caused by instrument failure or by overloading of the bath with hot workpieces.) Equipment manufactures should be consulted as to their recommendations regarding safe average bath temperatures.

2. Nitrate-nitrite mixtures are oxidising and should not be intermixed with materials that are readily oxidised. Cyanide and cyanide-bearing salts are especially incompatible with nitrate-nitrite salt mixtures. Austempering should never be done directly from an austenitising bath containing cyanide. When it is desirable to austemper cyanide or liquid carburised parts, they should first be immersed in a bath of neutral salt. This bath should be operated at the desired austenitising temperature and controlled to prevent the accumulation of more than 2 per cent sodium cyanide or potassium cyanide.
3. When austempering is carried out in the same area where cyanide-containing salts also are used, the salt pots and salt containers should be colour coded. Yellow, orange or red are commonly used for identification of nitrite and nitrite-containing salts. Cyanide-containing materials may be identified by white, blue or gray. Anyone handling these materials should be made aware that explosions can occur if they are intermixed.
4. Finely divided carbonaceous materials should not be used. For example, it is common practice to use graphite, coal or powdered coke as an insulating cover on lead baths. Under no circumstance should these be employed as coverings for nitrate-nitrite salt baths. Accidental contamination of austempering baths with accumulations of carbon black, which may build up at the discharge end of gas carburising or carbonitriding furnaces, must also be avoided. Where work is to be austempered directly from such furnaces, an opening for frequent cleanout should be provided.

ROLLING MILL

A rolling mill is a machine or factory for shaping metal by passing it between rollers. Rolling mills are often incorporated into integrated steelworks, but also exist as separate plants and can be used for other metals and other materials. Rolling mills historically have been of several kinds:

1. A slitting mill was used to cut flat bar iron into rods for nail-making.
2. A tinplate works normally contained at least two rolling mills, one for hot rolling and the other for cold rolling single plates, prior to tinning.
3. From the Industrial Revolution, puddled iron, after consolidation with a powered hammer (shingling), was rolled into bar iron using a rolling mill with grooved rolls. The grooves provided were progressively smaller, so that on successive passes through the rolls, the cross-section of the bar became smaller and the bar itself longer. By designing the rolls appropriately, it is possible to obtain iron or steel with various cross-sections, including I-shaped girders and railway rails. Fig. 12.6 showing a profile of rolling mill showin manufacture of plaes and wires.

ROLLING (METALWORKING)

Rolling is a fabricating process in which the metal, plastic, paper, glass, etc. is passed through a pair of rolls. There are two types of rolling process, flat and profile rolling. In flat rolling the final shape of the product is either classed as sheet (typically thickness less than 3 mm, also called 'strip') or plate (typically thickness more than 3 mm). In profile rolling, the final product may be a round rod or other shaped bar such as a structural section (beam, channel, joist etc). A profile of rolling mill showing manufacture of cone (Fig. 12.7). Rolling is also classified according to the temperature of the metal rolled. If the temperature of the metal is above its recrystallisation temperature then the process is termed as hot rolling. If the temperature of metal is below its recrystallisation temperature the process is termed as cold rolling.

Fig. 12.6. Plates and wires from rooling mill.

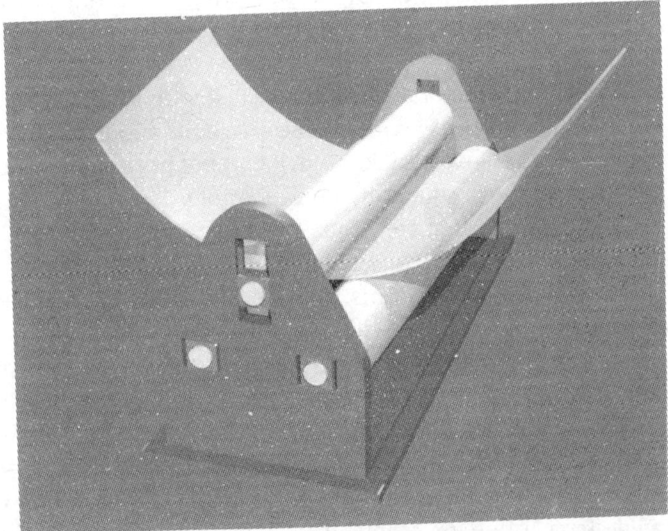

Fig. 12.7. Profile of rolling mill showing manufacture of a cone.

Other processes also termed as 'hot bending' are induction bending, whereby the section is heated in small sections and dragged into a required radius. Heavy plate tends to be formed using a press process and is termed forming, rather than rolling.

STEEL MILL

A steel mill is an industrial plant for the manufacture of steel. Steel is an alloy of iron and carbon. It is produced in a two-stage process. First, iron ore is reduced or smelted with coke and limestone in a blast furnace, producing molten iron which is either cast into pig iron or carried to the next stage as molten iron. In the second stage, known as steelmaking, impurities such as sulphur, phosphorus and excess carbon are removed and alloying elements such as manganese, nickel, chromium and vanadium added to produce the exact steel required. Steel mills turn molten steel into blooms, ingots, slabs and sheet through casting, hot rolling and cold rolling.

Integrated Plants

An integrated steel plant has all the functions for primary steel production:

1. Ironmaking (conversion of ore to liquid iron).
2. Steelmaking (conversion of pig iron to steel).
3. Bloom casting (production of large blocks of steel).
4. Roughing rolling/billet rolling (reducing size of blocks).
5. Product rolling (finished shapes).

The principal raw materials for an integrated plant are iron ore, limestone and coal (or coke). These materials are charged in batches into a blast furnace where the iron compounds in the ore give up excess oxygen and become liquid iron. At intervals of a few hours, the accumulated liquid iron is tapped from the blast furnace and either cast into pig iron or directed to other vessels for further steelmaking operations. Historically the Bessemer process was a major advancement in the production of economical steel, but it has now been entirely replaced by other processes such as the basic oxygen furnace.

Molten steel is cast into large blocks called 'blooms'. During the casting process various methods are used, such as addition of aluminum, so that impurities in the steel float to the surface where they can be cut off the finished bloom.

Because of the energy cost and structural stress associated with heating and cooling a blast furnace, typically these primary steelmaking vessels will operate on a continuous production campaign of several years duration. Even during periods of low steel demand, it may not be feasible to let the blast furnace grow cold, though some adjustment of the production rate is possible.

Integrated plants are large facilities that are typically only economical to build in 20,00,000 tonne per year annual capacity and up. Final products made by an integrated plant are usually large structural sections, heavy plate, strip, wire rod, railway rails, and occasionally long products such as bars and pipe.

A major environmental hazard associated with integrated steel mills is the pollution produced in the manufacture of coke, which is an essential intermediate product in the reduction of iron ore in a blast furnace.

Integrated plants may also adopt some of the processes used in mini-mills, such as arc furnaces and direct casting, to reduce production costs.

World integrated steel production capacity is at or close to world demand, so competition between suppliers results in only the most efficient producers remaining viable. However, due to the large employment of integrated plants, often governments will financially assist an obsolescent facility rather than take the risk of having thousands of workers thrown out of jobs. Such measures result in products then sold in international trade lead to allegations of dumping.

Mini Mills

A mini-mill is traditionally a secondary steel producer; however, Nucor (one of the world's largest steel producers), use mini-mills exclusively. Usually it obtains most of its iron from scrap steel, recycled from used automobiles and equipment or by-products of manufacturing. Direct reduced iron (DRI) is sometimes used with scrap, to help maintain desired chemistry of the steel, though usually DRI is too expensive to use as the primary raw steelmaking material. A typical mini-mill will have an electric arc furnace for scrap melting, a ladle furnace or vacuum furnace for precision control of chemistry, a strip or billet continuous caster for converting molten steel to solid form, a reheat furnace and a rolling mill.

Originally the mini-mill concept was adapted to production of bar products only, such as concrete reinforcing bar, flats, angles, channels, pipe, and light rails. Since the late 1980s, successful introduction of the direct strip casting process has made mini-mill production of strip feasible. Often a mini-mill will be constructed in an area with no other steel production, to take advantage of local resources and lower-cost labour. Mini-mill plants may specialise, for example, making coils of rod for wire-drawing use, or pipe, or in special sections for transportation and agriculture.

Since the electric arc furnace can be easily started and stopped on a regular basis, mini-mills can follow the market demand for their products easily, operating on 24 hours schedules when demand is high and cutting back production when sales are lower.

HOT ROLLING

The metallurgical process of Hot rolling, used mainly to produce sheet metal or simple cross sections from billets describes the method of when industrial metal is passed or deformed between a set of work rolls and the temperature of the metal is generally above its recrystallisation temperature, as opposed to cold rolling, which takes place below this temperature. Hot rolling permits large deformations of the metal to be achieved with a low number of rolling cycles.

Because the metal is worked before crystal structures have formed, this process does not itself affect its microstructural properties. Hot rolling is primarily concerned with manipulating material shape and geometry rather than mechanical properties. This is achieved by heating a component or material to its upper critical temperature and then applying controlled load which forms the material to a desired specification or size.

Mechanical properties of the material in its final 'as-rolled' form are a function of:

1. Material chemistry.
2. Reheat temperature.
3. Rate of temperature decrease during deformation.
4. Rate of deformation.
5. Heat of deformation.
6. Total reduction.
7. Recovery time.
8. Recrystallisation time.
9. Sub-sequent rate of cooling after deformation.

COLD ROLLING

Cold rolling is a metal working process in which metal is deformed by passing it through rollers at a temperature below its recrystallisation temperature. Cold rolling increases the yield strength and hardness of a metal by introducing defects into the metal's crystal structure. These defects prevent further slip and can reduce the grain size of the metal, resulting in Hall-Petch hardening. Cold rolling is most often used to decrease the thickness of plate and sheet metal. Fig. 12.8 Shows the cold rolling metal sheet like piece of brass.

Physical Metallurgy of Cold Rolling

Cold rolling is a method of cold working a metal. When a metal is cold worked, microscopic defects are nucleated through out the deformed area. These defects can be either point defects (a vacancy on the crystal lattice) or a line defect (an extra half plane of atoms jammed in a crystal). As defects accumulate through deformation, it becomes increasingly more difficult for slip or the movement of defects, to occur. This results in a hardening of the metal.

Fig. 12.8. Cold roll metal sheet like piece of brass.

If enough line defects accumulate, a grain may split into two or more grains in order to minimise the strain energy of the system. When large grains split into smaller grains, the alloy hardens as a result of the Hall-Petch relationship. If cold work is continued, the hardened metal may fracture.

During cold rolling, metal absorbs a great deal of energy, some of this energy is used to nucleate and move defects (and subsequently deform the metal). The remainder of the energy is released as heat.

A metal that has been hardened by cold rolling can be softened by annealing. Annealing will relieve stresses, allow grain growth, and restore the original properties of the alloy. After anealing, the metal may be further cold rolled with out fracturing.

Degree of Cold Work

Cold rolled metal is given a rating based on the degree it was cold worked. 'Skin-rolled' metal undergoes the least rolling, being compressed only 0.5–1 per cent to harden the surface of the metal and make it more easily workable for later processes. Higher ratings are 'quarter hard', 'half hard' and 'full hard'; in the last of these, the thickness of the metal is reduced by 50 per cent. The ridges and top edge of this can were formed by cold rolling.

Cold Rolling as a Manufacturing Process

Cold rolling is a common manufacturing process. It is often used to form sheet metal. Beverage cans are closed by rolling, and steel food cans are strengt`hened by rolling ribs into their sides. Rolling mills are commonly used to precisely reduce the thickness of strip and sheet metals.

COLD TREATING OF STEEL

Cold treating of steel is a supplemental treatment that may be used for such purposes as enhancing transformation of austenite to martensite and improving stress relief of castings and machined parts. Generally, 1 hour of cold treatment of adequate for each inch of cross section.

Hardening and Retained Austenite

Whenever hardening is to be done during heat treating, complete transformation from austenite to martensite generally is desired prior to tempering. From a practical standpoint, however, conditions vary widely and 100 per cent transformation rarely, if ever, occurs. Cold treating may be useful, in many instances, for improving the percentage of transformation and thus enhancing properties.

During hardening, martensite develops as a continuous process from start (M_s) to finish (M_f) through the martensite-formation range. Except in a few highly alloyed steels, martensite starts to form at well above room temperature. In many instances, transformation is essentially complete at room temperature. Retained austenite tends to be present in varying amounts, however and when considered excessive for a particular application, must be transformed to martensite and then tempered.

Cold treating vs tempering

Immediate cold treating without delays at room temperature or at other temperatures during quenching offers the best opportunity for maximum transformation to martensite. In some instances, however, there is a risk that this will cause cracking of parts. Therefore, it is important to ensure that the grade of steel and the product design will tolerate immediate cold treating rather than immediate tempering. Some steels must be transferred to a tempering furnace when still warm to the touch, to minimise the likelihood of cracking. Design features such as sharp corners and abrupt changes in section create stress concentrations and promote cracking. In most instances, cold treating is not done before tempering.

In several types of industrial applications, tempering is followed by deep freezing and retempering without delay. For example, such parts as gauges machineways, arbors, mandrils, cylinders, pistons and ball and roller bearings are treated in this manner for dimensional stability. Several freeze-draw cycles are used for critical applications. Cold treating is also used to improve wear resistance in such materials as tool steels, high-carbon martensitic stainless steels and carburised alloy steels for applications where the presence of retained austenite may result in excessive wear. Transformation in service may cause cracking and/or dimensional changes that can promote failure. In some instances, more than 50 per cent retained austenite has been observed. In such cases, no delay in tempering after cold treatment is permitted or cracking can develop readily.

Process limitations

In some applications in which explicit amounts of retained austenite are considered beneficial, cold treating might be detrimental. Moreover, multiple tempering, rather than alternate freeze-draw cycling, is generally more practical for transforming retained austenite in high speed and high-carbon/high-chromium steels.

Hardness testing

Lower than expected HRC readings may indicate excessive retained austenite. Significant increases in these readings as a result of cold treatment indicate conversion of austenite to martensite. Superficial hardness readings, such as HR15N, may show even more significant changes.

Precipitation-hardening steels

Specifications for precipitation-hardening steels may include a mandatory deep freeze after solution treatment and prior to ageing.

Shrink fits

Cooling the inner member of complex part to below ambient temperature may be a useful way of providing an interference fit. Care must be taken, however, to avoid the brittle cracking that may develop when the inner member is made of heat treated steel with high amounts of retained austenite, which converts to martensite on subzero cooling.

Stress Relief

Residual stresses often contribute to part failure and residual stresses frequently are the result of temperature changes that produce thermal expansion and phase changes—and consequently, volume changes.

Under normal conditions, temperature gradients produce nonuniform dimensional and volume changes. In castings, for example, compressive stresses develop in lower-volume areas, which cool first and tensile stresses develop in areas of greater volume, which are last to cool. Shear stresses develop between the two areas. Even in large castings and machined parts of relatively uniform thickness, the surface cools first and the core last. In such cases, stresses develop as a result of the phase (volume) change between those layers that transform first and the centre portion, which transforms last.

When both volume and phase changes occur in pieces of uneven cross section, normal contractions due to cooling are opposed by transformation expansion. The resulting residual stresses will remain until a means of relief is applied. This type of stress develops most frequently in steels during quenching. The surface becomes martensitic before the interior does. Although the inner austenite can be strained to match this surface change, subsequent interior expansions place the surface martensite under tension when the inner austenite transforms. Cracks in high-carbon steels arise from such stresses.

The use of cold treating has proved beneficial in stress relief of castings and machined parts of even or nonuniform cross section. The following are features of the treatment:

1. Transformation of all layers is accomplished when the material reaches –84°C (–120°F).
2. The increase in volume of the outer martensite is somewhat counteracted by the initial contraction due to chilling.
3. Re-warm time is more easily controlled than cooling time, thus allowing equipment flexibility.
4. The expansion of the inner core due to transformation is somewhat balanced by the expansion of the outer shell.
5. The chilled parts are more easily handled.
6. The surface is unaffected by low temperature.
7. Parts that contain various alloying elements and that are of different sizes and weights can be chilled simultaneously.

Advantages of Cold Treating

Unlike heat treating, which requires that temperature be precisely controlled to avoid reversal, successful transformation through cold treating depends only on the attainment of the minimum low temperature and is not affected by lower temperatures. As long as the material is chilled to –84°C (–120°F), transformation will occur; additional chilling will not cause reversal.

Time at temperature

After thorough chilling, additional exposure has no adverse effect. When heat is used, holding time and temperature are critical. In cold treatment, materials of different compositions and of different configurations may be chilled at the same time even though each may have a different high-temperature transformation point. Moreover, the warm-up rate of a chilled material is not critical as long as uniformity is maintained and gross temperature-gradient variations are avoided.

The cooling rate of a heated piece, however, has a definite influence on the end product. Formation of martensite during solution heat treating assumes immediate quenching to ensure that austenitic decomposition will not result in the formation of bainite and cementite. In large pieces comprising both thick and thin sections, not all areas will cool at the same rate. As a result, surface areas and thin sections may be highly martensitic and the slower cooling core may contain as much as 30 to 50 per cent retained austenite. In addition to incomplete transformation, subsequent natural ageing induces stress and also results in additional growth after machining.

Aside from transformation, no other metallurgical change takes place as a result of chilling. The surface of the material needs no additional treatment. The use of heat frequently causes scale and other surface deformations that must be removed.

Equipment for Cold Treating

A simple home-type deep freezer can be used for transformation of austenite to martensite. Temperature will be approximately –18°C (0°F). In some instances, hardness tests can be used to determine if this type of cold treating will be helpful. Dry ice placed on top of the work in a closed, insulated container also is commonly used for cold treating. The dry ice surface temperature is –78°C (–109°F), but the chamber temperature normally is about –60°C (–75°F).

Mechanical refrigeration units with circulating air at approximately –87°C (–125°F) are commercially available. A typical unit will have the following dimensions and operational features: chamber volume, up to 2.7 m^3 (95 ft^3); temperature range, 5 to –95°C (40 to –140°F); load capacity, 11.3 to 163 kg/hr. (25 to 360 lb/hr.); and thermal capacity, up to 8870 kJ/hr. (8400 Btu/hr.).

Although liquid nitrogen at –195°C (–320°F) may be employed, it is used less frequently than any of the above methods because of its cost.

Chapter 13

Martempering of Steel

INTRODUCTION

Martempering is a term used to describe an elevated-temperature quenching procedure aimed at reducing cracks, distortion or residual stresses. It is not a tempering procedure, as the name implies and is more properly termed 'marquenching'.

Martempering of steel (and of cast iron) consists of: (i) quenching from the austenitising temperature into a hot fluid medium (hot oil, molten salt, molten metal or a fluidised particle bed) at a temperature usually above the martensitic range (M_s point); (ii) holding in the quenching medium until the temperature throughout the steel is substantially uniform; and then (iii) cooling (usually in air) at a moderate rate, to prevent large differences in temperature between the outside and the center of the section. Formation of martensite occurs fairly uniformly throughout the workpiece during cooling to room temperature, thereby avoiding formation of excessive amounts of residual stress. The microstructure after martempering is essentially primary martensite, which is untempered and brittle for most applications. After being air cooled to room temperature, martempered parts are tempered in the same manner as if they had been conventionally quenched.

Fig. 13.1 shows the significant difference between conventional quenching (a) and martempering (b). The advantage of martempering lies in the reduced thermal gradient between surface and center as the part is quenched to the isothermal temperature and then air is cooled to room temperature. Residual stresses developed during martempering are lower than those developed during conventional quenched, because the greatest thermal variations occur while the steel is in the relatively plastic austenitic condition and because final transformation and thermal changes occur throughout the part at approximately the same time. Martempering also reduces or eliminates susceptibility to cracking.

Although martempering is used primarily to minimise distortion, eliminate cracking and minimise residual stresses, it also greatly reduces the problems of pollution and fire hazard provided that nitrate-nitrite salts are employed rather than martempering oils. This is especially true where nitrate-nitrite salts are recovered from wash water with systems that provide essentially no discharge of salts into drains. Any steel part or grade of steel responding to oil quenching can be martempered to provide similar physical properties. The quenching severity of molten salt is greatly enhanced by agitation and by water additions to the nitrate-salt bath. Both techniques are particularly beneficial in heat treating of carbon steels having limited hardenability. Table 13.1 compares the properties obtained in 1095 steel by martempering and tempering with those obtained by conventional quenching and tempering.

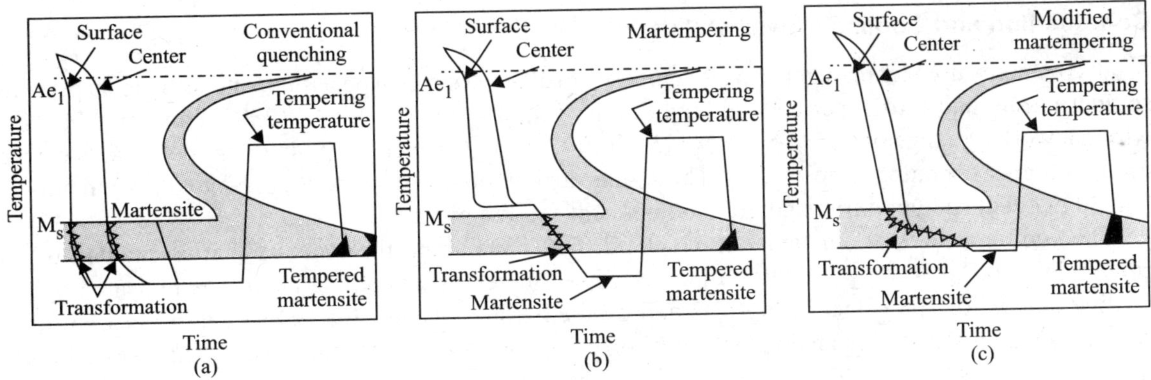

(a) Conventional process. (b) Martempering. (c) Modified martempering.

Fig. 13.1. Time-temperature transformation diagrams with superimposed cooling curves showing quenching and tempering.

Table 13.1 Mechanical properties of 1095 steel heat treated by two methods.

Specimen number	*Heat treatment*	*Hardness, HRC*	*Impact energy J*	*ft-lb*	*Elongation (a), %*
1	Water quench and temper	53.0	16	12	0
2	Water quench and temper	52.5	19	14	0
3	Martemper and temper	53.0	38	28	0
4	Martemper and temper	52.8	33	24	0

(a) In 25 mm or 1 inch.

In many instances, martempering eliminates the need for quenching fixtures that would be required for minimising distortion during conventional quenching, and thus reduces the cost of tooling and handling. However, changing from conventional quenching to martempering may require that dimensional variations in individual parts be studied before preheat-treatment dimensions are established.

Modified martempering differs from 'standard' martempering only in that the temperature of the quenching bath is below the M_s point Fig.13.1c. The lower temperature increases the severity of quenching. This is important for steels of lower hardenability that require faster cooling in order to harden to sufficient depth, or when the M_s is high and some bainite is detrimental to the finished part. Thus, modified martempering is applicable to a greater range of steel compositions than is the standard process.

Although hot oil is invariably the quenchant employed for modified martempering at 175°C (350°F) and lower, molten nitrate-nitrite salts (with water addition and agitation) are effective at temperatures as low as 175°C. Due to their higher heat-transfer coefficients, molten salts offer some metallurgical and operational advantages.

Martempering Media

Molten salt and hot oil are both widely used for martempering. Several factors must be considered when choosing between salt and oil. Operating temperature is the most common deciding factor. Oils are widely used for martempering at up to 205°C (400°F), and sometimes at temperatures as high as 230°C (450°F). Molten salt is used for martempering in the range from 160 to 400°C(320 to 750°F).

Composition and Cooling Power of Salt

A salt commonly used for martempering is composed of 50 to 60 per cent potassium nitrate, 37 to 50 per cent sodium nitrite, and 0 to10 per cent sodium nitrate. It melts at about 140°C (280°F) and may be used within a working range of 165 to 540°C (325 to1000°F). Higher-melting-point (less costly) salts can be used for higher operating temperatures. These salts are comprised of 40 to 50 per cent potassium nitrate, 0 to 30 per cent sodium nitrite and 20 to 60 per cent sodium nitrate.

The cooling power of agitated salt at 205°C (400°F) is about the same as that of agitated oil in conventional oil quenching. Addition of water to salt increases its cooling power, as indicated by the cooling curves for 1045 steel in Fig.13.2 and by the hardness values for 1046 steel in Fig.13.3 . The cooling power of salt is compared with the cooling power of water and of three types of oil, in Fig. 13.3.

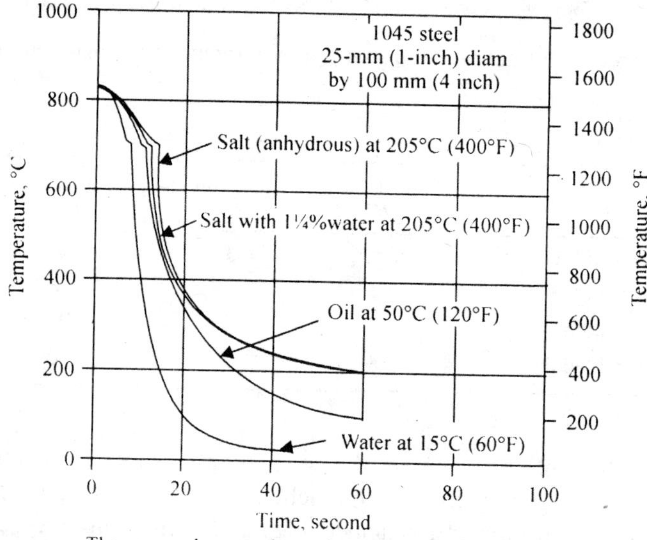

Fig. 13.2. Cooling curves for 1045 steel cylinders quenched in salt, water and oil.

Advantages of salt

Advantages of salt in comparison with oil for martempering are as follows:

1. The viscosity of salt changes only slightly over a wide temperature range.
2. Salt retains chemical stability, so that replenishment usually is required only to replace dragout loss. This is not always the case, however. In an installation for high-volume martempering of cylinder liners, for instance, breakdown of the salt did occur, forming a carbonate. This occurred on quenching from an endothermic atmosphere into an open-top salt quench bath at 245°C (475°F). The same operation conducted with a nitrogen atmosphere in the heating furnace does not produce such breakdown.
3. Salt has wide operating temperature range.
4. Salt is easily washed from the work with plain water.
5. Less time is required for workpieces to attain temperature equalisation in salt.
6. No unusual disposal problems attend the rinsing operation.

7. Salt is relatively easy to handle in powder form and easy to clean up if spilled; however, it is necessary to keep salts separated.

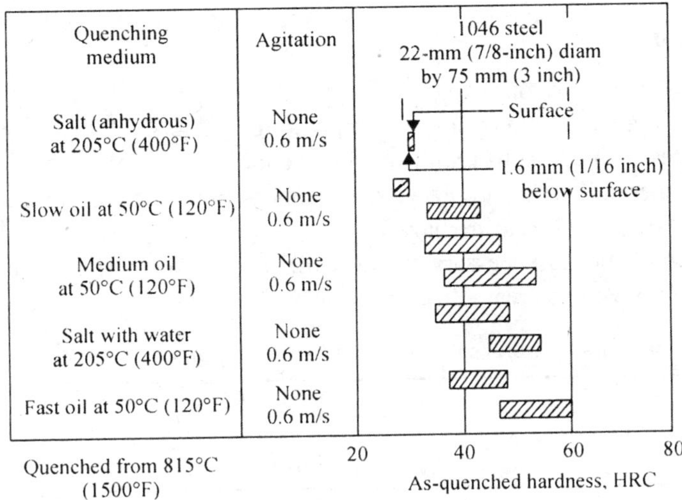

Fig. 13.3. Effects of quenchant and agitation on hardness of 1046 steel.

Disadvantages of salt

Disadvantages of salt in comparison with oil include the following:

1. The minimum operating temperature of salt is 160°C (320°F) .
2. Quenching from cyanide-base carburising salts is hazardous, because of possible explosion.
3. Explosion and splatter of high-temperature salt can occur if wet or oily parts are immersed.
4. Potentially explosive reactions also occur with sooty atmospheres in atmosphere furnaces connected to martempering salt quenches.
5. Quench salt can be contaminated by high-temperature neutral salt used for heating. Sludging operations are required to maintain quenching severity.

Oils for Martempering

Physical properties of two oils commonly used for martempering are listed in Table 13.2. Such oils are compounded especially for martempering and, in comparison with conventional quenching oils, offer higher rates of cooling during the initial stage of quenching.

Quenching oil requires special handling when used in the temperature range from 95 to 230°C (200 to 450°F). To prolong its life, the oil should be maintained under a protective atmosphere (reducing or neutral). Deterioration is accelerated when oil is exposed to air at elevated temperatures. For example, for every 10°C above 60°C (every 18°F above 140°F), the rate of oxidation is approximately doubled. This causes formation of acid and sludge, which may affect both hardness and colour of workpieces.

Advantages of oil

Advantages of oil in comparison with salt are as follows:

1. Oil can be used at lower temperatures.

2. Oil is easier to handle at room temperature.
3. Less dragout loss occurs with oil.
4. Oil is compatible with all austenitising salts.

Disadvantages of oil

Disadvantages of oil include the following:

1. Oil is limited to a maximum operating temperature of 230°C (450°F).
2. Oil deteriorates with use, thus requiring closer control.
3. When martempered in oil, workpieces require more time to attain temperature equalisation.
4. Oil, whether hot or cold, is a fire hazard.
5. Soap or an emulsifier is necessary to wash off the oil. Washers must be drained and refilled periodically. Oily wastes create disposal problems.

Table 13.2. Physical properties of two oils used for martempering of steel.

	Value, for oil with operating temperature of:	
Property	*95 to150°C (200 to 300°F)(a)*	*150 to 230°C (300 to 450°F)*
Flash point (min),°C (°F)	210 (410)	275 (525)
Fire point (min), °C (°F)	245 (470)	310 (595)
Viscosity, SUS, at:		
38°C (100°F)	235 to 575	–
100°C (210°F)	50.5 to 51	118 to122
150°C (300°F)	36.5 to 37.5	51 to 52
175°C (350°F)	–	42 to 43
205°C (400°F)	–	38 to 39
230°C (450°F)	–	35 to 36
Viscosity index (min)	95	95
Acid number	0.00	0.00
Fatty-oil content	None	None
Carbon residue	0.05	0.45
Colour	Optional	Optional

(a) Temperature range for modified martempering.

SAFETY PRECAUTIONS

In operation of a martempering installation (salt or oil), the precautions appropriate for operating any other hot liquid bath are applicable. The following operator safety precautions are recommended:

1. Operators of equipment should be thoroughly familiar with equipment manufacturer's instructions and with company safety recommendations before attempting to operate or service the equipment.
2. Operators should wear gloves, face shields and protective clothing, as required. Safety equipment contaminated with oil or nitrate-nitrite salts should be discarded.

Nitrate-Nitrite Salts

Several precautions should be taken in operating nitrate-nitrite salt baths:

1. All racks, fixtures and cleaning tools must be thoroughly washed clean of salts (hot water preferred), and thoroughly dried, before being reused; otherwise, the presence of salt could contaminate the austenitising furnace, and residual water could cause spattering of molten salt.
2. All salt bath furnaces should be labelled so that the proper salt is used. Cyanide-containing salts should not be combined with nitrate-nitrite salts, because such mixtures result in explosive reactions. Good housekeeping is essential to prevent accidents. Martempering salts must be stored in closed, well-marked containers. These salts are hygroscopic and will absorb water if exposed. Yellow, orange or red colouring is commonly used for identification of nitrate-nitrite salts.
3. When water is added to a nitrate-nitrite salt to increase its quenching severity, the water should be trickled or atomised onto the surface of the bath and should never be introduced below the surface or under pressure. Otherwise, spattering or eruption can occur. Equipment manufacturers should be consulted as to recommended procedures and hardware.
4. The salt bath should be equipped with a high-temperature limit controller to prevent the bath from exceeding 595°C (1100°F). At higher temperatures, nitrate salts decompose and may cause fire or explosion. Although nitrate-nitrite salts do not burn, they are oxidising and will support combustion. Therefore, oxidisable materials should not be stored near the bath, and combustible materials should not be introduced into the bath unless designed for the purpose. In the event of a fire, a carbon dioxide extinguisher should be used. Sand is also effective in smothering floor fires. To avoid explosive spread of the molten medium, water should never be used to extinguish a fire in or around a molten salt (or molten metal) bath.
5. Soot and carbonaceous materials should not be allowed to accumulate (on the surface of the molten bath).

Hot Oil

Hot oil also requires safety precautions. Among the hazards that accompany its use are: (i) fire; (ii) overflow, due to the increase in the volume of the oil during heating; (iii) explosion, when an atmosphere cover is used over the quench tank; and (iv) contamination by water, which can result in fire.

Equipment normally used for protection against these hazards includes:

1. Alarm system on temperature-control instrument.
2. An oil-level indicator.
3. If the oil is heated by gas-fired tubes, a safety control system to prevent firing when the air cooling system is off.
4. Expansion system to accommodate the change in volume that occurs when the oil is heated from room temperature to the operating temperature.
5. A safety control to prevent oil from being heated unless it is agitated.
6. A carbon dioxide or foam system, installed over the quench tank for fire protection.
7. A water-detection system, with automatic alarm, that monitors the oil for water content.

Caution should be used in installing ventilating systems and any other system that could cause addition of water to the hot oil.

SUITABILITY OF STEELS FOR MARTEMPERING

Alloy steels generally are more adaptable than carbon steels to martempering. In general, any steel that is normally quenched in oil can be martempered. Some carbon steels that are normally water quenched can be martempered at 205°C (400°F) in sections thinner than 5 mm (3/16 inch), using vigorous agitation of the martempering medium. In addition, thousands of gray cast iron parts are martempered on a routine basis.

The grades of steel that are commonly martempered to full hardness include 1090, 4130, 4140, 4150, 4340, 4640, 5140, 6150, 8630, 8640, 8740, 8745, SAE 1141 and SAE 52100. Carburising grades such as 3312, 4620, 5120, 8620 and 9310 also are commonly martempered after carburising. Occasionally, higher-alloy steels such as type 410 stainless are martempered, but this is not a common practice.

Success in martempering is based on a knowledge of the transformation characteristics (TTT curves) of the steel being considerd. The temperature range in which martensite forms is especially important. Fig. 13.4 shows the martensite temperature ranges for 14 carbon and low-alloy steels. Two trends may be observed in these data: (i) as carbon content increases, the martensite range widens and the martensite transformation temperature is lowered; and (ii) the martensite range of a triple-alloy (Ni-Cr-Mo) steel is usually lower than that of either a single-alloy or a double-alloy steel of similar carbon content.

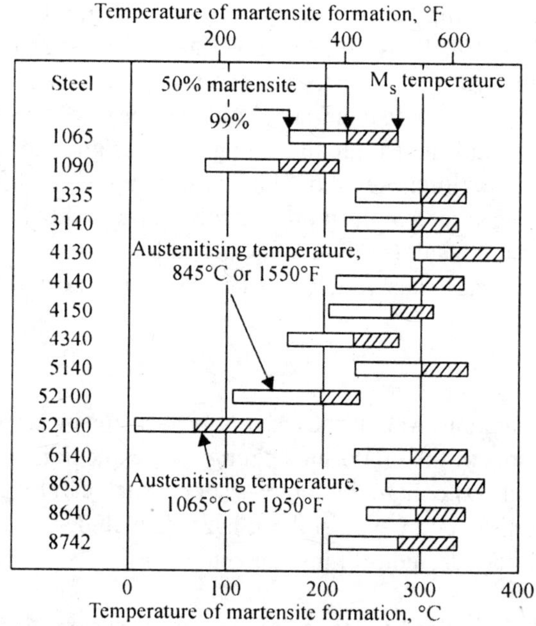

Fig. 13.4. Temperature ranges of martensite formation in 14 carbon and low-alloy steel.

Any steel that is to be martempered successfully must contain sufficient carbon or alloying additions to move the nose of the TTT curve to the right, thus permitting sufficient time for quenching of workpieces past the nose of TTT curve. The TTT diagrams for a hyperutectoid steel (1034) and a hypoeutectoid steel (1090) are shown in Fig.13.5. The diagram for 1090 steel is the simplest form of transformation diagram, because no proeutectoid constituents (free ferrite or free carbide) are involved in the transformation at temperatures above that corresponding to the nose of the curve.The speed of transformation at the nose is related to the hardenability of the steel; when nose of the TTT curve is far to the left on the diagram, the steel has lower hardenability; when the nose is to the right, the steel has higher hardenability. To achieve full hardening during quenching, the cooling curve of the steel must pass to the left of the curve farthest to the left on the diagram. In production, some loss in as-quenched hardness is usually accepted in order to achieve minimum distortion. A TTT diagram for a hypoeutectoid low-alloy steel (5140) suitable for martempering is shown in Fig.13.6. The chromium in this steel causes the characteristic shape of the TTT curve near 540°C (1000°F).

The TTT diagram for an extremely high-hardenability steel (4340) is also shown in Fig. 13.6. The combined effect of nickel, chromium and molybdenum on hardenability is illustrated in this diagram. These elements cause a double nose on the TTT curve. The nose that occurs at about 480°C (900°F) is more significant in martempering than the one occurring near 650°C (1200°F). Steels having such high hardenability are easily martempered to fully martensitic structures.

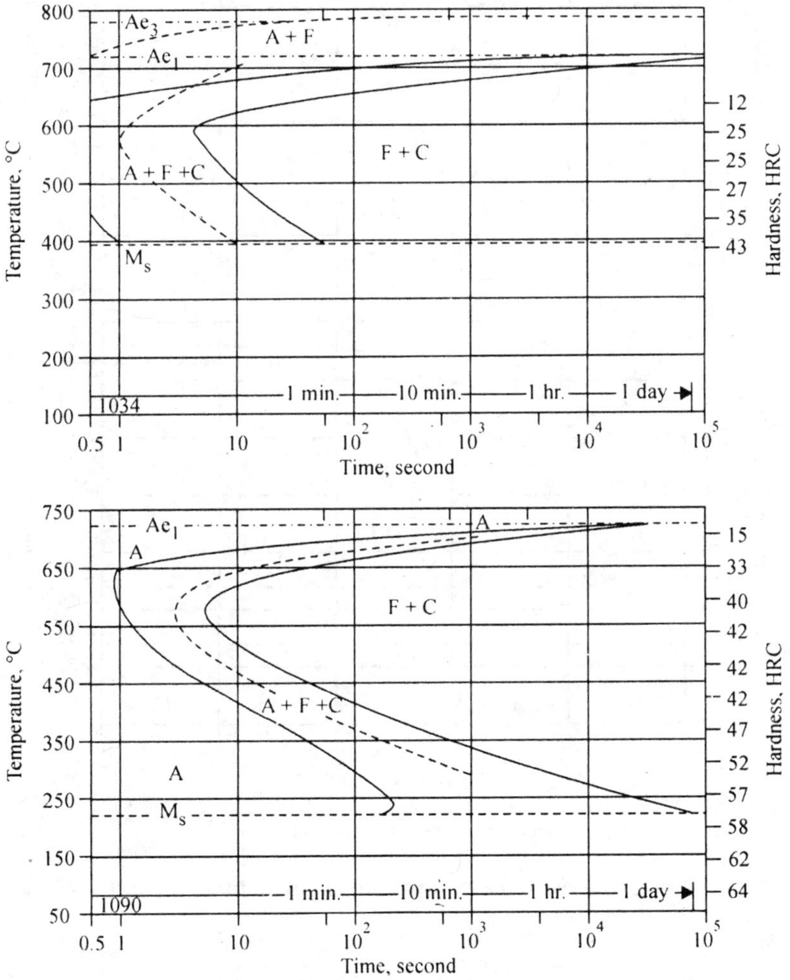

The 1090 steel was austenitised at 885°C (1625°F) and had a grain size of 4 to 5.

Fig. 13.5. Time-temperature transformation diagram for 1034 and 1090 steels.

Low-Carbon and Medium-Carbon Steels

Low-carbon and medium-carbon steels 1008 through 1040 are too low in hardenability to be successfully martempered, except when carburised. The TTT curve for the 1034 steel in Fig. 13.5 is characteristic of a steel that is unsuitable for martempering; except in sections only a few thousandths of an inch thick, it would be impossible to quench the steel in hot salt or oil without encountering upper transformation products.

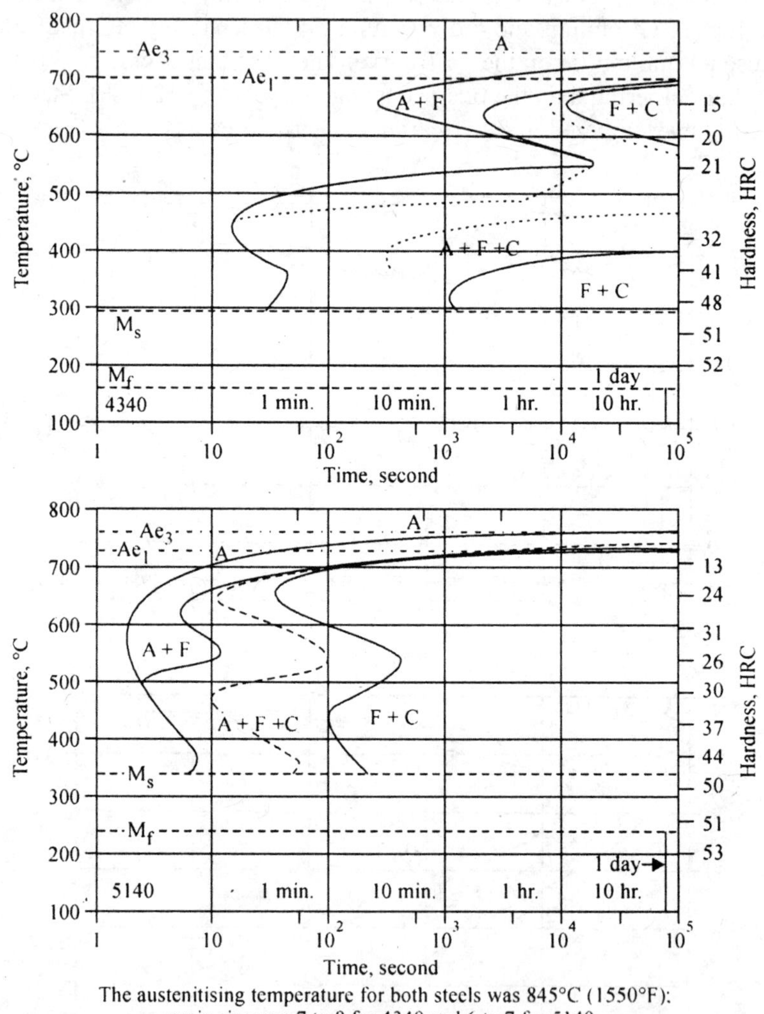

The austenitising temperature for both steels was 845°C (1550°F): grain size was 7 to 8 for 4340 and 6 to 7 for 5140.

Fig 13.6. Time-temperature transformation diagrams for 4340 and 5140 steels.

Borderline Grades

Some carbon steels higher in manganese content, such as 1041 and 1141, can be successfully martempered in thin-sections.

Low-alloy steels that have limited applications for successful martempering are listed below (the lower-carbon grades are carburised before martempering):

1330 to 1345	4520
4012 to 4042	5015 and 5046
4118 to 4137	6118 and 6120
4422 and 4427	8115

Most of the above alloy steels are suitable for martempering in section thicknesses of up to 16 or 19 mm (⅝ or ¾ inch). Martempering at temperatures below 205°C (400°F) will improve hardening response, although greater distortion may result in martempering at higher temperatures.

Effect of Mass

The limitation of section thickness or mass must be considered in martempering. With a given severity of quench, there is a limit to bar size beyond which the center of the bar will not cool fast enough to transform entirely to martensite. This is illustrated in Fig. 13.7, which compares the maximum diameter of bar that can be hardened by martempering, oil quenching and water quenching for 1045 steel and five alloy steels of various hardenabilities.

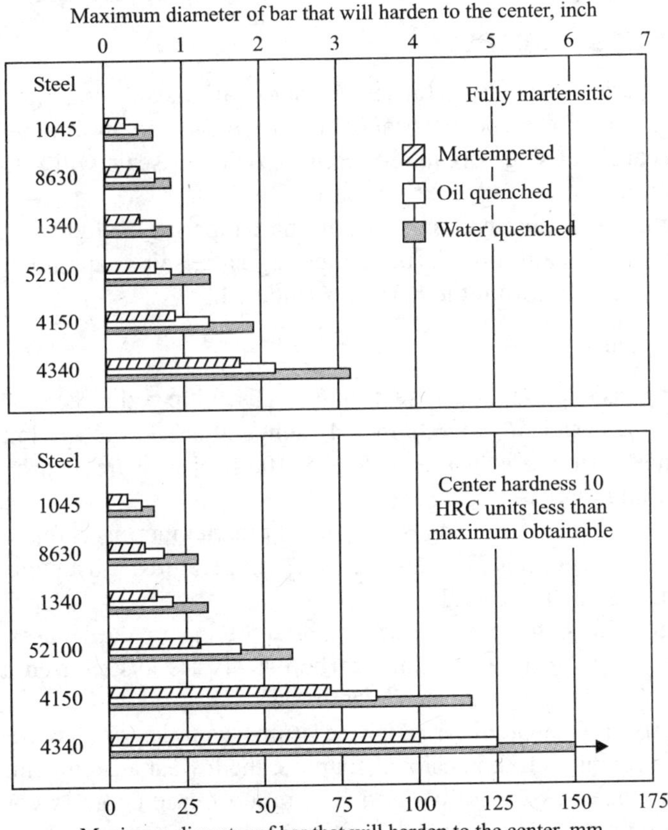

Fig. 13.7. Approximate maximum diameters of bars that are hardenable by martempering, oil quenching and water quenching.

For some applications, a fully martensitic structure is unnecessary and a center hardness 10 HRC units lower than the maximum obtainable value for a given carbon content may be acceptable. By this criterion, maximum bar diameter is 25 to 300 per cent greater than the maximum diameter that can be made fully martensitic (see lower graph in Fig.13.7). Nonmartensitic transformation products (pearlite, ferrite and bainite) were observed at the positions on end-quenched bars corresponding to this reduced hardness value, as follows:

1045	15 per cent pearlite
8630	10 per cent ferrite and bainite
1340	20 per cent ferrite and bainite
52100	50 per cent pearlite and bainite
4150	20 per cent bainite
4340	5 per cent bainite

The influence of mixed structures such as these on the mechanical properties of the steel would have to be determined for each application.

Steels selected for martempering must be judged on hardenability and section size. To form the same amount of martensite, for a given section size, the carbon content or alloy content, or both, must be somewhat higher for martempering than for conventional (uninterrupted) quenching.

CONTROL OF PROCESS VARIABLES

The success of martempering depends on close control of variables throughout the entire process. It is important that the prior structure of the material being austenitised be uniform. Also, use of a protective atmosphere (or salt) in austenitising is required, because oxide or scale will act as a barrier to uniform quenching in hot oil or salt.

The process variables that must be controlled in martempering include austenitising temperature, temperature of martempering bath, time in martempering bath, salt contamination, water additions to salt, agitation and rate of cooling from the martempering bath.

Austenitising Temperature

Austenitising temperature is important because it controls austenitic grain size, degree of homogenisation, and carbide solution, and because it affects the M_s temperatures, which is important in establishing martempering procedures. As indicated in Fig. 13.8 for 52100 steel, an increase in austenitising temperature lowers M_s temperature and increases grain size.

Temperature control during austenitising is the same for martempering as for conventional quenching: a tolerance of 8°C (15°F) is common. The austenitising temperatures most commonly used for several different steels are indicated in Table 13.3.

In most instances, austenitising temperatures for martempering are the same as those for conventional oil quenching. Occasionally, however, medium-carbon steels are austenitised at higher temperatures prior to martempering, to increase as-quenched hardness.

For carburised parts, low austenitising temperatures usually yield better size control during martempering. To obtain minimum dimensional changes, the lowest austenitising temperature that will yield satisfactory core properties should be used. The ratio of case depth to core, as well as the prior processing history of the steel (such as forging, rolling or drawing), can be controlling factors also, particularly for critical section shapes and sizes. Austenitising temperatures that have been used for several carburising steels are indicated in Table 13.3.

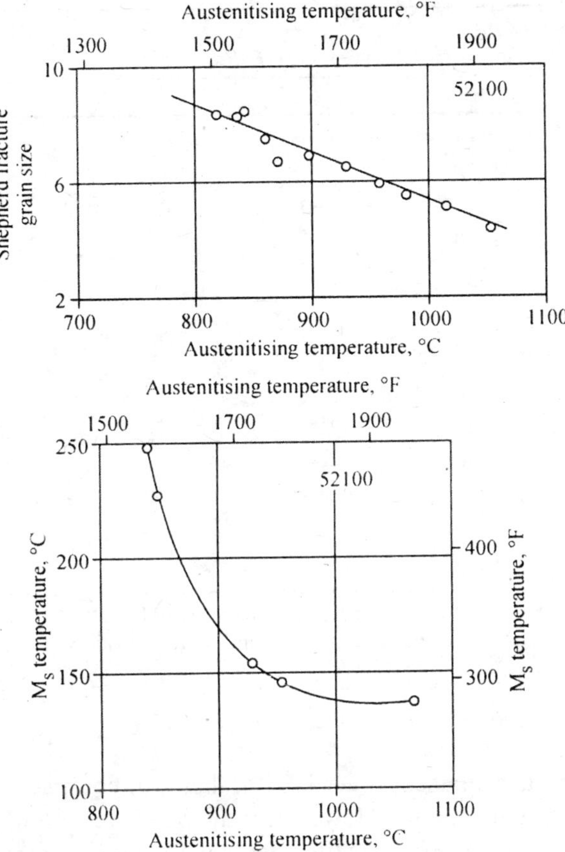

Fig. 13.8. Effects of austenitising temperature on grain size and M_s temperature of 52100 steel.

Table 13.3 Typical austenitising and martempering temperatures for various steels.

Grade	*Austenitising temperature*		*Martempering temperature*			
			Oil(a)		*Salt(b)*	
	°C	*°F*	*°C*	*°F*	*°C*	*°F*
Through-hardening steels						
1024	870	1600	135	275	–	–
1070	845	1550	175	350	–	–
1146	815	1500	175	350	–	–
1330	845	1550	175	350	–	–
4063	845	1550	175	350	–	–
4130	845	1550	–	–	205 to 260	400 to 500
4140	845	1550	150	300	–	–
4140	830	1525	–	–	230 to 275	450 to 525

(Contd...)

Grade	*Austenitising temperature*		*Martempering temperature*			
			Oil(a)		*Salt(b)*	
	°C	*°F*	*°C*	*°F*	*°C*	*°F*
4340, 4350	815	1500	–	–	230 to 275	450 to 525
52100	855	1575	190	375	–	–
52100	845	1550	–	–	175 to 245	350 to 475
8740	830	1525	–	–	230 to 275	450 to 525
Carburising steels						
3312	815	1500	–	–	175 to 190	350 to 375
4320	830	1525	–	–	175 to 190	350 to 375
4615	955	1750	190	375	–	–
4720	845	1550	–	–	175 to 190	350 to 375
8617, 8620	925	1700	150	300	–	–
8620	855	1575	–	–	175 to 190	350 to 375
9310	815	1500	–	–	175 to 190	350 to 375

a) Time in oil varies from 4 to 20 minute, depending on section thickness.(b) Martempering temperature depends on shape and mass of parts being quenched; higher temperatures in range (and sometimes above range) are used for thinner sections and more intricate parts.

Salt Contamination

When parts are carburised or austenitised in a salt bath, they can be directly quenched in an oil bath operating at the martempering temperature. However, if the parts are carburised or austenitised in salt containing cyanide, they must not be directly martempered in salt, because the two types of salts are not compatible, and explosions can occur if they are mixed. Instead, one of two procedures should be used: either: (i) air cool from the carburising bath, wash, reheat to the austenitising temperature for case and/ or core in a chloride bath, and then martemper; and (ii) quench from the cyanide-containing bath into a neutral chloride rinse bath maintained at the austenitising temperature and then martemper.

If the latter method is used, it is essential to control the amount of cyanide buildup in the neutral rinse. When tests indicate more than 5 per cent cyanide in the chloride rinse, part of the salt should be discarded and the remainder diluted with new salt.

All fixtures must be thoroughly cleaned after martempering, to prevent transfer of quenching salt to either cyanide baths or neutral chloride baths. Mixing of cyanide with nitrate-nitride salts will cause explosion. A chloride bath that is contaminated with nitrate-nitrite salts will produce pitting and decarburisation of parts immersed in it.

Temperature of the Martempering Bath

Temperature of the martempering bath varies considerably, depending on composition of workpieces, austenitising temperature, and desired results. In establishing procedures for new applications, many plants begin at 95°C (200°F) for oil quenching, or at about 175°C (350°F) for salt quenching, and progressively increase the temperature until the best combination of hardness and distortion is obtained. Martempering temperatures (for oil and salt) that represent the experience of several plants are listed in Table 13.3.

Time in the Martempering Bath

Time in the martempering bath depends on section thickness and on the type, temperature, and degree of agitation of the quenching medium. The effects of section thickness and of temperature and agitation of the quench bath on immersion time are indicated in Fig.13.9.

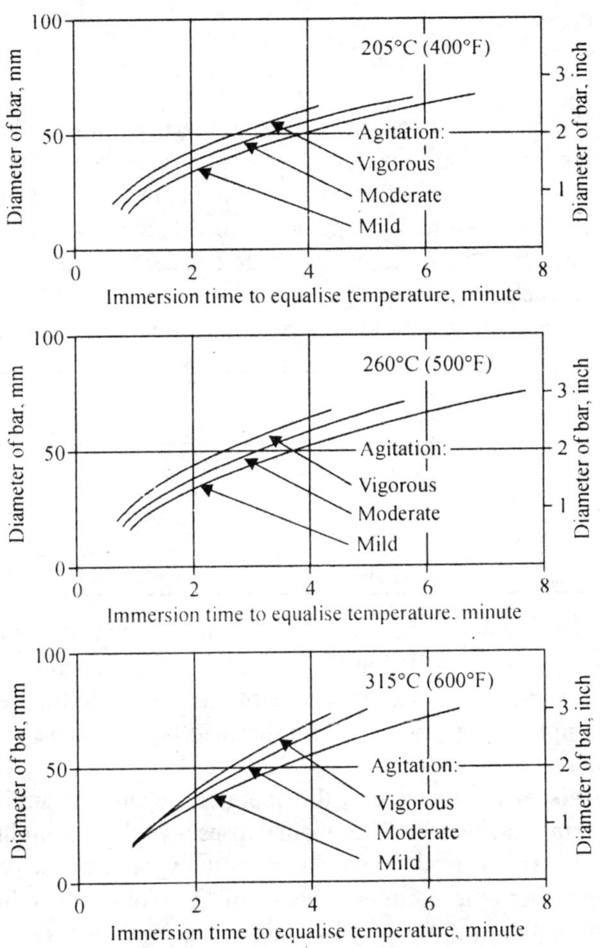

Effects of bar diameter and agitation of quench bath on time required for centers of 1045 steel bars to reach martempering temperature when quenched from a neutral chloride bath at 845°C (1550°F) into anhydrous nitrate-nitrite martempering salt at 205, 260 and 315°C (400, 500 and 600°F). Length of each bar was three times the diameter.

Fig. 13.9. Martempering time vs section size and agitation of quench bath for 1045 steel bars.

Because the object of martempering is to develop a martensitic structure with low thermal and transformation stresses, there is no need to hold the steel in the martempering bath for extended periods of time. Excessive holding lowers final hardness, because it permits transformation to products other than martensite. In addition, stabilisation of austenite may occur in medium-alloy steels that are held for extended periods of time at the martempering temperature.

The martempering time for temperature equalisation in oil is about four to five times that required in anhydrous salt at the same temperature. For example, for a 25-mm (1 inch) diam bar martempered in salt at 205°C (400°F) with moderate agitation, the customary immersion time is about 1½ to 2 minute, whereas about 8 to 10 minutes are required for the bar to attain temperature equalisation in oil at 205 to 220°C(400 to 425°F). The immersion time required in salt can be reduced by as much as 50 per cent by addition of 0.5 to 2 per cent water.

Water Additions to Salt

Addition of water increase the quenching severity of martempering salt usually is made by directing a fine spray of water onto the molten-salt surface. A protective shroud surrounds the water spray to prevent spattering. The turbulence of the salt carries the water into the bath without spattering or hazard to the operator. Water is continuously evaporating from the bath surface, and the rate of evaporation increases during quenching of hot work. Thus, it is necessary to add water periodically to maintain the water concentration and a uniform quenching severity. The amount of water to be added varies with the operating temperature of the salt, as indicated by the following recommended concentrations:

Temperature		*Water concentration,*
°C	°F	*per cent*
205	400	½ to 2
260	500	½ to 1
315	600	¼ to ½
370	700	¼

At present, there is no known means of automatically controlling the concentration of water in molten nitrate-nitrite salt. Water usually is controlled at the discretion of the operator, who will add water as needed. On the basis of experience, however, it is possible to anticipate the need for water; here, the addition of water may be simplified by use of timers that can be adjusted to time the frequency and duration of water additions.

The presence of water is visually detectable by the operator; because steam is released when the hot work is immersed into the nitrate-nitrite salt. The steam causes a visible mounding of the salt above the quench area, and there is a characteristic sizzling caused by the vapour phase.

Besides visual appearance, periodic hardness checks of the work will indicate the activity of the bath. A more accurate determination can be made by removing a small quantity of the salt and weighing it accurately before and after dehydrating it by heating to 370 to 425°C (700 to 800°F). Another method is to determine the freezing point of a small sample and then refer to a published curve that relates freezing point to water content for that specific salt.

Agitation

Agitation of the martempering salt or oil considerably increases the hardness obtainable for a given section thickness, in comparison to that obtainable in still quenching. This is demonstrated in Fig.13.10, which presents data for 52100 steel that relate section thickness, hardness and agitation. In some instances, the rapid cooling produced by the most vigorous agitation increases distortion. Thus, mild agitation often is used in combination with water addition to obtain minimum distortion without sacrificing hardness.

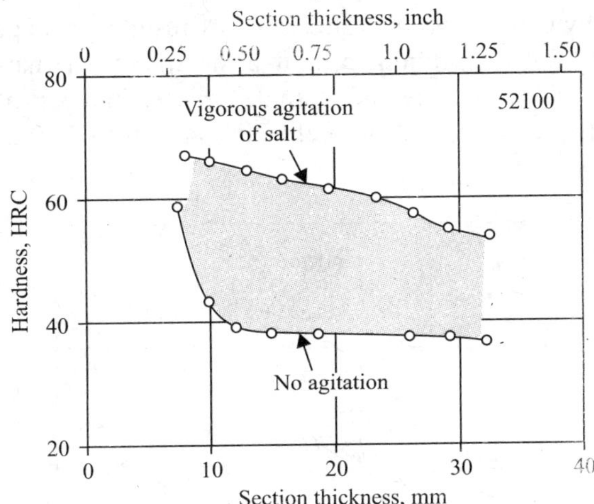

Fig. 13.10. Influence of agitation on surface hardness of 52100 steel in various section thicknesses martempered in hot salt.

Cooling from the Martempering Bath

Cooling from the martempering bath ordinarily is done in still air, to avoid large differences in temperature between the surface and the interior of the steel. Forced-air cooling by means of fans is occasionally used on sections over 19 mm (¾ inch) thick, but caution is required if the part varies in section thickness or has more surface exposed on one section, such as on threads or serrations, because objectionable amounts of distortion can occur on rapid cooling through the martensite range. Generally, cooling of workpieces in cool oil or water after removal from the martempering bath is considered undesirable, because cooling can re-establish thermal gradients and unequal stress patterns that can increase distortion.

Cooling time varies with the mass and density of the charge, the maximum section thickness of the workpiece, and the ambient air temperature. Usually, production loads of 365 to 815 kg (800 to 1800 lb) from either continuous or batch-type furnaces will require 2½ to 5 hours to reach room temperature. Fig. 13.11 shows the effect of section thickness. As indicated, the 25 mm (1 inch) diam sections were completely cooled in about half the time required for complete cooling of the 75 mm (3 inch) diam section.

After being cooled to room temperature, martempered parts usually can be held at room temperature for several hours, and sometimes days, without risk of cracking, because residual stresses are low compared to those in conventionally quenched parts. Holding of parts at room temperature also permits more nearly complete transformation of steels in which transformation is sluggish.

DIMENSIONAL CONTROL

In most instances, distortion is significantly less in martempered parts than in parts hardened by uninterrupted (conventional) quenching. However, prior processing often has a significant effect on distortion, regardless of the heat treating method used. Thus, for some applications, martempering may fail to solve distortion problems because excessive dimensional changes occurred during heating prior to martempering. Occasionally, it is necessary to consider the effects of fabricating stresses that occur during forging, stamping, rolling and machining. When the workpiece is heated, these stresses can cause a significant amount of distortion. A process anneal at 650° to 705°C (1200 to 1300°F) after rough machining or forming usually will relieve such stresses. Any resulting change in the size or shape of the part can then be corrected by finish machining prior to austenitising and martempering. During heating, distortion also can occur as the result of temperature differences in a part having both light and heavy sections. This condition often is corrected by preheating at 650 to 705°C(1200 to 1300°F) prior to austenitising.

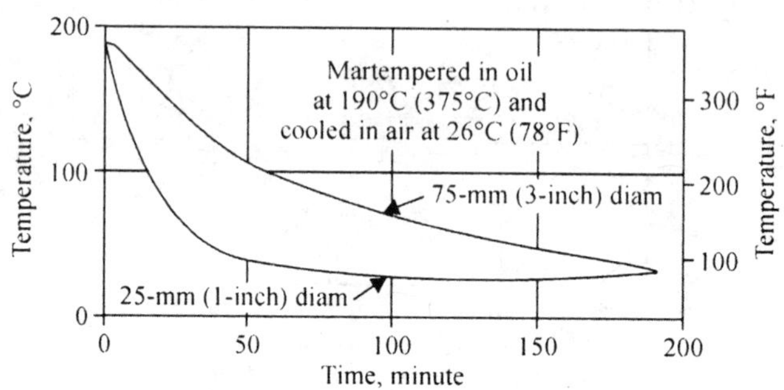

Temperature measurements were made at the surface. Times can be reduced by approximately 30 per cent by forced fan cooling.

Fig. 13.11. Effect of section thickness on time required for air cooling of steel after martempering at 190°C (375°F).

Relatively large parts that require extreme flatness often must be press quenched. For example, a gear 178 mm (7 inch) in diameter, 13 mm (½ inch) thick at the rim and with a 6.4 mm (¼ inch) web could not be martempered with an acceptable degree of flatness. A large ring gear with internal teeth and a thin wall is another part for which martempering has been replaced by press quenching to obtain dimensions with in acceptable limits.

Straightening After Martempering

Parts such as shafts, chain-saw guides, washers and springs require straightening or re-forming after heat treatment. This is sometimes not feasible after normal quenching and tempering. However, after such parts are removed from a martempering bath, straightening is readily accomplished, either by hand pressure or in a die press, while the parts are still essentially austenitic.

When parts are clamped between plates held at approximately 150°C (300°F), the parts cool to the temperature of the plate. At this temperature, depending on the M_f temperature of the steel, the transformation to martensite may be only partial, but it is usually near enough to completion for the part to set permanently before it is removed from the clamped position. At this temperature, the film of salt (melting point, about 145°C or 290°F) adhering to the part will not freeze and so removal of the part from the clamps or die can be accomplished without difficulty. The transformation of martensite is completed during subsequent air cooling to room temperature, and the part is within dimensional tolerance.

Forming After Martempering

Difficult-to-form materials, such as hot work die steels (H11) and some martensitic stainless steels used for missiles, rockets and high-speed aircraft, can be accurately formed and hardened by austenitising, martempering and hot forming to shape immediately after they are extracted from the quench bath, if the martempering temperature is above the martensite transformation temperature for the specific alloy being treated. During hot forming by rolling, forging, drawing or extruding, the metal consists of metastable austenite, which transforms to martensite on subsequent cooling to room temperature.

Parts of both simple and complex shape have been formed in this manner; after being air cooled to room temperature in or out of the forming die, the parts will accurately maintain their as-formed dimensions and yet be in the fully heat treated condition, requiring only a subsequent tempering operation.

This method of heat treating and forming has been successfully applied to sheet and plate that require nearly perfect flatness. It also has been used for forming type 420 stainless steel sheet into cups that require close dimensional tolerances and hardness values of 55 to 58 HRC. Working of metastable H11 that results in deformation of 58 to 94 per cent at 480°C (900°F) produces a 19 to 32 per cent increase in tensile strength.

APPLICATIONS

Typical applications of martempering in salt and oil are indicated in Table 13.4 and 13.5, which list and describe steel parts commonly treated and give details of martempering procedures and hardness requirements. From these tables, it is evident that martempering is used for parts of diverse shapes, weights, section sizes and steel compositions.

Although about a third of the applications in Table13.4 and most of the applications in Table13.5 are carburised parts, it is believed that in current industrial practice the tonnage of martempered through-hardening steels markedly exceeds that of martempered carburised steels. However, martempering is especially appropriate for carburised parts (particularly for splined shafts, cams and gears), because these parts generally are more difficult to grind and more costly to fabricate, and are made to closer dimensions, than are parts made of through-hardening steels.

Special adaptations of martempering are sometimes employed to achieve desired characteristics and to solve specific problems. Such modified techniques, however, usually require that all conditions be closely controlled, lest still greater problems result.

Table 13.4 Typical applications of martempering steel parts in salt.

						Martempering conditions			
		Maximum section thickness		*Weight*		*Temperature of salt*			*Required hardness,*
Part	*Grade*	*mm*	*inch*	*kg*	*lb*	*°C*	*°F*	*Minimum time in salt, minute*	*HRC*
Compliant tube	4130	0.8	0.03	0.11	0.25	160 (a)	320 (a)	5	50 (b)
Thrust washer	8740	5.1	0.20	0.05	0.1	230	450	1	52 min (b)
Chain link	1045	5.6	0.22	0.11	0.25	205 (c)	400 (c)	1	45 to 50 (b)
Cotton-picker spindle(d)	Type 410	6.4	0.25	0.05	0.12	315	600(c)	1½	44 to 48 (b)
Accessory drive shaft	9310 (e)	6.4	0.25	0.45	1.0	190	375	2½	90 (15N scale)
Clutch-adjustment	8740	7.6	0.30	0.14	0.3	230	450	2	52 min (b)
Seal ring	52100	7.6	0.30	0.18	0.4	190	375	10	65 (b)
Spur pinion	3312 (e)	7.6	0.30	0.23	0.5	175	350	1½	90(15N scale)
Internal gear	4350	8.9	0.35	0.36	0.8	245	475	2	54 minute (b)

(Contd...)

Part	Grade	Maximum section thickness mm	inch	Weight kg	lb	Martempering conditions: Temperature of salt °C	°F	Minimum time in salt, minute	Required hardness, HRC
Dual gear (f)	4815 (e)	9.4	0.37	2.13	4.7	260	500	2	62 to 63 (b)
Drive coupling	4340	10.2	0.40	0.27	0.6	230	450	2½	52 min (b)
Spline shaft	8720 (e)	10.2	0.40	0.50	1.1	190	375	2½	90(15N scale)
Arbor sleeve	1117L(e)	10.2	0.40	0.59	1.3	205	400	3	–
Screw machine spindle	8620 (e)	10.2	0.40	6.35	14.0	205	400	3	–
Driving barrel	4350	12.7	0.50	0.45	1.0	245	475	3	48 to 52 (g)
Bearing race (h)	52100	12.7	0.50	13.2	29.2	220	425	2½	63 to 64 (b)
Hog knive	9260	15.2	0.60	8.16	18.0	175 (j)	350 (j)	15	62 (b)
Landing-gear spring	6150	19.1	0.75	14.7	32.5	260	500	2¾	56 to57 (b)
Internal gear	1117L (e)	25.4	1.00	1.36	3.0	205	400	3	–
Spur pinion gear (k)	4047	25.4	1.00	16.4	36.2	230 (c)	450 (c)	3	50 to 52 (m)
Screw-machine sprocket	8620 (e)	38.1	1.50	9.07	20.0	205	400	3	–

(a) Salt contained 1½ per cent water; (b) As quenched; (c) Salt contained water; (d) 6.4-mm (¼-inch) diam by 203 mm (8 inch) long; (e) Carburised; (f) 124-mm (4 ⅞ inch) OD by 32-mm (1¼-inch) ID by 102 mm (4 inch); (g) Final; (h) 224 mm (8 13⁄16 inch) ID by 251 mm (9 ⅞ inch) OD; (j) Salt contained 1 per cent water; and (k) 19 mm (¾ inch) OD by 92-mm (3 ⅝ inch) ID by 140 mm (5½ inch). As-quenched hardness of teeth.

SELECTION OF AUSTENITISING EQUIPMENT

Austenitising of steel prior to martempering may be done in virtually any standard furnace. This phase of the operation has been successfully accomplished in furnaces ranging from small, simple box furnaces to large, fully automated, high-production installations. Both atmosphere-controlled furnaces and molten salt baths are widely used. Fluid beds are also being used to austenitise loads prior to martempering. The choice of austenitising equipment depends mainly on availability, shape and size of workpieces, production requirements, and permissible distortion.

Work that is austenitised in a gaseous atmosphere may be oxidised while being transferred to an oil or salt martempering furnace. However, no special considerations are necessary when salt media are used in both the austenitising and martempering furnaces.

Chloride salts are preferred for austenitising before martempering in salt, because chloride salts are compatible with, and are easily separated from, the martempering bath. The composition and characteristics of a typical chloride salt medium are as follows: composition, 45 to 55 per cent sodium chloride and 45 to 55 per cent potassium chloride: melting range, 650 to 675°C (1200 to 1250°F); working range, 705 to 900°C (1300 to 1650°F). If it is necessary to austenitise in a bath that contains cyanide, such as a liquid carburising bath, the work should be transferred to a neutral salt (chloride type) rinse at the austenitising temperature prior to martempering in salt. Direct quenching from a noncyanide-type liquid carburising bath is permissible. Salt bath furnaces for austenitising (and neutral rinsing, if used) are most commonly of the submerged-or immersed-electrode type, although externally heated pots also are satisfactory.

Table 13.5. Typical applications of martempering steel parts in oil.

hardness, Part	Grade	Maximum section thickness mm	inch	Outside diameter mm	inch	Weight kg	lb	Carburising temp., °C	μm	Depth of case 0.001 inch	temp., °C	Temp. of martem- Quenching oil (a)°C	Surface pering HRC
Sleeve	52100	3.2	0.125	–	–	0.1	¼	–	–	–	790	165	58–59
Spacer plate	1065	3.2	0.125	–	–	0.1	¼	–	–	–	790	165	56–57
Bushing	1117	4.8	0.1875	51.0	2.009	0.2	½	910	1015–1220	40–48	910	190	58–62
Bushing	1117	6.4	0.25	76.3	3.0034	0.6	1¼	910	1015–1220	40–48	910	190	55–60
Shifter rail	1018	9.5	0.375	–	–	1.0	2⅛	845(b)	255–455	10–18	845	165	55–60
Shifter rail	1018	9.5	0.375	–	–	1.4	3⅛	845(b)	355–610	14–24	845	165	55–60
Spur gear	8620	12.7	0.5	320.6	12.620	12.7	28	925	1145–1525	45–60	845	150	55–60
Helical gear	4620 H	19.1	0.75	331.5	13.050	16.9	37.2	925	760–1015	30–40	845	150	58–63
Herringbone gear	4820	19.1	0.75	283.2	11.150	16.3	36	925	1145–1525	45–60	845	150	55–61
Shifter rail	1141	25.4	1.0	25.4	1.0	0.9	1⅞	885(b)	455–660	18–26	885	165	45–50
Spiral bevel gear	4620	25.4	1.0	210.6	8.29	5.1	11.25	925	1015–1270	40–50	845	150	55 min
Helical pinion	8617H	25.4	1.0	35.8	1.409	0.4	0.9	925	510–710	20–28	845	150	58–63
Spur gear	8625	31.8	1.250	83.8	3.300	4.3	9⅜	925	1525–1725	60–68	925	190	58–62
Spur gear	4817H	34.0	1.340	186.7	7.350	8.6	19	925	1400–1780	55–70	845	150	58–63
Spur gear	8625	38.1	1.500	165.1	6.500	2.5	5½	925	1525–1725	60–68	925	190	58–62
Splined shaft	8625	39.7	1.564	39.7	1.564	2.7	5⅞	925	1400–1980	70–78	925	190	58–62
Spur gear	8625	44.4	1.750	108.0	4.250	3.5	7¾	925	1525–1725	60–68	925	165	58–62
Splined shaft	8625	44.4	1.750	44.4	1.750	2.0	4½	925	1525–1725	60–68	925	190	58–62
Splined shaft	8620	50.8	2.000	50.8	2.000	5.1	11¼	925	1525–1725	60–68	925	165	58–62
Splined shaft	8625	65.0	2.559	65.0	2.559	6.8	15	925	1525–1725	60–68	925	165	58–62
Spur gear	8625	84.7	3.3343	245.5	9.667	11.9	26¼	925	1525–1725	60–68	925	190	58–62

(a) Minimum time in oil, 5 min; and (b) carbonitriding temperature.

SELECTION OF MARTEMPERING EQUIPMENT

The furnace used for martempering is essentially a heat exchanger. Its basic functions are: (i) to absorb heat from the work being quenched; and then (ii) to dissipate this heat to the surroundings, to maintain a constant temperature.

In its simplest form, the martempering furnace consists of a steel pot that contains the oil or nitrate-nitrite salt and that is heated internally or externally. Such a simple furnace can be successfully used for martempering in limited production quantities.

For continuous production, more complex equipment is needed to maintain optimum quenching conditions.

Fuel-fired (usually gas-fired) immersion tubes, electrodes or immersion heaters, located across the back wall and sides of the furnace, are also used for internal heating. Occasionally, furnaces are externally heated by fuel or electricity, but such furnaces are limited to relatively small installations because they are difficult to control.

With internal heating, furnace size is unlimited and is based on production requirements. Sizes may range from 0.06 m^3 (2 ft^3) to lengths of more than 18 m (60 ft) and depths of more than 14 m (45 ft).

The operating temperature range for martempering usually is 165 to 595°C (325 to 1100°F). Temperature is measured by one or more thermocouples (depending on the size of the furnace), which are connected to controlling pyrometers that automatically control temperature within ±3°C (±5°F) by actuating the heating or cooling systems as required.

In continuous production, heat input from the work usually exceeds heat losses by radiation. Therefore, arrangements for cooling as well as for heating the martempering medium are required. To supplement heat losses by radiation from the surface of the bath, the exterior surface of the pot may be designed with cooling fins so that additional heat can be removed by forcing air through the cooling chamber between the pot and the casing walls. To increase heat dissipation, atomised water may be added to the stream of air. The mixture is then passed through a heat exchanger that is placed in the bath.

Agitation of the molten salt greatly improves the rate at which heat is extracted from the work. The furnace can be provided with a propeller-type pump that delivers the molten salt to the quench header into which the hot work is placed for quenching. By directing the flow of salt upward or downward through this quench header, effective control of the quenching severity of the salt is maintained, particularly if the speed of the pump can be varied. The salt also may be agitated by propeller mixers, centrifugal pumps or air bubblers located to produce effective agitation in the quench area. The use of air bubblers is not recommended, because they are not efficient and may cause carbonate buildup in the bath.

The furnace may have a second chamber in which the contaminating chlorides of barium, sodium and potassium that may be carried over from the austenitising salt bath are separated by gravity from the nitrate-nitrite martempering salt. Contaminated salt is continuously circulated through the separating chamber, and, with a drop in salt temperature, more chlorides are precipitated from solution and settle to the bottom of the separating chamber. Because clarification is continuous, a uniform quenching condition is provided at all times. When work is austenitised in an atmosphere-controlled furnace, this chamber is not required.

The equipment required for martempering in hot oil is essentially the same as that for martempering in salt. Although the operating temperature range of oil baths is lower (95 to 230°C, or 200 to 450°F), the problem of maintaining a constant bath temperature is the same as for salt.

Example 13.1. The equipment required for martempering miscellaneous carburised parts (70 gram to 1.0 kg, or 0.15 to 2.2 lb, per piece) made from 4024 and 4028 steels, at a rate of 455 kg/hr. (1000 lb/hr.), is listed in Table 13.6. Oil at 190°C (375°F) was the martempering medium used in this installation.

The parts were carburised at 915 to 925°C (1680 to 1700°F) to a depth of 0.5 to 0.75 mm (0.020 to 0.030 inch) in a radiant-tube, gas-fired, three-row continuous pusher furnace with automatic quenching facilities. Parts were quenched from the final zone at 895 to 905°C(1640 to 1660°F). The carburising atmosphere consisted of endothermic gas provided by a gas generator and 4.8 m^3 hr. (170 ft^3/hr.) of natural gas.

Table 13.6 Equipment requirements for oil martempering carburised parts made of 4024 and 4028 steels (Example 1).

Production requirements	
Production rate	455 kg/h (1000 lb/h)
Weight of each piece	70 g to 1.0 kg (0.15 to 2.2 lb)
No. of pieces per hour	Variable
Equipment requirements	
Capacity of quench tank	18925 1 (5000 gal)
Type of oil	Mineral oil (viscosity at 99°C or 210°F, 110 SUS)
Temperature of oil	190°C (375°F)
Agitation	High and low, as required

Example 13.2. In the operation for which equipment requirements are detailed in Table 13.7, oil at 150°C (300°F) was used as the martempering medium. In this operation, a 455-kg (1000-lb) load of 1.5 kg (3.3-lb) parts made of 8617 steel was quenched from an automatic batch-type furnace.

These parts were carburised to a depth of 1.0 mm (0.040 inch) in a radiant tube, gas-fired batch furnace with automatic quenching facilities. The parts were carburised at 925°C (1700°F) and cooled in the furnace to a quenching temperature of 845°C (1550°F). The carburising atmosphere consisted of endothermic gas and natural gas.

Table 13.7. Equipment requirements for oil martempering carburised parts made of 8617 steel (Example 2).

Production requirements	
Weight of load	455 kg (1000 lb) net
Weight of each piece	1.5 kg (3.3 lb)
No. of pieces treated per hour	75
Equipment requirements	
Capacity of quench tank	7570 1 (2000 gal)
Type of oil	Mineral oil with additives (viscosity at 38°C (100°F), 250 SUS)
Temperature of oil	150°C (300°F)
Agitation	Direct flow (a)

(a) Agitation provided by two 3.7-kW (5-hp) motors driving 455-mm (18-inch) propellers at 370 rpm, causing the oil to flow at a rate of 915 mm/s (36 inch/s).

Example 13.3. Table 13.8 lists details of equipment required for salt martempering 1.1 kg (2.5 lb) transmission shafts, made of 5040 steel, at a rate of 170 kg/h (375 lb/h). The complete treatment was as follows:

1. Austenitise for 35 minute in a neutral chloride salt at 845°C (1550°F) .
2. Martemper for 5 minute at 260°C (500°F).
3. Air cool (30 minute) to 65 to 95°C (150 to 200°F).

4. Temper for 45 minute at 425°C (800°F) .
5. Air cool (5 minute) to 95 to 120°C (200 to 250°F).
6. Wash, rinse and dry.

Hardness after tempering and cooling to room temperature was 40 to 42 HRC (required hardness was 38 to 42 HRC).

Table 13.8 Equipment requirements for salt martempering automotive transmission shafts of 5040 steel (Example 3).

Production requirements	
Weight of each piece	1.1 kg (2.5 lb)
Pieces per fixture	14
Production per hour:	
No. of loaded fixtures	10.7
No. of pieces	150
Weight of pieces	170 kg (375 lb)
Equipment requirements	
Martempering furnace	Steel salt pot with resistance immersion heaters (70 kW)
Size of chamber (a)	1.22 m by 510 mm by 560 mm (48 by 20 by 22 inch)
Size of chloride-separating chamber	380 by 815 by 940 mm (15 by 32 by 37 inch)
Capacity of salt pot	3630 kg (8000 lb)
Type of salt	Nitrate-nitrite
Operating temperature	260 ±3°C (500 ±5°F)
Agitation	One 2.2-kW(3-hp) 150-mm (6-inch) propeller pump
Cooling system (b)	One 0.25-kW (⅓ hp) blower (25.5 m^3/minute or 900 ft^3/minute)

(a) Total depth of salt was 940 mm (37 inch). (b) Cooling capacity of system was 215 kg/h or 475 lb/hr. (gross) from 845 to 230°C (1550 to 450°F) without exceeding 230°C.

Example 13.4. Details of salt martempering furnace used in a commercial heat treating plant for quenching up to 180 kg/h (400 lb/h) are given in Table 13.9. This salt bath is capable of being operated at up to 400°C (750°F) and therefore can also be used for austempering.

Table 13.9. Equipment requirements for salt martempering α variety of steel parts (Example 4).

Martempering furnace	Steel salt pot
Method of heating	100-mm (4-inch) immersion tube fired by natural gas (a)
Rated heat input	38.4 kW (131 000 Btu/hr.)
Operating temperature range (b)	205 to 400°C (400 to 750°F)
Capacity of salt pot	1725 kg (3800 lb)
Type of salt	Nitrate-nitrite
Size of chloride-separating chamber (c)	205 mm by 1.07 m (8 by 42 inch)
Agitation method	0.19-kW (¼-hp) propeller mixer
Cooling method (b)	Air through immersion tube

(a) Gas rated at 39.12 MJ/m^3 (1050 Btu/ft^3). (b) Temperature automatically controlled to ±3°C(±5°F). (c) Depth of salt, 760 mm (30 inch). (d) Cooling capacity, 180 kg/hr. or 400 lb/hr. (gross) from 845 to 260°C (1550 to 500°F) without exceeding 260°C.

MAINTENANCE OF EQUIPMENT

Lack of an established maintenance schedule may result in loss of process control or damage to the equipment, or both.

Salt System

Because martempering baths vary widely in design, shape, size and method of operation, it is not feasible to set forth a standard maintenance schedule. Manufacturers' recommendations for specific equipment should be followed; however, typical schedules for maintaining a salt bath are as follows:

Each 8-hour shift

1. Check instruments and thermocouples against a standard.
2. Check neutrality of austenitising bath (if salt containing cyanide is used for austenitising, the cyanide content should be less than 2 per cent).
3. Remove sludge from martempering bath; mechanical separators (filter baskets or pans) eliminate the need for this operation.
4. Check salt level.
5. Check agitation of bath, and adjust as needed.

Weekly

1. Lubricate all moving parts.
2. Remove sludge or contamination from the surface of the immersion heaters (radiant tubes or electrodes) and from the walls, bottom, and top of the furnace.

Monthly

1. Check operation of all moving parts, such as blowers and pumps; adjust belt tension and alignment of shafts.
2. Check electrical contacts of all contactors and relays, and repair as required; examine all electrical devices for proper operation.
3. Remove fallen parts or debris from the quench header or quench area, to avoid fouling.
4. Check heating and cooling facilities.

Semiannually or annually

1. Remove the salt from the furnace and check the condition of the pot, pumps and heating system.
2. Clean and repair all electrical parts, such as contactors, relays, motors and motor starters (it is especially important to remove condensed salt from all terminals and transformers).
3. Clean and repair all moving parts, and lubricate as required.

Oil System

Following is a typical procedure for maintaining a high-temperature (175 to 205°C or 350 to 400°F) oil-quench system:

Daily

1. Observe oil-temperature indicators every hour; verify the indicated temperature with a potentiometer once a day.

2. Check oil level in sight gauge, to ensure proper level and function of the automatic make-up unit.
3. Check closed system frequently for proper pressure of the atmosphere blanket over the quenching oil; pressure of this blanket should be equal to furnace-atmosphere pressure.
4. Check oil agitation by observation through sight doors and by noting if the pump shaft is operating at the proper speed or, preferably, by monitoring the load on the pump motor.
5. Check condition of oil on parts emerging from quench. Undue discolouration or varnishing may indicate deterioration of oil.
6. Visually check the performance of gas-fired immersion-heating tubes.

Weekly

1. Check condition of oil visually and by viscosity testing; record findings on graph to note trends.
2. Check speed of pump shaft with tachometer, to ensure consitstent oil flow.
3. Check performance of temperature control devices through 'on-off' range, to ensure postitve control.
4. Check thermocouple.
5. Check for proper operation and elevations of quenching mechanism or elevator.
6. Check and clean pilot lights.
7. Check and clean electrodes in ignition system of gas burners.
8. Check safety control on gas lines for heating tubes.
9. Check make-up oil supply.
10. Check motor-operated venting systems for proper operation, and remove carbon buildup to prevent jamming.

Semi-annually

1. Drain oil from, and clean, quench tank.
2. Operate, and inspect functioning of, mechanical elements, such as elevator, oil pumps and adjustable oil deflectors.
3. Check gas-fired heating tubes.
4. Inspect V-belts for pump drive and agitators; replace if necessary.
5. Check condition of temperature-measuring system.
6. Determine condition of oil by physical and chemical testing of various properties.

RACKING AND HANDLING

The techniques for handling parts to be martempered may be similar to those for conventional oil quenching. However, racking and fixturing can be simplified for some martempering applications, because distortion is less.

Example 13.5. Extensive fixturing was required during conventional oil quenching of shaftlike parts made of 52100 steel. These parts were about 180 mm (7 inch) long and had a major diameter of 25 mm (1 inch). The fixtures weighed about the same as the work load.

A change from conventional quenching to martempering for 5 minute in salt at 245°C (475°F) eliminated the need for the heavy, expensive fixtures and made it possible to hold distortion within the required limits. For martempering, the parts were placed vertically in simple fabricated baskets. This practice also resulted in a greater 'payload'.

The manner in which workpieces enter the bath generally is less critical in martempering than in conventional oil quenching. For example, large flat parts do not dish as much when martempered. However, the shape of each part must be considered individually, and some trial and error often is necessary before an optimum handling technique can be developed.

Nesting of small parts can be a problem, and development of a handling method that will result in uniform quenching often requires experimentation.

WASHING THE WORK

Regardless of the martempering medium used, the work load should not be washed until transformation is complete (all portions of workpieces should be near room temperature).

Martempering Salts

Martempering salts are completely water-soluble and any hot-water soaking tank or spray washer will remove all salt from accessible areas. Cold water can be used, but its washing action is much slower.

The speed of washing will depend on how much hot water unsaturated with salt flows over the salt-coated surface. Therefore, agitation will increase the washing action and a device such as an open-impeller sump pump can be used in a soaking tank to direct a stream of hot water into recessed area. Steam jets may be used as supplementary equipment to remove salt from sections of intricately shaped parts with difficult-to-reach recesses for which conventional washing equipment is inadequate.

Quenching Oils

Quenching oils often present washing problems. Martempering oil adheres more tenaciously to workpieces than does conventional quenching oil, because of its higher viscosity—as high as 1200 Saybolt Universal Seconds (SUS) at 38°C (100°F), compared with 100 SUS at 38°C for conventional oil.

Basically, the washing equipment is the same for both types of quenching oils, but a heavy-duty cleaner must be used to remove martempering oil. Several proprietary heavy-duty silicate-alkaline cleaners are available for cleaning parts martempered in hot oil. Vapour degreasing and steam cleaning without detergent, may be used in special applications.

Chapter 14

Hardenability of Steel

INTRODUCTION

Hardness is defined as resistance to plastic deformation or penetration. Hardenability is defined as the ease with which hardness may be attained by quenching. It (hardenability) is also defined as the ability to develop maximum hardness by quenching. It is also defined to have a hardened layer of martensite after quenching and also to have high hardness at the same given depth.

The hardenability is affected by: (i) method of manufacturing of steel; (ii) chemical composition of steel; (iii) critical cooling rate; (iv) quenching medium and method of quenching; and (v) size of piece, smaller pieces can be more uniformly hardened.

If the actual cooling rate at the core centre of the piece is more than the critical cooling rate, the whole cross-section of the piece will acquire martensite structure. In case the actual critical rate at the core is less than the critical cooling rate, the work is hardened only upto certain depth (from the surface towards the centre). Martensite structure will be produced only upto this (above mentioned) certain depth. At the core the austenite will decompose into ferrite + carbide structure i.e., troosite, sorbite or pearlite etc.

The hardenability determines the depth of hardening obtained on quenching, which is usually specified as the distance below the surface where the amount of martensite has been reduced to 50 per cent or more precisely to 50 per cent martensite and bainite. The characteristic property of a steel possessing high hardenability is that it shows a large depth of hardening or that it hardens through entirely in heavy sections. Without electron microscopy it may be difficult to distinguish martensite from lower bainite.

Since the depth of hardening is of great importance to tools and constructional parts it is customary to indicate this property by means of diagrams of the type shown in Fig. 14.1. When the depth of hardening is measured in this way the cooling medium must also be stated. In this instance the plain carbon steel, W1, was quenched in water and the other steels in oil. The heavier the section to be hardened the smaller is the depth of hardening and the lower is the core hardness, which is illustrated in Fig. 14.2. The reason why a steel is harder at the surface than at the centre is explained by referring to a continuous-cooling-transformation (CCT) diagram. By studying the schematic CCT diagram in Fig 14.3 is obvious that since the surface cools at a considerably faster rate than the centre, the cooling curve representing the surface will pass in front of the ferrite and bainite noses and as a result only martensite is formed. At the centre which cools more slowly some bainite will be formed, as may be inferred from the figure, and this will result in a lower hardness in the core.

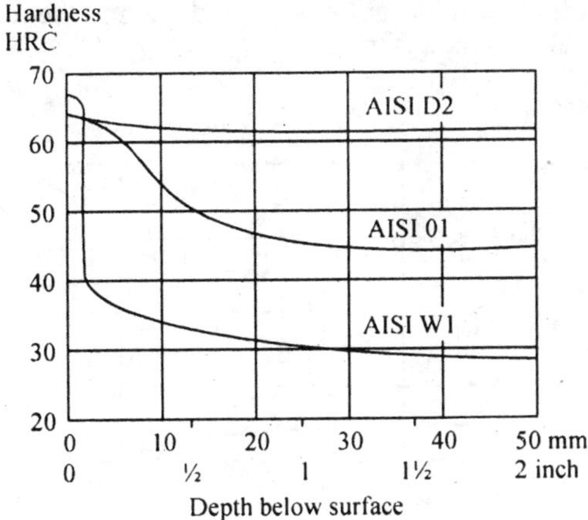

Fig. 14.1. Depth of hardening for various grades of steel. Bars 100 mm diameter. Steel W1 water quenched, the rest oil quenched.

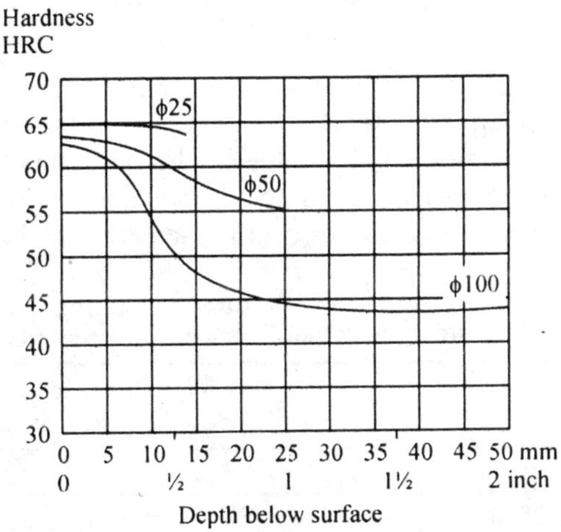

Fig. 14.2. Depth of hardening in different dimensions after oil quenching (AISI 01). Test-piece 25 mm diameter hardened from 800°C in oil. Test-piece 50 mm diameter hardened from 820°C in oil. Test-piece 100 mm diameter hardened from 840°C in oil.

As the dimensions of the steel increase, the rate of cooling decreases and the core hardness will be still further reduced owing to the formation of ferrite and pearlite. The hardness will also decrease when the cooling curve is so displaced as to be to the right of the critical cooling curve.

Fundamental hardenability data are of considerable use of the steel consumer and heat treater and therefore a number of simple methods have been developed whereby hardenability can be determined. Some of the best known methods are described below.

THE GROSSMANN HARDENABILITY TEST

To determine hardenability according to Grossmann's method, a number of cylindrical steel bars of different diameters are hardened in a given cooling medium. By means of metallographic examination the bar that has 50 per cent martensite at its centre is singled out and the diameter of this bar is designated as the critical diameter (D_0), the unit generally being inches.

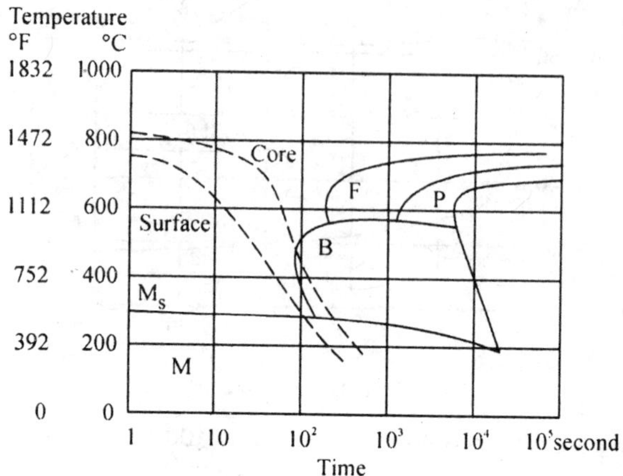

Fig. 14.3. Schematic illustration of the cooling curves for the surface and core, respectively, of an oil-quenched bar, 95 mm diameter. The surface is wholly martensitic; the core contains some upper bainite.

The cooling intensities of the different cooling media have been determined and are called the H-factors. The values of H are given in Table 14.1. Using the appropriate value of the H-coefficient of the cooling medium under consideration, the D_0-value can be converted to the ideal diameter D_i which is defined as the bar diameter which, when the surface is cooled at an infinitely rapid rate ($H = \infty$), will yield a structure, at the centre, containing 50 per cent martensite. The diagrams correlating D_0 and D_i are shown in Fig. 14.4.

Table 14.1. Cooling intensities of different cooling media, known as H-factors.

	Coefficient of severity of quench H		
	Cooling medium		
Agitation	*Oil*	*Water*	*Brine*
None	0.25–0.30	0.9–1.0	2.0
Mild	0.30–0.35	1.0–1.1	2.0–2.2
Moderate	0.35–0.40	1.2–1.3	–
Good	0.4–0.5	1.4–1.5	–
Strong	0.5–0.8	1.6–2.0	–
Violent	0.8–1.1	4.0	5.0

The value of D_i obtained is hence a measure of the hardenability of the steel and is independent of the cooling medium. In practice the D_i-values are used to determine the values of D_0 for bars quenched in various cooling media, using the diagrams in Fig. 14.4.

Example

By subjecting a steel whose D_i-values is 2.0 in to an oil quench, the H-coefficient of which is 0.4, it would yield a D_0-value of 0.8 inch. Familiarity with concept of hardenability enables a good indication of the hardenability of a steel to be obtained from its D_i-value since this is a useful figure for comparison purposes.

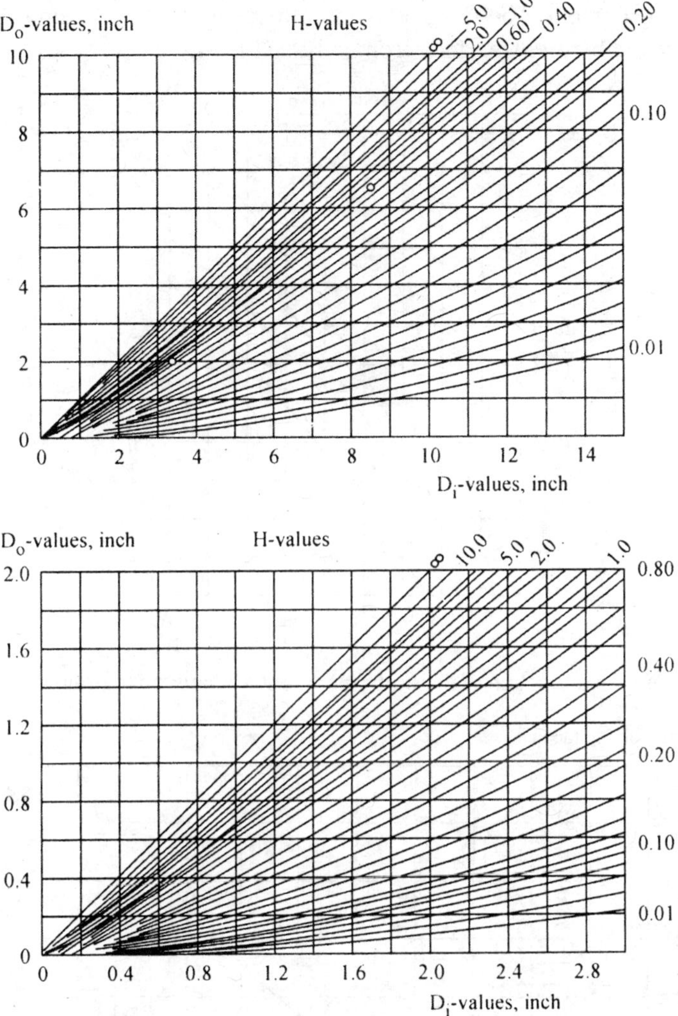

Fig. 14.4. Charts showing the correlation between critical diameter D_o, ideal dimeter D_i and H-value. The lower diagram is an enlargement of the lower, left-hand portion of the upper diagram.

Calculation of D_i Values from the Chemical Composition

The hardenability may be calculated from the composition of low-alloy and medium-alloy steels, taking into account only the amount of each element in solution at the austenitising temperature. The austenite grain size must also be taken into consideration.

The smaller the grain size the lower is the hardenability. This is due to the fact that the surface area of the grain boundaries increases as the grain size decreases, thereby in turn giving rise to an increasing number of nuclei which serve as initiating points for pearlite information. The computation starts from the C content and the grain size. By means of Fig. 14.5 a 'base' hardenability characteristic for D_i is obtained. For the other alloying elements the curves in Fig. 14.6 indicate the multiplying factor that corresponds to each alloy content. The factors given in this diagram have been selected from a number of test results and approved by the American Iron and Steel Institute.

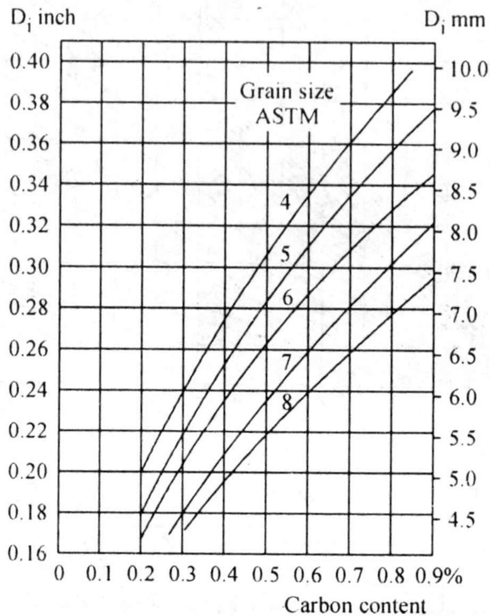

Fig. 14.5. The ideal diameter as a function of the carbon content and austenite grain size for plain carbon steels.

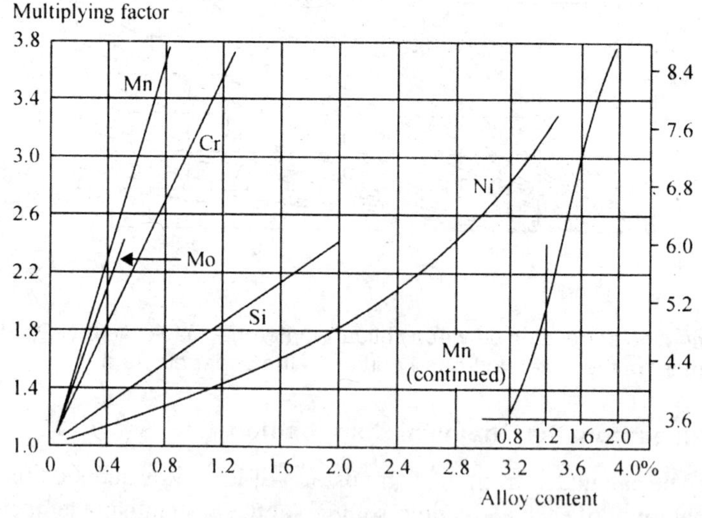

Fig. 14.6. Multiplying factors for different alloying elements for hardenability calculations.

Fig. 14.5 is applicable to C contents above 0.8 per cent, but only on the assumption that all carbides are in solution at the austenitising temperature. However, this is generally not the case since an unnecessarily high temperature would then have to be employed. Further, complete dissolution of the carbides might result in deleterious grain growth effects and high retained austenite content in the steel. Consequently if conventional hardening temperatures are used for low-alloy steels lying in the higher

C ranges of the equilibrium diagram, a falling-off in the hardenability must be expected when the C exceeds 0.8 per cent. This is because the 'excess carbon' combines with carbide and hardenability-inducing elements such as Cr and Mo. In spite of this reduction in hardenability, steels are still alloyed with about 1 per cent C and more, but under these circumstances the carbides are beneficial in increasing the water resistance of the steel.

Examples of hardenability calculations

Steel SS 2225, ASTM grain size 7, has the following composition (In per cent):

C	Si	Mn	Cr	Mo
0.25	0.3	0.7	1.1	0.2

From Fig. 14.5 the base value of D_i is 0.17 inch. On multiplying this value with the appropriate factors we obtain:

$$D_i = 0.17 \times 1.2 \times 3.3 \times 3.4 \times 1.6 = 3.7 \text{ inch}$$

Steel SS 2541 (BS 816M40), ASTM grain size 6, the following composition (In per cent):

C	Si	Mn	Cr	Ni	Mo
0.35	0.3	0.7	1.4	1.4	0.2

$$D_i = 0.22 \times 1.2 \times 3.3 \times 4.0 \times 1.5 \times 1.6 = 8.4 \text{ inch}$$

The D_i-values obtained may be converted to D_0 by means of diagrams in Fig. 14.4 as described above. For example, by quenching in oil with moderate agitation (H = 0.40) the critical diameter of steel SS 2225 is D_0= 2 inch and of steel SS 2541, D_0= 6.4 inch. These values are marked in the diagram, Fig. 14.4. Values of D_i, calculated as above, are only approximate but they are useful as a means of comparing the hardenability of different grades or heats.

The hardenability factors proposed by *Grossman* have been re-examined by *Moser* and *Legat* and a more precise relationship between grain size and C content has been obtained and this is presented in Fig. 14.7. The revised multiplying factors by *Moser* and *Legat* are shown in Fig 14.8.

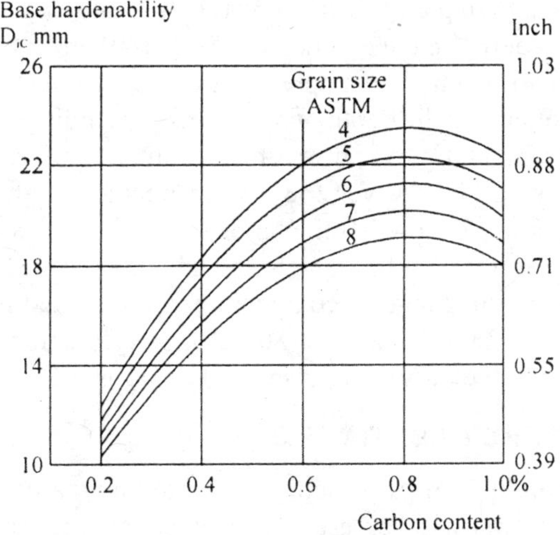

Fig. 14.7. Relationship between base hardenability, carbon content and grain size as obtaining in actual practice.

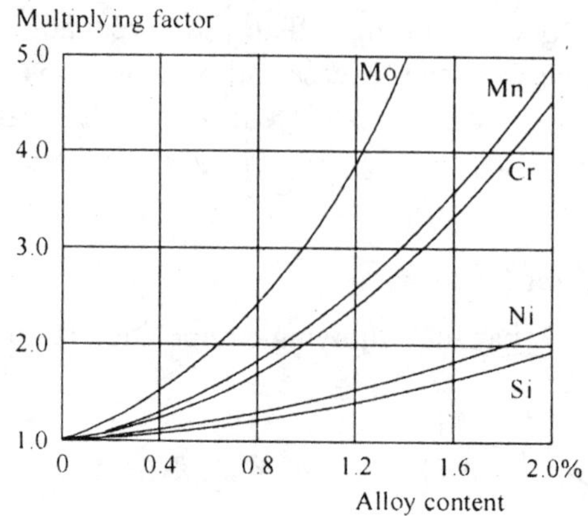

Fig. 14.8. Multiplying factors for calculating the hardenability as influenced by Mo, Mn, Cr, Si and Ni.

Using the results obtained by *Moser* and *Legat* the hardenability may be calculated from the following expression:

$$D_i = D_iC \times 2.21^{\text{Per cent Mn}} \times 1.40^{\text{Per cent Si}} \times 2.13^{\text{Per cent Cr}} \times 3.275^{\text{Per cent Mo}} \times 1.47^{\text{Per cent Ni}}$$

Recalculated values of D_i for the steels in the examples are as follows. (Grossmann's values are given in brackets).

SS 2225 $D_i = 2.5$ inch
SS 2541 $D_i = 8.0$ inch

Kramer, *Siegel* and *Brooks* have also examined Grossmann's equations and have published a diagram that is practically identical with the one in Fig. 14.7. *Jatczak* has extended this diagram by adding multiplying factors for other alloying elements. Fairly close agreement with the values of *Moser* and *Legat* were obtained in the case of low and medium-carbon steels. *Jatczak* has also published multiplying factors for steels containing C between 0.90 per cent and 1.10 per cent and which take account of the hardening temperatures. For these cases calculations can proceed straight from the composition of the steel without the necessity of applying any corrections for such alloying elements as are not in solution in the austenite. A comparison between the diagrams in Figs. 14.9 and 14.10 shows the effect of raising the hardening temperature. As a practical example illustrating the effect produced by the C content in the range mentioned, it was found that the depth of hardening in grindingmill rods, made from a shallow-hardening steel, increased from 11 to 15 mm as the C content was reduced from 0.90 per cent to 0.85 per cent. *Jatczak's* factor for Si, however, are not in agreement with the results obtained at Bofors where it has been shown beyond question that the hardenability of steels containing about 1 per cent C first falls as the Si content increases to about 1 per cent and then rises as the Si increases to about 2 per cent. The changing influence of Si in this range of composition is illustrated in Fig. 14.11 which shows the hardenability, denoted by the Jominy distance to 500 HV, as a function of the Si content. As the Mn content increases, the troughs of the curves are displaced towards higher Si contents.

JOMINY END-QUENCH HARDENABILITY TEST

On account of the high cost of Grossmann's method it is used now-a-days only to a very small extent, albeit it must be regarded as the most exact test. The most commonly used method at present has been developed by *Jominy*. For this test a round bar specimen is used, 25 mm in diameter and 100 mm in length.

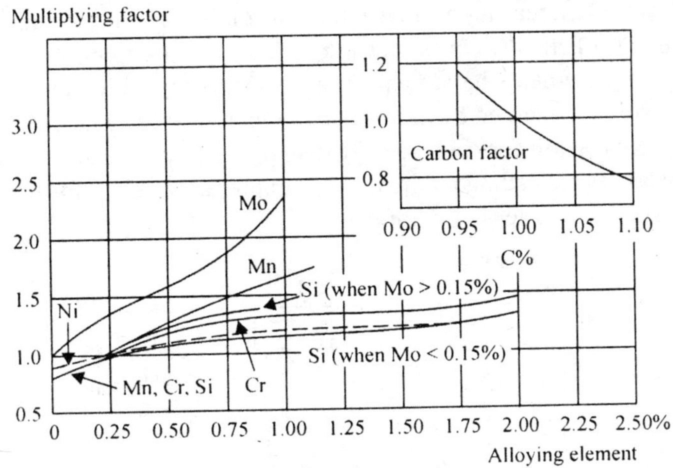

Fig. 14.9. This chart is used for determining the multiplying factors for high-carbon steels austenitised at 830°C (1525°F (base D_i is 1.13).

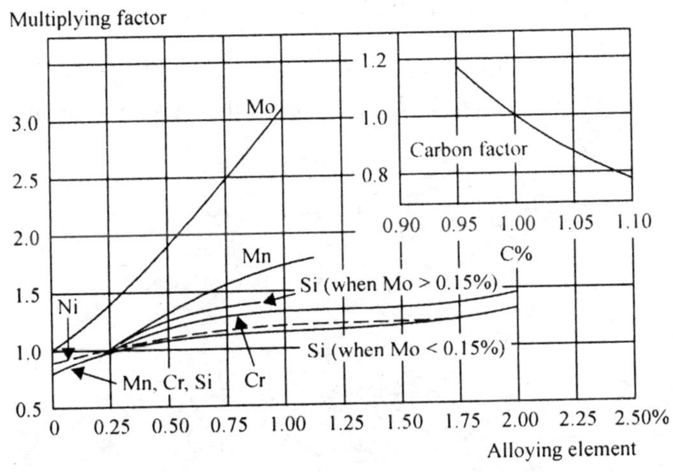

Fig. 14.10. This chart is used for determining the multiplying factors for high-carbon steels austenitised at 860°C (1575°F) (base D_i is 1.35).

The specimen is heated to the hardening temperature of the steel with holding time of usually 20 minutes. One end-face of the specimen is quenched by spraying it with a jet of water, as illustrated in Fig. 14.12. Hereby the rate of cooling decreases progressively from the quenched end along the length of the bar. When it is cool, two diametrically opposite flats, 0.4 mm deep and parallel to the axis of the bar are ground and the hardness is measured along the flats. The hardness values are plotted in a diagram against their distances from the quenched end. Jominy curves for some steel grades are shown in Figs. 14.13 to 14.22. The upper curve represents the maximum hardness values corresponding to the upper composition limit of the steel and the lower curve the minimum hardness values corresponding to the lower limit of the composition range. Together the curves form what is called a Jominy or hardenability band. Using regression analysis and proceeding from the chemical composition, just has derived expressions that enable the hardness at different Jominy distances to be calculated direct. It is found that all alloying elements have an increasing influence up to a Jominy distance of about 10 mm, but beyond this their influence remains practically constant. Carbon starts off at a Jominy distance zero with a factor of 50 but all other alloying elements have factor 0 at distance zero. This implies that the hardness at Jominy distance zero is solely governed by the carbon content. Using the derived factors it is possible to present a hardenability expression for each Jominy distance. Since the factors vary by only negligible amounts for Jominy distances exceeding 6 mm, a single expression can be formulated for the whole test specimen except for distances shorter that 6 mm.

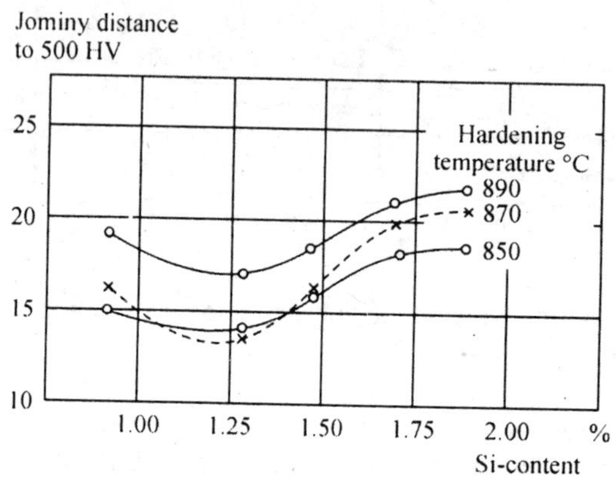

Fig. 14.11. Variations of hardenability with Si content in a steel with nominal composition 0.90 per cent C, 0.75 per cent Mn and 1.0 per cent Cr.

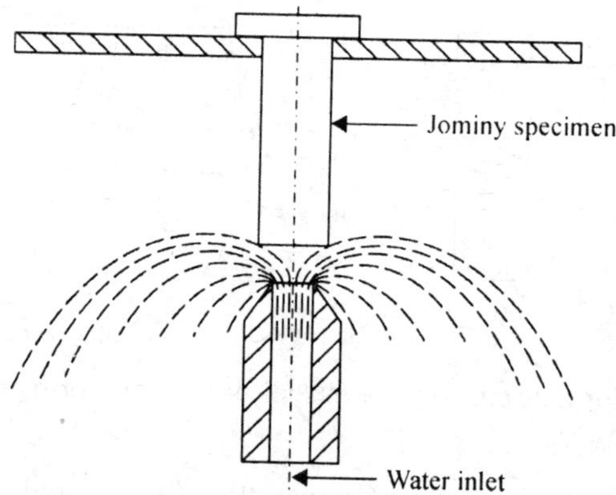

Fig. 14.12. Hardening of Jominy end-quench specimen.

The equation given below is formulated by *Just.*

$$J_{6\text{-}80} = 95\sqrt{C} - 0.0028\, s^2\sqrt{C} + 20\, Cr + 38Mo + 14Mn + 6Ni + 6Si + 39V + 96P - 0.8K - 12\sqrt{s} + 0.9\, s - 13\ \text{HRC}$$

where,

J = Jominy hardness, HRC.

s = Jominy distance, mm.

K = ASTM grain size.

The equation is valid for Jominy distances from 6 to 80 mm. The following limits are specified for the alloying elements:

C < 0.6 per cent , Cr < 2 per cent, μn < 2 per cent, Ni < 4 per cent, Mo < 0.5 per cent, V < 0.2 per cent. For Jominy distances shorter than 6 mm the influence of alloying elements other than C on the hardness is negligible. Hence the following equation holds for the end-face of the Jominy specimen:

$$J_o = 60 \times \sqrt{C} + 20 \text{ HRC } (C < 0.6 \text{ per cent})$$

It is found that the C content on the one hand and the rest of the alloying elements on the other interact mutually on the hardness. For instance, Cr acts more powerfully in medium-carbon steels than in low-carbon, i.e., case-hardening steels. The section already referred to, contains a number of expressions based on available Jominy curves. These expressions are to a great extent dependent on the origins of these curves. Thus, for steels specified by SAE, USS-Altas and MPI-Atlas, the expressions have different forms. Medium-carbon (steels for hardening and tempering) and low-carbon (case-hardening) steels have been given different expressions for hardenability, as seen below:

$J_{6\text{-}40}$ (case-hardening steel)

$$= 74\sqrt{C} + 14Cr + 5.4Ni + 29Mo + 16Mn - 16.8\sqrt{s} + 1.386s + 7 \text{ HRC}$$

$J_{6\text{-}40}$ (steel for hardening and tempering)

$$= 102\sqrt{C} + 22Cr + 21Mn + 7Ni + 33Mo - 15.47\sqrt{s} + 1.102s - 16 \text{ HRC}$$

In order to assess the accuracy of the first-mentioned expression a calculation has been carried out for two case-hardening steels and two steels for hardening and tempering, for which Jominy curves have been determined according to SS. As may be seen from Figs. 14.23 to 14.26, in some instances the conformity is reasonably satisfactory.

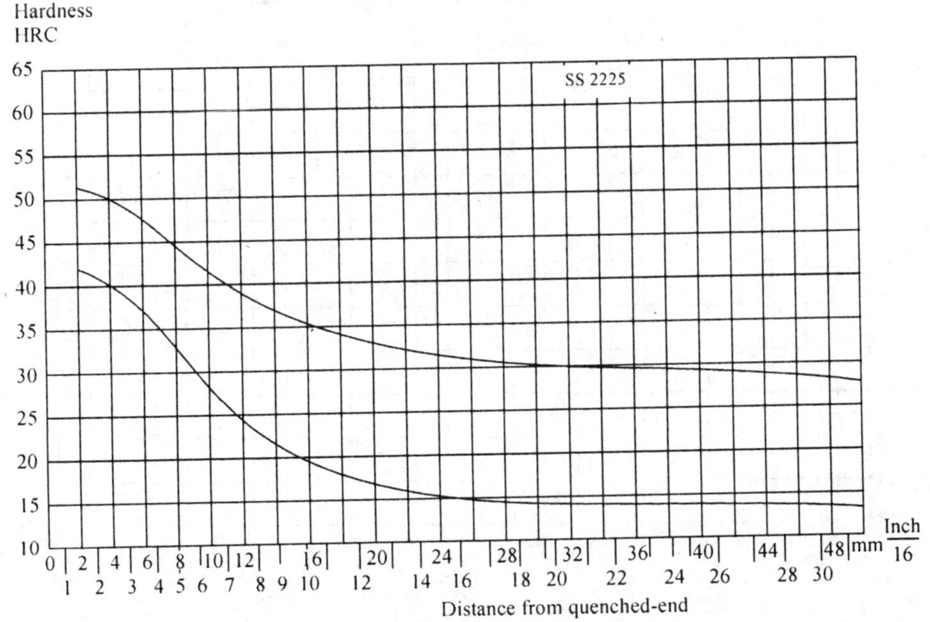

Fig. 14.13. Hardness values plotted against their distance from quench end.

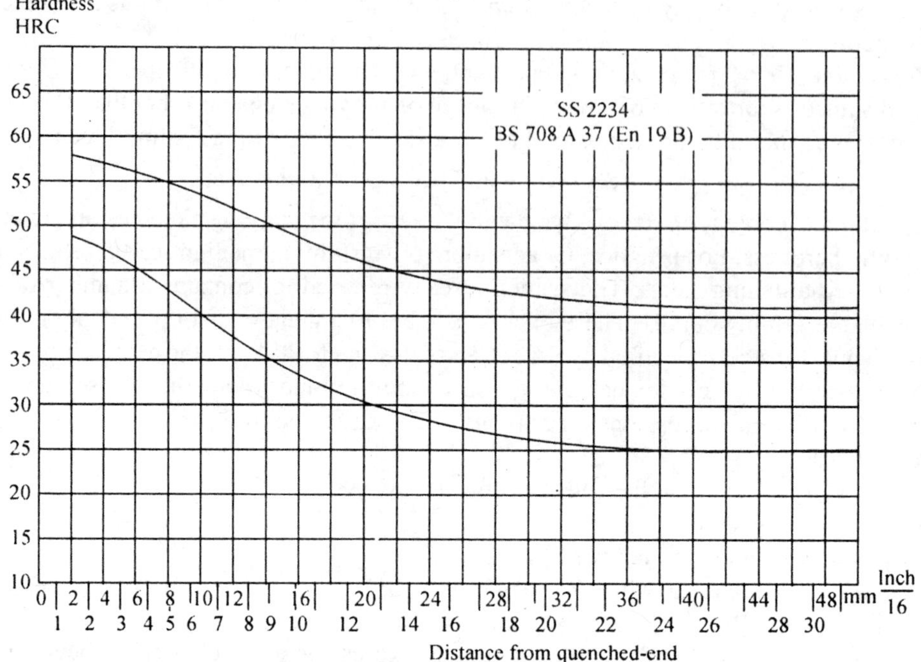

Fig. 14.14. Hardness values plotted against their distance from quench end.

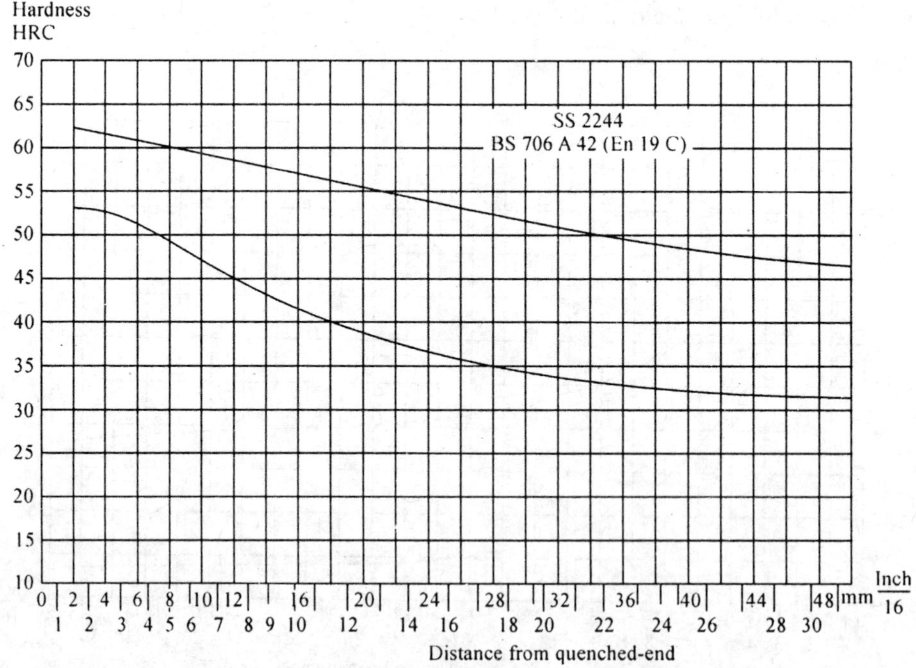

Fig. 14.15. Hardness values plotted against their distance from quench end.

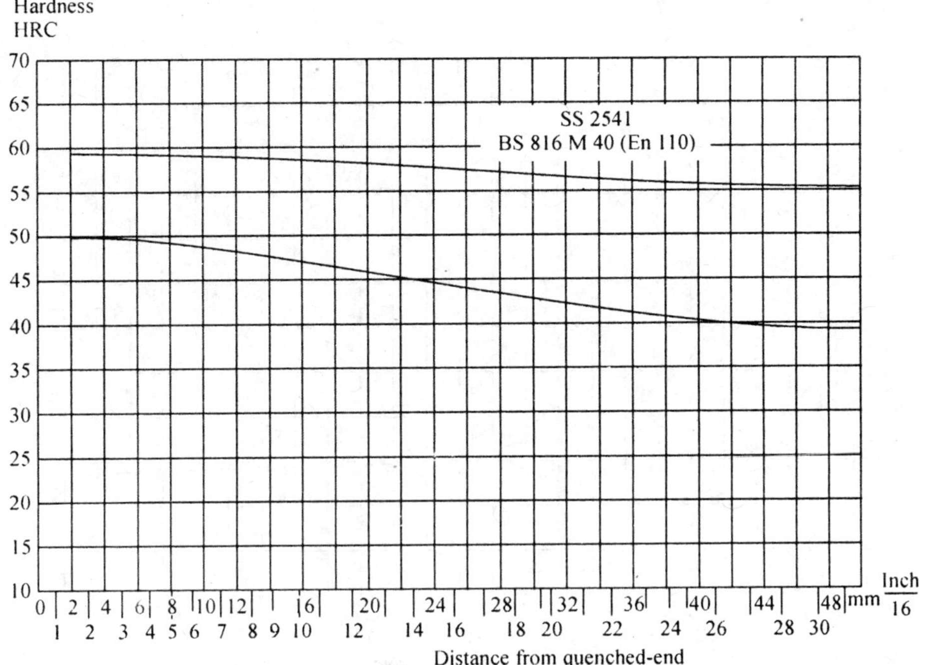

Fig. 14.16. Hardness values plotted against their distance from quench end.

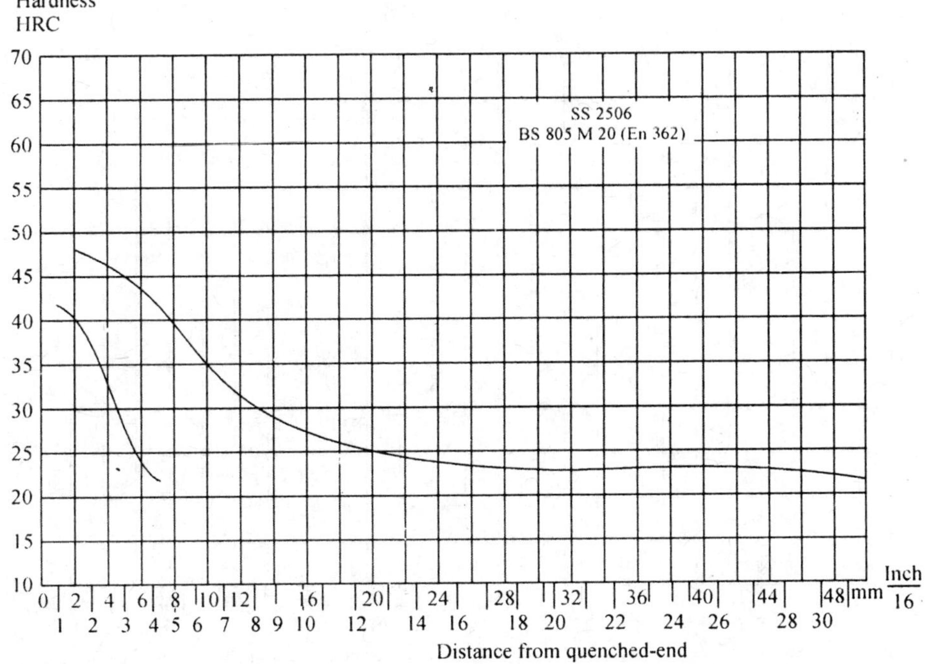

Fig. 14.17. Hardness values plotted against their distance from quench end.

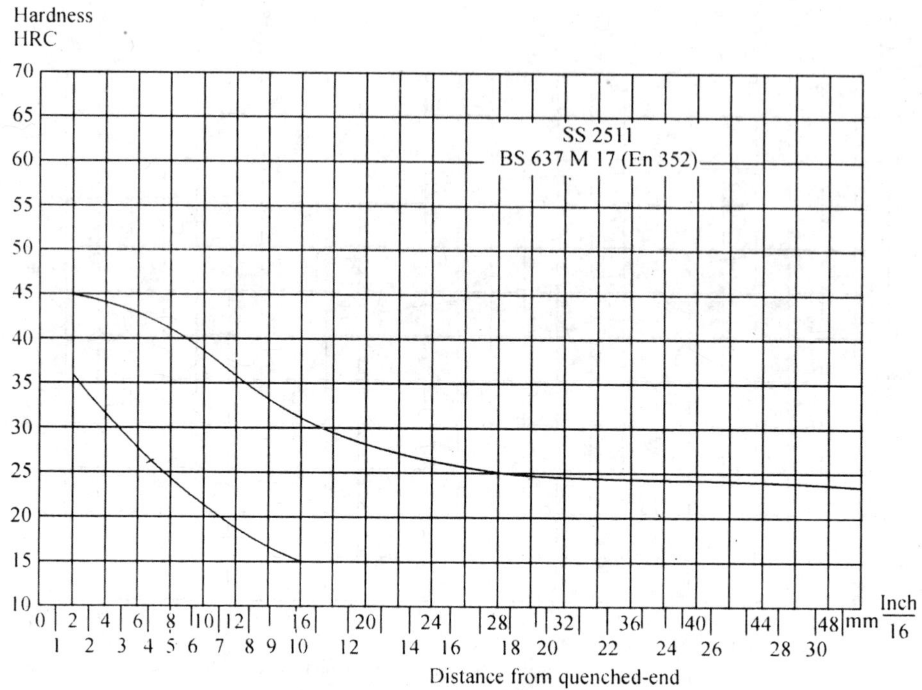

Fig. 14.18. Hardness values plotted against their distance from quench end.

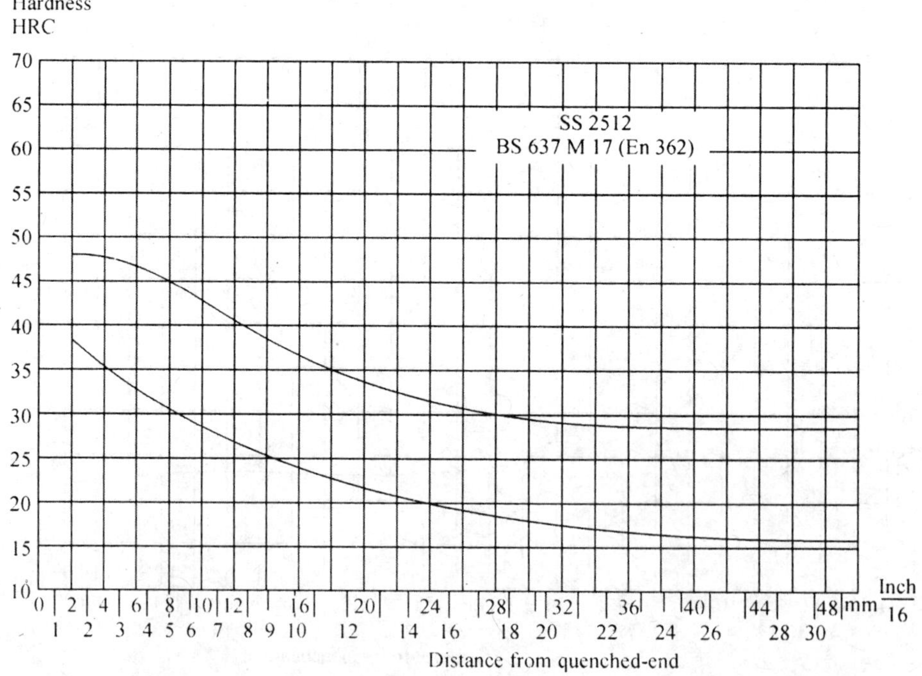

Fig. 14.19. Hardness values plotted against their distance from quench end.

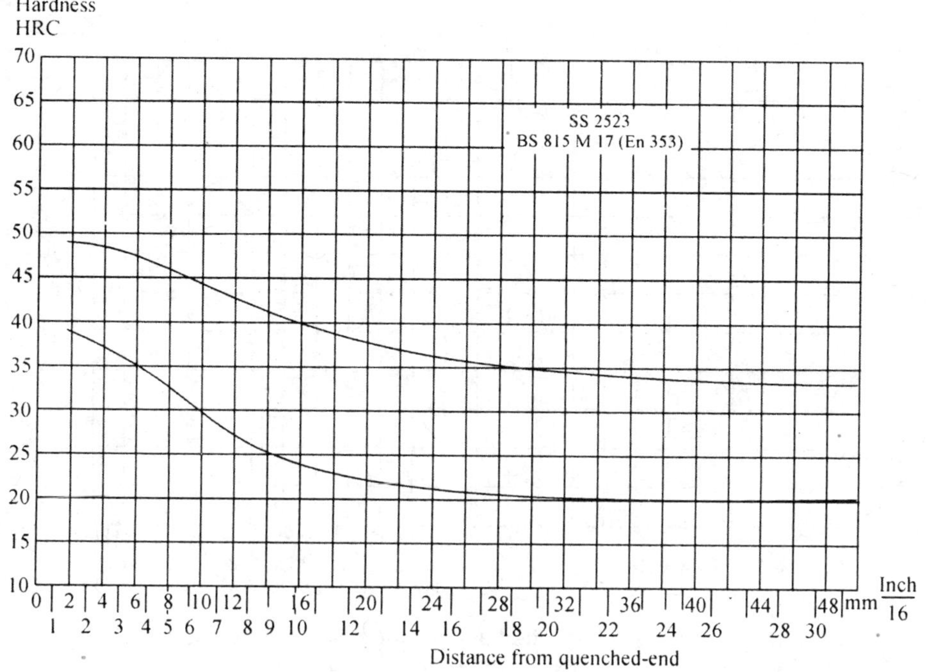

Fig. 14.20. Hardness values plotted against their distance from quench end.

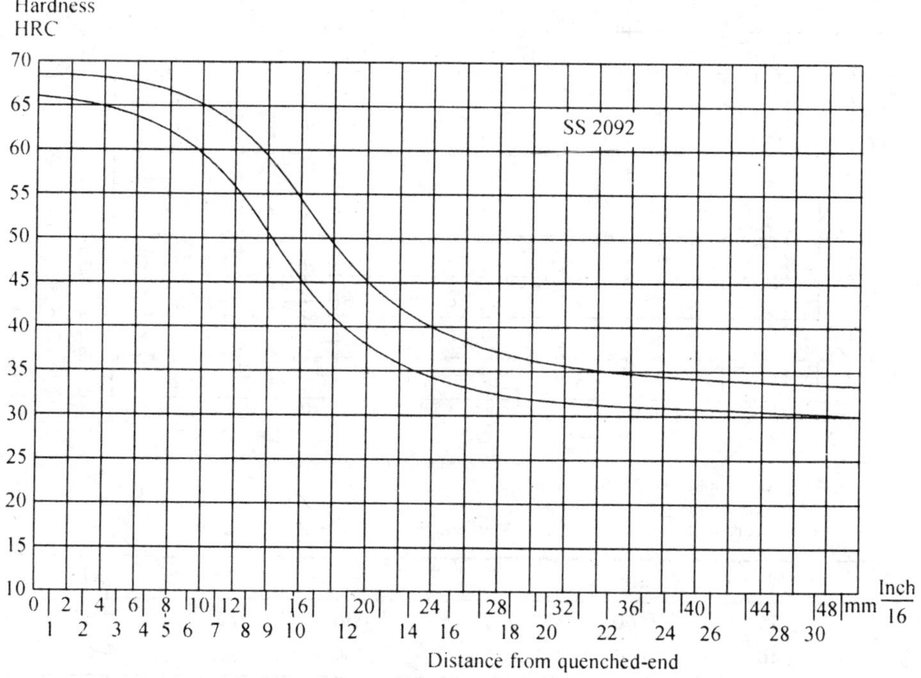

Fig. 14.21. Hardness values plotted against their distance from quench end.

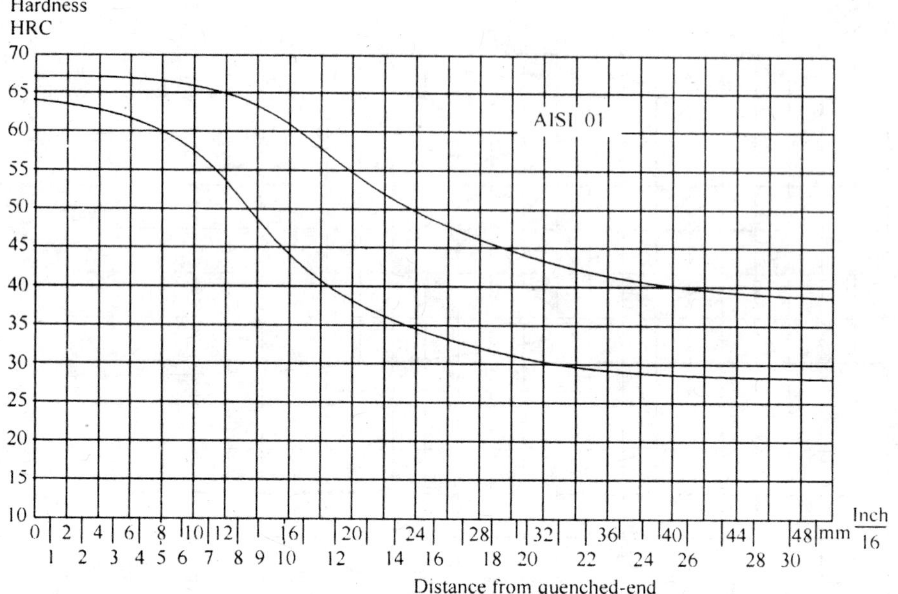

Fig. 14.22. Hardness values plotted against their distance from quench end.

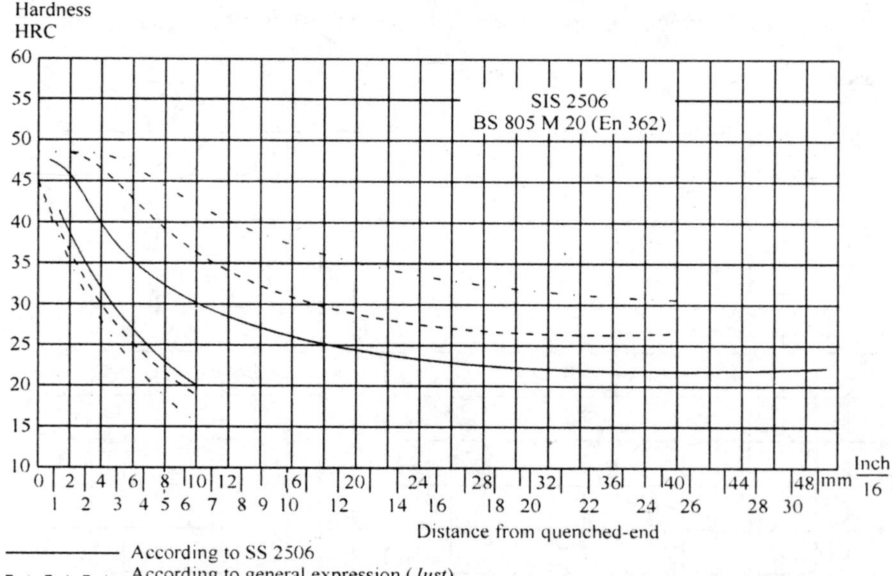

	C	*Si*	*Mn*	*P*	*S*	*Cr*	*Ni*	*Mo*
Min.	0.17	0.15	0.60	–	0.030	0.35	0.35	0.15
Max.	0.23	0.40	0.95	0.035	0.050	0.65	0.75	0.25

Fig. 14.23. Jominy curve for SS 2506 (BS 805M20). Comparison with curves calculated from composition of steel.

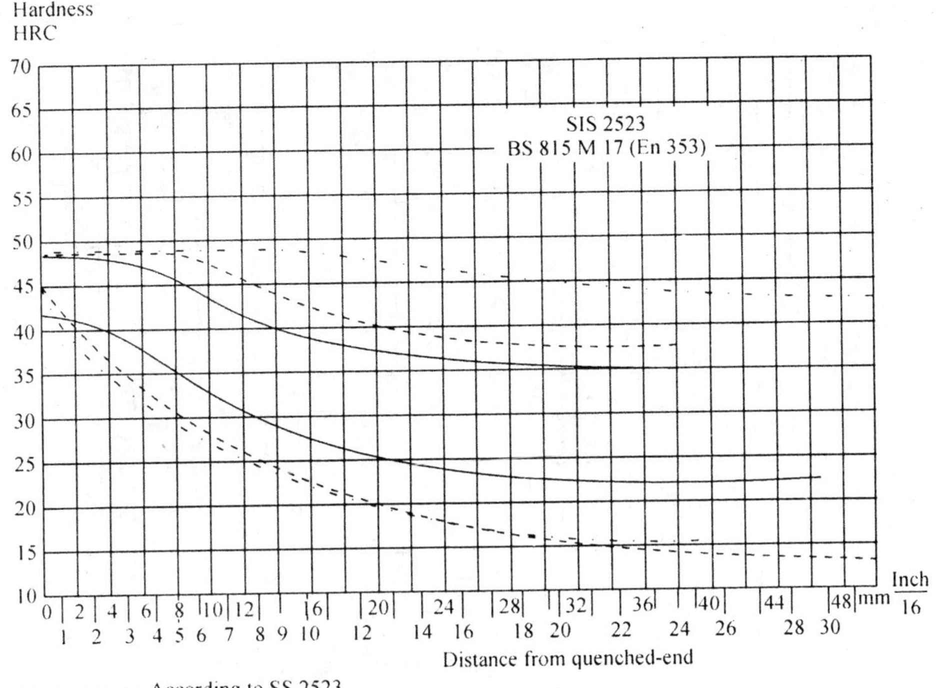

	C	Si	Mn	P	S	Cr	Ni	Mo
Min.	0.17	0.15	0.70	–	0.030	0.80	1.00	0.08
Max.	0.23	0.40	0.10	0.035	0.050	1.20	1.40	0.16

Fig. 14.24. Jominy curve for SS 2523 (BS 815 M17). Comparison with curves calculated from composition of steel *Just*.

Kirdaldy has developed a method for calculating Jominy diagrams by means of computers. He starts from the assumption that the nucleation of ferrite, pearlite and bainite takes place instantaneously; the problem is then to calculate the rate of growth *v(T)*. He had derived, analytically, an expression for the cooling rate at various Jominy distances. He assumes that the cooling rate producing 0.1 per cent of transformed phase at the temperature corresponding to the maximum *v(T)* also corresponds to the Jominy inflection point at which the martensite content is 50 per cent (Fig. 14.27).

By proceeding from the carbon content that gives the maximum hardness and Grossmann's D_I-value, which is an expression for the hardenability, and finally from the location of the point of inflection, it is possible to compute the whole Jominy curve. According to *Kirkaldy* the result obtained from such an exercise is shown in Fig. 14.28.

Ericsson had tested Kirkaldy's formulae by applying them to, *inter alia*, 25 steel grades from three different steel mills. One result that aims at giving the hardnesses at $\frac{1}{16}$ th inch Jominy distances is shown in Fig 14.29. It may be regarded as satisfactory but can be improved further by making use of correction factors that have been adapted to current steel-making processes.

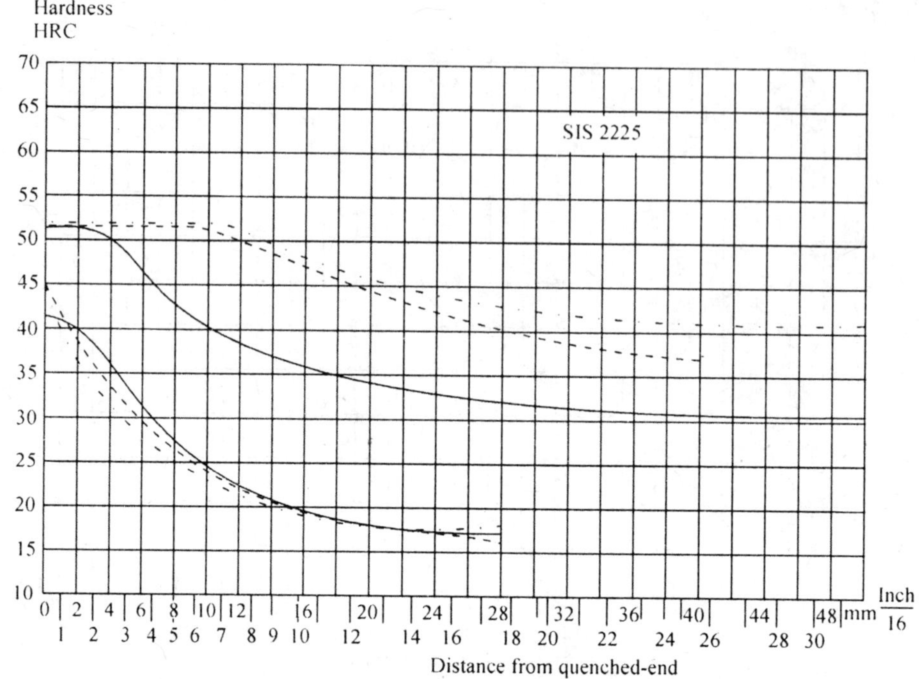

——— According to SS 2225
- · - · - According to general expression (*Just*)
- - - - - According to expression for steel for hardening and tempering (*Just*)

	C	*Si*	*Mn*	*P*	*S*	*Cr*	*Ni*	*Mo*
Min.	0.22	0.10	0.50	–	–	0.90	–	0.15
Max.	0.29	0.40	0.80	0.025	0.020	1.20	0.30	0.25

Fig. 14.25. Jominy curve for SS 2225 (ISO/R 683/2–1968). Comparison with curves calculated from composition of steel.

From practical experience it was found that although they may be of the same type, each individual melting furnace is able to influence in some way the hardenability of the steel even if the chemical composition-as determined by ordinary standard methods—is the same. The difference in the results between the various mills, as may be seen in Fig. 14.29 is, in fact, quite appreciable.

Practical Applications of Jominy Curves

Jominy curves may be used in the first instance to compare the hardenabilities of different heats of the same grades of steel, hereby serving as a valuable method of quality control. For more than 30 years there have been available in the USA special so-called H-steels for which there are specifications defining their hardenability bands (BS 970: Part 2 : 1970 and BS 970 : Part 3 :1971 also contain H-steels).

The curves may also be used to predict the expected hardness distribution obtained on hardening steel bars of different dimensions by cooling in various quenching media. The rates of cooling prevailing at different distances in the Jominy specimen may be compared with the cooling rates prevailing in bars of different diameters cooled in various quenching media.

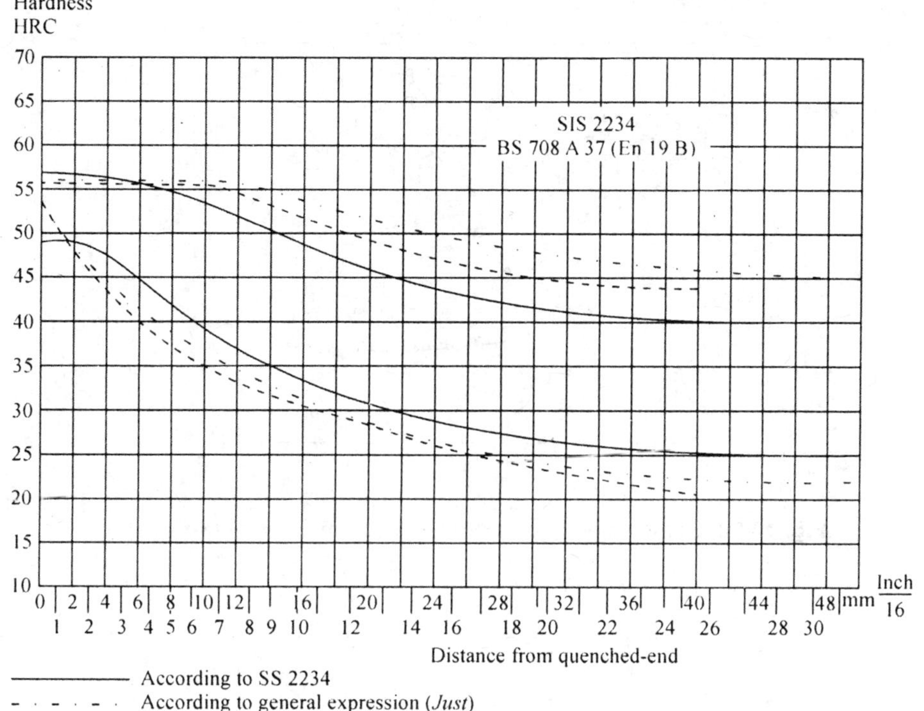

	C	*Si*	*Mn*	*P*	*S*	*Cr*	*Mo*
Min.	0.30	0.10	0.50	–	–	0.90	0.15
Max.	0.37	0.40	0.80	0.035	0.035	1.20	0.25

Fig. 14.26. Jominy curve for SS 2234 (BS 708 A 37). Comparison with curves calculated from composition of steel.

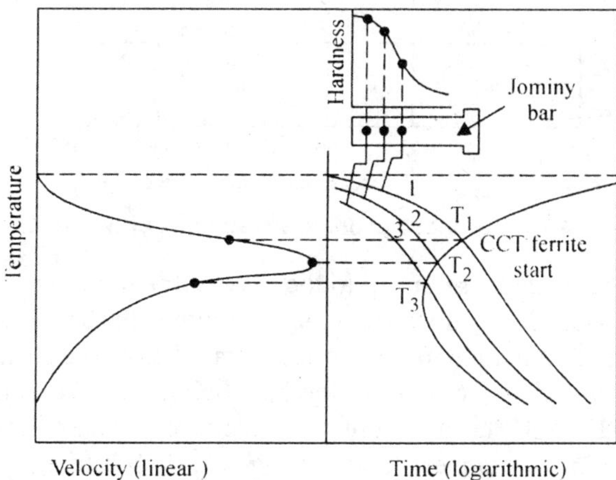

Fig. 14.27. Construction for locating the inflection point of a Jominy curve.

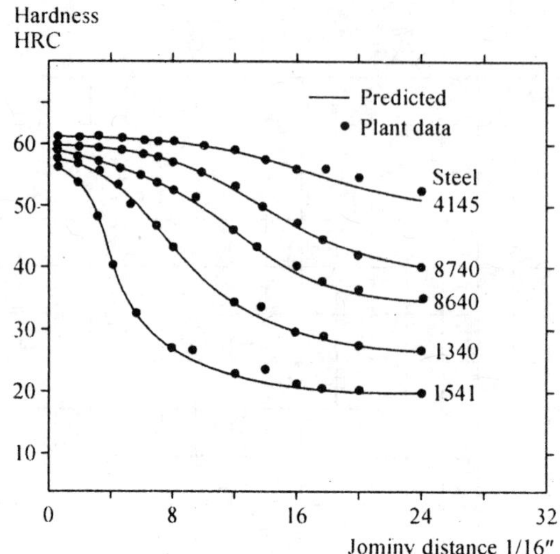

Fig. 14.28. Comparison of predicted and observed Jominy curve.

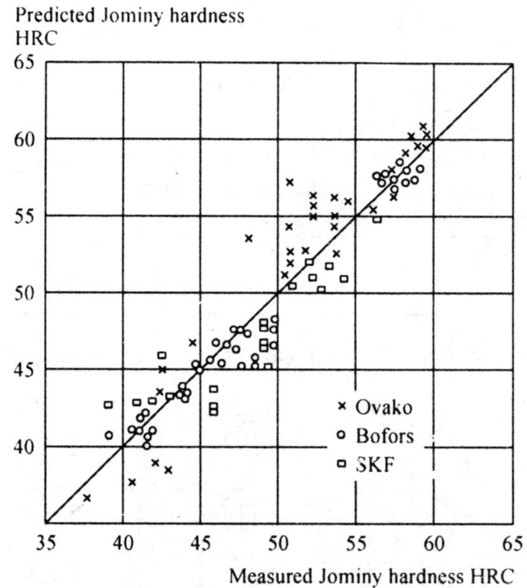

Fig. 14.29. Hardness at Jominy distances of $\frac{1}{16}$ inch.

Such a comparison may be made by referring to the curves drawn in Fig 14.30. From this it may be seen that the rate of cooling at a Jominy distance of, say, 14 mm is the same as that prevalent at a point 2 mm below the surface of a 75 mm diameter bar, or at 10 mm below the surface of a 50 mm diameter bar, or at the centre of a 38 mm diameter bar; all the bars being quenched in moderately agitated oil. Using the Jominy curves and Fig 4.31 it is therefore possible to construct a diagram showing the hardness distribution, after hardening, in the manner shown in Fig. 14.31. If a tempering curve for the Jominy specimen is available, a diagram showing the transverse hardness of a hardened and tempered bar may be constructed in similar way. The form such Jominy tempering diagrams is shown in Fig. 14.32. Fig. 14.31 applies to one rate of cooling only.

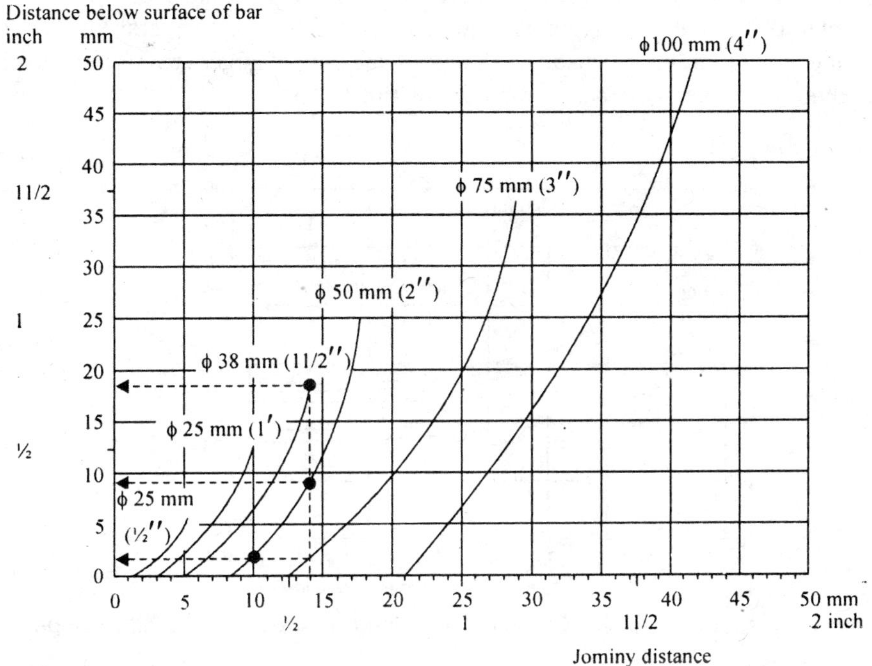

Fig. 14.30. Chart showing relationship between rate of cooling at different Jominy distances and rate of cooling in moderately agitated oil. Curves apply to round bars from 12.5 to 100mm (1/2-4 in) diameter (Diagram based on data taken from ASTM Standards).

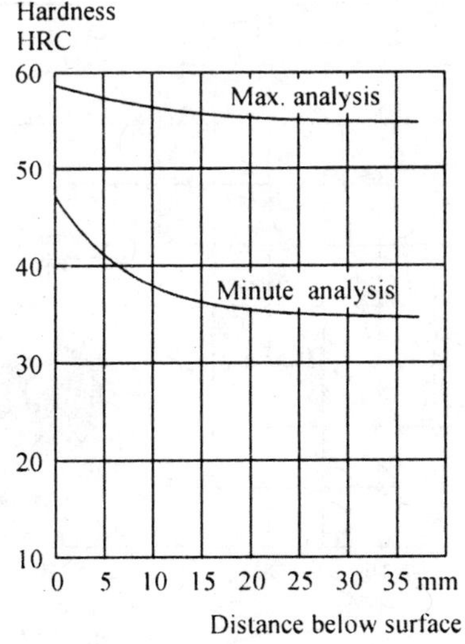

Fig. 14.31. Transverse hardness in a 75 mm diameter test-piece, steel BS 708A42. Construction from Jominy curves and Fig. 4.30.

The diagram may be supplemented to include various rates of cooling, the values of which correspond to those given in Table 14.1. Fig. 4.31 as well as Figs.14.33 to 14.38 are valid for Jominy distances up to 50 mm only. Therefore, the diagrams can be used to assess the core hardness of bars only to about 150 mm in diameter. Of course, the diagrams may be extended but their general limit is about 80 mm which is usually the maximum Jominy distance. The evaluate bars of larger cross-sections other methods must be used and these will be described in the next section.

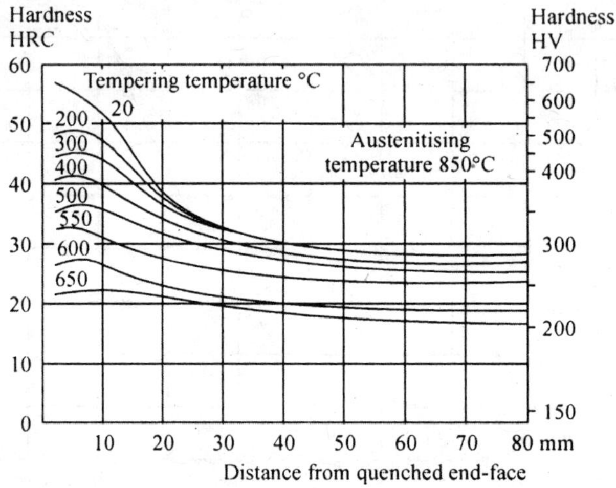

Fig. 14.32. Jominy curves for steel DIN 42Cr Mo4, corresponding to BS 708A42 after tempering at different temperatures.

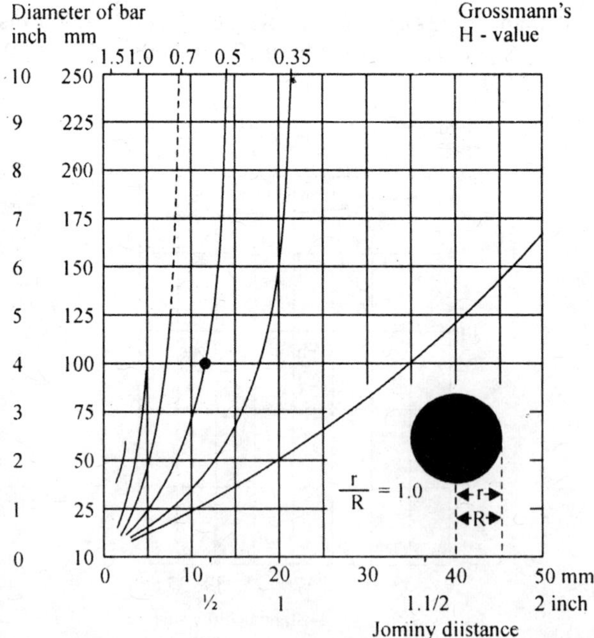

Fig. 14.33. Curves showing correlation between rates of cooling in the Jominy Specimen and rates of bars cooled in various quenching media.

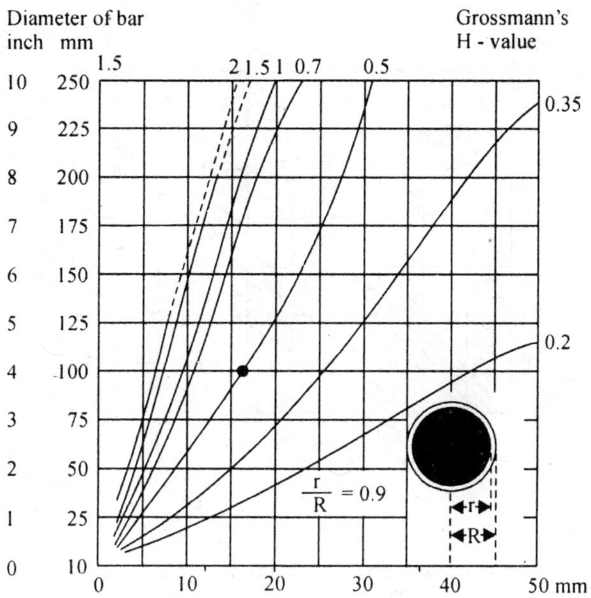

Fig. 14.34. Curves showing correlation between rates of cooling in the Jominy specimen and rates of bars cooled in various quenching media. The diagrams apply to the surface of the bar and to points situated 0·9R from the centre of the bar respectively (after *Lamont*).

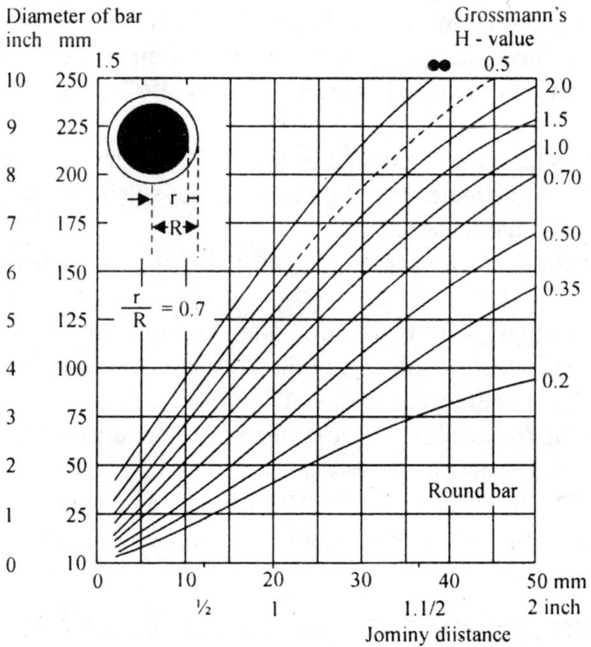

Fig. 14.35. Curves showing correlation between rates of cooling in the Jominy Specimen and rates of bars cooled in various quenching media.

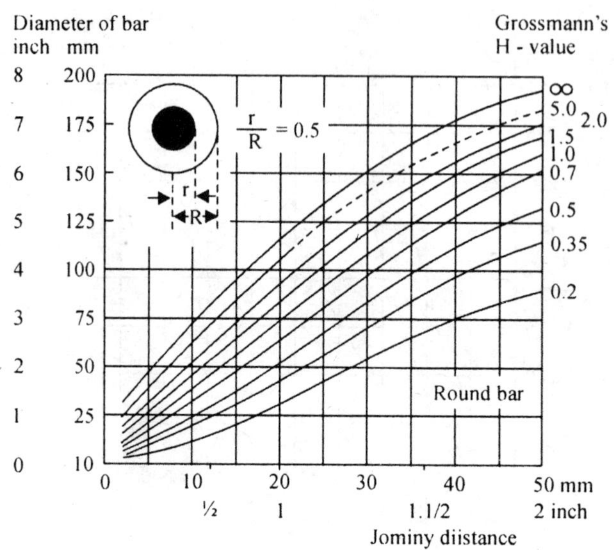

Fig. 14.36. Curves showing correlation between rates of cooling in the Jominy specimen and rates of cooling in bars cooled in various quenching media. The diagrams apply to points situated at distances of 0.7R and 0.5R from the centre of the bar respectively (after *Lamont*).

It is possible to combine the Jominy diagrams direct with the information contained in Fig. 14.30 and Figs 14.33 to 14.38, thereby enabling a rapid assessment to be made as to whether it is advisable to use a given type of steel or a steel with a given Jominy diagram in a certain dimension. Such combined diagrams may be look like the one shown in Fig. 14.39. From this diagram, which also holds for various cooling conditions, it may deduced, for example, that a bar, 50 mm in diameter, after being quenched in strongly agitated oil, will probably obtain a hardness of 40–49 HRC at the centre and 48–52 HRC at the surface.

If the quenching were to take place in moderately agitated oil an unsatisfactory hardness would most likely to be the result. A 75 mm diameter bar of this steel is not suited for oil quenching whereas water quenching will give acceptable hardness results.

If a Jominy diagram for the actual heat itself is available the assessment will obviously be more reliable.

The information given in Figs 14.33 to 14.38 applies to round bars only. When bars of square or rectangular cross-section are hardened these sections must be converted by estimation to equivalent round cross-sections for which diagrams are available.

In the compilation of an international steel grades standard a conversion diagram has been constructed, by means of which square and rectangular cross-sections may be converted to equivalent round ones, Fig. 14.40. For example, a 39 mm square and a 25 × 100 mm rectangular cross-section are each equivalent to a 40 mm diameter round section; a 60 × 100 mm cross-section is equivalent to an 80 mm diameter round one.

PRACTICAL APPLICATION OF THE TTE AND THE CCT DIAGRAMS

The familiar TTT diagrams, which show the progress of the transformation at constant temperature (isothermal transformation) are useful when isothermal processes are concerned such as isothermal annealing, austempering or martermpering. However, TTT diagrams of this type are suitable only for an approximate assessment of the hardenability of the various steels.

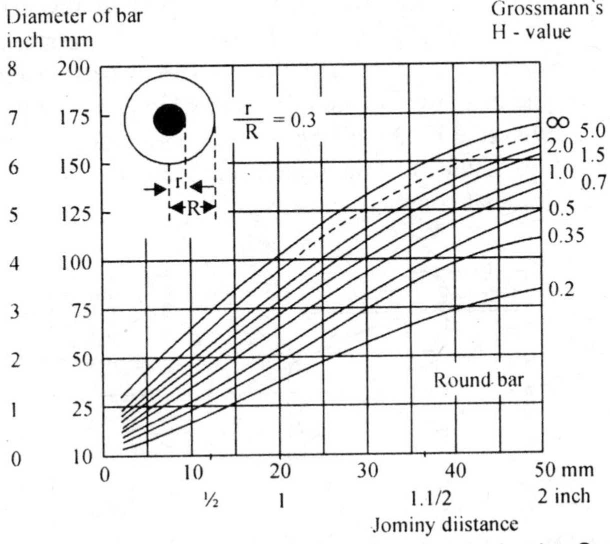

Fig. 14.37. Curves showing correlation between rates of cooling in Jominy Specimen with round bar.

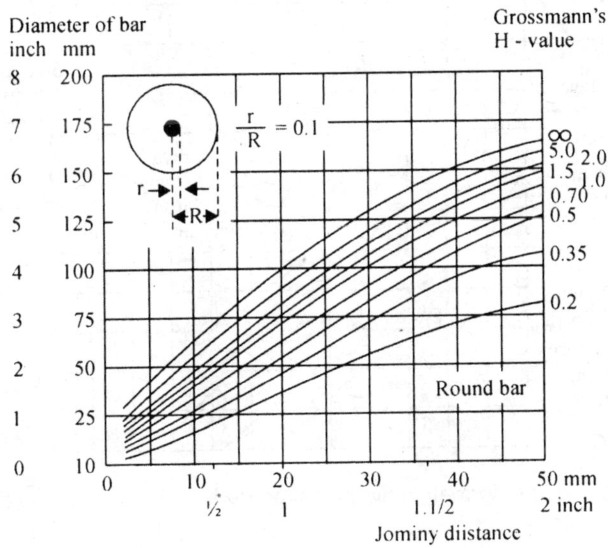

Fig. 14.38. Curves showing correlation between rates of cooling in the Jominy specimen and rates of cooling in various quenching media. The diagrams apply to points situated at distances of 0.3R and 0.1R from the centre of the bar respectively.

Various Types of CCT Diagram

Continuous-cooling-transformation diagrams (also called the CCT diagrams) are so constructed as to enable the personal dealing with heat treatment to predict which constituents will be produced as the rate of cooling varies. Provided that each individual point on the Jominy specimen corresponds to a known rate of cooling, it has been possible, by using data from the Jominy test, to construct CCT diagrams on which are superimposed cooling curves corresponding to the rates of cooling prevalent at various Jominy distances. Such a diagram is shown in Fig. 14.41. The rate of cooling at each point on the Jominy specimen is equivalent to some definite rate of cooling in a steel bar-see Fig. 14.30-and hence it is possible to predict from this type of transformation diagram which structural constituents will form in the steel bar, as well as the transverse hardness, the latter being computed from the Jominy diagram. However, it is not quite correct to assume that the same cooling curves apply to all grades of steel, since different steels have different thermal conductivities.

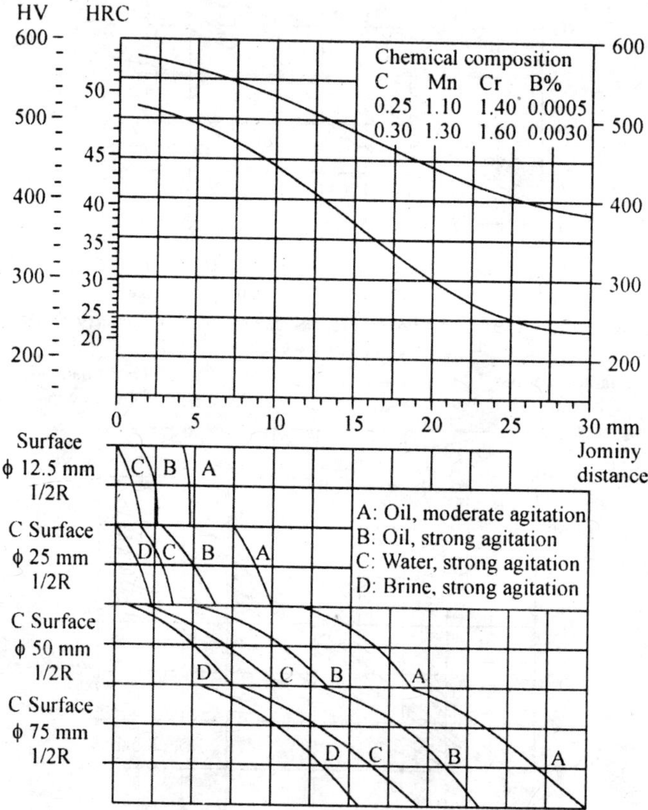

Fig. 14.39. Jominy diagrams and transformation diagrams for predicting the hardness of bar steel after hardening by means of various media.

Mechanisms that Control Hardenability

Influence of nucleation and growth

As a basis for the understanding of TTT and CCT diagrams there is in a schematic Fig. 14.42 that serves to explain simply what happens during isothermal transformation. It pre-supposes that the rate of nucleation is due mainly to the degree of supercooling, whereas growth is due mainly to diffusion.

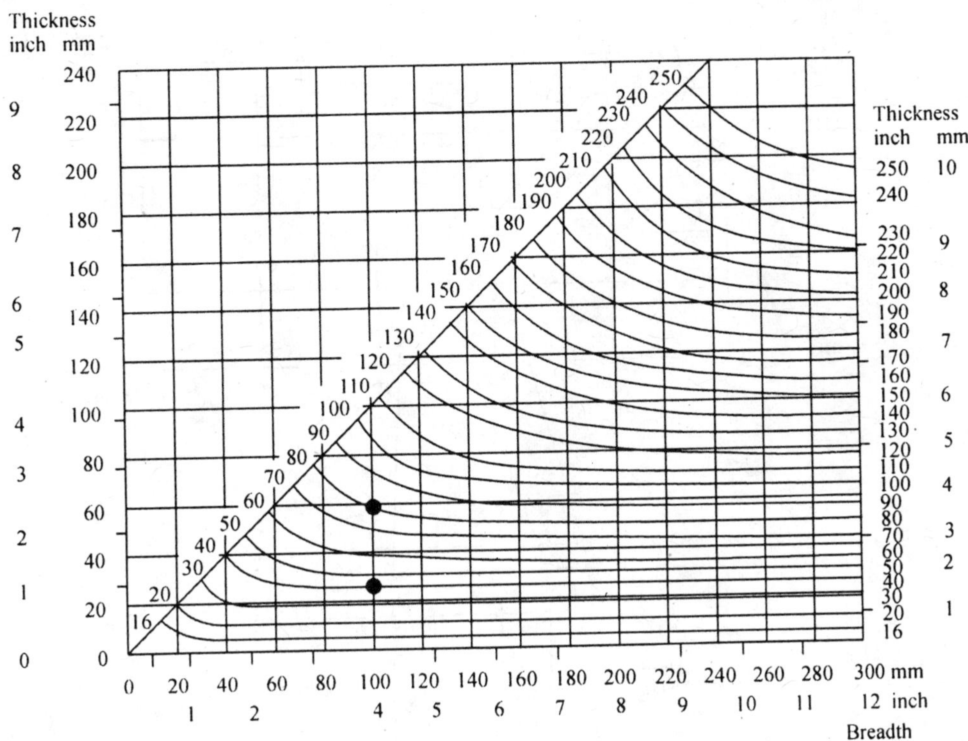

Fig. 14.40. Correlation between rectangular cross-sections and their equivalent round sections (according to ISO)

If cooling is rapid down to the temperature range of the nose and the specimen then held there (cooling mode 1) nucleation and growth both take place quite fast and the transformation proceeds as shown in the diagram. Should the cooling be taken to a lower temperature, nucleation will start after a shorter interval. If the temperature is now raised (mode 2) growth will be activated and the time taken for complete transformation will be shorter than if the transformation had followed mode1. When the TTE and CCT diagrams for the same steel are compared it is quite obvious that when the steady rate of cooling is interrupted in the temperature range in which transformation by diffusion takes place the progress of the transformation is stimulated by the interruption. On the other hand, an uninterrupted steady cooling is unfavourable to the transformation. This observation is illustrated by a test carried out with the spring steel En 45 (55 Si 7). Two continuous rates of cooling were simulated and superimposed on the TTE diagram for this steel (Fig. 14.43). The first cooling rate gave 98 per cent of martensite and the other one, 75 per cent. When the cooling was interrupted at 600°C for about 30 seconds and then continued, only 6 per cent of martensite resulted and the hardness was 341 HV, i.e., about the same is indicated on the TTT diagram at 600°C. Fig. 14.44 also shows how the hardness is affected when at first the cooling rate is constant down to the bainitic range and then when the cooling rate is varied. The highest cooling rate resulted in a hardness of 495 HV, the lowest, 340 HV.

If the cooling through the upper metastable austenite range is slow nucleation proceeds slowly, which enhances hardenability. Should the slow cooling continue till just before the nucleation range and then change over to rapid cooling, the steel passes into a temperature range where any nuclei previously formed will grow slowly. Such a cooling mode, mode 3 in Fig. 14.45, should give a high hardenability.

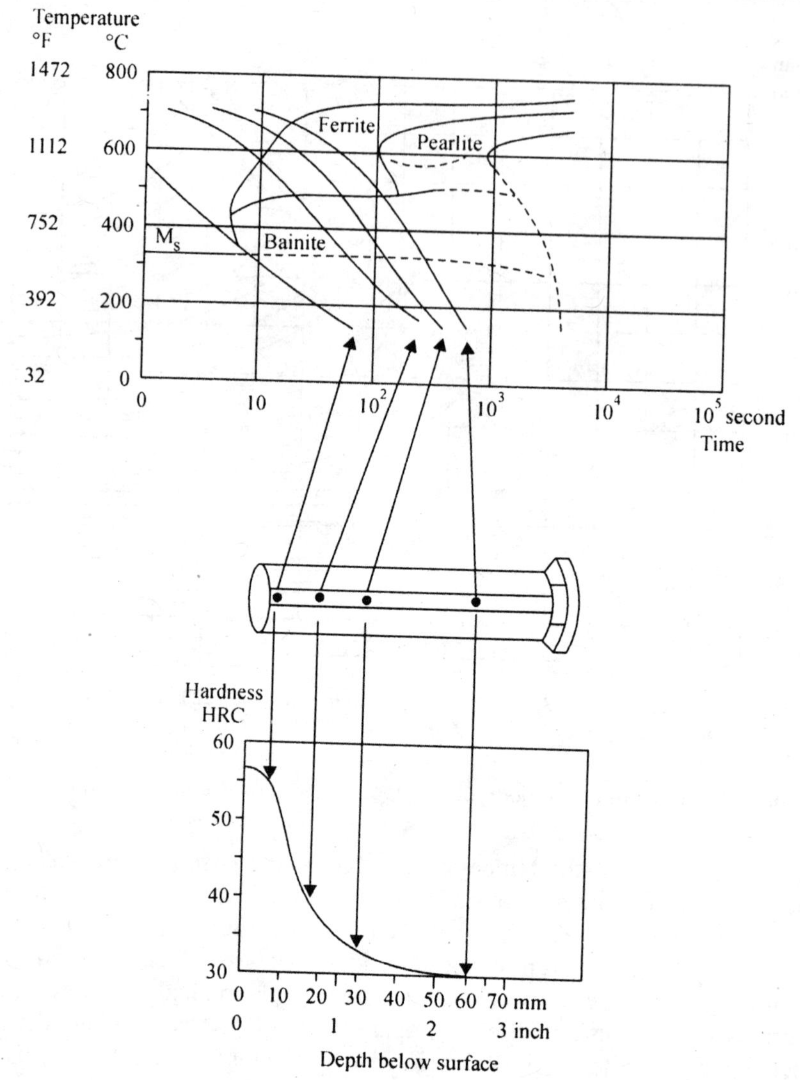

Fig. 14.41. CCT diagram with some cooling curves for corresponding Jominy distances. The lower diagram shows the hardness values along a Jominy test specimen.

In order to verify this hypothesis Fig. 14.45 was constructed in which various rates of cooling (dashed lines) down to 700°C were simulated. From this temperature the cooling progressed at a constant rate. These curves were superimposed on a CCT diagram, constructed from linear cooling data (solid lines). It was found that the slower the cooling proceeded in the first stage the higher became the hardness of the steel; an observation that may appear surprising.

Effect of internal stresses

It has been known for a long time that the formation of martensite is favoured by tensile stresses and retarded by compressive stresses.

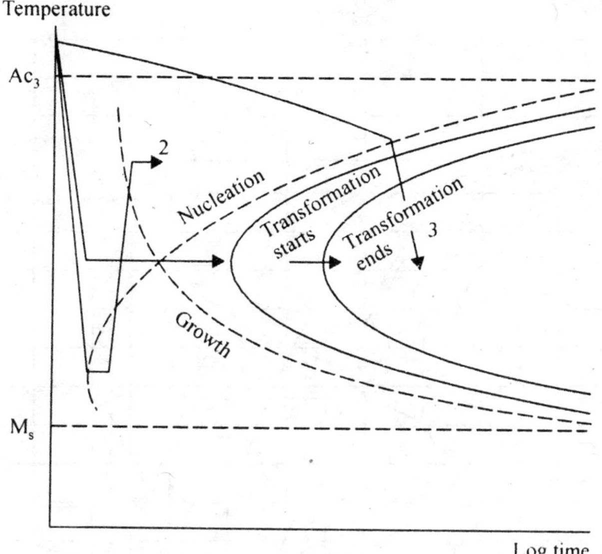

1. Conventional mode of cooling for TTT diagrams.
2. Mode of cooling that increases rate of transformation (reduces hardenability).
3. Mode of cooling that reduces rate of transformation (enhances hardenability).

Fig. 14.42. Schematic representation of nucleation and growth during isothermal transformation in an eutectoid carbon steel. The above modes of cooling which give rise to different rates of transformation are indicated in the diagram.

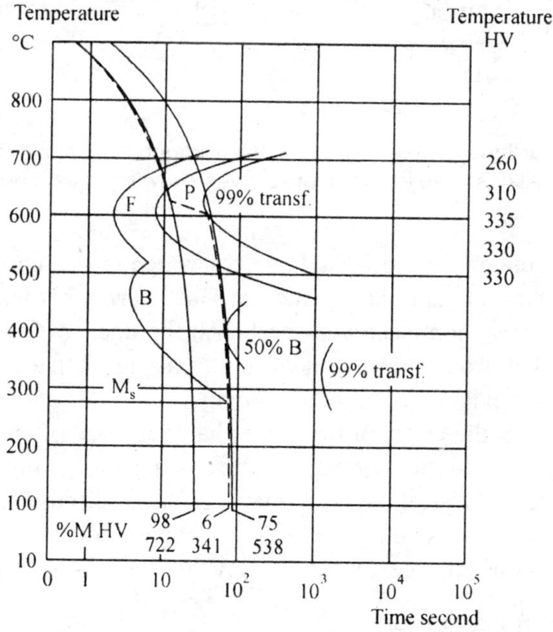

Fig. 14.43. TTT diagram for the spring steel En 45 (55 Si 7) also including curves for continuous cooling and interrupted mode of cooling.

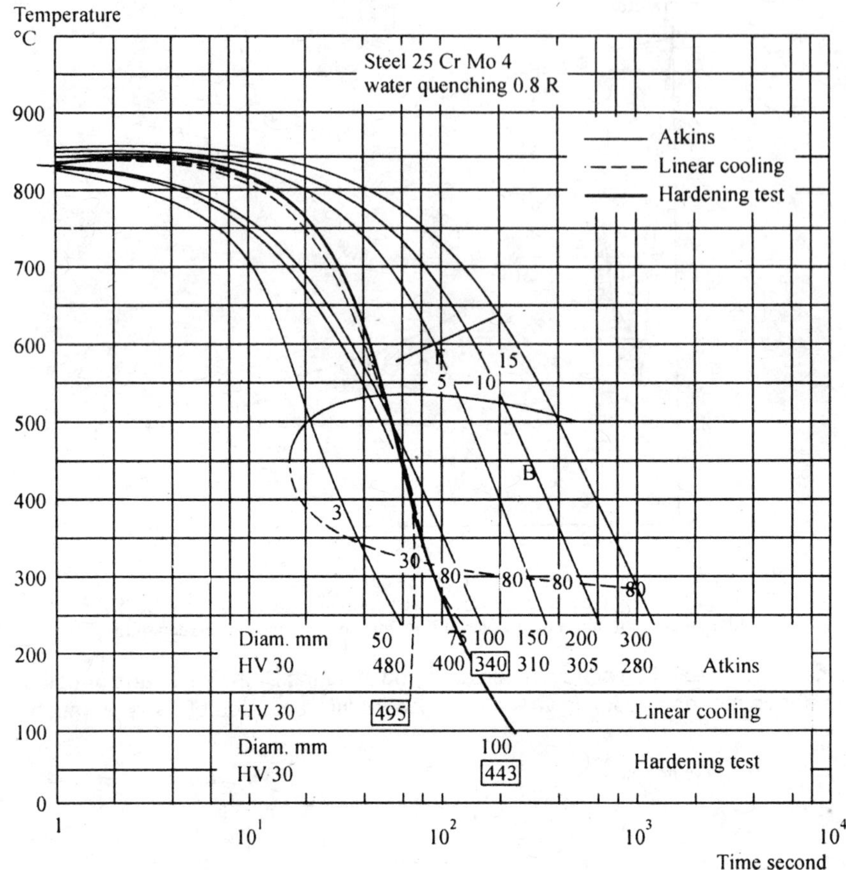

Fig. 14.44. The influence on hardness of the rate of cooling through the bainite region. Simulated curves for water quenching, according to Atkins: also included are curves showing linear cooling and hardening test with steel SS 2225 (25 CrMo 4)

In a dilatometirc test a specimen was subjected to a controlled rate of cooling simulating that taking place at the centre of a 250 mm diameter bar, grade SS 2541, during hardening. It was found that the hardness at the centre of the bar was appreciably lower than the hardness of the dilatometric test specimen, both samples having been taken from the same heat. It was assumed that this difference is due to the influence of the stress situation existing during the cooling.

An acceptable explanation of the cause of this lower hardness was offered with the aid of a simple model for assessing the stress situation prevaling during the hardening operation. According to Fig. 14.46 which is applicable to the steel and specimen size under discussion, there are tensile stresses acting at the centre while the bainite is being formed. This concurrence serves to facilitate the progress of bainite formation and as a consequence a relatively low hardness is the result.

Influence of the heat of transformation

When diffusion-dependent transformation products are being formed during the process of cooling, some heat is liberated which may influence the subsequent cooling process in various ways.

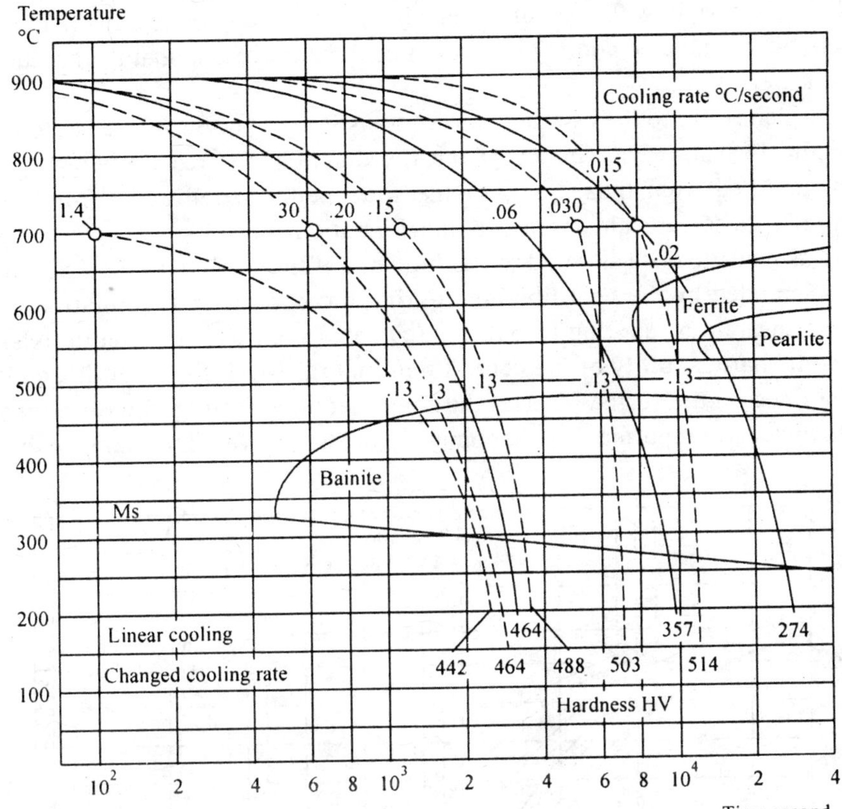

Fig. 14.45. Steel SS 2541. CCT diagram derived from linear cooling. Curves representing prolonged cooling above 700°C are drawn in the diagram. Austenitised at 850°C for 10 minutes.

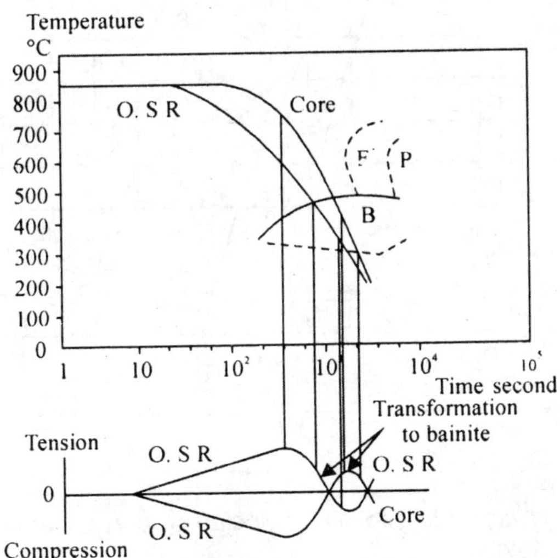

Fig. 14.46. Hypothetical stress patterns existing in a 300 mm diameter sample of steel SS 2541, on quenching in oil.

In Fig. 14.47 it is shown how the rate of cooling in oil, as actually measured on a 50 mm diameter test-piece, differs from simulated cooling in oil for this dimension; also that the hardness of the test-piece takes on a lower value.

This difference, which is due to the liberated heat of transformation, sets in even before the predicted start of the bainite formation, as deduced from the diagram, the reason being that the heat of transformation liberated from those parts of the specimen situated away from the area under consideration affects those parts already in the transformation zone.

If a steel has its pearlite nose situated rather high and if the cooling rate of the surface layer of the specimen is impeded by the heat liberated during the pearlite transformation it may happen that at the centre the rate of cooling through the zone of metastable austenite is slowed down, which enhances the hardenability of the central region. Fig. 14.48 shows both a cooling rate that gives a lower hardness at the surface than at the centre and the cooling rate for normal hardening. *Schimizu* and *Tamura* have reported this phenomenon and have given it the name inverse hardening.

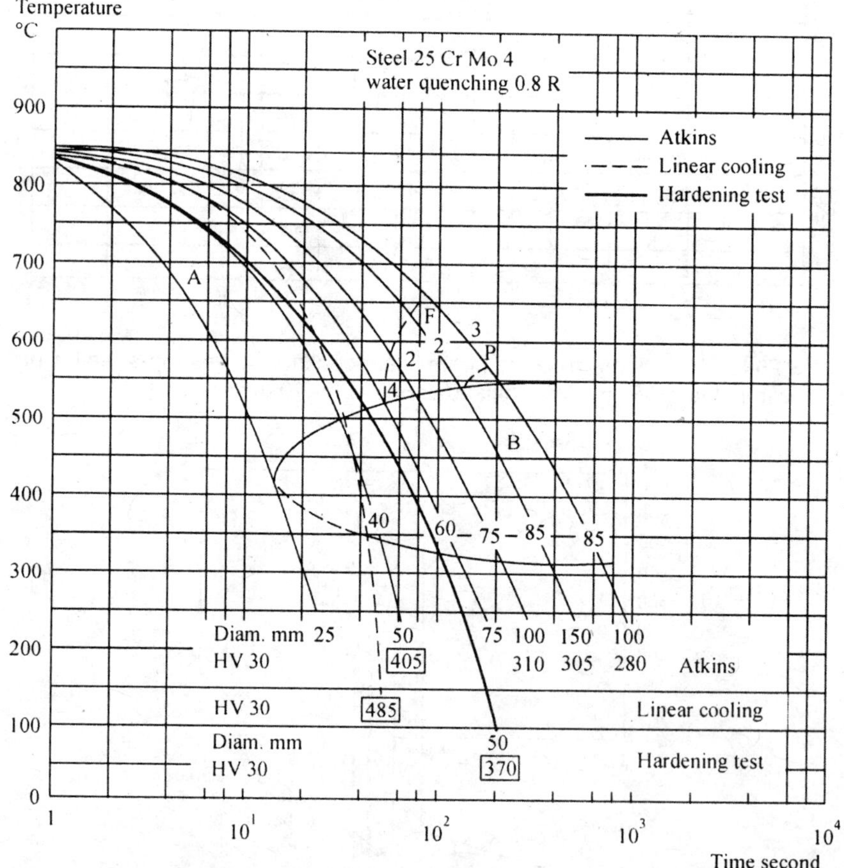

Fig. 14.47. Simulated quenching, according to Atkins, including curves for linear cooling and of hardening tests with steel SS 2225 (25 CrMo 4).

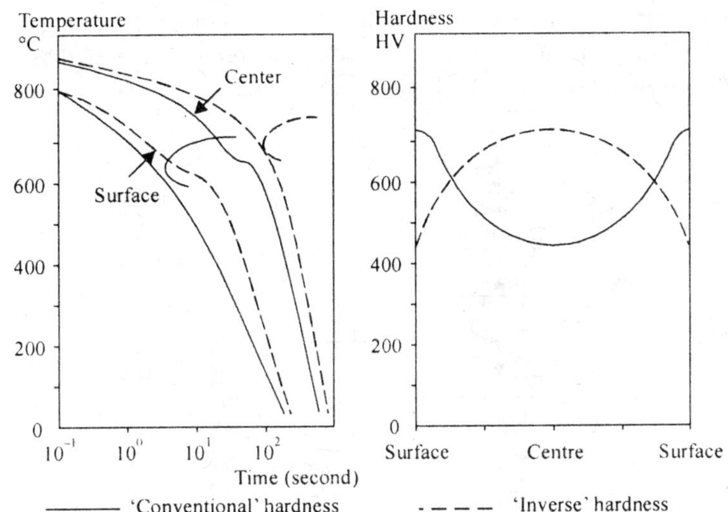

Fig. 14.48. The influence of the heat of transformation on the progress of cooling and transformation and on the hardness.

Computing CCT Diagrams from TTT Diagrams

TTT diagrams are quite easy to construct. Hence several attempts have been made to compute CCT diagrams by mathematical methods from TTT diagrams. Very successful work along these lines has been accomplished by *Hildenwall* who, by applying his computer program has succeeded in computing the temperature trend, phase transformations and hardness. Fig. 14.49 shows the flow sequence for the program and how it functions during the course of a straightforward computation.

The computer program consists in the main of three substages at which the following is worked out:

1. The temperature distribution in the specimen as a function of time and depth below the surface.
2. The transformation of austenite to martensite, ferrite, pearlite and bainite or cementite as a function of time and depth below the surface.
3. The build-up of internal stresses as a function of time and depth below the surface.

The cooling time is divided up into intervals and the specimen into small segments. For each time interval and specimen segment the temperature, the amount of newly formed phase and the internal stress are calculated. A number of numerical checks are added to the program. For each time interval and for each combination of temperature and phase content all pieces of information that have any bearing on subsequent computations are updated. When the ultimate state has been computed it is possible to work out the hardness profile for the specimen.

Temperature distribution is computed by solving the general equation for heat conduction adapted to those geometrical shapes that lend themselves to such treatment. The input data consist of such physical data as are functions of the temperature, i.e., the rate of heat flow between steel and the cooling medium, the specific heat capacity and heat conductivity along with the heat of transformation of austenite.

The rate of heat flow must be determined by cooling tests. When this has been found the temperature at all points of the specimen may be computed as a function of the time.

Computation of phase transformation is carried out according to the principle of dividing the cooling curve into a number of steps. By starting off from the mean temperature of each step the transformation taking place within the length of time set by the time interval is computed, using Avrami's exponential equation as a basis for the computation. The transformation increment is treated as if the transformation of austenite to ferrite, pearlite and bainite took place isothermally.

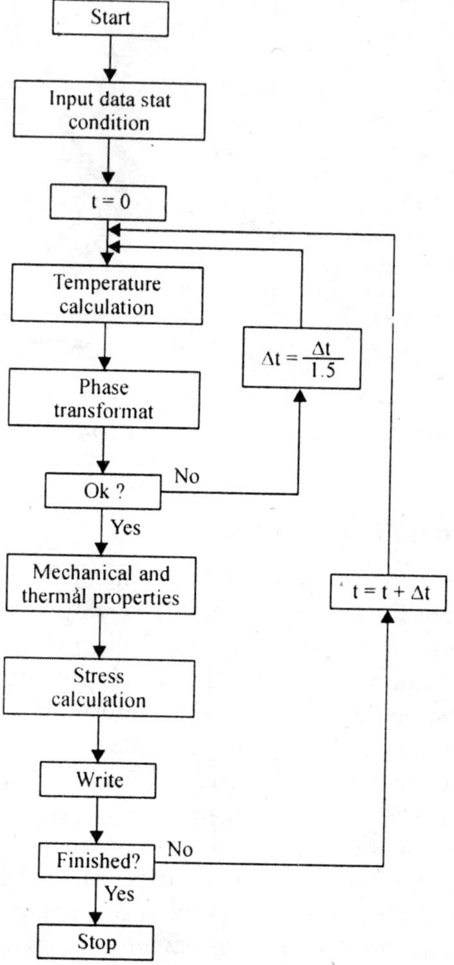

Fig. 14.49. Schematic flowchart of the computer program.

The transformation to martensite is computed according to an equation developed by Koistinen-*Marburger*. The input data required, besides the cooling curve, is a TTT diagram for the steel being investigated. It is possible to compute the amount of each phase for each segment of the specimen as a function of the time. By knowing which phases are present at every point in the body of the steel specimen it is possible to estimate the hardness of each phase in turn, given the ally content of the steel and the rate of cooling. The total hardness may then be calculated from the expression below:

$$H_{tot} = V_m \cdot H_m + V_{p+f} \cdot H_{p+f} + V_b \cdot H_b$$

where,

H = hardness (HV)
V = volume fraction

m = martensite
b = bainite
p + f = pearlite + ferrite

Internal stresses are computed by resolving the elastic-plastic problem. The effective elastic strain in each phase contributes to the total work hardening. It is assumed that the condition of flow, as stated by *von Mises*, holds as well as an associated flow law, propounded by *Prandtl Reuss*. For each structural component the mechanical properties, i.e., modulus of elasticity, yield point and rate of strain hardening and their dependence on carbon content and temperature, are obvious components of the input data.

PRACTICAL APPLICATION OF HARDENABILITY

There are two depth-of-hardening limits between which the hardenability should lie in order that the steel, after hardening, may be used to good technical and economic advantage. The upper limit is set by the hardenability that will result in an approximately 90 per cent martensitic structure at the centre. A steel having a higher hardenability, i.e., a steel that in a heavier section would give the same hardness as in a smaller cross-section would be regarded as uneconomic when used in the smaller dimension, provided that only the price factor is considered and that price is equated with hardenability. For every tool and every constructional part there is a rather narrow range of hardenability that will give optimum results, but it is usually for highly stressed parts only that such results are required. From the point of view of stock-keeping and safety it often happens that steels having a hardenability higher than necessary are used. The lower limit, applicable mainly to unalloyed and low alloyed steels, is set by that hardenability that will give a depth of hardening of about 3 mm, this hardened zone having at least a 90 per cent martensitic structure derived from the heat treatment and the mass effect of the cross-section. (This does not apply to case-hardening and other thermochemical surface treatment processes.)

Which properties are concomitant with high hardenability and which with low hardenability, respectively?

High Hardenability

Where steels for hardening and tempering in large cross-sections are concerned, the benefits of high hardenability are obvious. Such steels can be used in heavy dimensions for parts that are later to be machined to various cross sectional areas and for which strength requirements are equally high in the smaller and in the heavier sections. Unfortunately the test methods as stipulated by both SS and ISO are concerned only with a zone extending to about 20 mm below the surface.

Even if the hardness is the same at the surface as at the centre in a hardened and tempered bar it must be accepted generally that the impart strength of the central regions of a bar are often lower than at the surface. This is because the surface zones, as a rule, consist of tempered martensite or lower bainite whereas at the centre the structure is composed mainly of upper bainite or pearlite. These two latter constituents have considerably lower impact strength and higher transition temperature ranges than tempered martensite or lower bainite.

Another instance when high hardenability is required, concerns hardened and tempered forged blooms for drop-forging dies or pressure diecasting dies. These forgings are engraved subsequent to the heat treatment. Tools subjected to high pressures must also be made from steel that hardness through entirely, otherwise the softer core might be deformed. In such cases the surface layers will be exposed to execessive stresses which are liable to cause this layer to rupture.

Low Hardenability

In many cases a low hardenability is to be preferred, both for machine parts and for tools. A shallow-hardening steel, such a W1, can give a remarkably good performance. Long experience tells us that such steels can stand up to impact stresses extremely well and it is said they can do so because of the presence of hard case and a tough core. Tools with varying cross-sectional areas often need to have an even depth of hardness on all working faces. For such steel SS 2092 has shown itself to be eminently serviceable under working conditions. This has been subsequently verified by metallographic investigations and practical heat treatments. These investigations, which are of interest since they tend to verify the deductions drawn from data obtained from Jominy diagrams are described below.

The steels used for cold pilger rolls are SS 2092 (~90 CrSi 5) and AISI O 1. In order to harden a pilger roll made from O 1, outer diameter 340 mm, cross-sectional area 145 × 82 mm², it was quenched in oil from 830°C. The hardness at the corners was 55 HRC but on the cylindrical surfaces and in the groove the hardness was only 45 HRC. Yet another ring roll was hardened, a temperature of 890°C being used this time. Full harness of 62 HRC was realised all over the roll but it cracked in service after a short time. Metallographic examination revealed that the roll had a very coarse fracture, viz. fracture No. 5 according to the JK scale, and it had hardened throughout the whole section. It was apparent that to find the ideal hardening temperature for this tool was quite a problem. Examination of a similar roll made from steel SS 2092 and hardened from 860°C revealed the fracture. The hardened zone has approximately the same depth all the way round. Rolls made in this steel give satisfactory service. In order to obtain a proper insight into the behaviour of these various steels a systematic investigation was initiated which involved the hardening of Jominy test-pieces from different hardening temperatures. The results are shown in Figs. 14.50 and 14.51. From these curves we see that steel SS 2092 has the lowest hardenability and that the depth of hardening and grain size are least influenced by the hardening temperature.

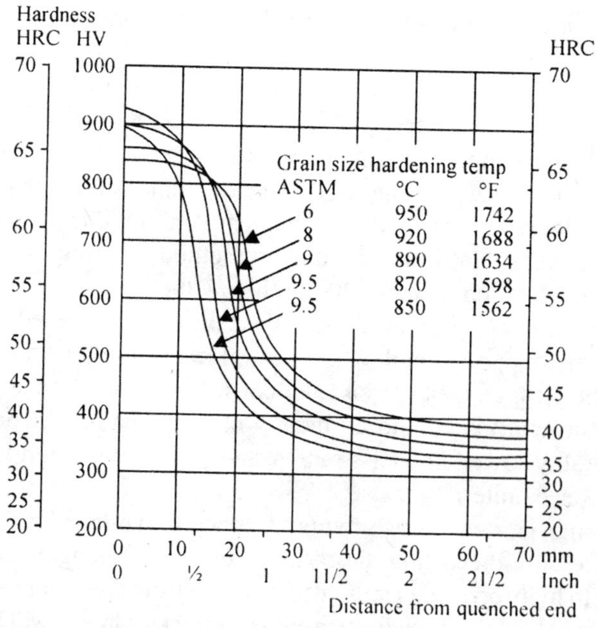

Fig. 14.50. Steel SS 2092 Jominy curves and ASTM grain size obtained with different hardening temperatures.

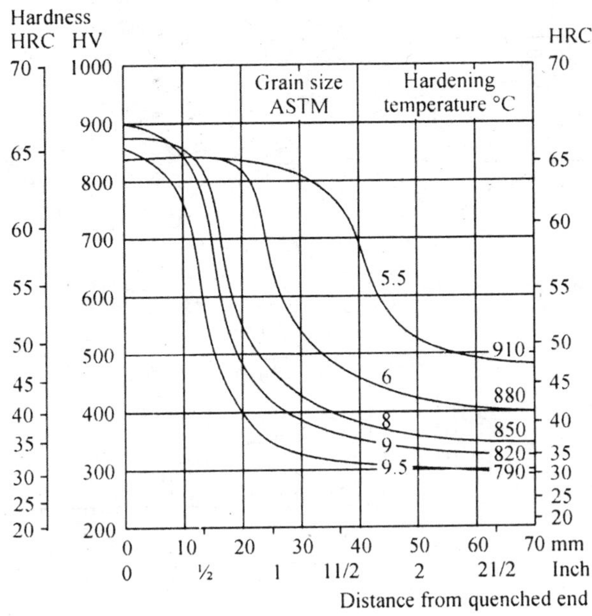

Fig. 14.51. Steel O 1. Jominy curves and ASTM grain size obtained with different hardening temperatures.

INFLUENCE OF THE DEPTH OF HARDENING ON THE STRESS PATTERN

In the preceeding section we have discussed the advantages attendant upon a favourable depth of hardening. In addition, an instance has been quoted illustrating the benefits accruing to a steel possessing a limited hardenability. However, this is only one aspect of the matter. The residual stresses that are created as a consequence of the hardening operation are of such importance that they must be taken into consideration when a steel and its heat treatment are being selected.

As a result of the hardening operation, thermal and transformation stresses are set up in the steel. If we regard the thermal stresses only and reckon them as tangential stresses, we obtain a stress distribution as shown if Fig. 14.52, which applies to a bar 100 mm in diameter and quenched in water from 850°C. At the point of time T the temperature difference between the surface and the centre is greatest, i.e., some 500°C. This corresponds to linear differential of about 0.6 per cent and gives rise to a stress that exceeds 1000 N/mm^2. If the body of the material were elastic enough to accommodate this stress, the stress diagram for the surface layer would take the shape of curve a. After temperature equalisation between the surface and the centre the stress should fall to zero. Since the yield point of steel is considerably lower at elevated temperatures than at room temperature the material will flow plastically, resulting in a surface stress diagram, such as curve b.

At the centre the stresses will be such as indicated by curve c. When time T has elapsed, contraction in the core will exceed that at the surface. This implies that a time O there is complete stress neutralisation and consequently at room temperature there will be compressive stresses in surface and tensile at the centre. When the bar has cooled to room temperature, the stress distribution will be as shown in the right-hand figure, i.e., tensile stresses at the centre and compressive stresses in the surface. The more effective the quenching medium and the larger dimensions of the bar, the greater will be the stresses.

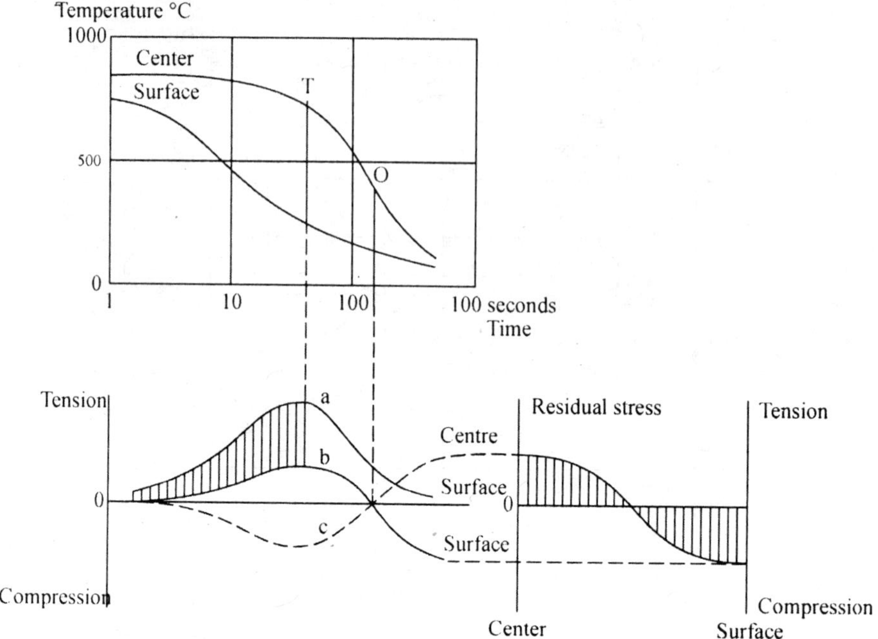

Fig. 14.52. Development of thermal stresses on cooling. Transformation stresses not taken into account.

The phenomena involved when steel hardens produce additional stresses due to the increase in volume that takes place as austenite transforms to other structures. The following increment of volume and linear expansion, Table 14.2, apply to a 1 per cent plain carbon steel. If the main part of the transformation takes place in the centre before the surface starts to transform, which is the course of events in shallow-hardening steels, the same stress pattern as has been described in Fig. 14.52 is set up. If the transformation to, say, pearlite has proceeded to 100 per cent completion at the centre before martensite has started to form in the surface, the compressive stresses increase considerably when martensite formation starts. Such as instance is shown schematically in Fig. 14.53. Up till time t_1 tensile stresses are created in the surface and compressive stresses at the centre.

Table 14.2. Volume and linear expansion.

Transformation	*Change in volume %*	*Change in length %*
Austenite to martensite	+4.2	+1.40
Austenite to lower bainite	+3.2	+1.07
Austenite to pearlite	+2.4	+0.80

When pearlite formation starts after time t_1 there is an incremental volume increase in the core at the same time as the fall in temperature produces an incremental volume decrease. Which of these phenomena is the overriding one is quite immaterial since the yield strength of the surface layer determines the maximum stress there. After time t_2 in this hypothetical case, pearlite formation at the centre has been completed just as the martensite starts to form in the outer layers. The volume increase in the surface and the contraction of the centre both combine to produce a rapid stress reversal. The compressive stresses in the martensitic outer layers may assume very high values. The yield strength of the central pearlitic regions is comparatively low and as the thickness of the outer martensitic layers increases, the tensile stresses at the centre increase; in fact they may gradually reach such a critical value that fracture takes place, often explosively.

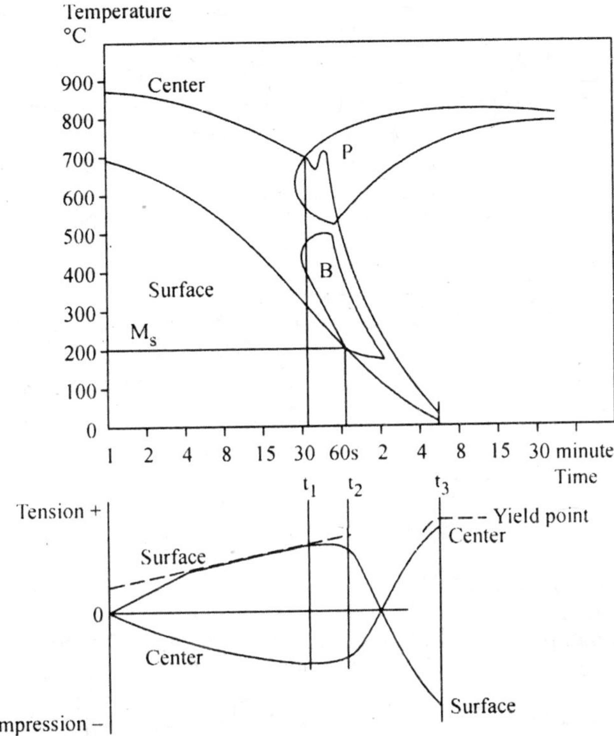

Fig. 14.53. Hypothetical stress pattern at the surface and center of a shallow hardening steel, 100 mm in diameter, quenched in water.

If the steel through-hardens the stress situation, as it is schematically illustrated in Fig. 14.54 may quite well develop. As they cool down to M_s the outer layers contract, thereby building up tensile stresses and creating correspondingly compressive stresses at the centre. When, after time t_1 the martensite transformation starts in the outer layers there is an increase in volume which reduces the tensile stresses somewhat. By the time t_2 is reached there are compressive stresses in the outer layers and tensile stresses at the centre. The fall in temperature at the centre, being greater there than in the surface layers, aggravates this stress situation.

A critical state is thus created for the central region but since the latter has an austenitic structure its ductility is good and owing to plastic deformation some degree of stress relief takes place. For this reason crack formation in central regions is a rare occurrence at this stage. When the temperature in this region has reached M_s a volume expansion sets in and as the temperature continues to fall the outer layers are subjected to tensile stresses which may become so severe that the steel cracks, often longitudinally. The risk that such cracking will occur increases with increasing section size and hardenability. Therefore the greater the hardenability of the steel the milder should be the quenching medium in order to avoid hardening cracks in through-hardening steels.

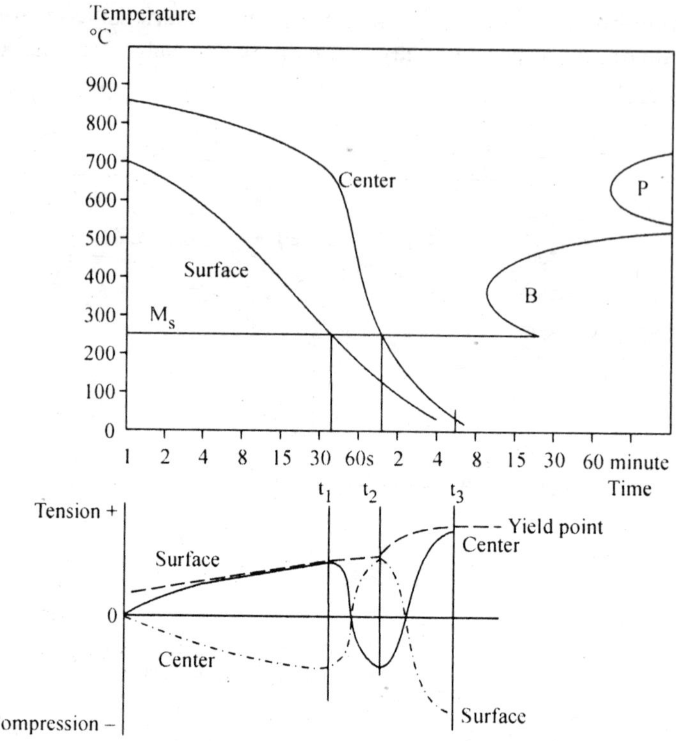

Fig. 14.54. Hypothetical stress pattern at the surface and center of a through-hardening steel, 100 mm in diameter, quenched in water.

From the above discussion it follows that there are two size groups that are most critical from the hardening point of view. One such size group among which most cracking actually occurs is the maximum section size in which the steel through-hardens. In this instance tensile stresses exceeding the tensile strength of the steel may build up in the outer layers. Cracking due to such causes is favoured by stress raisers on the surface and also by a high cooling rate. The other size group is the one where cracks start at the centre. This takes place when sufficiently high compressive stresses have been built up in the outer layers and correspondingly high tensile stress at the centre. In addition, this type of cracking is favoured by flaws, specially in the central regions of the steel and also by a high rate of cooling. Fig. 14.55 shows in diagrammatic form the two critical section size groups discussed, the type of crack formation that may occur and the prevailing stress patterns.

Rose and *Buhler* have carried out measurements on different specimens. One such series of measurements, shown in Fig. 14.56, deals with bars of various diameters, quenched in water. We can see how the dimension is instrumental in shaping the stress pattern.

The stress distribution in the 100 mm bar corresponds roughly to that due solely to thermal stresses. If the nose of the pearlite curve had been located farther to the left an additional increment of stress would have resulted.

With regard to the 30 mm diameter bar the start of the bainite formation at the center coincides approximately with that of the martensite in the surface. Compressive stresses were measured both in the surface and at the center and tensile stresses in the intervening zones.

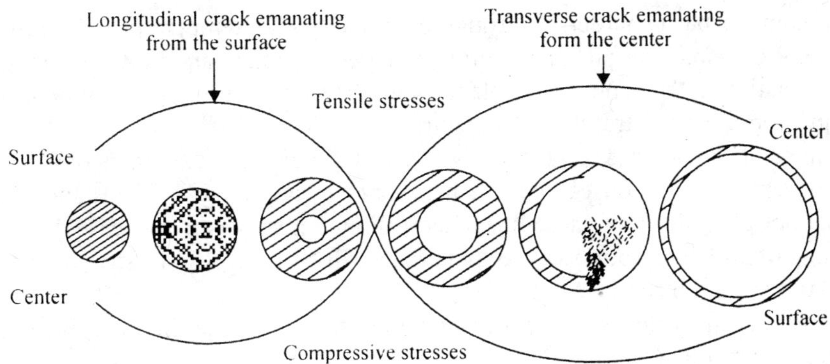

Fig. 14.55. Diagrammatic representation of the two critical size groups in which cracking may occur when the steel is being fully heated and quenched. Hatched areas represent martensite.

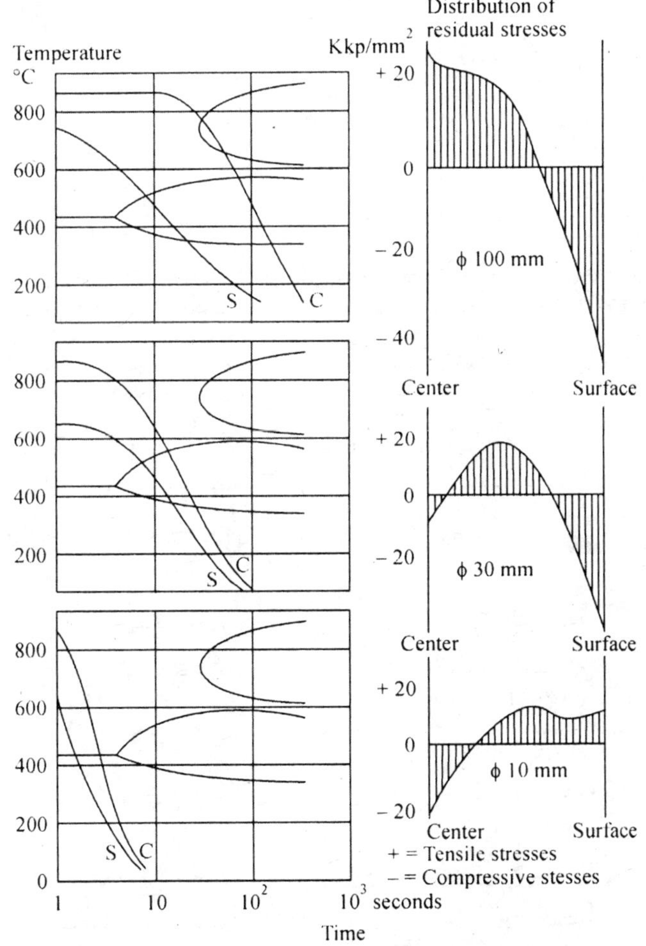

Fig. 14.56. Stress pattern for cooling, transformation and residual stresses in round test bars of different dimensions in steel DIN 22 CrMO 44.

The 10 mm diameter bar is hardened through completely. The temperature difference between the surface and the centre is smaller than that obtaining in layer-diameter bars. Furthermore, the main part of the martensite formation in the centre takes place when the surface-zone the creation of tensile stresses in the surface and compressive stresses at the centre.

The development of similar stress patterns may be envisaged if the alloy content of the steel is increased or if the hardening temperature is raised, the dimensions of the bars being kept constant (Fig. 14.57). There are endless variations possible along these lines. Let us limit our studies to two cases only.

1. Hardening of a plain-carbon steel so that result will be compressive stresses in the surface and tensile stresses at the centre.
2. Hardening of an alloy steel so that the result will be tensile stresses in the surface and compressive stresses at the centre.

In the first case the steel hardens at the surface and in the other case it hardens through entirely. The stress distribution is shown in Fig. 14.58.

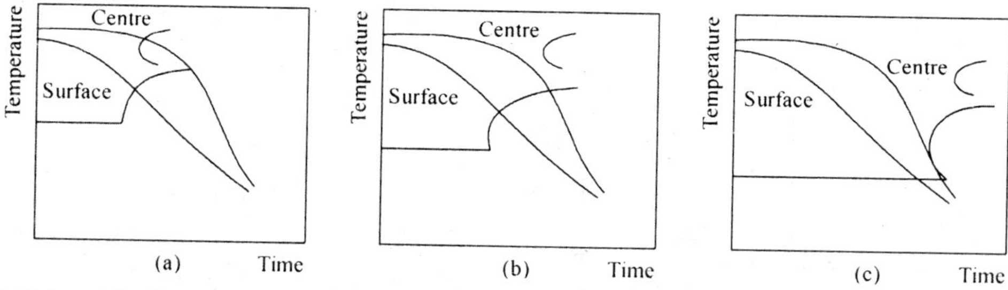

Fig. 14.57. Schematic illustrations showing how inherent hardenability influences the structures formed, on hardening, in the surface and at the center: (a), (b) and (c) represent increasing hardenabilities.

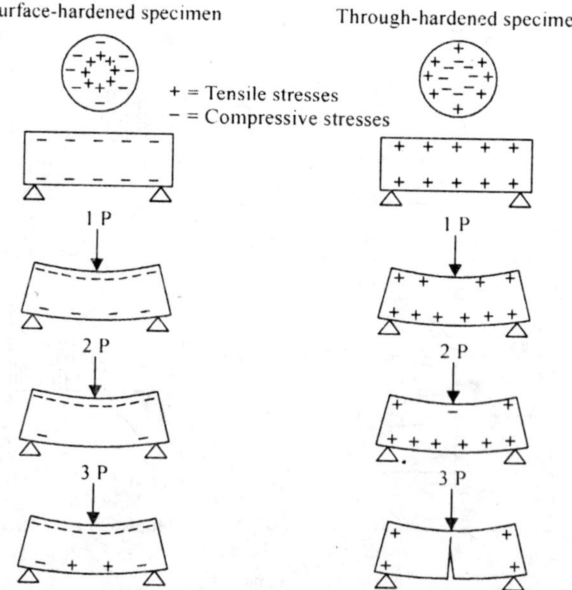

Fig. 14.58. Schematic sketch showing stress distribution, produced on bending, in surface-hardened and through-hardened test bars.

We shall discuss these cases with regard to the incidence of crack initiation on bending, assuming that the strength of the surface layer is the same for both bars and the hardening depth is at least 10 per cent of the bar diameter for the surface-hardened bar. On subjecting the bars to bending forces, as illustrated in Fig. 14.59 tensile stresses are induced in the lower layers and compressive stresses in the upper ones. A crack is more easily initiated in the layer subjected to tensile stresses. In order to balance the compressive stresses in the underside of the surface-hardended bar a definite bending force must be applied. This force, if applied to the through-hardened bar, will simply cause the tensile stresses on the underside to increase. On increasing this bending force additional tensile stresses are induced in the underside of this bar which will therefore be the first one to fracture, always assuming that there is an ample depth of hardening in the surface-hardened bar. What has been said above holds in a higher degree for fatigue and impact stresses.

In practice it has often been observed that surface hardening is an excellent means of increasing the endurance limit, whether by making use of shallow-hardening steel or by special methods of surface hardening, such as induction hardening or case hardening. Similarly, a long service life has been given by surface-hardened steels when used as tools subjected to impact stresses. A necessary condition is, however, that all working faces have been hardened.

To sum up, it may be said that whenever practicable from the point of view of the heat-treatment process, surface-hardened or alternatively not completely through-hardened tools and machine parts are to be preferred, due consideration of course being given to the service that the parts are subsequently to be subjected to.

Chapter 15

Heat Treatment of Ultrahigh-Strength Steel

INTRODUCTION

Ultrahigh-strength steels are heat treated by use of equipment and techniques similar to those employed for heat treating constructional alloy steels. The ultrahigh-strength steels ordinarily are quenched and tempered to specified hardnesses, but for critical applications it may be necessary to pull tensile specimens to ensure that a required combination of strength and ductility has been achieved. In still other instances, it may be necessary to conduct impact or fracture-toughness tests to ensure that a required level of resistance to brittle fracture has been attained.

In this chapter, only those commercial structural steels capable of a minimum yield strength of 1380 MPa (200 ksi) are considered; and general characteristics, heat treatment and properties are presented for three types of alloys from among the broad class of ultrahigh-strength constructional steels. These three alloy types are: (i) medium-carbon low-alloy steels; (ii) medium-alloy air-hardening steels; and (iii) high-alloy hardenable steel.

The present chapter also gives a brief description of the heat treatment and associated properties of the same types of steels. Chemical compositions are given in Table 15.1.

MEDIUM-CARBON LOW-ALLOY STEELS

The medium-carbon low-alloy steels considered here are of types 4130, 4140, 4340, 6150, 8640 and two modifications of 4340—namely, 300M and D-6a. These steels are readily hot forged. To avoid stress cracks resulting from air hardening, the forged part should be slowly cooled in a furnace or in an insulating medium. Prior to machining, usual practice is to normalise at 870 to 925°C (1600 to 1700°F) and temper at 650 to 675°C (1200 to 1250°F) or, if the steel is a deep air-hardening grade, to anneal by furnace cooling from 815 to 845°C (1500 to 1550°F) to about 540°C (1000°F). These treatments impart moderately hard microstructures suitable for machining.

A very soft spheroidised structure can be obtained by full annealing. Such a structure is less well suited for machining than the normalised-and-tempered structure. However, for severe very cold forming operations such as spinning, deep drawing, and wiredrawing, the soft and ductile spheroidised structure is preferred. If blanks for parts are produced by flame cutting, they are annealed before being formed or machined. Welded parts, especially if complex, are stress relieved, or hardened and tempered, immediately after welding.

Table 15.1. Compositions of ultrahigh-strength steels.

Designation or trade name	*Composition (a), %*							
	C	*Mn*	*Si*	*Cr*	*Ni*	*Mo*	*V*	*Co*
Medium-carbon low-alloy steels								
4130	0.28–0.33	0.40–0.60	0.20–0.35	0.80–1.10	–	0.15–0.25	–	–
4140	0.38–0.43	0.75–1.00	0.20–0.35	0.80–1.10	–	0.15–0.25	–	–
4340	0.38–0.43	0.60–0.80	0.20–0.35	0.70–0.90	1.65–2.00	0.20–0.30	–	–
300M	0.40–0.46	0.65–0.90	1.45–1.80	0.70–0.95	1.65–2.00	0.30–0.45	0.05 min	–
D–6a	0.42–0.48	0.60–0.90	0.15–0.30	0.90–1.20	0.40–0.70	0.90–1.10	0.05–0.10	–
6150	0.48–0.53	0.70–0.90	0.20–0.35	0.80–1.10	–	–	0.15–0.25	–
8640	0.38–0.43	0.75–1.00	0.20–0.35	0.40–0.60	0.40–0.70	0.15–0.25	–	–
Medium-alloy air-harding steels(b)								
H11 Mod	0.37–0.43	0.20–0.40	0.80–1.00	4.75–5.25	–	1.20–1.40	0.40–0.60	–
H13	0.32–0.45	0.20–0.50	0.80–1.20	4.75–5.50	–	1.10–1.75	0.80–1.20	–
9Ni-4Co steels								
HP 9–4–20	0.16–0.23	0.20–.040	0.20 max	0.65–0.85	8.50–9.50	0.90–1.10	0.06–0.12	4.25–4.75 HP
9–4–30	0.29–0.34	0.10–0.35	0.20 max	0.90–1.10	7.0–8.0	0.90–1.10	0.06–0.12	4.25–4.75

(a) Phosphorus and sulphur contents may vary with steelmaking practice. Usually, these steels contain no more than 0.035 P and 0.040 S; 9Ni-4Co steels are specified to have 0.10 maximum P and 0.10 maximum S. (b) ASTM A681; composition ranges utilised by some producers are narrower.

4130

Type 4130 is a water-hardening alloy steel of low-to-intermediate hardenability.

Heat treatments

The following standard heat treatments apply to type 4130 steel:

Normalise

Heat to 870 to 925°C (1600 to 1700°F) and hold for a period that depends on section thickness; air cool. Tempering at 480°C (900°F) or above is often done after normalising to increase yield strength.

Anneal

Heat to 830 to 860°C (1525 to 1575°F) and hold for period that depends on section thickness or furnace load; furnace cool.

Harden

Heat to 845 to 870°C (1550 to 1600°F) and hold, then water quench; or heat to 860 to 885°C (1575 to 1625°F) and hold, then oil quench. Holding time depends on section thickness.

Temper

At least ½ hr. at 200 to 700°C (400 to 1300°F); air cool or water quench. Tempering temperature and time at temperature depend mainly on desired hardness or strength.

Spheroidise

Heat to 760 to 775°C (1400 to 1425°F) and hold 4 to 12 hours, cool slowly.

Properties

Table 15.2 summarises the typical properties obtained by tempering water-quenched and oil-quenched 4130 steel bars at various temperatures. Because 4130 steel has low hardenability, section thickness must be considered when heat treating to high hardness or strength. Effects of mass on typical properties of heat treated 4130 steel are indicated in Table 15.3.

Table 15.2. Typical mechanical properties of heat treated 4130 steel.

Tempering temperature		*Tensile strength*		*Yield strength*		*Elongation in 50 mm*	*Reduction in*	*Hardness, HB*	*Izod impact energy*	
°C	*°F*	*MPa*	*ksi*	*MPa*	*ksi*	*or 2 inch,%*	*area,%*		*J*	*ft-lb*
Water quenched and tempered (a)										
205	400	1765	256	1520	220	10.0	33.0	475	18	13
260	500	1670	242	1430	208	11.5	37.0	455	14	10
315	600	1570	228	1340	195	13.0	41.0	425	14	10
370	700	1475	214	1250	182	15.0	45.0	400	20	15
425	800	1380	200	1170	170	16.5	49.0	375	34	25
540	1000	1170	170	1000	145	20.0	56.0	325	81	60
650	1200	965	140	830	120	22.0	63.0	270	135	100
Oli quenched and tempered (b)										
205	400	1550	225	1340	195	11.0	38.0	450	–	–
260	500	1500	218	1275	185	11.5	40.0	440	–	–
315	600	1420	206	1210	175	12.5	43.0	418	–	–
370	700	1320	192	1120	162	14.5	48.0	385	–	–
425	800	1230	178	1030	150	16.5	54.0	360	–	–
540	1000	1030	150	840	122	20.0	60.0	305	–	–
650	1200	830	120	670	97	24.0	67.0	250	–	–

(a) 25 mm (1 inch) diam round bars quenched from 845 to 870°C (1550 to 1600°F); and (b) 25 mm (1 inch) diam round bars quenched from 860°C (1575°F).

Table 15.3. Effects of mass on typical properties of heat treated 4130 steel.

Round bars oil quenched from 845°C (1550°F) and tempered at 540°C (1000°F); 12.83 mm (0.505 inch) diam tensile specimens cut from centre of 25 mm diam bar, and from mid-radius of 50 and 75 mm diam bars.

Bar size		*Tensil strength*		*Yield strength*		*Elongation in 50 mm*	*Reduction*	*Surface hardness,*
mm	*in*	*MPa*	*ksi*	*MPa*	*ksi*	*or 2 inch,%*	*in area,%*	*HB*
25	1	1040	151	880	128	18.0	55.0	307
50	2	740	107	570	83	20.0	58.0	223
75	3	710	103	540	78	22.0	60.0	217

4140

Type 4140 steel is similar in composition to 4130 except for a higher carbon content, which imparts greater hardenblility and strength. When any ultrahigh-strength steel is heat treated to high strength levels, it is subject to hydrogen embrittlement during operations such as acid pickling and cadmium or chromium electroplating. Ductility can be restored in thin sections by baking for 2 to 4 hours at 190°C (375°F) and in sections more than 38 mm (1½ inch) thick by baking for 23 hours at 190°C.

Heat treatments

The following standard heat treatments apply to 4140 steel:

Normalise

Heat to 845 to 900°C (1550 to 1650°F) and hold for period that depends on section thickness; air cool.

Anneal

Heat to 845 to 870°C (1550 to 1600°F) and hold for a period that depends on section thickness or furnace load; furnace cool.

Harden

Heat to 830 to 870°C (1525 to 1600°F) and hold; oil quench. (For water quencning, which is rarely used, hardening temperatures are 815 to 845°C, or 1500 to 1550°F). Holding time depends on section thickness.

Temper

At least ½ hr., at 175 to 230°C (350 to 450°F) or 370 to 675°C (700 to 1250°F); air cool or water quench. Tempering temperature and time at temperature depend mainly on desired hardness. To avoid blue brittleness, 4140 usually is not tempered between 230 and 370°C (450 and 700°F).

Spheroidise

Heat to 760 to 775°C (1400 to 1425°F) and hold 4 to 12 hours, cool slowly.

Properties

Table 15.4 summarises the mechanical properties obtained by tempering oil-quenched 4140 steel at various temperatures. Because 4140 is not a deep-hardening steel, section size should be considered, especially when specifying heat treatment for high strength levels. The effects of mass on hardness and tensile properties of 4140 steel are shown in Table 15.5.

Table 15.4. Typical mechanical properties of heat treated 4140 steel.

12.7 mm (½ inch) diam round bars, oil quenched from 845°C (1550°F).

Tempering temperature		*Tensile strength*		*Yield strength*		*Elongation in 50 mm or 2 in.,%*	*Reduction in area, %*	*Hardness, HB*	*Izod impact energy*	
°C	*°F*	*MPa*	*ksi*	*MPa*	*ksi*				*J*	*ft-lb*
205	400	1965	285	1740	252	11.0	42	578	15	11
260	500	1860	270	1650	240	11.0	44	534	11	8
315	600	1720	250	1570	228	11.5	46	495	9	7
370	700	1590	231	1460	212	12.5	48	461	15	11
425	800	1450	210	1340	195	15.0	50	429	28	21
480	900	1300	188	1210	175	16.0	52	388	46	34
540	1000	1150	167	1050	152	17.5	55	341	65	48

(Contd...)

Tempering temperature °C	Tempering temperature °F	Tensile strength MPa	Tensile strength ksi	Yield strength MPa	Yield strength ksi	Elongation in 50 mm or 2 in.,%	Reduction in area, %	Hardness, HB	Izod impact energy, J	Izod impact energy, ft-lb
595	1100	1020	148	910	132	19.0	58	311	93	69
650	1200	900	130	790	114	21.0	61	277	112	83
705	1300	810	117	690	100	23.0	65	235	136	100

Table 15.5. Effects of mass on typical properties of heat treated 4140 steel.

Round bars oil quenched from 845°C (1550°F) and tempered at 540°C (1000°F); 12.83 mm (0.505 inch) diam tensile specimens cut from center of 25 mm diam bars, and from mid-radius of 50 and 75 mm diam bars.

Diameter of bar mm	Diameter of bar inch	Tensile strength MPa	ksi	Yield strength MPa	ksi	Elongation in 50 mm or 2 inch, %	Reduction in area, %	Surface hardness, HB
25	1	1140	165	985	143	15	50	335
50	2	920	133	750	109	18	55	202
75	3	860	125	655	95	19	55	29

4340

Type 4340, the most popular steel in this class, is deep-hardening steel. In this sections, the steel is air hardening, although in practice it is usually oil quenched. When 4340 is heat treated to tensile strengths greater than about 1400 MPa (about 200 ksi), it is subject to hydrogen embrittlement. Parts exposed to hydrogen, such as in pickling or electroplating, should be baked at 185 to 195°C (365 to 385°F) for at least 8 hours and for 23 hrs. if thicker than 38 mm (1½ inch), as soon as possible after the pickling or plating operation.

Heat treatments

The following standard heat treatments apply to 4340 steel:

Normalise

Heat to 845 to 900°C (1550 to 1650°F) and hold for period that depends on section thickness; air cool.

Anneal

Heat to 830 to 860°C (1525 to 1575°F) and hold for a period that depends on section thickness or furnace load; furnace cool.

Harden

Heat to 800 to 845°C (1475 to 1550°F) and hold 15 minutes for each 25 mm (1 inch) of thickness (minimum, 15 minutes); oil quench to below 65°C (150°F) or quench into fused salt at 200 to 210°C (390 to 410°F), hold 10 minutes, then air cool to below 65°C (150°F).

Temper

At least ½ h at 200 to 650°C (400 to 1200°F); air cool. Temperature and time at temperature depend mainly on desired strength or hardness.

Spheroidise

Pre-heat to 690°C (1275°F) and hold 2 hours, increase temperature to 750°C (1375°F) and hold 2 hours., cool to 650°C (1200°F) and hold 6 hours, furnace cool to about 600°C (1100°F) and finally air cool to room temperature. An alternative schedule is to heat to 730 to 750°C (1350 to 1375°F) and hold several hours, then furnace cool to room temperature.

Stress relieve

After straightening, forming or machining, parts may be stress relieved at 650 to 675°C (1200 to 1250°F).

Properties

Through hardening of 4340 steel can be achieved by oil quenching round sections up to 75 mm (3 inch) in diameter, and by water quenching larger sections (up to the limit of hardenability). The influence of section size on tensile properties of oil-quenched and water-quenched 4340 is indicated by the data in Table 15.6.

Table 15.6. Effects of mass on mechanical properties of 4340 steel.

Data from alloy digest Ultrahigh-Strength Structural Steels', STP 498, American Society for Testing and Materials.

Section diameter		*Tensile strength*		*Yield sterngth*		*Elongation in 50 mm or 2 inch,%*	*Reduction in area, %*	*Hardness, HB*
mm	*in*	*MPa*	*ksi*	*MPa*	*ksi*			
Oil quenched and tempered (a)								
13	½	1460	212	1380	200	13	51	–
38	1½	1450	210	1365	198	11	45	–
75	3	1420	206	1325	192	10	38	–
Water quenched and tempered (b)								
75	3	1055	153	930	135	18	52	340
100	4	1035	150	895	130	17	50	330
150	6	1000	145	850	123	16	44	322

(a) Austenitised at 845 °C (1550 °F); tempered at 425°C (800 °F). (b) 75 mm (3 inch) diam bar austenitised at 800°C; 100 and 150 mm (4 and 6 inch) diam bars austenitised at 815°C (1500°F). All sizes tempered at 650°C (1200°F). Test specimens taken at mid-radius

Hardness of type 4340 as a function of tempering temperature is plotted in Fig.15.1. Typical mechanical properties of oil-quenched 4340 are given in Table 15.7. Additional data on mechanical properties (notch toughness and fracture toughness) of this steel tempered to different hardnesses are given in Table 15.8.

300M

Alloy 300M is basically a silicon-modified (1.6 per cent Si) 4340 steel, but is slightly higher in carbon and molybdenum and also contains vanadium. The steel exhibits deep hardenability. Because of its high silicon and molydenum contents, 300M is more prone to decarburisation than the steels so far described; and during heat treating, care should be exercised to avoid decarburisation. When heat treated to strength levels higher than 1380 MPa (200 ksi), 300M is susceptible to hydrogen embrittlement, and thus parts should be properly baked after plating (same baking procedure as that used for type 4340).

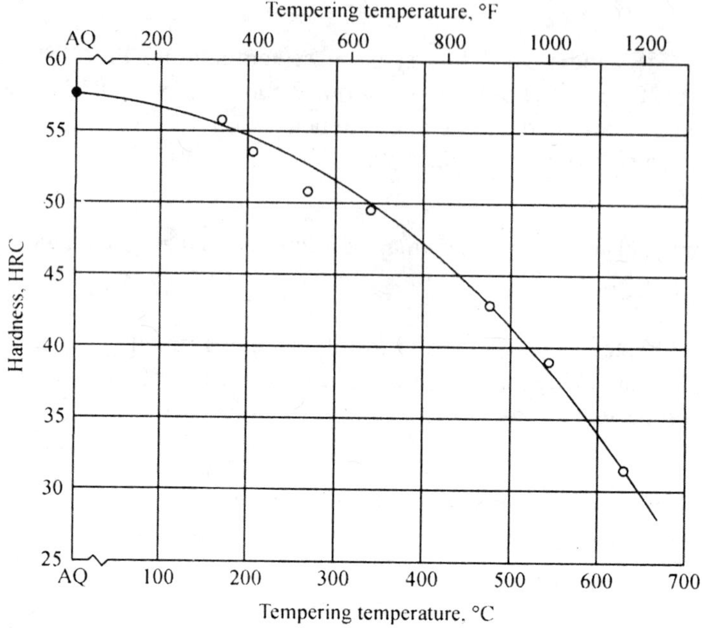

All specimens oil quenched from 845°C (1550°F) and tempered 2 hours at temperature

Fig 15.1. Variation for hardness with tempering temperature of 4340 steel.

Table 15.7. Typical mechanical properties of 4340 steel.

Oil quenched from 845 °C (1550°F) and tempered of various temperatures.

Tempering temperature		*Tensile strength*		*Yield strength*		*Elongation in 50 mm or2 inch, %*	*Reduction in area, %*	*Hardness*		*Izod impact energy*	
°C	°F	MPa	ksi	MPa	ksi			HB	HRC	J	ft-lb
205	400	1980	287	1860	270	11	39	520	53	20	15
315	600	1760	255	1620	235	12	44	490	49.5	14	10
425	800	1500	217	1365	198	14	48	440	46	16	12
540	1000	1240	180	1160	168	17	53	360	39	47	35
650	1200	1020	148	860	125	20	60	290	31	100	74
705	1300	860	125	740	108	23	63	250	24	102	75

Table 15.8. Notch toughness and fracture toughness of 4340 steel tempered to different hardnesses.

Hardness, HB	*Equivalent tensile strength(a)*		*Charpy V-notch impact energy*		*Plane- strain fracture toughness*	
	MPa	ksi	J	ft-lb	$MPa\sqrt{m}$	$ksi\sqrt{inch}$
550	2040	296	19	14	53	48
430	1520	220	30	22	75	68
380	1290	187	42	31	110	110

(a) Estimated from hardness.

Heat treatments

The following standard heat treatments apply to 300M steel:

Normalise

Heat to 915 to 940°C (1675 to 1725°F) and hold for a period that depends on section thickness; air cool. If normalising is to enhance machinability, recharge into a tempering furnace at 650 to 675°C (1200 to 1250°F) before the steel reaches room temperature.

Harden

Austenitise at 855 to 885°C (1575 to 1625°F). Oil quench to below 70°C (160°F); or quench in salt at 200 to 210°C (390 to 410°F) and hold 10 minute, then air cool to 70°C or below.

Temper

Two to four hours at 260 to 315°C (500 to 600°F); double tempering recommended. This tempering procedure produces the best combination of high yield strength and high impact properties. Tempering outside the range given above results in severe deterioration of properties.

Spheroidise

Heat to about 775°C (1430°F) and hold for a period that depends on section thickness or furnace load. Cool to 650°C (1200°F) at a rate no faster than 5.5°C/hr. (10°F/hr.), then cool to 480°C (900°F) no faster than 11°C/hr. (20°F/hr.), and finally air cool to room temperature. The same schedule is recommended for annealing.

Properties

Variations in hardness and mechanical properties of 300M with tempering temperature are presented in Table 15.9. This steel has deep hardenability, so that heat treated bars 75 mm (3 inch) in diameter have essentially the same tensile properties as bars 25 mm (1 inch) in diameter. Reductions in tensile strength, ductility and impact strength, however, are observed in heat treated bars 145 mm (5¾ inch) in diameter. Variations in properties of 300M with section size are presented in Table 15.10.

Table 15.9. Typical mechanical properties of 300 M steel.

Round bars, 25 mm (1 inch) in diameter, oil quenched from 860°C (1575°F) and tempered at various temperatures.

Tempering temperature		*Tensile strength*		*Yield strength*		*Elongation in 50 mm or 2 inch, %*	*Reduction in area, %*	*Charpy V-notch impact energy*		*Hardness, HRC*
°C	°F	MPa	ksi	MPa	ksi			J	ft. lb	
90	200	2340	340	1240	180	6.0	10.0	17.6	13.0	56.0
205	400	2140	310	1650	240	7.0	27.0	21.7	16.0	54.5
260	500	2050	297	1670	242	8.0	32.0	24.4	18.0	54.0
315	600	1990	289	1690	245	9.5	34.0	29.8	22.0	53.0
370	700	1930	280	1620	235	9.0	32.0	23.7	17.5	51.0
425	800	1790	260	1480	215	8.5	23.0	13.6	10.0	45.5

Table 15.10 Effects of mass on tensile and impact properties of 300M steel.

Round bars, normalised at 900°C (1650°F), oil quenched from 860°C (1575°F) and tempered at 315°C (600°F).

Bar diameter		*Tensile strength*		*Yield strength*		*Elongation in 50 mm or 2 inch,%*	*Reduction in area, %*	*Charpy V-notch impact energy when tested at*					
								+21°C (+70°F)		*−46°C (−50°F)*		*−73°C (−100°F)*	
mm	*inch*	*MPa*	*ksi*	*MPa*	*ksi*			*J*	*ft-lb*	*J*	*ft-lb*	*J*	*ft-lb*
25	1	1990	289	1690	245	9.5	34.1	30	22	26	19	24	18
75	3	1940	281	1630	236	9.5	35.0	26	19	19	14	12	9
150	5¾	2120	308	1800	261	7.3	22.3	12	9	9	7	7	5

D-6a and D-6ac

Ladish D-6a was designed primarily for use at room-temperature tensile strengths of 1800 to 2000 MPa (260 to 290 ksi). It is deeper hardening than 4340. The alloy is called D-6a when produced by air melting in an electric furnace and D-6ac when produced by air melting followed by vacuum arc remelting. Except for some improvements in mechanical properties of D-6ac due to melting practice, the characteristics of the two steels are identical.

Heat treatments

The following standard heat treatments apply to D-6a and D-6ac steels:

Normalise

Heat to 870 to 955°C (1600 to 1750°F) and hold for a period that depends on section thickness; air cool.

Anneal

Heat to 815 to 845°C (1500 to 1550°F) and hold for a period that depends on section thickness or furnace load; furnace cool to 540°C (1000°F) at a rate no faster than 28°C/hrs. (50°F/hrs.), then air cool to room temperature. Hot forgings may be annealed as follows for maximum machinability: Immediately after forging, charge the parts into a 650°C (1200°F) furnace and hold 12 hours, increase temperature to 900°C (1650°F) and hold for a period that depends on section size, cool to 650°C (1200°F) and hold 10 hours and finally air cool to room temperature.

Harden

Austenitise at 845 to 900°C (1550 to 1650°F for ½ to 2 hours. Sections no larger than 25 mm (1 inch) in thickness or diameter may be air cooled. Larger sections may be oil quenched to 65°C (150°F); or salt quenched to 205°C (400°F) and then air cooled. For optimum dimensional stability, 'aus-bay' quench into a furnace or salt bath at 525°C (975°F), equalise the temperature, then quench in an oil bath held at 60°C (140°F), or quench in 205°C (400°F) salt and air cool. The cooling rate during quenching significantly affects fracture toughness. For high fracture toughness, especially in heavy sections, austenitise at 925°C (1700°F), equalise temperature, and oil quench to 60°C (140°F).

Temper

Immediately after hardening, temper 2 to 4 hours in the range 200 to 700°C (400 to 1300°F), depending on desired strength or hardness.

Spherodise

Heat to 730°C (1350°F) and hold 5 hours, increase temperature to 760°C(1400°F) and hold 1 hour; furnace cool to 690°C (1275°F) and hold 10 hours; furnace cool to 690°C (1275°F) and hold 10 hours; furnace cool to 650°C (1200°F) and hold 8 hours; air cool room temperature.

Stress relieve

Heat to a temperature from 540 to 675°C (1000 to 1250°F) and hold for 1 to 2 hours; air cool.

Properties

Typical room temperature hardness of D-6a steel bar as a function of tempering temperature is plotted in Fig. 15.2; other typical mechanical properties of D-6a bar are given in Table 15.11. Tensile properties of heat treated D-6ac billet material are given in Table 15.12.

Table 15.11. Typical mechanical properties of D-6a steel bar.

Normalised at 900°C (1650°F), oil quenched from 845°C (1550°F) and tempered at various temperatures.

Tempering temperature		*Tensile strength*		*Yield strength*		*Elongation in 50 mm or 2 in.,%*	*Reduction in area,%*	*Charpy V-notch impact energy*	
°C	°F	MPa	ksi	MPa	ksi			J	ft-lb
150	300	2060	299	1450	211	8.5	19.0	14	10
205	400	2000	290	1620	235	8.9	25.7	15	11
315	600	1840	267	1700	247	8.1	30.0	16	12
425	800	1630	236	1570	228	9.6	36.8	16	12
540	1000	1450	210	1410	204	13.0	45.5	26	19
650	1200	1030	150	970	141	18.4	60.8	41	30

Table 15.12 Typical tensile properties of double tempered D-6ac billet.

Ausenitised 1 hr. at 900°C (1650°F), quenched in fused salt at 205°C (400°F) and held 5 minute, then air cooled to room temperature. Tempered 1 hour at 205°C; second temper, 4 hours at indicated temperature.

Second tempering temperature		*Tensile strength*		*Yield strength*		*Elongation in 50 mm or 2 inch,%*	*Reduction in area,%*
°C	°F	MPa	ksi	MPa	ksi		
480	900	1686.5	244.6	1540.3	223.4	11.1	40.0
510	950	1652.7	239.7	1519.7	220.4	13.2	44.1
540	1000	1613.4	234.0	1483.8	215.2	13.7	47.2

6150

Type 6150 can be considered an ultrahigh-strength steel, although as a constructional steel it is not as popular as the other steels in this class. It is a shallow-hardening steel. Parts made of 6150 can be readily welded; after welding, parts should be normalised, then hardened and tempered to the desired hardness.

Heat Treatments

The following heat treatments apply to 6150 steel:

Normalise

Heat to 870 to 955°C (1600 to 1750°F) and hold for a period that depends on section thickness; air cool.

Anneal

Heat to 845 to 900°C (1550 to 1650°F) and hold for a period that depends on section thickness or furnace load; furnace cool.

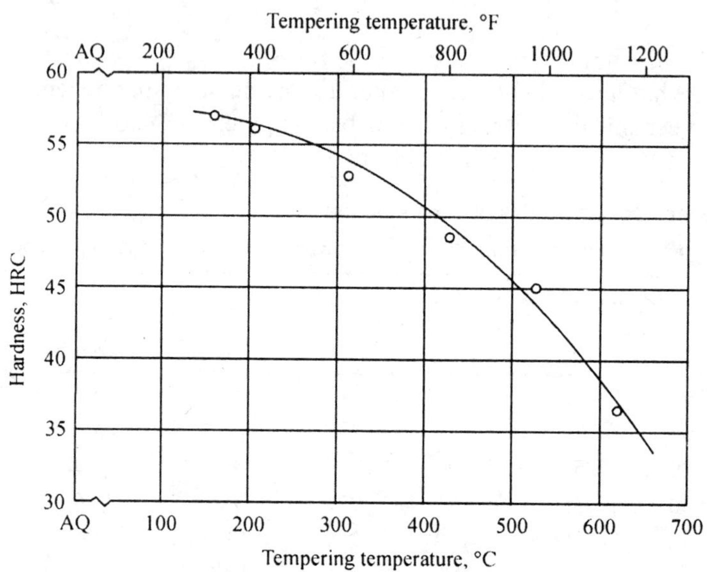

Fig. 15.2. Variation for hardness with tempering temperature for D-6a steel.

Harden

Austenitise at 845 to 900°C (1550 to 1650°F); oil quench.

Temper

At least ½ hr. at 200 to 650°C (400 to 1200°F). Tempering temperature and time at temperature depend on desired final hardness or strength.

Austemper

Austenitize in a salt bath at 845 to 900°C (1550 to 1650°F); quench in a salt bath at 230 to 315°C (450 to 600°F), hold 20 to 30 min, then oil quench or air cool to room temperature.

Martemper

Austenitize in a salt bath at 845 to 870°C (1550 to 1600°F); quench in a salt bath at 230 to 260°C (450 to 500°F), equalise, then air cool or quench to room temperature. Temper to desired hardness.

Spheroidise

Heat to 800 to 830°C (1475 to 1525°F), hold until heated through, furnace cool to 650°C (1200°F) and hold several hours, then cool slowly to room temperature.

Properties

Typical mechanical properties of small-diameter round sections of 6150 tempered at various temperatures are given in Table 15.13. Hardness and Izod impact energy as functions of tempering temperature are plotted in Fig.15.3. The effects of section size on tensile properties and hardness are given in Table 15.14.

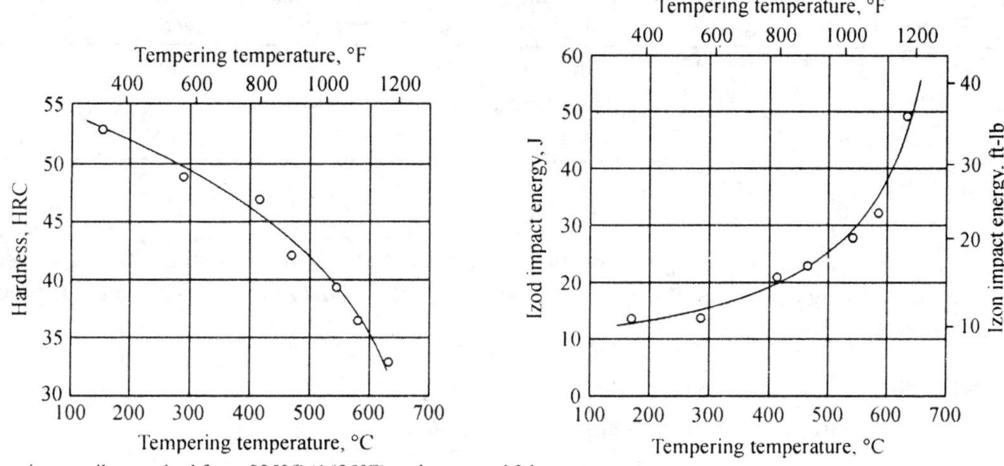

Specimens oil quenched from 885°C (1625°F) and tempered 2 hours at temperature

Fig 15.3. Variation for hardness and impact energy with tempering temperature for 6150 steel.

Table 15.13 Typical room-temperature tensile properties of heat treated 6150 steel.

Tempeirng temperature		*Tensile strength*		*Yield strength*		*Elongation in 50 mm or 2 inch,%*	*Reduction in area, %*	*Hardness, HB*	*Izod impact energy*	
°C	*°F*	*MPa*	*ksi*	*MPa*	*ksi*				*J*	*ft-ib*
Round bars, 14 mm (0.55 inch) in diameter(a)										
205	400	2050	298	1810	263	1	5	610	–	–
260	500	2070	300	1810	263	4	12	570	–	–
315	600	1950	283	1720	250	7	27	540	–	–
370	700	1770	257	1620	235	10	37	505	9	7
425	800	1585	230	1490	216	11	42	470	14	10
480	900	1410	204	1340	195	12	44	420	16	12
540	1000	1250	182	1210	175	13	46	380	20	15
595	1100	1150	167	1080	157	16	47	350	28	21
Round bars, 25 mm (1 inch) diameter (b)										
425	800	1570	228	1450	210	10	37	461	–	–
480	900	1360	197	1210	175	11	41	401	–	–
540	1000	1180	171	1030	150	12	45	341	–	–
595	1100	1030	150	875	127	15	50	302	–	–
650	1200	920	133	760	110	19	55	262	–	–
705	1300	810	118	660	96	23	61	235	–	–

(a) Normalised at 870°C (1600°F), oil quenched from 860°C (1575°F) and tempered at various temperatures. (b) Oil quenched from 860°C and tempered at various temperatures.

Table 15.14. Effects of mass on typical properties of heat treated 6150 steel.

Round bars, oil quenched from 830°C (1525°F) and tempered at 540°C (1000°F); 12.83 mm (0.505 inch) diam tensile specimens taken from center of 25 mm bars and from mid-radius of 50 and 75 mm diam bars.

Bar size		*Tensile strength*		*Yield strength*		*Elongation in 50 mm*	*Reduction*	*Hardness,*
mm	*inch*	*MPa*	*ksi*	*MPa*	*ksi*	*or 2 inch,%*	*in area, %*	*HB*
25	1	1185	172	1040	151	14	45	341
50	2	1170	170	1030	149	13	48	341
75	3	1090	158	950	138	13	47	331

8640

Type 8640 steel was especially designed for maximum hardenability and best combination of properties with minimum alloying additions. It is an oil-hardening steel, but may be water hardened if precautions are taken to prevent cracking.

Heat treatments

The following standard heat treatments apply to 8640 steel:

Normalise

Heat to 870 to 925°C (1600 to 1700°F) and hold for a period that depends on section thickness; air cool.

Anneal

Heat to 845 to 870°C (1550 to 1600°F) and hold for a period that depends on section thickness or furnace load; furnace cool.

Harden

Austenitise at 815 to 845°C (1500 to 1550°F); quench in oil or water.

Temper

At least ½ hour at 200 to 650°C (400 to 1200°F). Temperature and time at temperature depend on desired hardness.

Spheroidise

Heat to 705 to 720°C (1300 to 1325°F) and hold several hours; furnace cool.

Properties

Variations in typical properties of heat treated round sections of 8640 with tempering temperature are given in Table 15.15. Variations in properties with section size (mass effect) are given in Table 15.16.

MEDIUM ALLOY AIR-HARDENING STEELS

Heat treatments for the ultrahigh-strength steels H11 Modified (H11 Mod) and H13, which are also known as 5 per cent Cr hot work die steels, are discussed in this section. These steels are similar in composition, heat treatment and many properties. They have deep hardenability and can be hardened through in large sections by air cooling. Air hardening results in minimal residual stresses after hardening. Both H11 Mod and H13 are secondary-hardening steels, and thus develop optimum properties when tempered at temperatures above the secondary-hardening peaks at about 510°C (950°F). These high tempering temperatures provide substantial stress relief and stabilisation of properties so that the steels can be used advantageously at elevated temperatures. This also enables heat treated parts to be warm worked, or pre-heated for welding, at temperatures as high as 55°C (100°F) below the prior tempering temperature. Both H11 Mod and H13 steels are subject to hydrogen embrittlement and part should be baked after any exposure to environments where hydrogen may be absorbed. Baking for 24 hours or longer at 190°C (375°F) or higher is recommended.

Table 15.15. Typical room-temperature mechanical properties of 8640 steel.

Tempering temperature		*Tensile strength*		*Yield strength*		*Elongation in 50 mm or 2 inch,%*	*Reduction in area,%*	*Impact energy*		*Hardness*	
°C	*°F*	*MPa*	*ksi*	*MPa*	*ksi*			*J*	*ft-lb*	*HB*	*HRC*
Round bars, 13.5 mm (0.53 inch) in diameter (a)											
205	400	1810	263	1670	242	8.0	25.8	11.5(b)	8.5(b)	555	55
315	600	1585	230	1430	208	9.0	37.3	15.6(b)	11.5(b)	461	48
425	800	1380	200	1230	179	10.5	46.3	27.8(b)	20.5(b)	415	44
540	1000	1170	170	1050	152	14.0	53.3	56.3(b)	41.5(b)	341	37
650	1200	870	126	760	110	20.5	61.0	96.9(b)	71.5(b)	269	28
Round bars, 25mm (1 inch) in diameter (a)											
425	800	1382	200.5	1230	179	10	46	27(c)	20(c)	415	44
480	900	1250	181	1120	162	13	51	41(c)	30(c)	388	42
540	1000	1070	155	940	137	17	56	54(c)	40(c)	331	36
595	1100	1020	148	910	132	16	57	73(c)	54(c)	302	32
650	1200	865	125.5	760	110.5	20	61	83(c)	61(c)	269	28

(a) Oil quenched from 830°C (1525°F) and tempered at indicated temperature; (b) Izod. (c) Charpy V-notch.

Table 15.16. Effects of mass on typical properties of heat treated 8640 steel.

Oil quenched from 830°C (1525°F) and tempered at 540°C (1000°F).

Bar size		*Tensile strength*		*Yield strength*		*Elongation in 50 mm or2 in.,%*	*Reduction in area, %*	*Surface hardness, HB*
mm	*in.*	*MPa*	*ksi*	*MPa*	*ksi*			
25	1	1070	155	940	137	17	56	331
50	2	910	132	770	112	18	57	293
75	3	860	125	710	103	19	58	277

Because H11 Mod and H13 are air-hardening steels, forged parts must be cooled slowly after forging to prevent stress cracking. After forging, parts should be charged into a furnace at about 790°C (1450°F), soaked until the temperature is uniform, and then slowly cooled, either in the furnace or in an insulating medium such as ashes, lime, mica or silocel. When the forgings have cooled, they should be spheroidise annealed. Weldements, especiallly heavy-section weldments, should be cooled slowly in a funace heated to pre-heating temperature, or in an insulating medium, immediately after welding. After being cooled, weldments should be given a full spheroidising anneal.

H11 Mod

This steel is a modification of the martensitic hot work die steel H11, the significant difference being that H11 Mod has slightly higher carbon content.

Heat treatments

The following standard heat treatments apply to H11 Mod steel:

Normalise

Generally not recommended. For effective homogenisation, heat to about 1065°C (1950°F), soak 1 hr. for each 25 mm (1 inch) of thickness, air cool. Anneal immediately after the steel reaches room temperature. Note: There is a possibility that H11 Mod may crack during this treatment, especially if the surface is significantly decarburised.

Anneal

Heat to 845 to 885°C (1550 to 1625°F) in a furnace, preferably one with controlled atmosphere, and hold to equalise temperature; cool very slowly in the furnace to about 480°C (900°F), then more rapidly to room temperature. This treatment should produce a fully spheroidised microstructure free of grain-boundary carbide networks.

Harden

Pre-heat to 760 to 815°C (1400 to 1500°F), then raise the temperature to 995 to 1025°C (1825 to 1875°F) and hold 20 minute plus 5 minutes for each 25 mm (1 inch) of thickness (minimum, 25 minutes); air cool in still air. This can be conveniently done in a neutral salt bath or a controlled-atmosphere furnace. For a few applications, oil quenching from the low end of the hardening temperature range may be done. Air cooling, which produces less distortion than oil quenching, is more commonly employed.

Temper

At the secondary hardening peak temperature of about 510°C (950°F) for maximum hardness and strength, or preferably above the secondary-hardening peak to temper back to a lower hardness or strength with improved ductility and toughness. A minimum of 1 hour at temperature should be allowed, but parts preferably should be double tempered (2 hours at temperature, cool to room temperature, then 2 hours more at temperature). Triple tempering is more desirable, especially for critical parts. For high-temperature applications, parts should be tempered at a temperature above the maximum service temperature to guard against unwanted changes in properties during service.

Stress relieve

Heat to 650 to 675°C (1200 to 1250°F); cool slowly to room temperature. This treatment is often used to achieve greater dimensional accuracy in heat treated parts by stress relieving rough-machined parts, then finish machining, and finally heat treating to the desired hardness.

Nitride

For increased wear resistance, finish machined and heat treated parts may be nitrided. The nitriding operation can be considered as the second temper of a double tempering operation.The parts should be gas or liquid nitrided at about 525°C (980°F). The nitriding case depth depends on time at temperature. For example, gas nitriding in 20 to 30 per cent dissociated ammonia for 8 to 48 hours normally produces a case depth of about 0.2 to 0.35 mm (0.008 to 0.014 inch).

Properties

Variation of hardness with tempering temperature for H11 Mod is plotted in Fig.15.4. Variations in typical room-temperature longitudinal mechanical properties with tempering temperature are given in Table 15.17. As an indication of the deep air hardenability of this steel (to depths greater than 305 mm, or 12 inch), the transverse tensile strength and ductility obtained in large billets of air melted and vacuum arc-remelted H11 Mod are given in Table 15.18, which also shows the improvement in ductility that results from vacuum arc re-melting. Each value is the average of four tests, two from the top and two from the bottom of the ingot.

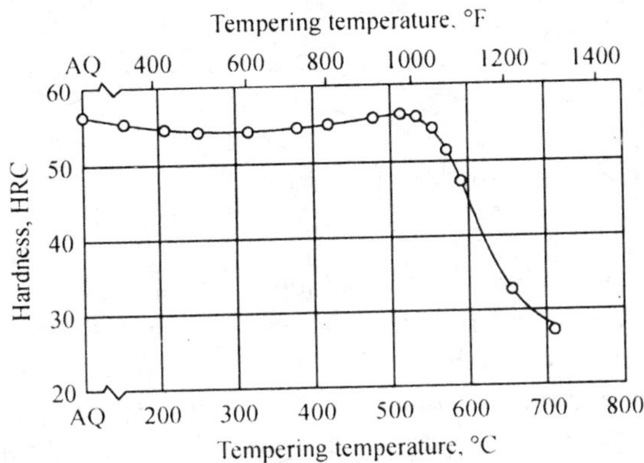

All specimens air cooled from 1010°C (1850°F) and double tempered, 2 + 2 hours at temperature

Fig. 15.4. Variation for hardness with tempering temperature for H11 Mod steel.

Table 15.17. Typical longitudinal mechanical properties of H11 Mod steel.

Air cooled from 1010°C (1850°F); double tempered, 2 + 2 hours at indicated temperature

Tempering temperature		*Tensile strength*		*Yield strength*		*Elongation in 50 mm or 2 inch, %*	*Reduction in area, % area, %*	*Charpy V-notch impact energy*		*Hardness, HRC*
°C	*°F*	*MPa*	*ksi*	*MPa*	*ksi*			*J*	*ft-lb*	
510	950	2120	308	1710	248	5.9	29.5	13.6	10.0	56.5
540	1000	2010	291	1675	243	9.6	30.6	21.0	15.5	56.0
565	1050	1850	269	1565	227	11.0	34.5	26.4	19.5	52.0
595	1100	1540	223	1320	192	13.1	39.3	31.2	23.0	45.0
650	1200	1060	154	850	124	14.1	41.2	40.0	29.5	33.0
705	1300	940	136	700	101	16.4	42.2	90.6	66.8	29.0

H13

The main difference in composition between H11 Mod and H13 is the higher vanadium content of the latter (Table 15.1); this leads to a greater dispersion of hard vanadium carbides, which results in higher wear resistance. H13 parts may be nitrided for additional wear resistance. Also, H13 has a slightly wider range of carbon content than H11 Mod. Depending on the producer, the carbon content of H13 may be near the high or low side of the accepted range and for a given heat treatment the properties will vary correspondingly.

Table 15.18. Effect of billet size and melting method on typical transverse strength and ductility of H11 Mod steel.

Air cooled from 1010°C (1850°F); triple tempered, 2 + 2 + 2 hours at 540°C (1000°F).

Billet size	*Melting method*	*Tensile strength*		*Reduction in area, %*
		MPa	*ksi*	
150 by 150 mm (6 by 6 inch)	Air	1965	285	16.1
	VAR	1985	288	25.7
300 by 300 mm (12 by12 inch)	Air	1972	286	7.2
	VAR	2013	292	19.7

Heat treatments

The following standard heat treatments apply to H13 steel:

Normalise

Not recommended for H13. Some improvement in homogeneity can be obtained by pre-heating to about 790°C (1450°F), heating slowly and uniformly to 1040 to 1065°C (1900 to 1950°F) and holding 1 hour for each 25 mm (1 inch) of thickness, and then air cooling. Just before or just as the steel reaches room temperature, it should be recharged into a furnace and given a full spheroidising anneal.
Note: There is a risk of cracking during this treatment, especially if done in a furnace where the atmosphere is not controlled to prevent surface decarburisation.

Anneal

Heat uniformly to 860 to 900°C (1575 to 1650°F) in a controlled-atmosphere furnace, or with the part in a neutral compound, to prevent decarburisation, and hold to equalise temperature; cool very slowly in the furnace to about 480°C (900°F), then cool more rapidly to room temperature. This treatment should result in a fully spheroidised microstructure free from grain-boundary carbide networks.

Harden

Pre-heat to 790 to 815°C (1450 to 1500°F), then raise the temperature uniformly to 995 to 1025°C (1825 to 1875°F) and soak 20 minutes plus 5 minutes for each 25 mm (1 inch) of thickness (minimum, 25 minutes); air cool in still air. For a few applications, oil quenching from the low side of the hardening temperature may be done, but at the risk of distortion or cracking. Air cooling is preferred, and usually is done from the high side of the hardening temperature range.

Temper

At the secondary-hardening peak of about 510°C (950°F) for maximum hardness and strength, but preferably at a higher temperature to temper back to a lower level of hardness or strength with improved toughness and ductility. Double tempering—2 hours at temperature, air cool, then 2 hours more at temperature—is recommended; for critical parts, triple tempering may be desirable.

Stress relieve

Heat to 650 to 675°C (1200 to 1250°F) and soak 1 hr. or more; cool slowly to room temperature. This treatment often is used to achieve greater dimensional accuracy in heat treated parts by stress relieving rough-machined parts, then finish machining and finally heat treating to the desired hardness.

Nitride

Finish-machined and heat treated parts may be nitrided. Because it is carried out at about the normal tempering temperature, nitriding can serve as the second temper in a double tempering treatment. The nitrided case depth depends on time at temperature. For example, gas nitriding at 510°C (950°F) for 10 to 12 hours produces a case depth of 0.10 to 0.13 mm (0.004 to 0.005 inch), where as gas nitriding for 40 to 50 hours results in a case depth of about 0.3 to 0.4 mm (0.012 to 0.016 inch). For selective nitriding, copper plating is preferred for stopping off areas that are not to be nitrided; stop-off containing lead should be avoided, because lead has been found to embrittle H13 steel.

Properties

The properties presented in this section are for H13 with a carbon content in the middle of the composition range (for composition range, see Table 15.1). Somewhat different properties should be expected when the carbon content is near the high end or the low end of the range.

Variation of hardness with tempering temperature for H13 is plotted in Fig.15.5. Typical room-temperature longitudinal mechanical properties of bars tempered to different hardness levels are given in Table 15.19. H13 has deep hardenability, although its hardenability is slightly less than that of H11 Mod. For example, an H13 bar 330 mm (13 inch) in diameter and 2743 mm (108 inch) long, when fast air cooled from 1010°C (1850°F), had an as-quenched hardness of 45 HRC.

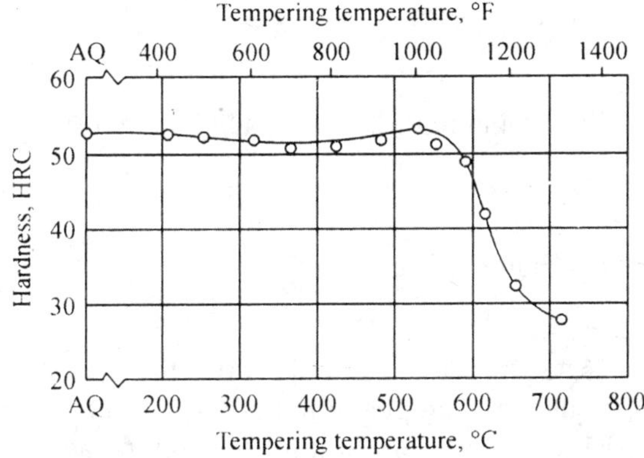

All specimens air cooled from 1025°C (1875°F) and tempered, 2 hours at temperature

Fig 15.5. Variation for hardness with tempering temperature for H13 steel.

Table 15.19. Typical room-temperature longitudinal mechanical properties of H13 steel.

Round bars, oil quenched from 1010°C (1850°F) and double tempered, 2 + 2 hours at indicated temperature.

Tempering temperature		*Tensile strength*		*Yield strength*		*Elongation in 4D, %*	*Reduction in area, %*	*Charpy V-notch impact energy*		*Hardness, HRC*
°C	*°F*	*MPa*	*ksi*	*MPa*	*ksi*			*J*	*ft-lb*	
527	980	1960	284	1570	228	13.0	46.2	16	12	52
555	1030	1835	266	1530	222	13.1	50.1	24	18	50
575	1065	1730	251	1470	213	13.5	52.4	27	20	48
593	1100	1580	229	1365	198	14.4	53.7	28.5	21	46
605	1120	1495	217	1290	187	15.4	54.0	30	22	44

9Ni-4Co STEELS

During the 1960's, a family of four HP 9-4 (9Ni-4Co) steels having high fracture toughness was developed, which when heat treated to very high strength levels. Among these, HP 9-4-20 and HP 9-4-30 are commercially available. They nominally contain 0.20 and 0.30 per cent C, respectively (Table 15.1 for chemical compositions). With the increase in carbon content, the attainable strength incresases but with corresponding decreases in toughness and weldability. The high nickel content of 9 per cent provides deep hardenability and toughness, and the 4 per cent Co prevents retention of excessive ausatenite in heat treated parts. The carbide-forming elements, chromium and molybdenum, also impart hardenability, but the amounts of these carbide formers are adjusted to provide a fairly flat response to tempering without pronounced secondary hardening and its attendant reduction in toughness.

HP 9-4-20, although it has good weldability and fracture toughness, cannot develop a yield strength of 1380 MPa (200 ksi). Therefore, only HP 9-4-30 is discussed below.

HP 9-4-30

HP 9-4-30 steel has deep hardenablilty and can be fully hardened to martensite in sections up to 1500 mm (6 inch) thick. Heat treated parts can be readily welded without any pre-heat or post-heat treatment. After welding, parts may be stress relieved at about 540°C (1000°F) for 24 hours.

Heat treatments

The following heat treatments apply to HP 9-4-30 steel:

Normalise

Heat to 870 to 925°C (1600 to 1700°F) and hold 1 hr. for each 25 mm (1 inch) of thickness (minimum, 1 hour); air cool.

Anneal

Heat to 620°C (1150°F) and hold 24 hours air cool.

Harden

Austenitise at 830 to 860°C (1525 to 1575°F) and hold 1 hour for each 25 mm (1 inch) of thickness (minimum, 1 hour); water or oil quench. Complete the martensitic transformation by refrigerating at least 1 hour at –87 to –60°C (–125 to –75°F); let warm to room temperature.

Temper

At 200 to 600°C (400 to 1100°F), depending on desired strength; double tempering preferred. The most widely used tempering treatment is double tempering (2 hours at temperature, air cool, then 2 hours more at temperature) at a temperature from 540 to 575°C (1000 to 1075°F).

Stress relieve

Usually required only after welding of restrained sections. Heat to 540°C (1000°F) and hold 24 hours; air cool to room temperature.

Properties

Room-temperature mechanical properties of HP 9-4-30 double tempered at three different temperatures are presented in Table 15.20. The data for material double tempered at 540°C (1000°F) represent the minimum mechanical properties for this condition; properties listed for the other conditions may be considered typical.

Table 15.20. Room-temperature mechanical properties of HP 9-4-30 steel.

Property	*Typical value for hardness of* 49–53 HRC(a)	44–48 HRC(b)	*Minimum value(C)*
Tensile strength	1650–1790 MPa (240–260 ksi)	1520–1650 MPa (220–240 ksi)	1520 MPa (220 ksi)
Yield strength(d)	1380–1450 MPa (200–210 ksi)	1310–1380 MPa (190–200 ksi)	1310 MPa (190 ksi)
Elongation in 4 D	8–12%	12–16%	10%
Reduction in area	25–35%	35–50%	35%
Charpy V-notch impact energy	20–27 J (15–20 ft.lb)	24–34 J (18–25 ft.lb)	24 J (18 ft.lb)
Fracture toughness (K_{Ic})	66–99 MPa $\sqrt{m}$ (60–90 ksi $\sqrt{inch}$)	99–115 MPa $\sqrt{m}$ (90–105 ksi $\sqrt{inch}$)	–

(a) Oil quenched from 845°C (1550°F), refrigerated to –73°C (–100°F) and double tempered at 205°C (400°F). (b) Same heat treatment as (a) except double tempered at 550°C (1025°F); and (c) For sections forged quenched to 75 mm (3 inch) or less in thickness (or to less than 0.016 m^2 or 25 inch^2, in total cross-sectional area), quenched to martensite and double tempered at 540°C (1000°F).

HEAT TREATING OF MARAGING STEELS

Maraging steels are ultrahigh-strength steel that differ from cnventional steels in that they are not hardened by carbon content. Instead these steels are strengthened by precipitation of intermetallic compounds produced by age hardening a matrix of very-low-carbon martensite. Carbon, in fact, is an impurity in maraging steels and is kept at the lowest possible concentration.

Compositions of the common grades of maraging steel are given in Table 15.21. Grades have been developed that provide specific levels of yield strength ranging from 1030 to 3450 MPa (150 to 500 ksi).

The absence of carbon and the use of intermetallic precipitation to achieve hardening produce several unique characteristics of maraging steels that set them apart from conventional steels. Hardenability is of no concern. The low-carbon martensite formed after annealing is relatively soft, about 30 to35 HRC. During age hardening there is only a very slight dimensional change. Thus, fairly intricate shapes can be machined in the soft condition and then hardened with a minimum of distortion.

Physical Metallurgy

Phase transformations

Phase transformations in maraging steels are illustrated in Fig.15.6. Plotted on the metastable diagram (Fig.15.6a) are the austenite-to-martensite transformation on cooling and the martensite-to-austenite reversion on heating. The equilibrium diagram (Fig.15.6b) shows that at higher nickel contents the equilibrium phases at low temperatures are austenite and ferrite.

In the metastable diagram, no phases transformations occur until the M_s temperature is reached and martensite is formed. Even very slow cooling of heavy sections produces only martensite, with no problems of low hardenability due to the large section size. The remaining alloying elements in the steel do, of course, alter the M_s temperatures significantly from those shown in Fig.15.6, but the characteristic independence of cooling rate is not altered. For most grades of maraging steel, M_s temperatures are on the order of 200 to 300°C (390 to 570°F) and the steel is fully martensitic at room temperature.

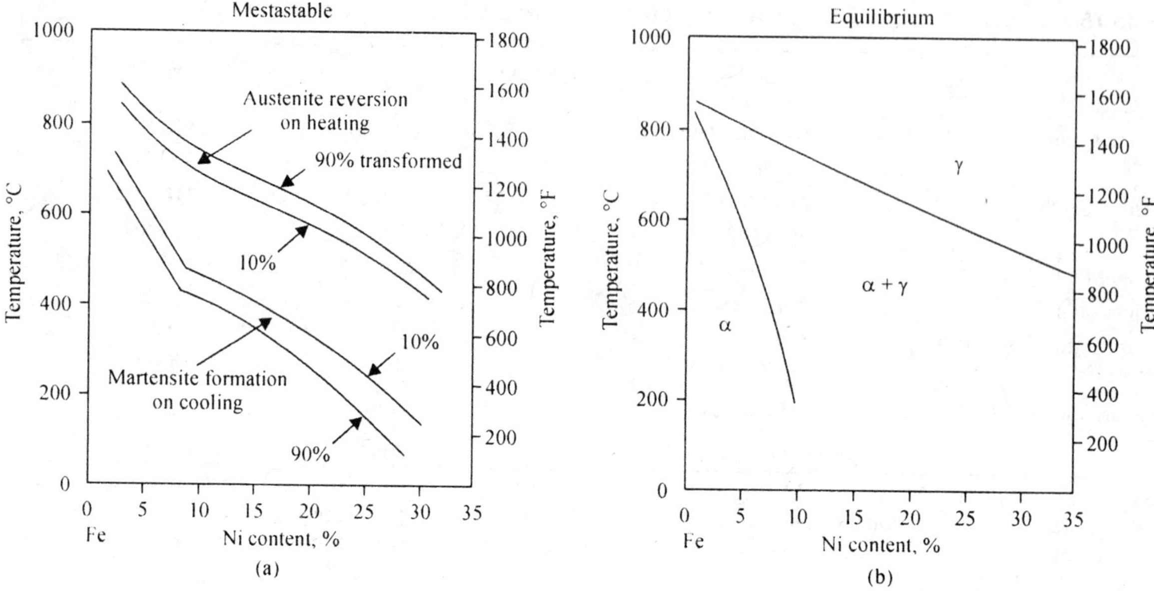

Fig. 15.6. Metastable and equilibrium phase relationships in the Fe-Ni system.

Age hardening

Age hardening is produced by heat treating for several hours at temperatures typically on the order of 480°C (900°F). During this stage the equilibrium-phase diagram becomes important. That is, with prolonged holding at 480°C, the structure tends toward the equilibrium phases, primarily ferrite and austenite. Fortunately, the precipitation reactions that cause hardening proceed much more rapidly than the reversion reactions that produce austenite and ferrite. Thus, very substantial increases in strength are achieved before reversion takes place. With longer aging times or higher temperatures, hardness will reach a maximum and then start to drop.

Table 15.21. Nominal compositions of commercial maraging steels.

	Composition(a), %				
Grade	*Ni*	*Mo*	*Co*	*Ti*	*Al*
18Ni (200)	18	3.3	8.5	0.2	0.1
18Ni (250)	18	5.0	7.75	0.4	0.1
18Ni (300)	18	5.0	9.0	0.65	0.1
18Ni (350)	18	4.2(b)	12.5	1.6(b)	0.1
18Ni (Cast)	17	4.6	10.0	0.3	0.1

(a) For all grades, carbon content is no more than 0.03 per cent. (b) Some producers use a combination of 4.8 per cent Mo and 1.4 per cent Ti, nominal.

Overageing

Softening in aged maraging steels usually is the result of: (i) overaging by coarsening of the precipitate particles and (ii) reversion to austenite. Very substantial amounts of austenite, of the order of 50 per cent, can eventually be formed.

Maraging steels normally contain little or no austenite after standard heat treatments, but sometimes austenite is deliberately formed. For example, if maraging steel is to be used in an application where overageing in service is expected, as in tooling for die casting of aluminium, the meterial will be slightly overaged beforehand. This minimises overageing in service and resultant surface tensile stresses. Extreme overageing to form large amounts of reverted austenite also has been employed as an intermediate treatment to enhance response to cold working or to minimise the effects of thermal gradients during hot working and subsequent storage of extraordinarily heavy sections.

Overaged maraging steels are expected to show good resistance to brittle fracture and stress-corrosion cracking. Unfortunately, there appears to be considerable heat-to-heat variability in these characteristics when overageing heat treatments are used. Therefore, it is difficult to recommend specific overageing treatments that will produce consistent mechanical properties. In general, if a specific yield strength is required, it is better to use a maraging steel designed to produce that yield strength by conventional ageing than to use an overaged higher-strength alloy.

Heat Treating

Conventional heat treatments for the standard grades of maraging steel are given in Table 15.22. Alloys with higher titanium contents are susceptible to formation of TiC films at austenite grain boundaries after holding at temperatures of the order of 900 to 1100°C (1650 to 2000°F). These films can severely embrittle the alloy when it is subsequently age hardened, leading to low-energy fractures along prior austenite grain boundaries. Prolonged annealing in this temperature range should be avoided for all compositions.

Table 15.22. Heat treatments and typical mechanical properties of standard 18Ni maraging steels.

Grade	Heat treatment (a)	Tensile strength MPa	Tensile strength ksi	Yield strength Mpa	Yield strength ksi	Elongation in 50 mm or 2 in.,%	Reduction in area, %	Fracture toughness MPa $\sqrt{m}$	Fracture toughness ksi $\sqrt{inch}$
18Ni (200)	A	1500	218	1400	203	10	60	155–200	140–220
18Ni (250)	A	1800	260	1700	247	8	55	120	110
18Ni (300)	A	2050	297	2000	290	7	40	80	73
18Ni (350)	B	2450	355	2400	348	6	25	35–50	32–45
18Ni (Cast)	C	1750	255	1650	240	8	35	105	95

(a) Treatment A: solution treat 1 hr. at 820°C (1500°F), then age 3 hrs. at 480°C (900°F).Treatment B: solution treat 1 hr. at 820°C (1500°F), then age 12 hours at 480°C (900°F). Treatment C: anneal 1 hour at 1150°C (1200°F), age 1 hour at 595°C (1100°F), solution treat 1 hour at 820°C (1500°F) and age 3 hours at 480°C (900°F).

Solution treatment

Maraging steels normally are solution annealed (austenitised) one hour for each 25 mm (1 inch) of section size. Atmosphere control may be necessary to minimise surface damage. Ordinarily, dry hydrogen or dissociated ammonia atmospheres are used. The cooling rate after annealing is of little consequence because it has no effect on either microstructure or properties. It is essential, however, that the steel be cooled to room temperature before it is age hardened. If this is not done, the steel may contain untransformed (retained) austenite and may be much softer than expected.

Age Hardening

Age hardening normally done at 455 to 510°C (850 to 950°F) for 3 to 12 hrs. In typical treatments at 480°C (900°F), grades 18Ni(200), 18Ni(250) and 18Ni(300) are held 3 to 6 hrs. and grade 18Ni(350) is held 6 to 12 hrs. The 350 grade also is aged for 3 to 6 hrs. at 495 to 510°C (925 to 950°F). For applications such as die casting tooling, ageing at temperatures of the order of 530°C (985°F) is employed.

The standard age-hardening heat treatments listed in Table 15.22 produce 0.04 per cent contraction in length in the 18Ni(200) grade 0.06 per cent contraction in the 18Ni (250 grade and 0.08 per cent contraction in both the 18Ni(300) and 18Ni(350) grades. These very small dimesional changes during hardening allow many maraging steel components to be finish machined in the annealed condition. The finished parts then can be hardened without further machining. When greater dimensional accuracy is required, an allowance for contraction is readily made.

Cleaning after Heat Treatment

For removal of oxide films formed by heat treatment, grit blasting is the most efficient technique. Maraging steels can be chemically cleaned by pickling in sulphuric acid or by duplex pickling, first in hydrochloric acid and then in nitric acid-plus-hydrofluoric acid. As with conventional steels, care must be taken to avoid overpickling.

Chapter 16

Quenching of Steel

INTRODUCTION

Quenching of steel is the rapid cooling of steel from a suitable elevated temperature. This generally is accomplished by immersion in water, oil, polymer solution or salt, although forced air is sometimes used. As a result of quenching, production parts must develop an acceptable as-quenched microstructure and in critical areas, mechanical properties that will meet minimum specifications after the parts are tempered. The effectiveness of quenching depends on the cooling characteristics of the quenching medium as related to the ability of the steel to harden.

Thus, results may be varied by changing the steel composition or the agitation, temperature and type of quenching medium. The design of the quenching system and the thoroughness with which the system is maintained contribute to the success of the process. The design of the part also contributes to the mechanical properties and the distortion that will result from a particular quench.

The rate of heat extraction attributable to a quenching medium is greatly modified by the manner or condition in which the quenching medium is used. These modifications have resulted in the assignment of specific names to various quenching methods, such as direct quenching, time quenching, selective quenching, spray, fog and interrupted quenching.

QUENCHING METHODS

Direct Quenching

Direct quenching is the most widely used method of treating steel. When carburised work is quenched from the carburising temperature, or from a slightly lower temperature, the term 'direct quenching' is used to distinguish this method from the more indirect practice of carburising, slow cooling, re-heating and quenching. Direct quenching practice is relatively simple and economical and distortion of carburised parts is usually less frequent with direct quenching than with re-heating and quenching, particularly on smaller parts.

Time Quenching

This form of quenching is used when the cooling rate of the part being quenched has to be changed abruptly at some time during the cooling cycle. The change in cooling rate may comprise either an increase or a decrease, depending on which is needed to attain desired results. The usual practice is to lower the temperature of the part by quenching in a medium (for instance, water) for a short time until the part has cooled below the nose of the time-temperature transformation curve and then to remove the part and quench it in a second medium (for instance, oil), so that it cools more slowly through the martensite transformation range. In many applications, the second medium is still air.

Time quenching is most often used for minimising distortion, cracking and dimensional change. It should be used with caution because successful application depends greatly on the skill of the operator.

Selective Quenching

This process is used when areas of a part are selected to remain relatively unaffected by the quenching medium. This can be accomplished by insulating the area to be protected or by allowing the quenchant to contact only those areas of the part that are to be quenched.

Spray Quenching

Streams of quenching liquid are directed at high pressure, up to 825 kPa (120 psi), to local areas of the workpiece in spray quenching practice. The cooling rate is fast and uniform over the entire temperature range of the quenching cycle because of the large volume of coolant used and because all of the coolant makes direct contact with the part being quenched. The velocity of the stream removes any vapour bubbles and creates spray droplets, which are available for heat transfer. However, low pressure spraying, in effect a flood-type flow, is preferred with certain polymer quenchants.

Fog Quenching

This procedure utilises a fine fog or mist of liquid droplets and the gas carrier as cooling agents. Although similar to spray quenching, fog quenching is less effective, because the quenching mist or fog is not readily adapted to rapid removal or replacement by the cooler fog or mist once it has been heated by contact with the part being quenched.

MECHANISM OF QUENCHING

Several factors are involved in the mechanism of quenching: (i) internal conditions of the workpiece that affect the supply of heat to the surface; (ii) surface and other external conditions that affect the removal of heat; (iii) the heat-extracting potential of the quenching fluid in the quiescent state at normal fluid temperatures and pressures ('standard' conditions); and (iv) changes in the heat-extracting potential of the fluid brought about by 'non-standard' conditions of agitation, temperature, or pressure. These factors are illustrated in Fig. 16.1 for a furnace heated gear that is quenched edgewise in a quiescent volatile liquid. The effect of irregular configuration of the gear on the flow of heat from within the gear to the quenching area is demonstrated in part A of Fig. 16.1. High temperature persists near the surface at the roots of the teeth where large vapour bubbles are trapped. If the gear were induction or flame heated (and thus has a uniformly thin heated layer conforming to the irregular contour of the gear), heat supply to the quenching area would be more consistent and quenching would progress more rapidly because heat also would flow simultaneously to the cold metal underlying the heated exterior.

A quiescent liquid experiences unavoidable movement as a result of the action of immersion, the turbulence of boiling and convection currents. This minimum agitation eventually dissipates the accumulated heat to the surrounding large body of liquid, but local volumes of liquid become heated excessively, or may even vapourise and this may affect the quenching action.

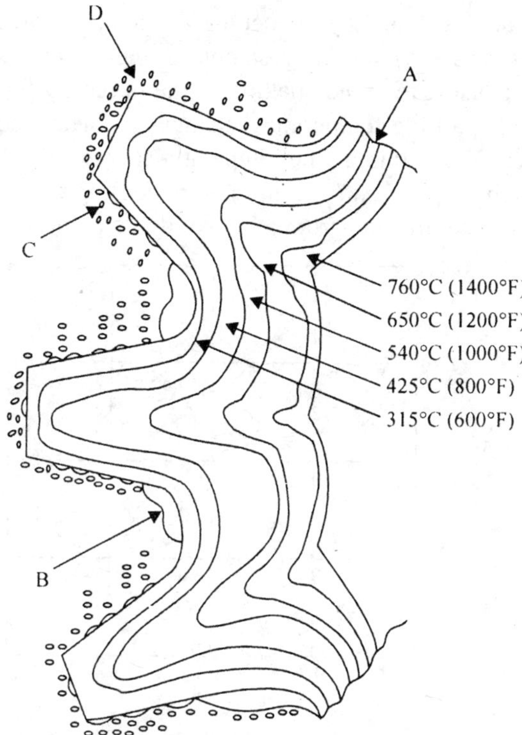

A, flow of heat from hot core of gear–temperature and flow rate vary with time; B, vapour blanket stage still exists due to large source of heat and poor agitation; C, trapped vapour bubbles condensing slowly; and D, vapour bubbles escaping and condensing. The gear was quenched edgewise in a quiescent volatile liquid.

Fig. 16.1. Temperature gradients and other major factors affecting the quenching of a gear.

Volatile quenching liquids produce some vapour at all operating temperatures. Above the boiling point, the supply of vapour becomes so plentiful that an envelope of gas is formed around the surface of the workpiece. This envelope, or 'vapour blanket', is maintained by radiated heat for as long as the heat is available (part B in Fig. 16.1). The temperature above which a total vapour blanket is maintained is called the 'characteristic temperature' of the liquid.

At lower temperatures, the vapour consists of bubbles, which vary in size depending on the relation of boundary tension of the liquid, the gas and the solid at the temperatures concerned. In one liquid, numerous small, easily detached bubbles may be formed (part D in Fig. 16.1), whereas large, adherent bubbles, fewer in number, may be formed in another (part C in Fig. 16.1); this is referred to as the 'bubble size characteristics' of the liquid. For any volatile liquid, the mechanical trapping of vapour bubbles (part C) will greatly retard the transfer of heat at the affected location.

COOLING CURVES

The most useful way of accurately describing the complex mechanism of quenching is to develop a 'cooling curve' for the quenching fluid under controlled conditions. A cooling curve test is sensitive to the factors that may affect the cooling ability of the quenchant, because the test simulates conditions of actual practice.

Cooling curves may be developed by quenching, from an elevated temperature, a test piece of the same steel of which the parts are made, in a sample of the quenching fluid. Sometimes an austenitic stainless steel specimen is used, to avoid soaking or the necessity for a protective atmosphere. A high speed recorder is used for plotting temperature changes as measured by one or more thermocouples embedded in the test piece. The resulting time-temperature curve indicates the heat transfer characteristics of the quenching fluid. The typical surface and centre cooling curves shown in Fig. 16.2 graphically describe the four stages of heat transfer from a hot solid to a cold liquid.

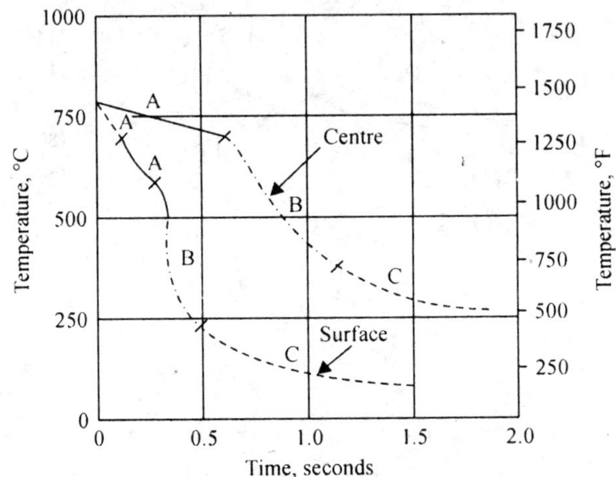

Fig. 16.2. Typical surface and centre cooling curves, indicating the stages of heat transfer from a hot solid to a cold liquid.

Stage A′

Stage A′ in Fig. 16.2 illustrates the first effects of immersion. Sometimes called the 'initial liquid contact stage', this stage is characterised by the formation of vapour bubbles that precedes the establishment of an enveloping vapour blanket. Stage A′ lasts for only about 0.1 second and is relatively unimportant in the evaluation of heat transfer characteristics. It is detectable only when extremely sensitive equipment is used and it cannot be detected when the liquid is viscous or contains entrained gases, or when the bath is operated at a temperature near the boiling point of the liquid.

Stage A

Stage A, called the 'vapour blanket cooling stage', is characterised by the Leidenfrost phenomenon—namely, the formation of an unbroken vapour blanket that surrounds the test piece. It occurs when the supply of heat from the surface of the test piece exceeds the amount of heat needed to form the maximum vapour per unit area of the piece. This stage is one of slow cooling, because the vapour envelope acts as an insulator and cooling occurs principally by radiation through the vapour film. This stage is not detectable in aqueous solutions of non-volatile solutes (at about 5 per cent concentration) such as potassium chloride, lithium chloride, sodium hydroxide or sulphuric acid. Cooling curves for these solutions start immediately with stage B.

When: (i) saturated solutions of barium hydroxide, calcium hydroxide or other slightly soluble materials; (ii) solutions containing finely dispersed solids; and (iii) colloidal solutions in water are used, films are deposited on the test piece during the A stage, which results in the prolongation of both the A and C stages. This condition usually causes a more violent action in stage B.

Solutions of some colloids or gels, such as polyvinyl alcohol, gelatine, soap and starch (but not water glass), form an envelope of gel outside the vapour blanket formed in stage A. The presence of this gel envelope prolongs the A stage and succeeding stages.

Stage B

Stage B, the 'vapour transport cooling stage', which produces the highest rate of heat transfer, begins when the temperature of the surface metal has been reduced somewhat and the continuous vapour film collapses; violet boiling of the quenching liquid then occurs and heat is removed from the metal at a very rapid rate, largely as heat of vapourisation. The boiling point of the quenchant determines the conclusion of this stage. Size and shape of the vapour bubbles are important in controlling duration of stage B, as well as the cooling rate developed within it.

Stage C

Stage C is called the 'liquid cooling stage'; the cooling rate in this stage is slower than that developed in stage B. Stage C begins when the temperature of the metal surface is reduced to the boiling point (or boiling range) of the quenching liquid. Below this temperature, boiling stops and slow cooling takes place thereafter by conduction and convection. The difference in temperature between the boiling point of the liquid and the bath temperature is a major factor influencing the rate of heat transfer. Viscosity also affects cooling rate in stage C.

Significance of Cooling Curves

The same mechanism is involved in cooling a test piece to evaluate a quenchant as is involved in quenching an actual part in a heat treating operation. For all or any selected portions of the curve, cooling curve information can be translated into cooling rate (in degrees per second), if desired.

Although a cooling curve relates only to the size and material of the test piece, the thermocouple location and conditions of the quenching liquid under which a test was performed, cooling curve data developed under one set of conditions can be translated to other conditions, as well as to H-values, by the application of heat transfer formulas.

Agitation

Agitation is externally produced movement of the quenching liquid relative to the part, either by stirring the liquid or moving the part, or both in combination. This activity has an extremely important influence on the heat transfer characteristics of the quenching liquid. It causes an earlier mechanical disruption of the vapour blanket in stage A and produces smaller, more frequently detached vapour bubbles during the vapour transport cooling stage (stage B). It mechanically disrupts or dislodges gels and solids, whether they are on the surface of the test piece or suspended at the edge of the vapour blanket, thus producing faster heat transfer in liquid cooling (stage C). In addition to the above effects, agitation also brings cool liquid to replace heat-laden liquid.

Temperature of Quenchant

The temperature of the liquid may markedly affect its ability to extract heat. As an example, water temperature is very important because it loses its cooling power as it approaches its boiling point. In oil, this effect is not as pronounced because oil becomes less viscous as the temperature is increased. This 'thinning' of the oil offsets the temperature rise by a substantial amount.

Workpiece Temperature

Increasing the temperature of the test piece has relatively little effect on its ability to transfer heat to the quenchant. The rate of heat transfer may be increased simply because a greater temperature difference exists. The most noticeable change in ability to transfer heat probably comes from the more rapid oxidation of the surface of the test piece at higher temperatures. This can either increase or decrease the heat transfer ability, depending on the thickness of the oxide developed.

METALLURGICAL ASPECTS

Steel is quenched to control the transformation of austenite to desired micro-constituents. The microstructures that may be secured are indicated in Fig. 16.3. Martensite is the as-quenched microstructure usually desired. As indicated by curve A in Fig. 16.3, to obtain the maximum amount of martensite, the cooling rate must be fast enough to avoid the nose of the time-temperature transformation (TTT) curve of the steel being quenched. If the cooling rate is not fast enough to miss the nose of the TTT curve (curves B, C and D in Fig. 16.3), some transformation to bainite, pearlite or ferrite will take place, with a corresponding decrease in the amount of martensite formed and the hardness developed.

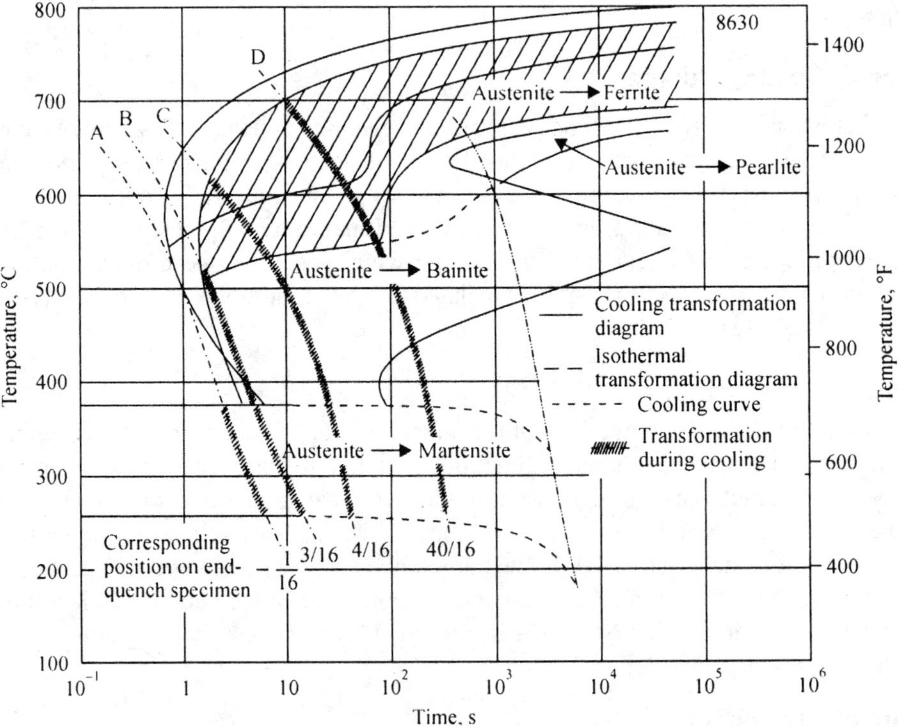

Fig. 16.3. Transformation diagrams and cooling curves for 8630 steel, indicating the transformation of austenite to other constituents as a function of cooling rate.

Carbon Content and Hardenability

The maximum hardness obtainable in a steel quenched at a sufficient rate to avoid the nose of the TTT curve depends on the carbon content. The cooling rate (quenching efficiency) necessary to obtain a fully martensitic structure depends on the hardenability of the steel and the section thickness of the part. The relation of carbon content and percentage martensite to hardness is shown in Fig. 16.4.

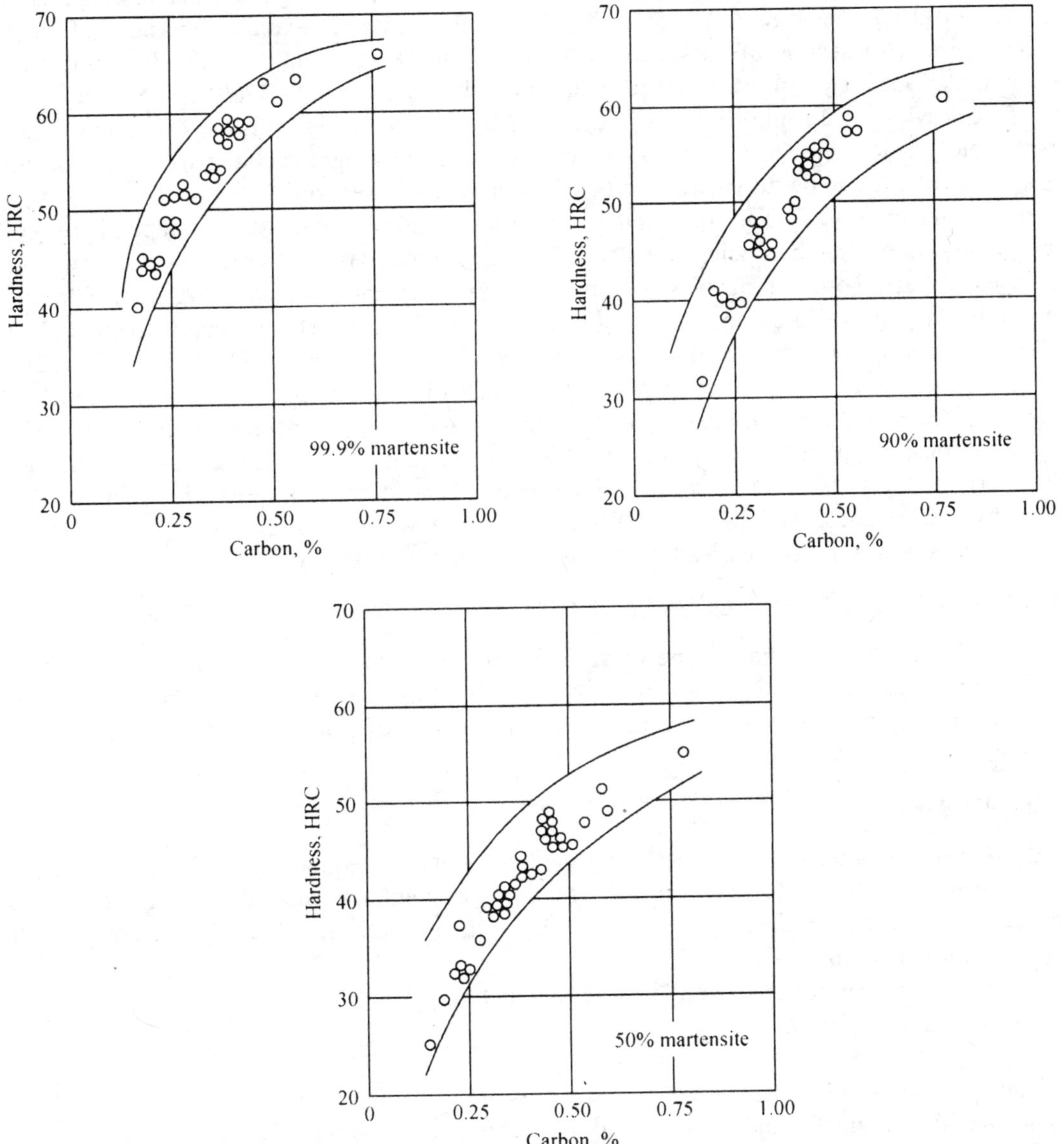

Fig. 16.4. Relation of carbon content and percentage martensite to HRC hardness.

Depending on the carbon content and hardenability of the steel, the cooling rate should be fast enough so that a high percentage of martensite will be produced in critically stressed areas of the part. Lower percentages of martensite are often acceptable in areas subject to lower stresses in service. Higher percentages of martensite in the as-quenched structure will produce higher fatigue and impact properties after tempering.

Cooling Rates

When carbon steel is quenched from the austenitising temperature, a cooling rate equal to or greater than about 55°C/s (100°F/s), measured at 705°C (1300°F), is necessary for avoiding the nose of the TTT curve. The entire cross-section of the piece must cool at this rate to obtain the maximum amount of martensite. Under ideal conditions, water provides a cooling rate of about 2760°C/s (5000°F/s) at the surface of steel cylinders 13 mm (½ inch) in diameter by 100 mm (4 inch) long. This rate of cooling decreases rapidly below the surface. Thus, for carbon steel, only light sections with a high ratio of surface area to volume can be fully hardened throughout the cross-section.

When water or brine is used for quenching, a high temperature gradient is developed between the surface and the centre of the part being quenched. This temperature gradient produces greater distortion and increases the hazard of cracking in all but simple, symmetrical shapes. Under favourable conditions, still quenching in a fast oil provides a cooling rate at the surface of a stainless steel cylinder 13 mm (½ inch) in diameter by 100 mm (4 inch) long of 2040°C/s (3700°F/s) between 830 and 545°C (1530 and 1010°F), compared with 2760°C/s (5000°F/s) for the surface of similar specimen in water. Quenching in oil results in a much lower temperature gradient from surface to centre, decreased distortion and less probability of cracking. With heavy sections, the cooling rate is limited by the rate of heat conduction from the interior to the surface. Rapid cooling of the centre of an extremely large section is impossible by any quenching method, because of the 'mass effect'. Therefore, when deep hardening of a heavy section is mandatory, it is necessary to use an alloy steel with higher hardenability.

TESTS AND EVALUATION OF QUENCHING MEDIA

The testing of quenching media can be classified into two categories: tests of hardening power (metallurgical response) and tests of cooling power (thermal response). The two properties are not the same, because hardening power also is related to the size, microstructure and composition of the steel being quenched.

Hardening-Power Tests

The final criterion for selection of a quenching medium is its hardening power, that is, its ability to develop a specified hardness in a given material-section size combination, under certain conditions of quenchant, agitation and temperature. The major factors influencing the hardness obtained on quenching can be summarised as follows:

1. Nature and concentration (if applicable) of quenchant.
2. Temperature of quenchant.
3. Agitation and mass flow of quenchant.
4. Material composition, structure and thermal history of the metal component.
5. Section size, geometry and surface condition of the metal component.

Jominy End-Quench Test (ASTM A255)

This test is widely used to determine hardenability of steels. It consists of heating a specimen 25 mm (1 inch) round by 100 mm (4 inch) long to the desired hardening temperature and then withdrawing heat from one end of the bar by the use of a water jet quench. The bar is then ground along its length to 0.4 mm (0.015 inch) below the surface and hardness readings of the specimen are taken 1.6 mm (1/16 inch) apart. Hardenability is expressed in terms of the distance from the quenched end to a given hardness.

Immersion Quench Test

Because of the importance of agitation in quenching, the evaluation of hardening power of quenchants may be achieved using an immersion quench of the test piece. The hardness readings taken directly indicate the quenching capability and test results of end quenching and immersion quenching can be easily correlated.

Cooling Power Tests

Because hardening power tests are time consuming and relatively difficult to perform, industry has developed simple, reproducible cooling power tests for the evaluation of quenching media. The four tests most commonly used are:

1. Cooling curve test (or basic thermocouple test).
2. The magnetic test.
3. The hot wire test.
4. The interval method, 5-second test.

Cooling curve test

The most useful method to determine the cooling power of a quenchant is by the basic cooling curve test, wherein cooling rates are established throughout the complete quenching process. The test is performed by quenching specimens, with thermocouples embedded at various points and following the cooling process by using a temperature measuring device. Specimens utilised can be made of stainless steel, silver, nickel or carbon steel. Cylindrical or spherical shapes are most common. A specimen of relatively small mass may be desired for slow quenching fluids, whereas faster quenching fluids may require larger mass, or a recorder having a very high response rate. The use of a silver probe is believed to have an advantage over the other metals because it is easy to prepare, free from transformation effects and resistant to corrosion. Its high conductivity averages out localised surface thermal conditions while high sensitivity is retained.

The latest test equipment is capable of the simultaneous recording of temperature versus time and temperature versus cooling rate curves. Depending on the quenching media, the cooling curves developed depict two or three stages of quenching as well as the point of recalescence in hardenable steels. Molten salts and metals have only two stages of quenching, that is, they have no vapour phase.

When analysing a cooling curve, the analyst should recognise that the heat transfer in metals is measured by the expression kps, where 'k' is the conductivity, 'p' is the density an s is the specific heat of the metal. The value of kps is 0.972 for silver, 0.49 for nickel and 0.308 for steel. Therefore, the temperature gradient in the test piece is much steeper in steel than in nickel and it is steeper in nickel than in silver. Correspondingly, the location of the thermocouple must be more precise in a steel test piece than in a silver test specimen and instruments must be capable of a faster response for silver than is required for steel.

A surface-located thermocouple is capable of disclosing peculiarities of the cooling curve that are obscured by a centre-located thermocouple. Moreover, the time factor is lengthened and the temperature of occurrence of characteristic points on the curve appears to be higher when measurement is made at the centre rather than the surface. If a surface thermocouple is used with a 10 m (0.4 inch) diam silver test piece, the recorder must be capable of accurately registering rates of temperature change of at least 540°C (1000°F) in 0.1 second throughout the temperature range being explored.

The 'characteristic temperature' (the lowest temperature at which a vapour envelope can be maintained) and the boiling point are unaffected by the temperature, the heat conductivity, or the mass of the test piece when the cooling curve is taken at the surface of the piece. This is not true when the curve is taken at any point below the surface. Surface thermocouple locations should always be used to obtain fundamental data, although other, more convenient, locations may be used for control purposes. The most common deviations from the above 'standard' conditions of performing the cooling curve test are variations in temperature of the test piece or the quenching medium and in the degree of agitation. Overall, the cooling curve test is an excellent research and quality control tool and provides for a great degree of comparative accuracy. Standardisation of the test would facilitate the comparison of results from different sources.

Magnetic test

This test makes use of the properties of metals that lose their magnetism when heated above a Curie point and regain their magnetism when cooled below this temperature. The magnetic test was devised to provide a means of comparing the heat extraction rates of oil, molten salt, water or other quenching media. The test method involves the heating of a 22 mm (7/8 inch) diam nickel sphere, weighing approximately 50 gram (1.8 oz) to 835°C (1625°F) in either air or a controlled atmosphere. (A chromium-diffused nickel sphere can be used to eliminate the need for a protective atmosphere to avoid marked alteration of surface finish). After uniformity of temperature is attained, the sphere is quenched in a 200 ml sample of quenchant located within a magnetic field. As the nickel sphere cools through the Curie temperature, it becomes magnetic and is attracted to the magnetic field. The time required for the nickel sphere to cool from 835°C (1625°F) to the Curie temperature, which is 355°C (670°F), is a measure of the cooling power of the quenching medium. The Curie temperature of nickel is below the nose of the isothermal transformation diagrams for most steels. Thus, with this method, quenching media can be compared for steels as well as for other metals. The faster the cooling power of the medium, the shorter the time-required for the nickel to regain its magnetism.

Other alloys with different Curie points can be used to establish additional points on a cooling curve. A modification of the magnetic quench test has been used to study the influence of circulation and heat on the cooling power of quenching oils. A device known as the electronic quenchometer is installed directly in the quenching system so that results are obtained from the actual quenching process conditions of media, temperature and agitation. To compare the results of the magnetic test with the actual quenching power for certain steels, tests were made by quenching AISI 1046 steel bars in various oils.

The test bars, 22 mm (7/8 inch) diam by 75 mm (3 inch) long, were quenched from an endothermic gas atmosphere after austenitising at 815°C (1500°F). Transverse hardness tests correlated favourably with magnetic test results.

Hot wire test

This method of quenchant evaluation consists of heating a nichrome or cupron wire (of standard gauge and electrical resistance) by means of an electrical current, in a small quantity (100 to 200 ml) of quenchant. The quenchant tested usually is at quenching temperature and the wire is supported by two copper or brass electrodes. Heating of the wire is accomplished by steadily increasing the current by means of a rheostat. The cooling power of the quenchant is indicated by the maximum current reading, as measured by an ammeter. Quenchants capable of extracting heat faster permit the passage of higher currents through the wire and thus register higher current values. This test resembles the 5 second test in that in compares quenchants only in the higher temperature range of quenching.

Interval test

This method, also known as the 5-second test, is a rapid method of comparing the cooling power of quenchants. In this test, a 2 litre (0.5 gal) sample of an oil is placed in an insulated container and the oil temperature is noted. A bar of metal (usually, stainless steel) weighing about 250 gram (8.8 oz) is heated to 815°C (1500°F) and quenched for 5 seconds. The oil sample is then stirred to ensure temperature equalisation throughout the bath and the temperature rise is noted to the nearest tenth of a degree Celsius. This process is repeated for a series of test bars. Finally, a bar of the same metal and size is fully quenched in a second 2 litre sample of oil and the rise in temperature is noted. Quenching power of the oil is computed according to the following equation:

$$\frac{A}{B} \times 100 = \text{Per cent quench speed}$$

where, A represents the average rise in oil temperature for the 5-second quench bars and B represents the maximum temperature rise of the oil sample for the fully quenched specimen.

The 5-second test is used for determining gross changes in quenching oil because it is expedient and requires no special equipment. A duration of 5 second in quenching, however, constitutes only a comparison in the higher temperature region of the quench and this may be misleading because it encompasses only a portion of the cooling curve—namely, the vapour blanket and vapour transport stages.

EVALUATION OF SEVERITY FOR QUENCHING OF STEELS

The ability of a quenching medium to extract heat from a hot steel workpiece can be expressed in terms of the 'H-value' (the severity of quench). If the H-value of still water is taken as 1.0, the H-values of oil, water and brine are as presented in Table 16.1.

Table 16.1. Effect of agitation upon the effectiveness of quenching.

Circulation of agitation	*H-value or quenching power* Oil	Water	*Caustic soda or brine*
None	0.25 to 0.30	0.9 to 1.0	2
Mild	0.30 to 0.35	1.0 to 1.1	2 to 2.2
Moderate	0.35 to 0.40	1.2 to 1.3	–
Good	0.4 to 0.5	1.4 to 1.5	–
Strong	0.5 to 0.8	1.6 to 2.0	–
Violent	0.8 to 1.1	4	5

The relations between bar diameter, inherent hardenability and H-value may be plotted as shown in Fig. 16.5, which is referred to as a 'Grossman' chart. The H-values of a material are plotted as such: if a 25 mm (1 inch) diam part is quenched in oil with 'good agitation' (0.50 H-value), the chart shows that, reading across the 25 mm (1 inch) diam horizontal line to the 0.50 H-value and then down, a value of 8 mm (5/16 inch) on the end-quench hardenability specimen is obtained. This means that the centre of the bar will have the same hardness as that shown at 8 mm (5/16 inch) from the quenched end of a standard hardenability specimen made from the steel being quenched.

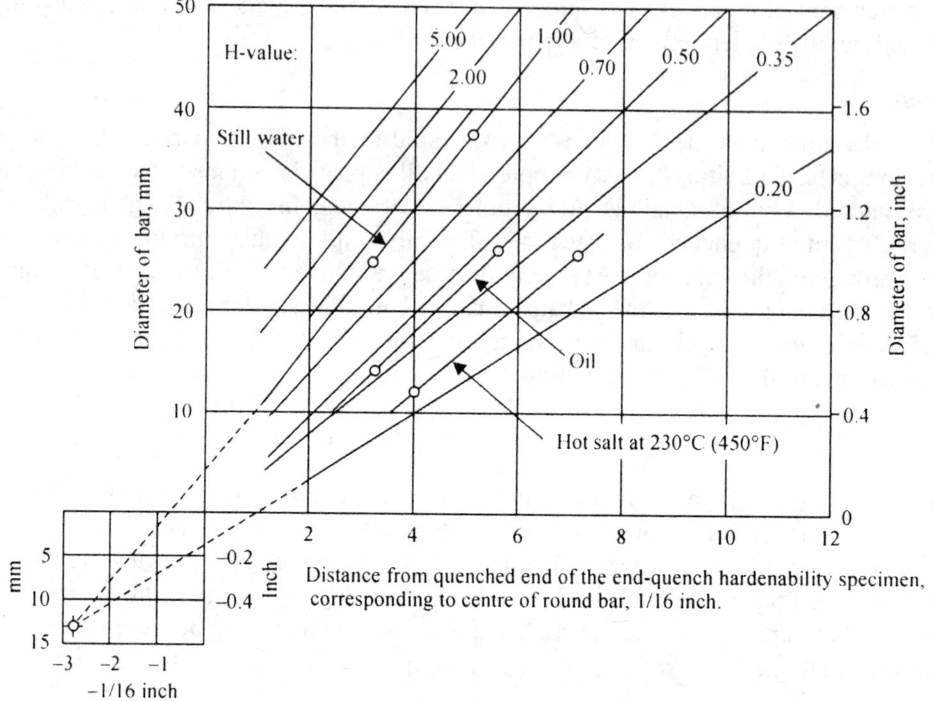

H-value (severity of quench): 5.00, strong brine quench, violent agitation; 2.00, poor brine quench, no agitation; 1.00, poor water quench, no agitation; 0.70, strong oil quench, good agitation; 0.50, good oil quench, good agitation; 0.35, good oil quench, moderate agitation; and 0.20, poor oil quench, no agitation.

Fig. 16.5. Grossmann chart relating bar diameter, hardenability of steel and severity of quench.

The application of the chart is limited because the quenchants involved are described in general terms (for example, 'very good oil quench—good agitation)'. An added complication is the fact that what appears to be good agitation in a quenching system that has no parts in it may be poor agitation when a load of parts is immersed in the bath.

A simple test for evaluating the cooling power of a quench bath under any condition of loading is based on the use of Fig. 16.5. If the hardenability of the steel and the bar diameter are known, it is possible to plot one point on the quench-severity curve. For example, if the 8 mm (5/16 inch) point on the end-quench specimen shows a hardness of 45 HRC and a 25 mm (1 inch) diam bar of the same steel quenched in the bath to be evaluated has a centre hardness of 45 HRC, the result indicates a quenched severity of 0.50. Regardless of the hardness obtained at the centre of the 25 mm diam test bar, it can be plotted on the chart of Fig. 16.5 at its correct location. For example, if the centre hardness had been 48 HRC and if the hardenability curve for the bar of steel tested had shown 48 HRC at 6 mm (¼ inch), the 6 mm line and the 25 mm diameter line would intersect at a quench severity of 0.70.

As shown in Fig. 16.5, for oil quenches the various severity lines of the Grossmann chart converge at a point outside the normal range of the chart. This point of intersection can be used as a second point in drawing any new quench-severity lines. An actual example is illustrated in Fig. 16.5 by the line labelled 'oil'. The two experimental points on this line were obtained by quenching a bar of steel of known hardenability having two different diameters, 25 and 12.5 mm (1 and ½ inch). The use of a bar with two diameters offers an excellent means of checking results because it provides three points on the line. The results of three different operating commercial quenches, as determined with stepped test bars or bars having two different diameters, are plotted in Fig. 16.5.

The principal advantage of this test is that the specimen can be treated along with other work just as if it were a production part; thus, the quench is evaluated under the conditions of temperature and agitation that actually prevail in the production quenching of a load of parts. The test bar should be of such diameter that the As-quenched hardness at the centre of the bar will fall on the sloping portion of the hardenability curve for the steel being quenched. The test can be performed most conveniently with a shallow-hardening steel because this type of steel permits the use of smaller diameter test bars, which are easier to prepare and to section after being quenched. The use of shallow-hardening steel in no way detracts from the validity of the test results, because severity of quench is an inherent characteristic of the bath and is not affected by part size or hardenability.

QUENCHING MEDIA

Many different media have been used for quenching. Most are included in the list that follows, but some of the these are used only to a very limited extent.

1. Water.
2. Brine solutions (aqueous).
3. Caustic solutions.
4. Oils.
5. Polymer solutions.
6. Molten salts.
7. Molten metals.
8. Gases, including still or moving.
9. Fog quenching.
10. Dry dies, commonly water cooled.

Water

As a quenching medium, plain water approaches the maximum cooling rate attainable in a liquid. Its other advantages are that it is inexpensive and readily available, is easily disposed of without attendant problems of pollution of health hazard and is an effective means of breaking scale from the surface of steel parts that are quenched from furnaces in which protective atmospheres have not been employed.

Water, therefore, is used wherever it is practical, that is, where the drastic quench afforded by water does not result in excessive distortion, or cracking of the workpiece. Water is widely used for quenching non-ferrous metals, austenitic stainless steel and other metals that have been given a solution treatment at elevated temperature. Plain water also is used to a considerable extent for rapid cooling of quench-hardenable steels where formation of vapour pockets with resulting soft spots is not objectionable.

One disadvantage of plain water as a quenchant is that its rapid cooling rate persists throughout the lower temperature range, in which distortion or cracking is likely to occur. Consequently, water usually is restricted to the quenching of simple, symmetrical parts made of the shallower-hardening grades of steel (plain carbon or low-alloy). Another disadvantage of using plain water is that its vapour blanket stage may be prolonged. This prolongation, which varies with the degree to which the complexity of the part being quenched encourages vapour entrapment and with the temperature of the quench water, results in uneven hardness and unfavourable distribution of stress. This may cause distortion, cracking or soft spots. Water quenched steel parts may rust unless immediately treated with a rust preventive.

To obtain reproducible results by water quenching, the temperature, agitation and contamination must be controlled. Water at a temperature of 15 to 25°C (55 to 75°F) will provide uniform quenching speed and reproducible results. As indicated in Fig. 16.6, the surface cooling power of water decreases rapidly as water temperature increases. (This behaviour also is demonstrated in Fig. 16.7). Hot water has a low cooling power because, as the boiling point is approached, the cooling action resembles that of steam.

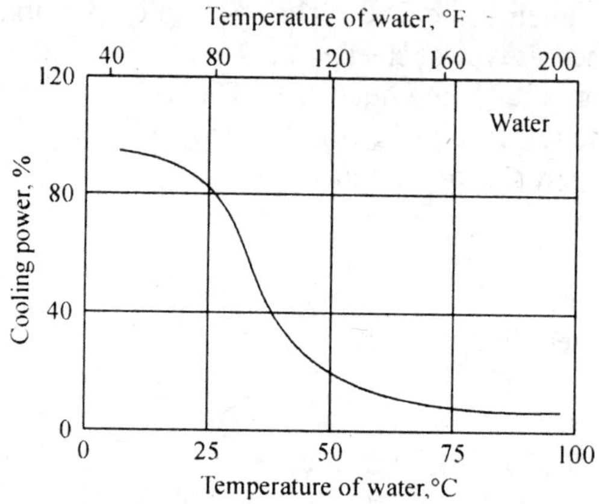

Fig. 16.6. Relation of surface cooling power of water with moderate agitation and water temperature.

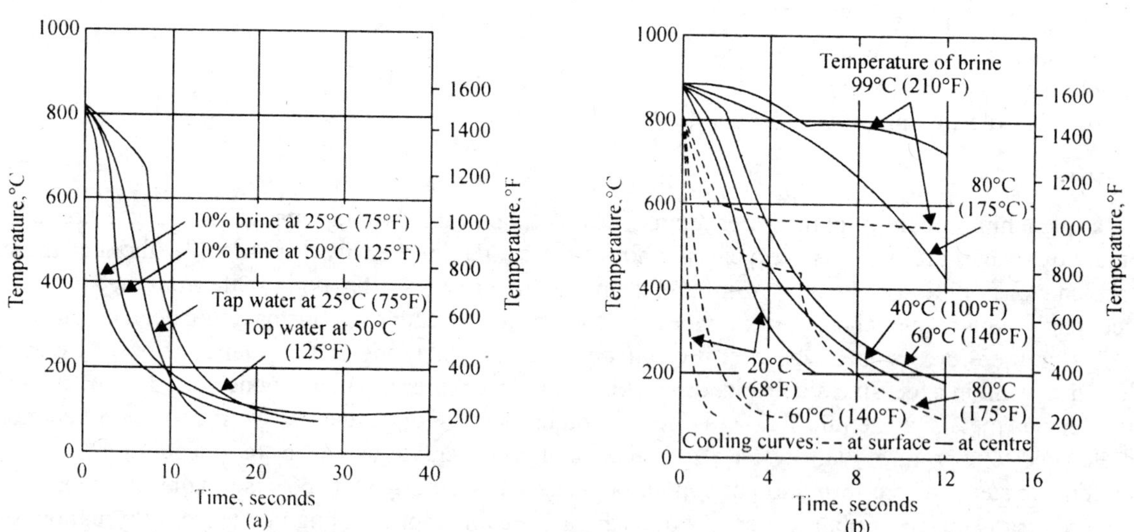

(a) Centre cooling curves for water and 10% NaCl solution in quenching 18–8 stainless steel specimens 13 mm (½ inch) diam by 64 mm (2½ inch). No agitation.

(b) Centre and surface cooling curves for 5% NaCl solution at 0.9 mm/s (3 ft/s) in quenching 0.95% C steel specimens 13 mm (½ inch) diam by 50 mm (2 inch).

Fig. 16.7. Cooling curves showing effect of temperature on cooling power.

Agitation

Agitation is especially important in water quenching because it disperses vapour bubbles from parts and directs cooler water against the workpiece.

Contamination

Contamination of water quench baths by dissolved heat treating salts is likely to increase the cooling rate, because salts effectively reduce the duration of the vapour blanket stage. However, contaminants such as soaps, algae, slimes or emulsion-formers reduce the cooling rate by trapping the steam or vapour film and thus prevent cooler water from contacting parts quickly enough to produce uniform results.

Insoluble salts that are carried over from activated salt baths into the quench bath may interfere with the operation of a closed pumping system or reduce the volume of available water in the quench tank. Periodic desludging or replacement of the quench water is therefore recommended.

Brine Solutions

The term 'brine', as applied to quenching, refers to aqueous solutions containing various percentages of salt (such as sodium chloride or calcium chloride).

Advantages and disadvantages

Brine offers the following advantages compared with plain water or oil, for quenching.

1. Cooling rate is faster than that of water for the same degree of agitation, or less agitation is required for a given cooling rate.
2. Temperature is less critical than for water, thus requiring less control (note cooling curves in Fig. 16.7).
3. Possibility of soft spots from steam pockets is less than in water quenching.
4. Distortion is less severe than in water quenching.
5. Heat exchangers are needed less often for the cooling of brine baths than they are for water or oil quench baths.
6. Complex baffling patterns in the quench tank and propeller agitation, both of which are often necessary with oil or water as quenchants, are not always required with a brine quench, but may be required for steels with very low hardenability.

Ordinarily, the disadvantages of brine quenching will not prevent its use, because brine quenching is used only where oil or water quenchants do not provide desired results. Some of these disadvantages are:

1. The corrosive nature of brine requires that, for reasonable service life, the quench tank, pumps, conveyors and other equipment in constant contact with the brine solution either be protected from corrosion by coating, plating or sheathing; or they must be made of corrosion resistant metals (such as copper-bearing or stainless steels).
2. A hood may be needed to carry off the corrosive fumes that emanate from brine baths, to protect nearby machinery or delicate equipment from attack.
3. Cost is higher than for water, largely because of the cost of the corrosion inhibitors that must be used after cleaning.
4. Labour cost is increased because of the necessity for testing to control solutions and for providing a safe environment.

Cooling rates

At 20°C (70°F) and moving at 0.9 m/s (3 ft/s), a solution containing 5 wt. per cent sodium hydroxide can cool the surface of a cylinder 12.5 mm (½ inch) in diameter and 50 mm (2 inch) in long from 870 to 200°C (1600 to 390°F) in 0.31 second. Under the same conditions, a 5 per cent solution of sodium chloride requires 0.48 second and plain water requires 1.2 second. However, a sodium chloride solution is safer, less costly and easier to handle than a sodium hydroxide solution and may be preferred even though the latter provides faster cooling rates. Brine is more widely used for small tool hardening applications, whereas caustic quenches are used for mass production processes.

Effect of concentration

In conventional quenching, it has been found that a 10 per cent NaCl solution is the most effective from the viewpoint of hardening. The concentration of the brine can be determined in several fashions, as in Table 16.2.

Table 16.2. Relation of brine density to brine concentration.

Salt, %	*Specific gravity*	*Degrees Baumé*	*Salt*	
			g/l	*lb/gal*
Sodium chloride solutions				
4	1.0268	3.8	41.1	0.343
6	1.0413	5.8	62.4	0.521
8	1.0559	7.7	84.5	0.705
9	1.0633	8.7	95.9	0.800
10	1.0707	9.6	107.1	0.894
12	1.0857	11.5	130.3	1.087
Sodium hydroxide solutions				
1	1.0095	1.4	10.1	0.0842
2	1.0207	2.9	20.4	0.1704
3	1.0318	4.5	31.0	0.2583
4	1.0428	6.0	41.7	0.3481
5	1.0538	7.4	52.7	0.4397

Although brine solutions (NaCl) of up to 24 per cent concentration are optimum for reducing the vapour stage, such high concentrations generally are impractical. The value of the 10 per cent NaCl solution in developing maximum surface hardness in conventional quenching is illustrated in Fig. 16.8. It should be noted that, although end quench values are used in this figure, the specimen was not quenched as in the standard end-quench test but was quenched in still water with only the end face contacting the water to simulate commercial quenching practice.

Effect of temperature

The cooling power of brine solutions is not critically affected by small variations in operating temperature. Although such solutions can be used at or near the temperature of boiling water, they provide maximum cooling power at about 20°C (70°F). The relatively small effect of temperature on cooling power of brine compared with water is demonstrated in Fig. 16.9.

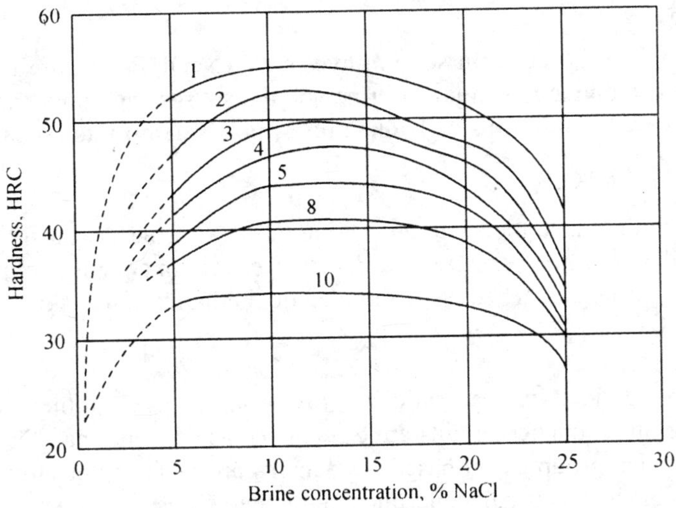

Fig. 16.8. Relation of hardness and brine concentration when still-quenching end-quench specimens in 99°C (210°F).

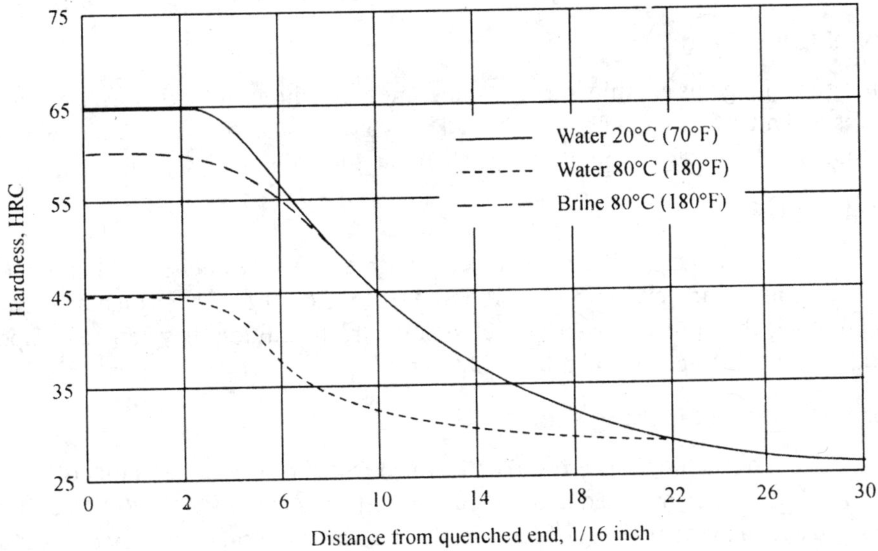

Fig. 16.9. Relation of hardness and distance from end of specimen quenched in water and brine.

Effect of contamination

Sludge and scale should be removed periodically, because both act to clog pumps and recirculating systems as well as to reduce the volume and cooling rate of the brine quench bath. Excess water reduces the solution strength and quenching power of the brine.

Caustic Solutions

Aqueous solutions of 5 to 10 per cent sodium hydroxide (NaOH) often are used for quenching. The performance of such solutions is similar to brine solutions—sharing about the same advantages as a quenchant. Its main shortcoming is its high alkalinity which is harmful to human skin.

PROPRIETARY SALT MIXES

Proprietary salt mixes that have the same quenching effect as brine or caustic solutions have become commercially available recently. When used at 5 to 15 per cent in aqueous solutions, they provide rust protection for the unwashed quenched parts and they do not harm human skin.

OIL QUENCHING

Quenching oils can be divided into several distinctive groups. Based on their composition, quenching effect and use temperature, quenching oils are categorised as conventional, fast, martempering, or hot quenching. An additional group of oil-based quenchants are oil-in-water emulsions. These find only limited use because of their unpredictable quenching characteristics, which are completely different from the characteristics of water-free quenching oils. All modern quenching oils are based only on mineral oils, usually paraffinebase and do not contain any fatty oils formerly used as additives to quenching oils or as quenching oils themselves. One reason for this is the much better ageing stability of mineral oils; another is the fact that certain well-suited fatty oils are now unavailable.

Conventional Quenching Oils

Conventional quenching oils are mineral oils, sometimes containing antioxidants; however, they are free from additives that alter their quenching effects. The typical viscosity of conventional quenching oils is 100 to 110 SUS at 40°C (100°F), but it can reach about 200 SUS at 40°C (100°F).

Fast Quenching Oils

Fast quenching oils are mineral oil blends, usually with a viscosity between 50 and 110 SUS at 40°C (100°F), but for the most part between 85 and 105 SUS at 40°C (100°F). They contain especially developed proprietary additives that provide faster quenching effects; in addition they can be compounded further with antioxidants, wetting agents and other additives.

Martempering or Hot Quenching Oils

Martempering or hot quenching oils are solvent-refined paraffin-type mineral oils of very good thermal and oxidation stability. They are used at temperatures between about 95°C (200°F) and 230°C (450°F) for modified and actual martempering of ferrous metals. Martempering oils may contain antioxidants to improve their ageing stability. They are also available with comparatively high quenching effects, even at high use temperatures, because of the additions of very effective speed improvers.

Emulsions

Emulsions of soluble oils, usually employed as coolants for metal-working such as grinding, cutting and sometimes forming, are used for quenching with concentrations of 3 to about 15 per cent. Such quenching emulsions are similar in characteristics to water-based quenchants, particularly when compared to synthetic polymer solutions.

Typical physical and chemical properties of several commercially available quenching oils are listed in Table 16.3, which also show the appropriate ASTM test methods for the determination of the properties listed. These tests are used by vendors for quality control on new quenching oils. They may be applied also to the routine evaluation of used quenching oils after a certain time-span of usage. Oils from different producers may differ in general properties and quenching effects. These differences results from variations in the compositions of base oils and the additives used.

Table 16.3. Typical properties of commercially available quenching and martempering oils.

Type of quenching oil	*No.*	*API gravity*	*Flash point °C*	*Flash point °F*	*Pour point °C*	*Pour point °F*	*Viscosity at 40°C (100°F), SUS*	*Sponi-fication*	*Ash, %*	*Water, %*
Conventional,	1	33	155	315	−12	10	107	0.0	0.01	0.0
no additives	2	27	185	365	−9	15	111	0.0	0.03	0.0
Fast, with speed	3	33.5	190	370	−12	10	95	0.0	0.05	0.0
improvers	4	35	160	320	−4	25	60	0.0	0.20	0.0
Martempering,	5	31.1	235	455	−9	15	329	0.0	0.02	0.0
without speed	6	28.4	245	475	−9	15	719	0.0	0.05	0.0
improvers	7	26.6	300	575	−7	20	2550	0.0	0.10	0.0
Martempering,	8	28.4	230	450	−9	15	337	2.0	1.1	0.0
with speed	9	27.8	245	475	−9	15	713	2.2	1.1	0.0
improvers	10	25.5	300	570	−7	20	2450	2.5	1.4	0.0
ASTM test		D287	D92	D92	D97	D97	D445, D2161	D94	D482	D95, D1533

Cooling Characteristics

The ideal quenchant would have a higher initial quenching effect during vapour-phase and boiling-range periods but would cool slowly through the final convection range (liquid-cooling phase). Cold water and especially the aqueous solutions of inorganic salts, show the highest initial quenching speeds, but they also quench very fast at the end of the quenching process, that is, during the convection phase. Thus, they are restricted to quenching simple shapes and steels of comparatively low hardenability. For other workpieces, they would cause either intolerable degrees of distortion, or warpage and high quench crack rate. All quenching oils have considerably lower quenching effects than water or aqueous inorganic salt solutions. However, the heat extraction is more uniform and particularly slow at the end of the cooling cycle, that is, during the convection range. As a result, the dangers of distortion or cracking are diminished decisively.

Conventional quenching oils exhibit a comparatively long vapour-phase period, during which period the quenching speed is very low. The rate of cooling becomes higher during the boiling range, followed again by very slow cooling in the convection range. Thus, the quenching power of conventional quenching oils is far less than that of water and often inadequate for steel of lower hardenability.

Fast quenching oils show a high initial quenching speed, in some situations approaching the initial speed of water, followed by fast cooling during the boiling range. The cooling rate in the convection range is usually about the same for fast and conventional quenching oils. However, some fast quenching oils contain special additives that cause a faster quench speed in the convection range that results in a much better through-hardening effect than attainable with regular fast quenching oils.

Fig. 16.10 shows the spread of cooling curves of eight different compounded (speed-improving additive-containing) fast quenching oils as a shaded band and gives a curve for one typical conventional additive-free 100 SUS at 40°C (100°F) mineral quenching oil. These cooling curves, which show clearly the difference in quenching effects of conventional and fast quenching oils, were recorded without agitation and using a cylindrical austenitic stainless steel test probe of 13 mm (0.5 inch) diameter by 64 mm (2.5 inch) length, containing a thermocouple in its geometrical centre. All oils were at a temperature of 50°C (125°F).

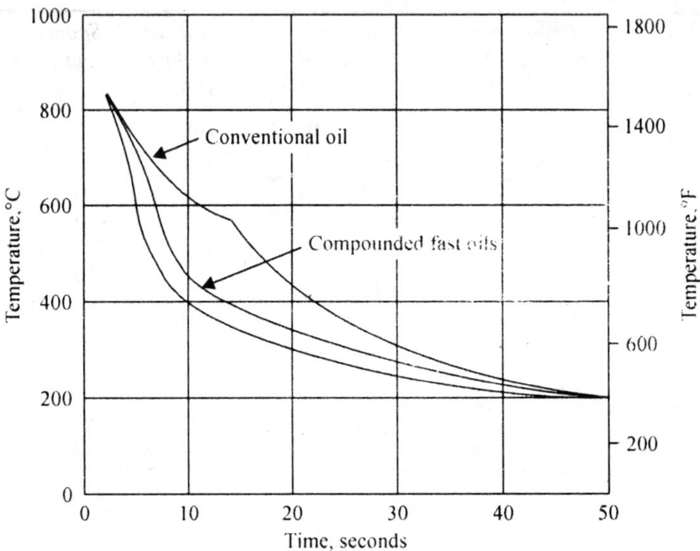

Centre cooling curves for austenitic stainless steel quenched in conventional oil and compounded oils with fast quenching capacities. All oils at 52°C (125°F). Specimens were 13 mm (0.5 inch) in diam by 64 mm (2.5 inch) long.

Fig. 16.10. Centre cooling curves in oil.

Table 16.4 shows the results of magnetic quenchometer tests in quenching durations, in seconds from 885°C (1625°F) to 355° (670°F), of 22 mm (7/8 inch) diameter nickel balls (48 to 52 gram). These results are listed for commercially available conventional and fast quenching oils as well as for martempering oils without and with additions of speed improvers. Conventional and fast quenching oils usually are tested at 25°C (80°F); the permissible range is 20 to 25°C (70 to 80°F). Martempering oils, which are used at higher temperatures, are commonly tested at 120°C (250°F). The test balls are made of either pure nickel or chromised pure nickel. The latter type was developed to approach the conditions and results obtained with pure nickel balls heated to 885°C (1625°F) under reducing atmospheres. This magnetic test, using the chromised nickel balls, is specified in ASTM D 3520, but it is so cumbersome that the much simpler plain nickel ball test usually is preferred.

Table 16.4 compares test results of the two magnetic quenchometer test methods. The already discussed difference in quenching effect between conventional and fast quenching oils is clearly indicated by either variant of the test. Table 16.4 also shows that martempering oils with speed improvers have, at 120°C (250°F), quenching effects as fast as the slower-quenching, low-temperature fast oils. As shown in Table 16.5, the martempering oils quench as fast as conventional 100 SUS quenching oils when used at about 175°C (350°F). Thus, these modern fast quenching martempering oils permit modified or actual martempering of steel parts that are either thicker or lower in hardenability than previously permissible when using the slower quenching martempering oils without speed-improving additives.

Table 16.4. Comparison of the cooling power of commercially available quenching and martempering oils according to magnetic quenchometer test results.

Type of quenching oil	*Oil sample*	*Viscosity at 40°C (100°F) SUS*	*Flash point °C*	*°F*	*Quenching duration from 885°C (1625°F) to 355°C (670°F) At 27°C (80°F) Ni-ball*	*Chromised Ni-ball*	*At 120°C (250°F) Ni-ball*	*Chromised Ni-ball*
Conventional	1	102	190	375	22.5	27.2	–	–
	2	105	195	380	17.8	27.9	–	–
	3	107	170	340	16.0	24.8	–	–
Fast	4	50	145	290	7.0	(a)	–	–
	5	94	170	335	9.0	15.0	–	–
	6	107	190	375	10.8	17.0	–	–
	7	110	185	370	12.7	19.6	–	–
	8	120	190	375	13.3	17.8	–	–
Martempering, without speed improvers	9	329	235	455	19.2	27.6	18.4	22.1
	10	719	245	475	26.9	29.0	25.1	30.4
	11	2550	300	575	31.0	32.0	31.7	32.8
Martempering, with speed improvers	12	337	230	450	15.3	(a)	12.8	(a)
	13	713	245	475	16.4	17.9	14.0	15.6
	14	2450	300	570	19.7	17.0	15.1	15.4

(a) Not available.

The typical ranges for the magnetic quenchometer speeds of the different quenching oil groups are listed in Table 16.5 as a function of their actual use temperature. Table 16.5 clearly shows the well-known fact that the quenching effect of quenching oils does not change much with changing oil temperature. Significant losses of quenching power occur only at temperatures over about 150°C (300°F), where martempering oils are preferred.

Table 16.5. Comparison of the ranges of cooling power of commercially available quenching and martempering oils according to magnetic quenchometer test results using pure nickel balls.

Type of quenching oil	*Quenching duration from 885°C (1625°F) to 355°C(670°F) At 65° (150°F)*	*At 120°C(250°F)*	*At 175°C(350°F)*	*At 230°C(450°F)*
Conventional	14 to 22	14 to 22	–	–
Fast	7 to 14	7 to 14	–	–
Martempering, without speed improvers	18 to 34	18 to 34	22 to 38	Approx 47
Martempering with speed improvers	14 to 20	13 to 18	16 to 22	Approx 33

The insignificant effect of temperature changes and the influence of increased agitation on the cooling power of one conventional and one fast oil, as determined by a modification of the magnetic quenchometer test, are shown in Fig. 16.11.

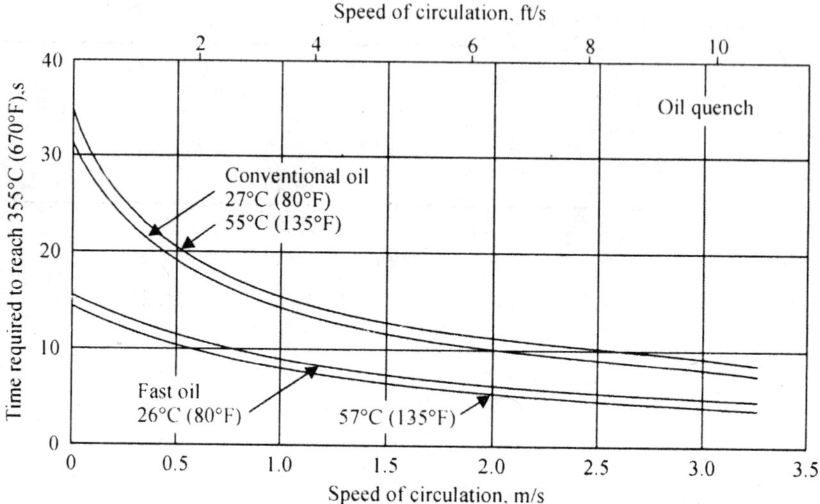

Effects of circulation and temperature of quenchant on the heat extraction rate of conventional and fast oils. A 25 mm (1 inch) nickel ball was heated in a reducing atmosphere and quenched to the Curie temperature 355°C (670°F).

Fig. 16.11. Comparison of cooling power of conventional and fast quenching oils.

The cooling characteristics of quenching oils are not changed over a wide range of operating temperatures. The only differences are found in the last stage of cooling, when the temperature of the quenched part approaches the oil temperature. During this stage, when heat is extracted only by convention, the cooling speed decreases with higher quenching oil temperatures. These differences can be important when quenching critical parts. Slower cooling speeds cause reduced temperature gradients between surface and core and result in lower internal stresses. This is especially beneficial while the steel part is passing through the temperature range of martensite formation, below 260°C (500°F), because it minimises distortion and lowers the probability of quench cracking.

As indicated by the comparison of cooling curves presented in Fig. 16.12, emulsions of water and soluble oil combine the poorest features of water and oil quenchants. The centre-cooling curves of an oil-in-water emulsion, consisting of 10 per cent soluble oil and 90 per cent water at 25°C (75°F) and 50°C (125°F), show that extremely stable vapour envelopes are formed at the beginning of the quenching cycle. Thus, at the higher bath temperature, the initial severity of the quench is reduced to below that of a conventional oil. Cooling in the latter stage is essentially the same as in water. Water-in-oil emulsions containing 90 per cent soluble oil and 10 per cent water have much lower cooling rates than conventional oil. For these reasons, oil and water emulsions generally are inferior to other quenchants. They are rarely used for quenching, although sometimes they may be used for spray quenching of induction or flame-heated parts. In these instances, the better-suited aqueous polymer quenchants would be more desirable.

The only legitimate use for oil-in-water emulsions is the final cooling of quenched and tempered fasteners to provide a black surface and rust protection. Such emulsions commonly contain about 3 to 10 per cent of the soluble oil and are used at temperature between about 30°C (85°F) and 80°C (175°F).

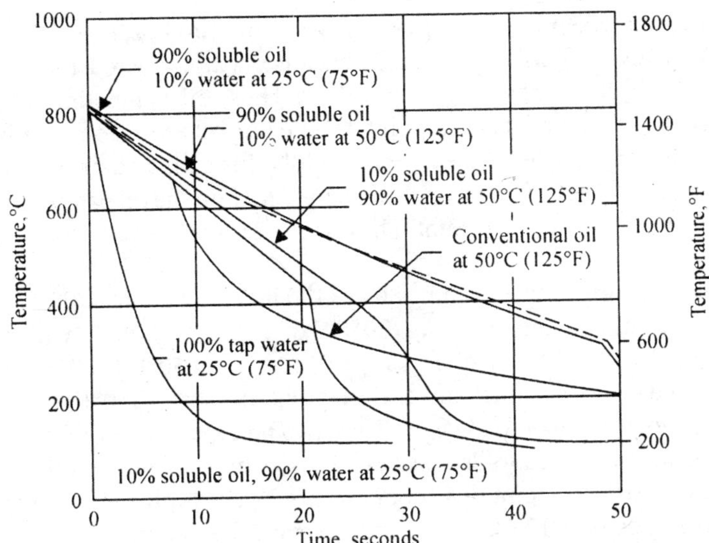

Centre cooling curves for still-quenched 18–8 stainless steel specimens 13 mm (0.5 inch) in diameter by 64 mm (2.5 inch) long, indicating comparative cooling characteristics of plain water and soluble oil, at temperatures of 25 and 50°C (75 and 125°F).

Fig. 16.12. Comparison of water, oil and emulsions of water and soluble oil.

Oil Temperature

As shown in the previous section, the quenching effects of conventional and fast quenching oils are almost independent of their use temperatures. For practical reasons, these oils are generally maintained at temperatures between 40 and 95°C (100 and 200°F), most often at 50 to 70°C (120 to 160°F). Higher or lower oil temperatures are possible without any change in the as-quenched hardness of the hardened work, but are rarely utilised. Higher temperatures will cause faster ageing of the oil and can lead to increased fuming. For safety purposes, the maximum use temperature should be at least 50°C (120°F) below the flash point of the oil. Lower oil temperatures may result in more distortion because of the faster quenching effect of the oil during its convection range and may increase the fire hazard because of the higher viscosity of the cold oil. Fig. 16.13 shows the advantage of a more fluid quenching oil. Any given oil will have a lower viscosity at higher temperature and higher viscosity at lower temperatures.

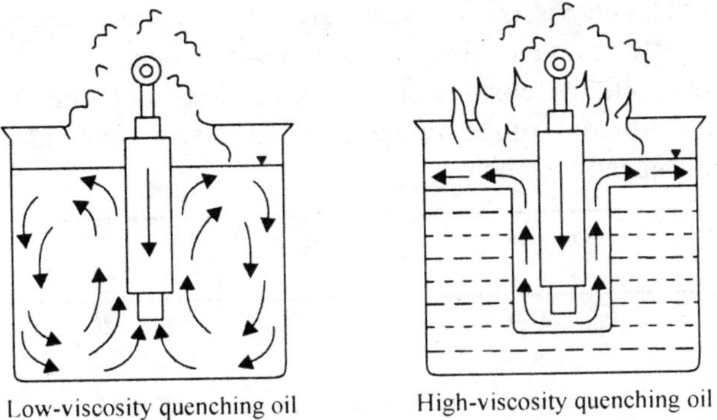

Fig. 16.13. Advantage of low-viscosity quenching oil over high-viscosity oil.

When a hot part is immersed into a low-viscosity oil, the fire hazard is less even if the oil's flash point is comparatively lower. This effect occurs because heat extracted from the part is distributed rapidly through the oil by the 'thermosyphon' effect. Thus, oil contacting the hot metal at the air interface does not reach and exceed its flash point. Quenching oils with comparatively high viscosities—for example, cold oils—do not distribute heat as rapidly. Instead, a comparatively thin layer of oil is heated at the part surface and gets hotter as it flows upward. Thus, the oil can reach the air-oil interface at a temperature that is higher than its flash point. The oil temperature may even pass the fire point, resulting in immediate ignition.

In one test, two similar parts, each weighing 80 kg (175 lb) and at about 870°C(1600°F), were partially immersed in separated baths containing 180 kg (395 lb) of oil at 20°C (70°F) with no circulation. One oil had a flash point of 205°C (400°F) with a viscosity of 120 SUS at 50°C (120°F); it ignited immediately. The other tank contained oil with a much lower flash point of 120°C (250°F) but a lower viscosity of 46 SUS at 50°C (120°F); it took 20 seconds to ignite. Thus, two precepts are apparent: first, it is recommended that cold, viscous oils be pre-heated to decrease their viscosities, thereby diminishing the likelihood of ignition; and second higher-flash-point oils should be preferred. Typical conventional or fast quenching oils have minimum flash points of 125°C (260°F) and average flash points of 175°C (350°F).

Martempering oils are designated to be used at higher temperatures permitting modified or actual martempering of ferrous metals. The typical use temperature ranges for commercially available martempering oils are as follows:

		Temperature range, °C	
Viscosity(a)	*Flash point(b) (°C)*	*In open air*	*Under protective atmosphere*
250–550	220	95–150	95–175
700–1500	250	120–175	120–205
2000–2800	290	150–205	150–230

(a) At 40°C (100°F) SUS; and (b) minimum.

Martempering oils, when covered by a protective (inert, neutral or reducing) atmosphere in an integral quenching chamber of a furnace with controlled atmosphere, are largely protected from oxidation by air. Thus, they may be used at higher temperatures and closer to their flash points than similar oils used in open air. The dragout loss of a quenching oil depends mainly on its viscosity, which is dependent on its temperature. Because oils of higher viscosities usually are used at higher temperatures, the actual viscosities and the dragout losses of different quenching oils do not vary much.

The following table shows the viscosities of one conventional and one fast quenching oil as well as two martempering oils as a function of the temperature. The recommended-use temperature ranges for the four oils appear in bold-faced type below:

Oil temperature, °C	*Conventional*	*Viscosity of oil, SUS fast*	*Martempering*	
40	100	95	720	2630
65	55	60	190	555
95	40	40	85	180

(Contd...)

Oil temperature, °C	Conventional	Viscosity of oil, SUS fast	Martempering	
120	35	35	55	90
150	30	35	45	60
175	–	–	40	45
205	–	–	35	40
230	–	–	–	35

When these oils are used at the middle of their use temperature ranges, their viscosities vary only from 40 to 60 SUS.

Quench Loads

Quench loads will determine the temperature rise of the quenching oil, but uniform hardening usually can be maintained despite large temperature increases. However, the danger of fire must be considered when the oil temperatures approach the flash point. Parts with high ratio of surface area to volume increase the fire hazard because they transfer heat more rapidly to the oil, thus increasing the fire hazard. Such parts require lower quenching oil temperature, larger oil volumes or oils of higher flash point.

The recommended capacity of an oil quench tank can be estimated by either of the following generalised rules:

1. Volume of oil in litres equals weight of quench load in kg ÷ 10.
2. Volume of oil in gallons equals weight of quench load in pounds.

The two variations of the rule deliver somewhat different results, but the difference is only 17 per cent. Thus, either of the rules can be used for a first approximation.

For very large, massive parts, the oil volume can be much lower than estimated using above rules. Such parts transfer the heat comparatively slowly to the oil. However, it is necessary to install larger coolers and provide fast circulation from tank to cooler in order to maintain the oil temperature at a safe, low level.

To minimise oxidation, all quenching oils should have as little contact with air as possible. It is recommended that the surface of open-air quench baths be kept as small as possible. For oil baths under controlled atmospheres, external exposure to the air should be as short as possible.

Quenching oils should be agitated (i) to distribute heat quickly and uniformly; and (ii) to maintain a uniform temperature, which provides a uniform quenching effect throughout the tank. Only mechanical agitation should be used (pumps or impellers). Strong agitation should be used only when work is actually quenched.

Note: Never bubble compressed air through oil for agitation.

Heating elements used for pre-heating quenching oils should have less than 1.5 W/cm^2 (10.0 W/inch^2) to prevent local overheating, which would cause not only unnecessary ageing of the oil (polymerisation) but also the formation of an insulating layer of oil-coke on the heater with the possible result of heating element burn-outs.

The quenching system, particularly heating elements and coolers, should not comprise copper or copper alloys. These metals act as catalysts, accelerating the oxidation and polymerisation (ageing) of mineral oils. Use of steel, ferrous castings, stainless steel, nickel-plated or tin-plated materials is recommended instead. Existing copper or brass equipment should be tin plated.

Water Contamination

Water in quenching oils is very dangerous, particularly in martempering oils used at temperatures above 100°C (212°F). Not only does it causes non-uniform or insufficient as-quenched hardness of the work, but also it creates heavy foaming and increased fire hazard because oil foam catches fire very easily. Oil foam that develops in an integral-quenching chamber can cause an explosion if the foam level reaches the hot furnace chamber.

The effect of water in quenching oils is shown in Fig. 16.14 in the form of centre cooling curves. These results were recorded by quenching a cylindrical stainless steel probe in a fast quenching oil free of water and after water additions of 0.06 to 2.0 per cent. At water concentrations below 0.12 per cent, the initial quenching speed was increased, whereas higher water concentration caused a more stable vapour blanket around the probe resulting in a lower initial quenching speed.

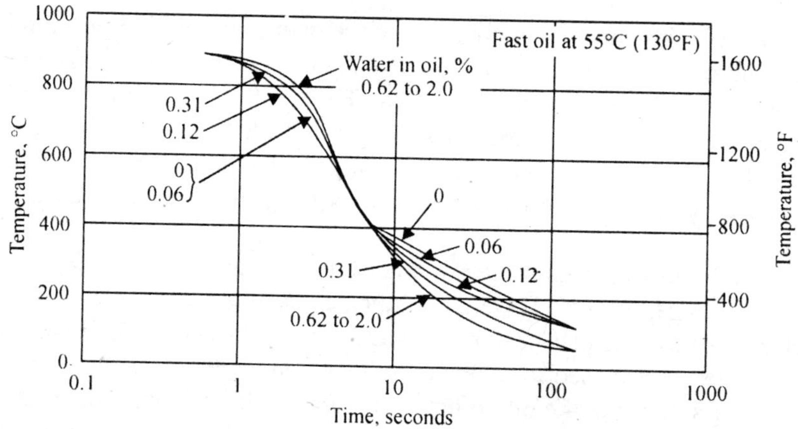

Effects of water contamination on the quenching power of fast oil at 55°C (130°F) in quenching type 304 stainless steel specimens 13 mm (0.5 inch) in diam by 100 mm (4 inch) long. Oil was not circulated, and specimens were not agitated. Thermocouples were located at the geometric centre of specimens.

Fig. 16.14. Effects of water contamination.

All water-containing oil samples quenched faster than the water-free fast oil during boiling and convection ranges. Because of the higher cooling speeds in the convection range, which overlaps with the martensite formation range of most common steels, increased distortion and an increased likelihood of cracking can be expected when using water-contaminated quenching oils.

Entrapment of water in localised sections of an oil bath can result in an explosion from the sudden formation of large volumes of steam. However, except where oil and water are distributed in stratified layers under stagnant conditions, there is little danger of explosion. With normal circulation of the bath, water content can be high without danger of explosion.

Water can be removed from an oil bath by several methods: (i) raising the temperature above the boiling point of the water (Fig. 16.15); (ii) allowing the water to settle and draining it off; and (iii) passing the bath through a centrifuge. However, used quenching oils may sometimes emulsify the water content. Such water-containing oils cannot be treated by using the method described in (ii). The other two methods will work.

Surface Conditions of Quenched Work

In order to attain clean quench hardening, the quenching oil must retain its non-staining features. This is best accomplished by maintaining low quenching oil temperature to minimise oxidation or oil degradation. On the other hand, higher temperatures usually will minimise distortion. Carbonaceous deposits on quenched parts (sludging) are symptomatic of oil breakdown. Normally, an oil does not reach this stage of degradation without detection, except in unprotected systems. Carbonaceous deposits are objectionable because they are difficult and costly to remove. However, they do not adversely affect hardening of the steel because they deposit after parts have reached a temperature below the range where rapid cooling is required.

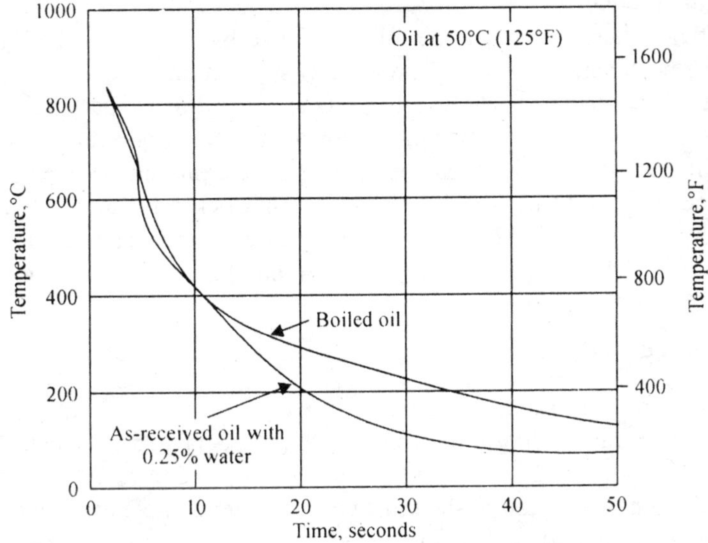

Centre cooling curves for austenitic stainless steel specimens 13 mm (0.5 inch) in diam by 64 mm (2.5 inch) long, showing the effects of quenching in As-received oil containing 0.25% water and in the same oil after it had been freed of water contamination by being boiled at 120°C (250°F) for 5 hrs. Quenching temperature of oil in both conditions was 52°C (125°F).

Fig. 16.15. Effects of boiling off water in oil.

When sludging occurs in installations that employ carbonaceous-protective atmospheres, the problem is complicated because carbon fallout may be the initial source of carbonaceous deposition. When quenching oil is contaminated by carbon fallout, it is sometimes difficult to determine other possible causes of carbon deposition.

Control of Quenching Oils

Routine analyses to determine the state of a used quenching oil should be performed about every three months. Used quenching oils are usually checked as follows:

1. Crackle test (water, qualitative).
2. Water content (quantitative, only if above test is positive) by ASTM test method D95.
3. Viscosity at 40°C (100°F) by ASTM test methods D445, D2161.
4. Sludge test by ASTM test method D2273.
5. Magnetic quenchometer speed, using plain nickel balls, at 25°C (80°F) for conventional and fast quenching oils; at 120°C (250°F) for martempering oils.

Selection of Quenching Oil

The quenching of steel involves unsteady heat flow and kinetics of solid-state transformation; both of these are difficult to handle quantitatively. Therefore, the selection of oils for specific applications must be based largely on trial and error. In many plants, selection of quenching oil is based on experience with similar parts and steels. When such experience is not available, it will be necessary to utilise one or more of the test methods previously described.

The most important criterion in selecting a quenching oil is that it provide a cooling rate that will harden the part without cracking it. Other considerations are degree of distortion, cost for the finished part and microstructure produced.

If a steel has sufficient hardenability to be fully hardened by being quenched in a conventional oil, there will probably be no advantage in using the more costly fast oils.

In selecting a quenching oil, its contribution to the cost of the finished part must be considered. Aside from their initial cost (and the cost of replacing dragout), quenching oils can influence overall cost by: (i) limiting the type of steel of which parts can be made; (ii) determining the percentage of rejected parts that will require re-working; and (iii) determining the extent of cleaning after quenching.

Although it is more common to fit the oil to the steel being treated than to fit the steel to the oil, there have been instances in which fast quenching oils permitted changing to steels of lower hardenability, with accompanying reductions in cost.

POLYMER SOLUTIONS

The technology of heat treating has seen dramatic growth during the current century. In the past, the principal quenchants were water, perhaps modified by some special additive such as inorganic salts and naturally occurring oils. With the discovery of petroleum oil, hydrocarbon products became a major medium for quenching. It has been established that certain water-soluble organic polymers are useful in modifying the cooling characteristics of water.

Polyvinyl Alcohol (PVA)

Although PVA can be regarded as a polymer of vinyl alcohol, in reality all PVA resins are made by the hydrolysis of polyvinyl acetate.

Cooling characteristics

At concentrations of less than 0.01 per cent, the cooling characteristics at room temperature are only modestly different from those of water alone. With such small concentration variations, close control of PVA solutions is necessary. Control is complicated by the fact that quenched parts can become coated with an insoluble layer of resin, thus reducing the bath concentration. Maintaining an 'effective' concentration requires specific control measures.

Polyalkylene Glycols (PAG)

Polyalkylene glycols (PAG) or polyalkylene glycol ethers, are these materials are formulated by the random polymerisation of ethylene and propylene oxides (although higher alkylene oxides and/or aryl oxides might be used also). Although block polymerisation of these same oxides are possible, these derivatives are less attractive as quenchants. By varying the molecular weights and the ratio of oxides, polymers having broad applicability may be produced. Certain of the higher molecular-weight products were shown to have utility as metal quenchants when used in aqueous solution.

Proper selection of the polymer composition and its molecular weight, provides a PAG product that is completely soluble in water at room temperature. However, the selected PAG molecules exhibit the unique behaviour of inverse solubility in water—that is, water insolubility at elevated temperatures. This phenomenon provides the unique mechanism for cooling hot metal by surrounding the metal piece with a polymer-rich coating that serves to govern the rate of heat extraction into the surrounding aqueous solution. As the metal part approaches in temperature the temperature of the quenchant itself, the PAG polymer coating dissolves to again provide a uniform concentration in the quenchant bath.

Cooling characteristics

In the application of PAG quenchants for heat treating, three principal parameters are recognised to control the rate of cooling:

1. Quenchant concentration.
2. Quenchant temperature.
3. Quenchant agitation.

The influence of polymer concentration on cooling rates is illustrated by the cooling curves shown in Fig. 16.16. No specific concentrations are listed because the shape of the cooling curve, as well as the change in rate with concentration, will vary with the selection of the PAG quenchant. The slower rates of cooling achieved at the higher concentrations reflect the thickness of the polymer layer that surrounds the heated part during quenching. The polyglycol quenchants also are less sensitive to minor changes in polymer concentration, which is a recognised deficiency of polyvinyl alcohol and the other 'film-forming' polymer quenchants.

Just as water exhibits a marked decreases in cooling capability as its temperature is elevated (Fig. 16.6), this same loss is translated to the aqueous solutions of PAG quenchants. The curves shown in Fig. 16.17 are illustratire of the general trends that would occur with changes in bath temperature; more detailed data would require specific identification of the particular PAG quenchant employed.

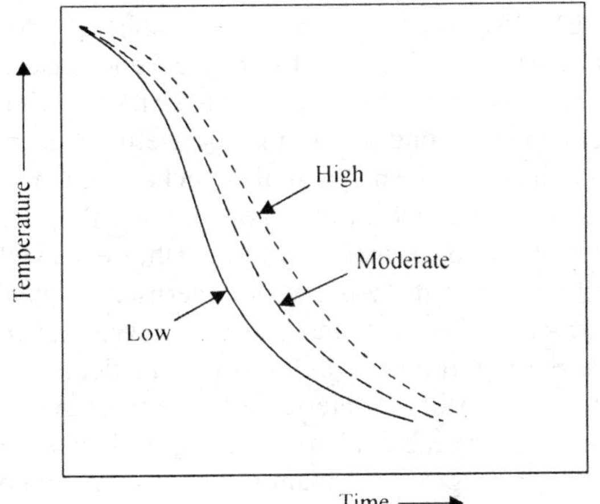

Fig. 16.16. Effect of concentration on cooling rate.

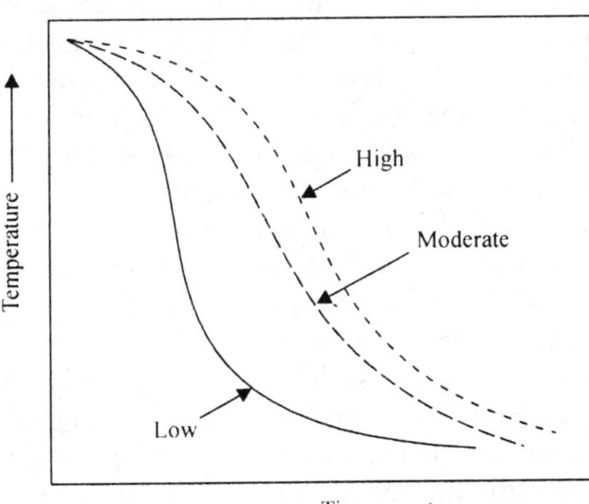

Fig. 16.17. Effect of temperature on cooling rate.

The use of no agitation with a polyglycol quenchant would be unusual. In general, low to moderate agitation is essential (i) to ensure that adequate replenishment of polymer occurs at the hot metal surface; and (ii) to provide uniform heat transfer from the hot part to the surrounding reservoir of cooler quenchant.

Vigorous agitation may be essential for achieving a rapid rate of cooling (for example, with a low hardenability steel) to avoid undesired transformation. Fig. 16.18 clearly illustrates that, as agitation is increased, the cooling curves shift to more rapid rates.

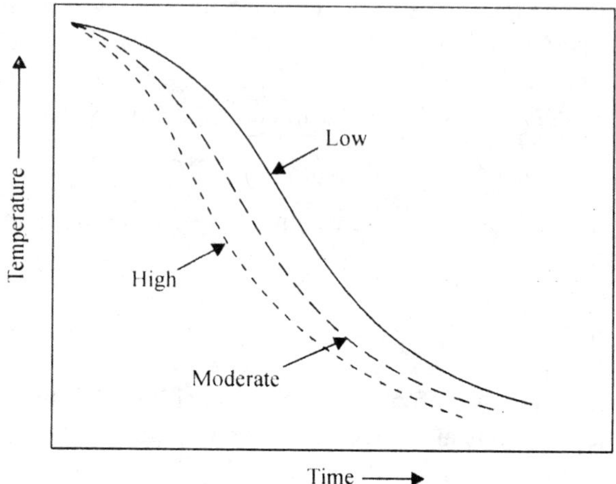

Fig. 16.18. Effect of agitation on cooling rate.

Control measures

The refractive index of oxyalkylene glycol polymer solutions (in the range employed for quenching) is essentially linear with concentration. Thus, the refractive index of a PAG quenchant solution serves as a measure of product concentration. Industrial model optical refractometers that employ an arbitrary scale may be calibrated. Whereas such instruments prove invaluable for day-to-day monitoring of the quenchant concentration, the refractometer also will register other water-soluble components that are introduced to the used quenchant. When the indicated refractometer reading begins to provide erroneous numbers, some other analytical test is required to define the 'effective' quenchant concentration. With polyglycol quenchants, kinematic viscosity measurements (which are correlated with concentration) have proven to be most useful.

As required, additional analytical tests for pH, inhibitor level and conductance may be useful adjuncts to a successful monitoring programme. If the level of contaminants in the polyglycol quenchant becomes excessive—where these contaminants may be, in part, the same undesirable constituents that are detrimental to water alone, or oil—quenchant recovery can be effected thermally. By heating the quenchant solution (in whole or in part) above the separation temperature, a more-dense polymer-rich layer is obtained. Much of the water-soluble contamination can be withdrawn with the supernatant water layer. Solid contaminants such as scale or carbon would require settling, filtration and/or centrifugation.

Because PAG quenchants are highly bioresistant, the addition of a bactericide to the as-supplied quenchant is not required. Further, biochemical activity in use is traceable not to the polyglycol polymer itself but to the introduction of nutrient contaminants. Microbiological treatment such as is employed with other aqueous metal working fluids generally will keep under control this foreign biological activity.

Polyvinylpyrrolidone (PVP)

Polyvinylpyrrolidone (PVP) is derived from the polymerisation of N-vinyl-2-pyrrolidone. PVP is a water-soluble polymer characterised by its unusual complexing and colloidal properties and by its physiological inertness. PVP has been available as a white, free-flowing powder, manufactured in four molecular-weight grades.

Quenchant variables

As with other polymer-type quenchants, concentration, bath temperature and agitation all play a role in establishing the cooling characteristics. By comparison, the quenching rates tend to be faster during the stable film and nucleate and boiling stages, but are slower during the convection stage. Because PVP does not have inverse solubility in water, only very small amounts of polymer film are retained on quenched parts at quenching temperatures from 30°C (85°F) to near boiling. Thus, a broader working range of temperatures for quenching can be employed.

Optical refractometer readings will provide initial control of concentration, but back-up with viscosity measurements is strongly recommended.

Polyacrylates

The most recent addition to commercially available polymer quenchants in the US is a product comprising an aqueous solution of sodium polyacrylate. The following reaction illustrates that the polymer may be achieved from the direct polymerisation of sodium acrylate or the alkaline hydrolysis of some polyacrylate ester.

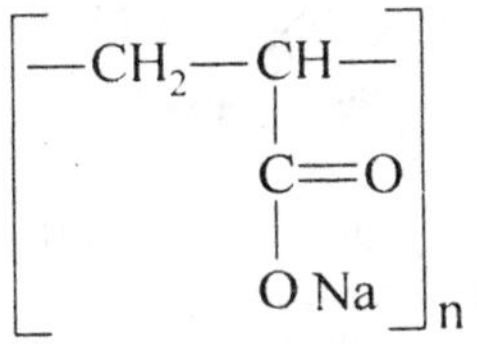

$$\left[-CH_2-\underset{\underset{\displaystyle O\,Na}{|}}{\underset{\displaystyle C=O}{\underset{|}{CH}}}- \right]_n$$

By utilising the salt of an alkali metal, in this instance sodium, the polymer is provided solubility in water. Polyacrylate quenchants represent a class of quenchants completely different from the PVA, PAG or PVP types. The latter polymers fall into the type characterised as non-ionic, that is, not ionisable or neutral. Polyacrylate quenchants are considered anionic, which is negatively charged. The charged character of the polymer imparts another dimension to the quenchants, strong polarity. The strong polarity provides water solubility but is also suspected of causing the polymer to operate by a different mechanism of heat extraction.

Unlike other polymer quenchants, polyacrylate solutions do not split on heating and do not form plastic films on the surface of the hot work. Their slower rate of cooling is based on the molecular weight of the polyacrylate and the subsequent viscosity of its solutions. By varying the molecular weights of the polymer, a whole family of quenchants can be designed covering a full range of applications from the fast quenching of water to the slow cooling of oils.

The quenching effect of the polyacrylate quenchants is a function of the three basic parameters: (i) polymer concentration; (ii) bath temperature; and (iii) bath agitation.

The cooling rate of the polyacrylate solutions are distinctly slower than those of any other polymer quenchants. This illustrates their advantages as quenchants in particular for applications requiring a slow quenching effect.

With increasing polymer concentration and bath temperature, the cooling rate of the solutions can be slowed to the extent that many ferrous metals do not transform to martensite at all, but form bainite or pearlite. This 'non-martensite' quenching can be utilised for new and unique heat treating procedures which are beneficial in many respects—for example, cost, energy, safety and environmental control.

Some applications of the polyacrylate solutions for non-martensitic quenching are as follows:

1. Deep carburised parts, such as bearing races, balls and rollers, are usually double hardened in oil to obtain grain refinement. The first quenching can be done in a polyacrylate solution.
2. Direct quenching of plain carbon steel with a high carbon content in polyacrylate solutions to obtain similar mechanical properties as obtained by quench and temper or austempering.
3. Rod or wire patenting in polyacrylate solutions, instead of the commonly used lead or salt bath process at 510 to 565°C (950 to 1050°C).
4. Direct quenching of hot formed parts in polyacrylate solutions to obtain good machinability without the usual necessary tempering process subsequent to hardening, in particular for steels of low hardenability, such as low-alloy carburising steels.
5. Cooling of hot formed parts in polyacrylate solutions to prevent excessive scaling and decarburisation during slow cooling in air, whereby practically the same microstructure is obtained as during air cooling.
6. Large concast steel slabs are water quenched to shorten cooling duration, thus permitting inspection shortly after casting. This is possible only for slabs of plain carbon steel with up to 0.2 per cent carbon. Alloy steels and steels with higher carbon content have to be air cooled, which requires several days. They would crack when water is quenched. Now, the addition of polyacrylates to the slab quench water allows faster cooling of high carbon and alloy steel slabs.

The polyacrylate quenchants also can be used for quenching aluminium alloys after solution treatment to minimise distortion and warpage.

As any other polymer quenchants, the polyacrylates solutions require the use of agitation, the degree of which depends on the specifics of applications. As a general rule, a high degree of agitation is recommended for hardening operations, whereas minimal agitation is usually sufficient for non-martensitic quenching. The concentration control of the polyacrylate quenchants is based on their kinematic viscosities. To take into account the influence of contaminations, they also should be periodically checked with other laboratory methods including cooling curves analysis. This service usually is provided by a quenchant supplier.

GAS QUENCHING

Gas quenching is used to provide a cooling rate that is faster than the obtained in still air and slower than that obtained in oil.

In gas quenching, the austenitised workpiece is placed directly into the gas quenching zone or chamber and heat is rapidly extracted from the metal by a fast-moving stream of gas. The cooling rate of the metal is related to the surface area and mass of the part and to the type and velocity of the cooling gas. The cooling rate can be adjusted and controlled by altering the last two of these variables, thus providing a degree of flexibility that can be obtained with very few other quenching media.

Recirculation

During the quench period, large volumes of relatively cold gases are directed through nozzles or vanes to achieve the impingement of high-velocity gas on the surface of the work load. After absorbing heat from the material being processed, the gases are cooled by being passed over water cooled or refrigerated coils. Recirculating fans return the chilled gases to the high-velocity nozzles, through which they are again directed at the work to absorb more heat.

Gas quenching unit

Gas quenching unit may be designed for either batch or continuous processing and it is gastight, to prevent infiltration of air or loss of gas pressure. Various gases, ranging from air to complex mixtures, may be used for cooling, depending on process requirements. Protective atmospheres are commonly used, not only to produce bright work but also to increase the heat transfer rate between the gas and the work, thus bringing about an increase in the rate of cooling.

Applications

In some applications, quenching in still air is too slow and oil quenching is undesirable because of distortion, cost factors, handling problems or the final hardness obtained. In these applications, gas quenching constitutes a useful compromise between two extremes.

Hardening tool steel

Gas quenching has been used to obtain maximum hardness in various grades of aır-hardening tool steel.

Gas quenching also has been employed to develop the desired mechanical properties in other steel products, such as thin sheet and small-diameter wire.

FOG QUENCHING

In fog quenching, heat is rapidly extracted from the metal by a fast-moving stream of gas that contains water droplets. The cooling capacity of the 'fog' is derived from both the absorption of heat by the gas and the heat of vapourisation of the water. The addition of water droplets (or fog) to an air stream can increase its cooling capacity by a factor of 4½. Fog quenching is most effective in the lower temperature ranges. An outgrowth of gas quenching, fog quenching normally is substituted for liquid quenching in an effort to minimise distortion. Although less severe than a liquid quench, fog quenching is more severe than a plain gas quench of equivalent velocity. In general, the types of parts that are fog quenched are similar to those that are considered suitable for gas quenching.

Agitation

The most important influences on the cooling rate of steel in various quenching media are:

1. Agitation (flow rate) of the quenching liquid.
2. Surface oxidation of the workpiece.
3. Mass and section size of the workpiece.

The various quenching media available do not automatically supply all the cooling rates desired, particularly in the rage of quenching speeds between those obtainable in water and in still oil. The rate of cooling by oil quenching can be increased by altering the rate of oil flow past the work or by agitation of the work in the oil.

Factors Controlling Agitation

In all agitated quenching baths, the degree and character of agitation vary from point to point within the bath; these variations are even more pronounced in spray quenching. Although accurate description or measurement of agitation is difficult, the principal factors that control agitation are well known. These include the general shape of the bath, location of the work, direction of flow currents, type of agitator, flow rate and power consumed. In spray or jet quenching, additional factors are encountered, such as the shape, arrangement and placement of the spray head in relation to the work, the pressure, velocity and size of the jets and the total volume of quenching fluid used per unit of time.

SURFACE OXIDATION

The effects of surface oxidation have been evaluated by magnetic testing, observation of high-speed motion pictures and cooling curve measurements.

Magnetic Testing

Magnetic testing is used to determine the effect of surface oxidation on the rate of heat transfer from nickel specimens to the quenching medium. Specifically, the heat transfer rate of nickel balls heated to 885°C (1625°F) in an oxidising atmosphere was compared to that of similar balls heated to the same temperature in a protective, reducing atmosphere. Rates of heat transfer from 915 to 355°C (1675 to 670°F) as shown in Tables 16.4 and 16.5 indicate that the balls heated in an oxidising atmosphere cooled faster than the balls heated in a reducing atmosphere.

High-speed Motion-Picture Techniques

High-speed motion-picture techniques have been used to reveal the influence of oxide coatings on the quenching rate of steel in water containing polyvinyl alcohol, trisodium phosphate or carboxyl methylcellulose. When a steel specimen that had been heated in a protective atmosphere was submerged in the liquid, it was noted that a visible envelope, or cocoon, surrounded the specimen, preventing contact between the liquid and the steel surface. Finally, after a delay of many seconds, the envelope burst and quenching action commenced. The total quenching time in a 0.3 per cent solution of polyvinyl alcohol in water was 37 seconds, a slower quenching rate than was obtained with conventional quenching oils. When an identical specimen that has been heated in an oxidising atmosphere was quenched in the same polyvinyl alcohol solution, immediate contact was made between the solution and the specimen surface and the total quenching time was only 2.3 seconds. This is comparable to the rate obtained in plain water.

QUENCHING SYSTEMS

Equipment requirements for quenching may vary widely. A small plant making machine parts might each day require the hardening of only a few simple carbon steel parts weighing about 1.4 kg (3 lb) a piece. For such an application, the quenching 'system' would comprise a barrel of water, with piping to the water supply and a drain to the sewer. Handling equipment would consists simply of pair of tongs. As quantity of work to be quenched and complexity of workpieces increase, however, various other items of equipment must, of course, be added to the quenching system.

For a complete quenching system, the following functional equipment usually is required and installed: (i) work tank or machines; (ii) facilities for handling the parts quenched; (iii) quenching medium; (iv) equipment for agitation; (v) coolers; (vi) heaters; (vii) pumps and strainers or filters; (viii) quenchant supply tank; (ix) equipment for ventilation and for protection against hazards; and (x) equipment for automatic removal of scale from tanks.

Continuous Quenching

A tank such as that illustrated schematically in Fig. 16.19 is used commonly for continuous quenching of similar parts that can be dropped into the quenching liquid. In this installation, the parts fall from the furnace belt through a chute into the quenching medium and onto another belt that conveys them out of the quench. Agitation is provided by an impeller directed at the lower end of the quench chute. If the mass of work entering the tank is large, a means of controlling the temperature of the quenchant would be added. This design of quench system is usually limited to symmetric and simple shapes because free fall from the furnace belt can result in distortion and non-uniform quenching.

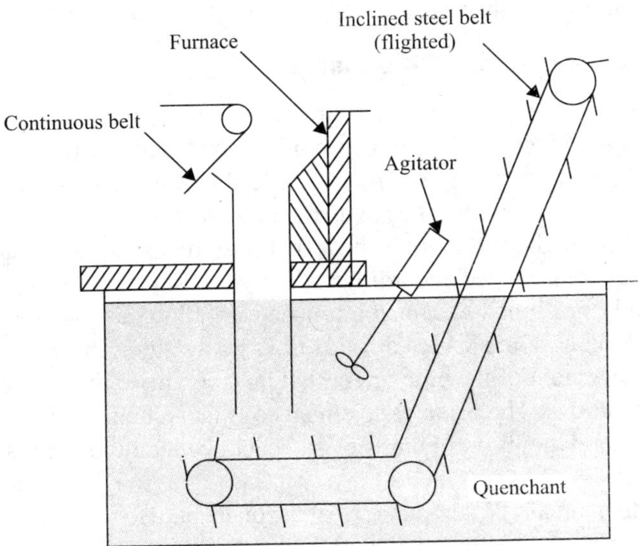

Schematic illustration of tank for quenching parts of similar shape from a continuous hardening furnace. Parts fall from the furnace belt through the quenchant and onto a conveyor belt that transports them from the quench tank. Quenchant may be water or oil.

Fig. 16.19. Tank for continuous quenching.

Water Quenching System

A simple arrangement for water quenching, incorporating a system for controlling water temperature, is shown in Fig. 16.20. In this system, a supply of fresh water is continuously introduced directly into the bath and is allowed to overflow to the sewer.

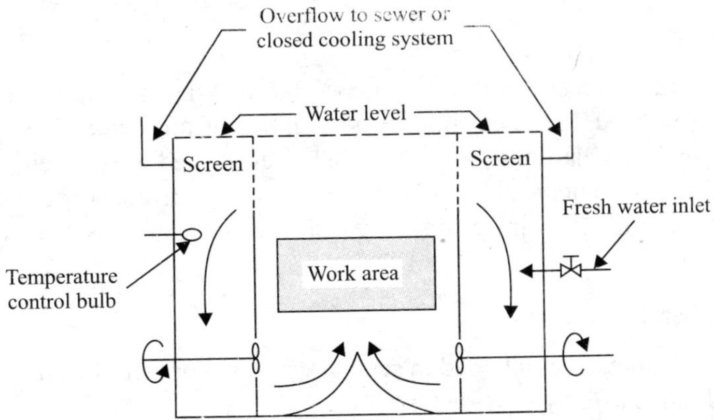

Fig. 16.20. Temperature-controlled overflow tank for water quenching.

Temperature control of the water is provided by a self-operated temperature regulator (which opens and closes at pre-determined settings, these depending on the volume of parts that are being processed and on the water temperature desired). The system is intended for use where water is plentiful and inexpensive. Where water is scarce, however, this system could be connected to a central or closed water-cooling system that would condition the water for re-use.

Oil Quenching Systems for Continuous Carburisers

Additional controls at the quenching operation are required for mass production of parts such as finished gears through a continuous carburiser. In a large automotive plant manufacturing rear and front axle assemblies, the quenching of gears, stem pinions and differential side gears and pinions requires special loading fixtures for batch-quenched parts and special quenching presses for ring gears. Because low cost per unit must be maintained throughout the entire heat treating process, quenching is integrated with heating and handling facilities are arranged accordingly.

The clean, soft gears, fully machined except for finished grinding, are charged into a continuous pusher-type gas carburiser. Workpieces are loaded on alloy trays with fixtures that permit direct quenching.

Fig. 16.21 shows the arrangement of the equipment that may comprisc the quenching system for an application such as described above. The principal elements of this system are:

1. Quenchant storage-supply tanks and pumps of sufficient capacity to maintain an efficient heat transfer from the gears.
2. Coolers and heaters to maintain the temperature of the quenchant for consistent quenching results.
3. Filters in the system to minimise free carbon and other foreign elements that may affect the rate of heat transfer or interfere with the efficient operation of quenching presses.
4. Agitation of the quenchant to obtain uniform quenching and minimise distortion.
5. Special alloy fixtures and handling facilities to transfer workpieces to the quench.
6. Quenching oil selected to provide the desired cooling rate.
7. Quenching presses for processing ring gears.

A typical batch-quenching tank that can be used for high-volume production of work on trays. The pressure of the atmosphere gas must be greater than that of the outside air; otherwise, air will enter the chain drive unit opposite the agitator motors and the revolving chain will carry air into the oil and cause oxidation, especially when hot oil is being used.

Special Techniques

Special techniques for quenching are often expensive, but they are sometimes required in order that specified results can be attained. Non-uniform hardness and occasional excessive out-of-roundness were encountered. The non-uniform hardness was attributed to entrapped air in each part, which prevented uniform quenching. This was remedied by installing a vent pipe in the quenching tank. By quenching the parts individually over this pipe, the air was displaced and the quenchant reached all areas uniformly and uniform hardness was obtained.

AGITATION EQUIPMENT

The agitation of a quenchant can be obtained in several ways. In conventional quench tanks, circulation of the quenching medium is usually provided by:

1. Pumps.
2. Passage of the workpiece through the quenching medium.

3. Manual or mechanical movement of the workpiece.
4. Mechanical propellers.

The selection of any type of agitation method is dependent on the tank design, type of quenchant, volume of quenchant, the part design and the severity of quench required.

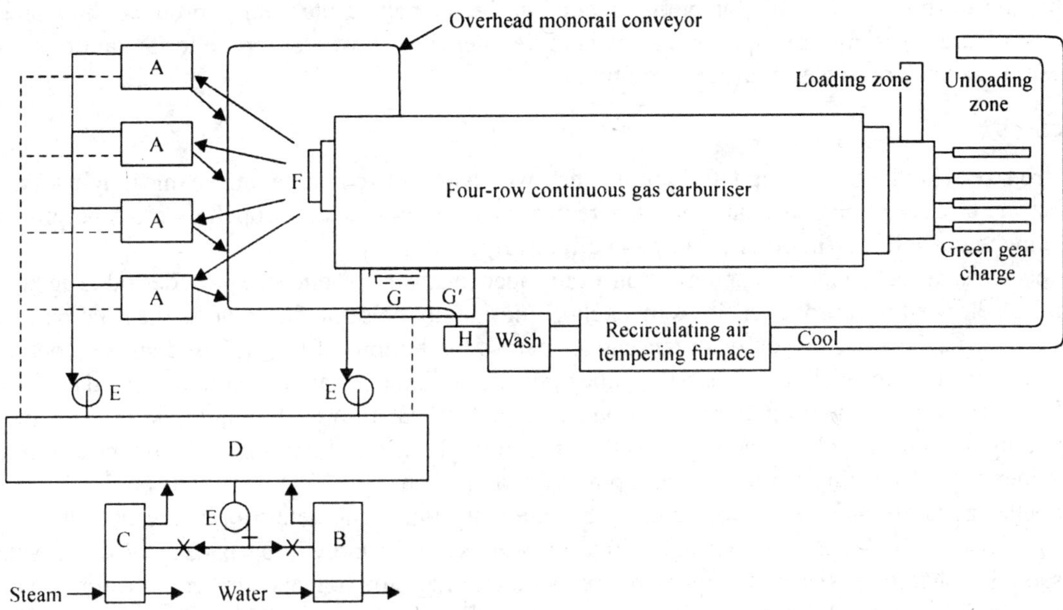

A, press quenches; B, cooler; C, heater; D, storage tank; E, pumps, strainers; F, slot door; G, atmosphere-hooded quench; G', open portion of G; and H, washer loading platform.

Fig. 16.21. Layout showing the elements of a quenching system used with a four-row continuous gas carburising furnace.

Pumps

Pumps are commonly employed because they provide a controllable means for directing the quenchant. Also, the flow of the quenchant can be readily divided to provide circulation to more than one location in the tank. When the quenching medium is oil and a cooling system is employed, the pump used to recirculate oil to the cooling system also may be used to provide agitation. Recirculating pumps may be preferred for use in a 'floating fountain' quench that is capable of removing heat quickly from an internal cavity of workpiece.

Gravity Fall

Gravity fall of the workpiece through the quenchant usually is employed for quenching lightweight parts that have a large ratio of area to unit weight. If pumps or mechanical agitators were used, such workpieces probably would drift in the tank and it would be difficult to handle the load with a conventional elevator-conveyor discharging quench tank.

Movement of the Workpiece

During the quenching of steel, it is usually is desirable to obtain the most rapid cooling when the steel temperature is above about 540°C (1000°F). In this temperature range, the part is usually enveloped in the vapour blanket and the cooling rate is slowest. To accelerate cooling in this temperature range and to remove undesirable scale, rapid relative motion of the part in the quenchant is necessary. On small, low-production items, this can be accomplished by moving the part, or a small basket or tray of parts, through the quenchant by hand. Workpieces may also be driven mechanically with respect to the quenching medium. For example, shafts are sometimes rotated in the quenching medium to produce the effect of agitation. Where a variety of shapes and sizes is to be quenched, however, propeller agitation is most desirable, because it produces a turbulent motion.

Propellers

Aside from providing effective agitation, propellers, which are self-contained mechanical agitators, are compact, require no piping and can be easily removed for maintenance. Propellers must be properly located in the quench tank in order to function effectively.

Propellers are usually placed near the bottom of the quench tank to produce the most desirable agitation. A liquid accelerated by a propeller flows in a helical motion, in the same direction as the rotation of the propeller blades. This high-velocity stream moves across the bottom of the tank and spreads out as it moves away from the propeller. Upon striking the opposite wall, the stream is diverted upward and in the direction of rotation of the propeller. This produces a general rotation of the liquid, which is partially disrupted by the return cycle of the liquid to the propeller. The general motion of the quenchant in the tank is, therefore, a swirling motion and an up-and-down motion.

Propeller agitators may be either top-entry or side-entry units. Side-entry units are usually placed below the floor level, to conserve floor space. Top-entry units require more floor space but less excavation for installation. After the size of the quench tank is determined, the horsepower required for agitation may be calculated from the data shown in Table 16.6. These power requirements are minimums to produce a velocity of about 15 m/min (50 ft/min). Higher horsepower is required for increased agitation.

Table 16.6. Power requirements for propeller agitation of quenching liquids.

Volume of liquid in tank		*Power required for velocity of approx. 15 m/min (50 ft/min)*			
		Conventional oil		*Water or brine*	
l	*gal*	*W/l*	*hp/gal*	*W/l*	*hp/gal*
190 to 3000	50 to 800	0.985	0.005	0.788	0.004
3000 to 7600	800 to 2000	1.182	0.006	0.788	0.004
7600 to 11400	2000 to 3000	1.182	0.006	0.985	0.005
Over 11400	Over 3000	1.380	0.007	0.985	0.005

VARIABLES AFFECTING AGITATION

Effect of Velocity

Tests have shown that quenching oil moving at a velocity of 23 m/min (75 ft/min) will increase the depth of hardening over that obtained at lesser velocities in sections up to 100 mm (4 inch) in diameter. In heavier section, it is possible that no advantage can be gained by increasing the rate of agitation, because the rate of heat transfer through the workpiece becomes the limiting factors. However, the practical limit on agitation is that which produces foaming on the surface of the oil. Foaming is undesirable and indicates that air is being entrapped.

Number of Agitators

The number of units required depends on the total horsepower required and the horsepower of available units. Two or more smaller units provide more uniform agitation and more versatility than one large unit. The velocity of the quenchant can be varied by using two-speed motors or variable-pitch propellers, or a combination of both.

Relation of Tank Design to Agitation

A properly designed tank and two or more propellers usually will provide uniform agitation throughout the working area. Sometimes baffles are required to direct the quenchant flow uniformly throughout the tank. On top-entry units, guide vanes usually are installed in the draft tube to prevent the quenchant from vortexing.

DESIGN OF QUENCH TANKS

To conform with present-day requirements, quenching systems are integrated into heat treating line and the automated lines are normally used to process many types of parts. It is seldom economically feasible to design a quench tank for only one type of parts. One of the major goals of design is to obtain as much flexibility in the quenching system as possible without appreciably affecting the cost of the unit.

The design of a quench tank is primarily based on such factors as the kilograms (pounds) of the steel quenched per hour and size of the parts, as well as its shape, weight, section thickness, grade of steel and propertise needed. A design based solely on kilograms (pounds) per hour may lead to miscalculations and unacceptable results.

Practical Suggestion for Design of Quench Tanks

Some practical suggestions for the design of quench tanks are as follows:

1. Time necessary to 'quench out' the part treated should be measured and allowance should be made for this cycle when the hourly rating or capacity of the work tank is determined. When this factor is considered and the mechanism determined that provides for the desired quenching technique, the size of the enclosure of the work tank follows. This volume of medium then is used in other calculations concerning equipment and in the selection of size; for example, the volume of medium influences the selection of the size of cooler and whether a supply tank is necessary.
2. Adequate space should be provided around the workpiece to get the full benefit of circulation and maximum heat absorption from the quenched parts.
3. At best, it is difficult to control distortion caused by thermal strains when quenching. Therefore, the hot and plastic workpiece should not be mutilated and deformed by allowing it to fall on the chute or quench conveyor until sufficient quenching has taken place to form a hard crust that will protect the part. On the other hand, if the part is large and roughly machined so that deformation in handling is not serious, premature destruction of the chute and convey or mechanism is likely to result.
4. The work tank should be accessible to facilities for maintenance and cleaning.
5. If removal of scale from the tank is a serious problem, the use of a controlled atmosphere furnace should be considered.
6. Adequate ventilation should be provided to protect workmen from fumes and to improve working conditions.

7. If the quenching medium is oil, protection against fire hazard should be provided.
8. Use of clean-out plates below the solution level of the work tank should be avoided, if possible. Gaskets are difficult to maintain and will leak unless many studs are used on closely spaced centres.
9. Mechanisms for use in tanks for oil quenching may be of almost any material.
10. Special selection of materials for mechanisms is usually necessary if brine or caustic is used. A high cost of maintenance will usually result.
11. Alloy trays and fixtures of the 25Cr-12Ni or 35Ni-15Cr type can be quenched in oil quite satisfactorily, but not in water, caustic or brine.
12. If water, caustic or brine must be used, the use of special materials for pumps, coolers and other equipment should be considered.
13. Wherever possible, the design of the work tank should provide for flexibility and control of such conditions as: (i) time cycle of quench; (ii) volume of circulation; and (iii) uniformity of bath conditions.
14. It should not be expected that agitation will solve all problems. Excessive agitation that will blow light parts off the conveyor and thereby lose them, should not be employed. A medium that has a higher speed of quenching should be used instead.
15. A quenching bath should not be agitated so violently that foaming occurs on the bath surface. This means that air pockets are forming and that the maximum benefit of circulation is not being obtained.

FIXTURES

Fixtures are used to provide support or restraint, as discussed below:

Support Fixtures

Support fixtures are widely used to minimise distortion during quenching. They may vary in design from a simple tray or rack to complex compartmented baskets and special holders. Distortion of parts such as shafts is minimised if they are hung or supported vertically during heating and quenching.

Circular parts, such as large rings, may be supported on flat surfaces during heating and quenching. However, this practice may cause non-uniform hardness, because it restricts the flow of quenchant. A preferred practice is to support such members on fixtures that have radial ribs, machined or ground flat, to permit free flow of the quenching medium to all areas.

Distortion of thin-wall circular members is sometimes corrected by inserting pins inside the ring to force the small diameter outward prior to tempering. The pins are a part of a turn-buckle assembly that permits adjustment. To ensure acceptable roundness after tempering and removal of the pin assembly, an over-correction of about 50 per cent is usually required. However, corrective methods such as this are practical only for small-scale production.

Restraint Fixturing

Restraint fixturing is costly and is used primarliy for highly specialised applications. Notable examples of parts that require restraint during heating and quenching are rocket and missile casings, or other large components with thin wall sections. For such components, two or more external restraining bands may be used. An articulated internal fixture also may be required. It is built up of numerous cast or stamped pieces pinned together to provide many points of support and a free flow of quenchant to the inner surface of the workpiece. Articulated fixtures can accommodate workpieces of various diameters and lengths. The whole assembly is hung in the furnace and then lowered into the oil quench tank.

COLD DIE QUENCHING

Thin disks, long slender rods and other delicate parts that distort excessively when they are quenched in conventional liquid media, often can be quenched between cold dies with virtually no distortion.
Example 16.1. Large, thin thrust washers, which were blanked from cold rolled steel and contained machined oil grooves, developed considerable distortion as a result of blanking and machining stresses. A dimensional variations of 1.3 mm (0.050 inch) was common and even large variations sometimes occurred. Nevertheless, these washers had to be flat to within 0.13 mm (0.005 inch) after hardening.

To ensure the required flatness, the washers were squeezed between a pair of water-cooled die blocks immediately after they left the hardening furnace. The die blocks provided the necessary quenching action while maintaining flatness. When water-cooled beryllium copper die blocks were used in this application, the rate of cooling approached that obtained in water quenching.

Cold die quenching is limited to parts with a large surface area and a relatively small mass, such as washers, rods of small diameter and thin blades.

PRESS QUENCHING

Commercial quenching presses are designed of controlled quenching of ring gears and other round, flat or cylindrical parts, to permit heat treating with minimum distortion.

COOLING SYSTEMS

As quenching proceeds in a given medium, heat is removed from the workpieces and unless a cooling system is provided, the temperature of the quenching medium rises. Uniform results in the hardening operation depend in part on the control of quenchant temperature. Accordingly, coolers (and sometimes heaters) are required to achieve the desired temperature control.

Many types of heat exchangers are used for cooling quenching liquids and each offers certain advantages. The rate of heat transfer through the cooling unit always depends on the velocity of the cooling medium passing through it. As the velocity of the cooling medium increases, the rate of heat transfer increases. Therefore, the efficiency of the cooling unit depends on an adequate velocity of flow, which in turn requires an adequate pumping facility.

The maintenance of cooling equipment is related to its design. Thus, if the cooling unit is capable of maintaining a uniform quenching oil temperature, the possibility of fouling is minimised. To facilitate maintenance, cooling equipment should always be designed for easy cleaning.

Shell-and-Tube Heat Exchanger

Shell-and-tube heat exchanger is a type of oil cooler that consists of a series of copper or steel tubes enclosed in a steel shell. The quenching liquid is circulated through either the tube section or the shell section, with the remaining section reserved for circulating the cooling liquid. City water or well water is often used as the coolant; with larger heat loads, an evaporative water-cooling unit may be employed to conserve water.

In general, disposable water is used when the total quenching load is less than 910 kg/hr. (2000 lb/hr.). However, the use of disposable water will depend on the water-consumption costs as compared to the initial and operating cost of the evaporative unit. Usually, the evaporative unit does not provide water at as low a temperature as can be expected from well water or the normal city source. Therefore, it normally requires a larger heat exchanger.

The shell-and-tube heat exchanger requires very little maintenance. If properly designed, the tubes may be removed for cleaning. The heat exchanger is flexible in its application and can easily be designed to comply with various specified conditions of velocity.

Maintenance Costs

Maintenance costs are related to the number of mechanical units involved. Pumps are essential for water circulation and in some type of towers, motor-driven fans, used for air circulation, are another maintenance item.

Selection of Cooling System

To select the optimum cooling system for the quenching circuit, the maximum demand or heat load must be determined from the following data:

1. Weight of steel quenched during a specified period.
2. Specific heat of steel.
3. Temperature of steel entering the quench.
4. Temperature of steel leaving the quench.
5. Temperature of quenching liquid entering the work tank.
6. Temperature of quenching liquid leaving the work tank.
7. Characteristics of the quenching medium such as specific heat, specific gravity, viscosity, flash point, boiling point and point of vapourisation of toxic solutions.

From the above data and with full knowledge of costs and acceptable metallurgical results, proper cooling equipment can be selected for the quenching system.

STORAGE OR SUPPLY TANKS

For uniform hardness and minimum distortion in batch quenching, experience has indicated that the rise in temperature of the quenchant should be maintained within as narrow a range as possible—preferably within 5°C (10°F). This requires a large volume of quench liquid to limit the instantaneous temperature rise in the quenching tank; this volume can be supplied by an auxiliary storage tank in the system. In addition to providing sufficient quenching liquid, the storage tank is an economical means of linking equipment such as pumps, coolers, heaters and filters and it provides a means for the continual re-use of quenching liquids.

Design

Design of the storage tank depends on individual requirements and conditions. Fundamentally, the tank should be of the closed type and sufficiently large to supply the necessary quench capacity for optimum temperature control when three-quarters full. The tank should be equipped with one or more manhole covers and numerous other opening on the top for drainage connections from the quench tanks. The ends of the tank should be provided with capped valve outlets for tapping the liquid supply. These outlets should conform in size with pipe outlets of normal size. A sight gage and a drain valve also should be attached to the storage tank.

There are advantages in having tanks positioned above the floor of a well-ventilated basement room that has the capacity to house pumps, coolers, filters or other equipment required in large quenching systems. This will permit gravity feed of the quench liquid to the pumps, eliminate any possibility of loss of pump prime and allow insulation of the storage tank to minimise heat loss where temperature control of the liquid is desirable. However, there also are advantages in using underground storage tanks (notably, saving of space).

HEATERS

Storage tanks containing water or brine may be heated economically by steam or hot water pipes or by radiant tubes inserted in the tanks. The optimum storage temperature range for water or brine is 13 to 25°C (55 to 75°F). When steam or hot water is used for heating quenching oils, there is always danger of contamination that will change the quenching characteristics of the oils, cause foaming and create danger of explosion.

When fuel-fired radiant tubes are used they should have the characteristic of burning deep in the tube, thus providing a cold leg to a distance well below the height of the quenching liquid. The products of combustion must be well vented. For oil storage tanks (or quenching tanks) the preferred temperature range is about 50 to 70°C (120 to 160°F), depending upon the type of quenching oil.

Immersion-type electrical heating elements also are used for heating the oil in either quenching or storage tanks. A choice between immersion electric heaters and radiant-tube burners is a matter of economics and convenience.

PUMPS

Primarily, pumps are installed for one or both for the two reasons:

1. It is necessary to maintain a specified flow of the quenching liquid through the heat exchanger; usually, this is a continuous flow and need be only enough to compensate for heat transfer.
2. Agitation is necessary in the quenching tanks; quite often, a second pump is installed for this purpose.

Centrifugal pumps are used where ever possible, because wearing is minimised and the initial cost usually is lower. However, where a storage tank is employed, it usually is necessary to raise the oil from the storage tank to the quenching tank.

Because a centrifugal pump is not self-priming, it is customary to install a small positive-displacement pump for priming the centrifugal pump when necessary. Because a continuous-duty pump is usually small, this is often a positive-displacement pump and is used both for continuous circulation when necessary and for priming the centrifugal pump used in agitation.

By the installation of valves, it is usually feasible to connect the pump in such a way that it can be used for emptying or filling the quenching tank or tanks. Valves also should be installed to simplify the repair of pumps. Centrifugal pumps are not designed for great suction-lift operation and therefore, should be installed as near the oil supply as possible. Consideration should be given also to the resistance to flow set up by the piping system, so that the pump itself may be properly sized. If resistance of the piping is considerably less than anticipated, the motor of the pump is likely to become overloaded.

MECHANICAL CONDITIONING OF QUENCHING OILS

Mechanical conditioning of quenching oils involves the removal of the following contaminants:

1. Scale.
2. Carbonaceous materials, which may be: (i) products of oil oxidation (sludge); and (ii) carbon fallout, encountered in protective-atmosphere installations.

3. Other insoluble solids, such as sand.
4. Water.
5. Soluble compounds, such as carbon dioxide.

These contaminants can be removed by filtering, evaporation or draining. The solids can best be removed by appropriate by-pass filters. The choice of filtering medium for removing solids is important. The most commonly used filtering media are the waste-pack, mineral-wool and cellulose types that must be replaced after their filtering ability has been exhausted. Clay filtering media are more expensive than the above types but can be re-used after exhaustion, by suitable regeneration. However, the regeneration will not remove scale or sand. Clay media should be carefully selected when fast quenching oils are to be filtered, because it is possible to remove the additive along with the undesirable carbonaceous materials.

Magnetic filters or traps and strainers are useful in removing scale and other foreign materials. These types of filters can be easily cleaned and returned to service. They are especially helpful for preventing premature filter clogging and for protecting pumps.

Water can be removed by filtering or centrifuging, but these methods are expensive and are rarely used. Usually, bulk water is removed by draining and suspended water is removed by heating. Carbon dioxide is removed by heating.

MAINTENANCE OF QUENCHING INSTALLATIONS

Because quenching baths vary widely in design, shape, size and method of operation, it is not feasible to establish standard procedures for maintenance. However, typical schedules for maintaining large quenching installations are as follows:

Oil Quenching

Daily

1. Check oil level in quench tanks.
2. Check oil temperature.
3. Check oil filter pressure.
4. Check oil pumps and oil flow.

Weekly

1. Check quenching rate of oil in production system.
2. If oil filters are not included in oil system, check for solids in oil.
3. Check oil temperature controller and control setting.

Monthly

Drain quench tanks and remove sludge.

Semi-annually

1. Check heat exchanger coils, pipes and pumps.
2. Replace oil filters when necessary.
3. Check screens ahead of oil filter.
4. Check oil storage tank for sludge, water leaks and general condition.
5. Calibrate oil-temperature gauge.
6. Check oil for contamination.

Water Quenching

Daily

1. Check water temperature.
2. Check water pressure.
3. Check water circulation.

Weekly

1. Drain quench tanks and remove sludge.
2. If water is recirculated, make necessary chemical addition to prevent calcium compounds from building up in tubes.

Brine Quenching

Daily

1. Check brine temperature.
2. Check brine concentration and adjust as required.

Weekly

1. Drain tanks and remove sludge.
2. Check pumps and tank condition.
3. Check quenching fixtures for signs of deterioration.

QUENCHING OF INDUCTION HARDENED PARTS

A great many induction hardening applications employ water as the quenching medium. Other media, such as conventional quenching oil, soluble oil, compressed air, water-base solutions of polyvinyl alcohol or brine, are occasionally used. Because water provides greatest ease in handling, minimum installation and maintenance of equipment and greatest safety, it is used unless metallurgical considerations indicate the necessity for one of the other quenching media.

QUENCHING OF FLAME HARDENED PARTS

The proper application of a suitable quench in flame hardening is as important as proper heating. The quench must remove heat at a rate that will produce the desired structure and assist in controlling the depth of the hardened case.

SAFETY PRECAUTIONS

Fire hazards always attend the use of quenching oils. The causes of quenching-oil fires and the methods of extinguishing them should be thoroughly understood.

One of the most common causes of fires in open quenching tanks—and one of the which operating personnel should be made aware—is ignition of the oil surface from a partially submerged load. Some aids to the prevention of such fires are:

1. Careful design of hoists or conveyors used to lower the work load into the quench.
2. Periodic inspection of chains, sprockets and other components that may fail.
3. Having available an alternative power supply for use in the even that a power failure occurs during a quenching cycle.

The ignition of oil overflowing from a tank can produce a most damaging type of fire. When an oil quench tank is located adjacent to a furnace or any other source of ignition, special precautions are required, such as:

1. Providing inside overflow drains, to prevent overfilling and outside drains, to prevent the spread of burning oil.
2. Providing a system for detecting the presence of water in the oil (several detectors are commercially available). If the oil temperature approaches 120°C (250°F) with water present, foaming can result in overflow that exceeds the drain capacity.

A third hazard arises when oil is heated by the work load to a temperature above its flash point. A temperature less than 10°C (50°F) below the flash point is considered dangerous. To minimise the temperature rise, several methods may be employed, including the use of: (i) cooling coils in the tanks; (ii) external heat exchangers; (iii) more agitation; (iv) a larger tank; and (v) oil with a higher flash point.

Extinguishing Oil Fires

A planned programme for extinguishing quenching oil fires should include:

1. A quick method of extinguishing the fire without contaminating the oil, such as by smothering it with a tank cover or by the use of a carbon dioxide system.
2. An auxiliary method giving longer-lasting protection, such as foam, dry chemical, or draining the tank.
3. Periodic training of personnel in fire prevention and fire extinguishing.

Facilities to drain oil from the tank to a safe location constitute one method of preventing serious fires. Tank covers or lids can be used effectively to smother a fire in a small tank. These covers can be actuated by heat or, from a safe distance, manually.

There are two general types of carbon dioxide systems: (i) high-pressure systems, in which the gas is stored at room temperature; and (ii) low-pressure systems, in which the gas is stored refrigerated to –20°C (0°F).

The primary effectiveness of carbon dioxide lies in its ability to reduce the supply of oxygen at the surface of the oil to the point at which it cannot support combustion. In some systems, carbon dioxide has a cooling effect as it sublimes from a solid 'snow' to a gas.

The advantages of carbon dioxide are that it does not contaminate the quenching oil or require clean-up. Disadvantages are the short duration of the protection afforded and the storage costs for a large installation.

Fire-fighting foam is a mass of fine, heat-resisting bubbles. It smothers an oil fire by floating over the oil surface and setting up into a stiff, long-lasting blanket. There are two general types of foam: chemical and mechanical (or air); both are equally effective.

An advantage of foam is its long-lasting protection. When a quench load has only partially submerged, or when the fire has heated surrounding metal, the protection must last until these sources of re-ignition are eliminated. Some disadvantages of foam are the clean-up problem after use and that it can cause oil over 120°C (250°F) to boil or foam unless the extinguisher is discharged several feet above the tank surface. The dry chemical extinguisher uses mainly sodium bicarbonate discharged through heads with high-pressure nitrogen. The extinguishing action, advantages and disadvantages are similar to those of the foam type. Dry powders have been known to contaminate quenching oil.

SECTION III

Chemical Heat Treatment of Steel

Chapter 17

Basic Concepts of Chemical Heat Treatment of Steel

INTRODUCTION

The saturation of the surface of steel with a certain element by diffusion of this element from the surrounding medium at a high temperature is called chemical heat treatment or, more simply, case-hardening.

Diffusion in a crystalline solid is based on the fact that atoms can change their equilibrium positions within the crystal lattice.

Each temperature of a solid corresponds to a definite energy level associated with its thermal vibrations. Energy is not uniformly distributed among the individual atoms of a crystal lattice. The great majority of atoms possess a certain mean energy corresponding to the temperature to which the solid is heated. On the other hand, due to energy fluctuations, there are always sufficient number of atoms possessing either more or less energy than the mean value.

THEORY OF CHEMICAL HEAT TREATMENT

In chemical heat treatment, parts are usually placed in a medium rich in the desired element which penetrates the metal by diffusion.

With chemical heat treatment in a gaseous phase (the commonnest case), the treatment is based on the three elementary processes.

Dissociation

Dissociation takes place in the gaseous phase and consists in the decomposition of molecules and formation of active atoms of the diffusing element. For instance, molecules of carbon dioxide or ammonia can dissociate by the reactions:

$$2CO \rightarrow CO_2 + C$$

or

$$NH_3 \rightarrow 3H + N$$

and form active atoms of carbon and nitrogen which are capable of dissolving in the metal; the degree of decomposition of gas molecules is called the degree of dissociation and is measured in per cent.

Absorption

Absorption occurs at the gas-metal interface and consists in dissolution of free atoms in the surface. The process is possible only when the diffusing element B is soluble in the base metal A.

Diffusion

Diffusion is penetration of the saturating element into the depth of metal.

All these processes result in the formation of a diffusion layer in which the concentration of the diffusing element varies from the highest value in the surface to zero at a depth y (Fig. 17.1). Such is the case if the diffusing element forms a continuous series of solid solutions with the base metal. If the saturating element B and the base metal A form a system of alloys with limited solubility or with chemical compounds (Fig. 17.2a), the structure of the layer will be determined by the isothermal section through the constitutional diagram of that system at the temperature of diffusion saturation.

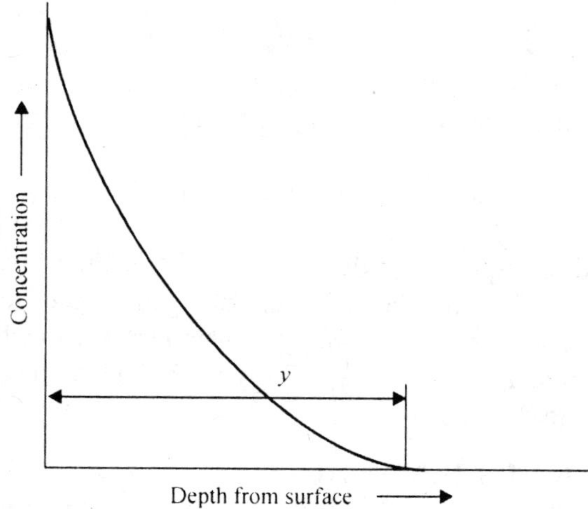

Fig. 17.1. Variation of the concentration of diffusing element in depth.

Suppose that the system (metal A and diffusing element B) is described by the constitutional diagram shown in Fig. 17.2 and that the saturation takes place at temperature t_1. If the processes of dissociation, absorption and diffusion have enough time to occur actively, a layer of the B (A) solid solution of a variable concentration will form in the surface (Fig. 17.2b); beneath it there will be a layer of the solid solution of A_nB_m compound, also of a variable concentration and finally, a layer of A (B) solid solution of a concentration diminishing from the saturation limit (at the given temperature) to zero. The concentration at the boundaries between the layers varies jump-wise in accordance with the conditions of equilibrium co-existence of the phases, as follows from the constitutional diagram in Fig. 17.2a.

If the process of saturation is not active enough, the concentration in the surface will not reach 100 per cent B.

Types of Diffusion Movements of Atoms in Metals

There are two types of diffusion movement of atoms in metals:

1. Self-diffusion, is the movement of atoms of the base metal in its own crystal lattice.
2. Heterodiffusion, is the movement of foreign (dissolved) atoms in the base metal crystal lattice.

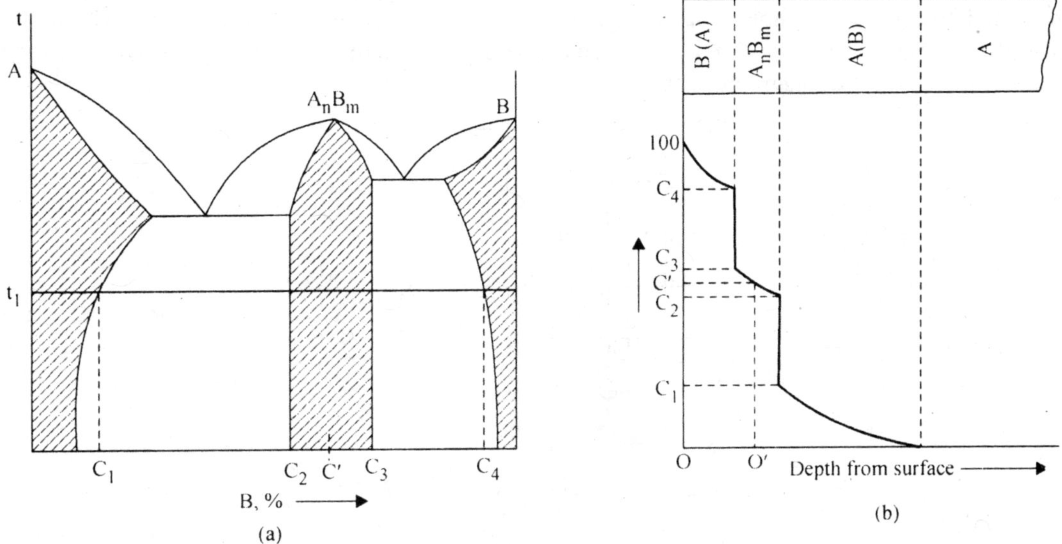

Fig. 17.2. A system of alloys with limited solubility and compounds. (a) Constitutional diagram (single-phase regions shown hatched); and (b) distribution of concentration of diffusing element and structure of diffusion layer.

In the former case, atoms of the base metal are in a chaotic thermal motion and some of them may change places in the crystal lattice by jumping from one site to another. This process of displacement of like atoms occurs continually and chaotically in direction and does not alter the concentration.

In the latter case, the movement of foreign atoms occurs in the direction from places of a high to those of a lower concentration; the process is spontaneous, since a state of a non-uniform concentration has a higher free energy than that having a uniform distribution of solute atoms.

If the concentration of the solution becomes uniform, heterodiffusion, like self-diffusion, is directed chaotically. Cases are known of what is called ascending diffusion in which the atoms move from places of a lower concentration to those of a higher concentration. In each case the process is caused by some factors (usually a non-uniform distribution of a third, slowly diffusing, element or a non-uniform distribution of stresses) and is based on the tendency of the system to diminish its free energy.

Diffusion Displacement of Atoms

As already discussed, all substances undergo thermal fluctuations owing to which some atoms can acquire an appreciably higher energy than the mean energy characteristic of the given temperature of the body. These atoms escape from their equilibrium positions in the lattice sites and move in interstices, thus leaving some lattice points empty.

An atom positioned in a lattice interstice is called a displaced atom (Fig. 17.3) and a crystal lattice site not occupied by an atom is called a vacancy. A vacancy or a displaced atom disturbs the regular arrangement of the adjacent atoms (Fig. 17.3) and thus a field of elastic distortions of the crystal lattice appears around the vacancy or displaced atom.

The hypothetical mechanism of heterodiffusion, which is analogous to the mechanism of self-diffusion is now generally recognised. If there is a vacancy (hole) near an atom A (Fig. 17.3), the latter can move easily from its place into that hole; an atom B will occupy the place of atom A, an atom C, the place of atom B, and so on, i.e., the hole displaces, as it were, together with displacing atoms. The process of heterodiffusion can be more properly described as the displacement of foreign atoms and the process of self-diffusion, as the displacement of 'holes'

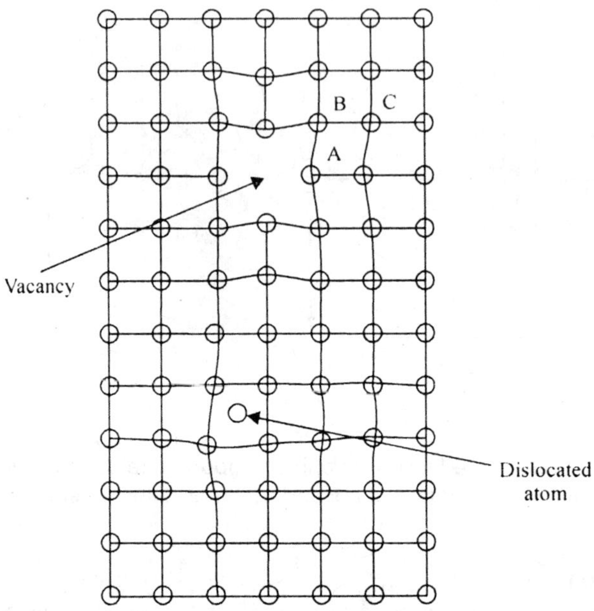

Fig. 17.3. Lattice interstice.

To move from the initial position to a neighbouring 'hole', atom A should first occupy an intermediate position in the interstice. The work required to displace an atom from its equilibrium position is called the activation energy, which is the most important characteristic of the capacity of atoms to move. The activation energy is determined by the nature of the substance and is independent of temperature.

Apart from the 'hole' mechanism, other diffusion processes are also possible. For instance, the motion of a displaced atom from one interstice to another (until it finds a 'hole' and is quieted) or an exchange of places between two neighbouring atoms. The 'hole' mechanism is effected most easily. For instance, calculations of self-diffusion in copper have given the following values of activation energy: 64 kcal/g-atom for the 'hole' mechanism, 230 kcal/g-atom for the motion of a displaced atom and 400 kcal/g-atom for the exchange mechanism. Because of these large differences, real diffusion occurs only by the 'hole' mechanism and the role of the other mechanisms is negligible.

The effect of temperature, which is known to appreciably accelerate the diffusion processes, can be explained by the fact that it increases thermal fluctuations and thus forms conditions for appearing more holes in a lattice.

The process of diffusion can be characterised quantitatively by what is called the coefficient of diffusion D which is numerically equal to the quantity of a substance that can diffuse through an area of 1 cm^2 for 1 second at a concentration gradient of unity at the two sides of the area (0 and 100 per cent of substance B); its dimension is cm^2/second.

The effect of temperature on the diffusion coefficient can be described by the formula:

$$D = Ae^{-Q/RT}$$

where, A is a coefficient depending on the type of crystal lattice; Q is the activation energy; e is the base of natural logarithms; R is the gas constant; and T is the absolute temperature. The value of constants for this equation for specific substances are given in Table 17.1.

Table 17.1. Values of the constants in the diffusion coefficient equation.

Matrix metal	*Diffusing element*	D_o, cm^2/sec	*Q, calories per gram-atom*
γ-iron	Carbon	0.04 + 0.08 (per cent C*)	31,400 ± 800
	Nitrogen	0.335×10^{-2}	34,660
	Aluminum	–	44,000
	Chromium	–	80,000
α-iron	Carbon	0.02	20,000
	Nitrogen	4.67×10^{-4}	17,950

* Per cent C is the carbon content of the austenite.

Since R and Q are independent of temperature, the diffusion coefficient of a given substance substantially increases with temperature (Fig. 17.4a); in the log D–1/T co-ordinates, this relationship is expressed by a straight line (Fig. 17.4b).

As follows from the formula above, the coefficient of diffusion of various substances depends quite appreciably on the activation energy, being substantially lower at greater values of Q.

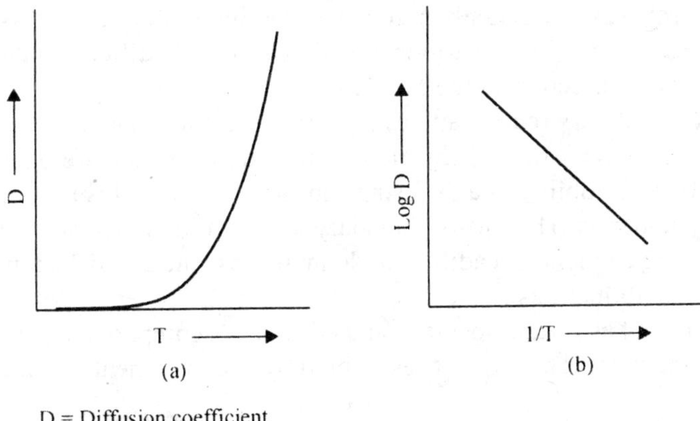

Fig. 17.4. Effect of temperature on diffusion coefficient.

In interstitial solid solutions, the process of diffusion is facilitated by that it is not needed to displace solvent atom (ion) into an irregular position, therefore, the activation energy is lower than in the formation of substitutional solid solutions. For instance, the activation energy in the diffusion of carbon in γ-iron is about 30 kcal/g-atom, compared to about 60 kcal/g-atom for the diffusion of some metals (substitutional solid solutions in γ-iron).

The coefficient of diffusion in these two cases may differ in magnitude thousands or even tens of thousands of times. Thus, for a steel with 0.2 per cent C at 1100°C, $D = 6 \times 10^{-7}$ for the diffusion of carbon and $D = 6 \times 10^{-11}$ for the diffusion of molybdenum.

The time of a process of chemical heat treatment is determined by the desired depth of diffusion layer. With all other parameters of the process (temperature, etc.) being the same, the depth of diffusion layer (y) varies with time (τ) by a parabolic law (Fig. 17.5).

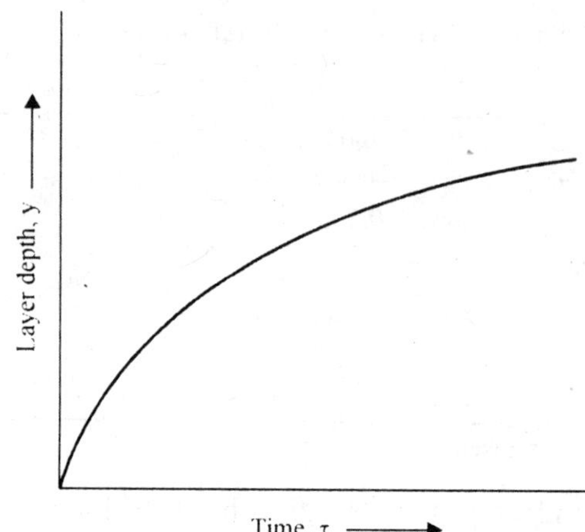

Fig. 17.5. Effect of process time on the depth of diffusion layer.

The thicker the layer already obtained, the less it grows in the same period of time. Some specific features of diffusion in metals can be explained by their crystal structure.

The anisotropy of properties of crystals may have a definite effect on diffusion. For instance, the diffusion of copper in hexagonal zinc in various directions occurs at different rates: more quickly in the base plane and slowly in the direction of the main axis.

In lattices of a larger symmetry (cubic lattices), the rate of diffusion is less dependent on orientation.

The diffusion processes occur more easily at grain boundaries which are places of concentration of crystal lattice defects. If the solubility of a diffusing substance in a metal is low, diffusion can often take place mostly at grain boundaries. The role of boundary layers of increased solubility is less appreciable in cases when the diffusing element is readily soluble in the base metal. Diffusion processes occur more quickly during phase transformations.

Chemical heat treatment has certain specifics and advantages compared to surface hardening, though is less productive than the latter. The advantages offered by chemical heat treatment may be formulated as follows:

1. The process is independent of the shape of article and produces a hardened layer of a uniform depth over the whole surface of both simple and complicated parts, whereas surface hardening is largely dependent on the shape of article and is inapplicable with many parts of intricate shape.
2. The process produces a greater difference between the properties of the metal in the surface and core as that obtained by surface hardening. Here the properties are determined by the structure and composition of the metal while in surface hardening, by the structure only.
3. The effects of surface overheating can be eliminated by subsequent heat treatment, whereas in surface hardening, which is usually the final procedure, it cannot be eliminated.

Chapter 18

Carburising of Steel

INTRODUCTION

The carburising process essentially consists in the saturation of the surface layers of steel with carbon, i.e., the formation of a high-carbon surface layer. Since the original metal is low in carbon, the core remains soft and tough, even after the carburised surface layer is subjected to hardening.

In accordance with the carbon source, carburisation is classified as: (i) pack carburising with solid carbonaceous mixtures (carburisers); (ii) gas carburising; and (iii) liquid carburising.

The main purpose of the carburising process is to obtain a hard and wear-resistant surface on machine parts by enrichment of the surface layer with carbon to a concentration from 0.75 to 1.2 per cent and subsequent quenching. Steel which has been carburised and quenched (case-hardened) has a higher fatigue limit.

Low carbon steels, containing from 0.1 to 0.18 per cent C, may be subjected to carburising. Steels with a higher carbon content (0.2 to 0.3 per cent C) may be employed for large components. Recently, there has been a tendency to use carburised steels, of higher carbon content for medium and small machine components as well. Steel with an increased carbon content (0.25 to 0.35 per cent C) has a stronger core. This enables the depth of the carburised case to be reduced thus simplifying subsequent heat treatment.

In some countries, low-carbon steels of various grades are extensively used as carburised steels. Alloy steels are used for highly stressed components, especially of large size, in applications when the core must possess high mechanical properties and the case must be highly wear resistant. Such alloy steels include chromium steels, chromium-manganese-titanium steels, chromium-nickel steels etc.,

The use of alloy steels, as mentioned above, reduces distortion in quenching since oil is used instead of water as a quenching medium. This enables complex machine parts to be successfully carburised and subsequently heat treated.

Machine parts are carburised directly after machining or they may have an allowance of 0.05 to 0.1 mm for finish grinding. Selective carburising is frequently applied. Here, only certain portions of the surface of the work are carburised. The remaining surface is protected against the carburising effects by a thin layer of copper (0.03–0.04 mm) applied by electrolytic plating, or by coating with special pastes. Sometimes, an allowance is provided and these already-carburised sections of the surface are machined off before hardening.

An anti-carburising paste called linite (talc and fine ground white clay mixed in water glass) is often employed. It must be noted that in drying and carburising, the paste may crack easily and so does not provide reliable protection. Copper plating is the most reliable procedure. Numerous industrial processes of chemical heat treatment are shown in Table 18.1.

Table 18.1. Types of chemical heat treatment.

Diffusion saturation with non-metals	*Diffusion saturation with metals*	*Diffusion removal of elements*
Carbourising (cementation)	Alitising	Dehydrogenation
Nitriding	Chromising	Deoxygenation
Cyaniding	Chrome-alitising	Decarbonisation
Nitrocementation	Zincing	Complex removal of impurities
Boriding	Copperising	
Silicification	Titanising	
Sulphidising	Berrylising	
Sulpho-cyaniding	Vanadising	
Oxygenising		

HARDENING TREATMENT

The hardening treatment may be carried out as follows:

1. Direct, i.e., quenching straight from the carburising medium.
2. Single quenching, i.e., heating and quenching the carburised parts after first allowing them to cool to room temperature from the carburising treatment.
3. Double quenching which usually consists of a direct quench and then a re-quench from a lower temperature. Double quenching may also imply that after carburisation the parts are allowed to cool to room temperature and are then subjected to their first quenching or normalising treatment from 850–900°C. The second quenching is carried out in the same way as in the single quenching, usually from a temperature range between 780°C and 820°C, which is the proper quenching range for the case.

Double quenching, when carried out according to the latter method is applied to steels that have not been fine-grain treated. The other methods, however, are applied mainly to fine-grained steels. The various methods of quenching and tempering will be described in more detail in further section of this chapter.

By the term depth of carburisation is meant either the distance below the surface to a definite *carbon concentration*, or the total depth of *carbon penetration*. In addition to time and temperature, the depth of carburisation depends on the carbon potential of the carburising medium and on the composition of the steel. The higher the carbon potential the higher the carbon concentration at the surface of the steel, when equilibrium has been established and the deeper the carburising depth.

The *carbon potential* may be defined as that carbon content which a specimen of a carbon-steel foil acquires when equilibrium conditions have been established between the carbon potential of the carburising medium and the carbon content of the foil. The dependence of the depth of carburisation on the carbon potential and carburising time is shown in Fig. 18.1.

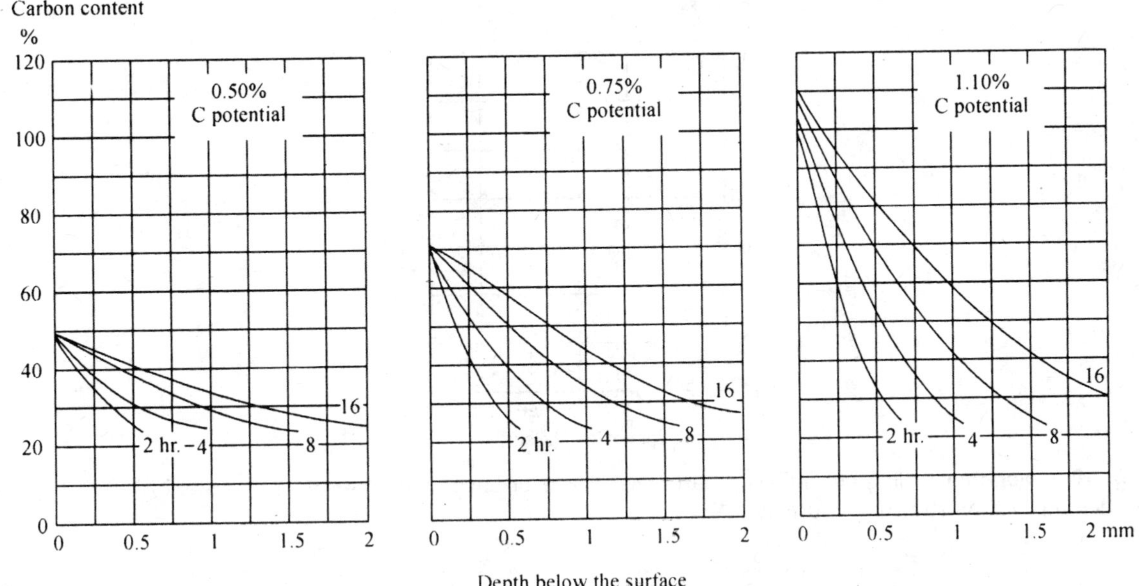

Fig. 18.1. Carbon profile of carburised case versus carbon potential and duration of carburisation for a plain-carbon steel carburised at 920°C.

A mathematical model for calculating the carburised concentration profile for carbon has been advanced by *Collin, Gunnarson* and *Thulin*. The highest carbon content that the austenite of plain carbon steel can hold may be deduced from the iron-carbon equilibrium diagram. If the carbon concentration is higher, carbides will have formed. This formation is facilitated if the steel contains carbide-forming elements or if it has a composition that favours carbide formation.

The depth of case hardening, DC, is defined, according to SS 11 70 08, as the distance from the surface to plane at which the hardness is 550 HV. This definition is also adopted by ISO (ISO 2639–1973). These Standards cover depth of hardening of 0.3 mm and more. For smaller depth the Standards SS 11 70 10 or ISO 4970–1979 apply. The depth of case hardening is determined by HV measurements, using a load of 1 kp, on a section through a case-hardened surface and at right angles to it. Figure 18.2 shows the hardness profile across a section of a case-hardened steel. With the aid of a magnifying lens the depth of case hardening, viewed at a magnification of about 10 times, may be approximately assessed on samples or on actual parts fractured after case hardening.

There exists a fair measure of agreement between the determined and the assessed depths of case hardening and the latter is widely used for control and check of the caurburising process. Should an exact measure of the depth of case hardening be required, hardness measurements must be resorted to.

Figure 18.3 has been constructed to serve as an aid to heat-treatment operators who handle steel parts of various grades and dimensions. The curves are based on a large number of case-hardening tests and are sufficiently accurate for practical carburising work when using a solid compound, a salt bath or a gas. The curves apply to alloy case-hardening steels in dimensions up to about 50 mm in thickness, oil quenched; also to unalloyed case-hardening steels in sections up to about 100 mm, water quenched. The diagram is in good agreement with documented data when the depth of case hardening has been determined in the same way as that described in ISO 2639–1973.

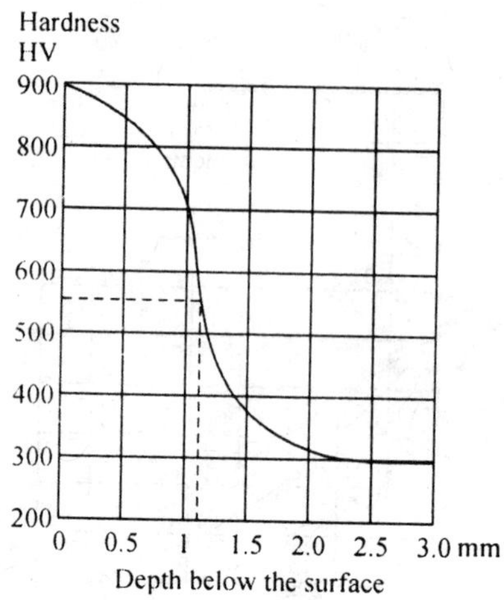

Fig. 18.2. Schematic diagram showing hardness profile of case in a case-hardened steel. Depth of case hardening. DC, as shown in figure, is 1.15 mm.

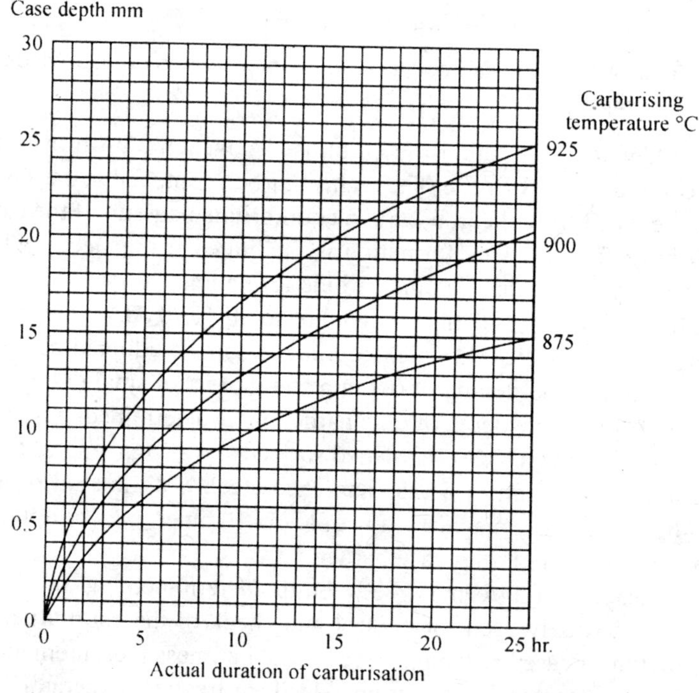

Fig. 18.3. Case depth versus duration and temperature of carburisation.

The validity of the simple diffusion equation $x = k\sqrt{t}$ has been checked against the diagrams and the following values have been obtained for the constant k when x = DC in mm and t = time in hours.

Temperature °C	875	900	925
Constant K	0.34	0.41	0.52

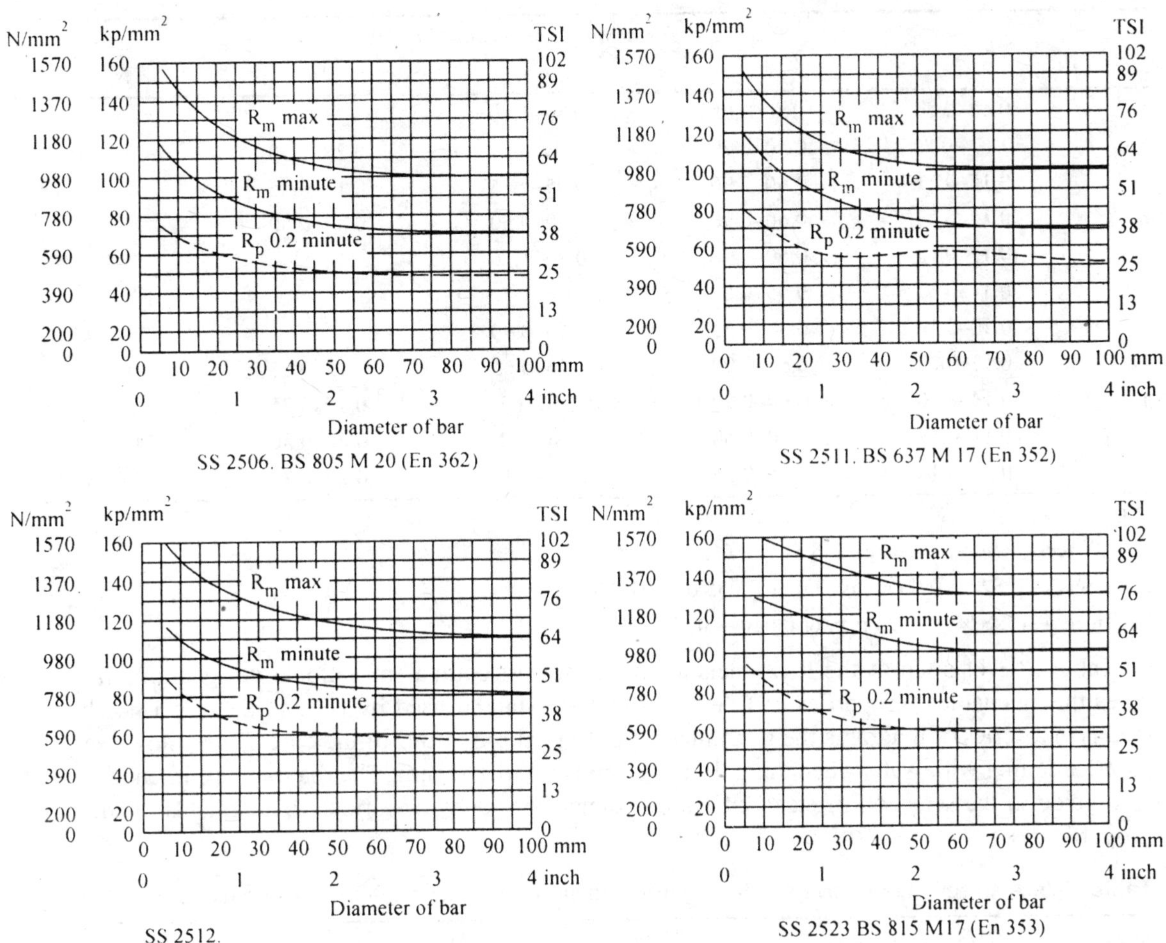

Fig. 18.4. Four standardised case-hardening steels showing relation between bar. Diameter and core strength after case-hardening with water quenching.

Grades of Steel

ISO has issued a recommendation, ISO R/683/XI-1970, covering 15 case-hardening steels, the composition of each being given in Table 18.2.

Table 18.2 Chemical composition of steels as covered by ISO/R683/XI-1970.

	Chemical composition%				
Steel no.	*C*	*Mn*	*Cr*	*Mo*	*Ni*
1	0.07–0.13	0.30–0.60	–	–	–
2	0.12–0.18	0.30–0.60	–	–	–
3	0.12–0.18	0.60–0.90	–	–	–
4	0.17–0.23	0.60–0.90	0.70–1.00	–	–

	Chemical composition%				
Steel no.	*C*	*Mn*	*Cr*	*Mo*	*Ni*
5	0.13–0.19	1.0–1.30	0.80–1.10	–	–
6	0.17–.0.23	0.40–0.70	–	0.20–0.30	1.60–2.00
7	0.15–0.21	0.60–0.90	0.85–1.15	0.15–0.25	–
8	0.17–0.23	0.60–0.90	0.30–0.50	0.40–0.50	–
9	0.12–0.18	0.60–0.90	0.80–1.10	–	1.3–1.7
10	0.11–0.17	0.35–0.65	1.4–1.7	–	1.3–1.7
11	0.10–0.16	0.35–0.65	0.60–0.90	–	2.75–3.25
12	0.17–0.23	0.60–0.90	0.35–0.65	0.15–0.25	0.40–0.70
13	0.14–0.20	0.60–0.90	0.80–1.1	0.15–0.25	1.2–1.6
14	0.11–0.17	0.30–0.60	0.80–1.1	0.20–0.30	3.0–3.5
15	0.12–0.18	0.25–0.55	1.1–1.4	0.20–0.30	3.8–4.3

N.B.

For all steels: P % = 0.035 max and Si % = 0.15–0.40.

For steels 1–3: S % = 0.035 max. 0.020–0.035 or 0.030–0.050.

For steels 4–15: S % = 0.035 max or 0.020–0.035.

Table 18.3 lists ten of the ISO steels and their nearest equivalent national designation. Table 18.4 gives the mechanical properties of five steels that are standardised by BS and SS. Fig 18.4 shows by means of a series of diagrams the guaranteed mechanical properties of the core of the alloy steels, after case-hardening with water quenching, in dimensions up to 100 mm. The diagram are intended to serve as a guide in the selection of steels of various dimensions and for various requirements of mechanical properties.

Table 18.3. A selection of equivalent case-hardening steels with their national designations.

ISO	*AISI*	*BS*	*En*	*DIN*	*NF*	*SS*
2	1015	–	–	C 15	(CC10)	–
3	1016	080M15	32C	–	XC18	1370
4	5120	527M20	–	(15 Cr 3)	–	–
5	(5115)	–	–	16 MnCr 5	16MC6	–
8	–	–	–	20 MoCr 4	–	–
(9)	–	637M17	352	–	16NCD6	2511
10	–	–	–	15 CrNi 6	–	–
11	–	655M13	36A	–	–	(2514)
12	8620	805M20	362	–	20NCD2	2506
(13)	–	815M17	353	–	18NCD6	2523

METHODS OF CARBURISING

Regardless of what method is used, carburisation always takes place via a gaseous phase. However, each method has its own intrinsic characteristics which produce differing case-hardening results.

Table 18.4. Guaranteed properties of case-hardened steels. The mechanical properties are valid for the test-bar dimensions recommended by ISO .The values refer to the material of the core.

			$R_{p0\cdot2}$ minute		R_m		A_5 %	Hardness	KCU
BS	SS	Dia mm	kp/mm^2	N/mm^2	kp/mm^2	N/mm^2	minimum	HB	kpm/cm^2
080M15	1370[1]	11	40	390	70–110	690–1080	9	205–330	5
		30	25	250	50–70	490–690	16	140–205	5
080M15	1370[2]	11	60	590	90–135	880–1320	7	270–405	5
		30	35	340	60–80	590–780	14	170–240	
805M20	2506[3]	11	65	640	100–140	980–1370	8	300–420	
		30	50	490	75–105	740–1030	11	220–315	6
		63	40	390	65–90	640–880	13	185–270	6
637M17	2511[3]	11	65	640	100–135	980–1320	8	300–405	
		30	50	490	75–110	740–1080	10	220–330	6
		63	45	440	70–100	690–980	11	200–300	6
–	2512[3]	11	75	740	110–150	1080–1470	8	330–450	
		30	55	540	85–125	830–1230	9	255–375	5
		63	50	490	75–105	740–1030	10	220–315	5
815M17	2523[3]	11	85	830	125–160	1230–1570	7	375–480	
		30	70	690	105–145	1030–1420	8	315–435	5
		63	60	590	90–125	880–1230	9	270–375	5

1. Water quenched from 780°C ± 10°C, tempered for 1 hr. at 180°C.
2. Water quenched from 890°C ± 10°C, tempered for 1 hr. at 180°C.
3. Oil quenched from 820°C±10°C, tempered for 1 hr. at 180°C.

Carburising by means of Solid Substances (Granular Compounds, Pack Carburising)

Carburisation with charcoal alone (without activating agents or energisers) can take place thanks to the presence of the atmospheric oxygen enclosed in the carburising box. The oxygen reacts with the charcoal and during the heating-up period a CO_2-rich mixture is produced which then continues to react with the charcoal as follows:

$$CO_2 + C \rightarrow 2CO \qquad \text{... (18.1)}$$

As the temperature rises the equilibrium is displaced to the right, i.e., the gas mixture becomes progressively richer in CO.

At the steel surface the CO breaks down as follows:

$$2CO \rightarrow CO_2 + C \qquad \text{... (18.2)}$$

The atomic or nascent carbon thus liberated is readily dissolved by the austenite phase of the steel and diffuses into the body of the steel, even to the extent of forming carbides (cementite). The CO_2 produced reacts again with the charcoal according to 18.1 above and the cycle of reactions is repeated.

Since the amount of atmospheric oxygen in a charcoal packing can vary and may be insufficient to produce the carburising gas it is current practice to mix the charcoal with an energiser, usually barium carbonate, which reacts during the heating-up period as follows:

$$BaCO_3 \rightarrow BaO + CO_2 \qquad \text{... (18.3)}$$

$$CO_2 + C \rightarrow 2CO \qquad \text{... (18.4)}$$

The CO_2 originally formed then reacts with the carbon in the charcoal, producing the active CO.

Since all the reactions mentioned above can proceed in both directions, i.e., they are reversible, the correct way of representing reaction (18.4) is

$$CO_2 + C \leftrightharpoons 2CO \qquad \text{... (18.5)}$$

If the temperature is increased but the pressure kept constant the reaction proceeds from left to right, i.e., more CO is produced. On lowering the temperature the proportion of CO_2 increases at the expense of the CO. Fig. 18.5 shows the equilibrium relationships existing between the constituents in the gaseous mixture expressed as percentages by volume, at a total pressure of one atmosphere. AT 900°C, which is a customary carburising temperature, equilibrium is at about 96 per cent CO and 4 per cent CO_2.

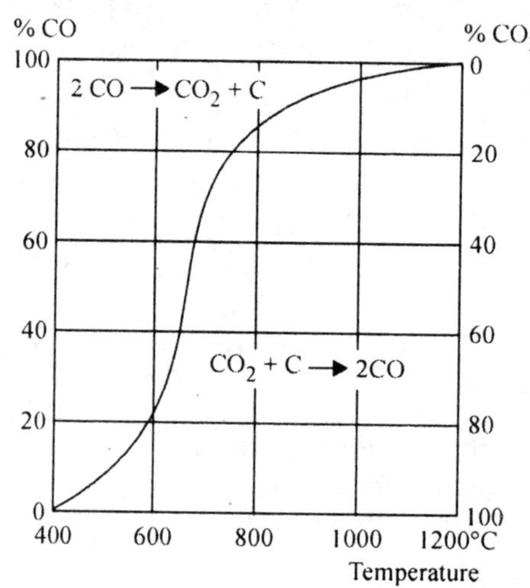

Fig. 18.5. Equilibrium diagram for reaction $2CO \rightleftarrows C + CO_2$ at pressure of one atmosphere.

This equilibrium curve is also included in Fig 18.6, which is part of the equilibrium diagram for the Fe–O–C system. Given that at 900°C the mixed gases are in equilibrium according to equation (18.5) the carbon concentration at the surface of the steel will adjust itself to some value between 0.70 per cent and 1.2 per cent, which is the equilibrium carbon concentration at the phase-field boundary A_{cm} at this temperature. If the CO_2 content falls to 2 per cent some Fe_3C may be precipitated; conversely, should for some reason the CO_2 content increase, owing to say, the admittance of air to the carburising compound, the carbon content of the steel will fall correspondingly. For example: at 900°C, 6 per cent CO_2 results in about 0.70 per cent C and 14 per cent CO_2 in about 0.30 per cent C.

When steel parts are being pack carburised in a carburising compound continuously or regularly, boxes should be made from heat-resisting steel, type 25 per cent Cr, 20 per cent Ni being suitable for this purpose. A box for occasional work may be made from ordinary mild-steel sheet.

After being cleaned the parts are packed into the boxes on the bottom of which is first laid a layer of the compound, about 25 mm thick. The purpose of this layer is to keep the parts from coming into contact with the bottom itself of the box since they might otherwise become deformed during the carburising process. When a layer of the parts has been placed in the box more compound is added to make a somewhat thinner layer, then more parts are packed and so on. When the top layer of the parts has been packed in the box there should be room at the top for a 50 mm layer of compound before the lid is placed in position. It is important to make the lid airtight with clay or similar material as the carburising process may be prevented by air ingress. During the heating-up period the clay dries and may crack and hence its sealing properties are not wholly reliable. A good way of preventing the ingress of air is to make a rimmed lid about 25 mm larger than the box. When the lid has been placed in position the box is turned upside-down and the space between it and the rim of the lid is sealed with refractory compound (Fig. 18.7). Round boxes made from heat-resisting steel sheet have adequate rigidity and there is generally no need to use any sealing material.

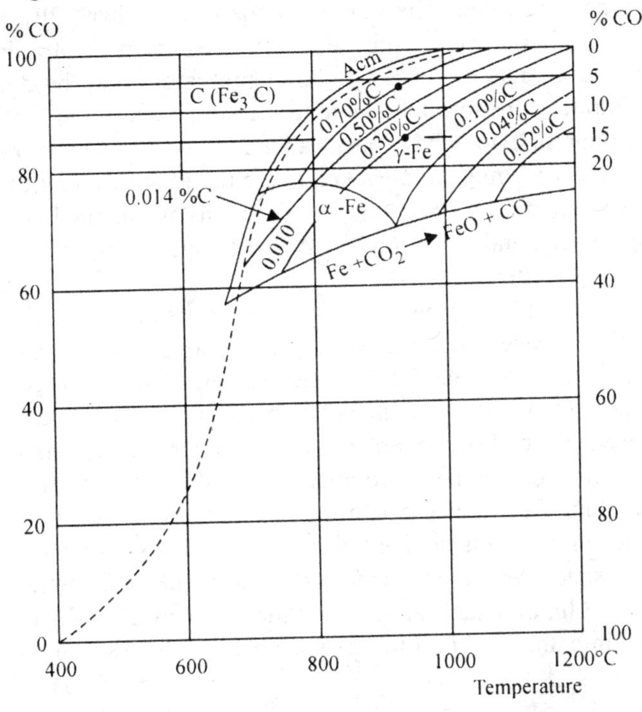

Fig. 18.6. Section of equilibrium diagram for Fe–O–C at pressure of one atmosphere.

Conventional carburising compounds contain between 6 and 20 per cent of energiser. This substance is usually evenly mixed with the charcoal and in principle, the carbon of the compound can be fully utilised. A batch of fresh carburising compound will shrink a good deal in volume after its first use, a reduction by as much as a fifth of its original volume being not unusual. Hence, before starting another treatment fresh compound is added to make up the original volume. In this way mixture is obtained which shrinks less and, more important, which shrinks by a known amount.

Carburising compounds have relatively poor heat conductivity. Consequently when large carburising boxes are used, workpieces lying in the middle of the box are carburised considerably less than those lying near the sides. The smaller the boxes used the more uniform is the carburising effect likely to be. The boxes themselves must not be placed so close together that heat radiation or the free circulation of the furnace atmosphere is impeded. It is good practice to make the boxes cylindrical whereby more space between them is automatically created. The service life of cylindrical boxes is also longer and the risk of cracking smaller. From the point of view of both production economics and uniformity of carburisation, optimum results are obtained if the diameter or width of the box is about 150 mm. Boxes having a diameter or width exceeding 300 mm should not be used unless the workpiece dictates this size.

The case depth is kept under control by inspecting test specimens or, if large series are concerned, by checking actual parts. The curves in Fig. 18.3 can serve as a guide when estimating the time necessary to obtain a stipulated case depth. Since the curves refer to holding times at the temperatures concerned, the heating-up time must be assessed. The dimensions of the boxes, the size of the charge and the heating capacity of the furnace are all important factors. A simple way of judging the temperature in the middle of the box is to insert a steel rod through a hole in the box and then pull it out when it is thought that the preset temperature has been reached. The colour of the red-hot bar will give an indication of the temperature inside the box. Alternatively, a tube closed at one end and protruding into the box may then be welded to one side and at right angles to it. The temperature may be determined by means of a thermocouple inserted into the tube through a hole in the box.

In order to assess heating time, tests have been carried out with boxes, 130 mm diameter × 320 mm. Specimens 10 mm diameter × 50 mm, were packed in the boxes at different distances from the wall. In one test they were placed immediately next to larger specimens as sketched in Fig. 18.8. Thermocouples were attached to the specimens and were connected to a temperature recorder. The furnace measured 378 × 500 × 800 mm and was rated at 24 kW.

During the heating-up period the temperature increased at the rate shown in the diagrams. Specimen No. 1 in box A reached the furnace temperature only slightly ahead of specimen No. 2. Specimen No. 3 required an additional 10 minute to reach 900°C. The specimens in box B required about twice as long to reach the pre-set temperature but they all reached it in about the same time. At first the temperature of specimen No. 6 increased more slowly than that of the other (see specimen No. 3). At the higher temperatures the rate of heating was faster compared with the other specimens. This was due to the radiation absorbed at the top of the box and also to the lower rate of heating of the larger specimens.

In the test just described it is estimated that the specimens in box A obtained about one hour's longer carburising time than the specimens in box B. Assuming a total furnace time of say, 5 hours at 900°C, this would imply 4 hr. and 3 hr. of actual carburising time, respectively, which, in turn, would result in a carburised depth of 0.85 mm and 0.70 mm for the specimens in boxes A and B respectively (according to Fig. 18.3).

Salt-bath Carburising

The active carburising agent in a salt bath is sodium cyanide (NaCN) or potassium cyanide (KCN). It is believed that the carburisation proceeds via a gaseous phase according to the following reaction:

$$2NaCN + O_2 \rightarrow 2\,NaCNO \quad \text{... (18.6)}$$

$$4NaCNO \rightarrow 2NaCN + Na_2CO_3 + CO + 2N \quad \text{... (18.7)}$$

$$3Fe + 2CO \rightarrow Fe_3C + CO_2 \quad \text{... (18.8)}$$

The first reaction takes place at the interface between the salt bath and the atmosphere; the other two reactions take place at the interface between the salt bath and the steel. Some of the nitrogen liberated by reaction (6.9) is also taken up by the steel.

The conventional cyanide baths that give a case depth up to 0.8 mm have a cyanide content (NaCN) of 40–50 per cent. When a fresh cyanide bath is being prepared it is customary to start with pure NaCN which is melted down in the furnace. NaCN decomposes quite rapidly at the beginning of the process whereby the cyanide content assumes the value stated above.

A salt bath containing about 20 per cent NaCN may be used to treat very thin parts requiring a case depth of only a few tenths of a millimetre. The parts are quenched direct into water after being carburised. Such a bath is prepared by melting together equal parts of sodium cyanide and anhydrous sodium carbonate.

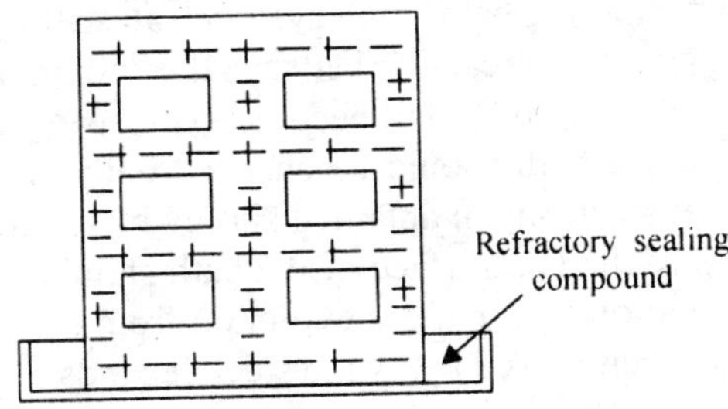

Fig. 18.7. Section through fully packed carburising box.

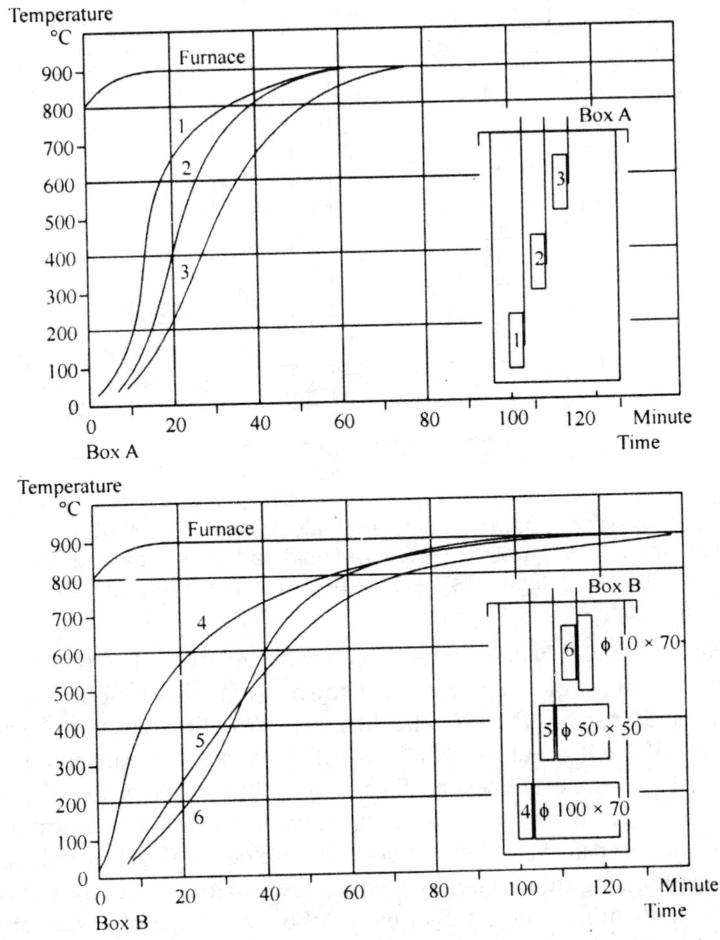

Fig. 18.8. Rate of temperature increase during heating-up period for boxes A and B and experimental carburising set-up in order to assess heating-up time at various distances from box walls. Box dimensions: 130 mm diameter × 320 mm.

To obtain case depths up to 1.5 mm, so-called activated baths are employed, the NaCN content of which is about 10 per cent. These baths are also used in conjunction with baths for the martempering of steel at about 200°C, which contain sodium nitrite and potassium nitrate. When the parts are transferred from the cyanide bath to the martempering bath some cyanide salt will be carried over. If the content of the bath is more than 10 per cent NaCN the reaction between the carbon in the cyanide salt being carried over and the oxygen in the martempering bath will become explosively violent.

The NaCN content of the cyanide bath should be checked daily and fresh salt added as required. Generally the NaCN content is satisfactorily maintained simply by replacing the volume of salt carried out with an equal amount of fresh salt. The surface of the bath should be kept covered with a layer of graphite fines which prevents radiation loss and gas evolution from the bath.

The amount of carbon and nitrogen picked up by the steel depends mainly on the cyanide content of the bath and on its temperature. Fig. 18.9 shows that the carbon concentration increases where as the nitrogen decreases as the cyanide content increases from 10 to 50 per cent.

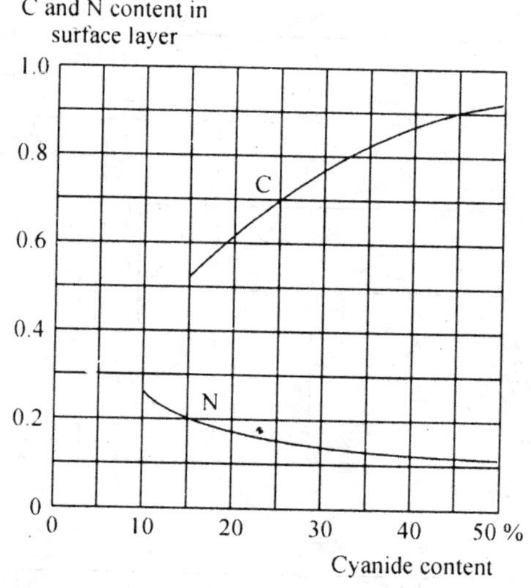

Fig. 18.9. Surface concentrations of carbon and nitrogen as functions of the NaCN content when carburising unalloyed case-hardening steel for 2.5 hr. at 950°C in a liquid carburising bath under a protective cover of graphite. The carbon and nitrogen estimations were carried out on a 0.075 mm thick surface layer on the specimen (after Waterfall).

With rising temperature, the NaCN content remaining constant at about 50 per cent, the concentration of carbon increases and that of nitrogen decreases as shown in Fig. 18.10. Both diagrams have common co-ordinates at 950°C and 50 per cent NaCN. The differences existing between the curves at this point are due to the somewhat different conditions under which the tests were carried out 37. Obtained from the same sources, Fig. 18.11 gives the hardness profiles of a mild steel, specimens of which were carburised in a cyanide bath at different temperatures, oil quenched, heated to 780°C and water quenched.

The diagram in Fig. 18.13 can also be used to assess the case depth. Leaflets and brochures describing the various liquid bath carburising salts generally give curves showing the total carburised depth but this information is usually of only minor interest. Salt-bath carburising is mainly used for small parts that in general require a case depth less than 0.5 mm. The smaller the case depth required, the greater is the economy achieved by using a salt bath since the rate of heating is very high compared with that in solid carburising media.

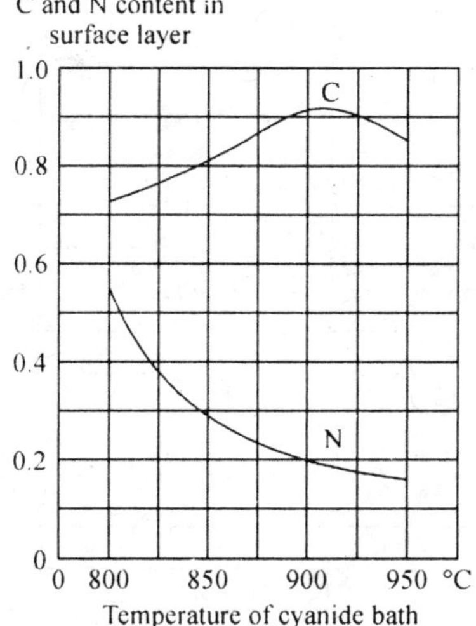

Fig. 18.10. Surface concentrations of carbon and nitrogen as functions of the carburising temperature when carburising unalloyed carbon steel for 2.5 hr. in a 50 per cent NaCN bath. The carbon and nitogen estimations were carried out on a 0.075 mm thick surface layer on the speimen (after Waterfall).

Before the parts are immersed in the salt bath they should be preheated to between 100°C and 400°C, partly to remove any traces of moisture and partly to utilise more efficiently the capacity of the salt bath. For further information on carburising salts the reader is recommended to contact the suppliers concerned.

It may happen than even when parts are being carburised in a cyanide bath the surface carbon concentration becomes too high, with the result that the quenched hardness may show somewhat low values on account of there being too much retained austenite present. The carbon content increases as the time of treatment increases and as the temperature rises. Therefore, when the parts are to be quenched direct from the cyanide bath a moderate carburising temperature should be chosen. The result of a trial hardening treatment, using BS 655M13 (En 36 A) is shown in Fig. 18.12. From these curves it can be seen how the thickness of the soft surface zone increases as the treatment time increases. The surface hardness of the specimens was also checked, using HRC and from this it could be seen that in this instance the case depth should be about 0.4 mm if the hardness is to exceed 60 HRC. By tempering the specimens at 180°C the surface hardness is reduced but the case-hardened depth remains largely the same (Figures 18.12 and 18.13).

If a case depth exceeding 0.5 mm is required it is more economical to use higher carburising temperatures, e.g., 900–925°C. When alloy steels are quenched direct the surface hardness is rather low and consequently they are usually rehardened by a requenching treatment from some lower temperature. If such double quenching is resorted to, the first quench takes place in oil or in a salt bath. If the parts are allowed to cool in air oxidation of the surface may occur.

Bungardt, *Brandis* and *Kroy* have studied the carburising process in a liquid salt bath at temperatures between 900°C and 1000°C in great detail and found that a normal NaCN salt bath will give a soft surface due to high content of retained austenite. There are, however, now available on the market salt baths that will build up a definite surface carbon concentration, say 0.80 per cent and a tolerable content of retained austenite.

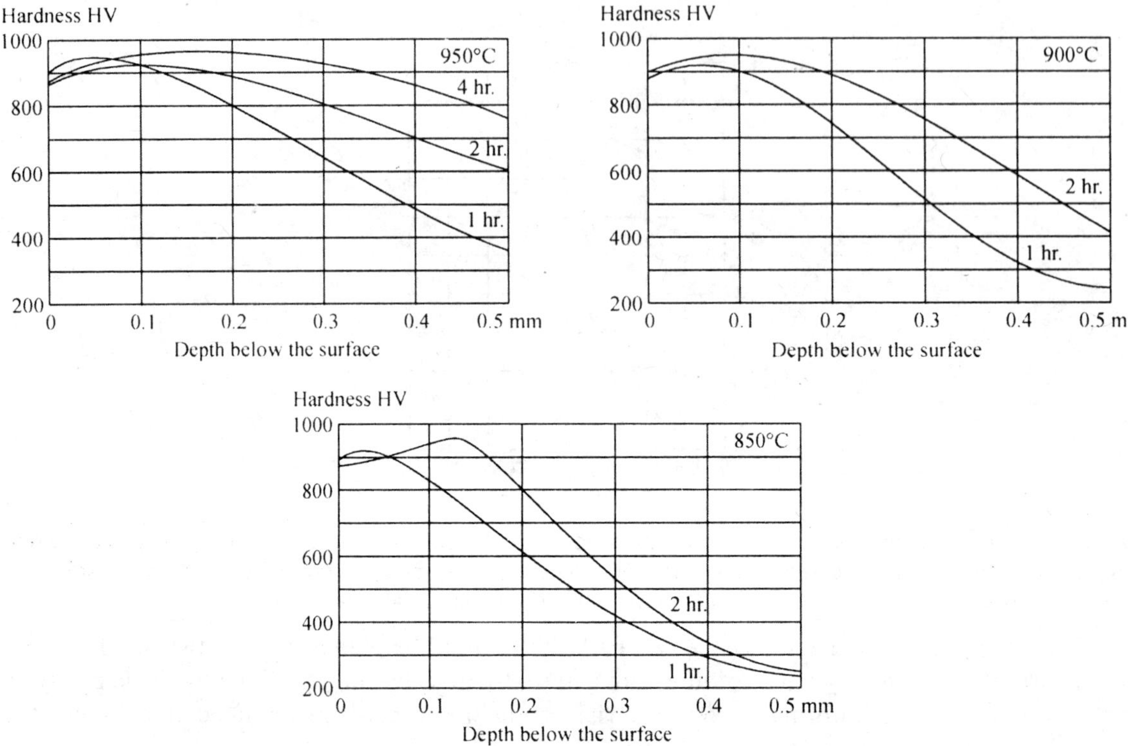

Fig. 18.11. Curves showing depth of hardening depth of hardening in mild-steel test bars, 14 mm diameter, carburised in a cyanide bath at various temperatures and oil quenched. Bars reheated to 780°C and water quenched (after Waterfall).

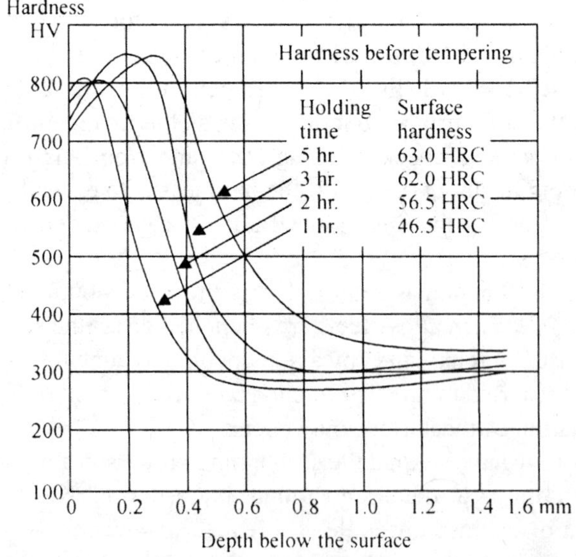

Fig. 18.12. Steel BS 655 M13 (En 36 A). Hardness as quenched after cyaniding at 870°C for various times. (NO tempering).

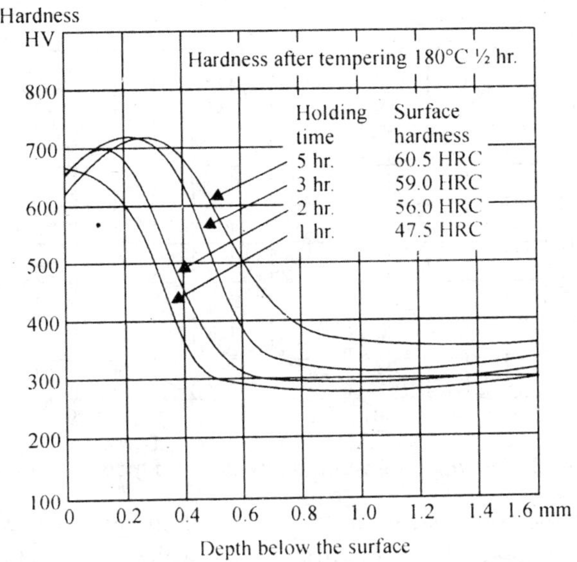

Fig. 18.13. Steel BS 655M13 (En 36 A). Hardness after cyanide hardening at 870°C for various times. Tempering for ½ hr. at 180°C.

N.B. Sodium cyanide is extremely poisonous if it gets into the human system. There are special instructions to be followed when work with cyanide salts is in progress and workers should make themselves familiar with these and the procedure to be adopted in case of accidents. Municipalities using bacteriological water purification, i.e., purifying sewage by means of bacteriological action, will require the installation of a detoxification unit since even very low cyanide concentrations will destroy the bacteria. Detoxification is usually carried out by 'neutralising' the salt by means of sodium hypochlorite. The disposal of waste salt must be rigorously controlled in conformity with official regulations.

There are now available on the market a number of cyanide-free carburising salt baths, the residues of which need not be detoxified.

Gas Carburising

During the last few decades carburising by means of gas has become the most popular method of case hardening. The furnaces are very dependable and relatively simple to operate. Since hydrocarbons enter into the process several reactions take place simultaneously. In addition to the reaction

$$2\,CO \rightleftarrows C + CO_2 \qquad \text{... (18.9)}$$

there is also the methane reaction

$$CH_4 \rightleftarrows C + H_2 \qquad \text{... (18.10)}$$

and the water-gas reaction

$$CO + H_2 \rightleftarrows C + H_2O \qquad \text{... (18.11)}$$

The composition of the furnace gas can be determined by analysis and hence the carbon potential may be estimated. The moisture content of the gas has a great influence on the carbon potential as indicated by reaction (18.11) and may be measured by simply determining the dew point of the gas. The dew point is that temperature at which water droplets (dew) are deposited from the gas (saturation). The relationship moisture content and dew point is shown in Fig. 18.14.

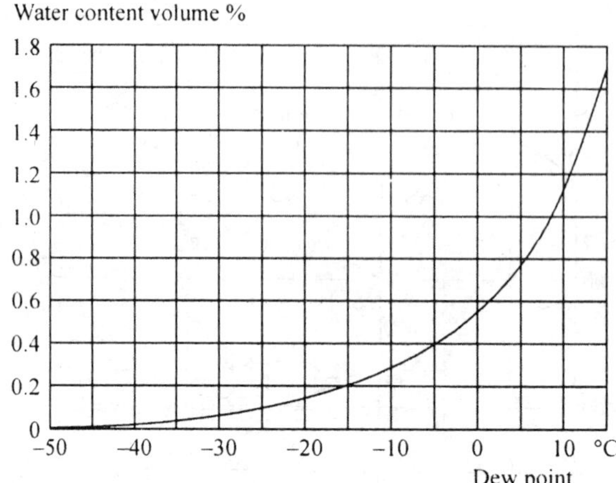

Fig. 18.14. Relation between dew point and moisture content.

There are a number of methods in existence for the continuous measurement and control of the carbon potential. One such method is based on infrared rays which are used to register the concentration of CO_2 in the gas. This value gives a measure of the CO concentration which, in turn, relates to the carbon potential. By means of a special measuring unit the resulting impulse can be converted direct to a numerical value of the carbon potential. The control may also be carried out by measuring the resistance of a steel wire which is calibrated for different carbon concentrations. The resistance measured is proportional to the carbon concentration. Indirect measurement of the carbon content can take place by means of an oxygen probe.

When gas carburising is being employed, a primary requirement is the availability of furnaces sufficiently gas tight to prevent air ingress, since this would interfere with the process.

In principle there are four different ways of producing the gas, viz:

1. Dripping carburising liquid into the furnace. This liquid generally contains ethyl alcohol and white spirit or only isopropyl alcohol. It drops on to a plate inside the furnace where it is dispersed and vapourised. By this method it is possible to employ a liquid that produces a carrier gas having a relatively low carbon potential. This potential can then be adjusted by dripping a more active liquid into the furnace. In practice, however, only one liquid is used and the carbon potential is adjusted by varying the volume of liquid added.
2. Admission of gas from a gas-producing unit. The gas is produced by endothermic combustion. The raw gas, usually propane, is mixed with air in carefully balanced proportions and burnt in the gas-producing unit to produce a carrier gas with a carbon potential of 0.35 to 0.50 per cent. Before it enters the furnace the carrier gas is given an extra addition of propane, there by adjusting the carbon potential to the required level which is usually around 0.8 per cent.
3. Direct admission of gas and air into the furnace. The gas actually doing the carburisation is produced in the furnace chamber by the reaction between the gas and air which are admitted in balanced amounts.
4. Carburising in vacuum furnace. Only hydrocarbons are admitted into the furnace from which air is evacuated. Since no oxygen takes part in the reaction there is no internal oxidation. The process can take place at a temperature higher than that normally used, thus allowing the carburisation to proceed more quickly.

INFLUENCE OF HEAT TREATMENT AND STEEL COMPOSITION ON CASE DEPTH, SURFACE HARDNESS, CORE HARDNESS AND MICROSTRUCTURE

As a result of the carburising treatment the steel has obtained a certain carbon content in the surface zone. On subjecting the steel to further heat treatment this zone can give it the required surface properties. Case hardening is, however, not merely a matter of surface hardness and depth of penetration. Stipulated requirements may call for a certain correlation between the hardness of the case and the core.

Depth of Hardness Penetration (Case Depth)

In the first instance the depth of hardness penetration depends on the carbon content of the carburised layer. Provided that only martensite is formed in the case on quenching, the depth of case hardening would be equivalent to a depth of carbon penetration down to 0.40 per cent C. This would agree well for small parts but as the section dimensions increase, the rate of cooling decreases and hence the conditions necessary for the formation of martensite are changed. This implies that steel components of different sizes but which have the same depth of carbon penetration obtain a case-hardening depth that is dependent on the dimensions of the piece.

The depth of case hardening is also dependent on the quenching medium. Water quenching gives a greater depth of case hardening than oil quenching, particularly in heavy sections. The differences in case-hardening depths for different dimension are evened out if the quenching takes place in water.

Since the hardenability is influenced by the quenching temperature the latter is also a factor affecting the depth of case hardening.

The course of the carburising process may conveniently be followed by means of test bars of case-hardening steel which are taken out of the furnace and quenched in water. The bars are fractured and the fractured surface is inspected at a magnification of about × 10 to assess the depth of case hardening.

The fact is often lost sight of that the chemical composition of the steel and hence its hardenability, does exert a great influence on the case-hardened depth; also that with low-alloy case-hardening steels in heavy dimensions it is not possible as a rule to obtain large depths of case hardening on quenching in oil even if the carburising times have been very long. It should also be borne in mind that since the carbide-forming elements tend to bind the carbon, hardenability may be reduced.

Surface Hardness

As before, the carbon dissolved in the austenite is the decisive factor determining surface hardness. When the carbon concentration at the surface of conventional alloy case-hardening steels exceeds 0.70 per cent the M_s temperature falls steeply, the amount of retained austenite after quenching increases and the hardness decreases. If the surface layer of the steel contains the appropriate concentration of carbon for maximum hardness the quenching temperature is of minor importance to the hardness provided that the grain size is not altered. By varying the quenching temperature of an 'over-carburised' steel it is possible to control the amount of carbon going into solution and hence the amount of retained austenite which, in turn, affects the hardness.

When pack carburisation is used the carbon concentration in the case can be controlled to a certain extent by adjusting the activity of the carburising compound and carburising temperature. A low carbon activity in the compound and a low carburising temperature result in a low carbon content in the surface.

When carburising in gas it is possible to control the carbon content and keep it at the desired level, generally 0.7–0.8 per cent C. The quenching temperature is not so decisive a factor in this instance. For practical reasons the temperature in the furnace is lowered to about 830°C before quenching. At this low temperature carbide is formed and hence the amount of carbon dissolved in the austenite is somewhat less. In addition the lower quenching temperature reduces distortion.

It is difficult to decide how much retained austenite will produce the best result in each individual instance. For some machine components, such as gears, it has been found, in practice, that retained austenite in larger amounts than previously considered advisable can have beneficial effects.

The surface hardness is obviously influenced by the rate of cooling. When quenching into a martempering bath the hardness falls as the quenching bath temperature is raised.

It is sometimes difficult to obtain a high surface hardness in alloy case-hardening steels and a subzero treatment is thus commonly employed. Even unalloyed steels (which are water quenched) may contain appreciable amounts of retained austenite in the surface layers, particularly when the parts are quenched direct after a deep carburising treatment.

On account of the presence of oxygen or oxygen-containing compounds in most carburising furnace atmospheres it often happens that certain alloying elements in the surface of the steel are oxidised. This occurrence is called internal oxidation and as the oxygen proceeds to bind the alloying elements the hardenability tends to be reduced resulting in incomplete martensite formation in the outer zone. The oxygen diffuses into the steel, mainly along the grain boundaries and reacts with those elements that have a greater affinity for oxygen than iron has. Silicon, manganese and chromium are susceptible to oxidation where as molybdenum and nickel are not.

The degree of internal oxidation is dependent on the CO_2–content and on the time and temperature of the carburisation.

Internal oxidation gives rise to only a small hardness reduction in the surface zone, but, more importantly, it causes a reduction in the fatigue strength of the material.

When carrying out conventional case hardening it is not possible to avoid some internal oxidation [46]. Since this phenomenon is dependent on both time and temperature the degree of internal oxidation may be kept to a minimum by making the case depth as small as permissible.

Core Hardness

When a gear, for example, is being designed with regard to its flexural strength the mechanical properties of the material of the core form the basis for the calculations.

The increase in hardness just below the carburised layer, particularly noticeable in light sections, may give rise to machining difficulties when the material of the core is being machined after the hardening treatment. In such cases the core hardness can be controlled by adjusting the temperature of quenching.

FATIGUE STRENGTH OF CASE-HARDENED STEEL

By and large the increase in fatigue strength brought about by case hardening is proportional to the hardness increase in the surface zone. It is also dependent on the compressive stresses in this zone. The maximum fatigue strength value is obtained when the carbon is about 0.6 per cent. Most fatigue strength investigations of case-hardening steels show that optimum properties are obtained when the surface carbon content is 0.6–0.8 per cent and the surface hardness at least 700 HV.

Fatigue fractures that occur in unnotched highly polished test specimens do not start at the surface but often originate in the boundary area between the core and the hardened case. This phenomenon may be explained by means of the schematic stress diagram shown in Fig. 18.15 which represents a cross-section through the plane of the surface and the neutral axis. H represents the strength of the material of the core and of the hardened case. If OP represents the stress acting across the section the strength, H, of the steel is not exceeded. A larger stress, OP', exceeds the strength of the material in the outer zone of the core (see the black area) and hence a crack appears here. To avoid this a higher core strength H^1 or a larger case depth should be chosen. The fatigue crack initiated in the boundary area is quickly propagated through the body of the core thus inducing a progressively increasing stress in the case. Cracks begin to appear here and ultimately the test specimen fractures suddenly. From this we see that the strength of the core has a decisive influence on the fatigue limit owing to the fact that the cracks are usually initiated in the core, the fatigue limit of which is lower than that of the carburised case.

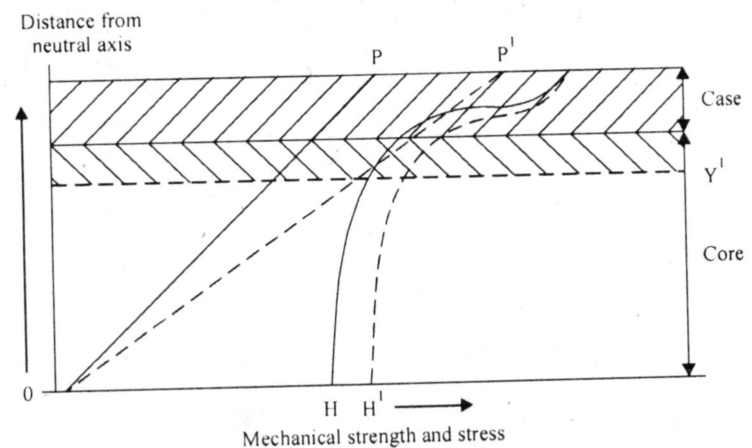

Fig. 18.15. Schematic stress diagram for section between surface and neutral axis.

There is, however, an optimum value of the core strength which, if it should be exceeded, will cause a reduction in the fatigue limit. In step with increasing contents of the alloying elements—carbon in particular—there is a successive lowering of the transformation temperature range of the austenite during its cooling from the hardening temperature of the steel. The strength of the core is thereby increased but the longer the time lapse before the transformation takes place-in the core the smaller will be the compression stresses in the case. This implies a lower fatigue limit and a greater notch sensitivity of the latter. For maximum resistance to fatigue fracture the optimum tensile strength of the core after case hardening is usually in the region of 1300 N/mm^2 corresponding to approximately 400 HV.

This point is illustrated in Fig 18.16, which shows the fatigue life of case-hardened gears with a gear modulus of 3, as a function of the core hardness.

RECOMMENDATIONS FOR CASE HARDENING

From practical experience gained in this field the following general heat-treatment recommendations are suggested for case-hardening steels. The treatment is designed to give a surface hardness of at least 60 HRC or 710 HV after quenching.

Carburising in Solid Compound (Pack Carburising)

Temperature: 900–925°C

Depth of case hardening mm	Method of hardening
< 0.50	Direct quench

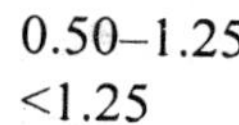

0.50–1.25	Single quench from 800 to 820°C
<1.25	Double quench

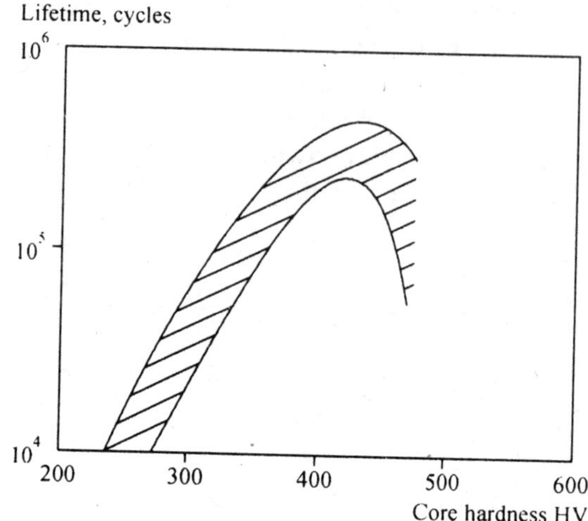

Fig. 18.16. Fatigue life of case-hardened gears, gear modulus 3, as a function of core hardness.

When direct quenching is employed a carburising temperature of 900°C is recommended. If a case-hardening depth of more than 1.25 mm is aimed at it might be good practice to perform the traditional practice of a double quench, i.e., a first quench from about 880°C and then a final quench. During the first heating for quenching, the carbon concentration is reduced by diffusion. In addition it breaks up and dissolves the carbide network that usually forms when pack carburising is used to give a deep carbon penetration.

Carburising in a Salt Bath (Liquid Bath Carburising)

Temperature: 850–900°C

Depth of case hardening mm	Method of hardening
<0.05	Direct quench
0.05–1.0	Requench from 800–820°C (Double quench)

For direct quenching, a carburising temperature of 850–870°C is recommended; for double quenching, 900°C may be used.

Carburising in Gas (Gas Carburising)

Temperature: 900–940°C

The carburisation is carried out to the required depth of case hardening and to a surface carbon concentration of 0.70–0.80 per cent .When the process is completed the temperature in the furnace is lowered to 830°C and the part is then quenched in a suitable medium.

Tempering

Case-hardened steels are tempered at temperatures generally around 160–200°C. Temperatures below 160°C should not be used, particularly if a grinding operation is to follow, since grinding cracks develop very easily. Tempering is not necessary after a cyaniding treatment that gives a case depth of only a few tenths of a millimetre.

The hardness falls quite rapidly when the steel is tempered between 160°C and 200°C. If a hardness of 60 HRC is required the tempering temperature should not be higher than 180°C.

CASE HARDENING OF TOOL STEELS

The wear resistance of tools made from certain medium-carbon steels may be increased by case hardening. This treatment is very often applied to simple punches, for instance, made from S 1. When such tools are being hardened they are heated in a cyanide bath at 900°C, i.e., the normal hardening temperature for the steel, for some 10–30 min longer than the usual holding time and then quenched in the customary way in water or oil.

Hot-work tool steels are case hardened for certain applications. First, the tools are carburised at about 900°C, generally in a solid compound and then allowed to cool in the compound. If the tool is to be put to hot-work service it is quenched from 1000°C. For cold-work applications it is better to use a lower quenching temperature, 980°C for example. This gives the steel a somewhat higher impact strength.

The heat treatment just discussed, involving low quenching temperatures, is applicable if the tool is to work at moderate temperatures. If high working temperatures are intended the tool, after carburisation, should be rehardened by quenching from 1050–1100°C.

PROTECTING AGAINST CARBURISATION (SELECTIVE CARBURISING)

For various reason it may be necessary to perform some machining operation after the carburising treatment. Hence restricted carburisation may be carried out according to some of the methods outlined below.

An Extra Machining Allowance

An extra machining allowance greater than the total depth of carbon penetration and usually about three times DC, is allowed to remain on those surfaces that are to be left soft. If the parts are pack carburised they should be allowed to cool slowly in the box. After this they are generally quite machinable. If the parts are too hard they may be tempered at 650°C. Should they have been quenched from the carburising temperature they must be tempered. After machining the parts are hardened in the usual way.

Protective Pastes

Protective pastes of 'NO CASE' type are applied to those surfaces that are to remain soft. These pastes are made up of a powder that contains mainly copper suspended in a varnish-type binder. The pastes are not suitable for use with gas or salt-bath carburising. Specially compounded pastes are available for gas carburising.

Electrolytic Copper Plating

Electrolytic copper plating to a thickness of 30–50 μm. It is important to ensure that the parts are very carefully cleaned before being copper plated and that the copper bath itself is clean. An alkaline bath gives a more compact and even coating than an acidic one. Any pores in the copper deposit may be detected by means of a reagent consisting of 10 parts $K_4Fe(CN)_6$, 10 parts of NaCI and 20 parts of gelatine dissolved in distilled water. A filter paper is moistened with the solution and placed on the copper deposit. If there are any pores blue spots will appear on the paper in 3–5 minute.

Those surfaces that are to be carburised can be laid bare in different ways:

1. By machining off the copper deposit. In this case machining allowance equal to the depth of cut must be planned beforehand. Great care must be exercised so that the copper coating is not damaged on those surface that are to be protected by it.
2. By locally coating the surface with an insulating varnish or by sticking on a special tape before copper plating.
3. By locally dissolving the copper deposit by chemical means. This method consists of painting the surfaces that are to remain soft with a special varnish (Bonosol B 40) after the copper plating is completed. A red pigment is added to the varnish for detection purpose. When the varnish has dried the copper deposit is dissolved from the unprotected areas by immersing the parts in a solution containing 5 liter water, 2.4 kg chromic acid and 125 ml concentrated sulphuric acid. Before commencing the carburising treatment the protective varnish is dissolved by exposing it to trichlorethylene vapour for 2 minute.

Electrolytic copper plating also serves to protect a surface against carburisation by a cyanide bath of the 10 per cent NaCN type. Higher concentrations of NaCN will cause the copper deposit to dissolve fairly quickly. Even a low concentration of cyanide will gradually dissolve copper. Regular carburisation of copper-plated parts in a cyanide bath should be avoided since the copper that is dissolved in the salt will cause the bath gradually to lose its carburising power.

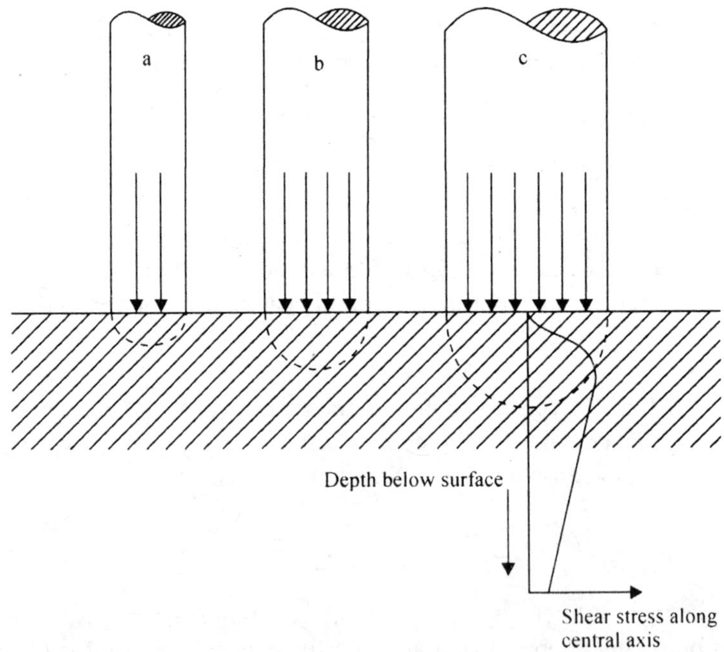

Fig. 18.17. Schematic sketch showing distribution of maximum shear stresses created by a force pressing on a surface. Largest value of shear stress exists at points under the surface along an arc of a semicircle that runs from the edges of the contact surfaces. The unit pressure (in N/mm^2) is the same in all three cases. In (c) are drawn the shear stress variations along the central axis.

CHOICE OF CASE-HARDENING DEPTH

When steel is case hardened it develops increased surface hardness and wear resistance, greater ability to withstand flexural and torsional forces and enhanced fatigue strength. The depth of case hardening is of decisive importance to several of these properties. If the surface of the steel is subjected to adhesive wear at low loads a case depth of some tenths of a millimetre is generally sufficient. As the load increases the case depth must also be increased.

The pressure applied to a case-hardened surface is propagated into the material. The most dangerous of the stresses thus created is considered to be the shear stress. In a section through a stressed surface the largest shear stresses act along a semicircle emanating from the edges of the surface to which the load is applied. If the same unit stress (N/mm^2) is applied, the distance down to the greatest shear stress increases as the area of the surface under load increases (see Fig. 18.17). If the shear stress directly below the hardened case exceeds the yield strength in shear of the material the latter will deform plastically and the case will have to support a greater load which may cause cracking. Hence the choice lies between, increasing the depth of case hardening or increasing the core strength of the steel.

The choice of case depth is often a matter of experience and is rarely the result of theoretical considerations. In actual practice, calculations are made complicated by the fact that the load is seldom evenly distributed across the surfaces of contact between two bodies. For case-hardened gears the design load is proportional to the gear module. The depth of case hardening is usually taken to be 0.15–0.18 times the module. Case depths of more than 2.0 mm are rarely encountered in gears.

Chapter 19

Cyaniding and Carbonitriding of Steel

INTRODUCTION

Cyaniding is a process in which both carbon and nitrogen are added to the surface layer of steel to increase its hardness and wear resistance. These methods are also effective for increasing the fatigue limit, particularly of medium and small size machine parts (gears, shafts, wrist pins, etc.). Cyaniding involves heating the steel in liquid or solid media; if the process is performed in a gas atmosphere it is called carbonitriding.

Cyaniding can be carried out in solid, liquid and gaseous media, and acccordingly, is classified into solid, liquid and gaseous cyaniding (the last process is often called carbonitriding).

Solid cyaniding is similar to carburising, with the carburiser containing additionally cyanide salts[1]. The process is appreciably less productive than liquid or gaseous cyaniding and cannot be recommended for wide application.

Liquid cyaniding which is made in molten cyanide salts is the most popular cyaniding process. The process is based on the decomposition of cyanide salts with the formation of free atoms of carbon and nitrogen which diffuse into the metal depth.

Gaseous cyaniding is effected in a mixture of carburising and nitriding gases (for instance, a mixture of lighting gas and ammonia).

The effect of cyaniding is determined by the depth of the difffusion layer and the concentration of carbon and nitrogen in the surface layer. The composition and properties of the cyanided layer substantially depend on the temperature of the process: at a higher temperature, the layer becomes richer in carbon and at a lower temperature, in nitrogen (Fig.19.1).

Since cyaniding is the simultaneous saturation of steel with carbon and nitrogen, as if a combined process of carburising and nitriding, we may say for simplicity that the process approaches carburising at higher temperatures and nitriding at lower temperatures.

This is why a distinction is made between high-temperature cyaniding (at 800–950°C) and low-temperature cyaniding (at 500–600°C).

[1]A typical carburiser for the process contains 30–40 per cent $K_4Fe(CN)_6$ (potassium ferrocyanide), 10 per cent Na_2CO_3 (soda) and charcoal is the balance.

LIQUID CYANIDING

In this process, the work is heated in molten salt baths containing various prussiates. During heating, carbon and nitrogen are introduced into the surface layer of the steel. As to techniques employed and results obtained, the liquid cyaniding of tool and structural steels may be classified into three chief groups.

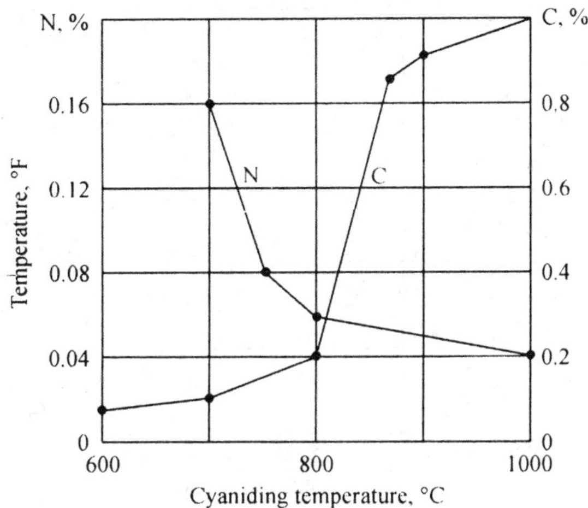

Fig. 19.1. Effect of cyaniding temperature on the content of carbon and nitrogen in surface layer.

Low-Temperature Cyaniding

Low-temperature cyaniding is employed to increase the life of high-speed steel tools. This process is conducted at 550–560°C in a bath containing 30 to 50 per cent NaCN, 20 to 45 per cent Na_2CO_3 and 10 to 20 per cent NaCl. The time required depends upon the depth of the case and is usually between 5 and 30 minutes. Cutting tools are cyanided after having undergone complete heat treatment, grinding and sharpening. No further heat treatment is required after cyaniding. A thin diffusion layer (0.02 to 0.04 mm) of high nitrogen content is produced by low-temperature cyaniding. This case is distinguished for its high hardness, wear resistance, and low coefficient of friction in metal cutting. These factors provide for high tool life in operation.

Medium-Temperature Cyaniding

Medium-temperature cyaniding is used to obtain a thin diffusion layer (0.075 to 0.25 mm) which is wear resistant under small specific loads. In this procedure, the steel is heated to 750–860°C in molten salts containing 20 to 35 per cent NaCN. Soda ash (Na_2CO_3) and sodium chloride (NaCl) are employed as the neutral salts. Cyaniding time is determined by the depth of hardened case required and varies from 5 to 90 minutes. The resulting case contains 0.6–0.8 per cent C and 0.4–0.5 per cent N. Cyaniding is followed by quenching direct from the salt bath and low-temperature tempering. The case has a hardness of R_c 56–60. This process is applied to bolts, screws, nuts, small gears, and other machine parts of low and medium-carbon steels.

High-Temperature (deep) Cyaniding

This process is accomplished in baths containing barium chloride, sodium chloride, potassium chloride and 6 to 10 per cent sodium cyanide (6 to 10 per cent NaCN, 80 to 84 per cent $BaCl_2$, and not over

10 per cent NaCl). The effect of cyaniding in a bath of the given composition, in the first place the depth of the layer, depends on the temperature and time of the process, as may be seen from Fig. 19.2.

The temperature and composition of the bath also determine the structure of the layer metal and its saturation with carbon and nitrogen.

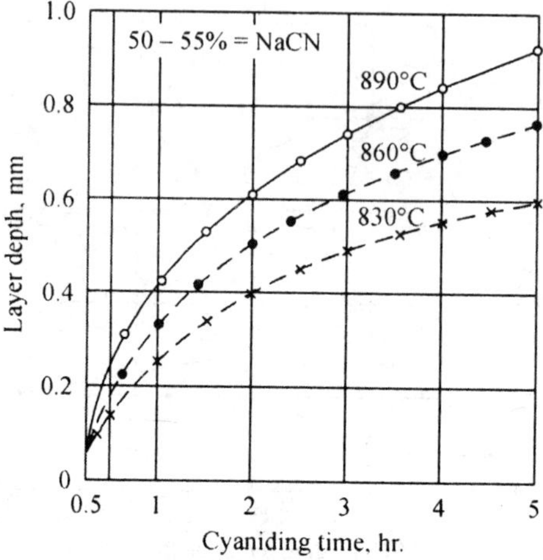

Fig. 19.2. Effect of temperature and time of cyaniding on the depth of cyanided layer.

Cyanided steel parts have a bright crust on the surface which is rich in nitrogen and consists of a carbonitride phase of the type of M_3C; then follows a layer of nitrous martensite (if cooling from the cyaniding temperature was made quickly). The presence of a bright surface crust is not always desirable, since it maker the cyanided layer excessively brittle.

As compared to carburised layers, cyanided layers have a higher wear resistance, higher hardness and higher corrosion resistance. Cyaniding also increases the fatigue strength, as any other case-hardening process resulting in the appearance of compressive residual stresses in the surface.

Since cyaniding is carried out at a lower temperature and for a shorter time than carburising, it causes no grain coarsening. Cyanided steel parts can be immediately hardened to form a high hardness in the surface. Cyanide salts are poisonous; this is an essential drawback of cyaniding. For that reason the process is carried out in special rooms and with strict observance of safety engineering rules.

Cyanided parts have an attractive mat surface and for that reason the cyaniding process is often used to improve the appearance of finished steel parts. To do this, parts are heated in a cyanide bath for a short time. The process is carried out at 900–960°C from one to six hours in accordance with the required depth of the cyanided case (Table 19.1).

Table 19.1. Relationship between case depth and cyaniding time.

Cyaniding time, hours	*1*	*2*	*3*	*4*	*5*
Depth of case, mm	0.5–0.6	0.8–0.9	1.0–1.1	1.2–1.3	1.4–1.5

Baths containing $BaCl_2$ must be covered with a layer of ash-free graphite to reduce the degree of depletion of the bath in cyanide salt and heat loss. To maintain normal conditions, the bath must be periodically refreshed by alternately adding $BaCl_2$ and NaCN in the proportion 4 : 1. Cyaniding is followed by quenching direct from the bath and low-temperature tempering.

High-temperature cyaniding will produce a case from 0.5 to 1.5 mm deep (Table 19.1) containing 1.0 to 1.2 per cent C and 0.2 to 0.3 per cent N.

The properties of a deep cyanided case are near to those of a carburised case.

Deep cyaniding is a relatively new process and possesses many advantages over solid and gas carburising. Principle advantages include the high speed of the process, the simultaneous saturation with both carbon and nitrogen which provide a higher wear resistance and the considerably less deformation of complex machine parts (gears, shafts, etc.):

Due to these features, deep cyaniding is being more and more extensively used in many plants.

The following main chemical reactions occur in the process of low and medium-temperature cyaniding.

The sodium cyanide in the bath reacts with the oxygen in the air and is oxidised according to the reaction:

$$2NaCN + O_2 \rightarrow 2NaCNO$$
$$2NaCNO + O_2 \rightarrow Na_2CO_3 + CO + 2N_{at}$$
$$2CO \rightarrow CO_2 + C_{at}$$

Carbon monoxide, formed by the decomposition of sodium cyanate (NaCNO), evolves atomic carbon, when it comes into contact with the iron. This atomic carbon is diffused in the iron. At the same time nitrogen, produced by the dissociation of sodium cyanate, is also introduced into the iron.

Somewhat different reactions occur in high-temperature liquid cyaniding. Here, the composition of the bath is selected so as to exclude the accumulation of a large amount of sodium carbonate which impedes the penetration of carbon into the steel. On the other hand, the bath should have a sufficiently high melting point since cyaniding is done at 900–950°C. Therefore, a large amount of barium chloride ($BaCl_2$) is included in the bath. This prevents the formation of Na_2CO_3 and activates the carburising of the steel.

In the presence of $BaCl_2$, the following reaction occurs:

$$2NaCN + BaCl_2 \rightarrow 2NaCl + Ba(CN)_2$$

The barium cyanide $Ba(CN)_2$ formed here is unstable and decomposes at high temperatures:

$$Ba(CN)_2 \rightarrow BaCN_2 + C_{at}$$

The carbon produced by this reaction diffuses into the iron. Thus, the gaseous phase CO is excluded in the production of atomic carbon. In the second stage of the reaction, barium cyanamide is oxidised to form CO and nitrogen:

$$BaCN_2 + O_2 \rightarrow BaO + CO + 2N_{at}$$

The different character of the reaction occurring in a high-temperature cyaniding bath leads to the penetration into the steel of mainly carbon and a limited amount of nitrogen.

PACK CYANIDING

In this procedure, the cyaniding is done in a mixture of 60 to 80 per cent charcoal and 40 to 20 per cent potassium ferrocyanide [$K_4Fe(CN)_6$]. Pack cyaniding is chiefly used at the present time to better the cutting properties of high-speed steel tools. The cyaniding temperature is from 540° to 560°C and the holding time is from 1.5 to 3 hours. Though pack cyaniding produces a very hard, wear-resistant case, it has many disadvantages. It takes more time than liquid cyaniding and does not ensure stable properties of the work. Various sections of the work may be cyanided selectively by oating them with certain pastes containing substances which ceasily evolve carbon and nitrogen.

Structures and Properties of the Cyanided Case

The cyanided case on structural steel contains both carbon and nitrogen as mentioned above. This case has a very high hardness, reaching 800–900 VHN. Its structure is similar to that of a carburised case but, at the surface, the high nitrogen content forms a thin (0.01–0.05 mm) unetched layer (ε-phase). The nitrogenous γ-phase may also be present, deeper in the case. After hardening, the case contains martensite in which carbon and nitrogen are dissolved.

In addition to increasing the hardness and wear resistance, cyaniding substantially increases the fatigue limit which is of great practical importance. The short time that the work is held in the cyaniding bath (15 to 30 minutes) enables the process to be combined with heating for hardening. Such heating in a cyaniding bath also excludes decarburising which sharply reduces the fatigue limit.

The principal disadvantage of cyaniding is the high cost and toxicity of the cyaniding salts and the necessity for corresponding measures to protect the personnel involved.

CARBONITRIDING

Carbonitriding is the treatment of parts made of various tool and structural steels in a gaseous mixture consisting of a carburising gas and 20 to 30 per cent ammonia (NH_3).

In carbonitriding, as in liquid cyaniding, both carbon and nitrogen simultaneously saturate the surface of the steel but the process is slower. Carbonitriding is classified as:

Low-Temperature Carbonitriding

Low-temperature carbonitriding which is carried out at 540–560°C. This procedure is applied for high-speed steel tools. The process requires from 30 minutes to 3 hours. For example, twist drills, reamers, and core drills are carbonitrided in from one to two hours. (depending on the diameter), taps in from 30 minutes to one hour milling cutters (plain, form, and end mills) in from one to two hours, etc.

This time is sufficient to produce a case from 0.02 to 0.04 mm deep possessing high hardness (950–1100 VHN) and wear resistance.

Medium-Temperature Carbonitriding

Medium-temperature carbonitriding is carriedout at a temperature of 840–860°C in an atmosphere of the carburising gas (natural gas, gas produced in the pyrolysis and cracking of kerosene, town gas, etc.) and ammonia in a 3:1 proportion. The time required may be determined from the following data:

Depth of case, mm	0.2–0.3	0.4–0.5	0.6–0.7	0.8–1.0
Carbonitriding time, hours	1.5–2.0	3.0–4.0	5.0–6.0	8.0–10

Carbonitriding is followed by quenching either direct from the furnace or after pre-cooling to 800–825°C or after re-heating. The final operation is tempering at 160–180°C. This type of carbonitriding is usually applied to complex machine parts that have a tendency to warp during heat treatment.

High-Temperature Carbonitriding

High-temperature carbonitriding is carried out at a temperature of 900–950°C. This procedure is used instead of gas carburising to save time and to prevent the deposition of soot and coke on the work. It also increases the wear resistance due to the introduction of nitrogen into the steel. The rate of gas carburising at 930°C is from 0.15 to 0.18 mm per hour. An addition of 15 to 25 per cent ammonia to the carburising gas enables the rate to be increased to 0.20–0.25 mm per hcur (to obtain a case 1.0 mm deep).

As in gas carburising, this carbonitriding procedure is followed by quenching and, finally, tempering at 160–180°C. Carbonitriding is, without doubt, a highly progressive method for the heat treatment of structural steels. It has a number of essential advantages as compared to gas carburising. These include the considerable acceleration of the process, the exceptionally high wear resistance of the case obtained, and the lower costs due, chiefly, to the lower processing temperature (medium-temperature carbonitriding). This lower temperature of the process increases the service life of the furnace and its accessories and also reduces the fuel consumption.

ABSORBTION OF CARBON AND NITROGEN

As discussed above during carbonitriding, C and N are absorbed simultaneously in the steel, the N increasing the hardness of the carburised layer. This process can be carried out in a salt bath or gas. The treatment temperature is normally 800–900°C, but both lower or higher temperatures can be employed. Basically, carbonitriding in a salt bath is the same as cyanide-bath hardening. In the section on case hardening we have already dicussed N pick-up during carburising in the cyanide bath. However, unless anything to the contrary is stated, by carbonitriding is meant carburising in gas with simultaneous N pick up. Carburising takes place in the same way as described previously. Nitrogen, in the form of gaseous ammonia, is supplied direct into the furnace or in the same feed pipe as is used by the carburising gases. The ammonia content is usually kept between 3 and 8 per cent. The normal value for the N content of the surface layer is about 0.4 per cent. After treatment the steel is hardened in the same way as after conventional gas carburising. In some cases ammonia is supplied only during the last 30–60 minutes of the process. The object of this procedure is to avoid the soft skin which can occur as a result of internal oxidation in the surface layer. As a result of such internal oxidation, the alloying elements are combined with oxygen and thus do not contribute to hardenability.

Theory of Carbon and Nitrogen Absorption

At the treatment temperatures normally applied the solubility of N in α-iron is approximately 0.1 per cent. Its rate of diffusion is roughly the same as that of C, as shown in Table 19.2. The diffusion of N in α- iron process ~50 times more quickly than it does in γ-iron. The corresponding figure for C is 40. However, the equilibrium diagram for iron–nitrogen resembles the iron–carbon diagram (Fig. 19.3) and at the temperatures used for carbonitriding, nitriding can be taken in principle as equivalent to carburising.

Table 19.2. The diffusion coefficients of nitrogen and carbon for diffusion in α-iron.

Temperature °C	*Diffusion coefficients $cm^2.s^{-1}$ in α-iron*	
	N	*C*
300	5.3×10^{-10}	4.3×10^{-10}
500	3.6×10^{-8}	4.1×10^{-8}
700	4.4×10^{-7}	6.1×10^{-7}
900 (extrapolated)	2.3×10^{-6}	3.6×10^{-6}

PRACTICAL RESULTS

Fischer and *Schaaber* studied the relationship between N content on the one hand and treatment temperature, the amount of ammonia supplied and the C content on the other. Fig 19.4 shows the effect of treatment temperature and the amount of ammonia supplied. The N content increases with increasing quantity of ammonia and decreases with increasing temperature. Figs. 19.5a and 19.6b show that the C content has very little influence on the N content with ammonia addition up to 5 per cent.

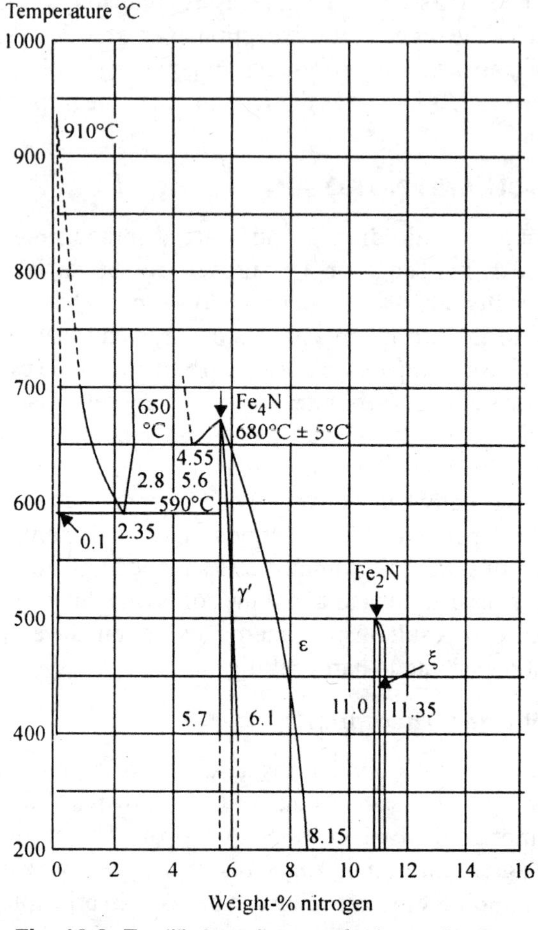

Fig. 19.3. Equilibrium diagram for iron–nitrogen.

The investigations were made on iron foils 0.05 mm in thickness. Parallel tests were also made using cylindrical test specimens made from plain-carbon and alloy steels, from which thin surface layers were machined off to enable the N content to be determined. In practically all cases the same N contents were found in the removed 0.1 mm outer layer, as were found in the foils. Fig 19.6 shows a typical distribution of C and N in the surface layer of a carbontirided mild carbon steel.

Prenosil made detailed studies of the effect of N on the hardenability of steel and establised that contents of around 0.20 per cent N have a favourable influence. This is illustrated in Fig. 19.7 which is a TTT diagram for a steel with 0.08 per cent C and 0.22 per cent N. For comparison purposes the transformation curves have been drawn for the same steel without N. Apart from a direct increase in hardenability, N results in a lowering of both A_1 and M_s. Contents higher than 0.20 per cent increase hardenability to a moderate extent, but lower A_1 and M_s considerably. The values in Table 19.3 were obtained on the same steel as above, with higher N contents. Since N lowers M_s substantially the amount of retained austenite in alloy steels can become high when carbonitriding.

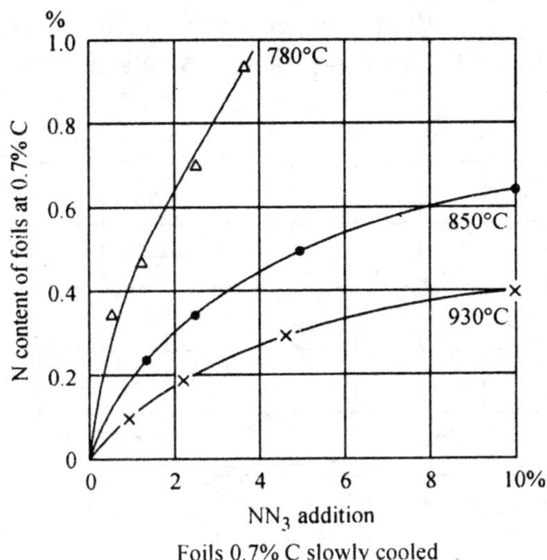

Fig. 19.4. Dependence of nitrogen potential at different temperatures on ammonia addition during carbonitriding.

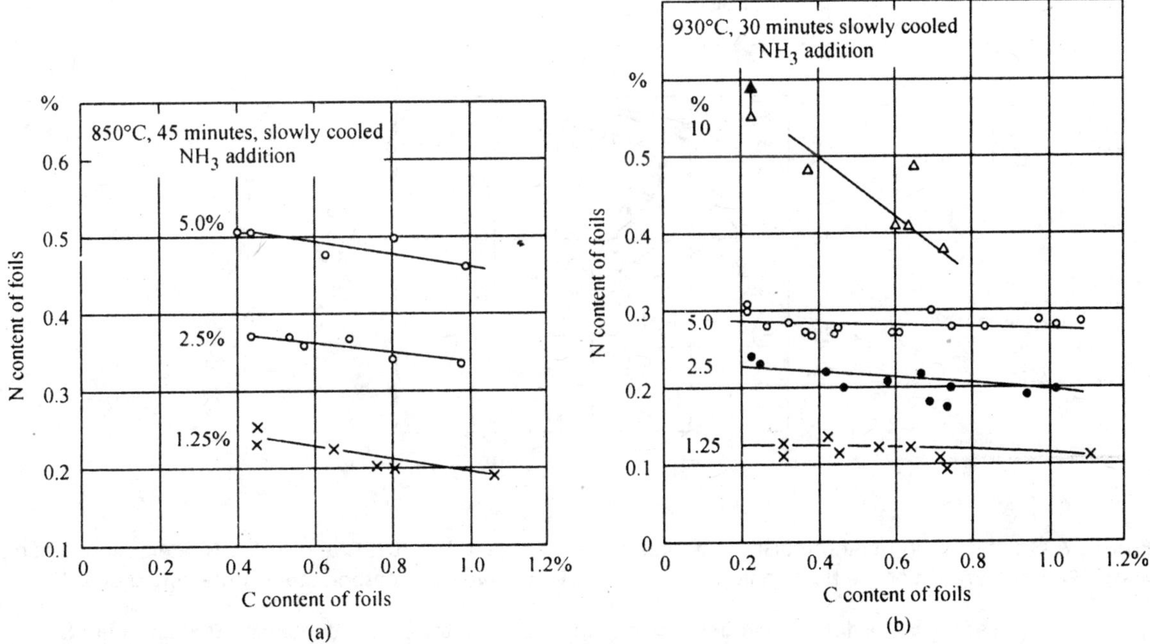

Fig. 19.5. Nitrogen content as a function of carbon content in foils which were carbonitrided with different ammonia additions at 850 and 930°C.

Tests have also been made using Jominy test bars which were carburised or carbonitrided using different N activities. The test bars were then austenitised at 820°C for 30 minutes and quenched in the conventional manner. The hardness was measured after the test bars had been ground down to depths between 0.1 and 0.5 mm below the surface. The contents of C and N in these layers were analysed. The results obtained are shown in Fig. 19.8. With N contents below 0.20 per cent there is hardly any tendency towards increased hardenability, but this tendency increases noticeably with higher N content.

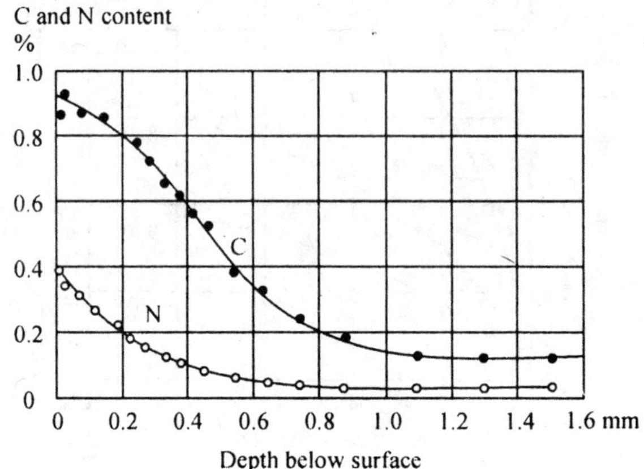

Fig. 19.6. Typical carbon and nitrogen concentraion in a mild carbon steel after carbonitriding at 860°C • = C, ∘ = N.

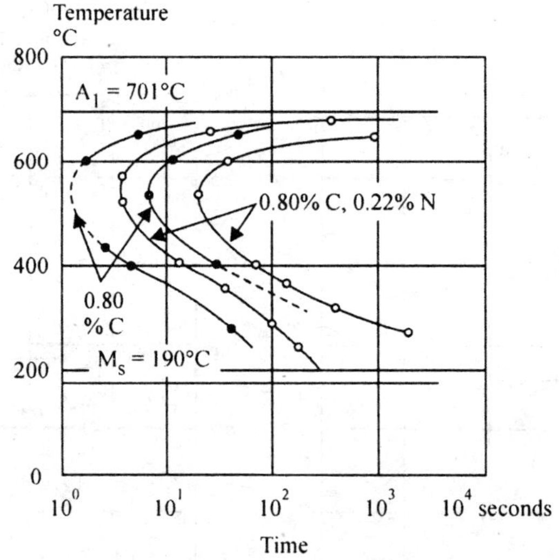

Fig. 19.7. Graph showing isothermal transformation of austenite with 0.80% C and 0.22% N after carbonitriding at 860°C. For comparison we have included transformation curves of a carbon steel containing 0.80% C.

Table 19.3. Effect of nitrogen content on A_1 and M_s for a steel containing 0.80 per cent C.

N%	A_l,°C	M_s°C
0.0	720	205
0.39	682	154
0.66	670	108

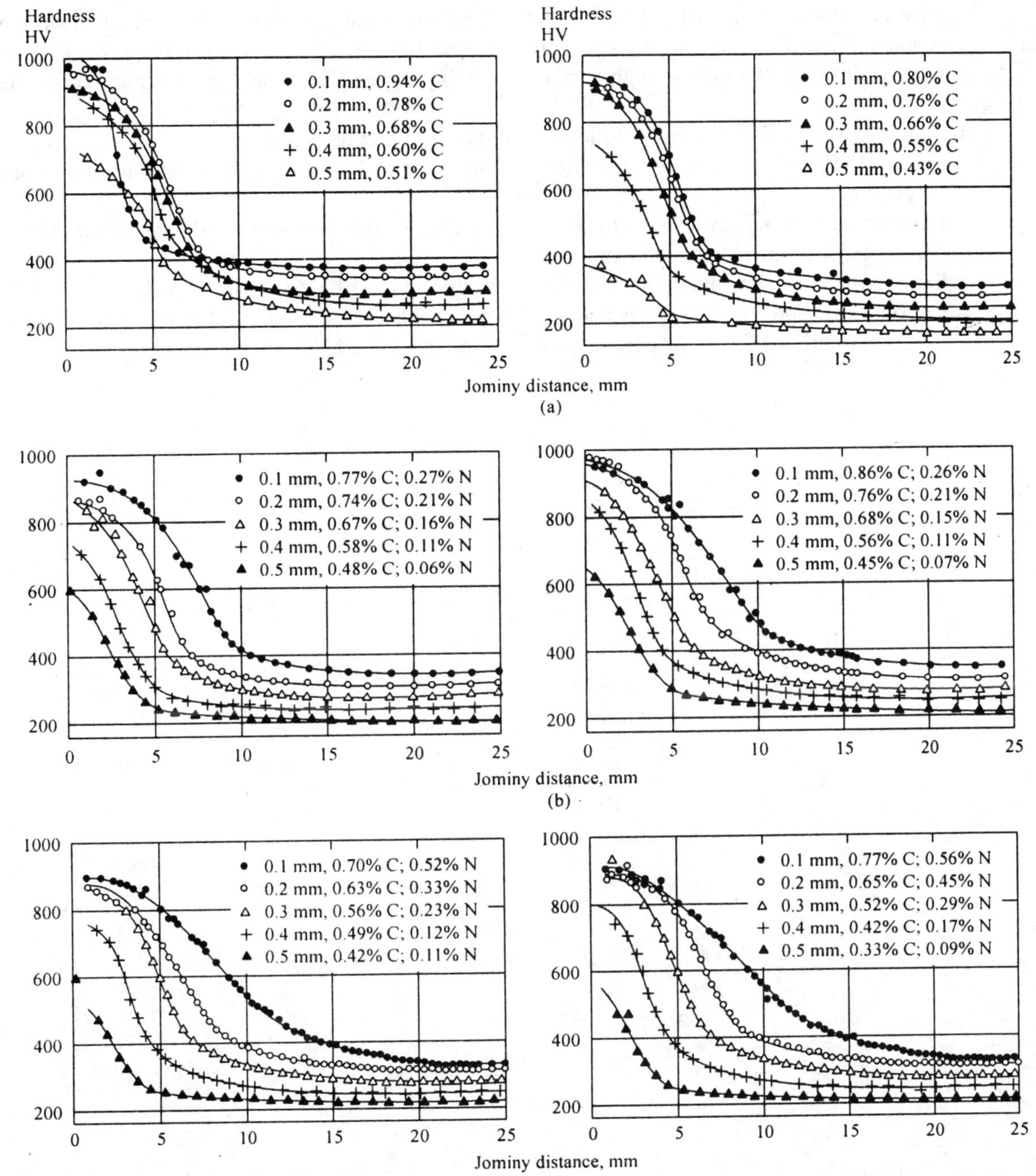

Fig. 19.8. Jominy diagrams. After either carburising or carbonitriding the Jominy test specimen is austenitised at 820°C for 30 minutes. Base material is 0.08% C; 0.19% Si; 0.40% Mn. The keys in the diagrams give the carbon and nitrogen content at different depths below the surface: (a) carbourised Jominy tests specimen; (b) carbonitrided Jominy test specimen (0.06–0.27 N); and (c) carbonitrided Jominy test specimen (0.09–0.56% N).

The increase in hardenability can be studied in the diagrams at Jominy distances around 10 mm. If the nitrogen content exceeds about 0.6 per cent there is a risk that pores may appear in the surface zone, which in turn will reduce fatigue strength very considerably, especially if the load tends to exceed the fatigue limit.

Thus from the above discussions the following conclusions can be drawn.

1. Carbonitriding gives increased hardenability and wear-resistance compared with ordinary case hardening.
2. Carbonitriding gives a hard and uniform layer or case to such plain-carbon steels as tend to have soft spots when quenched in water.
3. Plain-carbon and low-alloy steels which normally have to be quenched in water may, when carbonitrided, sometimes be quenched in oil. This is applicable to small parts only. The potentiality of nitrocarburising for larger parts must not be overestimated.

Chapter 20

Nitriding of Steel

INTRODUCTION

Nitriding is the process of saturating the surface of steel with nitrogen by holding it for a prolonged period at a temperature from 480° to 650°C in an atmosphere of ammonia (NH_3). Nitriding increases the hardness of the surface to a very high degree (high hardness is retained even when the steel is subsequently heated to 600–650°C). It also increases the wear resistance, fatigue limit and resistance to corrosion in such media as the atmosphere, water, steam, etc.

The carburising treatments as already dissussed are austenitic thermochemical treatments, in that they involve the diffusional addition of the interstitial alloying elements into the austenite phase and rely on the subsequent transformation of the austenite to martensite to produce a high surface hardness. The process of nitriding, however, is a ferritic thermochemical treatment and usually involves the introduction of atomic nitrogen into the ferrite phase in the temperature range 500–590°C and consequently no phase transformation occurs on cooling to room temperature. The method was first used at the end of the 1920s and since then its application has continuously spread, due among other things to the fact that the process has been further developed and can now be applied to a much larger number of steels than was originally thought possible.

The properties imparted to steel by nitriding can be summarised as follows:

1. High surface hardness and wear strength, together with reduced risk of galling.
2. High resistance to tempering and high-temperature hardness.
3. High fatigue strength and low fatigue notch sensitivity.
4. Improved corrosion resistance for non-stainless steels.
5. High dimensional stability.

These properties will be dealt with later in this chapter. The main factors which govern the depth of the nitride case are: treatment time, temperature, nitrogen activity and steel composition.

When studying the nitriding process we make use of the iron-nitrogen equilibrium diagram Fig. 20.1. At the nitriding temperatures customarily used, the nitrogen will dissolve in iron, but only up to a concentration of 0.1 per cent. When the nitrogen content exceeds this value, γ'-nitride is formed, the chemical formula of which is Fe_4N. If the nitrogen concentration exceeds about 6 per cent, the γ'-nitride starts to change into ε-nitride ($Fe_{2-3}N$). Below 500°C ζ-nitride begins to form. The nitrogen content of this phase is about 11 per cent and its chemical formula Fe_2N.

When making obsevations in the metallurgical microscope the γ' and ε-nitrides can be seen as a white surface layer, called the 'white layer' or 'compound layer'. Simultaneously with the increase in thickness of the white layer during nitriding, the nitrogen diffuses further into the steel. When the solubility limit is exceeded, nitrides are precipitated at the grain boundaries and along certain crystallographic planes.

Among the alloying elements, Al, Cr and Mo are used as nitride formers. Nitrogen together with carbon, forms carbonitrides. The amount of carbonitride can be increased by simultaneous pick-up of nitrogen and carbon by the steel.

The high surface hardness which is obtained after nitriding is due to the formation of finely dispersed nitrides and carbonitrides which distort the ferrite lattice. Basically, all steels can be nitrided.

DETERMINATION OF DEPTH OF NITRIDING

The depth of nitriding, designated DN, is determined in accordance with SS 11 70 09 by taking HV measurements on a cross-section through a nitrided surface. The limiting hardness is 400 HV. This figure is applicable only to constructional steel of the type included in Table 20.1. Other limiting values should be chosen for other steels. The Standard is applicable to nitriding depths of 0.3 mm and more. For depths smaller than 0.3 mm the Standard SS 11 70 10 applies, which is most nearly equivalent to ISO 4970-1979, *Steel-Determination of Total* or *Effective Thickness of Thin Surface-Hardened Layers*. However, this International Standard is based on somewhat different norms and specifies a method of measuring the total or effective thickness of thin surface-hardened layer having thicknesses less than or equal to 0.3 mm, obtained, for example, by mechanical (shot-blasting, shot-peening, etc.) or thermochemical (nitrocarburising, carburising and hardening, etc.) treatment. The two methods usually selected are the micrographic method and the microhardness measurement method.

The choice of the method and its accuracy depend on the nature of the thin layer and on its presumed thickness. Since the method used also affects the result obtained, the choice has to be made by prior agreement between parties concerned. Surface-hardened layers with thicknesses of more than 0.3 mm are covered by ISO 2639, Steel-Determination and *Verification of the Effective Depth of Carburised and Hardened Cases;* and ISO 3754, *Steel-Determination of Effective Depth of Hardening after Flame* or *Induction Hardening.*

The latter of the two methods consists in determining the changes in Vickers or Knoop microhardness under a 2.94 N (300 gf) load, measured from the circumference inwards to the centre of the product.

The total thickness of the thin surface layer is defined by the distance from the surface to the limit beyond which the hardness of the unaffected metal is reached.

The effective thickness is defined by the distance from the surface to the limit beyond which the required hardness (reference hardness) is reached.

The depth of nitriding can be assessed with the aide of a fractured surface in the same way as that employed for case-hardening steels. Fig. 20.1 shows the agreement obtained between the values assessed and determined at 400 HV for steel BS 905M39 (En 41 B).

METHODS OF NITRIDING

The four main methods of nitriding are gas nitriding, salt-bath nitriding, plasma nitriding and powder nitriding. There are variations within these main methods and the commonest variations will be described. Nitriding by means of ammonia gas, which involves the absorbtion of nitrogen only, is usually referred to simply as nitriding or gas nitriding. When besides nitrogen, carbon also is introduced the treatment is called nitrocarburising and this process may take place in a gaseous atmosphere or in a salt bath.

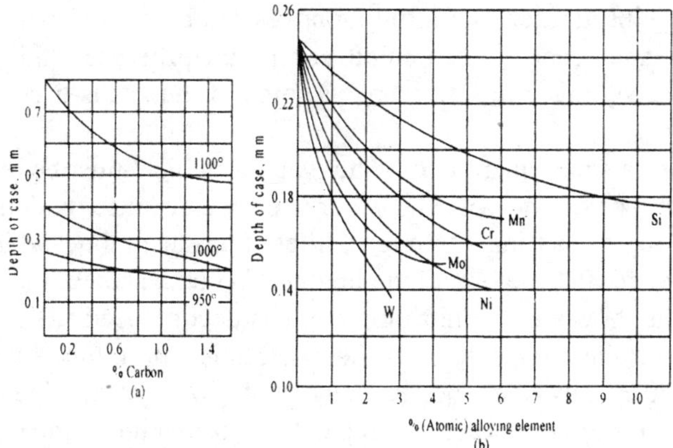

Fig. 20.1. Depth of nitriding for BS 905M39, En 41 B as function of nitriding period at 510°C.

Gas Nitirding with Ammonia

In the original method of gas nitriding, ammonia is allowed to flow over the parts to be hardened, normally at about 510°C. The ammonia dissociates in accordance with the equation:

$$2NH_3 \rightarrow 2N_{Fe} + 3H_2$$

At the instant of dissociation, nitrogen occurs in the atomic form and as such it can be absorbed by the steel. Fig. 20.2 gives a schematic representation of the dissociation and nitrogen pick-up in the steel.

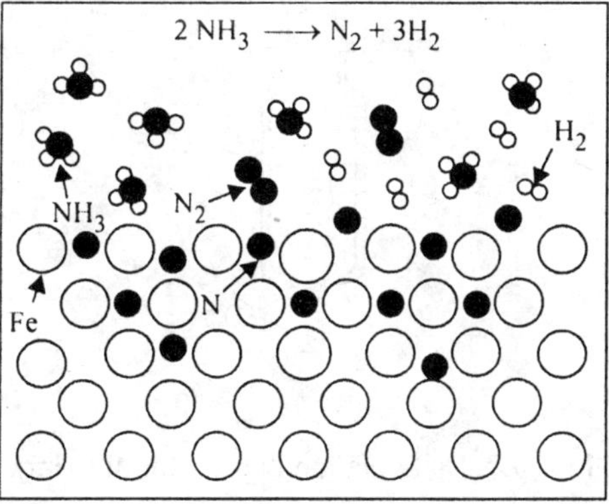

Fig. 20.2 Dissociation of ammonia and nitrogen pick-up in steel during gas nitriding.

Gas nitriding is used for parts that require a case depth between 0.2 and 0.7 mm. Nitriding is undertaken in an electric furnace equipped for precise temperature control. For large-scale nitriding pit furnaces with specially designed containers provided with tightly closing lids are used. For small-scale nitriding operations a conventional muffle furnace with a separate nitriding box, can be used. Aluminium sheet or asbestos yarn is used as the caulking material between the flange and cover of the box.

The boxes used for nitriding should be made from a material which does not react with the gases. Nickel, inconel and similar alloys are ideal, but heat-resisting steel with 25 per cent Cr and 20 per cent Ni, particularly, have also proved quite suitable for this purpose. If only a few nitriding operations are to be carried out in an *ad hoc* installation, the nitriding boxes can be made from iron sheet. However, the iron will react with the gas and hence it is difficult to control the processes taking place during the initial nitriding operations in a newly made nitriding box as some of the gas is used up in nitriding the inner surfaces of the box.

The gas always contains a small quantity of water vapour, which must be removed before the gas is introduced into the nitriding box, otherwise the water vapour can cause oxidation of the nitrided parts. This removal is best achieved by passing it through a filter of unslaked lime, which should be replaced or regenerated by heating it at 1000°C before all the lime has become slaked.

To check that the ammonia gas in the nitriding box has decomposed (dissociated) according to the requirements of the process, the composition of the exit gas is determined by means of a dissociation pipette. The sketch in Fig. 20.3 shows the general arrangement of one type of pipette customarily employed. During the course of the nitriding process or preparatory to making a measurement, the gas passes through the taps C and D. During the measurement these taps are closed after which tap E is opened. The water contained above at B is then sucked into the meter because the undissociated ammonia dissolves in the water, which then takes up precisely the volume occupied previously by the ammonia. The remaining gas, i.e., the dissociated ammonia, which consists of nitrogen and hydrogen, does not dissolve in the water. Fig. 20.4 shows some examples of height of the water column for different degrees of dissociation.

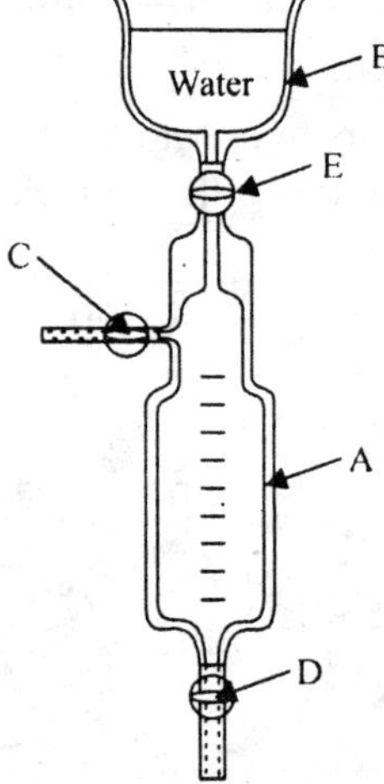

Fig. 20.3. Dissociation pipette for nitriding.

The parts to be nitrided should be well cleaned and degreased. Under no circumstances should there be any traces of rust or mill scale. If certain areas of the part are to remain soft, they can be given an electrolytic coating, e.g., of copper, tin or nickel, which prevents the nitrogen from diffusing into the steel. 'Stopping-off' agents containing tin are also available and these are painted on to the surfaces to be protected. Since the layer of tin is molten at the nitriding temperatures, it is most important that the coating should be so thin (not exceeding 0.01 mm) that it does not spread to surfaces that are to be nitrided.

The parts are placed inside the nitriding boxes in such a way that all surface come into contact with the gas. After the box has been closed, it is purged with ammonia gas until all the air has been expelled, after which it is placed inside the furnace. The composition of the exit gas is checked regularly, using the dissociation pipette and for the first 5–10 hours is kept at 15–20 per cent residual gas. If determinations indicate a lower value than this, the gas velocity is too high and should be reduced. If the converse should prevail, the gas velocity must be increased. When, after adjustment, the system has arrived at a steady-state condition, it is enough to check the gas composition two or three times per day, a residual gas content of about 50 per cent being regarded as a suitable value. Formerly a value of about 30 per cent residual gas was regarded as satisfactory, but this can result in a more brittle nitrided layer.

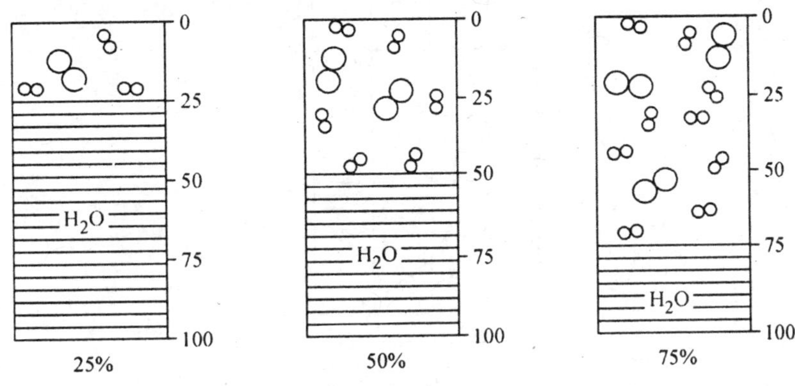

Fig. 20.4. Height of water column in meter for different degrees of disscciation.

When the nitriding is completed, the box is removed from the furnace without interrupting the flow of gas. After the charge has cooled to 200°C, the gas supply is cut off and the gases remaining in the box are most conveniently expelled with compressed air before the box is opened. The nitrided parts now normally exhibit a characteristic matt grey colour. If on occasions the colour has shades of yellow, blue or purple, it does not mean that the nitriding procedure has been at fault. The various tints derive from the presence of oxygen in the system, which may have originated from incomplete drying of the gas, or from some leak in the box or in the supply tubing.

Gas Nitriding with Ammonia and Nitogen or Hydrogen

Although this method has very seldom been used it is included here for the sake of completeness. *Sorokin* recommended a mixture of 20 per cent ammonia and 80 per cent nitrogen. The low ammonia content gives a low nitrogen activity and this results in a tougher case. However, similar results can be achieved by increasing the degree of dissociation when nitriding with ammonia only or by adding hydrogen.

Gas Nitriding with Ammonia and Hydrocarbons (Nitrocarburising)

When gas nitriding with ammonia, hydrocarbons can also be added, usually in the form of pure propane or endogas, produced from propane and air. Nitrocarburising takes place in a gaseous atmosphere by several different methods. These methods differ from one another according to the character and composition of the carbonaceous gases. Three of the most popular methods are referred below:

Designation	*Carbonaceous gas*
Nitemper	Endothermic gas
Triniding	Exothermic gas + natural gas
Nitroc	Exothermic gas with about 90 per cent N

Nitorocarburising is performed at somewhat higher temperature than normal gas nitriding, e.g., 570°C. The carbon diffuses into the steel at the same time as the nitrogen and together they form a carbonitride mainly of the ε-type. The composition and micro-hardness of the surface layer of pure iron which has been nitrided in an atmosphere of 50 per cent ammonia and 50 per cent propane at 580°C for 4 hours is shown in Fig. 20.5. The high carbon content in the surface layer gives improved wear resistance and less risk of scuffing.

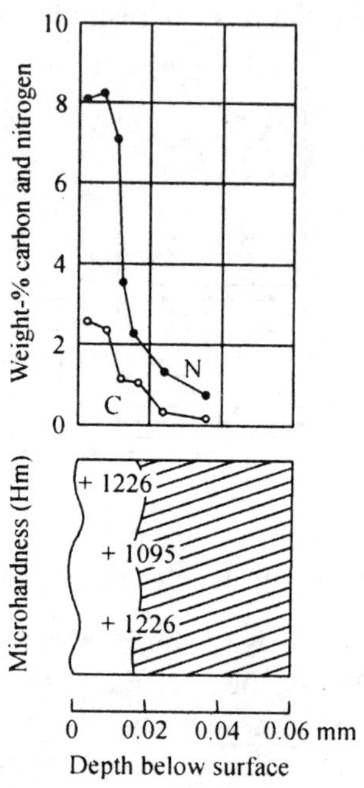

Fig. 20.5. Composition and microhardness of pure iron after nitriding for 4 hrs., at 580°C in a mixture of 50 pre cent ammonia and 50 per cent propane. Degree of dissociation of ammonia was 20 per cent.

X-ray structural analysis confirmed that the compound layer did consist mainly of the cph phase, ε-carbonitride, which has a variable composition in the range $Fe_3(CN)$ and $Fe_2(CN)$. The carbionitride phase is derived from the ε-nitride phase by the replacement of nitrogen atoms by carbon. A small amount of Fe_4N and traces of oxygen were also present in the compound layer.

Plasma or Glow-Discharge Nitriding

The glow-discharge or plasma nitriding process makes use of an ionised gas that serves as a medium for both heating and nitriding. The parts to be treated are charged into an airtight chamber, which constitutes the anode of the plasma nitriding unit. The parts making up the charge are so arranged that they are in electrical contact with the cathode, either by being placed on a bottom plate or suspended in a suitable manner. After being evacuated the chamber is filled with the process gas which, besides nitrogen, may contain hydrogen and methane. Thus, plasma nitriding may function both as conventional gas nitriding and as nitrocarburising. The working pressure lies between 0.1 and 10 mbar. On applying a potential difference of 500–1000 V the gas becomes ionised with the result that positive nitrogen ions having a high kinetic energy bombard the charge. Thereby the parts become heated to the required temperature and at the same time the nitrogen required for the nitriding process is introduced into the surface of the parts. All surfaces of the charge may be subjected to ion bombardment. This causes the formation of a blue-white glow discharge plasma around the components. The temperature is controlled by means of thermocouples located in the charge. The heating as well as the cooling after treatment may be controlled without difficulty.

The process may be operated at temperatures between 420 and 700°C which admits of a large degree of flexibility. Maraging steels, for example, may be simultaneously plasma nitrided and precipitation hardened at 420–480°C. As with gas nitriding, the time of plasma nitriding is governed by the required case depth. Plasma nitriding gives a nitriding depth that is about the same as that obtained by conventional gas nitriding at the same temperature. However, in actual practice for treatment times, up to approximately 30 hrs., the depth of hardening with plasma nitriding is greater than that of conventional gas nitriding since the heating up time for plasma nitriding is appreciably shorter.

The composition of the compound layer can be controlled so that either γ' or ε-nitride is formed or that both phases form simultaneously. This is the most usual case according to *Varchoschkow*. Other authors claim that the most usual situation involves the formation of the monophased γ', Fe_4N nitride. Stainless steels and other heat-resisting steels (chromium-alloyed) which generally are difficult to nitride by other methods are readily plasma nitrided.

The only factor restricting a more general application of the plasma nitriding process seems to be the high installation costs, but taking into account that it is possible to control the composition of the compound layer it should be accepted that at present this is, technically, the most advanced process.

Salt-Bath Nitriding

The nitriding of steel in a bath of molten salt is almost as old as the original gas nitriding method. The salt mixtures initially used contained 60–70 per cent by weight NaCN and 30–40 per cent by weight KCN. In addition there are a few per cent of carbonates (Na_2CO_3) and cyanates (NaCNO). By ageing the bath at 575°C for 12 hrs., the cyanate can be raised to the desired level, about 45 per cent. Steel parts should not be immersed in the bath during the ageing stage. The normal working temperature for salt-bath nitriding is 550–570°C and the holding time seldom exceeds 2 hours. The salt bath gives off carbon and nitrogen.

Up to the middle of the 1950s, nitriding salt baths were used only on a very restricted scale and then mainly for short-period nitriding (10–30 minutes) of twist drills for example. If longer nitriding periods were used, pitting occurred on the steel surfaces and in addition the results of the nitriding were irregular. Both of these phenomena were due to variations in the cyanate content at different levels in the bath and also to the solution of iron from the crucible, resulting in oxidation of the surface of the steel parts comprising the charge. During the last decades the method has become more popular because, by injecting air into the bath, it has become possible to obtain better control of the cyanate content. A further improvement has been achieved by employing a titanium crucible, thus avoiding the decomposition of the salt which took place previously as a result of contact with the iron crucibles hitherto used. The method which has been developed by DEGUSSA is called the Tufftride process. Extensive literature has been published on the Tufftride process.

Another variation of salt-bath nitriding is 'Sulfinuz treatment', in which the active constituent in the bath, apart from NaCN and NaCNO, is sodium sulphide (Na_2S). During conventional salt-bath nitriding, nitrogen and carbon are absorbed by the surface of the steel. During the Sulfinuz treatment, sulphur is present and this enhances still further the excellent anti-friction properties of the nitrided layer. Fig. 20.6 shows the composition of the surface layer of a carbon steel containing 0.20 per cent C after Sulfinuz treatment at 570°C for 2 hours.

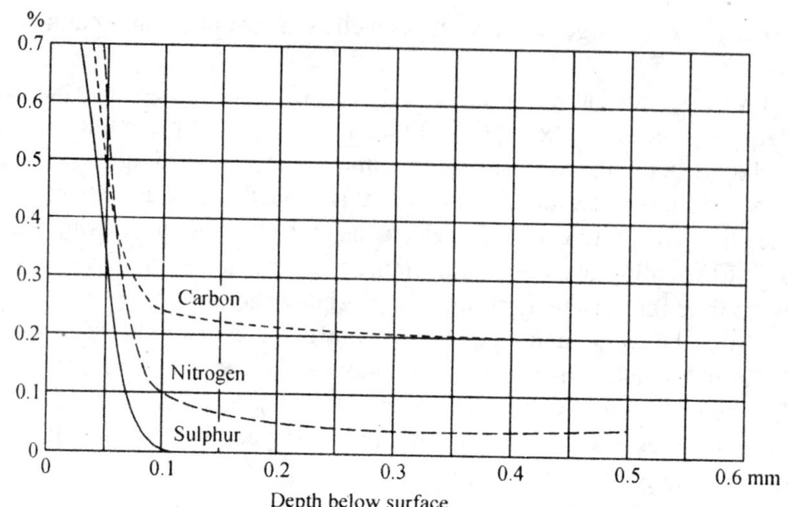

Fig. 20.6. Sulphur, nitrogen and carbon contents in surface layer of mild steel after Sulfinuz treatment at 570°C for 2 hours.

In some cases of Sulfinuz treatment it is possible for light corrosion of the steel surface to take place and this can amount to a loss of 0.005 mm from the surface. Furthermore a thin high-sulphur layer can be formed on the outside of the steel parts. This layer should not be ground off, because it has high anti-friction properties. Should it have to be removed for dimensional reasons, this is best done with steel wool. Normally, grinding should not be performed after salt-bath nitriding.

After Tufftride or Sulfinuz the best surface is obtained if the steel parts are quenched in warm water. This rapid cooling results in a supersaturated solid solution of nitrogen in α-iron, which adds an extra contribution to the increase in fatigue strength which is a characteristic feature of nitriding. However, the high internal stresses introduced at the same time reduce toughness. With regard to plain-carbon and low-alloy steels, if these are tempered at 200°C for 1 hour nitride precipitation takes place which reduces the fatigue strength and increases toughness. During slow-cooling, as with gas nitriding, this precipitation is obtained automatically. It is essential to make a daily check on the cyanate content of the salt bath (also the sulphur content for the Sulfinuz bath). The deposits which form in the bath and which sink to the bottom of the crucible should be collected and removed by means of a scoop. The entire bath should be renewed at regular intervals.

Powder Nitriding

In powder nitriding the parts to be treated are packed into boxes in roughly the same way as when pack carburising. About 15 per cent by weight of an energiser is first placed on the bottom of the box, after which is laid a layer of nitriding powder on which the parts are placed. (The parts must not come into contact with the energiser). More nitriding powder is then added, one layer of powder alternating with one layer of parts. The boxes are then closed as tightly as possible and are then placed in a muffle furnace at a temperature of 520–570°C. A nitriding period not exceeding 12 hours is recommended.

Nitriding tests on steels BS 905M39 (En 41 B), BS 817M40 (En 24) and H 13, gave roughly the same values for hardness and case depth as those obtained during gas nitriding for corresponding times and temperatures. After powder nitriding, the steel surfaces can exhibit some pitting and spalling.

COMPARISON BETWEEN GAS AND SALT-BATH NITRIDING

In the preceding section we have described gas and salt-bath nitriding without making any direct comparisons between these methods. To some extent the differences between the two methods have been exaggerated and many statements which must now be regarded as outmoded are still cited in literature references. In this instance gas nitriding implies nitriding by means of ammonia gas only. The statements usually made when comparing the two methods are as follows:

1. Gas nitriding requires 12–120 hours nitriding time whereas salt-bath nitriding requires 1–4 hours.
2. Gas nitriding makes the steel brittle, but this is not the case with salt-bath nitriding.
3. The reason why steel becomes brittle during gas nitriding is the presence of the brittle γ'-nitride which is not present after nitriding in the salt-bath.

The usual reason for the statement that gas nitriding makes steel brittle is that gas-nitrided aluminium-alloyed steel is compared with plain-carbon or low-alloy steel that has been nitrided in a salt bath.

In other contexts the same steel has been employed for purposes of comparison, in such tests the case thickness of the gas-nitrided steels is considerably greater.

To permit a closer study of the circumstances prevailing during gas and salt-bath nitriding, a number of test discs were made from quenched and tempered Cr-Ni-Mo-steel, BS 817M40 and then nitrided; one test was carried out in gas for 10–40 hours at 510°C, with 50 per cent degree of dissociation and a second test in a Tufftride salt bath, for 2 and 4 hours at 570°C.

Surface Hardness and Depth of Hardening

After nitriding, taper sections were ground on a plane specimen surface and the hardness was measured in HV, load 1 kp.

After gas nitriding at 510°C, the surface hardness was found to be about 675 HV. Nitriding at 550°C gave a lower surface hardness of about 575 HV, but on the other hand the depth of nitriding was greater. Nitriding for 20 hours at 550°C corresponds to 40 hours at 510°C.

The specimens after the Tufftride treatment gave somewhat differing results, depending on where they were treated. In one test the resulting difference in case depth between the 2 and 4 hours treatment time was negligible, but in another instance there was a marked difference. In both instances the surface hardness was about 650 HV.

As shown in Fig. 20.7, a comparison between gas-nitrided and Tufftrided specimens shows that 10 hours of gas nitriding at 510°C corresponds very closely to 4 hours nitriding in a salt bath at 570°C.

Testing for Brittleness

It is difficult to find a reliable method for assessing the brittleness of the white layer. In the present case the assessment was made by means of a metallographic study of a section through the white layer. In addition the hardness (HB 5/750) of the specimens was measured and the appearance of the indentations was studied at 50 × magnification. From this study the following facts emerged:

During gas nitriding the thickness of the white layer increased with increasing time and temperature. However, the character of the layer remained homogeneous throughout. On performing the Brinell hardness test on a specimen after it had been nitrided for 10 hours at 510°C, both radial and concentric cracks were observed. After nitriding for a longer period or at higher temperatures concentric cracks only were observed, the width of which decreased with increasing nitriding time. This is probably due to the enhanced load-bearing capacity of the nitrided layer.

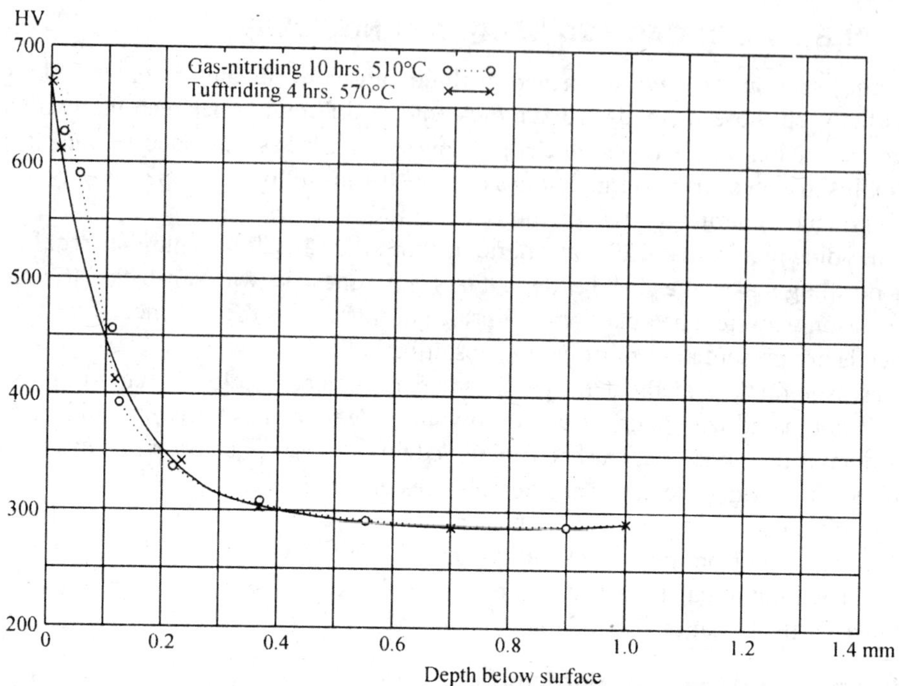

Fig. 20.7. Hardness profile through surface layer after gas nitriding and Tufftriding respectively, of BS 817 M40.

The Tufftride treatment also caused the thickness of the white layer to increase with treatment time. After a treatment of 4 hours the layer contained pores which could clearly be distinguished at 500 × magnification. According to *Finnern* this is a normal phenomenon that occurs even after a 90 minute treatment. Brinell hardness testing gave rise to radial and concentric cracks after both a 2 and 4 hours Tufftride treatment. On the basis of these test methods it is difficult to make any assertions as to which method of nitriding gives the toughest surface layer.

Nitride Formation

The composition of the nitride layer in the nitrided specimens was established by means of X-ray diffraction (Cu-radiation). After gas nitriding at 510°C for 20 and 40 hours no reflection was obtained for α-Fe. As a result of the treatment time there was a reduction in the amount of γ′-nitride and an increase in the amount of ε-nitride. Gas nitriding at 550°C for 10 and 20 hours gave similar results.

After the 2 hours Tufftride treatment at 570°C an indication of a α-Fe could be detected. Both γ′-nitride and ε-nitride were also present. As a result of a 4 hrs. treatment at 570°C, the α-Fe disappeared the γ-nitride disappeared almost completely, but the ε-nitride increased. The investigation was repeated using Cr-radiation and basically the same results were obtained.

These investigations have shown clearly that γ′-nitride is present both after gas nitriding and salt-bath nitriding at the temperatures employed. Nevertheless, the amount of γ′-nitride is greater after gas nitriding than after salt-bath nitriding. Summarising, the investigations made have shown that none of the three statements cited by way of introduction and which derive from the published literature; turns out to have any validity whatever. The opinion, often put forward, that low-alloy steel should be nitrided in a salt bath only and not gas-nitrided can thus be disregarded.

Advantages and Disadvantages

The following additional points can be made for the two methods.

Treatment time

In this respect, salt-bath nitriding has its obvious advantages when compared with conventional gas nitriding. This applies particularly if a nitriding depth of 0.1 mm (at 400 HV) is adequate, which is normally obtained after salt-bath nitriding for 2 hrs. with the steel investigated.

Cleanliness

With gas nitriding there is no soiling of the work premises or of the nitrided components. However, with salt-bath nitriding, it is quite easy to spill salt on the top of the furnace and on the surrounding floor area. Furthermore, parts treated in the salt bath must be cleaned by rinsing in hot water. The cyanide salt is toxic, but cases of poisoning are very rare provided that the correct procedure is followed. Since the rinsing water together with residues of salt can easily be flushed into the drainage system, we can expect that in future the nature conservancy authorities will impose more stringent regulations.

Even now, traces of cyanide must be detoxified in districts where the effluent water is subjected to bacteriological purification. The reason for this is that the cyanide destroys the bacteria that are used for purification purposes.

Depth of nitriding

As a result of the formation of pores, which increases with increasing treatment time during salt-bath nitriding, the nitriding period is restricted to 4 hours. No such restrictions are imposed on gas nitriding, but for practical reasons a nitriding time of 90 hours should normally not be exceeded. Often a greater depth of nitriding is required than that obtainable with salt-bath nitriding and this is when gas nitriding comes into its own.

Dimensional stability

For the same depth of nitriding, the resulting distortion is the same for both methods, if by this we mean the change in dimensions occurring as a result of the compressive stresses induced in the steel during nitriding. After gas nitriding, the box containing the nitrided parts is removed from the furnace and allowed to cool freely in air, whereas components which have been nitrided in a salt bath are quenched in water or oil after the treatment. As a result of the water quench thermal stresses are created which can result in changes of shape.

Wear resistance

During salt-bath nitriding a higher content of ε-carbonitride is formed than during conventional gas nitriding. In critical cases the higher content of ε-carbonitride provides increased wear resistance and reduced risk of scuffing. By adding hydrocarbons during gas nitriding (nitrocarburising) the content of ε-carbonitride can be raised to the same level as that obtained during salt-bath nitriding. Nitrocarburising and salt-bath nitriding yield results that in actual practice are very similar.

Toughness

As a result of the rapid cooling employed during salt-bath nitriding, the toughness of the nitrided layer is inferior to that obtained by gas nitriding, where cooling takes place slowly. The practical consequences of this have been observed, for instance, when comparisons were made between gas-nitrided and salt-bath nitrided moulds made from quenched and tempered steel. Several instances of spalling were observed after salt-bath nitriding.

What has been said above indicates that the shorter treatment time and the resultant lower costs represent the only definite advantage that salt-bath nitriding can offer *vis-à-vis* gas nitriding, which possesses many advantages within other spheres. Hence the heat treater as well as the user of nitrided material has to select which method is the most favourable one for his special or general case.

NITRIDABILLITY

The concept of 'nitridability' includes on the one hand the ability of the steel to absorb nitrogen and on the other the increase in hardness imparted by the nitrogen. The effect of the alloying elements has been described. For the sake of convenience the diagram shown previously is reproduced in Fig. 20.8. When deciding on the composition of a nitriding steel for which high surface hardness is required, the choice falls mainly on those elements that form nitrides, such as Al, Cr and Mo.

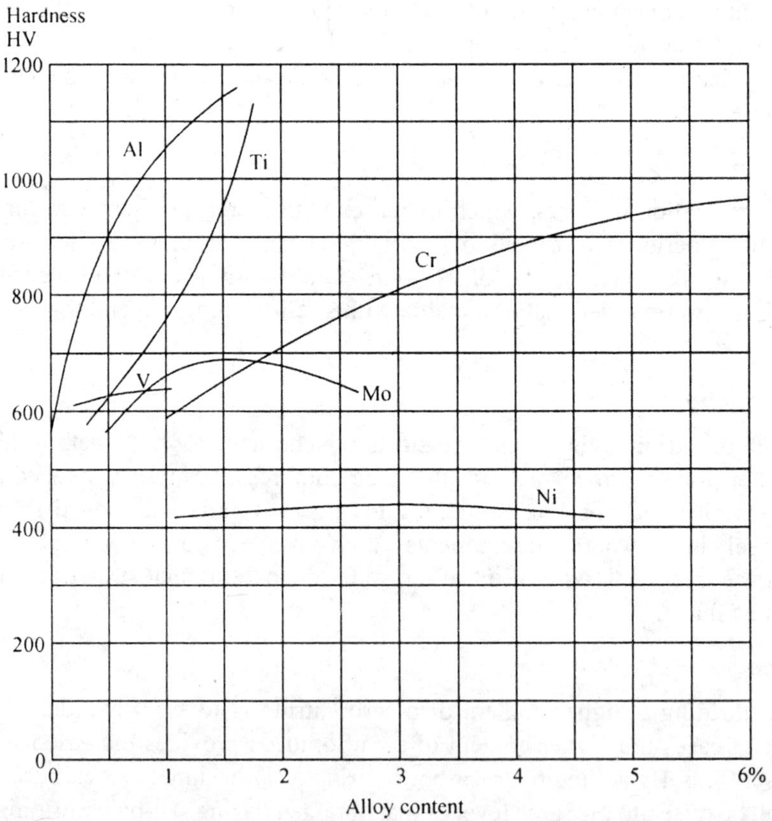

Fig. 20.8. Influence of alloying elements on hardness after nitriding. Base alloy 0.35 per cent C, 0.30 per cent Si, 0.70 per cent Mn.

The diagram shows how these elements along with a few others affect the surface hardness after nitriding. Al, closely followed by Ti, has the greatest effect and then come Cr, Mo and V. In the Ni steels the same surface hardness has been obtained as in plain-carbon steels (about 400 HV). In steel containing several alloying elements higher hardness values are obtainable than if the alloying elements are used separately.

The depth of nitriding decreases with increasing content of alloying element, as shown in Fig. 20.9. Al and Ti, which have the greatest effect in increasing the hardness, also have the strongest delaying influence on the diffusion of nitrogen into the steel as their contents increase. An optimum value for hardness and depth of nitriding is obtained at roughly 1 per cent Al and this is also the content normally used in Al-alloyed nitriding steels. The reason for the inhibiting effect of the alloying elements is that they bind the nitrogen as nitrides. Carbon, too, has a strong inhibiting effect on the diffusion of nitrogen.

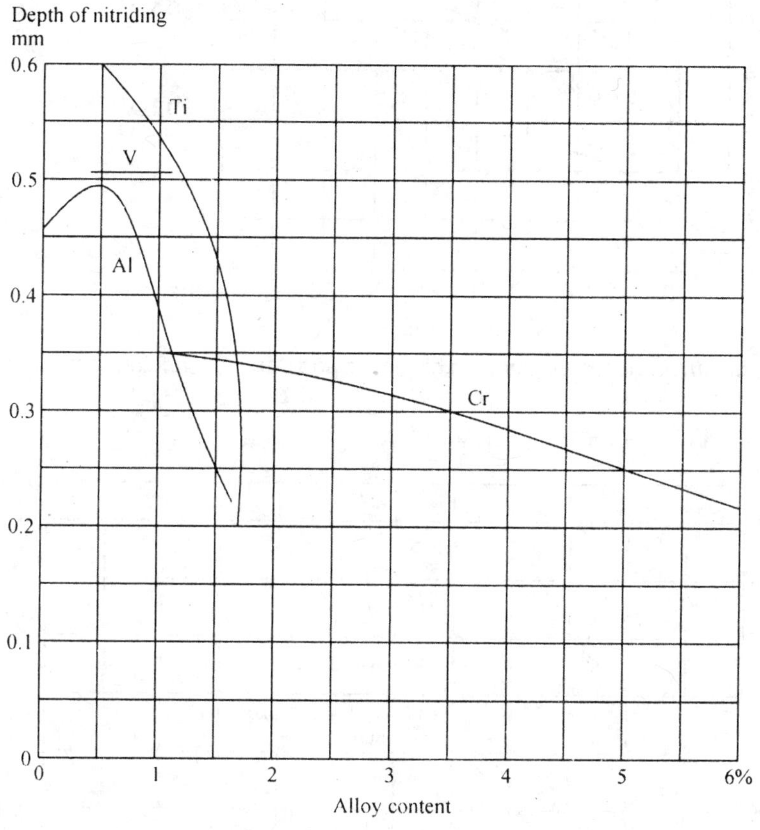

Fig. 20.9. Influence of alloying elements on depth of nitriding measured at 400 HV. Nitriding for 8 hours at 520°C.

The microstructure influences nitridability in the following two ways. A high content of free ferrite favours diffusion of nitrogen. A low carbide content in the structure favours both the diffusion of nitrogen and hardness. Usually alloy steels in the heat-treated state are employed for nitriding, i.e., they have been quenched and tempered at 500–650°C. Within this temperature range both precipitation and coagulation of carbides takes place. Since carbide precipitation starts preferentially at the grain boundaries, a strong barrier is formed at these points, where the diffusion-retarding effect is accentuated because diffusion—which normally takes place more rapidly at the grain boundaries than through the grains—is impeded by the nitrides and carbonitrides formed in the grain boundaries. The effect of the tempering temperature used during the heat treatment is illustrated in Fig. 20.10, which applies to steel En 29 B. Tests have shown that the nitridability of this steel after: (i) induction hardening and tempering at 180°C; and (ii) surface decarburisation, is better (Fig. 20.11)—than after quenching and tempering (tough hardening).

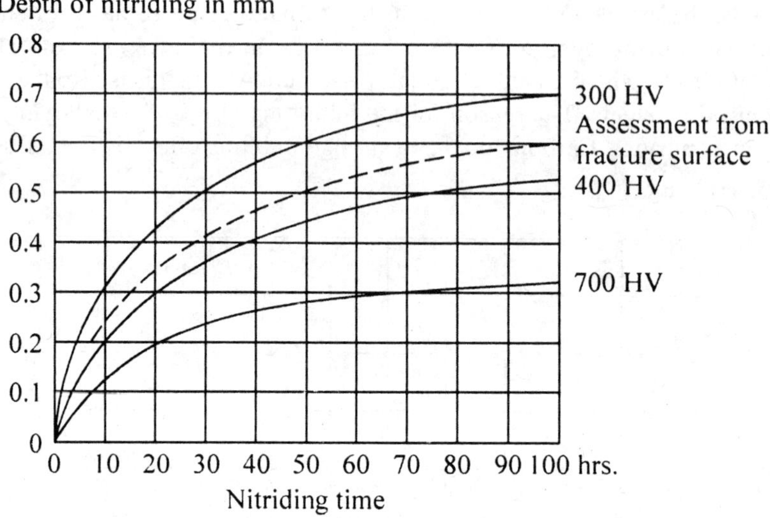

Fig. 20.10. Influence of tempering temperature on hardness and depth of nitriding of steel En 29 B. Nitrided for 60 hr at 510°C.

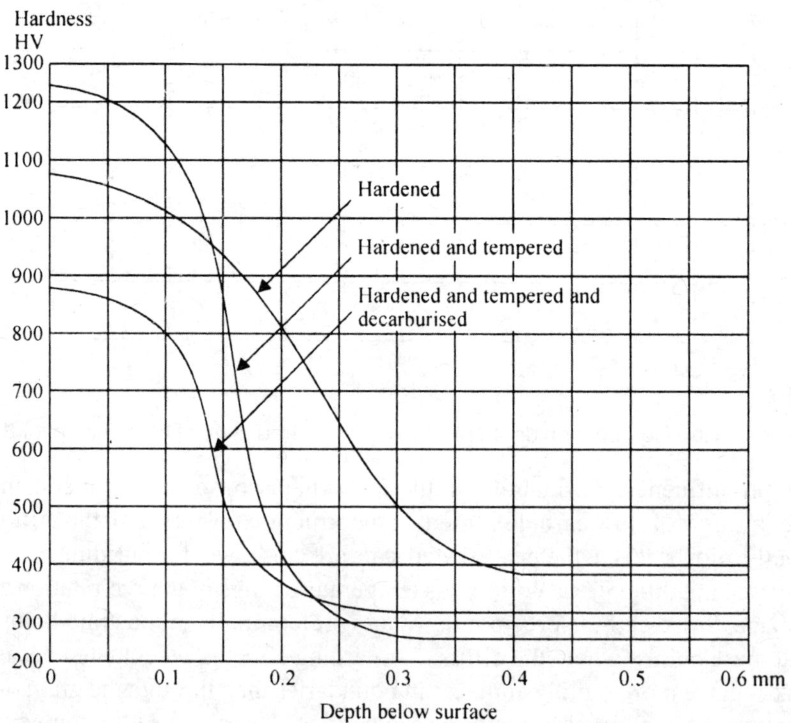

Fig. 20.11. Influence on surface hardness and depth of nitriding of the microstructure of En 29 B after glow discharge nitriding for 8 hrs. at 520°C.

Hodgson and *Boron* have made detailed tests on the nitriding of heat-treatable steels containing approximately 3 per cent Cr and 0.05 per cent Mo, with carbon contents 0.10–0.37 per cent and subjected to varying tempering temperatures.

The results are summarised in Fig. 20.12 which, among other things, shows that the response of the steel to elevated tempering temperatures is dependent on the carbon content.

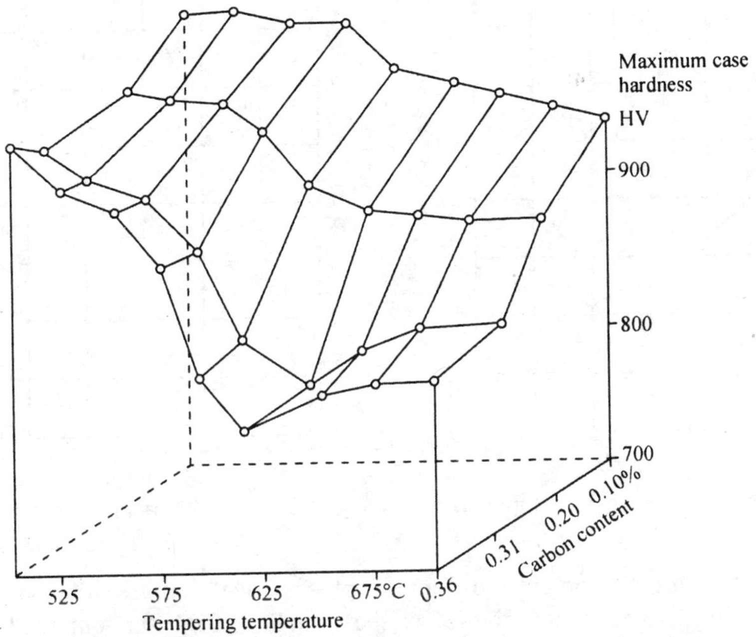

Fig. 20.12. Relationship between carbon content, tempering temperature before nitriding and hardness after nitriding. Tempering time 100 hrs. Base analysis 3 per cent Cr and 0.5 per cent Mo.

The grain size and carbide distribution in the steel influence the surface smoothness obtained after nitriding. With increasing grain size, the nitrides are precipitated in wide bands along the grain boundaries. If carbides have already been precipitated at the grain boundaries, the nitrogen forms carbonitrides with the carbides, which will hereby increase in size. As a result the carbonitrides can displace individual grains in the surface layer, so that the white layer will become deformed or crack (spall).

Measurements of surface smoothness on nitrided specimens of SS 2940 (ISO steel 3) show that there is a definite correlation between grain size and surface smoothness, as has been shown by measurements made on material taken from the surface and centre of hardened and tempered rods, roughly 150 mm diameter. As shown in Fig. 20.13, the surface smoothness of the nitrided specimens taken from the centre is inferior to the smoothness of those taken from the surface. The centre portions contain considerably more grain-boundary precipitates than the sections located near the surface, owing to the slower cooling of the centre during hardening. This investigation resulted in a modification of the composition of steel SS 2940 (ISO steel 3), so that its hardenability improved and the tendency to form grain-boundary precipitates in sections exceeding 100 mm was reduced (ISO steel 4).

During hot working or heat treatment of steel SS 2940, the surface layer may become decarburised and results in grain growth. On nitriding, this will have the consequences outlined above. If the machining allowance has been inadequate, spalling will occur during nitriding. If stress relieving takes place at too high a temperature, re-crystallisation with the formation of columnar grain may occur in the surface layer. This type of grain growth, also causes rough surfaces and spalling.

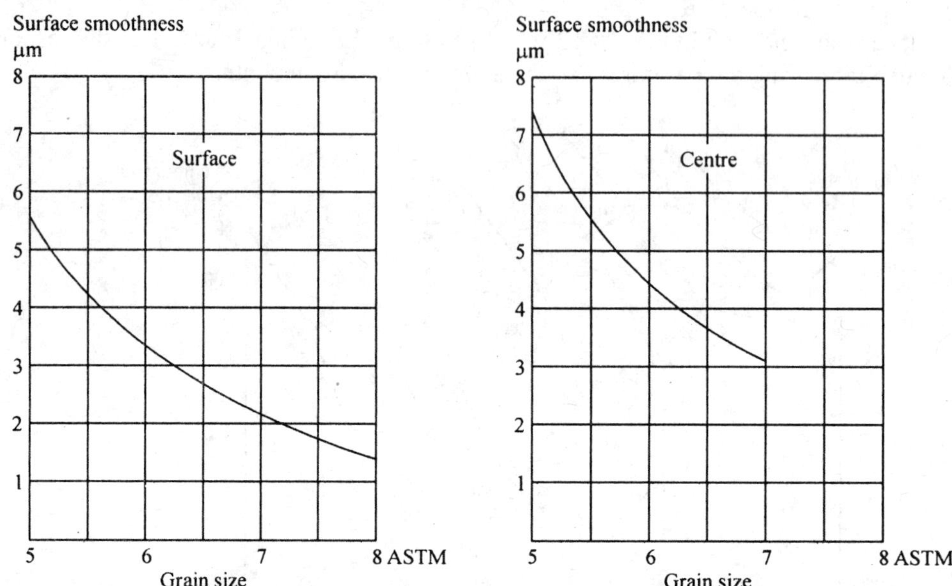

Fig. 20.13. Surface smoothness as a function of grain size of nitrided specimen of ISO steel 3 taken from surface and centre.

When studying the nitrogen content at different depths below the surface after salt-bath nitriding of a carbon steel containing 0.15 per cent C, the values shown in Fig. 20.14 were obtained. These confirm the validity of the equation shown above quite well, if the calculations assume a constant nitrogen content. Basically, the effect of the nitriding temperature is such that the lower the temperature the higher is the surface hardness but, at the same time, the smaller is the depth of nitriding. This applies particularly to the alloy steels and is shown by way of example in Figs. 20.15 and 20.16 which apply to steels En 41 B and En 29 B.

Nitrogen activity is another of the controlling factors during nitriding. In accordance with the laws governing diffusion, the degree of nitrogen penetration is governed by the temperature and the nitrogen content which can be built up in the outermost layer of the steel. In the nitriding salt bath, nitrogen activity can be controlled by means of the cyanate content which, as mentioned previously, should be maintained at about 45 per cent. During gas nitriding the nitrogen activity is controlled by the degree of dissociation and the flow rate of the gas and the following equation is assumed to apply:

$$A_N \sim a \times v$$

where,

A_N = activity of atomic nitrogen
a = degree of dissociation
v = flow rate.

Consequently the nitrogen activity is a function of the number of ammonia molecules dissociated at the steel surface per unit of time. At constant pressure and temperature the degree of dissociation is reduced as the flow rate increases but the product av increases. Therefore in actual practice the nitrogen content can be checked by means of the degree of dissociation and controlled by the flow rate.

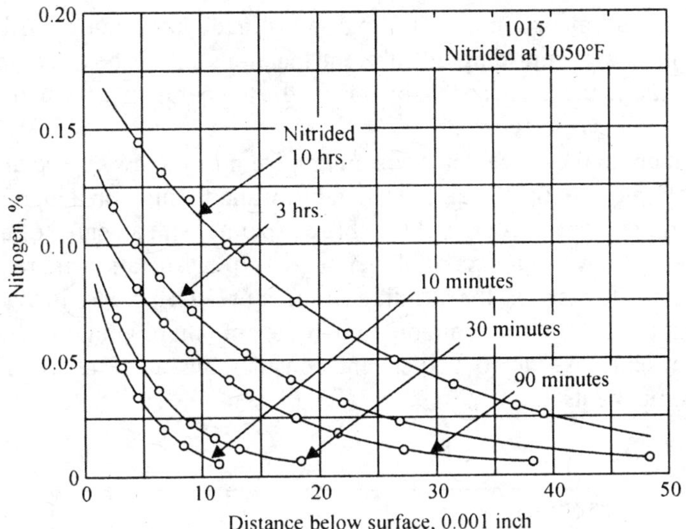

Fig. 20.14. Nitrogen gradients in 1015 steel as a function of time of nitriding at 1050°F (566°C), using the aerated bath process.

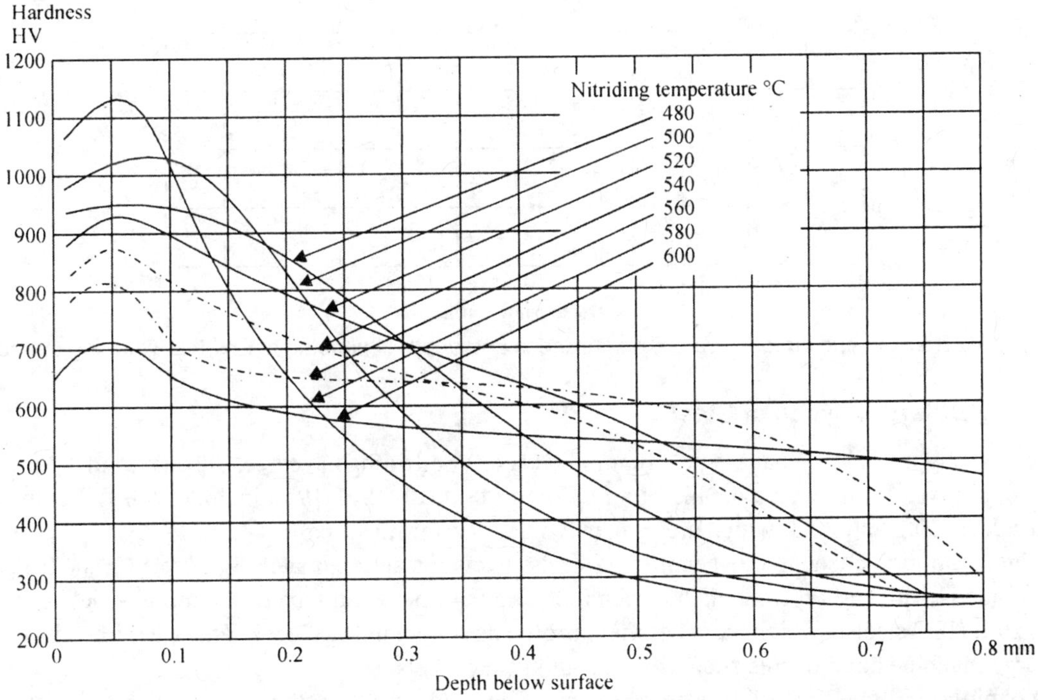

Fig. 20.15. Influence of nitriding temperature on hardness and depth of nitriding of BS 905M39, En 41 B. Nitriding period 60 hours

A high nitrogen activity is thus obtained with high flow rates, which give a low degree of dissociation. A low nitrogen activity is obtained with low flow rates, which give a high degree of dissociation. If the flow rate is too low, the turbulence in the nitriding box is inadequate unless the box is provided with a fan. This can result in uneven nitriding at different locations in the box. The degree of dissociation increases as the temperature increases so that by raising the temperature a constant degree of dissociation can be ensured with a high gas flow rate.

At the commencement of the process it is customary to employ a low degree of dissociation (high gas flow rate), the object being to expel the air present rapidly and to build up a high nitrogen content in the surface layer. In actual practice it has proved feasible to operate with a degree of dissociation of roughly 50 per cent even in furnaces without fans. If higher degrees of dissociation are required, it is essential to have furnaces with fans or to increase the nitriding temperature. An increase in the degree of dissociation from 30 to 50 per cent at 510°C implies a reduction in the amount of ammonia by *ca* 25 per cent. If the degree of dissociation is increased to 50–60 per cent, considerable increases are achievable in the service life of gas-nitrided components.

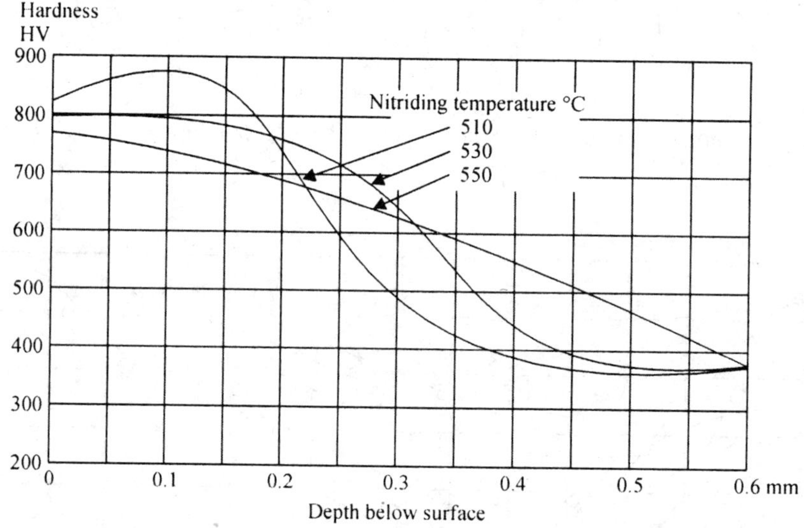

Fig. 20.16. Effect of nitriding temperature on hardness and depth of nitriding of En 29 B. Nitriding period 60 hr.

Nitriding Different Types of Steel

Some of the types of steel that are not commonly used for nitriding have been discussed in the preceding section, but we will now give details of a few more. ISO/R 683/X–1970 includes the following steels, Tables 20.1, 20.2 and 20.3, which are primarily intended for nitriding.

Chromium-molybdenum-alloyed nitriding steels containing approximately 3 per cent Cr and 0.5 per cent Mo are made with carbon contents between 0.20 and 0.40 per cent. One of the more widely used steels is En 29 B (ISO steel 1) which has a hardness profile as shown in Fig. 20.17. This steel finds its main use for large machine components such as shafts and gear wheels.

Aluminium-alloyed nitriding steels are manufactured with varying compositions, depending on hardenability requirements. One of the rather commonly occurring steel is BS 905M39 (En 41 B), which on nitriding achieves a surface a hardness of about 1100 HV and has the hardness profile shown in Fig. 20.18. A nitriding period shorter than 5 hours is not recommended if the very highest surface hardness is required.

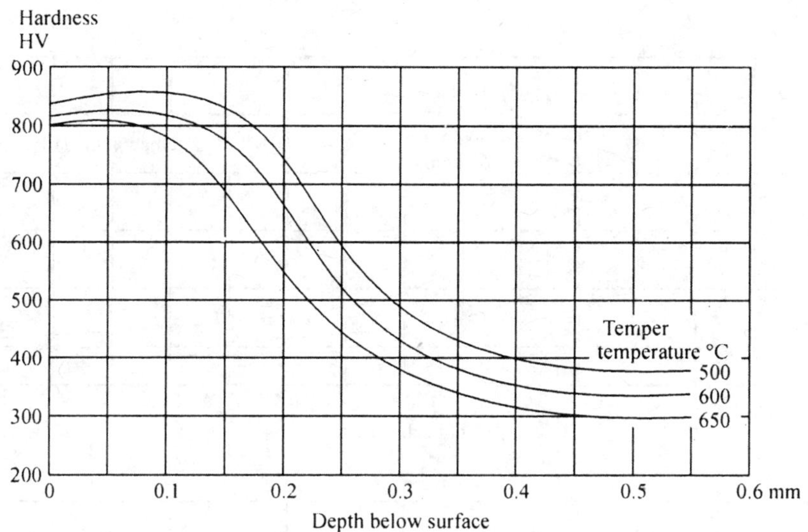

Fig. 20.17. Influence of tempering temperature on hardness and depth of nitriding of steel En 29 B. Nitrided for 60 hours at 510°C.

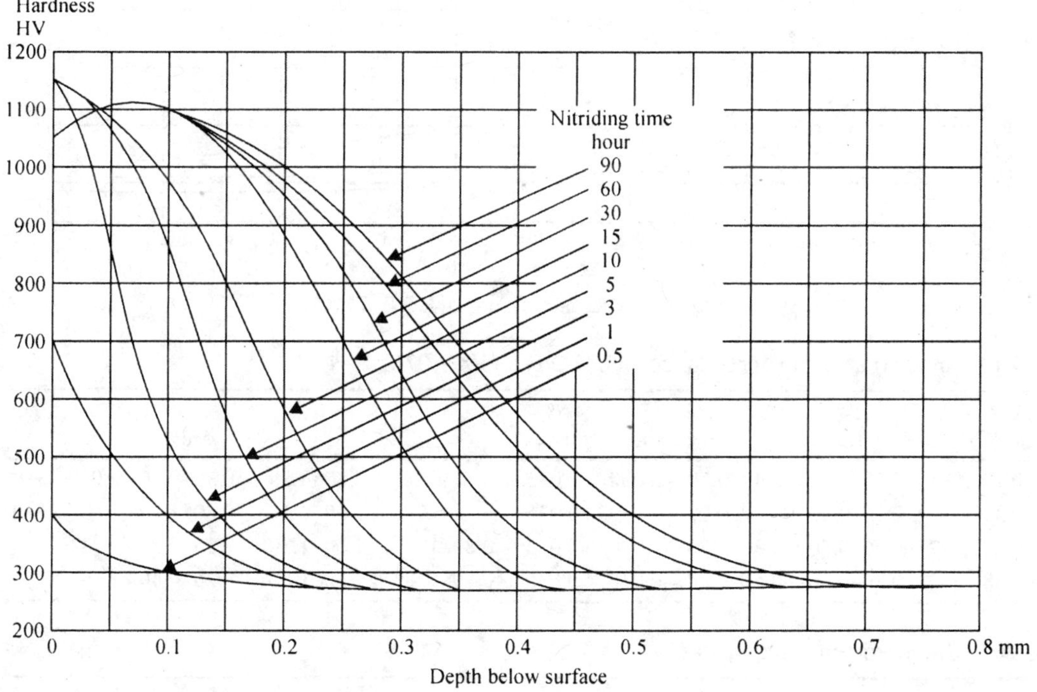

Fig. 20.18. Hardness profile through surface layer of BS 905M39 after gas nitriding for 0.5–90 hours at 510°C.

Alloy heat-treatable steels of the type BS 708A37 (En 14 B) and BS 817M40 (En 24) exhibit a surface hardness of around 650 HV after gas nitriding at 510°C, if their initial state was hardened and tempered. If the nitriding temperature is raised to 550°C, the hardness drops to around 550 HV. Basically the graph in Fig. 20.19 applies also to BS 768A37.

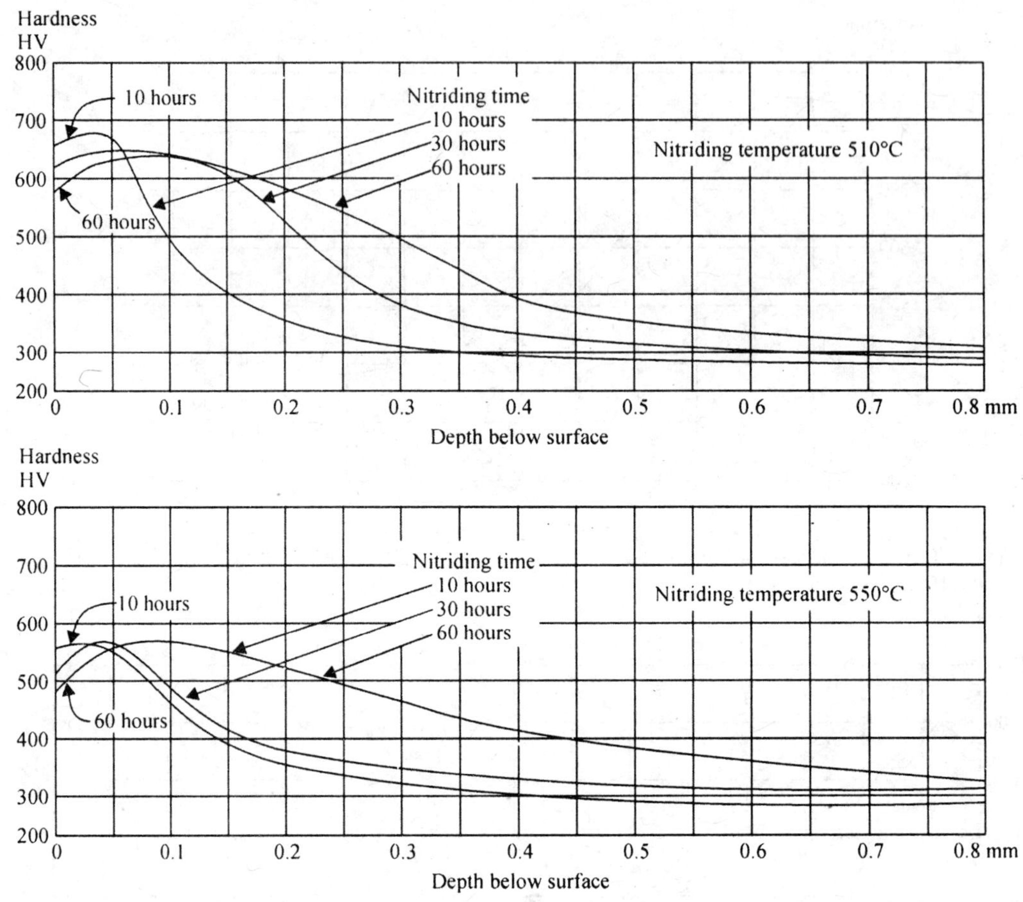

Fig. 20.19. Gas nitriding of BS 817M40.

Table 20.1. Nitriding steels in accordance with ISO/R 683/X–1970.

Type of steel	*C%*	*Si%*	*Mn%*	*P% max.*	*S% max.*	*Al%*	*Cr%*	*Mo%*	*Ni% max*	*V%*
1	0.28–0.35	0.15–0.40	0.40–0.70	0.030	0.035	–	2.80–3.30	0.30–0.50	0.30	–
2	0.35–0.42	0.15–0.40	0.40–0.70	0.030	0.035	–	3.00–3.50	0.80–1.10	–	0.15–0.25
3	0.30–0.37	0.20–0.50	0.50–0.80	0.030	0.035	0.80–1.20	1.00–1.30	0.15–0.25	–	–
4	0.38–0.45	0.20–0.50	0.50–0.80	0.030	0.035	0.80–1.20	1.50–1.80	0.25–0.40	–	–

Table 20.2. Standardised nitriding steels.

ISO	*Nitralloy*	*BS*	*En*	*DIN*	*SS*
1	–	(722M24)	29B	32 CrMo 12	2240
2	–	897M39	40C	32CrMoV 1210	–
3	G	–	–	34 CrAlMo 5	2940[1]
4	135M	905M39	41B	–	2940

1. Previous composition replaced by steel 4.

Alloy case-hardening steel, e.g., BS 637M17 can be nitrided in the annealed state, where by the hardness and depth of nitriding obtained are somewhat higher than that achieved with heat-treatable steels (see Fig. 20.20). Unalloyed case-hardening steels and plain low-carbon steels show, after gas nitriding the hardness profiles given in Fig. 20.21. After salt-bath nitriding or nitrocarburising the hardness distribution obtained is roughly the same. However, in the outermost layer the surface hardness is somewhat greater.

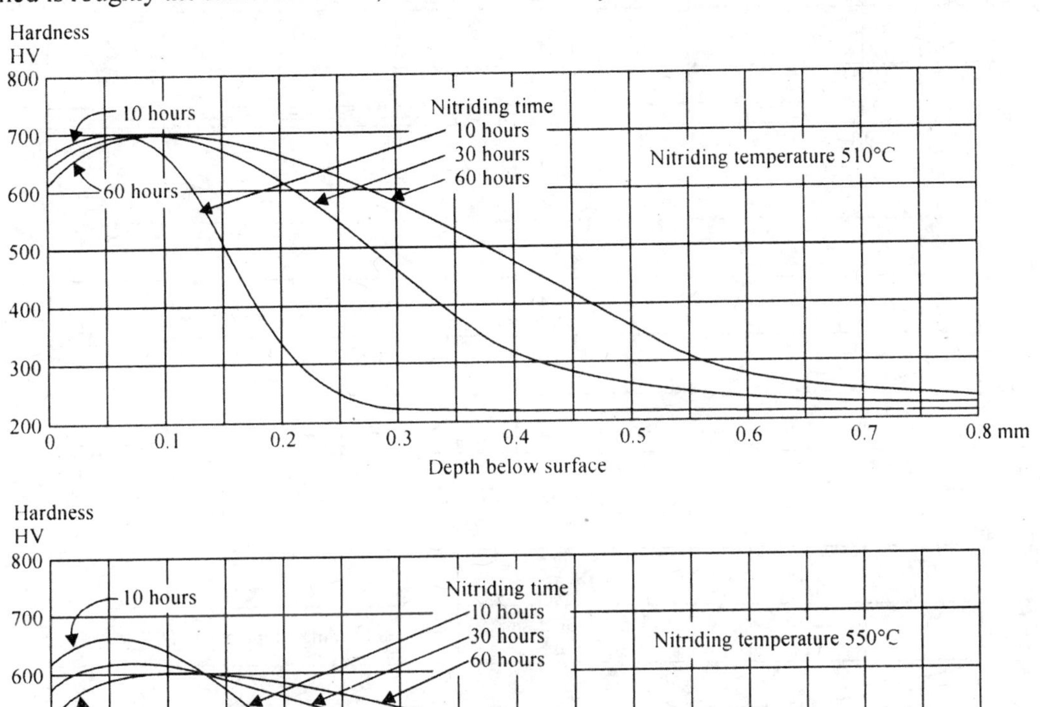

Fig. 20.20. Gas nitriding of annealed BS 637M17.

Low-alloy tool steels type NiCrMo can be nitrided either in the annealed or hardened and tempered state. The surface hardness becomes roughly 500 HV.

High-alloy tool steels containing 5–12 per cent Cr of type A 2, D 2 and D 6 follow a common pattern, which suggests that the highest hardness is obtained if, after hardening the steel is tempered at only 200°C. The surface hardness is 1100–1200 HV (Figs. 20.22 and 20.23).

If, for example, the hardening temperature is raised to 1050°C, secondary hardening takes place during nitriding. The hardness of the core obviously drops with increased nitriding time but it is considerably higher than if a normal hardening temperature had been used (Fig. 20.24).

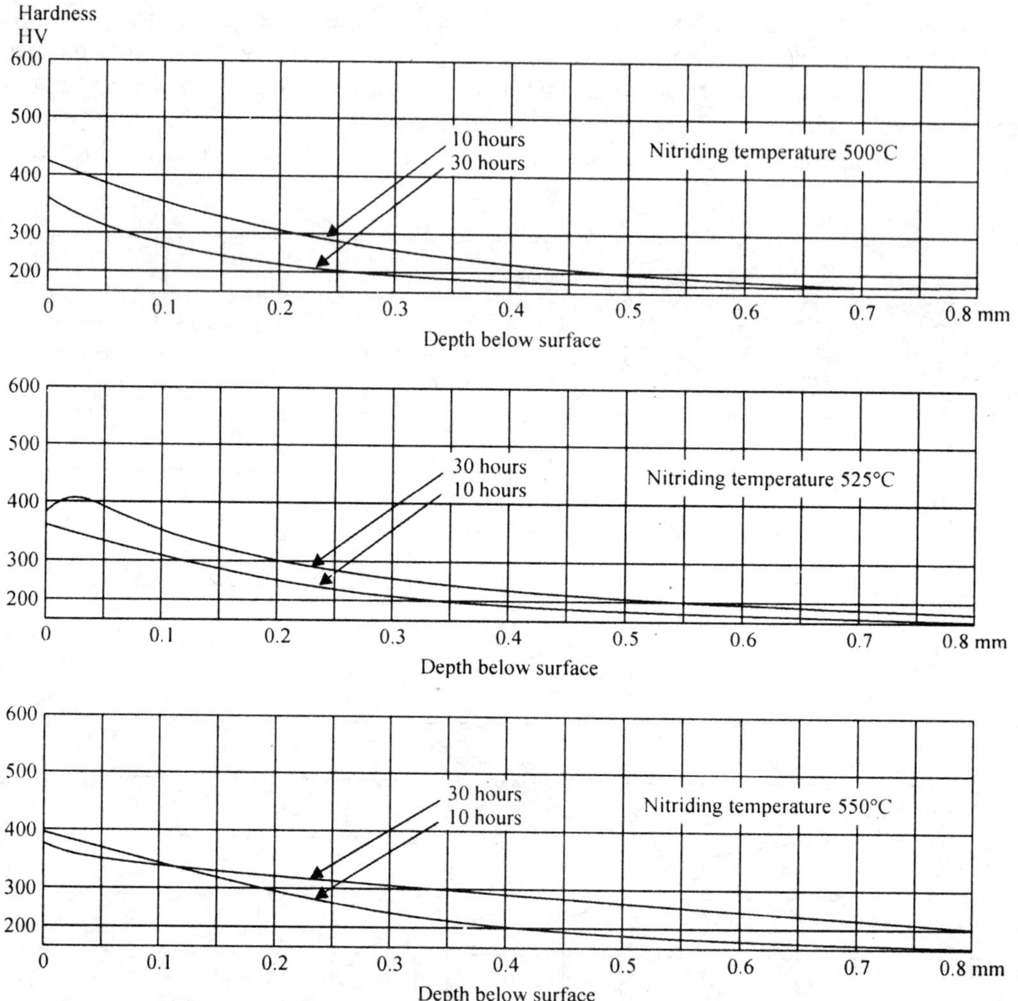

Fig. 20.21. Effect of nitriding temperature on the hardness and depth of nitriding of SS 1370 (0.15% C steel).

A tempering treatment at 510°C prior to nitriding has not affected the depth of nitriding or the surface hardness. Steel D 6, after hardening from 1050°C, behaves in precisely the same way during nitriding as steel D 2. Hot-work steels of the type normally employed exhibit surface hardness values after nitriding that lie between 1100 and 1300 HV. Owing to the high alloy content, the depth of nitriding for these steels is relatively small as is also the case for the high-alloy chromium steels. Since the steels are normally tempered at temperatures above the nitriding temperature, the hardness of the core is not reduced during nitriding. As shown in Figs. 20.25 and 20.26 the temperature employed during tempering has a marked effect on the depth of nitriding for steel H 13. A high initial hardness tends to have a favourable effect on the depth of nitriding. Nitriding can be carried out between 510 and 525°C without noticeably affecting the surface hardness. Nitriding at 550°C gives a surface hardness of about 1100 HV and a somewhat greater depth of hardening than that obtained at lower temperatures. Tools made from this steel are nitrided on quite a large scale for both cold and hot-work service.

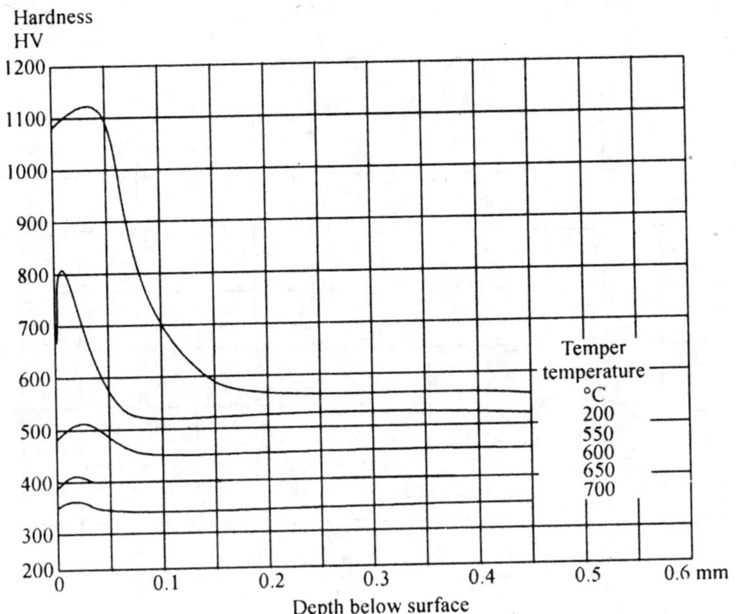

Fig. 20.22. Gas nitriding of D 6 for 10 hours at 510°C after hardening from 980°C followed by tempering at different temperatures.

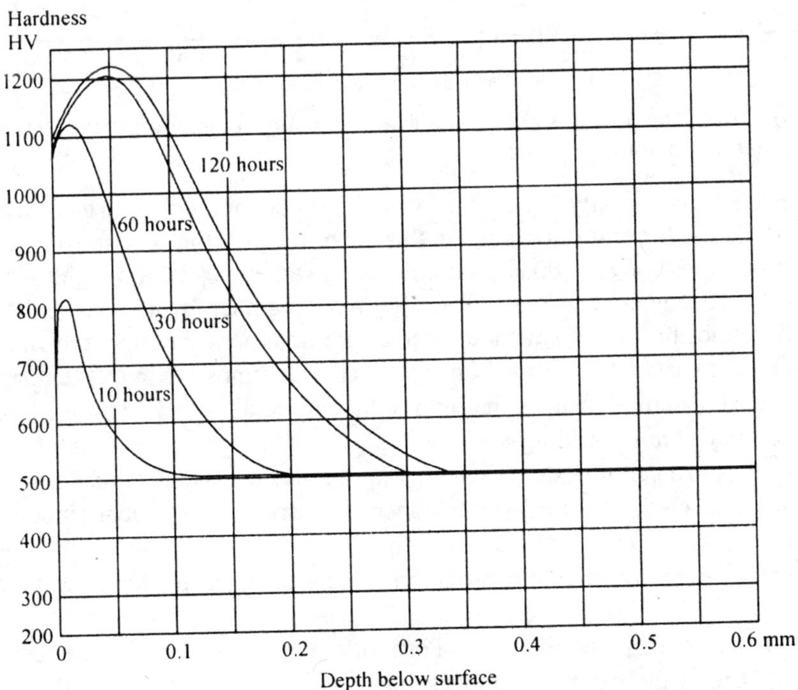

Fig. 20.23. Gas nitriding of D 6 for 10–120 hours at 510°C after hardening from 980°C followed by tempering at 550°C.

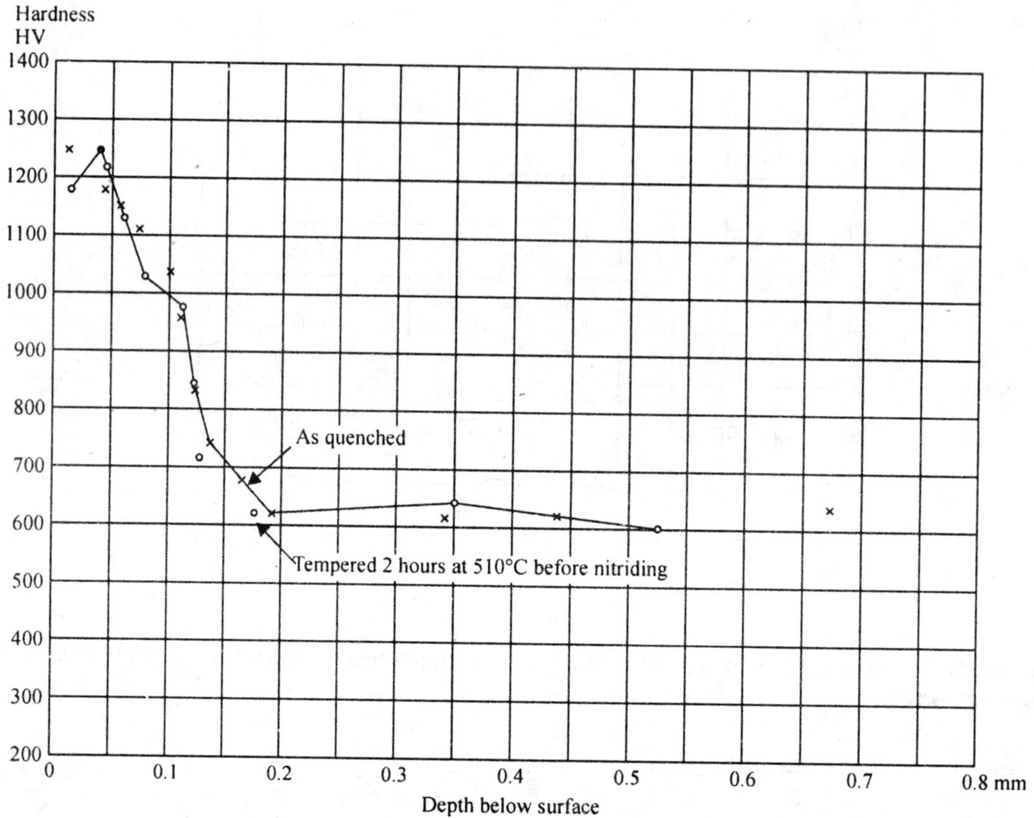

Fig. 20.24. Gas nitriding of D 2 for 30 hours at 510°C with and without prior tempering for 2 hours at 510°C. The specimens were first hardened from 1050°C in oil.

High-speed steel tools are normally salt-bath nitrided for 10–60 minutes. With gas nitriding the period should not exceed 10 hours. If periods longer than this are employed, the outermost layer becomes soft (Fig. 20.27). Hardness values above 1300 HV are obtainable which may be too high for many purposes. In such cases the tools are tempered at 550–600°C after nitriding, thereby increasing the toughness.

Staninless steels cannot be nitrided without special pre-treatment, because the film of chromium oxide which forms on the surface of the steel after contact with atmospheric oxygen prevents nitrogen absorption. For martensitic stainless steels the oxide film can be removed or thinned down by a light sand blasting immediately before nitriding.

The same also applies to a certain extent to the $^{18}/_{8}$ austenitic steels and similar steels. Better results are obtained with these steels if chemical pre-treatment is carried out in accordance with any of the following methods:

1. Pickling in 50 per cent by volume hydrochloric acid at 70°C, with subsequent thorough rinsing in water.
2. Pickling in 1 per cent by volume orthophosphoric acid containing 0.7 per cent primary zinc phosphate at 70°C without subsequent washing.

In both cases nitriding must be carried out directly after the pre-treatment. If, for practical reasons, this should bot be possible, the parts can be dealt with in accordance with any of the following methods:

3. Pickling in 10 per cent hot sulphuric acid, rinsing in water and copper-plating in a copper cyanide bath to give a coating of thickness 0.3 μm.
4. According to *Coppola* 12 per cent Cr stainless steel can be nitrided after lapping the steel surface followed by passivation in hot nitric acid, or heating in air at 510°C. Heating in air at 260–370°C prior to nitriding gives unsatisfactory results.

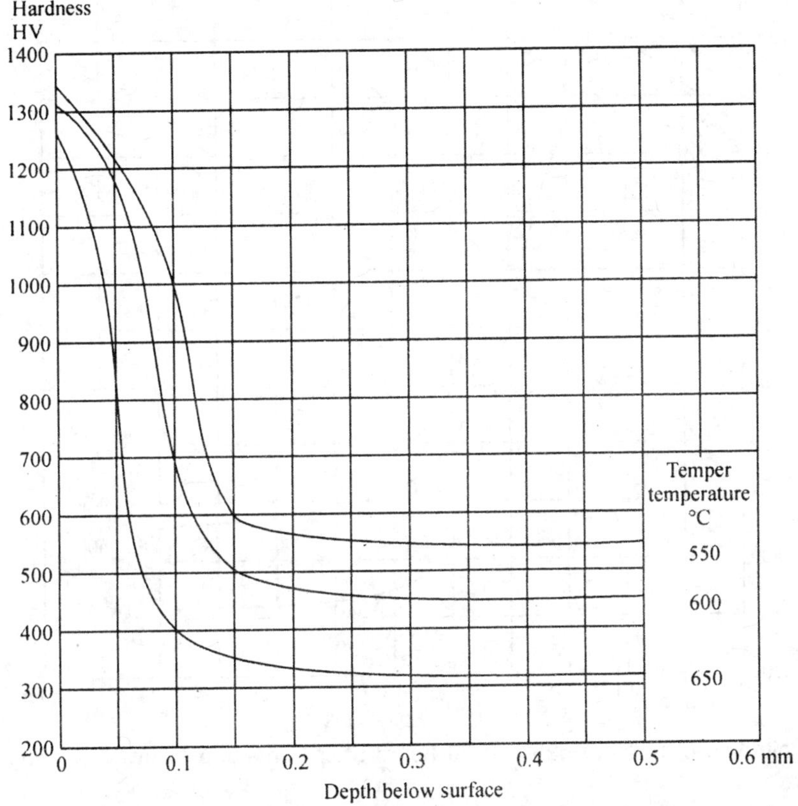

Fig. 20.25. Gas nitriding of H 13 at 525°C for 10 hours after hardening followed by tempering at different temperatures.

When plasma nitirding it is possible to remove the interfering oxide film by a simple sputtering operation carried out in the nitriding chamber immediately before the nitriding proper.

Stainless steels are nitrided at higher temperatures than those used for other steels. For martensite stainless steels 550–550°C is used and for the austenitic steels, 570–620°C. It is important that the air should be purged from the furnace before heating starts.

Fig. 20.28 shows results obtained from nitriding tests on creep-resisting 13 per cent Cr steel at temperatures between 520 and 600°C for 60 hrs. This graph can also be used for other martensite steels of the type AISI 430. Fig. 20.29 shows hardness curves which have been plotted for an $^{18}/_{8}$ steel and steel A 286 after nitridng for 60 hrs. at 600°C.

As a result of nitriding there is a considerable reduction in resistance to rusting and hence it is doubtful whether a stainless steel chosen for its corrosion-resisting properties, should be subjected to nitriding. Nitriding of steel A 286 which is employed as a creep-resisting steel can, on the other hand, be quite justifiable. Also, as a result of nitriding, austenitic steels can become slightly magnetic in the nitrided layer, partly because the iron nitride Fe_4N is ferromagnetic and partly because of partial transformation of austenite to ferrite in the nitrided layer.

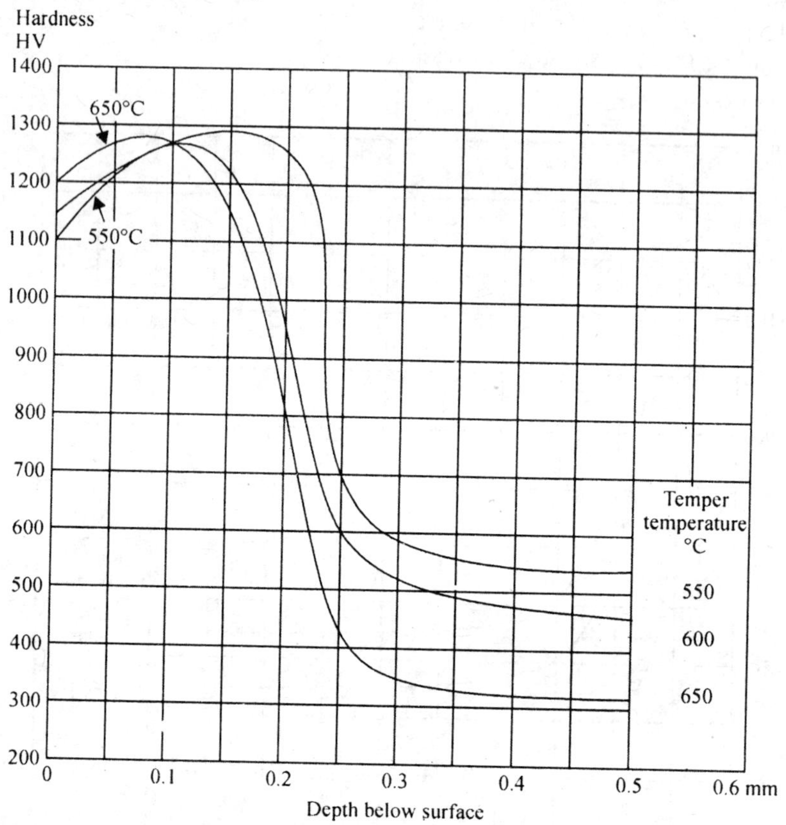

Fig. 20.26. Gas nitriding of H 13 at 525° for 30 hours after hardening followed by tempering at different temperatures.

Properties of Nitrided Steels

Wear resistance

The wear resistance of steels increases with increasing hardness. Since the concept of 'wear resistance' is difficult to define, a warning should be given against any generalisations. If the hardness of a cutting edge should exceed a certain value, small fragment of the edge break away; as machining continues, the edge becomes blunt and a pattern reminiscent of wear arises. The nitrided layer is very liable to chip in this way. In the case of abrasive wear of a scratching or eroding nature, it is doubtful whether nitriding will provide any long-term improvement in wear resistance. With gouging or grinding abrasion, nitriding is not to be recommended. Fig. 20.30 shows the results obtained from shot-blasting tests with steel grit on steels of different hardness, the least abrasion being reported for the nitrided steel.

It is for adhesive wear especially that the main advantages are achieved with nitriding, because the coefficient of friction is reduced, as is also the risk of scuffing. A further improvement is achieved by nitrocarburising and particularly by the Sulfinuz treatment, a fact confirmed by practical tests and laboratory investigations. During wear tests in an Amsler machine, in which two discs made from a 0.15 per cent carbon steel rotated against each other at a speed of 440 and 400 rpm respectively, the values shown in Fig. 20.31 were obtained. The specimen given the Sulfinuz treatment is superior to both the case-hardened and the nitrided specimens. Scuffing has started at the point where the curves noticeably change direction. In the initial stages the Sulfinuz-treated specimen suffered somewhat more abrasion than the case-hardened or the nitrided specimens because of the wearing away of the black surface coating.

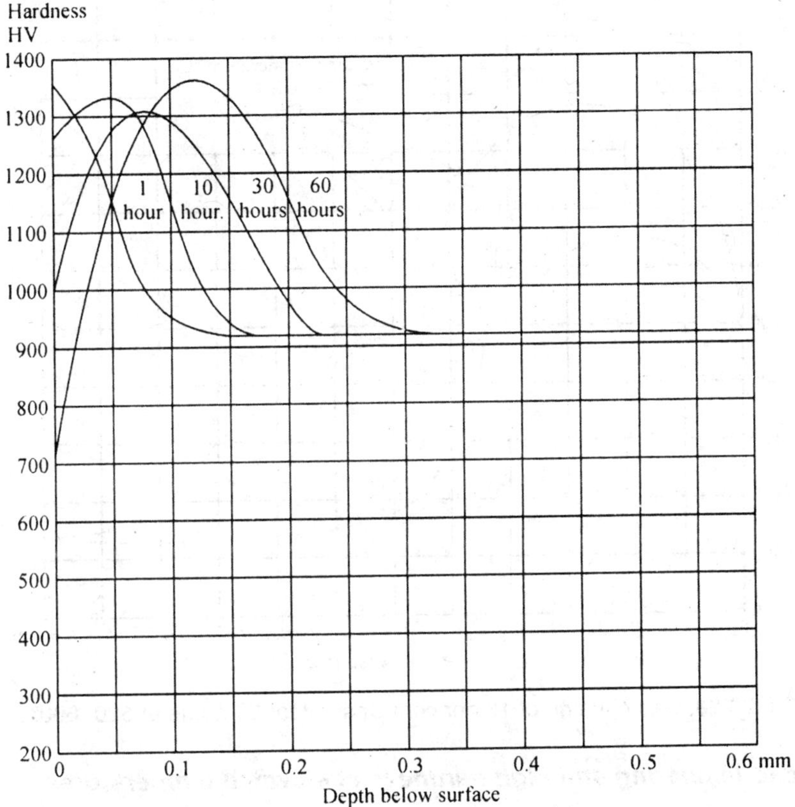

Fig. 20.27. Gas nitriding at 510°C for 1–16 hrs., of high-speed steel M 42. Degree of dissociation: 30 per cent.

Under certain conditions nitrided steels can function without lubrication, this being particularly marked with Sulfinuz-treated steels. This characteristic has been tested in a Faville-Levally machine, in which a test bar rotates at a speed of 300 rpm between two hardened V-jaws. The pressure is gradually raised until scuffing occurs. When tests were made with oil-lubricated specimens made from 0.15 per cent carbon steel, the untreated bar scuffed at a loading of 380 kp. The Sulfinuz-treated bar exhibited no sign or scuffing whatever at the maximum loading of the machine, i.e., 1140 kp (Fig. 20.32). Corresponding tests were made on unlubricated specimens and the untreated specimen scuffed at a loading of 250 kp. The test with the Sulfinuz-treated bar had to be discontinued at a loading of 630 kp because the specimen lengthened as a result of the frictional heat (Fig. 20.33).

Practical tests were performed involving the Sulfinuz treatment of case-hardened steel parts that exhibited a tendency to scuffing. As an example we can mention a bushing made from case-hardened Cr-

Ni steel which started scuffing against the shaft after 7 minutes at a speed of 4000 rpm. After Sulfinuz treatment at 550°C for 2 hours, the bushing could be run for 2 hours, without scuffing. Because of the heat of friction generated the machine then started to operate sluggishly. After being allowed to cool to room temperature the running test could be resumed.

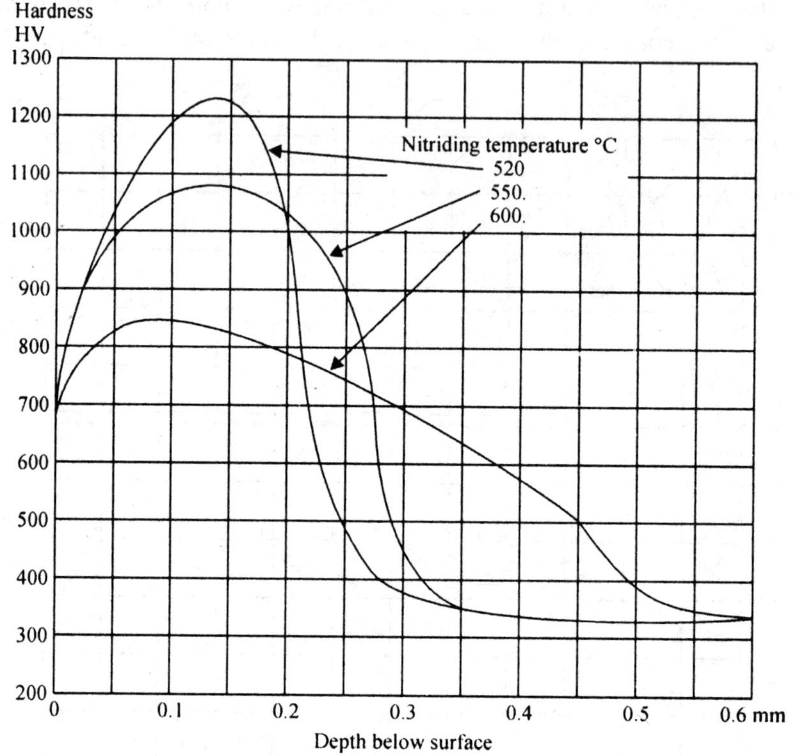

Fig. 20.28. Gas nitriding of 13 per cent Cr steel for 60 hours at 520–600°C.

High resistance to tempering and high hardness at elevated temperatures

These are valuable properties of nitrided steels. After nitriding, the steel can be heated up to the same temperature as that at which it was nitrided and in some cases to an even high temperature, without loss of hardness at room temperature (Fig.20.34). Fig. 20.35 shows that the hardness at elevated-temperatures is also relatively high. The excellent wear resistance and elevated-temperature hardness combined ensure that nitrided steels can be used with good results for hot-work dies. However, the choice of steel and nitriding time must be adapted to suit working conditions.

High fatigue strength and low notch sensitivity

Nitrided steels are characterised by high fatigue strength and low notch sensitivity. The increase in fatigue strength is highest for plain-carbon and low-alloy steels and this generally increases as the nitriding time increases. The standardised nitriding steels heat treated to an ultimate tensile strength of 900 N/mm^2 have a fatigue strength of roughly 450 N/mm^2, as determined during rotary bending fatigue tests on highly finished ground test bars, 10 mm diameter. By gas nitriding at 510°C for 20 hours the fatigue strength increases to roughly 600 N/mm^2. Nitriding for 60 hours gives a fatigue strength of around 700 N/mm^2 at 10^7 load cycles.

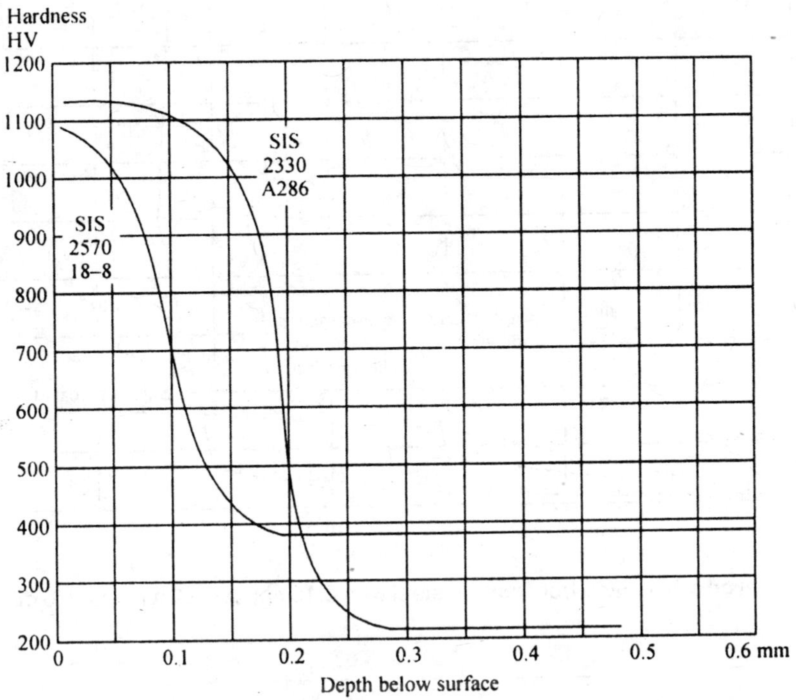

Fig. 20.29. Gas nitriding of $^{18}/_{8}$ steel and steel A 286 60 hr at 600°C.

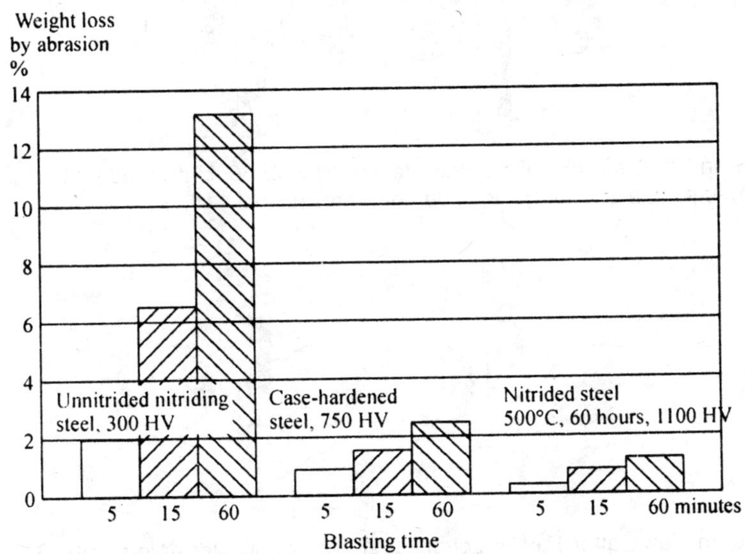

Fig. 20.30. Abrasion loss occurring during blasting with steel grit on unnitrided, case-hardened and nitrided test specimens respectively.

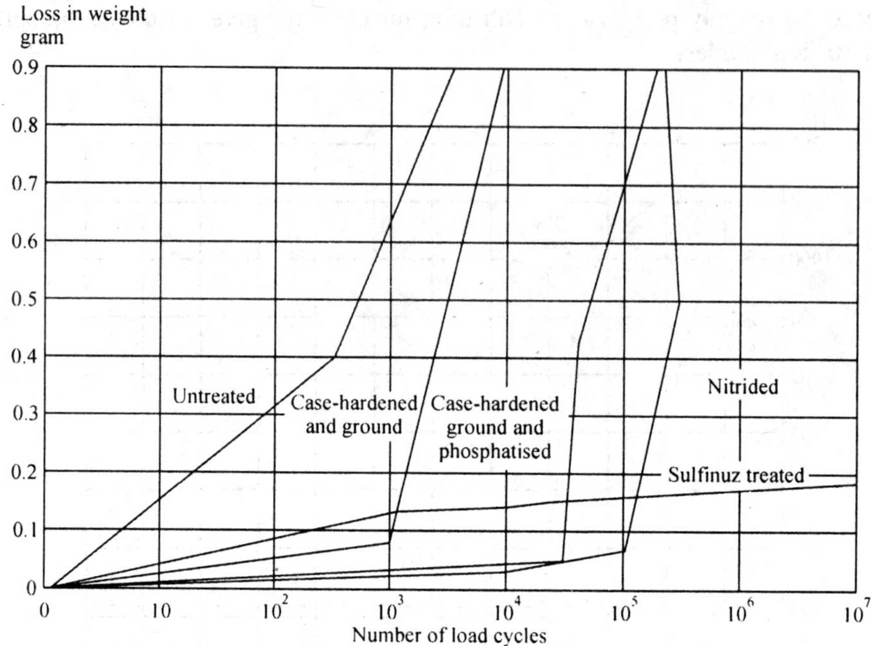

Fig. 20.31. Amsler test on discs made from carbon steel with 0.15 per cent C subjected to different treatments. Loading 20 kp.

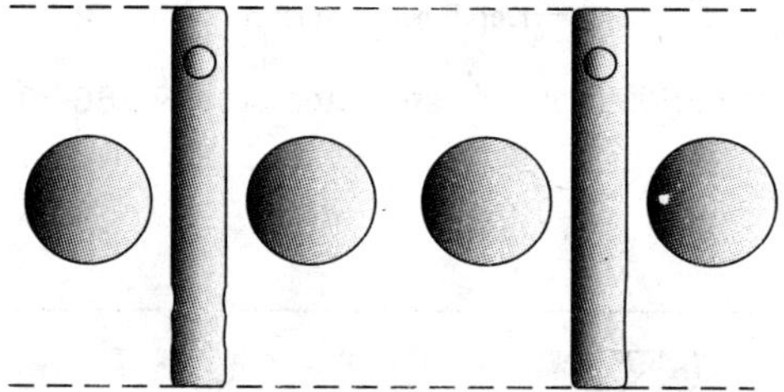

Fig. 20.32. Test bars and jaws after Faville-Levally test in oil SAE 30. Untreated (left): Scuffing at pressure of 380 kp. Sulfinuz treated specimen (right): No scuffing at pressure of 1140 kp.

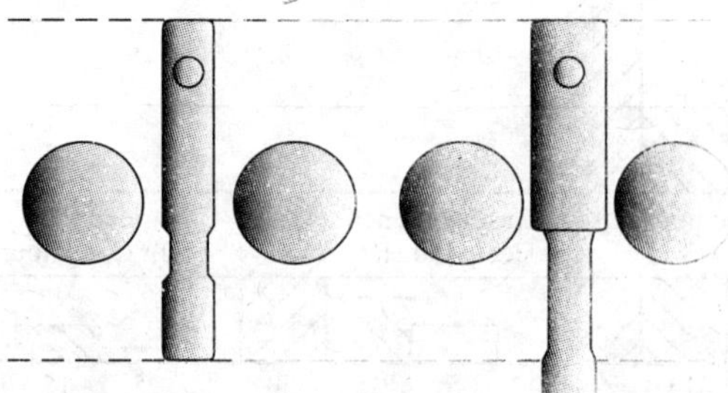

Fig. 20.33. Test bars and jaws after Faville-Levally test without oil. Untreated (left): Scuffing at pressure of 250 kp. Sulfinuz treated (right): No scuffing at pressure of 630 kp, but test bar lengthened as a result of plastic deformation.

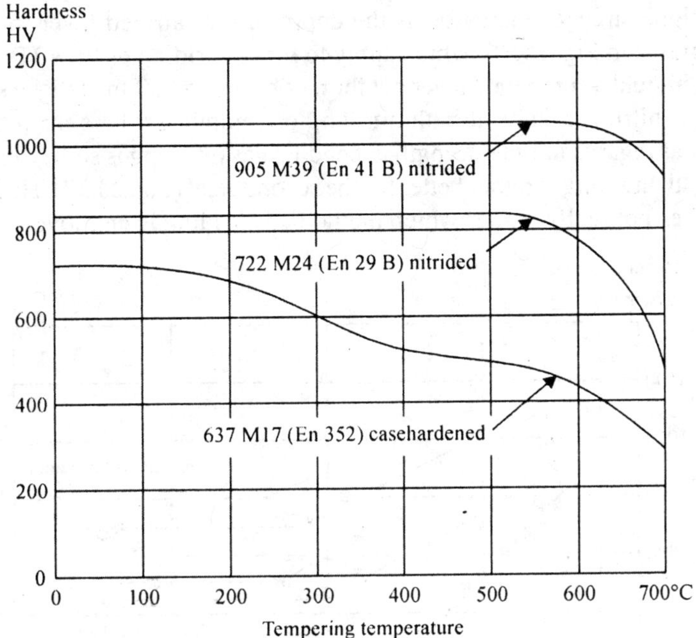

Fig. 20.34. Tempering diagram for nitrided and case-hardened steels respectively.

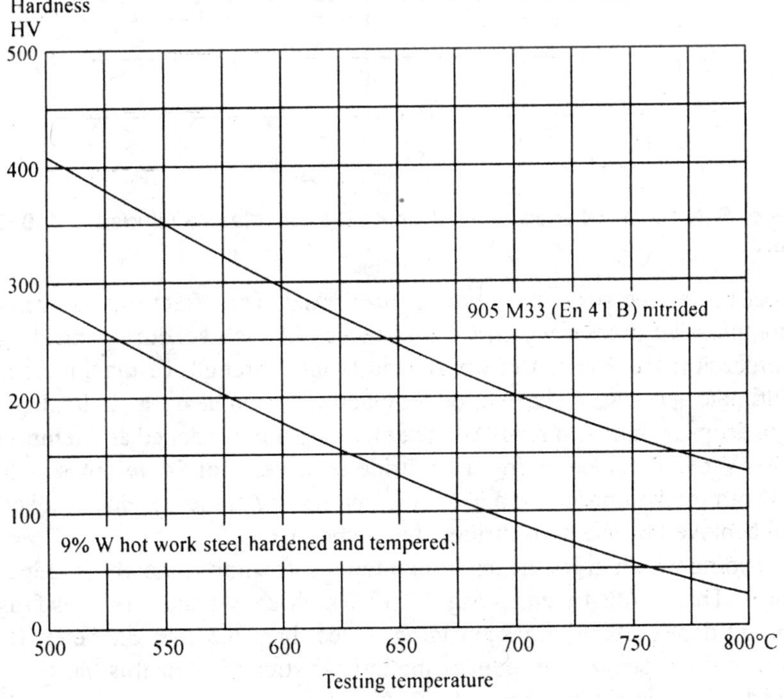

Fig. 20.35. Elevated temperature hardness of nitrided steel and hot-work steel.

In general the fatigue strength increases as the depth of the nitrided layer increases. This point is illustrated in the diagrams in Fig. 20.36, which apply to the ionitriding of En 40 B where the compound layer was 1 μm thick. Should a compound layer not form or be ground off the fatigue strength will be lower.

Tests involving gas nitriding and salt-bath nitriding of smooth test bars made from steel DIN C 35 and 34 Cr 4 show that as regards fatigue strength in bending gas nitriding is somewhat more advantageous for the first and salt-bath nitriding is rather better for the second steel (Fig. 20.37). However, the differences are so slight that it does not really matter which method of nitriding is employed.

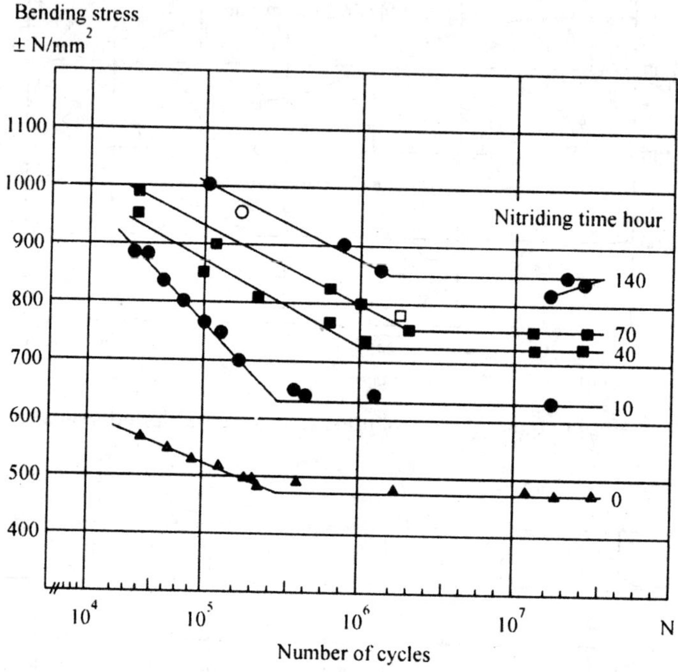

Fig. 20.36. Wöhler or S–N curves of specimens of En 40 B bright plasma nitrided at 480°C for 1,04,070 and 140 hours respectively.

The notch effect on fatigue strength is extremely marked. The effect of this unfavourable factor is lessened by the compressive stresses which are introduced into the steel during nitriding. Hence nitriding is a very useful process if the aim is to increase the fatigue strength of machine components which, because of their ultimate application, have to be provided with notches, e.g., bolts. Fig. 20.38 shows that the fatigue strength drops by half as a result of the notch effect in hardened and tempered test bars made from DIN 14 CrMoV 69. By undertaking a Tufftride treatment for 90 minutes at 570°C the fatigue strength increases both for the unnotched and for the notched bars, the nitrided notched bar reaching the same value as that achieved by the non-nitrided, unnotched bars.

The depth of nitriding which gives the maximum fatigue strength is, however, dependent on material thickness and shape. This is illustrated in Fig. 20.39 which shows the torsional fatigue strength as a function of nitriding depth after the gas nitriding of test bars 6.5 mm diameter. One test bar has a transverse hole 1 mm diameter, which reduces the fatigue strength. For this bar the maximum fatigue strength is achieved with a nitriding depth of 0.3 mm. In the case of the undrilled bar the maximum fatigue strength is obtained with a nitriding depth of 0.4 mm. With thin-walled parts, salt-bath nitriding is preferable in such cases as compared with gas nitriding because it is easier to control the depth of nitriding during bath nitriding when such shallow depths are required.

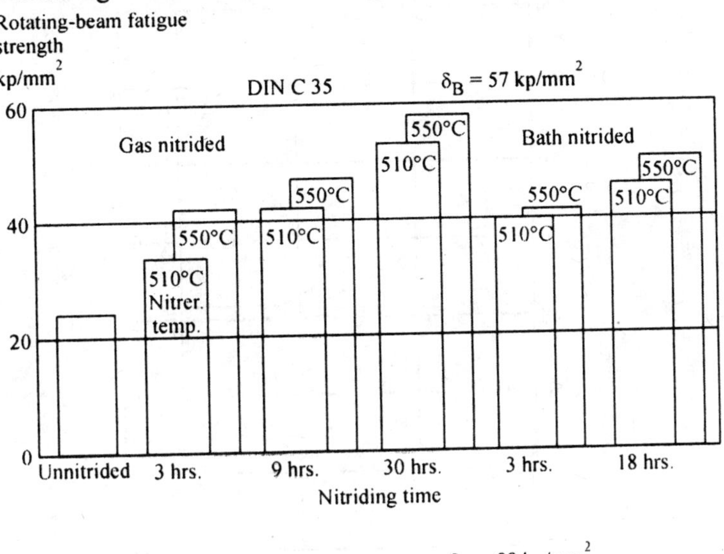

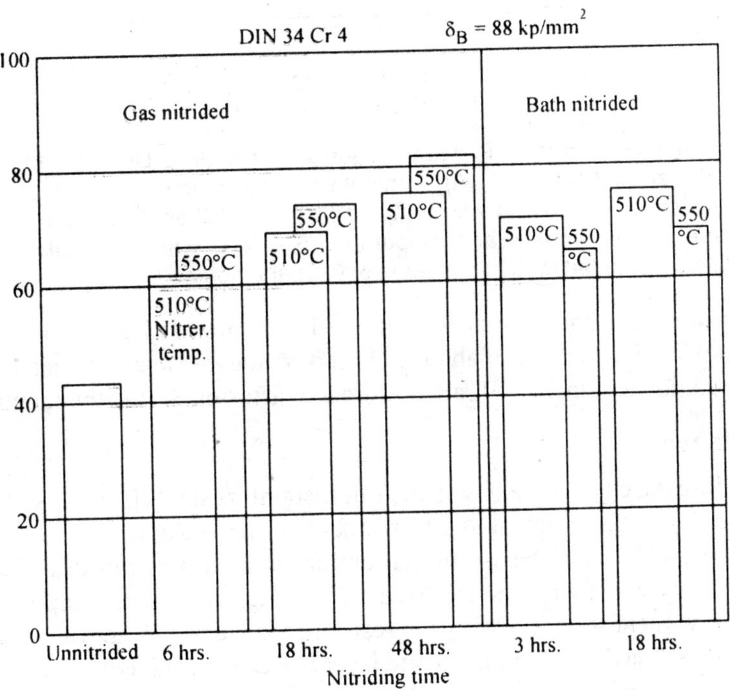

Fig. 20.37. Increase in fatigue strength of test bars 5.9 mm diameter made of hardened and tempered steel after gas and salt-bath nitriding at 510 and 550°C.

Should a nitrided part be subjected to such high loads that cracks form in the nitrided layer the fatigue strength may fall off drastically. Such parts as have pronounced stress raisers, where stress concentrations are high, are also particularly, susceptible to fatigue failure after nitriding.

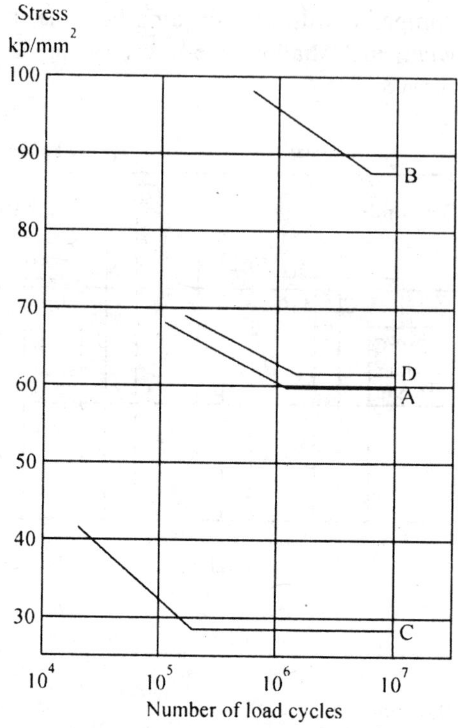

Fig. 20.38. Effect of Tufftriding on rotating-beam fatigue strength. Steel DIN 14 CrMoV 69 (0.14 per cent C, 1.5 per cent Cr, 0.09 per cent Mo + V). (A) = smooth test specimen, hardened and tempered at 570°C; (B) = smooth test specimen, hardened and tempered at 600°C then Tufftrided for 90 minute at 570°C; (C) = notched test specimen, hardened and tempered at 570°C, D = notched test specimen, hardened and tempered at 600°C then Tufftrided for 90 minutes at 570°C; and notched factor α_k = 2.

Fior and *Corredor* have shown that the fatigue life of nitrided SAE 7140 (Nitralloy 135H) is exceptionally short when frequencies of about 5 Hz and smaller are applied. This fact is probably due to longer times of relaxation and increased rate of nitrogen diffusion during the fatigue test.

Corrosion resistance

As a result of nitriding the corrosion resistance of non-stainless steels is increased. From the corrosion viewpoint the white layer is equivalent to a 13 per cent martensite chromium steel. In stainless steels, corrosion resistance is reduced because the chromium is bound as nitrides. The temperature used during nitriding may also render non-stabilised 18/8 stainless steels susceptible to intercrystalline corrosion during subsequent service. Corrosion resistance is greatest for nitrided and polished surfaces however by grinding off, say 0.03 mm, the white layer is removed and corrosion resistance is reduced. After grinding 0.03–0.06 mm steel BS 905M39 has better corrosion resistance than the conventional heat-treatable steels and BS 402S37. When tested in salt-water spray, the two last mentioned groups exhibited roughly the same corrosion resistance in the nitrided layer. From various nitriding tests on 18/8 steels, contradictory results have been obtained, but in any case we must expect a reduction in corrosion resistance, at least down to the same level as that shown by martensitic chromium steels in the hardened and tempered state.

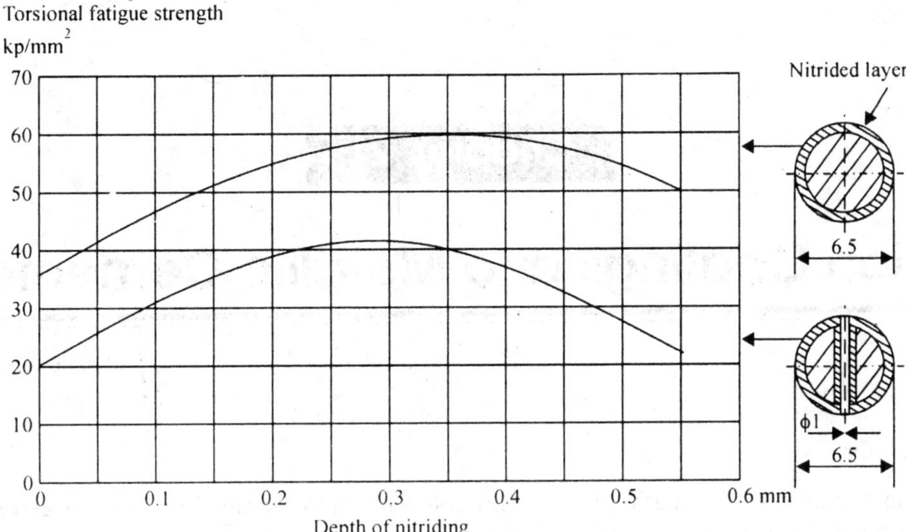

Fig. 20.39. Torsional fatigue strength as function of depth of nitriding of a smooth test specimen and a drilled test specimen, respectively, made from a Cr–Mo–V steel having the following composition: 0.3 per cent C, 2.5 per cent Cr, 0.2 per cent Mo, 0.25 per cent V; σ_B = 110 kp/mm^2.

Similar results were obtained during corrosion tests in tropical atmosphere testing chambers. In this instance test specimens were used having dimensions of 15 mm diameter × 50 mm. They were tested in the 'normal state'; after a Sulfinuz treatment for 1 hour at 570°C; after gas nitriding for 30 hours at 510°C; and after gas nitriding and then having approximately 0.05 mm on the diameter ground off. The matt surface on the specimens that were nitrided only is due to the film of oxide that forms during nitriding and that tends to reduce somewhat the corrosion resistance. The best results are obtained if, as mentioned above, polishing is undertaken subsequent to nitriding. After grinding, the specimen made from BS 905M39 has withstood the corrosion attack better than have the other grades of steel. The poor results obtained after Sulfinuz treatment were due in part to the fact that the specimens were not polished prior to corrosion testing.

Chapter 21

Diffusion Coatings and Metallic Cementation

INTRODUCTION

The art of impregnation, or penetration of elements into a metal by atomic diffusion processes, has been practiced for hundreds of years. Damascus steel was produced by the cementation process of diffusing carbon into iron. To be successful, an alloy system must possess a wide range of intersolubility or must produce intermetallic compounds of the two elements. The depth of penetration is a function of temperature and time at temperature, as in the case of carburising and nitriding.

Saturation of the surface of steel by aluminium, chromium, beryllium and other elements is called diffusion coating. Parts, whose surface is coated in this manner, acquire various valuable properties, including high heat resistance, corrosion resistance, increased wear resistance and hardness.

Diffusion coating of steel by various metals and also by silicon is done by embedding the parts in the corresponding powdered mixtures (usually ferroalloys), by immersing in the molten metal if the diffusing element has a comparatively low melting point (for example, zinc and aluminium), or by saturation from a gaseous medium.

Diffusion coating of iron (or steel) by metals by means of direct contact of the powdered medium with the work surface presents various difficulties due to the high temperatures and the long holding time required. These difficulties may be overcome to a great extent by saturation from a gaseous medium. Volatile chlorous compounds of the metals ($AlCl_3$, $CrCl_2$, $SiCl_4$, etc.,) are used as the medium. They are obtained by the action of chlorine (or hydrogen chloride) on the metals or their alloys with iron (ferroalloys) at high temperatures. These chlorides dissociate on the surface of the work and the metal produced in the atomic state diffuses into the iron.

The rate of diffusion depends to a considerable extent upon the composition of the steel. Increased carbon content impedes diffusion and a re-distribution of the carbon content is observed in the surface layers. In certain cases, the carbon atoms are crowded out of the diffusion layer by aluminium or silicon; in other cases, on the opposite, carbon diffuses intensively toward the surface (for example, in saturation with chromium).

Direct-contact impregnation may be accomplished by heating steel to a high temperature with the cementation medium in the form of a solid, liquid, or gas. A common method consists of packing the steel part with a metal powder and carrier, such as chlorine or a volatile chloride. Complex alloy coatings may be produced by impregnation with more than one metal.

Indirect methods of impregnation involve the diffusion by heat treatment of coatings previously applied by hot dipping, electroplating, spraying, cladding, etc.

Many binary systems of iron have been studied to determine which of the elements might be applied as a coating for steel by the impregnation process; however, relatively few of the elements have gained commercial acceptance. Those systems having present or potential value are shown in Table 21.1.

Table 21.1. Diffusion coatings on iron and steel.

Commercial processes		*Promising systems*	
Element	*Process*	*Element*	*Characteristics of coating*
Al	Calorising	B	Hard; resists HCl
C	Carburising	Be	Very hard; resists atmospheric corrosion
Cr	Chromising		
N	Nitriding	Cd	Similar to galvanising
Si	Siliconising	Co	Improved corrosion resistance
Sn	Stannising	Mn	Fair corrosion resistance
Zn	Sherardising	Mo	Very hard; fair corrosion resistance
		Ni	Good corrosion resistance
		Sb	Resists HCl and dilute H_2SO_4
		Ta	Hard; resists dilute HCl and H_2SO_4
		Ti	Hard
		V	Very hard; resists HNO_3
		W	Very hard; resists H_2SO_4
		Zr	–

One process of aluminium impregnation, or calorising, consists of heating steel parts in an airtight box with a mixture of powdered aluminium, aluminium oxide and aluminium chloride for a period of from 6 to 24 hours at a temperature between 1500 and 1800°F (815 and 980°C). Impregnation is designed to take place in the first 6 hours., followed by a homogenising anneal to promote diffusion of the coating. The diffused layer usually extends from 0.025 to 0.040 inch in depth, with the surface containing about 25 per cent aluminium. Calorised steel resists the attack of flue gases containing sulphur that are found at high temperatures in oil refineries. Other uses include bolts, salt pots, furnace parts, etc., for service up to about 1500°F (815°C).

CALORISING

Calorising is a procedure in which aluminium penetrates into the surface of steel to impart to it high heat resistance (up to 850–900°C). This property of calorised parts is due to the formation of a dense surface film of aluminium oxide (Al_2O_3) which provides reliable protective action against oxidation and corrosion.

Calorising Techniques

Several calorising techniques are employed which are discussed below.

Powder calorising

In power calorising the parts are packed in a box where they are embedded in a mixture of aluminium or ferroaluminium powder and a small amount of ammonium chloride (0.5 to 2 per cent). Sometimes clay, alumina, or chamotte are added to prevent caking. The boxes are held for several hours at a temperature from 950 to 1000°C. The time required for calorising low-carbon steel at a given temperature depends upon the specified depth of the diffusion layer. From 3 to 6 hours will be required to obtain a case 0.2–0.3 mm deep at 950°C; a case 0.4–0.5 mm deep will require from 12 to 16 hours.

Chemical processes that occur in powder calorising consist in dissociation of the ammonium chloride at a high temperature into ammonia and hydrogen chloride:

$$NH_4Cl \rightarrow NH_3 + HCl$$

Further, the hydrogen chloride reacts with the aluminium (or ferroaluminium) to form the volatile compound, aluminium chloride:

$$6HCl + 2Al \rightarrow 2AlCl_3 + 3H_2$$

Then the following reaction occurs upon contact with iron:

$$Fe + AlCl_3 \rightarrow FeCl_3 + Al_{at}$$

Atomic aluminium diffuses into the iron.

Calorising is sometimes followed by diffusion annealing at 900–1000°C to reduce the brittleness of the calorised layer which is due to heavy supersaturation of the surface zones in the diffusion layer with aluminium (40 to 50 per cent). Diffusion annealing reduces the aluminium concentration at the surface (by diffusion deeper into the steel) thus reducing its brittleness. This is the most extensively used calorising procedure.

Dip calorising

This process entails the immersion of steel parts into a bath of molten aluminium saturated with iron (6 to 8 per cent) to pervent the aluminium from corroding (dissolving) the steel part. The bath temperature is 750–800°C and the process takes from 45 to 90 minutes. This will provide a calorised case from 0.2 to 0.35 mm deep. Calorising is followed by diffusion annealing at 900–1000°C.

Disadvantages of this procedure are the embrittlement caused by the supersaturation of the layer with aluminium and the low service life of the crucibles in which the aluminium is melted.

In addition to the two calorising methods described above, there are several methods in which the steel is first simply mechanically coated with aluminium (by plating or metal spraying) which diffuses into the steel upon subsequent high-temperature annealing.

The calorised layer, obtained in powder mixtures or by the dip method, is a solid solution of aluminium in iron. The aluminium content at the surface of the case is about 30 per cent.

An increase in the carbon content or that of alloying elements in the steel impedes the diffusion of aluminium (Fig. 21.1).

Calorising is applied to the fire grates of gas generators, tubes of thermocouples, components of pouring ladles, and to other parts operating at high temperatures.

New methods of rapid calorising in molten aluminium under a layer of flux are of great practical interest. Work with a carefully cleaned surface is heated in a flux melt (for example, 40 per cent KCl, 40 per cent NaCl, 10 per cent Na_3AlF_6 and 10 per cent AlF_3) in a bath having a layer of molten aluminium at the bottom.

Heating in the flux continues for 4 or 5 minutes and then the part is lowered into the molten aluminium which has a temperature of 720–780°C and is held there from several seconds to 10 or 15 minutes depending on the type of the part being treated.

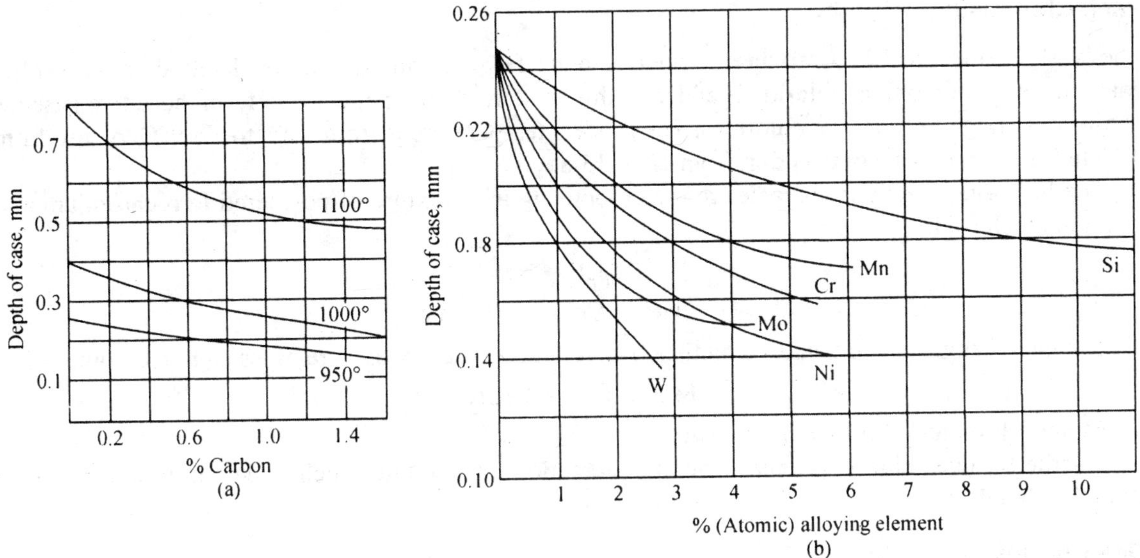

Fig. 21.1. Relationship between the depth of the calorised case and the content (a) of carbon; and (b) of alloying elements.

Finally, the part is blasted with hot air to remove adhering aluminium.

The case obtained by this method comprises two parts: (i) a layer of pure aluminium 0.04 to 0.05 mm thick; and (ii) a diffusion layer 0.015 to 0.05 mm thick consisting mainly of α solution.

A rapid calorising procedure by means of metallisation is being used in the USA. The work is heated to 200–250°C and aluminium is sprayed on its surface by an oxyacetylene spraying device. After this, the work is heated to 780–810°C and is held at this temperature for 10 to 15 seconds. Heating is accomplished either in a molten flux bath or by high-frequency induction. The layer obtained will be from 0.015 to 0.025 mm deep.

Rapid calorising is applied to automobile valves and also to sheets and wire, instead of galvanising, as an anticorrosive measure.

CHROMISING OF STEEL

Chromium impregnation, or chromising, may be accomplished by heating iron or low-carbon steel parts in a container with a mixture of powdered chromium and alumina in a hydrogen atmosphere for about 4 hours at a temperature in the range of 2375 to 2550°F (1300 to 1400°C). Gaseous methods have also been developed by the reaction of hydrogen and hydrogen chloride with chromium to form $CrCl_2$ as the active carrier. The process is relatively expensive and produces hard, brittle coatings varying from 0.0005 to 0.005 inch in depth with the surface containing about 10 or 30 per cent chromium. Chromising has been employed to protect turbine buckets from corrosion or erosion and to provide protection for steel parts from atmospheric corrosion and attack by nitric acid.

The saturation of the surface of steel machine parts by chromium provides a high resistance to corrosion by gases at temperatures up to 800°C and high anticorrosive properties in such media as fresh water, sea water and nitric acid. High-carbon steels are chromised to obtain hardness and wear resistance. There are three principal types of chromising techniques.

Pack Chromising

The work is embedded in a mixture of finely ground ferrochromium and powdered alumina, kaolin, chamotte, etc. Ammonium chloride is added to this mixture (1 to 5 per cent) to form the active gaseous phase consisting of chromium chlorides. The process temperature is from 1100 to 1200°C. A case from 0.15 to 0.20 mm may be obtained in from 12 to 15 hours.

The following reactions characterise the chromising in a mixture of powdered ferrochromium and ammonium chloride:

$$NH_4Cl \rightarrow NH_3 + HCl$$

$$2HCl + Cr \rightarrow H_2 + CrCl_2$$

The interchange reaction, caused by the contact with the iron (work), releases atomic chromium:

$$Fe + CrCl_2 \rightarrow FeCl_2 + Cr_{at}$$

Atomic chromium diffuses into the work.

Volatile ferrous chloride produced by the interaction of chromium dichloride and iron is removed from the chamber.

Gas Chromising

In this process, the work is heated to 950–1050°C in an atmosphere of chromium dichloride and chromic chloride vapours ($CrCl_2$ and $CrCl_3$) obtained by passing hydrogen chloride vapours over heated chromium or ferrochromium.

At the chromising temperature, the vapours of chromium dichloride and chromic chloride contact the surface of the work. This causes an interchange reaction in which atomic chromium is released to diffuse into the iron:

$$CrCl_3 + Fe \rightarrow FeCl_3 + Cr_{at}$$

$$CrCl_2 + Fe \rightarrow FeCl_2 + Cr_{at}$$

Liquid Chromising

Here, the work is held at 900–1000°C in the molten salts $BaCl_2$, $MgCl_2$, etc., to which is added either chromium dichloride ($CrCl_2$), in a proportion of 15 to 30 per cent by weight, or ferrochromium (20 to 25 per cent).

Liquid chromising is based on the following reaction:

$$CrCl_2 \text{ (in the molten salt)} + Fe \text{ (work)} \rightarrow FeCl_2 + Cr_{at}$$

When commercial iron is chromised, the diffusion layer consists of a solid solution of chromium in α-iron. In high-carbon steels the layer will consist almost completely of the chromium carbide $(FeCr)_7C_3$. The formation of a continuous carbide layer is due to the diffusion of carbon from the inner layer to meet the chromium. Carbon has a higher rate of diffusion than chromium and therefore, not all of it is used in forming the carbide layer; a portion forms a transition layer with a high carbon content. Due to the intensive carbon diffusion in low-carbon steels a transition layer consisting of chromium carbide may be formed to meet the chromium. The carbide $(FeCr)_7C_3$ formed at the surface of the diffusion layer considerably increases its hardness. The chromised case obtained on iron has a hardness of 250–300 VHN, its hardness on high-carbon steel is 1200–1300 VHN. The depth of the case does not ordinarily exceed 0.15–0.20 mm. Carbon in steel impedes the diffusion of chromium into the iron. The addition of tungsten, molybdenum, and silicon to steel accelerates diffusion while chromium, nickel, and manganese reduce its rate (Fig. 21.2).

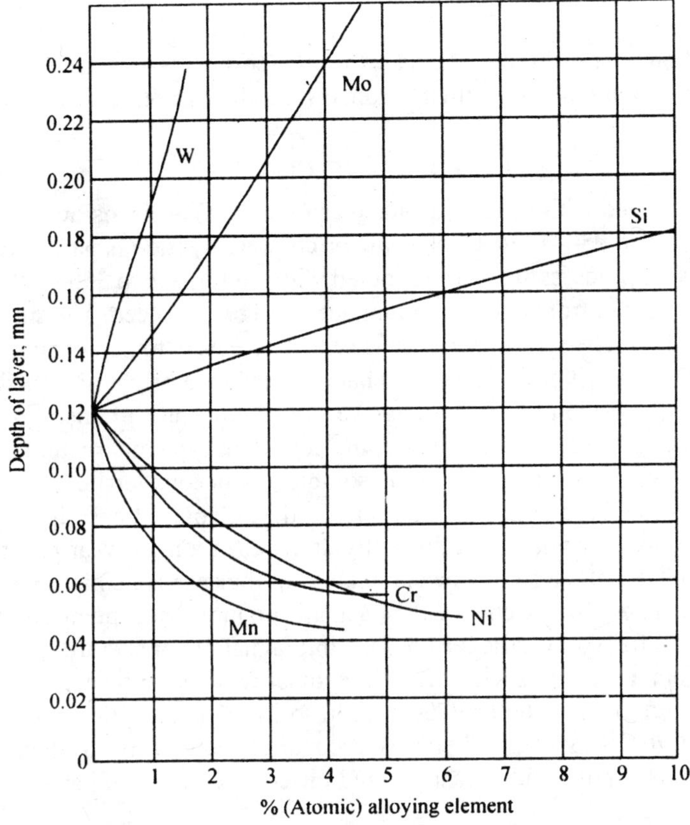

Fig. 21.2. Effect of alloying elements on the depth of the chromised case.

Chromising is applied in the manufacture of components for steam power equipment, water and steam fixtures, valves, connection pipes, and machine parts subject to wear and exposed to corrosive media.

SILICONISING

Siliconising is a process in which silicon diffuses from the surrounding medium to saturate the surface of steel (or malleable iron). Siliconised steel acquires a high resistance to the corrosive effects of sea water, as well as nitric, sulphuric, and hydrochloric acids, resistance to scaling at high temperatures (up to 750°C), and a somewhat higher wear resistance.

Methods of Siliconising

Two principal siliconsing methods are employed.

Pack siliconising

The work is embedded in a layer of ferrosilicon (50 to 95 per cent Si), ammonium chloride (2 to 5 per cent), and neutral components (chamotte and quartz sand). The process is carried out at 1100–1200°C and takes from 2 to 24 hours. The case obtained is from 0.2 to 0.8 mm deep. Siliconsing in powdered mixtures is rarely applied.

Gas siliconising

Gas siliconising is accomplished in retort furnaces where the steel is heated in an atmosphere of silicon tetrachloride ($SiCl_4$). Upon interaction with the iron, this produces atomic silicon which diffuses into the metal:

$$3SiCl_4 + 4Fe \rightarrow 4FeCl_3 + 3Si_{at}$$

The retort is first changed with silicon carbide (carborundum) or ferrosilicon and then with the part. After heating the furnace to 950–1050°C, chlorine or chlorine hydride is passed through the retort at a low rate. At the end of the process, the work is cooled with the furnace to 300–400°C. The time required is from 2 to 5 hours and a case from 0.5 to 1.25 mm deep will be produced. The siliconised case is a solid solution of silicon in α-iron. The surface zone of the case has a concentration of about 14 to 15 per cent Si.

The microstructure of the siliconised case is characterised by a bright unetched layer at the surface. This is a solid solution of silicon in α-iron. Pearlite segregations are frequently observed beneath the diffusion layer. Below this, the structure is that of the core. This type of structure is due to the expelling of carbon from the diffusion layer since it is little soluble in siliceous ferrite.

Notwithstanding its low hardness (200–300 VHN), the siliconised case is very brittle and may be machined by cutting tools only with great difficulty. It acquires a high wear resistance, however, after being saturated with oil at 170–200°C. A special feature of the case is its high porosity.

Components of machinery for the chemical, paper, and oil industries (pump shafts, pipelines, fittings, nuts, bolts, etc.) are most frequently siliconised. Zinc impregnation, or sherardising, involves the heating of iron and steel parts in a rotating sealed drum in contact with zinc powder for a period of from 3 to 12 hours at a temperature of from 660 to 700°F (350 to 370°C). A coating about 0.002 inch thick is obtained in 3 hours at 700°F (370°C). Sherardised coatings are uniform in thickness and particularly adapted to the protection of bolts, nuts, washers, and miscellaneous small parts.

BORON SATURATION

Steel, saturated with boron, acquires very high hardness, wear resistance, and corrosion resistance in various media. Steel may be saturated with boron by embedding in ferroboron powder and heating to 950–1050°C.

A more promising procedure involves the electrolysis of molten salts containing boron. The parts to be treated are immersed as cathodes in a bath of molten borax ($Na_2B_4O_7$). Both of molten chlorous salts (NaCl, $BaCl_2$), to which powdered ferroboron or boron carbide has been added, are also suitable for this process.Excellent results are obtained by a gaseous saturation method. Here steel is saturated with boron in a medium of diborane (B_2H_6) mixed with hydrogen (B_2H_6: H_2 = 1:50) or boron chloride (BCl_3). mixed with hydrogen, at a temperature of 850–900°C.

These boron treatments result in a case having a layer of iron monoboride FeB at the surface, followed by iron sub-boride Fe_2B and then a solid solution of boron in iron.

The borides in the diffusion layer form specific columnar crystallites. Only a part of the layer is continuous. The hardest component is iron monoboride FeB (2000 VHN); the hardness of iron sub-boride Fe_2B is somewhat lower (1600–1800 VHN). The more carbon and alloying elements in the steel, the less the depth of the case will be. High-alloy steels, including stainless and heat-resistant steels, give good results when they are saturated with boron.

SPRAYED, FACED AND CLAD COATINGS

The coating methods discussed in this chapter require parts that can be placed in a processing pot, tank, or retort. Large parts and heavy coatings are more economically prepared by special techniques such as metal spraying, facing, or cladding.

Metal spraying consists of melting metal wire or powder with an oxyhydrogen or oxyacetylene flame in a special metallising gun which sprays the metal particles by an air blast. Molten particles impinging on the previously prepared surface to be coated are flattened and interlocked to provide mechanical bonding rather than alloying with the base metal. The process requires roughening of the surface by sand blasting, rough turning, or other techniques. A subsequent diffusion heat treatment may be required to obtain the best bonding conditions. In general, coatings of from 0.001 to 0.015 inch thick are applied for the corrosion protection of steel. Somewhat heavier coatings are used to re-build worn or undersize parts. Almost any metal that is obtainable in wire or powder form may be employed, provided it can be melted in the oxyacetylene flame. Metal spraying is adapted to the coating of large structures, such as bridges, storage tanks, tank cars, etc., and the building up of worn pistons, shafts, bearings, etc.

Flame plating is process where in tungsten carbide particles are propelled from a special gun at supersonic velocity which is produced by the detonation of oxygen and acetylene. This action heats the particles so that they become plastic prior to becoming imbedded in the surface of the work. Coatings 0.002 to 0.010 inch thick may be applied, and the temperature of the workpiece seldom exceeds 400° F (210°C). The process finds application in increasing the wear resistance of such items as gauges.

Metal facing, or hard surfacing, is the operation of welding metal to the surface of a part in order to improve the abrasion, corrosion, or heat resistance of the surface, or to re-build worn or eroded parts. Facing materials include the high-carbon and high-alloy steels, cobalt and nickel-base alloys and carbides of tungsten, tantalum and boron mounted in suitable alloy binders to increase their toughness. Hard facing is applied to valve seats, metal-working dies, oil-well drilling tools, dredge-bucket teeth, etc.

When a surface coating exceeds about 3 per cent of the total mass of the steel base, the resultant product is known as a composite steel and the process is called cladding. Cladding may be accomplished in several ways, including casting the clad metal around or adjacent to a previously formed steel ingot, rolling composite plates to produce bonding by pressure welding, or building up a thick surface layer by fusion welding in a manner similar to hard facing.

Examples of composite or clad steels include high-carbon, alloy or high-speed steel bonded to mild steel for use in hacksaw blades, circular knives, broaches, blanking dies, jailbar, vault plate, etc. Stainless clad steels are produced with a 10 to 20 per cent thickness of stainless steel cladding on one or both sides of a mild steel backing for use in cooking vassels for the varnish, canning, and soap industries. The backing serves to increase the thermal conductivity of the stainless cladding and avoids localised hot spots. Nickel-clad and nickel-alloy clad steel plates are manufactured with coatings of from 5 to 25 per cent of the thickness for use in pressure vessels incorporating excellent corrosion resistance with high strength. Copper-clad steel wire is produced with high electrical conductivity on the surface, combined with a high-strength steel core.

NON-METALLIC COATINGS

Artificial black or blue-black oxide coatings may be produced on steel by heat alone or by contact with various solid, liquid, or gasesous oxidising agents. Room-temperature oxidation with chemical oxidisers or by electrolytic treatment has the advantage of avoiding possible tempering of hardened steel parts, but it does not, in general, produce the lasting finish that may be obtained at higher temperatures. A prerequisite for successful blueing and blackening of steel is a 'chemically clean' surface, free of oil and other contaminating agents. Oxide finishes are employed chiefly for appearance, with very limited protection against corrosion. Typical applications include gun parts, screws, nails, bicycle chain, blued sheet products, etc.

Phosphate coatings are applied to steel by dipping or spraying with a solution of phosphoric acid and manganese oxide to produce an insoluble crystalline coating of iron phosphate. One commercial process, known as parkerising, produces coatings of from 0.00015 to 0.003 inch thick. Phosphate coatings readily absorb oil, grease, wax, and dyes. They resist corrosion, aid in deep-drawing of sheet products, and reduce wear in such applications as piston rings, valve tappets, gears, camshafts, gun parts, etc.

Bonderising is a process of applying a thin phosphate coating to serve as a base for paint on sheet products, such as refrigerator cabinets, type-writer frames, automobile bodies, etc.

Vitreous or porcelain enameling involves the application of fused silicate glasses to the surface of steel. A 'frit' is prepared by melting refractories, fluxes, and opacifiers in a furnace and then pouring the charge into a tank of water. The resulting glass particles are ground in a ball mill with special chemicals, clay, pigments, water, etc., to form 'slip'. This material is usually applied to the metal surface by dipping, drying, and 'firing' in an enameling furnace at a temperature of from 1500 to 1600°F (815 to 870°C) to produce a ground coat. This coating adheres to the base metal and prevents reaction between the metal and one or more cover coats. The cover coats are applied by spraying and are fired in a similar manner to provide service protection and decorative qualities. Cover coats may be white or may contain suitable inorganic pigments.

In contrast to the organic enamels, paints, varnishes, etc., vitreous enamels possess durability, attractiveness, hardness, and resistance to heat, abrasion, and chemical attack. In general, heat resistance is satisfactory up to 600 to 1000°F (315 to 540°C) with some special coatings formulated for short-period service up to about 1700°F (925°C). Thermal shock resistance is influenced by the difference in thermal expansion between the coating and the base metal. Since the coating is actually a glassy phase, it has very poor resistance to mechanical shock. Sheet-metal coatings range between 0.007 and 0.015 inch with heavier coatings being used on castings. To minimise chipping, thin coatings are preferred. Chipping and cracking of the coating cannot be economically repaired.

Vitreous enamel resists the action of nearly all chemicals except hydrofluoric acid and hot caustic solutions. In fact, its resistance to chemical attack makes it uneconomical to strip defective coatings for the salvage of steel parts for subsequent re-coating.

Iron and steel enameling stock requires close control of composition. Low-carbon rimmed steel and ingot iron are generally used with manganese, phosphorus and sulphur closely controlled. Gas-forming elements, such as carbon, tend to react with the enameling frit during firing and to produce brown spots, known as 'copperheads', on the surface. Sheet-enameling stock also requires good surface finish, forming qualities, weldability, and resistance to sagging at the firing temperature.

The successful application of vitreous coatings to cast iron is complicated by the high carbon content and variations in foundry practice. Cast irons containing low combined carbon and high silicon are considered best for enameling. Vitreous enameled products include bathtubs, refrigerators, dishwashers, washing machines, water heaters, stove bodies, cooking utensils, sign panels, etc.

METALLIC CEMENTATION

Metallic cementation is a general name implying the saturation of the surface of steel with various metals by diffusion. The particular processes are called chromising (saturation with chromium), alitising (saturation with aluminium), siliconising (saturation with silicon), etc. If the steel is simultaneously saturated with two or more metals, the process is accordingly called chrome-alitising, chrome-tungstenising, etc.

The processes of diffusion saturation with chromium, aluminium and boron have been studied most thoroughly, whereas those of saturation with molybdenum, beryllium and other elements are known worse.

Types of Metallic Cementation

As other types of chemical heat treatment, metallic cementation can be carried out in solid, liquid and gaseous media.

Solid cementation

Solid cementation is performed in a respective ferro-alloy with an addition of ammonium chloride (NH_4Cl). The metalliser reacts with HCl or Cl_2 and forms a volatile metallic chloride ($AlCl_3$, $CrCl_2$, $SiCl_4$, etc.) which dissociates into free atoms on contact with the metal surface.

Liquid cementation

Liquid cementation is made by immersing steel parts into molten metal, for instance, aluminium.

Gaseous cementation

Gaseous cementation is carried out in gaseous chlorides of various metals.

The diffusion of chromium, aluminium and other metals occurs much more slowly than that of carbon or nitrogen, since the former form substitutional solutions with iron and the latter, interstitial solutions. For the same temperature and time of the process, the saturated layer in cementation is substantially thinner (by a factor of a few tens or hundreds) than a carburised layer.

Because of the low rate of diffusion, cementation processes are expensive and time-consuming and for that reason are not widely used in industry. The process is usually carried out at a high temperature (1000–1200°C). They have found application only in special cases when it is needed to form special properties in the surface layer or to save alloying elements.

One of the principal qualities of metal-saturated surfaces (chromised, alitised or siliconised) is a high heat resistance. For that reason steel parts for operation at temperatures to 1000–1100°C are made of plain carbon steels with subsequent alitising, chromising or siliconising.

An exceptionally high hardness (up to 2000 HV) and a high resistance to abrasion wear can be obtained by boriding, i.e., by the formation of hard iron borides FeB and Fe_2B in the surface; borided layers are however very brittle.

SECTION IV

Surface Hardening of Steel

Chapter 22

Surface Hardening of Steel: An Overview

INTRODUCTION

Aim of surface hardening is to increase the hardness only on the surface while the core does not show any structural change. It has good compression strength and significantly higher fatigue resistance with good ductility. It competes with thermo-chemical treatment which is much costlier. Steels containing 0.4–0.6 per cent carbon or low alloy steels are used for surface hardening. High alloy steels are not suitable for this treatment as their hardenability cannot be taken advantage of, besides, with quick and short time heating, alloy carbides do not dissolve and austenite obtained is practically low alloyed. Before surface hardening, the article may be subjected to toughening treatment which improves the strength and ductility properties all through. Surface hardening involves heating the surface at a fast rate to a temperature above Ac_3, superheating above Ac_3 can be based on Fig. 22.1. The superheat increases with increase in heating rate. Immediately after heating, it is quenched with water spray. The depth up to which austenite forms undergoes hardening. The faster the heating, the thinner is the hardened layer. This mainly depends on the heating method.

Many steel parts are heat treated so as to have a high hardness in the surface and a high toughness in the core, which ensures a high wear resistance and at the same time a high dynamic strength. This is done by methods of surface hardening and chemical heat treatment.

Surface hardening is advantageous over chemical heat treatment as it requires appreciably less time.

DIVERSITY OF SURFACE HARDENING METHODS

Despite the large diversity of methods of surface hardening, all of them essentially consist in heating and hardening of the surface layers only. The heating can be done by various processes: (i) in molten metals or salts; (ii) in the flame of oxy-acetylene or gas burners (called flame hardening); (iii) in electrolytic solutions; and (iv) by electric currents induced in the surface layers (induction, or high-frequency, hardening in which high-frequency currents are induced in the surface layers of steel parts).

The last process is now increasingly employed in the machine building industry.

Process of Surface Hardening

The essence of any process of surface hardening consists in rapid heating of the surface layers above the critical points, which forms a sharp temperature gradient across the cross-section (Fig. 22.1). If the heating is interrupted and the part is cooled quickly, the layer that has been heated above Ac_3 (I) will be full-hardened; the layer heated above Ac_1 but below Ac_3 (II) will be hardened only partially; and the core (III) will be not hardened at all, since it is not heated or is heated only below Ac_1.

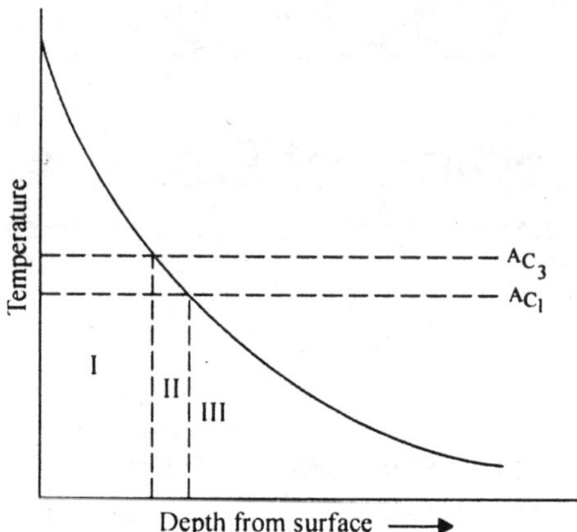

Fig. 22.1. Temperature distribution from surface to depth in surface hardening.

Heating of surface layers appreciably above Ac_3 is a common case in all processes of surface hardening, but it does not always means that the metal is overheated and its structure is impaired. When we considered the pearlite-to-austenite transformation, we have noted that with a higher heating rate, the transformation takes place at a higher temperature; since the time of heating is short, the microstructure of the final austenite is less coarsened and the mechanical properties are not worsened much.

Experience shows that each heating rate has a corresponding temperature interval in which a fine-grained structure is obtained.

Fig. 22.2 shows the temperature region of correct heating (I), depending on heating rate (without holding at the heating temperature). As may be seen, the temperature of heating for hardening increases with the heating rate. Temperatures below the correct temperature interval are insufficient for completion of the austenitic transformation, while higher temperatures cause overheating and intensive grain coarsening. It is also seen from the diagram that the temperature interval of hardening for slow heating in the furnace (II) is appreciably lower than that for high-speed heating.

As has been indicated, the temperature of hardening for a particular steel is assigned according to the Fe-C constitutional diagram. This is only true, however, for slow heating in the furnace in which the transformations occur without an appreciable overheating. With high-speed heating, the temperature of hardening should be selected by using diagrams of the type shown in Fig. 22.2.

GROUPS OF SURFACE-HARDENING METHODS

Group 1–Surface Heat Treatment

These methods can be used only on steels with sufficient carbon to harden by quenching. In one method (flame hardening), the steel surface is heated with a flame and subsequently quenched in a suitable medium. Another method, induction hardening, utilises eddy currents, induced in the steel surface by a high-frequency coil. The depth of their penetration is inversely proportional to the frequency of the current in the coil. In this method, the temperature of the affected zone rises rapidly, reaching austenitisation levels within seconds. The part is quenched by dropping into a tank or by spraying the surface with a quenching medium. A third method, laser heating, provides very fast localised heating that can be applied to larger surfaces by appropriate scanning movement of the laser gun. Quenching occurs through rapid heat transfer to the cold substrate.

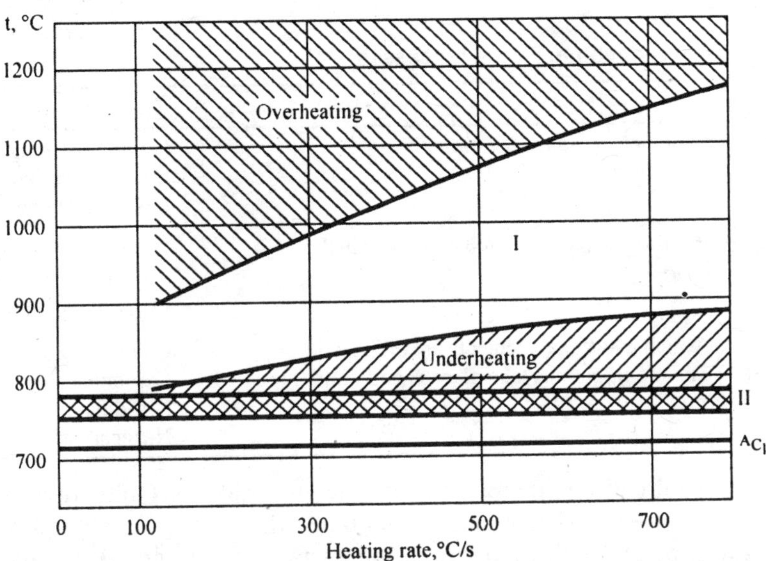

Fig. 22.2. Diagram to select heating temperature. I—high-speed electric heating; II—slow heating in furnace.

Group 2—Hard Coating

The oldest known representative method belonging to this group is electroplating, mainly chrome. The disadvantage of wet plating methods is waste disposal and neutralisation in order to meet today's high ecological standards. Modern, high-tech methods include chemical vapour deposition (CVD) and physical vapour deposition (PVD). The result of such deposition is referred to as a plating or coating.

In chemical vapour deposition, hard coatings are produced by chemical reactions at elevated temperatures, 800–1100°C (1472–2012°F) from gaseous media. The disadvantage is high temperature, which may cause transformations in the core and distortion. Low-temperature CVD coatings for wear protection are successfully carried out by some providers but, generally, this is not yet a mass-scale technology.

In physical vapour deposition, metal vapour is produced and directly deposited on the steel substrate, in the presence of a reactive gas, to form a hard coating. The process is characterised by low substrate temperatures, down to 200°C (392°F), preventing loss of core hardness. A typical PVD application is the titanium nitride (TiN) coating, used for diverse cutting tools and components. It provides an attractive golden-yellow appearance. Limiting factors of this technique are high capital equipment and operational costs. Gaining popularity are multilayered, so-called duplex coatings, which comprise hard PVD or CVD layers over a nitrided substrate. The nitrided layer provides a high hardness support for the thin (1–3/μm) coating, resulting in exceptional wear resistance. Utilisation of duplex coatings has opened new field of application for nitriding.

Group 3—Diffusion Layers

In these processes, various elements may be released by an active medium, absorbed by the surface and diffused into the material of the part. The resultant surface modifications are termed hardened *cases* or *layers*. The most popular processes are carburising, nitriding, and their combinations, i.e., carbonitriding or nitrocarburising, depending on processing conditions. The main groups of surface hardening steel are shown in Table 22.1.

Table 22.1. Main groups of surface-hardening processes.

Group	*Description*	*Examples*
1	Methods that do not introduce additional elements into surface	Flame hardening Induction hardening Laser hardening
2	Methods depositing a coating on top of the original surface of a part	Plating, e.g., chrome Chemical vapour deposition (CVD) Physical vapour deposition (PVD)
3	Methods in which additional elements are diffused into the surface and modifications take place at and under the original surface	Carburising Carbonitriding Nitriding Nitrocarburising

But many products require different properties at different locations. Quite frequently, this variation takes the form of a hard, wear-resistant surface coupled with a tough, fracture-resistant core. The methods developed to produce such properties can generally be classified into three basic groups: selective heating of the surface, altered surface chemistry, and deposition of an additional surface layer.

Selective Heating Techniques

If a steel has sufficient carbon content to attain the desired surface hardness, the different properties can be obtained by simply varying the thermal histories of the various regions. Selective heating for surface hardness generally requires steels of at least 0.3 per cent carbon. Maximum hardness depends on the carbon content, whereas the depth of hardness depends on the material's hardenability.

Flame hardening

Flame hardening employs a high-intensity oxyacetylene flame to raise the surface temperature high enough to reform austenite, which is then quenched and tempered to the desired surface properties. Heat input is quite rapid and is concentrated on the surface. Slow heat transfer and short times leave the interior at low temperature and therefore free from any significant changes.

Considerable flexibility is provided since the rate and depth of heating can be easily varied. Depth of hardening can range from thin skins to over ¼ inch. Flame hardening is often used on large objects, as alternative methods tend to be limited by both size and shape. Equipment varies from crude, hand-held torches to fully automated and computerised units.

Induction heating

Induction heating is particularly well suited to surface hardening, the rate and depth of heating being controlled by the amperage and frequency of the generator. The steel part is placed inside a wound coil, which is then subjected to alternating current. The changing magnetic field induces a current in the steel that flows through the surface layers and heats by electrical resistance. Heating rates are extremely rapid and efficiency is high. The process can be adapted to special shapes and offers the benefits of excellent reproducibility, good quality control, and the possibility of automation.

Laser beam hardening

Laser beam hardening has been used to produce hardened surfaces on uneven geometries. An absorptive coating, such as zinc or manganese phosphate, is first applied to the steel to improve the efficiency of converting light energy into heat. The surface is then scanned with the laser, where beam size, beam intensity, and scanning speed (often as high as 100 inches per minute) have been selected to obtain the desired results. While some heat is removed through transfer into the cool, underlying metal, a water or oil quench is often used on the surface. In addition, the process operates at high speeds, produces little distortion, induces residual compressive stresses on the surface and can be used to harden selected surface areas while leaving the remaining surfaces unaffected. Computer control and automation can be used to control the process parameters, and conventional mirrors and optics can be used to shape and manipulate the laser beam.

Electron beam hardening

Electron beam hardening is very much similar to laser beam hardening. Here, the heat source is a beam of high-energy electrons, rather than a beam of light. The charged particles can be focused and directed by means of electromagnetic controls. Like laser beam treating, the process can be readily automated and production equipment can perform multiple operations with efficiencies often greater than 90 per cent. Electrons cannot travel in air, however, so the entire operation must be performed in a hard vacuum and this requirement provides the major limitation.

Still other selective heating techniques employ immersion in a lead pot or salt bath as the means of heating the surface.

Techniques Using Altered Surface Chemistry

When steels contain insufficient carbon to attain the desired surface properties by selective heating, an alternative approach is to alter the surface chemistry. Carburising is the most common technique within this category. In the pack-carburising process, components are packed in a high-carbon solid medium, enclosed in a gas-tight box and heated in a furnace for 6 to 72 hours at roughly 1650°F (900°C). The hot carburising compound produces CO gas, which reacts with the metal, releasing carbon. This, in turn, is readily absorbed into the austenite, which is the stable steel structure at these temperatures. The boxes are then taken from the furnace and the parts are removed and thermally processed. Direct quenching from the carburisation treatment can produce the different surface and core properties due to the different carbon contents of the steel at these locations. A slow cool from carburising, reheat and quench or a duplex core-case heat treatment are alternative thermal processes that can produce superior properties. The carbon content of the surface varies from 0.7 per cent to 1.2 per cent depending on the details of the process. Case depth may range from a few thousandths of an inch to over $\frac{3}{8}$ inch, but cases over 0.06 inch (1.5 mm) are seldom employed. Several problems are encountered in pack carburising. Heating is inefficient; temperature uniformity is questionable; handling is often difficult; and the process is not readily adaptable to continuous operation. Gas carburising overcomes these difficulties by replacing the solid carburising compound with a carbon-providing gas, usually containing an excess of CO. Although the mechanisms and processing are the same, the operation is faster and more easily controlled. Accuracy and uniformity are increased and continuous operation is possible. Special types of furnaces are required, however.

Liquid carburising or cyaniding

In liquid carburising or cyaniding, the carbon is supplied by immersing the part in a molten carbon-containing bath. Safety and environmental concerns associated with toxic cyanide fumes have severely limited the use of molten cyanide salts, but nontoxic liquid carburising compounds have been developed to preserve the technique. Most applications involve the production of thin cases on small parts.

Nitriding

Nitriding produces hardening by the production of alloy nitrides in the surface layers of special steels that contain nitride-forming elements such as aluminium, chromium, molybdenum or vanadium. The parts are heat-treated and tempered at 1000–1250°F (525–675°C) before nitriding. After cleaning and removal of any decarburised surface material, they are heated in a dissociated ammonia atmosphere (containing nitrogen and hydrogen) for 10 to 40 hours at 950–1150°F (500–625°C). Nitrogen diffusing into the steel then forms alloy nitrides, hardening the metal to a depth of about 0.025 inch (0.65 mm). Very hard cases are formed and distortion is low. No subsequent thermal processing is required. In fact, it should be avoided because the differential thermal expansions and contractions will crack the nitrided case. Finish grinding should also be avoided, if possible, because of the exceptionally thin case. Thus, although the surface hardness is higher than for other hardening methods, the long times at elevated temperatures, coupled with the thin case, restrict the application of nitriding to the production of high-quality cases.

Ionitriding

Ionitriding is a plasma process that has emerged as an attractive alternative to the conventional methods. Parts to be treated are placed in an evacuated 'furnace' and a direct current potential of 500 to 1000 volts is applied between the parts and the furnace walls. Low-pressure nitrogen gas is introduced into the chamber and becomes ionised. The ions are accelerated toward the product surface, impact and generate sufficient heat to promote inward diffusion. This is the only heat associated with the process; the 'furnace' acts only as a vacuum container and electrode. The advantages of the process include shorter cycle times, reduced consumption of gases, significantly reduced energy costs, reduced space requirements, and the possibility of total automation. Product quality is improved over that of conventional nitriding and the process is applicable to a wider range of materials.

Ion carburising

In ion carburising, a low-pressure methane plasma is created, producing atomic carbon and transferring the carbon to the surface. Production time is generally 30 per cent less than with gas carburising and the product is oxide-free because of the near-vacuum conditions. Ion plating and ion implantation are other new technologies in surface modification.

Chapter 23

Flame Hardening

INTRODUCTION

Flame hardening and induction hardening are two methods which, in many respects, give equivalent results. The advantages of induction hardening enumerated in the preceding section also apply, by and large, to flame hardening. The capital investment costs for flame hardening are lower than for induction hardening, but the operating costs are higher. Flame hardening in its simplest form consists of heating the steel to the hardening temperature by a flame from a welding torch and then quenching it in water or oil.

The grades of steel which were recommended for induction hardening are also suitable for flame hardening, which, moreover, can be used for the partial hardening of tools from alloy tool steels. If the steels exhibit sufficiently high hardenability, the requisite hardness can be achieved by cooling in air. In certain cases flame hardening is preferred to induction hardening, because it is easier to direct the heat to selected surface areas.

Flame hardening is a heat treating process in which a thin surface shell of a steel part is rapidly heated to a temperature above the critical point of the steel. After the grain structure of the shell has become austenitic, the part is quickly quenched, transforming the austenite to martensite while leaving the core of the part in its original state. In contrast, slow cooling causes transformation, as the temperature passes through the corresponding ranges, to pearlite, bainite and martensite, with the final structure being a combination of the three. The result is a relatively soft and ductile steel. To achieve hardness, therefore, the steel must be cooled rapidly so that it by-passes the first two transformation phases and transforms directly from austenite to martensite.

Flame hardening employs direct impingement of a high temperature flame or of high velocity combustion-product gases; the part is then cooled at a rate that will produce the desired levels of hardness and other properties. The high temperature flame is obtained by combustion of a mixture of fuel gas with oxygen or air; flame heads are used for burning the mixture. Depths of hardening from about 0.8 to 6.4 mm ($1/32$ to $1/4$ inch) or more can be obtained, depending on the fuels used, the design of the flame head, the duration of heating, the hardenability of the work material and the quenching medium and method of quenching used. The process can be used for through hardening of work 75 mm (3 inch) or less in cross-section, depending on the hardenability of the steel.

Hardening by flame differs from true case hardening, because the hardenability necessary to attain high levels of hardness is already contained in the steel and hardening is obtained by localised heating.

Although flame hardening is used mainly to develop high levels of hardness for wear resistance, the process also improves bending and torsional strength and fatigue life. One of the major advantages of flame hardening is the ability to satisfy stringent engineering requirements with carbon steels.

SCOPE AND APPLICATION

Flame hardening is applied to a wide diversity of workpieces and ferrous materials (when required mechanical properties can be provided by selective or localised hardening) for one or more of the following reasons:

1. Because parts are so large as to make conventional furnace heating and quenching impracticable or uneconomical. Typical examples include large gears, machineways and large dies and rolls.
2. Because only a small segment, section or area of a part requires heat treatment, or because heat treating all over would be detrimental to the function of the part. Typical examples include the ends of valve stems and pushrods and the wearing surfaces of cams and levers.
3. Because dimensional accuracy of a part is impracticable or difficult to attain or control by furnace heating and quenching. A typical example is a large gear of complex design, for which flame hardening of the teeth would not disturb the dimensions of the gear.
4. Because the use of flame hardening permits a part to be made from a less costly material, thus effecting an overall cost saving in comparison with other technically acceptable methods. The process gives inexpensive steels the wear properties of alloyed steels and parts can be hardened without scaling or decarburisation, thereby eliminating costly cleaning operations. For example, a large carburised low-carbon alloy steel part might be made at less cost from a flame-hardened plain carbon steel.

METHODS OF FLAME HARDENING

The versatility of flame-hardening equipment and the wide range of heating conditions obtainable with gas burners often permit flame hardening to be done by a variety of methods, of which the principal ones are:

1. Spot or stationary.
2. Progressive.
3. Spinning.
4. Combination progressive-spinning.

Selection of the appropriate method depends on the shape, size and composition of the workpiece; the area to be hardened; the depth of case required; and the number of pieces to be hardened. In many instances, more than one method will be capable of providing the desired result; the choice will then depend on comparative costs. Principle illustrating progressive flame hardening is shown in Fig. 23.1.

Spot (Stationary) Method

The spot (stationary) method, consists of locally heating selected areas with a suitable flame head and subsequently quenching. The heating head may be of either single-orifice or multiple-orifice design, depending on the extent of the area to be hardened. The heat input must be balanced to obtain a uniform temperature over the entire selected area. After being heated, the parts usually are immersion quenched; however, in some mechanised operations, a spray quench may be used.

Basically, the spot method requires no elaborate equipment (except, perhaps, fixtures and timing devices to ensure uniform processing of each piece). However, the operation may be automated by indexing the heated parts into either a spray quench or a suitable quench bath.

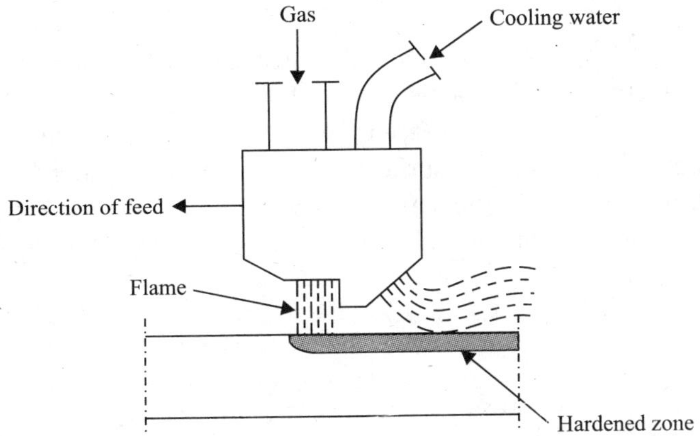

Fig. 23.1. Principle illustrating progressive flame hardening.

Progressive Method

The progressive method is used to harden large areas that are beyond the scope of the spot method. The size and shape of the workpiece, as well as the volume of oxygen and fuel gas required to heat the specified area, are factors in selection of this method. In progressive hardening, the flame head is usually of the multiple-orifice type and quenching facilities may be either integrated with the flame head or separate from it. The flame head progressively heats a narrow band that is subsequently quenched as the head and quench traverse the workpiece.

The equipment needed for flame hardening by the progressive method consists of one or more flame heads and a quenching means mounted on a movable carriage that runs on a track at a regulated speed (flame-cutting machines are adaptable to this type of flame hardening). Workpieces mounted on a turntable or in a lathe can be hardened readily by the progressive method; either the flame head or the workpiece may move. There is no practical limit on the length of parts that can be hardened by this method, because it is easy to lengthen the track over which the flame head travels. Single passes as wide as 1.5 meter (60 inch) can be made; wider areas must be hardened in more than one pass.

When more than one pass is required to cover a flat surface or when cylindrical surfaces are hardened progressively, such surfaces will exhibit soft bands because of overlapping or underlapping of the heated zones. These soft bands can be minimised, however, by closely controlling the extent of the overlapping. (Wherever overlapping occurs, the possibility of severe thermal upset and cracking should be anticipated. Tests should be conducted to determine whether overlapping will cause cracking or other harmful effects). Simple curved surfaces may be hardened progressively by means of contoured flame heads and some irregular surfaces may be traversed by the use of tracer-template methods.

The rate of travel of the flame head over the surface is governed mainly by the heating capacity of the head, the depth of case required, the composition and shape of the work and the type of quench employed. Speeds ranging from 0.8 to 5.1 mm/s (2 to 12 inch/minute) are typical with oxyacetylene heating heads. Ordinarily, water at ambient temperature is used as a quenchant, although air sometimes is used when a less severe quench is indicated; under special conditions (particularly for quenching alloy steels), warm or hot water or a polymer-base synthetic quenchant also may be employed.

Spinning Method

The spinning method (Fig. 23.2) is applied to round or semi-round parts such as wheels, cams or gears. In its simplest form, the method employs a mechanism for rotating or spinning the workpiece, in either a horizontal or a vertical plane, while the surface is being heated by the flame head. One or more water-cooled heating heads equal in width to the surface to be heated are employed. The speed of rotation is relatively unimportant, provided uniform heating is obtained. After the surface has been heated to the desired temperature, the flame is extinguished or withdrawn and the work is quenched by immersion or spray or a combination of both.

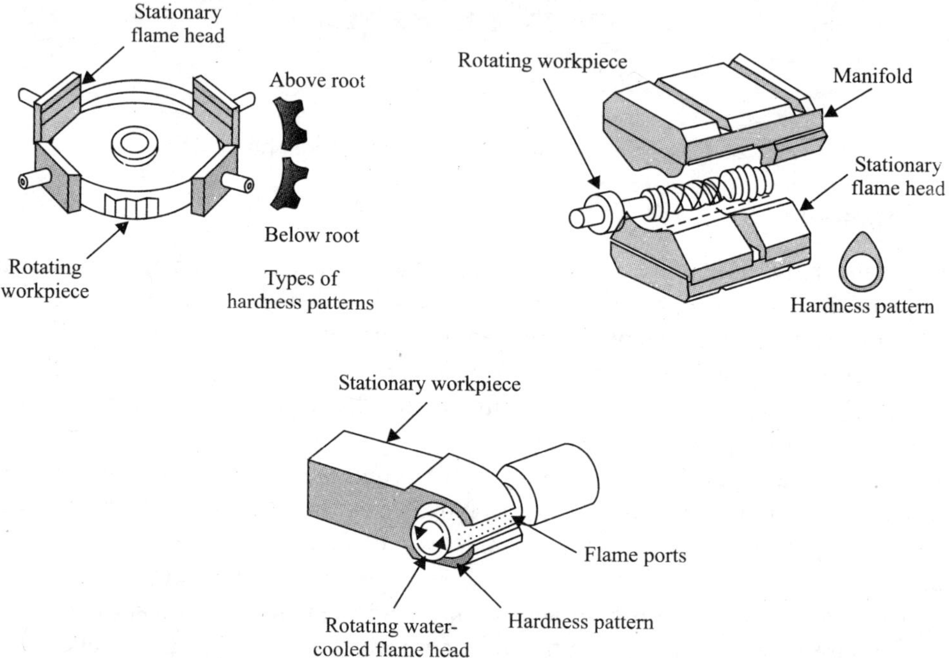

Fig. 23.2. Spinning methods of flame hardening.

The spinning method is particularly adaptable to extensive mechanisation and automation; this makes it possible, for example for all the cams on a camshaft to be hardened at the same time.

Today, fully automatic flame-heating equipment is available that can treat round components upto 1.5 meter (59 inch) in diameter and upto 2 tonne in weight. Much of it has been designed to treat gear wheels of all types.

Commercial machines have been built that provide automatic control of timing, temperature and quenching, as well as accurate control of gas flow, so that close metallurgical specifications can be met consistently. Frequently, when production is sufficient, the spinning method can be set up so that parts either are loaded manually and unloaded automatically, or are both loaded and unloaded automatically.

This method has been extended to components of irregular cross-section and mass distribution. Typical are large drive wheels for tracked vehicles cams and camshafts for marine diesel engines and crane travelling wheels. Speed, deep hardness penetration, localised hardness zone and uniformity of hardness pattern are the main advantages.

In one technique, a rotating flame head is used for internal spin hardening of odd-shaped parts that would present handling problems if the parts themselves were rotated. Each part is positioned by a simple handling device and the flame head rotates inside the part.

In contract to the progressive method, in which acetylene usually is used (because of its high flame temperature and rapid heating rates), satisfactory results can be obtained in spin hardening with natural gas, propane or manufactured gas. The choice of gas depends on the shape, size and composition of the workpiece and on the depth of case required, as well as on the relative cost and availability of each gas.

A wide choice of quenchants also is possible in the spin hardening method. Because the flame is extinguished or withdrawn before the part is quenched, any appropriate quenchant may be used for immersion quenching. In spray quenching, the quenchant is usually water, a water-base liquid such as soluble oil or a simulated oil in the form of a polymer-base quenchant; air also has been used.

Combination Progressive-Spinning Method

The combination progressive-spinning method (Fig. 23.3), as the name implies, combines the progressive and spinning methods for hardening long parts such as shafts and rolls. The workpiece is rotated as in the spinning method; but in addition, the heating heads traverse the roll or shaft from one end to the other. Only a narrow circumferential band is heated progressively as the flame head moves from one end of the work to the other. The quench follows immediately behind the heating head, either as an integral part of the head or as a separate quench ring.

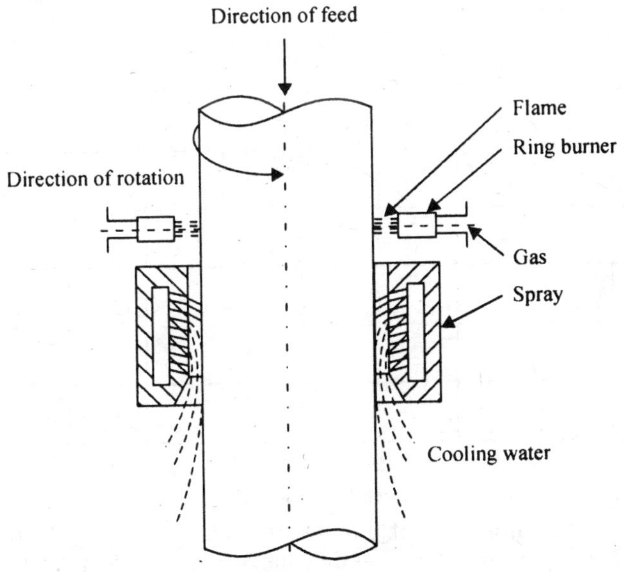

Fig. 23.3. Principle illustrating progressive spin hardening.

This method provides a means of hardening large surface areas with relatively low gas flows. Progressive spinning units designed to handle a broad range of diameters and lengths are available commercially.

Fig. 23.4 shows the hardness distribution in test bars made from carbon steel C50 after trial hardening when using a feed rate of 50 mm/min and different burner distances.

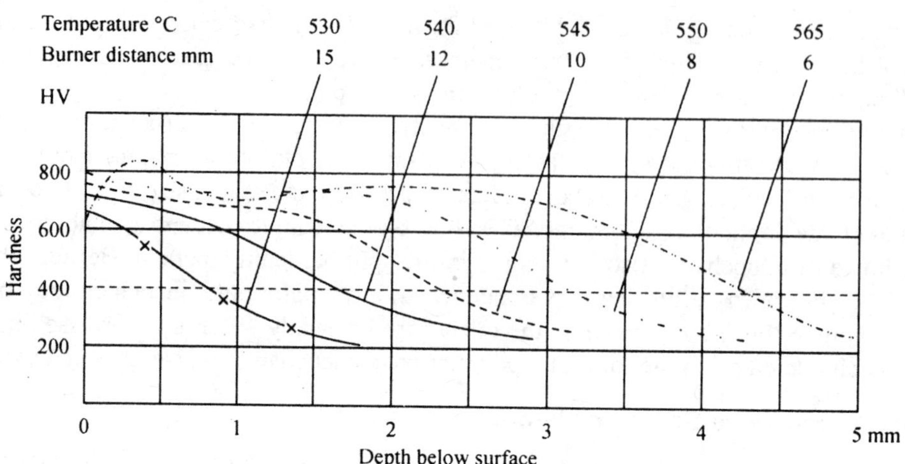

Fig. 23.4. Relationship between hardness profile and burner distance during flame hardening of a carbon steel with 0.50% C. Test specimen dimension 25 x 75 x 100 mm. Feed 50 mm/min. Water spray quenching. Temperature measured 10 mm below the surface.

It is also possible for tool steels of type O1 and D2 to be flame hardened without cracking. With these steels it is sufficient to cool them in air in order to obtain surface hardness values of around 800 HV. Fig. 23.5 shows result from a trial steel O1. During a hardening run, with the minimum burner distance, the steel became coarse grained in the surface layer.

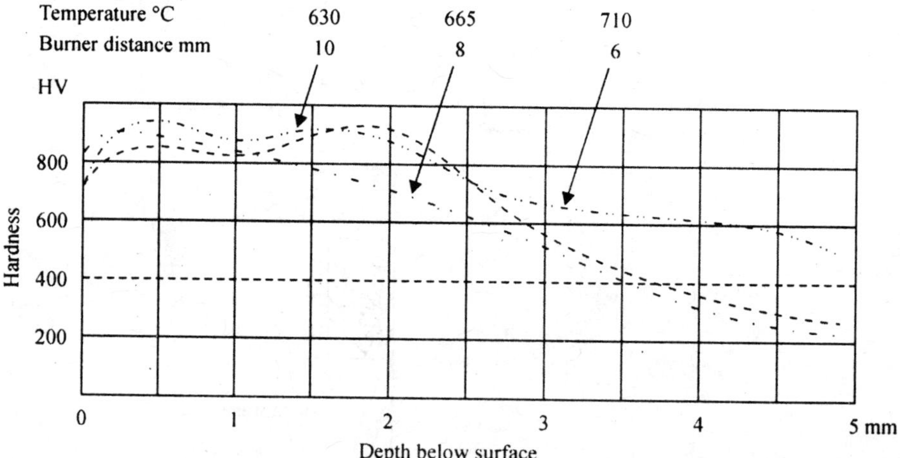

Fig. 23.5. Relationship between hardness profile and burner distance during flame hardening of steel O1. Test specimen dimensions: 25 x 75 x 100 mm. Feed 50 mm/min. Air cooling, temperature measured 10 mm below surface.

Steel D2 is unaffected by overheating as regards grain size, but it is liable to contain retained austenite and be prone to melting. During the tests it was clear that this effect was due, among other things, to the rather low heat conductivity of the steel, which resulted in a steep temperature gradient. Fig. 23.6 also shows that the temperatures measured 10 mm below the surface are considerably higher than in the preceding case. When hardening the edges only of the test bars and with only a slight change in the feed rate, both normal and overheated structures were obtained.

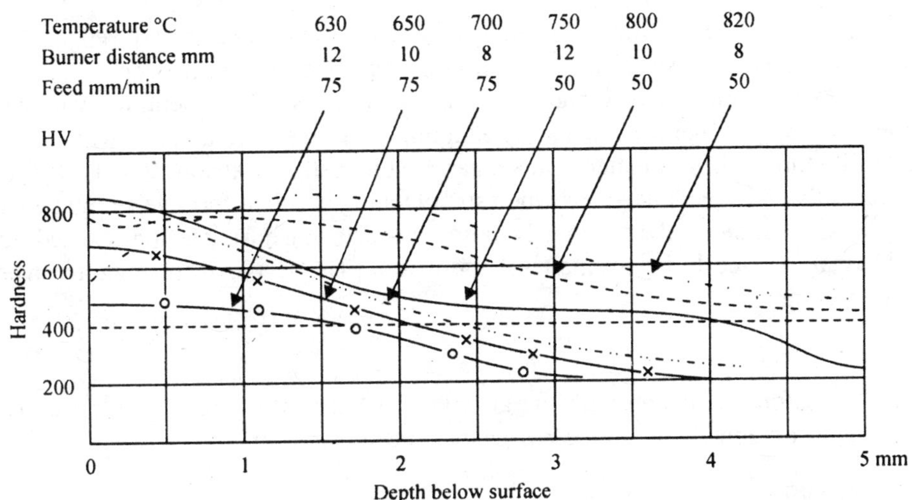

Fig. 23.6. Dependence of hardness profile on burner distance and feed during flame hardening of steel D2. Test specimen dimensions: 25 × 75 × 100 mm. Air cooling. Temperature measured 10 mm below surface.

EXAMPLES OF FLAME-HARDENED MACHINE COMPONENTS AND TOOLS

Gears can be flame-hardened by several methods. Up to module 5, spin hardening is used. If only the tooth tips are to be hardened, gears of larger modules can be spin hardened. With gears of module 4 or more, one tooth at a time can be progressively hardened. This can be undertaken in such a way that either the flanks on one tooth are hardened simultaneously (flank hardening) or the flanks on two teeth are hardened simultaneously (tooth-gap hardening). From module 6 and up, it is also possible to harden the space between the teeth (tooth bases), so that the fatigue strength is increased.

Progressive hardening can be carried out with a quenching spray directly after heating, this being essential for gears made from carbon steel. Each tooth is heated progressively, after which the gear wheel is rotated 15–60° and the tooth allowed to cool in air. The other parts of the gear are kept cool by means of cooling water.

A similar method is employed for spur gears and sprocket wheels which are placed vertically, with roughly one quarter of the wheel immersed in water. When the tooth located at the higher position has been heated to the hardening temperature, the wheel is rotated through 180°, so that the heated tooth is quenched. For single-tooth hardening pre-indexed machines are used which give symmetrical stresses in the gear wheel and thereby only slight deformation is created.

Long Objects

Long objects, such as lathe guides, are hardened progressively, whereby a flame-cutting machine can be used for the feed. The depth of hardening is regulated by the amount of heat supplied (gas pressure) and the feed rate. Lathes guides have been hardened by using the same gas-pressure setting but different feed rates. One special case where a cam shaft for a printing press is hardened simultaneously along four sections, length about 100 mm each. The larger parts of the cam shaft is immersed in water, which provides sufficient cooling to ensure that satisfactory hardening is achieved. A similar procedure can be used when hardening long edges, for example. The object is placed in water with the edge just level with the surface. As the flame sweeps towards the edge, the water is blown away and the edge is exposed and heated. As the burner travels onwards, the water flows back and quenches the edge.

During the flame hardening of long and relatively thin objects it is difficult to avoid deformation. When one side of a straight edge is being hardened it might be presumed that this will become convex, since the martensite formed has a greater volume than the other structural constituents. This does in fact occur quite frequently and is advantageous when straightening is called for. The straight edge may also become concave, which is due to the fact that the upsetting occurring during heating exceeds the increase in volume that has taken place during the martensite formation. The degree of contraction can be regulated by means of the feed rate.

Plates

Plates can be successfully hardened with the aid of wide burners. Large plates can be hardened in several passes, but a soft zone between the passes must be allowed for.

Simple Blanking Tools

Simple blanking tools (punching tools), as mentioned previously, can be given a wear-resistant cutting edge by means of flame hardening. This method is used mainly for tools employed in blanking large plates. Such a tool can either be built up of shear plates which are bolted firmly to a parent tool or it can be manufactured from plate in which the desired contour has been cut to shape by flame cutting and then machined. In the former instance the shear plates can be dismantled and flame hardened separately if so required. In the latter case hardening must be carried out using a torch which is guided around the entire cutting edge. When the torch has returned to the point which was hardened first, a narrow, softer zone forms, the location of which should be at a point where minimum wear is anticipated.

Carbon steel containing 0.50 per cent C, as well as steel W1, can be used for simple tools. Quenching is carried out by water spraying. It may also suffice to keep the tools immersed in water, with the cutting edge just above the surface of the water.

Moulds for Plastics

Moulds for plastics are largely made from steel P20 hardened and tempered. If the plastic is very abrasive or if the pressure at the mating surfaces is extremely high, flame hardening of these mating surfaces can impart a considerable improvement in wear resistance. After flame hardening and cooling in air the hardness of P20 is 50–53 HRC.

General Recommendations

The following recommendations are given for flame hardening the edges of moulds and other tools made from alloy steels when using a conventional welding torch or burner which is passed over the edges of the tool as shown in Fig. 23.7.

1. The area to be hardened should be clean and free from scale.
2. An oxy-acetylene flame is preferable, initially with slightly more acetylene than oxygen—a 'blue or green cone' will be noticed in the flame.
3. Oxygen should be increased until the 'blue or green cone' disappears.
4. A slight pre-heat by movement of the flame over the beginning of the area to be hardened should be given.

5. The tip of the core flame should then be held about 8 mm (¼ inch) back from the edge of the material, approximately 8 mm (¼ inch) above the material and at an angle of 60–90° to the horizontal plane and 15–30° to the vertical.
6. Once a dull-red hot spot becomes visible, the flame must be moved slowly along, so that the red spot follows the flame at a constant speed and remains a constant size. Speed is approximately 50 mm (2 inch) per minute.
7. The flame should not be allowed to dwell or the process to discontinue and then be re-started. If the flame dwells, sparks may appear, which indicates burning of the metal. If the process is discontinued it is virtually impossible to start where the flame left off and the localised uneven heating gives rise to stressed or soft areas.
8. Tempering at 180–200°C is recommended, but not necessary, after flame hardening. Local heating with the flame is used.

A suitable nozzle for flame hardening is AGA welding nozzle No. 4. Alternatively BOC nozzle no. 18. Saffire 3 range.

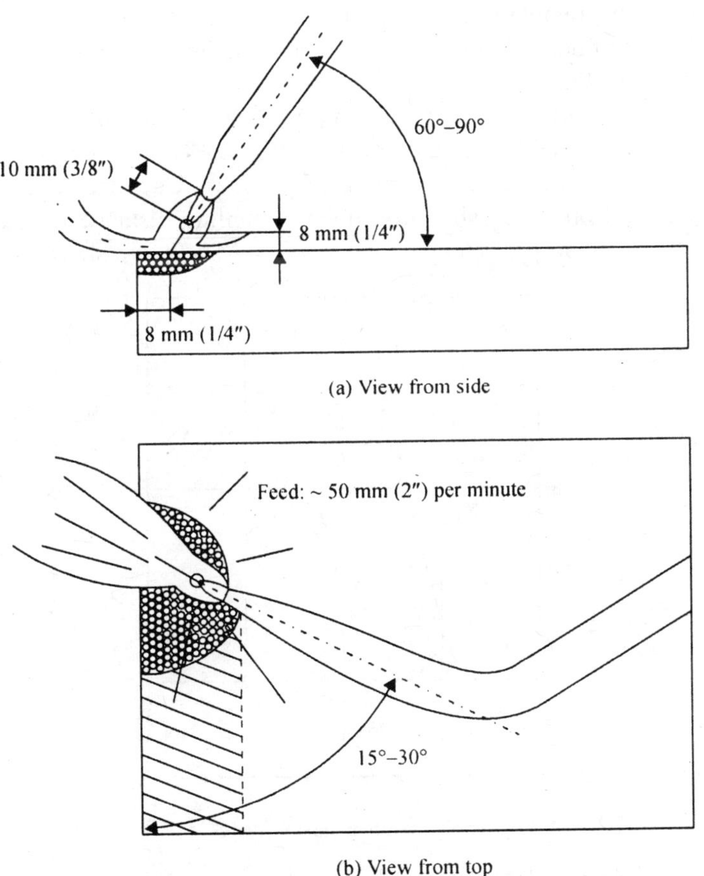

Fig. 23.7. Flame hardening of the edge of a tool made from steel P20.

FUEL GASES

Several different fuel gases are used in flame hardening. In selecting a fuel gas for a given application, the required rate of heating and the cost of the gas must be considered along with the initial cost of equipment and maintenance.

Flame hardening does not alter the composition of the base metal if done properly. Carburising, neutral and oxidising flames can be used. Oxidising flames have high oxygen ratios and can be detrimental because they produce extremely hot temperatures that can cause decarburisation and overheating. A carburising flame can prevent some decarburisation but also can introduce unwanted carbon into the surface. For best results, neutral or slightly carburising flames should be used.

A comparison of the rates of heating of fuel gases can be made when certain fundamental properties of usable mixtures with oxygen are known. A parameter that correlates well with actual heating speed is 'combustion intensity' or 'specific flame output'. This is the product of the normal velocity of burning multiplied by the net heating value of the mixture of oxygen and fuel gas.

Thus, a knowledge of these two properties often permits selection of the most suitable fuel gas for a specific hardening speed and depth of case. The fuels of greatest commercial interest are ranked by combustion intensity (at metallurgically suitable ratios of mixture with oxygen) in the following order: acetylene, propane, methane. Values of normal burning velocity and the heating values of metallurgically suitable mixtures are listed in Table 23.1.

The time required for heat penetration is another good criterion for judging the heating qualities of a fuel provided that all other variables remain constant. Fig. 23.8 shows comparative heating times for stabilised MAPP which is a registered trademark of the Dow Chemical Company (methylacetylene propadiene), acetylene and propane using an efficient coupling distance for each fuel. These curves show that a greater depth of hardness can be obtained with MAPP in a shorter length of time.

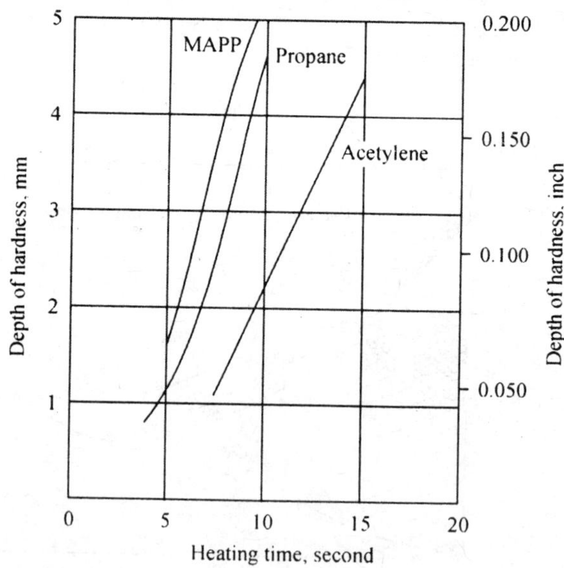

Flame velocity, 170 m/s (550 ft/s); port size, No. 69 drill (0.74 mm or 0.0292 inch); coupling distance, 9.5 mm (3/8 inch); material 1036 steel. Oxygen-to-fuel ratios: MAPP, 5.0; acetylene, 1.33; propane, 4.5

Fig. 23.8. Comparison of heating times for MAPP, acetylene and propane.

Table 23.1. Fuel gases used for flame hardening.

Gas	Usual Heating value MJ/m³	Usual Heating value Btu/ft³	Flame temperature With oxygen °C	Flame temperature With oxygen °F	Flame temperature With air °C	Flame temperature With air °F	Usual ratio of oxygen to fuel gas	Heating value of oxy-fuel gas mixture MJ/m³	Heating value of oxy-fuel gas mixture Btu/ft³	Normal velocity of burning mm/s	Normal velocity of burning inch/s	Combustion intensity(a) mm/s × MJ/m³	Combustion intensity(a) in./s × Btu/ft³	Combustion ratio of air to fuel gas
Acetylene	53.4	1433	3105	5620	2325	4215	1.0	26.7	716	535	21	14 284	15 036	–
City gas	11.2–33.5	300–900	2540	4600	1985	3605	(b)	(b)	(b)	(b)	(b)	(b)	(b)	(b)
Natural gas (methane)	37.3	1000	2705	4900	1875	3405	1.75	13.6	364	280	11	3 808	4 004	9.0
Propane	93.9	2520	2635	4775	1925	3495	4.0	18.8	504	305	12	5 734	6 048	25.0

(a) Product of normal velocity of burning multiplied by heating value of oxy-fuel gas mixture; and (b) varies with heating value and composition.

The ratio of oxygen-to-fuel is very important in obtaining maximum heating efficiency from the fuel. However, oxygen-to-fuel ratios should not be confused with oxygen and fuel consumption rates, which vary with flame velocity, port size and heating time. Stoichiometrically, acetylene requires 2½ mols of oxygen per mol of gas for complete combustion and MAPP and propane require 4 and 5 mols, respectively. With acetylene, however, only 1 to 1.5 volumes of oxygen is provided directly and the remainder is drawn from the surrounding atmosphere. Neither MAPP nor propane has sufficient heat for flame hardening unless more oxygen is supplied, normally at a rate of 4 parts of oxygen per part of fuel. MAPP burns over a rather wide range of oxygen-to-fuel ratios, however and thus permits a wider range of heat output while still providing high heat generation when necessary. Use of MAPP increases oxygen consumption but decreases fuel consumption and fuel is usually more expensive than oxygen.

Bulk systems of supply for oxygen and fuel gases greatly reduce their cost, but of greater importance is the elimination of cylinder handling and of the residual losses usually attending the use of gases in cylinders. Bulk systems also provide a more nearly constant supply of gas at uniform pressure.

Depth of Heating

Shallow hardness patterns (less than 3.2 mm or 0.125 inch deep) can be attained only with oxy-gas fuels. The high-temperature flames obtained with oxy-gas fuels provide the fast heat transfer necessary for effective localisation of the heat pattern. Deeper hardness patterns permit the use of either oxy-gas fuels or air-gas fuels. Oxy-gas fuels will localise the heat, but care is required in their application to avoid overheating the surface during development of the deeper-seated heat. Air-gas fuels, with their slower rates of heat transfer (lower flame temperatures), minimise or eliminate surface overheating but generally extend the heat pattern beyond the desired hardness pattern. For this reason, air-gas flame hardening is generally limited to steels of shallow hardenability. In this manner, the hardness pattern is controlled by the quench rather than by the heating. The deeper-seated heat produced by air-gas flames may preclude the use of air-gas mixtures, because excessive distortion may occur. In consideration of these factors, the use of air-gas heating will depend primarily on the shape of the part insofar as the configuration favours heat localisation and a lower rate of heat transfer.

Gas Consumption Time and Speeds

Gas consumption in flame hardening varies with the thickness of the case to be obtained; increasing or decreasing the depth of hardening increases or decreases the amount of gas used. Massive parts increase gas consumption because of their greater internal cooling effect. In order to take advantage of the maximum flame temperature from the oxy-fuel gas flame, the distance from the end of the inner cone of the flame to the work should be 1.6 mm ($^1/_{16}$ inch).

The speed of travel of the flame head over the work in the progressive method, as well as the time of heating for the spot and spinning methods, will speeds usually vary between 0.8 and 5.1 mm/s (2 and 12 inch/minutes) for most applications, although very thin parts may require speeds of 42 mm/s (100 inch/minute) or more, to avoid overheating or burning. Because of the intense heat involved, the necessity for accurate control of the rate of travel in the progressive and progressive-spinning methods cannot be overemphasised.

Time-temperature-depth relations for various fuel gases employed in the spot (stationary), spinning and progressive methods of flame hardening are shown in Fig. 23.9. The family of curves for the spot method (Fig. 23.9a) was obtained analytically by considering the flow of heat in three dimension away from a heated spot on the surface of a rocker-arm pad. The calculations were based on heat sources of different strengths, which were varied on the basis of combustion intensity of the gases considered (acetylene, propane and natural gas). It is understood that strength of heat source will be affected also by such factors as size of tip, distance from tip to work, total gas flow and ratio of oxygen-to-fuel gas; hence, these curves are intended to indicate trends in time-temperature-depth relations rather than to provide operational values for particular applications.

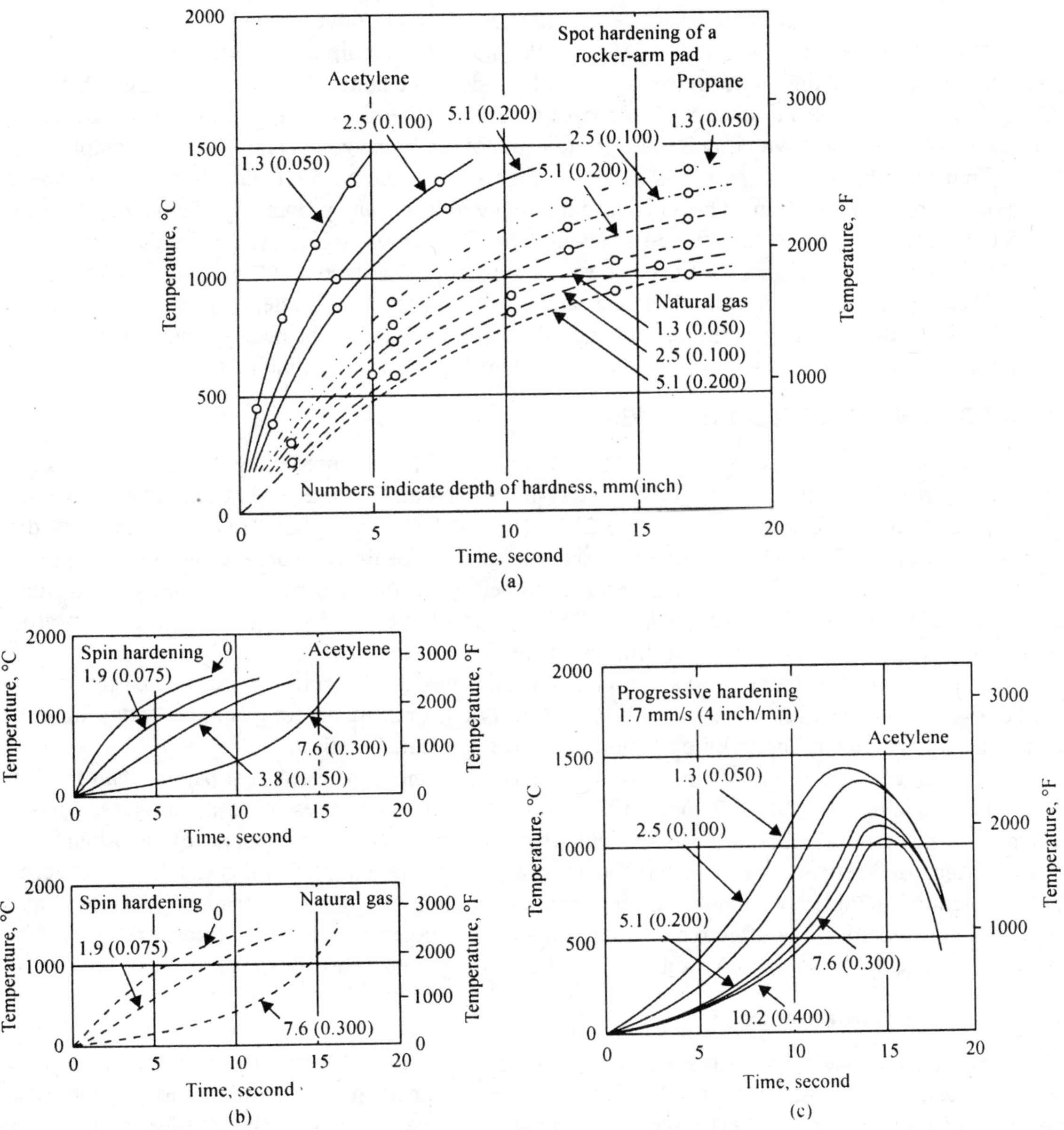

Fig. 23.9. Calculated time-temperature-depth relations for spot (stationary), spinning and progressive flame hardening.

The curves for the time-temperature-depth relationships of fuel gases used in the spinning method, shown in Fig. 23.9b, were obtained analytically by considering the flow of heat into a cylindrical body from a heat source supplying heat uniformly over the outer cylindrical surface. It was assumed that the temperature at the axis would not be raised appreciably during surface heating and that, in spin hardening, the cylinder would be rotated rapidly enough to give the effect of substantially uniform surface heating. No temperature decrease at high values of time is indicated, because in this type of process the body is quenched immediately when the surface temperature reaches a pre-determined value.

The curves for the progressive method (Fig. 23.9c) also were obtained analytically by considering heat flow into a body from a line source moving along a flat side of the body. The heat source strength and body configuration were chosen to be representative of progressive flame hardening. In this instance, the time variable can be correlated with travel speed if the width of the flame zone is known. For example a flame zone 25 mm (1 inch) wide will pass over a point on the work surface in 15 second when the travel speed of the tip is 1.7 mm/s (4 inch/min). This width of flame zone indicates the probability of a multirow tip and if the heat source is as strong as assumed for computing the curves, will undoubtedly damage the work surface. Consequently, a higher travel speed would be used—for example, 2.5 mm/s (6 inch/min), which would heat a point on the work for 10 seconds and result in hardening to a depth of about 2.5 mm (0.100 inch). The rapid decrease in temperature at large values of time is due to the mass quenching effect, which, in practice, would be augmented by use of water spraying or other quenching means.

BURNERS AND RELATED EQUIPMENT

Burners are basic components of equipment for all methods of flame hardening. Burners vary in design, depending on whether they are fired by an oxy-fuel or an air-fuel gas mixture. Flame temperatures obtained by oxy-fuel gas combustion are 2540°C (4600°F) and higher. Heat transfer is by direct impingement of the flame on the surface of the workpiece. For this reason, oxy-gas burners are more commonly referred to as 'flame heads'. Flame temperatures obtained by air-fuel gas combustion are considerably lower (Table 23.1) and heat transfer is by impingement of high-velocity combustion-product gases (no direct flame) or by radiation from an incandescent refractory surface.

There is no universal flame head and no flame head is designed specifically for one particular gas. A well designed flame head can be used with MAPP, acetylene or propane, for example. Better flame-head design usually results in improved operation and lower gas consumption.

In general, a large number of small ports will produce a more efficient heat pattern than few larger ports. Port spacing of 2.3 to 3.2 mm (0.090 to 0.125 inch) on centres is, in most cases, advisable. Counterboring permits higher flame velocities and frequently is advantageous or necessary when propane is used. Because acetylene has a higher flame-propagation rate, counterboring usually is unnecessary and frequently is undesirable. Counterboring reduces coupling distance and permits a stabilised flame at higher flame velocities. For effective operation, the ratio of counterbore area to throat area should be of the order of 2 to 1. Ratios as high as 4 to 1 may be used in some cases, however.

Oxy-Fuel Gas Flame Heads

Oxy-fuel gas combustion develops flame temperatures above those at which useful metals and refractory materials can survive. Accordingly, the flame head is designed in such a way as to provide a flame pattern that avoids any direct heating of its metal parts. Generally, the flame head consists of a tube or a shell with one or more orifices drilled into it; the number and arrangement of orifices depend on the required area of heat coverage. Flame heads for use with oxy-fuel gas are shown in Fig. 23.10.

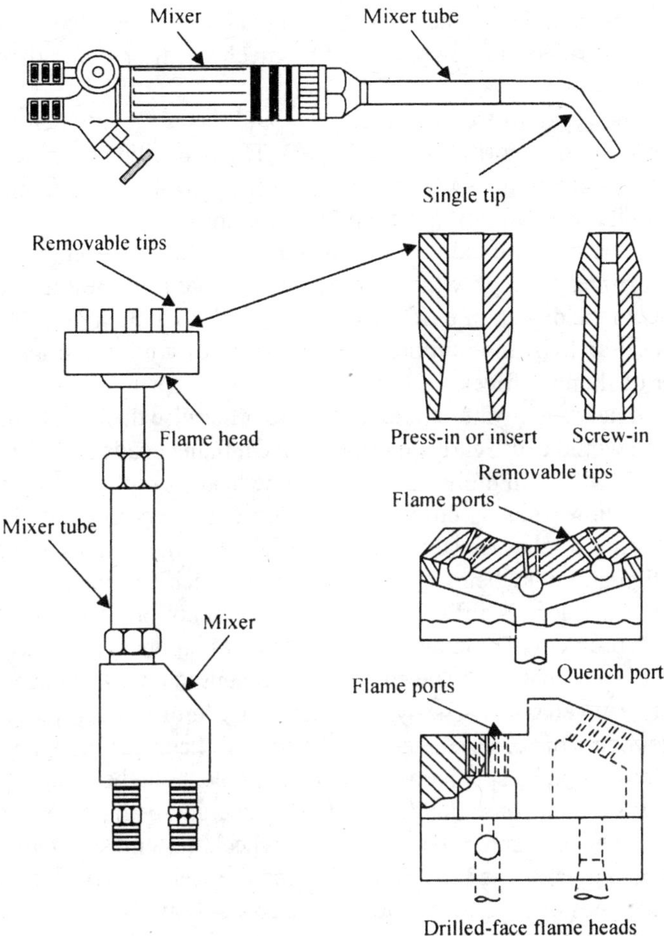

Fig. 23.10. Flame heads for use with oxy-fuel gas.

The drilled-face flame head has a limited range of application and usually is designed to meet the requirements of one specific part to be flame hardened in an appreciable quantity. For other applications, the flame head may be fitted with removable tips of the screw-in or insert type.

The screw-in type of removable tip shown in Fig. 23.10 is used widely enough so that it is available in off-the-shelf commercial flame heads or supplied as a standard items by manufacturers of flame-hardening equipment; it is also simple enough that is can be made to specification by plants applying the flame-hardening treatment.

The insert, or press-in, type of tip shown in Fig. 23.10 is smaller than the screw-in type and permits closer spacing of orifices, approaching that of a drilled flame head. Flame heads with removable and replaceable tips can be used over a wider range of applications by removing one or more tips and replacing them with plugs.

The heat output of the flame head is governed by the number and size of the orifices, when all other factors are equal. The individual orifices range from No. 73 to No. 51 drill size (0.6 to 1.7 mm or 0.024 to 0.067 inch, in diameter). Such small holes can readily become plugged and when this occurs the flame head will not function properly. Flame heads with removable tips have an advantage over fixed-tip heads in that oversize or out-of-round holes, caused by mechanical or flame damage, can be corrected without replacing the head.

Integral parts of the flame system are the mixer block and mixer tube, which mix the component fuel gases and convey them through the orifices (Fig. 23.10). The capacities of mixer block and mixer tube must match the number and size of the orifices; if the mixer is too small the flame will flash back and if the mixer is too large the flame head will not function efficiently.

To ensure identical velocity of mixed gases at all flame orifices, it is common practice to design a flame head with baffle orifices through which the gases must pass before they are burned at the flame orifices. Two rules apply to the design of baffle orifices: (i) their total area must be 1.25 to 1.50 times the area of the flame orifices; and (ii) the number of baffle orifices within a single baffle should be one-fourth the total number of flame orifices.

Multiple-orifice flame heads are water cooled, because otherwise the high temperatures developed at and around the flame head would cause early deterioration. On flame heads used for progressive hardening, the quench water cools the head. On multiple-orifice flame heads used in the spinning and progressive-spinning methods, the cooling water is circulated through chambers integrated into the head. Single-orifice flame heads (welding torches, for example) generally are not water cooled.

New equipment configurations are being designed to handle specific problems. With conventional flame-hardening equipment, for example, it is difficult to obtain a zone of uniform temperature across the surface of the tooth gap in gears. The problem of obtaining uniform depth of hardness, not only from tooth to tooth but also across the entire face width, has been solved by development of a two-chamber burner with individual control of energy input to each chamber. The system ensures uniform heating, particularly in the tooth root and incorporates separate control of nozzles for tooth-root and tooth-flank hardening. With this system, complex components also can be processed by spin hardening. For example, both the straight gear and bevel gear of a double pinion shaft can be simultaneously hardened, if the quenching-bath immersion depth is adequate. By distributing heating power between spur wheel and bevel wheel, austenitising temperature can be reached under the tooth roots of both gears at the same time. This requires selection of different heating-up times for the two gears and different rates of oxygen flow to the two sets of burners.

Air-Fuel Gas Burners

Air-fuel gas combustion develops lower flame temperatures that are compatible with available refractory materials. Thus, burners are designed with the aim of completely utilising the heat generated. The burners, incorporating heat resistant refractory liners, are of two types, generally designated as either the radiant type or the high-velocity convection type.

Radiant Type Burner

Radiant type burner is essentially a refractory cup in a protective metal casing. Air-gas pre-mixture is supplied through the pipe at the back and passes through an accurately moulded ceramic tip that is screwed into the pipe and located at the bottom of the cup. With numerous narrow slots moulded into its periphery, the tip functions essentially as the distributing head of a multiport burner. The many small flames wash the inner surface of the cup, making it highly incandescent for rapid heat transfer by radiation. Because combustion is completed with in the cup, the burner may be positioned close to the work with no flame impingement.

The standard radiant burner employed in flame hardening is approximately 75 mm (3 inch) in cup diameter. It is particularly effective for spin hardening the teeth of large gears. A single row of burners may be arranged in a ring surrounding a gear, as shown in Fig. 23.9(a), or multiple rows of burners may be arranged to cover completely the surface to be hardened.

High-Velocity Convection Burner

The high-velocity convection burner is basically a miniature refractory-lined furnaces in which heat is released at rates as high as 414 MJ/m^3.s (4×10^7 Btu/ft^3.hr.). The air-gas pre-mixture is supplied through a pipe connection and flows through the orifices of the ceramic port plate. The burner design is such that the burning gases heat the chamber lining to a temperature approaching the theoretical flame temperature. This permits pre-heating of the reacting gases and accelerates combustion. In this manner, gases at approximately 1650°C (3000°F) are discharged through a restricted slot opening to impinge on the workpiece at velocities up to 760 m/s (2500 ft/s). High-velocity convection burners are well adapted to localised heating of parts in spin-hardening operations.

Related Equipment

Burners for both the oxy-fuel gas and the air-fuel gas methods of flame hardening are implemented by pressure regulators, valves, flowmeters and protection devices. For the air-fuel gas method, a separate mixer and compressor are utilised, because the mixing function is not incorporated within the burners (Fig. 23.11).

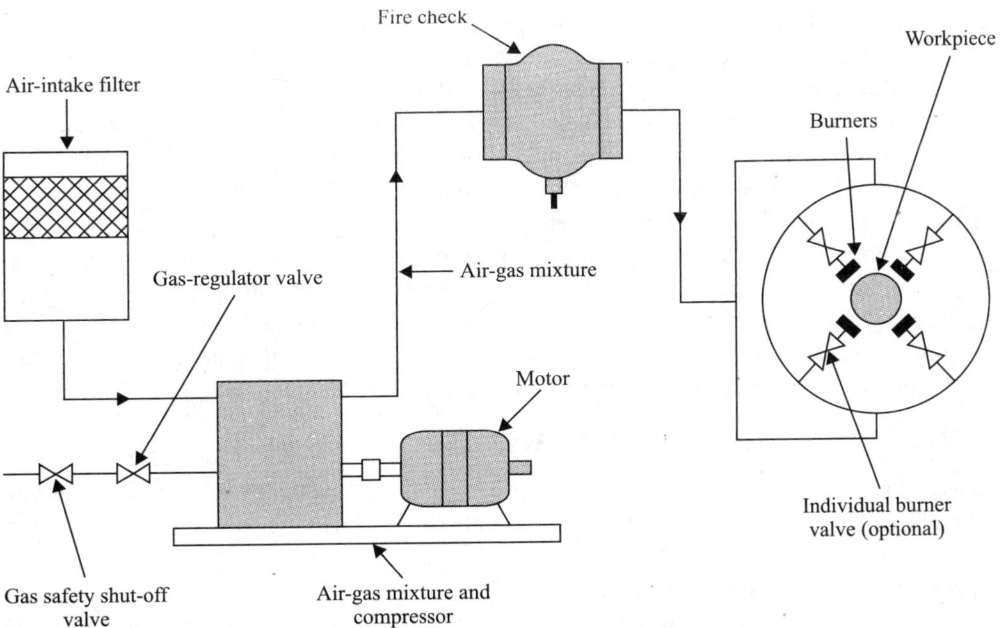

Fig. 23.11. Mixer-burner system for use with air-fuel gas.

Materials of Construction for Burners

Metallic components of burners may be of a variety of materials, depending on the type of burner and the service to which it is subjected. The flame heads for oxy-fuel gas flame heating are in nearly all instances made of copper or a dense grade of lead-free brass. These metals are relatively inexpensive, have excellent thermal conductivity and are readily machined. Because flame velocity is relatively low, there is little likelihood that the metal flame head will overheat and deteriorate as a result of the backwash of hot gases. Normally, the flow of gases within the head provides sufficient cooling to maintain a safe temperature. If additiional cooling is required, the head may be water cooled. Under no circumstances should the flame head be permitted to overheat to an extent that will cause damage to the burner metal, brazed burner joints or drilled flame ports.

Life of oxy-fuel gas flame heads depends largely on the conditions of a particular application, such as containment of the heat resulting from the configuration of the piece being heated and the degree of backwash of flame or hot gases impinging on the burner. Therefore, it is difficult to predict the exact life to be expected for a new application. Brass heads used for progressive hardening of machineways have shown an average life of 1000 to 2000 hours in continuous service.

Because the melting points and ranges of heat resistance of available materials are not greatly different (compared with the large differences between flame temperatures and metal melting points), life expectancy is a relatively unimportant basis for selection of a particular metal. Availability, cost and ease of fabrication are likely to be more important factors in material selection. A metal frequently used for flame heads in Muntz metal (60–40 lead-free brass); when cooling is a problem, however, copper is preferable.

Screw-in tips usually are made of copper, but when higher heat resistance is required, because of secondary flame, Monel K-500 can be used. Insert or press-in tips usually are made of copper, but when high wear resistance of the tip bore is desired, tips made of sintered carbide are employed.

Cast iron, stainless steel and heat-resisting alloys are used for the casings that surround the refractory cups of radiation-type air-fuel gas burners. Heat-resisting alloys may provide reasonable life at temperatures up to 1150°C (2100°F). Casings of high-velocity convection burners generally are constructed of stainless steel, but may be of Inconel for more severe service. For spin hardening of gears, either type of burner may have a useful life of 10 to 15 40 hours weeks of continuous operation. In many applications, longer life of casings has resulted when the operator has exercised care in setting up and operating burners. Also, water cooling of the burner casing can markedly extend its life and may be employed provided that the added cost is justifiable and that space for the cooling arrangement is available in the heating set-up.

The refractory parts of all types of air-gas burners are formed in steel dies and prefered in kilns and are capable of withstanding the most extreme temperatures generated in the burners. These parts generally fail from thermal or mechanical shock; hence, service life is unpredictable.

OPERATING PROCEDURES AND CONTROL

The success of many flame-hardening applications depends largely on the skill of the operator. This is especially true when the volume of work is so small or so varied that the cost of automatic-control equipment is not justified.

The principal operation variables are:

1. Distance from inner cone of oxy-fuel gas flames or from air-fuel gas burner, to work-surface.
2. Flame velocities and oxygen-to-fuel ratios.
3. Rate of travel of flame head or work.
4. Type, volume and angle of quench.

These variables must be closely controlled to ensure duplication of desired surface hardness and depth of hardness. It is highly desirable to develop a specific procedure for each item to be flame hardened. The procedure is developed by preliminary tests on the production piece itself, if warranted, or on mock-up sections of approximately the same cross-section as the production piece. After the desired contour and depth of hardened zone have been developed, the procedure is applied to production pieces and when established, is made a part of the heat treating specification.

Gas Pressures

Oxy-fuel gas and air-fuel gas pressures should be controlled closely for uniform input of heat. Flat oxy-fuel gas flame heads are somewhat less efficient when used on circular or curved surfaces, because each cone of flame is at a different distance from the work. Overheating causes cracking.

Oxygen-to-Fuel Ratio

Oxygen-to-fuel ratio is a key factor in determining flame temperature. For example, propane produces flame temperatures of 2700°C (4900°F) at a 5-to-1 ratio, 2540°C (4600°F) at a 4-to-1 ratio and 2340°C (4300°F) at a 3-to-1 ratio.

Flame Velocity

Flame velocity is one of the most important variables, because when balanced with other variables it is the main determinant of case depth. In flame hardening 1045 steel, the basic flame velocity required for establishing a case depth of 3.2 to 4.8 mm ($\frac{1}{8}$ to $\frac{3}{16}$ inch) is 152 m/s (500 ft/s).

In flame hardening cast irons, where high surface temperatures are undesirable, use of lower flame velocities is distinctly advantageous. In this regard, both MAPP and propane are easy to control and are effective over a wide range of velocities. The ability to control the flame over a wide velocity range not only provides flexibility but also makes the operation much safer and results in better heat-pattern control.

Coupling Distance

Coupling distance is another key parameter in flame hardening. The relations between coupling distance and depth of hardening for three fuel gases are illustrated in Fig. 23.12. In general, the coupling distance for MAPP is equal to or very slightly greater than that for acetylene. The effect of counterboring is a shortening of the coupling distance. By the same token, coupling distance increases as flame velocity increases. Efficient coupling distances for MAPP generally range from 6.4 to 9.5 mm (½ to $\frac{3}{8}$ inch) depending on gas velocity and port size. A counterbored port of No. 69 drill size (0.74 mm or 0.0292 inch) will operate efficiently at a coupling distance of 4.8 mm ($\frac{3}{16}$ inch).

Hardening Temperatures

Hardening temperatures can be judged by competent operators, but to the inexperienced the heated metal will appear colder than it actually is, because of the light from the burning gases, the consequent tendency is to overheat unless the operator is equipped with didymium-tinted glasses. Radiation or optical pyrometers are often used to judge more accurately the temperatures developed. Radiation-pyrometer systems of fast response are used extensively to control work temperatures and heating times. Metallurgical examination is the best method for establishing operating conditions. Overheated spots may appear under individual flames, but with properly designed flame heads and good scanning technique, this effect is minimised and not readily detectable by microscopic examination.

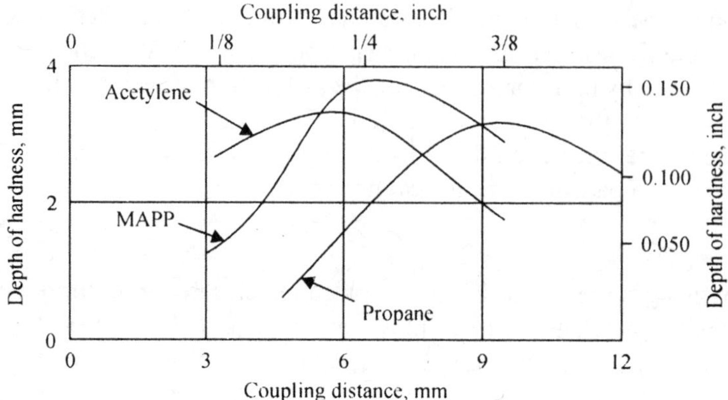

Flame velocity, 170 m/s (550 ft/s); port size, No.69 drill (0.74 mm or 0.0292 inch); material, 1052 steel. Oxygen-to-fuel ratios: MAPP, 4.5; acetylene, 1.33; propane, 4.5. Heating times: MAPP, 6 seconds; acetylene and propane, 8 seconds

Fig. 23.12. Effect of coupling distance for MAPP, acetylene and propane.

PRE-HEATING

In flame hardening parts of large cross-section, difficulty in obtaining the desired surface hardness and hardness penetration often can be overcome by pre-heating. When the available power or heat input is limited, the hardened depth can be increased by pre-heating.

The hardness data in Fig. 23.13 illustrate the effectiveness of pre-heating in developing hardness penetration in the ring gear.

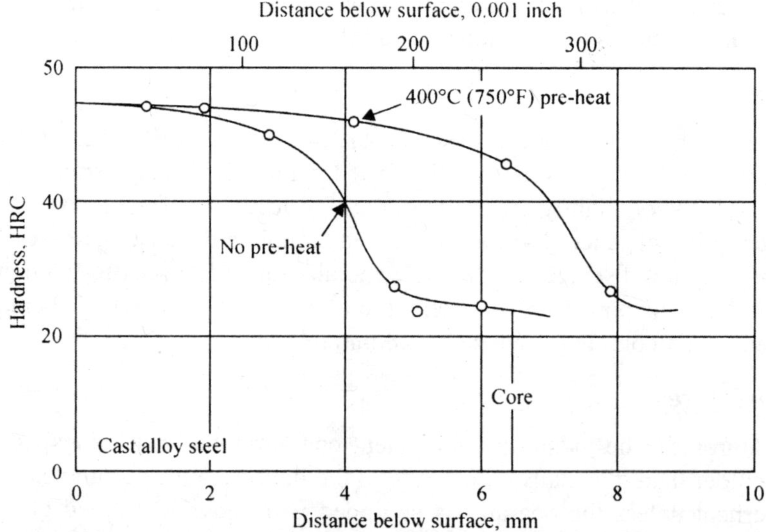

Fig. 23.13. Effect of preheating on hardness gradient in a ring gear.

Flame hardening of pre-hardened and tempered steels, especially some of the alloy steels, requires careful control of heating to avoid cracking. Pre-heating of the part may be advisable to minimise cracking in such steels. Hardenable cast iron also is susceptible to cracking. In one application, cast iron crane wheels had to be pre-heated to 480°C (900°F) to prevent rupture of the spokes caused by non-uniform expansion during spin heating of the tread area.

In another application, the difficulty of heat treating the internal teeth of a planetary-gear housing was overcome by spin hardening. The component had an irregular mass distribution, which led to large distortions when heat treating was done by ordinary flame-hardening or induction-hardening equipment. Case hardening of teeth also failed. The problem was solved by using a specially designed gear-wheel spin-hardening machine with close control over heating time and power and by pre-heating the part with the largest mass and cross-section to an accurately pre-determined temperature.

DEPTH AND PATTERN OF HARDNESS

In some instances, the flame hardening procedure results in a greater depth of hardness than desired. Fig. 23.9, for example, shows that spot hardening a rocker-arm pad after heating 4 seconds at 870°C (1600°F) with oxyacetylene flame produces a depth of hardness of 5.1 mm (0.200 inch). If excessive, this depth may be reduced by heating the steel to the same temperature but in a shorter time. Thus, with a reduction in heating time to 3.2 seconds at 870°C the depth of hardening will be 2.5 mm (0.100 inch) (Fig. 23.8a). Because of the shorter heating period, the cost of hardening each pad also will be reduced.

The problem of excessive depth resulting from a spin-hardening operation may be resolved similarly, as shown in Fig. 23.9b. In this operation, heating with an oxyacetylene flame for 13.5 seconds at 870°C (1600°F) produces a hardness depth of 7.6 mm (0.300 inch). Heating at the same temperature for a period of 8 seconds results in a hardened depth of about 3.8 mm (0.150 inch). In a progressive-hardening operation with a rate of travel of 1.7 mm/s (4 inch/min), a hardness depth of 5.1 mm (0.200 inch) will be produced when the heating time is 12 second at 870°C (1600°F), as shown in Fig. 23.9c. The flame head contains two rows of flame holes that produce a heating zone 20 mm (0.8 inch) wide. If the rate of travel is increased 20 per cent to 2 mm/s (4.8 inch/min), the heating time is proportionately reduceds to 10 second. This results in a hardened depth of about 2.5 mm (0.100 inch).

Variation in Depth of Hardness

Variation in depth of hardness may occur in steels of the same nominal composition but from different heats. This is demonstrated in Fig. 23.14, which shows the variation in depth of hardness obtained in flame hardening hubs made of 1062 steel from three different heats. The inside diameter of these hubs were flame hardened for 12 seconds to produce a minimum hardness of 59 HRC at a depth of 1.9 mm (0.075 inch).

Surface Hardness Pattern

The hardness pattern of any portion of a flame-hardened surface can be determined by lightly blasting the area with fine sand. The hardened portion of the surface will be less affected than the unheated area by the cutting action of the sand. This procedure also may be employed to indicate soft spots when the entire surface has been subjected to flame hardening. (Soft spots may result from non-uniform heating or the interference of scale on the surface).

Another procedure for determining surface hardness pattern is etching of the area with a 10 per cent nitric acid solution. The hardened area will appear darker than the unhardened area. These procedures are applicable also to cross-sections of hardened areas for an indication of heat penetration and they are useful as preliminary checks on the performance of the flame-hardening set-up. Variations in the hardness of the surface are exemplified by the data shown in Fig. 23.15. These hardness readings taken on the surface of races made of 4063 steel cover a one-month period. These races were heated for 9 seconds to produce a minimum hardness of 59 HRC at 1.3 mm (0.050 inch) below the surface.

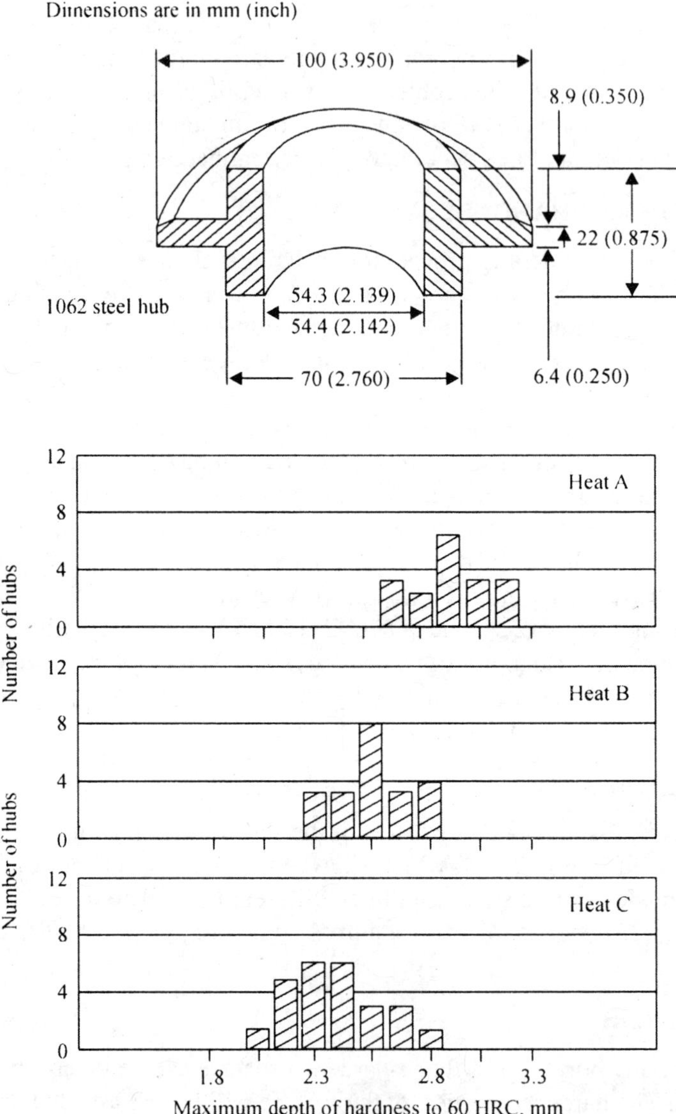

Variations in depth of hardness among three heats of 1062 steel hubs flame hardened on the ID to 59 HRC min at 1.9 mm (0.075 inch) below the surface. Hubs were heated 12 second and quenched in oil. Hardness was measured on cross-sections of heated area.

Fig. 23.14. Heat-to-heat variations in depth of hardness.

Thus, hardness of the case in flame hardening is a function of the carbon content of the steel and will range up to 65 HRC. Medium-carbon steels with 0.40 to 0.50 per cent carbon are ideal for flame hardening, but steels with carbon contents as high as 1.50 per cent also can be flame hardened with special care.

Normally, hardening depth ranges from 1.3 to 6.4 mm (0.05 to 0.25 inch). Heavier sections, such as large rolls and wheels, can have case depths of up to 13 mm (0.5 inch). Manganese-bearing alloys aid in the depth of hardening by decreasing the critical cooling rate which contributes to deep hardening. Therefore, manganese and free-machining grades of steel are considered excellent for flame hardening.

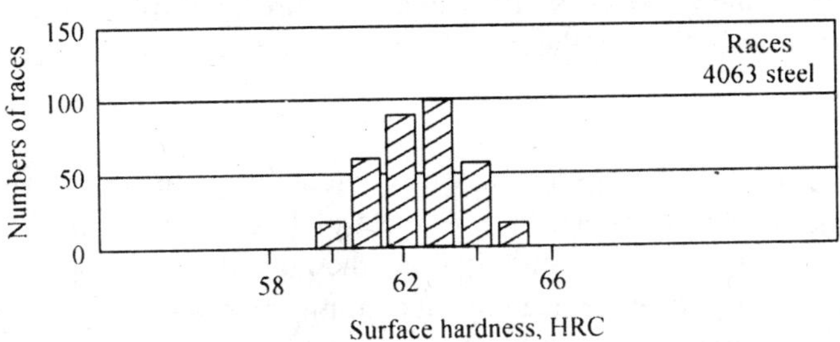

Results obtained during one month of testing flame-hardened races (4063 steel) that were heated for a total of 9 seconds to produce a hardness of 59 HRC min at a depth of 1.3 mm (0.050 inch)

Fig. 23.15. Distribution of surface hardness.

When hardening depths are required beyond the capabilities of ordinary carbon steels (0.60 to 0.90 per cent Mn), elevated manganese ranges such as 0.80 to 1.10 per cent, 1.00 to 1.30 per cent or 1.10 to 1.40 per cent can be efficiently used. Wear resistance in many cases is not the only critical design criterion. Under high compressive loading, the hardened layer must be deep enough not only to provide the required wear life of the part, but also to contribute to the support of heavy contact loads. The case must be fully martensitic and the material supporting the hardened layer must be of sufficient strength. However, increased hardenability may lead to cracking problems at least with water quenching.

MAINTENANCE OF EQUIPMENT

Flame Heads (Nonferrous) of the Oxy-Fuel Gas Type

Experience with non-ferrous flame heads of three types of construction (drilled face, screw-in tips and press-in tips) has shown carbon deposit, lack of proper cooling, erosion and corrosion to be the main causes of maintenance problems common to all.

Carbon Deposit

The intermittent igniting and extinguishing of the flame causes a small deposit of carbonaceous material to build up on the sidewalls of the port, from the retrogression of the burning fuel below the orifice of the port as the flame is extinguished. Although small for each cycle, the deposit builds up gradually until it causes a restriction in the port and thus a variation in the velocity of the gas and therefore in the heating performance. Several thousand cycles may be completed before it is necessary to clean the head.

A slight amount of cleaning can be done without removing the head from the mixer tube, by pushing through the ports a wire of the proper size or a drill one size smaller than the ports. However, care must be exercised to prevent bell-mouthing or increasing the size of the port. Cleaning by this method deposits loose carbonaceous material in the fuel chamber and this material may be moved by the velocity of the oxygen and fuel gas mixture and become lodged in other ports. Flame heads operating on volume production should be removed from the machine and thoroughly cleaned after 40 to 120 hour of operation; the interval depends on the installation, the oxygen and fuel gas mixture and the heating cycle time.

Drilled-port and insert-port heads, when removed, can be cleaned by inserting in the ports a wire of the proper size or a drill one size smaller than the ports and then immersing the heads in a suitable solvent or cleaner that will loosen carbonaceous deposits. Again, care must be exercised to prevent bell-mounting or enlarging of the port. Also, the cleaner or solvent must not attack the head itself. After cleaning and rinsing in clean water, all traces of dirt and moisture should be blown from the ports, fuel chamber and water lines. Screw-in tips and heads may be cleaned in the same manner.

Erosion and Corrosion

All types of heads are subjected to the by-products of combustion, which may attack the face of the burner or the tips, causing erosion and corrosion. A flash plate of chromium on the burner face has proved helpful in reducing erosion and corrosion. In time, the face of a drilled-port burner may erode to a point at which it is necessary to re-face the head by machining and re-working the counterbore. If there is enough space between ports on drilled-port heads, the port can be repaired by drilling oversize and pressing in an insert. These inserts can be removed when necessary and replaced.

Because the oxy-fuel gas type of flame head is water cooled, it may be necessary, in localities where extremely hard water is supplied, to softened the water to keep scale from forming in the cooling passages; commercial water softeners of the zeolite or poly-phosphate type may be used. Recirculation systems with heat exchangers are sometimes required.

Burners of the Air-Gas Type

Failures of air-gas burner liners occur by spalling, erosion and cracking of the refractory cup or tunnel and outlet. These liners should be inspected periodically, at intervals determined by experience with the individual installation; when conditions are severe, inspection may be required as often as once every shift.

The causes of deterioration of burner liners are thermal shock from repeated and rapid heating or cooling (or both) and mechanical shock from work striking the burners. The burner casings deteriorate mainly from prolonged exposure to escaping hot gases that heat the casings locally, causing more or less progressive oxidation, cracking, growth and infrequently, burning. If installation space permits the use of water or air cooling, these usually can be expected to increase the life of burners. Shielding, designed to conduct hot gases away from the burners and other parts of the machine, also is valuable in reducing maintenance costs.

Mechanical Components

Because both oxy-fuel and air-fuel gas systems employ pressure gauges, pressure regulators, valves, flowmeters and safety devices, the maintenance problems associated with these accessories are common to both methods of flame hardening. Symptoms and probable causes of trouble with several mechanical components used in flame hardening are listed in Table 23.2.

Air-gas systems also have maintenance problems related to their compressors and blowers. Devices for protection against backfire and explosion should be serviced according to manufacturers' instructions.

Table 23.2. Maintenance problems in flame-hardening equipment.

Component	*Symptom of trouble*	*Probable causes*
Pressure gauge	Erratic behaviour	Mechanical damage to mechanism
Gas-supply pressure regulator	Failure to hold pressure setting at outlet	Diaphragm broken or hardened

(Contd...)

Component	Symptom of trouble	Probable causes
		Diaphragm too taut
		Sticking stem or valve parts
		Damaged or dirty valve seat or plugged orifices or vents
		Broken or damaged spring
Solenoid valve	Failure to close or open	Defective valve mechanism or wiring
Fuel valve	Flame lingers after shut-off	Dirty seat or stem; damaged seat
Oxygen valve	Flame popping after shut-off	Damaged seat

Electrical Components

Electrical contacts should be cleaned periodically; when the heating cycle is controlled by a timer, periodic checks of the timer are advisable. Generally, if the heating cycle is inaccurate, it will be evident in the inspection of the parts. If the heating cycle is controlled by a thermopile, several items must be checked: (i) excessive flow of cooling water through the housing will cause condensation on the lens; (ii) a deposit of any type on the lens will cause erratic results; and (iii) in high-production work, the lens should be cleaned at least twice each shift and the temperature recorder should be balanced once each shift.

Spindle and Movable Holding Fixtures

It is important that positioning of the parts and the flame head be consistent. Therefore, holding fixtures and spindles should be periodically cleaned, checked and lubricated. Any work part should be replaced before it becomes improperly located.

Piping

Types of piping and fittings recommended by underwriters should be used in installing the gas lines from the gas source to the site of use. In long gas lines, means should be provided for purging accumulated condensate from the line before the gas reaches the mixing chamber or torch.

PREVENTIVE MAINTENANCE

The following is a typical schedule of preventive maintenance for flame-hardening equipment:

Daily

1. Check all solenoids for leaks; clean, repair or replace, as necessary.
2. Replace damaged screw-in flame tips; clean dirty or clogged tips or ports.
3. Remove foreign material from surfaces of flame heads and movable machine parts.
4. Check for stable pilots.
5. Check for leakage of fuel gas, water, oxygen and air.
6. Check level, circulation and temperature of quench oil.
7. Clean lens of radiation pyrometer and balance recorder.
8. Check temperature of water-cooled parts.
9. Check oil level and flow of oil to spindle.

Weekly

1. Check liquid level in backflash arrestor.
2. Check relief valve on backflash arrestor.
3. Lubricate all movable parts as required.
4. Check all fuel-gas, oxygen and air connections for leaks (use soap solution).

Bimonthly

1. Remove and clean flame heads (frequency will vary with installation and use).

Semi-annually or Annually

1. Remove, clean and repair oxygen and fuel-gas regulators.
2. Remove, clean and repair all solenoids.
3. Check electrical contacts and wiring.
4. Clean heat exchanger and quench tank.
5. Check orifice in mixing blowpipe.
6. Re-pack motor bearings.

SAFETY PRECAUTIONS

All fuel gases are explosive when mixed with either air or oxygen within their flammable limits. However, they are widely used throughout industry and are safe when properly handled, stored and transported in compliance with established laws of safe practice. Stabilised methylacetylene-propadiene combines the safety and handling ease of propane with the high heat energy of acetylene.

Operators should be taught to keep equipment clean, free of grease and oil and in good condition and to avoid leaks of oxygen or fuel gas by regularly testing for them with soapy water free of oil and grease. They should open oxygen or fuel-gas valves slowly and be sure that the lines are purged before igniting. An air hose should be available for dispersing fuel gas that has been bled from the lines. Any spaces in and around the machine in which gases might accumulate should also be purged with the air hose. No flame or other source of ignition should be allowed in proximity to the area where fuel gas or oxygen is released into the air and good ventilation is of primary importance.

Operators should recognise backfires or backflashes immediately and shut-off the gases; the cause and extent of damage must be determined and corrections made, before the torch is re-ignited. To prevent major explosions, a backflash arrestor must be installed in the fuel-gas line.

Hand friction lighters should not be used to ignite the gases, except for very small flame heads of low gas capacity. A flame from an acetylene torch or a burner using some other gas should be used. The pilot flame should be very close to the heating head, to prevent accumulation of gas and oxygen before ignition. For automatic flame-hardening machines, the pilot light can be permanently mounted (in which case a flame detector and alarm system should be installed to give warning when the pilot light is extinguished accidentally) or electric-spark ignition can be used. When operations are stopped for any length of time, the valves in the main supply lines should be closed. These valves must not leak, because even a small leak can cause a dangerous accumulation of gas.

Oil or grease may ignite violently in the presence of oxygen under pressure and must be kept away from cylinders, regulators, hoses and other apparatus.

The cylinders in a manifold should be secured with clamps or chain to prevent them from tipping or falling. A fall or any severe bump can detonate acetylene, with disastrous consequences. Propane is not likely to detonate, but any gas under pressure is dangerous if the valve is damaged or knocked off the cylinder. In some instances, noise can be a problem in a gas-air system. Equipment should be designed to keep the noise level within generally accepted limits, to avoid injury to nearby personnel.

QUENCHING METHODS AND EQUIPMENT

The proper application of a suitable quench in flame hardening is as important as proper heating. The quench must remove heat at a rate that will produce the desired structure and assist in controlling the depth of the hardened case.

Method of quenching and type of quenchant vary with the method of flame hardening. In spot hardening, immersion quenching is generally used, but spray quenching may be used.

Quenching after Progressive Heating

Parts heated by the progressive method usually are quenched by a spray integrated into the flame head, although for steels of high hardenability or where it is desirable to vary or adjust the distance between the heated zone and the quench, a separate spray-quench head is sometimes used. The spray quench should issue from the head at an angle away from the flame head, to prevent interference with heating and must provide full coverage over the heated band. The integrated type commonly sprays the quenching medium on the work 19 to 32 mm (¾ to 1¼ inch) behind the last row of flames. When the surface to be hardened by the progressive method is vertical or overhead, additional cooling may be required after the usual spray quench.

Quenching after Spin Heating

Parts heated by the spinning method can be quenched by several different procedures. In one, the heated part is removed from the heating area and quenched by immersion in a separate quench tank. Another method integrates quenching by making the quench tank a part of the flame-hardening machine. Parts may be heated on the machine spindle or arbour and then be quenched by being dropped into a quench tank located immediately below the arbour. Parts that are either too heavy or too fragile to be dropped may be lowered into the quench by means of arms or may be lowered into the quench while still on the arbour. Parts may be removed manually from the quench or be removed automatically by means of a conveyor belt. A spray-quenching ring sometimes is submerged in the quench bath to increase cooling rate.

Parts heated by the spinning method may be quenched also with 'quench blocks' on the same plane as the flame heads. When the heating cycle is completed and the flames are extinguished, the quench is turned on. The quench blocks should cover the heated band and provide enough quenching liquid to obtain suitable and uniform hardness. There should be enough quench points around the periphery of the block to envelop the area completely with quenchant. Water or a polymer quench solution is usual for this type of quench, because of the hazard of contaminating the heating heads with oil. When oil quenching is used, the oil should not contact the heating head. This may be avoided by indexing the heated part to a separate quenching station or by retracting the heating heads into a shielded area.

When air-fuel gas mixtures are employed in spin-hardening operations, the heat pattern developed, in many instances, extends beyond the limits desired. Immersion quenching in such instances would potentially extend the hardening into areas requiring subsequent machining or contribute to an excessive amount of distortion. Localised quenching, such as that provided by quench rings, is a solution to this problem. For example, this quenching method was used in hardening the teeth of large sprockets weighing about 205 kg (450 lb). The entire tooth surface, including the root, was hardened to a depth of 4.8 mm

($\frac{3}{16}$ inch). The 10-minute heating cycle completely heated the teeth and back-up rim. When quenching was done by rotation of the sprocket in a ring of water-spray nozzles, hardening was confined to the desired surfaces and the back-up rim remained soft. A similar practice is employed for hardening of large gears. Parts heated by the combination progressive-spinning method usually are quenched by a spray integrated in the flame heads or by separate quench blocks located below the flame heads.

QUENCHING MEDIA

With spray quenching, either integrated in the flame head or by separate quench blocks, water or a dilute polymer solution is used as the quenching medium. Quenching oils should not be allowed to come in contact with oxygen or to contaminate equipment using it.

As with conventional hardening, alloy content of the steel determines the type of quench that should be used; the quenchant may be water, a brine solution, the glycol-base polymer quench or air.

By reducing the pressure of the quenchant, the rate of cooling by spray quenching can be reduced from the maximum for which the integrated or separate quench blocks are designed. Increasing the distance between the last row of flames and the point at which the quenchant impinges will allow the mass of metal below the area to be hardened to extract heat and thus will decrease the severity of quenching.

Quenching distance is another factor the operator must consider, because the quenchant must strike the heated area while it is still at the critical temperature to avoid formation of pearlite or other undesirable transformation products in the microstructure. If the spray head is angled too close to the flame, spotty hardness and blowouts can occur. If the angle of spray is too far away, the case may not be fully hardened.

Forced Air

In progressive hardening operations, forced air frequently is used as the quenchant for steels normally considered to be oil-quenching steels. Water is not used immediately after heating, because the fast cooling action would result in surface checks. Because most of these steels have relatively low Ar_3 transformation temperatures, the forced air quickly reduces the surface temperature to a point at which water can be applied without causing surface checks. The resulting hardness usually is close to that obtained with a direct oil quench. For example, 52100 steel quenched with forced air and then with water attains a surface hardness of 60 to 61 HRC.

Forced-air quenching is used also in applications where intermediate hardnesses are required. One example is the hardening of railroad rail ends to reduce 'end-batter' by the impact of car wheels. The rails are indexed under the burners in four pre-heat stations and one high-heat station (high-temperature high-velocity burner), which heats the rail to the hardening temperature of 870°C (1600°F) in 95 seconds. In a sixth station, air at 690 kPa (100 psi) is directed against the heated surface. This treatment results in a hardened structure of fine pearlite that provides wear resistance and sufficient ductility to withstand the impact of the moving wheels. Water is unsuitable in this application, because the high-carbon rail steel is susceptible to cracking. Rail lengths for curves, crossovers and switches are similarly hardened; however, for these, hardening is done progressively, utilising oxy-gas burners and air quenching.

Immersion Quenches

Immersion quenches vary in type in relation to the metal used, the hardness and depth desired and the mass, design and dimensional tolerances of the part. The quenching medium can be a caustic or brine solution, water, a soluble-oil emulsion or any of a large variety of oils or simulated oils such as polymer quenches.

Self-Quenching

During any type of flame hardening other than through hardening, the mass of cold metal underneath the heated layer aids quenching by withdrawing heat. Thus, cooling rates are very high compared with rates for conventional quenching. During progressive hardening of the teeth of gears made from medium carbon steels such as 4140, 4150, 4340 or 4640, for example, the combination of rapid heating and the temperature gradient between the surface and the interior of the gear results in a subsequential self-quench that is equivalent to quenching in oil. For uniform quenching, a coolant can be used at some distance from the tooth being heated. In one application involving hardening the teeth of 4150 steel gears in the size range of 4 to 12 pitch, one tooth at a time was heated and cooled by conduction. The coolant was directed at the body of the gear. With this procedure, the hardness of the teeth was 50 to 55 HRC after the gears had been tempered at 205°C (400°F).

FLAME HARDENING PROBLEMS AND THEIR CAUSES

Problems that may occur in flame hardening and their causes, are listed below.

Overheating

1. Controlling pyrometer set too high.
2. Millivolt compensator of controlling pyrometer set incorrectly.
3. Heating cycle too long.
4. Flame heads too close to work.
5. Oversize flame ports.
6. Excess oxygen in flame.
7. Excessive fuel-gas pressure.
8. Improper pattern of flame tip.

Hardness Below Minimum Required

1. Controlling pyrometer set too low.
2. Millivolt compensator of controlling pyrometer set incorrectly.
3. Heating cycle too short (underheating).
4. Severity of quench too low.
5. Delay before quenching too long.
6. Part not thoroughly quenched.
7. Material hardenability too low for quench.
8. Surface decarburised.

Spotty or Uneven Hardening

1. Non-uniform heating.
2. Time interval between heating and quenching too short.
3. Quenching medium not agitated enough.
4. Water in quenching oil.
5. Scale on work.
6. Improper quenching medium.
7. Surface decarburised.

Distortion

1. Shape of part or relation of portion to be hardened to remainder of section, not well adapted to flame hardening.
2. Metallurgically unsuitable prior structure.
3. Heating cycle too long.
4. Non-uniform heating.
5. Non-uniform quenching.
6. Excessive rate of quenching.
7. Material hardenability excessive.

Shallow Depth of Hardening

1. Material of low hardenability.
2. Excessive rate of gas flow; check for oxidising flame.
3. Flame-port velocity too high.
4. Controlling pyrometer set too low.
5. Millivolt compensator of controlling pyrometer set incorrectly.
6. Short heating cycle or excessive scanning speed.
7. Severity of quench too low.
8. Delay before quenching too long.

Excessive Depth of Hardening

1. Low flow rate of gas; check for reduction flame.
2. Excess fuel gas in flame.
3. Controlling-pyrometer temperature set too high.
4. Millivolt compensator of controlling pyrometer set incorrectly.
5. Flame-port velocity too low.

Excessive Scaling

1. Heating cycle too long.
2. Rates of gas flow too low.
3. Flame velocity too low.
4. Delay before quenching too long.
5. Improper arrangement of flame heads around periphery; overheating or banding.

TEMPERING OF FLAME-HARDENED PARTS

It is usually desirable to temper parts that have been flame hardened; the need for tempering of martensite is the same regardless of the heat treating method used to produce it. Flame-hardened steel will respond to a tempering treatment as it would if hardened to the same degree by any other method. Standard procedures, equipment and temperatures can be used. However, for work that is flame hardened because it is too large to be heated in a furnace, flame tempering may be the only feasible method of tempering available.

Flame Tempering

Large articles usually are hardened by the progressive method and can be tempered immediately by re-heating the hardened surface with a flame head placed a short distance behind the quench. The re-heating or tempering flame head must be designed correctly in regard to number and size of orifices or tips (flame ports) to produce the desired work temperature and temperature gradient in the flame-hardened zone at the flame-hardening speed.

Exact final adjustments can be made by varying the flow of gases through the flame head and by adjusting the distance between flame head and work surface. Tempering flame heads must have smaller heat outputs than hardening flame heads, because too-rapid heating of the hardened zone can cause cracking and because, for tempering the temperatures required are lower and temperature control usually is more critical.

Self-Tempering

On large parts flame hardened to depths of about 6.4 mm (¼ inch) or more, the residual heat present after quenching may be sufficient to accomplish satisfactory relief of hardening stresses and subsequent tempering in a separate operation may be unnecessary. Air-fuel gas heating, because of its lower rates of heat transfer, promotes development of the deep-seated heat required for self-tempering.

When residual heat cannot be utilised and it is desirable to eliminate the separate tempering operation, the use of a lower-carbon steel is suggested if hardness requirements permit. Pre-heating of heavy parts will increase the residual heat available for tempering. For this purpose, pre-heating may be accomplished by: (i) heating the part in a furnace, if feasible; and (ii) spinning the part in the burner ring with the burners firing at reduced or idling input. The choice will depend on the availability of equipment and the economics involved. The exact pre-heat temperature will vary, depending on the size and configuration of the part and on the degree of stress relief or temper desired. In each application, the exact schedule may be established on a trial basis as a preliminary step.

SURFACE CONDITIONS

For wrought and cast steel parts, the surface conditions likely to be detrimental to successful flame hardening are, in general, those that interfere with heating or quenching, cause localised overheating, initiate cracking or result in the presence of a soft surface skin after proper heating and quenching.

Table 23.3 summarises the more common defects or conditions, their origins and the detrimental effects to be expected when they are present on flame-hardened areas. The extent of these defects determines the amount of difficulty they may cause.

Table 23.3. Surface conditions detrimental to flame hardening of steel parts.

Defect or condition	*Probable origin of condition*	*Detrimental effects to be expected on flame-hardened areas*
Laps, seams, folds, fins (wrought parts)	Rolling mill or forging operations	Localised overheating (or, at worst, surface melting), with consequent grain growth, brittleness and greater hazard of cracking
Scale (adherent) (a)	Rolling or forging; prior heat	Insulating action against heating, with resulting treatment; flame cutting underheated areas and soft
spots		Localised retardation of quench causing soft spots
Rust, dirt(a)	Storage and handling of material or parts	Similar to scale as noted above. Severe rusting may result in surface pitting that will remain after hardening

(Contd...)

Defect or condition	*Probable origin of condition*	*Detrimental effects to be expected on flame-hardened areas*
Decarburisation	Present in as-received steel bar stock; heating for forging or prior heat treatment of parts or stock	In severely decarburised work, no hardening response will be found when parts are tested by file or other superficial means(b)
Pinholes, shrinkage (castings)	Casting defects	Localised overheating (or, at worst, surface melting), with consequent grain growth, brittleness and greater hazard of cracking
Coarse-grain gate areas (castings)	Casting gates located in areas to be flame hardened (avoid, if possible)	Increased cracking hazard during quenching, compared with non-gated areas; shrinkage defects also likely in these areas
Improper welds	Parts welded with an alloy dissimilar to base metal	Weld-zone reaction dissimilar to base-metal reaction. Weld may separate, requiring rewelding or scrapping of the part (c)

(a) In addition to having detrimental effects on flame-hardened surfaces, scale, rust or dirt in the path of the flame may become dislodged and cause fouling of oxy-fuel gas burners or react chemically with ceramic air-fuel gas burner parts (causing rapid deterioration). When such materials enter a closed quenching-system, they may clog strainers, plug quench orifices and cause excessive wear of pumps; (b) partial decarburisation lowers surface hardness as a direct function of actual carbon content of stock lost at surface, provided that steel was adequately heated and quenched; and (c) to avoid these and other problems, it is mandatory that the flame hardener be given accurate and complete information on any changes in composition and past processing of the part. For example, previously hardened parts should never be flame hardened unless they have been annealed; otherwise, cracking is inevitable.

DIMENSIONAL CONTROL

Because of its ability to heat specific areas of a part selectively, flame hardening in many applications affords greater control of dimensional stability than is obtainable by furnace heating and quenching. The magnitude of the dimensional changes that occur in flame hardening is influenced by such factors as size and shape of the part, area heated, depth of heating, hardenability and quenching medium.

SELECTION OF PROCESS

Flame hardening is basically adaptable to surface hardening and to through hardening of selected areas. Manufacturing specifications often call for heat treatment of an entire part, but in many cases this is not necessary. Crane wheels, mill rolls and gears are examples of items for which flame hardening can be used to harden only the working surfaces. Dies often require hardening only along parting lines or in forming areas.

Other hardening processes that may be applicable for accomplishing the same purposes include: (i) induction hardening for surface or through hardening; (ii) carburising, nitriding and other furnace processes in which surface composition is altered to permit hardening of the case; (iii) application of hard facing materials to surfaces; and (iv) through hardening of selected sections by partial immersion in molten lead or salt baths. Table 23.4 lists the different processes in use and describes the benefits realised from each.

Induction hardening and flame hardening are the most effective methods for selectively hardening portions of a part (gear teeth, for example) without affecting other aresa. A 4150 steel shaft, 760 mm (30 inch) long, with a 25 mm (1 inch) long gear at one end, can be core hardened to 26 to 32 HRC all over (a good machining hardness) and the gear can be hardened to 55 to 60 HRC by flame or induction hardening. If carburising or nitriding were used instead, the main shaft would have to be masked with copper, which would have to be stripped away after hardening.

The differences among these three basic types of surface hardening processes can be small enough for certain applications so that their feasibilities overlap. For example, a ring gear used in almost identical applications has been made for four customers using four different surface-heating procedures: (i) induction hardening a 4150 steel gear, with or without separate core hardening, depending on the specific application; (ii) flame hardening the teeth of a 1045 steel gear; (iii) carburising an 8620 or 9310 steel gear, then finish grinding; and (iv) core hardening a 4150 steel gear to 32 to 64 HRC, then nitriding.

Table 23.4. Relative benefits of five hardening processes.

Carburising	Hard, highly wear-resistant surface (medium case depths); excellent capacity for contact load; good bending fatigue strength; good resistance to seizure; excellent freedom from quench cracking; low-to-medium-cost steels required; high capital investment required.
Carbonitriding	Hard, highly wear-resistant surface (shallow case depths); fair capacity for contact load; good bending fatigue strength; good resistance to seizure; good dimensional control possible; excellent freedom from quench cracking; low-cost steels usually satisfactory; medium capital investment required.
Nitriding	Hard, highly wear-resistant surface (shallow case depths); fair capacity for contact load; good bending fatigue strength; excellent resistance to seizure; excellent dimensional control possible; good freedom from quench cracking (during pre-treatment); medium-to-high-cost steels required; medium capital investment required.
Induction hardening	Hard, highly wear-resistant surface (deep case depths); good capacity for contact load; good bending fatigue strength; fair resistance to seizure; fair dimensional control possible; fair freedom from quench cracking; low-cost steels usually satisfactory; medium capital investment required.
Flame hardening	Hard, highly wear-resistant surface (deep case depths); good capacity for contact load; good bending fatigue strength; fair resistance to seizure; fair dimensional control possible; fair freedom from quench cracking; low-cost steels usually satisfactory; low capital investment required.

Careful analysis, however, would no doubt show that one of the four variations would be most economical. Examples of false economy are common.

In selecting a flame-hardening process, consideration must be given to the method of applying the flame to the work and to the choice of the gas mixture. Oxy-gas equipment differs significantly from air-gas equipment in the design of the systems that control, deliver and burn the gas mixtures.

The heating characteristics of the combustion mixture largely determine the degree to which the hardening and heat-affected zones can be localised in the workpiece. Shallow hardness patterns (less than 3.2 mm or $\frac{1}{8}$ inch deep) will be obtained only with oxy-gas mixtures, because the higher-temperature flames produced provide sufficiently fast heat transfer to effectively localise the heat pattern. Deeper hardness patterns may permit the use of either oxy-gas or air-gas mixtures. With oxy-gas mixtures the heat is localise, but care is required in controlling the heat release rate to avoid overheating the surface during development of the deeper-seated heat.

Air-gas mixtures, because of the lower rate of heat transfer obtained with them, reduce the tendency toward surface overheating but generally extend the heat-affected zone well beyond the desired hardness pattern. For this reason, air-gas flame hardening is generally limited to those applications in which the portion to be hardened may be through heated or in which materials of shallow hardenability are employed. In the latter instance, the depth of the hardness pattern is primarily controlled by quenching rather than by heating intensity. Because of potential excessive distortion, the deeper-seated heat developed in a part by air-gas flames may preclude the use of this heating medium.

In view of these considerations, the success of air-gas flame hardening depends largely on the configuration of the part insofar as that configuration favours heat localisation with its lower rates of heat transfer. In selective through hardening of gear and sprocket teeth, flanges, ribs, edges and similar projections, air-gas flame heating may be applied. On the other hand, surface hardening of gear teeth, rolls, journals, shaft areas, machineways, wear areas of forming dies, inside and outside diameters of hubs and massive sections is usually done with oxy-fuel flame equipment. Small parts frequently require the very small, easily controllable flame characteristics obtainable with oxy-fuel burning equipment.

When more than one mode of flame application or combustion-gas mixture can be used for a flame-hardening operation, the selection of process and equipment becomes primarily an economic consideration. The equipment then may be chosen on the basis of expected cost to meet immediate and anticipated production requirements.

Whether or not a flame-hardening process is selected in preference to other hardening processes for a specific application is commonly determined on the basis of: (i) suitability of the process to produce the required results; (ii) control of distortion; and (iii) the obvious factor of cost.

Selection of Material

Application of flame hardening is limited to hardenable steels (wrought or cast) and cast irons. Typical hardnesses obtained for various grades of these materials by flame heating and quenching in air, oil or water are given in Table 23.5.

Table 23.5. Response of steels and cast irons to flame hardening.7

	Typical hardness, HRC, as affected by quenchant		
Material	*Air (a)*	*Oil (b)*	*Water (b)*
Plain carbon steels			
1025 to 1035	–	–	33 to 50
1040 to 1050	–	52 to 58	55 to 60
1055 to 1075	50 to 60	58 to 62	60 to 63
1080 to 1095	55 to 62	58 to 62	62 to 65
1125 to 1137	–	–	45 to 55
1138 to 1144	45 to 55	52 to 57(c)	55 to 62
1146 to 1151	50 to 55	55 to 60	58 to 64
Carburised grades of plain carbon steels(d)			
1010 to 1020	50 to 60	58 to 62	62 to 65
1108 to 1120	50 to 60	60 to 63	62 to 65
Alloy steels			
1340 to 1345	45 to 55	52 to 57(c)	55 to 62
3140 to 3145	50 to 60	55 to 60	60 to 64
3350	55 to 60	58 to 62	63 to 65
4063	55 to 60	61 to 63	63 to 65
4130 to 4135	–	50 to 55	55 to 60
4140 to 4145	52 to 56	52 to 56	55 to 60

(Contd...)

Material	Typical hardness, HRC, as affected by quenchant		
	Air (a)	*Oil (b)*	*Water (b)*
4147 to 4150	58 to 62	58 to 62	62 to 65
4337 to 4340	53 to 57	53 to 57	60 to 63
4347	56 to 60	56 to 60	62 to 65
4640	52 to 56	52 to 56	60 to 63
52100	55 to 60	55 to 60	62 to 64
6150	–	52 to 60	55 to 60
8630 to 8640	48 to 53	52 to 57	58 to 62
8642 to 8660	55 to 63	55 to 63	62 to 64
Carburised grades of alloy steels(d)			
3310	55 to 60	58 to 62	63 to 65
4615 to 4620	58 to 62	62 to 65	64 to 66
8615 to 8620	–	58 to 62	62 to 65
Martensitic stainless steels			
410 and 416	41 to 44	41 to 44	–
414 and 431	42 to 47	42 to 47	–
420	49 to 56	49 to 56	–
440 (typical)	55 to 59	55 to 59	–
Cast irons (ASTM classes)			
Class 30	–	43 to 48	43 to 48
Class 40	–	48 to 52	48 to 52
Class 45010	–	35 to 43	35 to 45
50007, 53004, 6003, 60003	–	52 to 56	55 to 60
Class 80002	52 to 56	56 to 59	56 to 61
Class 60-45-15	–	–	35 to 45
Class 80-60-03	–	52 to 56	55 to 60

(a) To obtain the hardness results indicated, those areas not directly heated must be kept relatively cool during the heating process; (b) thin sections are susceptible to cracking when quenched with oil or water; (c) hardness is slightly lower for material heated by spinning and combination progressive-spinning methods than for material heated by progressive or stationary methods; and (d) hardness values of carburised cases containing 0.90 to 1.10% C.

Maximum hardness is not the sole criterion used in selecting flame hardening as a heat treatment. Proper steel selection is essential to minimise distortion, for example. Plain carbon steels should be used, if possible instead of steels whose deep hardening characteristics are more likely to incur higher internal stresses.

Some flame hardeners feel it is important to stress relieve all alloy steels and other steel with more than 0.40 per cent carbon at 175 to 245°C (350 to 475°F), depending on customer specifications. This low-temperature tempering decreases hardness, but it also removes internal stress and restores toughness and ductility.

Selective heating has the disadvantage of developing residual tensile stresses in the surface. As one area of a piece of metal is heated while the remainder stays cold, the hot metal expands; if restraint is sufficient, the heated metal will upset itself.

Upon cooling, this upset metal becomes short. As it cools to room temperature, it will often stabilise in a state of tension, which can be high enough to crack the part. When a part is to be induction or flame hardened, the materials engineer should work closely with the designer to keep the level of hardness and necessary carbon, as low as possible while still meeting engineering requirements. Carbon content is the most important factor determining the level of hardness that can be attained in steels by induction or flame heating. It controls hardness level, the tendency of the part to crack and the magnitude of residual surface stresses. The practical level of minimum surface hardnesses attainable with water quenching for various carbon contents is shown in Fig. 23.16. The curve is applicable for induction hardening as well as for flame hardening. It applies also for alloy steels, except those containing stable carbide formers such as chromium and vanadium.

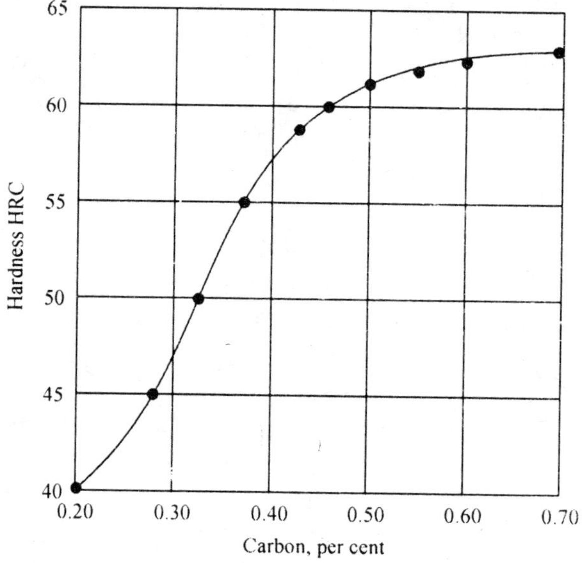

Practical minimum carbon contents can be determined from this curve. Because actual carbon content is the criterion, the check analysis tolerances shown in the AISI steel products manuals should be considered in selecting grades of steel.

Fig. 23.16. Relation of carbon content to minimum surface hardness attainable by flame or induction heating and water quenching.

For best results, steels to be induction or flame hardened should be as rolled, normalised (particularly from a high temperature) air-blast quenched or quenched and tempered. These preferred heat treatments result in microstructures conducive to rapid and complete austenitisation and full hardening. In selecting steels for either induction or flame hardening, it is important that the necessary steps be taken to ensure that the areas to be hardened are free of decarburisation. Depending on stock size, steel grade, producing mill and several other factors, the depth of decarburisation for as-rolled bar may run from near zero to 3.2 mm (0.125 inch). It should not be assumed that turned and polished bar is free of decarburisation unless it is specifically ordered with this requirement. Carbon-restored and cold finished bar is available from mills in various carbon and alloy grades.

When maximum resistance to fatigue is desired, the hardened surface should contain residual compressive stresses; a recommended level is 172 MPa (25 ksi). Because surfaces hardened to depths of less than 1.9 mm (0.075 inch) are commonly residually stressed in tension, it is suggested that depth of hardening be at least 2.7 mm (0.105 inch) to ensure that residual stresses are compressive. This depth is particularly appropriate for manufacturers not equipped with residual stress-measuring equipment. Further, microstructure should be at least 90 per cent martensite with no ferrite visible at a magnification of 500 ×.

Carbon Steels

Plain carbon steels in the range of 0.37 to 0.55 per cent carbon are the most widely used for flame-hardening applications. They can be through hardened in sections up to 13 mm (½ inch). This response permits the use of carbon steel for selectively flame-hardened small gears, shafts and other parts of small cross section in which uniform properties are needed throughout the section. These same steels can be used for larger parts in which hardness is necessary only to shallow depths from 0.8 to 6.4 mm (1/32 to ¼ inch). Carbon steels 1035 to 1053 are suitable for flame hardening; 1042 and 1045 are the most widely available and are recommended for all flame-hardening applications except where they would be incapable of meeting requirements, as in the following examples:

1. Failure of a 1045 steel part to harden with a given quench would necessitate the use of a steel of highes hardenability—for example, one with higher carbon or manganese or both or possibly an alloys steel.
2. If increased depth of hardening is required, 1042 or 1045 may be inadequate where heavy sections are progressively hardened; therefore, substitution of 1541, 1552 or an alloy steel would be necessary.
3. In applications in which wear resistance is of prime importance, it might be advisable to use a steel with 0.60 per cent carbon or more to produce maximum surface hardness. Steels this high in carbon content often are quenched in oil or simulated oil to avoid the hazard of cracking due to water quenching. Thus, greater hardenability may be needed with the higher carbon content.
4. When a severe quench in brine or caustic is required for hardening 1042 or 1045 steel and such quenching causes cracking, a steel of higher hardenability—either carbon or alloy—should be selected, which can be hardened by a less severe quench.

Alloy Steels

The use of alloy steels for flame-hardening applications is justified only for the following reasons:

1. Because high core strength is required (through heat treatment before flame hardening) and because carbon steels are inadequate to achieve this strength in the section sizes involved.
2. Because the mass and shape of a part, restrictions on distortion or the hazard of cracking preclude the use of carbon steel quenched in water.
3. Because certain alloy grades may be more readily obtainable than carbon grades appropriate for the application. This situation often precludes use of the highes-manganese carbon grade such as:

1036	1048	1137	1141
1037	1051	1138	1144
1039	1052	1139	1145
1041	1053	1140	1151

Steels such as 4135H, 4140H, 6150H, 8640H, 8642H and 4340H are typical of the more readily obtainable alloy steels.

Carbon and alloy steel castings are widely used for flame-hardening applications and selection of a specific composition or grade is made on much the same basis as for wrought carbon and alloy steels.

Cast Iron

Gray cast irons, ductile irons and pearlitic malleable irons having combined carbon contents of 0.35 to 0.80 per cent can be flame hardened and will respond the same as steel.

Cast irons having less than 0.35 per cent combined carbon will not respond readily to flame hardening because of the inability of austenite to dissolve graphite during the extremely rapid heating that occurs during flame hardening. Flame hardening of these irons produces a typical hardness of about 40 HRC. Malleable iron, in which all the carbon is in graphitic form, is not amenable to flame hardening for this reason.

Cast irons having combined carbon contents greater than 0.80 per cent are difficult to flame harden because of their inherent brittleness and susceptibility to cracking when heated and quenched rapidly. The low melting point of cast iron and the presence of graphite in the microstructure make cast iron susceptible to 'burning' or even localised melting during flame hardening; therefore, extreme care must be taken to reduce the rate of heating when cast iron is hardened using equipment designed specifically for flame hardening of steel. For example, the distance between the inner cone and the workpiece can be increased or the flame velocity can be decreased. Use of a flame head with smaller orifices also will decrease the heating rate.

Perhaps the most significant factor in the response of cast irons to flame hardening is prior microstructure.

Other Materials

Flame hardening can be applied to other hardenable ferrous materials—for example, alloy cast irons, martensitic stainless steels and tool steels. The nature of flame hardening, however—especially the relatively high temperature gradients and higher-than-normal surface temperatures may cause retention of excessive amounts of austenite in many of the highly alloyed materials, with possible low hardness and transformation to untempered martensite in service, which is accompanied by brittleness.

Carburised parts made of plain carbon or alloy steels can be flame hardened to provide a hard case of high carbon content. The depth of carburised case may vary from a few hundredths of a millimetre to 1.6 mm ($1/16$ inch) or more. The flame-hardening procedure is adjusted to heat the carburised case to its full depth for hardening. Because the core of low-carbon steel so treated does not harden substantially, the method provides a means of accurately controlling the depth of hardened case.

FLAME ANNEALING

The chemical reaction between oxygen and steel during flame cutting evolves intense heat, raising the temperature of the cut edge to the melting point on the surface and above the transformation temperature to a shallow depth. If the carbon or alloy content of the steel is high enough, hardening results when the heated zone is cooled rapidly.

Carbon steels containing 0.30 per cent carbon or less do not harden sufficiently to prevent the use of flame-cut pieces in structures or in subsequent fabrication where bending is involved. When cut by flame, steels containing alloying elements and steels containing more than 0.30 per cent carbon may harden sufficiently to prevent use of flame-cut pieces for some purposes.

When hardening is likely to occur along a cut edge, oxyacetylene flames, applied by suitable equipment, can be used either to prevent hardening or to soften an already hardened cut surface. The term 'flame annealing' is applied to this process and to selective flame softening of areas of hardened steel parts.

Tool steels and certain highly alloyed steels will crack during flame cutting or very soon after, unless they are heated to 205 to 425°C (400 to 800°F) before being cut.

OTHER HARDENING PROCESSES

Some other hardening processes of less importance are discussed below:

Hardening with Electric Contact Resistance Heating

This method consists in supplying alternating current of industrial frequency (50 cps) and at low voltage (2 to 6 V) to the workpiece through electrodes which have the form of rollers. The power is supplied from single-phase step down transformers (Fig. 23.17). The current has a density of 400 to 700 a per millimetre of roller width. The steel is rapidly heated at the point of contact with the roller. The workpiece is rotated and the roller is fed gradually along its length (Fig. 23.17). Due to these motions, the point of contact with the roller travels along the surface of the shaft (workpiece) in a helix and consecutively heats the shaft (band hardening). Cooling is done either by heat conduction into the metal or by quenching in the ordinary manner.

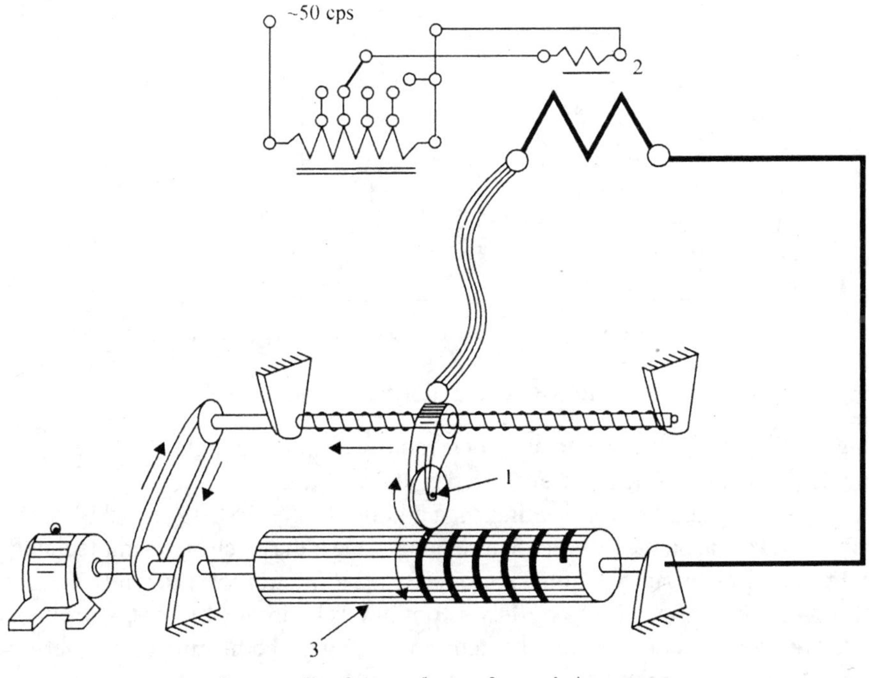

1— Contact roller, 2—transformer, 3—workpiece

Fig. 23.17. Hardening with electric contact heating.

The feed of the roller along the work is varied in a range from 2 to 15 mm per second in accordance with the depth of hardened layer required.

The roller is pressed against the surface of the workpiece with a pressure of 10 to 15 kg per millimetre of roller width. The chief disadvantage of this method is the helical 'tempered band' of lower hardness on the surface after hardening. The tempered band is obtained because, in heating each new turn of the helix, the roller partially heats and tempers the preceding, already-hardened turn. This hardening method is seldom employed at the present time.

The surface of steel hardened by this method usually has a lower hardness due to excessive cooling by the water-cooled copper rollers. The maximum hardness will lie at a depth of 0.05 to 0.1 mm below the surface.

Hardening with Electrolytic Heating

In this method the workpiece, which serves as the cathode, is immersed in an electrolyte (a 5 to 10 per cent solution of soda or potash, (Fig. 23.18). The shell of the bath is the anode. When direct current at a high voltage (250 to 300 V) is passed through the electrolyte, a hydrogen blanket is formed surrounding the cathode (i.e., workpiece).

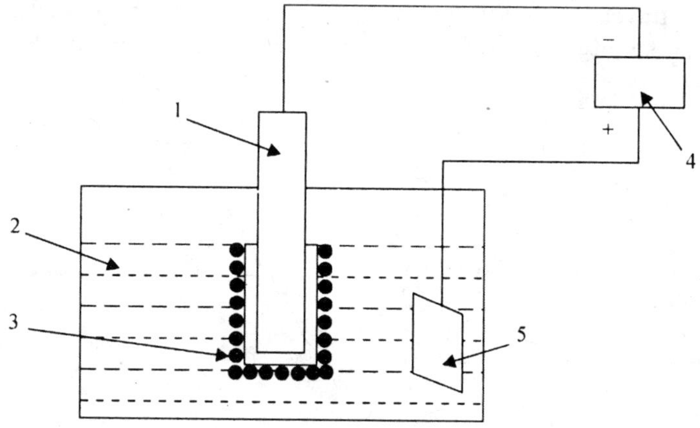

1. Workpiece (eathode), 2. Electrolyte, 3. Hydrogen blanket, 4. Current supply, 5. Anode

Fig. 23.18. Electrolytic heating.

Due to its high resistivity, the hydrogen blanket is rapidly heated to a high temperature (about 2000°C) and a considerable portion of its heat is transferred to the workpiece.

Two stages may be distinguished in heating the cathode. In the first stage, bubbles of hydrogen are intensively evolved and settle on the cathode. These bubbles heat the electrolyte, immediately surrounding the cathode and produce a vapour blanket. This stage takes a fraction of a second.

In the second stage, the cathode is covered by a continuous hydrogen blanket, surrounded, in turn, by a steam blanket. The cathode is intensively heated in this stage. The heating rate is varied by changing the voltage, current and heating time.

Heating time is varied in a range from 10 to 40 seconds to suit the type of workpiece being heated. The current density varies from 3 to 6 per sq cm.

The principal disadvantage of this method is that the surface is inevitably overheated if a comparatively shallow depth of hardening is required. This shortcoming, in conjunction with the high power consumption and difficulties encountered in mechanising the process, has limited the application of this method.

Chapter 24

Induction Hardening

INTRODUCTION

When electric current passes through a conductor, a magnetic field is created round it. If the conductor consists of a coil, a magnetic field is established inside the coil. This field persists even if a metal bar is inserted into the coil, as shown in Fig. 24.1. If the magnetic flux is created by a high-frequency alternating current, it gives rise to eddy currents in the surface of the metal bar, which consequently becomes heated (see Fig. 24.2). In iron, hysteresis losses also contribute to some extent to the temperature rise up to the Curie point (768°C), above which iron is non-magnetic.

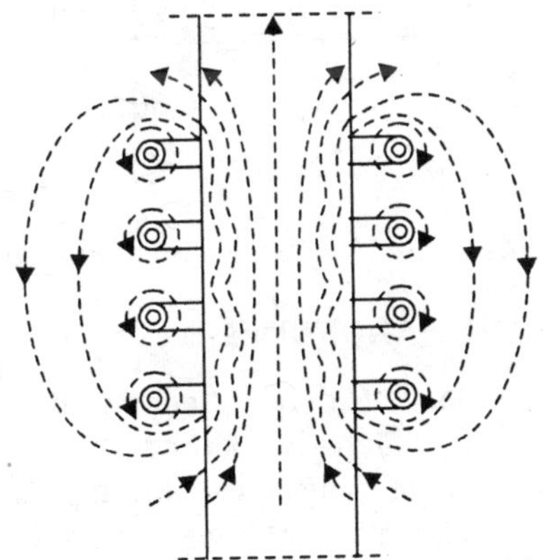

Fig. 24.1. Path of magnetic flux through a metal bar inserted in a coil through which an electric current is flowing.

FUNDAMENTAL PRINCIPLES

The depth of penetration of the heat is governed mainly by the power and frequency employed. The normal power input is 0.1–2 kW/cm^2 of the heated surface. The relationship between depth of penetration and frequency can be calculated approximately by using the following simplified expressions, which are valid for the temperature rise in steel up to the hardening temperature:

$$d_c = \frac{20}{\sqrt{f}} \text{ cold state (20°C)}$$

$$d_h = \frac{500}{\sqrt{f}} \text{ hot state (800°C)}$$

where d = depth of penetration in mm
f = frequency in cycles per s (Hz)

Owing to heat conduction in the material during heating, the overall depth of penetration is larger. It is possible to calculate the additional penetration due to heat conduction from the following expression.

$$d_l = 0.2\sqrt{t}$$

where d_l = depth of penetration in mm
t = time in second

The total depth of penetration is obviously d + d_t. It should be stressed that these expressions give only a rough estimate of the depth of penetration and they have been included here only to show the fundamental effects of frequency and time.

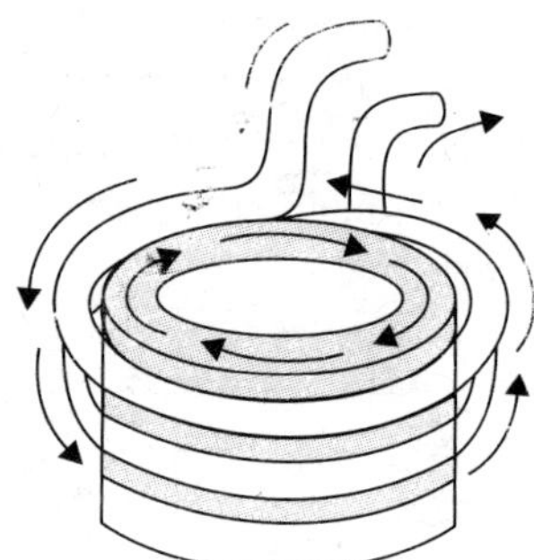

Fig. 24.2. Path of the electric current in the coil and metal bar during induction heating.

STEEL GRADES FOR INDUCTION HARDENING

By induction hardening we can achieve a considerable increase in surface hardness and steels are used having C contents of 0.30–0.50 per cent, which give hardness values of 50–60 HRC. The upper limit is governed by the C content that may risk hardening cracks. From practical tests it has been found that this C content is around 0.50 per cent. If the heat treatment is well controlled, higher C contents are permissible, e.g., for rolls, that are made from steels having 0.80 per cent C , 1.8 per cent Cr and optionally 0.25 per cent Mo. In accordance with ISO 683/XII–1972, the steels in Table. 24.1 are recommended for flame and induction hardening. Steel 3, with increased Mn content (ca 1.30 per cent) has also proved suitable for this purpose. Steel 9 corresponds to BS 108A42 (En 19 C). Cr-Mo steels with lower carbon contents are also widely used. The steels are normally quenched in water. In certain cases the alloy steels can be cooled by means of an oil emulsion. Steels 9–11 can also be oil quenched.

The steel, prior to induction hardening, should be in the hardened and tempered or normalised state and the hardness should be such as is acceptable for the unhardened sections. Fully annealed steel is not so suitable, because normally the time needed for dissolution of the carbides is longer than the normal heating-up time. Consequently induction hardening of fully annealed steel can result in inadequate and irregular values unless the heating time is matched to suit the fully annealed state.

Table 24.1. Steels for flame and induction hardening ISO 683/XII-1972.

Type of steel	*C%*	*Si%*	*Mn%*	*P% max*	*S% max*	*Cr%*	*Mo%*	*Ni%*
1	0.33–0.39	0.15–0.40	0.50–0.80	0.035	0.035	–	–	–
2	0.38–0.44	0.15–0.40	0.50–0.80	0.035	0.035	–	–	–
3	0.43–0.49	0.15–0.40	0.50–0.80	0.035	0.035	–	–	–
4	0.48–0.55	0.15–0.40	0.60–0.90	0.035	0.035	–	–	–
5	0.50–0.57	0.15–0.40	0.40–0.70	0.035	0.035	–	–	–
6	0.42–0.48	0.15–0.40	0.50–0.80	0.035	0.035	0.40–0.60	–	–
7	0.34–0.40	0.15–0.40	0.60–0.90	0.035	0.035	0.90–1.20	–	–
8	0.38–0.44	0.15–0.40	0.60–0.90	0.035	0.035	0.90–1.20	–	–
9	0.38–0.44	0.15–0.40	0.50–0.80	0.035	0.035	0.90–1.20	0.15–0.30	–
10	0.38–0.44	0.15–0.40	0.70–1.00	0.035	0.035	0.40–0.60	0.15–0.30	0.40–0.70
11	0.37–0.43	0.15–0.40	0.50–0.80	0.035	0.035	0.60–0.90	0.15–0.30	0.70–1.00

EQUIPMENT FOR INDUCTION HARDENING

As the frequency governs the depth of penetration (depth of hardening), equipment should be selected mainly on the basis of the required depth of hardening. Since the latter generally increases with the size of the hardened part, the consequence is that, for large parts, equipment operating at low frequencies and low power input per unit of surface is selected whereas for small parts the best results are obtained using equipment operating at high frequencies and with high power input per unit of surface (Table. 24.2).

Table 24.2. A guide to frequency selection.

Frequency range	*Power supply source*	*Frequency kHz*	*Minimum depth of hardening obtainable (mm)*	*Minimum diameter of workpiece (mm)*
Intermediate frequency	Rotary converter	0.5–10	1.5	ca 15
Intermediate frequency	Solid state	0.5–10	1.5	ca 15
High frequency	Valve generator	100–500	0.2	ca 1.5

As an example it can be stated that for the hardening of large rolls for cold rolling Bofors use three rotary converters operating at 500 Hz with a total power output of 900 kW and a power input per unit surface of 0.1 kW/cm^2, by means of which a hardness depth of up to 20 mm can be obtained. As may be seen in Table 24.2 rotary converters are used in the frequency range 0.5–10 kHz. The same frequency range is also covered by a new generation of convertors called solid state converters, the functioning of which is based in semi-conductor techniques. The efficiency of these units lies around 75 per cent as against approximately 60 per cent for the rotary converters. The solid state converters are very reliable in operation but they require servicing by highly qualified personnel. Rotary converters, besides requiring only simple maintenance, also give reliable and trouble-free service. For the hardening of small components, valve generators operating at frequencies around 5,00,000 Hz and power outputs up to 100 kW are employed.

WORKING COILS AND FIXTURES

The coils, also known as inductors, also become heated by the electric current and by thermal radiation from the heated steel. The coils are often made from copper tubing and are cooled during operation by internal water flow. Fig. 24.3 show three basic types of working coils designed for external and internal heating and for heating of a flat surface.

The internal diameter of the copper tube is matched to suit generator power output. For the power output normally required, around 50 kW, tubing with an internal diameter of around 5 mm is used. The coils are usually wound with a distance of 2–5 mm between turns and with roughly the same spacing between the coil and the workpiece. By varying this distance it is possible to influence the rate of heating to a very large extent.

Parts that are to be hardened must be held in a fixture of some sort so that they are correctly located in the inductor and small components must be firmly held so that they are not displaced by the magnetic field. A metallic fixture that comes near the coil may become heated by the induced current. To avoid this the fixture and centre washers are normally made from an insulating and heat-resistant material such as 'sindango' or a heat-resistant plastic.

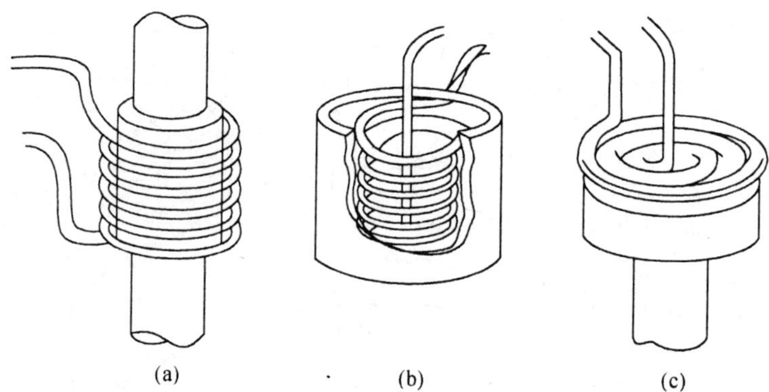

Fig. 24.3. Types of coil used for induction heating.

PROCEDURE DURING INDUCTION HARDENING

According to the grade employed, the steel, after it has been treated in the coil, is cooled with water or oil or, in certain cases, in air. Heating and cooling can be carried out in accordance with two main methods, these being single-shot hardening and progressive hardening.

Single-shot Hardening

During single-shot hardening the component is positioned and heated, after which it can be quenched in various ways as shown in Fig. 24.4.

1. The component can either be immersed by hand in the quenching liquid or it can be released from the fixture so that it drops into the liquid.
2. In hardening machines the heated steel parts can be fed automatically into a quenching spray when heating has terminated.
3. The coil can also be provided with a quenching spray which starts to operate when the high-frequency current is switched off.

The parts may rotate during heating and, if possible, during cooling. This method is employed for parts that have rotational symmetry. This gives a more uniform depth of hardness penetration, because it compensates for any irregularities in the coil.

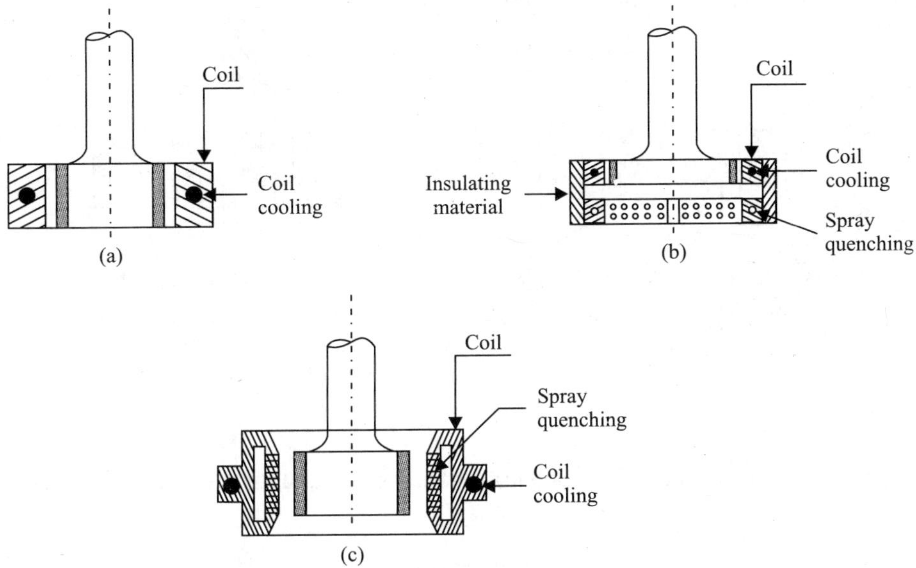

Fig. 24.4. Example of different methods of quenching when induction hardening.

Progressive Hardening

Progressive hardening is used if the power output is not adequate to permit single-shot hardening, or if a particularly shallow depth of hardness is required. The method is employed for very long parts and normally requires a cross-sectional area that is uniform along the entire length of the hardened surface. During hardening the component should be held in position, e.g., in a rotating chuck or between a pair of jaws. It is fed forward at a certain rate through the coil, so that it is heated and quenched progressively during its passage through the coil and spray. A separate spray as shown in Fig. 24.5a can be employed; alternatively the working coil can be combined with the spray into one unit, so that the spray ring itself also functions as the inductor, (Fig. 24.5b).

During hardening the spray coil is located underneath the inductor. Progressive hardening can also be carried out on parts that are not axially symmetrical, e.g., as shown in Fig. 24.6. Here cooling is carried out by means of air.

DETERMINATION OF DEPTH OF HARDENING

The method used for determining the depth of hardening is the same for both induction hardening and flame hardening and is specified according to SS 117007, i.e., by measuring the value of HV 1 on a cross-section through the hardened zone. The limiting hardness value is 400 HV. The letters DI and DF refer to induction hardening and flame hardening, respectively. The corresponding ISO Standard is: ISO 3754–1976, Steel—Determination of Effective Depth of Hardening After Flame or Induction Hardening. However, this specification is set out somewhat differently.

This depth of hardening, designated DS, applies to:

1. Hardened layers with a depth greater than 0.3 mm.
2. Parts which, in the surface-hardened condition, have at a distance 3 × DS from the surface a hardness less than the hardness limit (HV)-100.

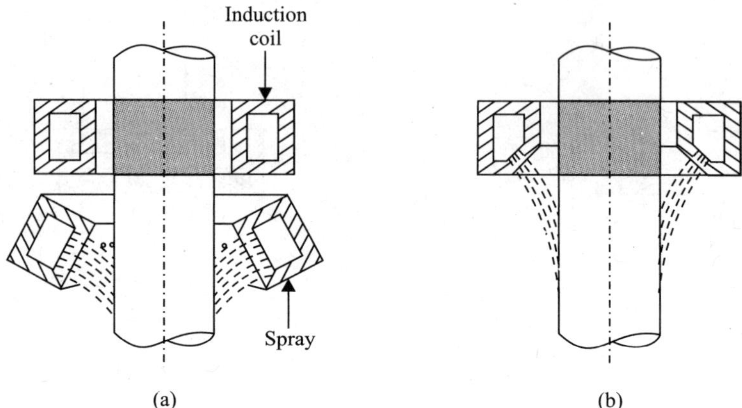

Fig. 24.5. Progressive hardening of shafts using (a) separate spray and (b) integrated working and spray coil.

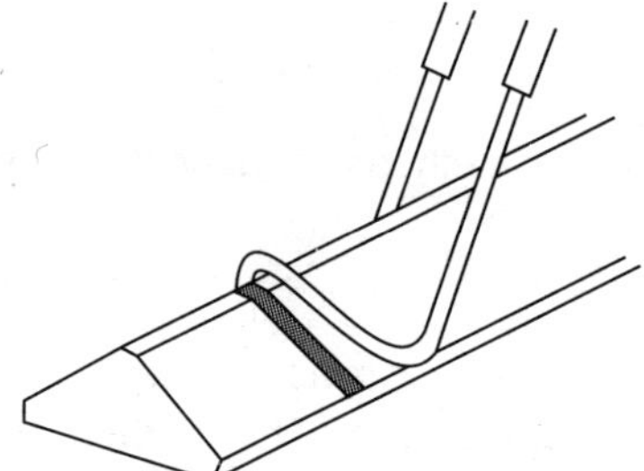

Fig. 24.6. Progressive hardening without rotating the workpiece. In this example, air cooling is used.

Where these conditions are not satisfied the effective depth of hardness after flame or induction hardening should be defined by agreement between the parties concerned. For steels where the hardness of a part at a distance 3 × DS from the surface is above the hardness limit (HV)–100, the criterion may still be used on condition that a higher hardness limit is chosen for the assessment of the effective depth of hardening.

The effective depth of hardening after flame or induction hardening is the distance between the surface of the product and the layer where the Vickers hardness (HV) under a load of 9.8 N (1 kgf) is equal to the value specified by the term 'hardness limit' which is a function of the minimum hardness required for the part, given by the following equation:

Hardness limit (HV) = 0.8 × minimum surface hardness (HV).

INFLUENCE OF VARIOUS FACTORS ON HARDNESS AND DEPTH OF HARDENING

As mentioned previously, the C content has a decisive influence on hardness after hardening. As the C content increased up to about 0.80 per cent the hardenability also increases, particularly when in combination with other alloying elements. Both the hardening temperature and holding time are decisive factors that govern the hardness and depth of hardening. For induction hardening with conditions otherwise remaining the same the depth of hardening is governed by the frequency of the inductive current. Since the hardening temperature, heating time and holding time can be controlled relatively easily during induction hardening, it is possible with this method to determine in advance the required hardness depth with a relatively high degree of certainty by using calculated and empirical values.

The hardening temperatures employed for induction hardening are normally some 50°C higher than those used for conventional hardening. The holding times are extremely short. In many cases there is no holding time, i.e., cooling starts directly when the desired temperature has been reached. Temperatures are normally controlled by means of optical pyrometers.

The pressure of the cooling medium and the angle of incidence of the jet as well as its width are of paramount importance to the intensity of cooling. Hence these factors influence the depth of hardening and the hardening itself. *Kegel* and *Pennekamp* have arrived at the following recommendations based on experimental tests and theoretical calculations of the coefficient of heat transfer through the vapour film:

Pressure	1–3 bar
Angle of incidence	20–30°
Width of slit	0.8–1.2 mm, preferably 0.9–1.0 mm

Usually the spray nozzle contains a number of holes, the diameters of which are taken as equivalent to the slit width of the above recommendations.

Figs 24.7 to 24.10 show examples of hardness and depth of hardening of steels having a base composition of 0.7 per cent Mn, 1.1 per cent Cr, 0.25 per cent Mo and varying C contents. Induction hardening was carried out at 1,00,000 Hz. In this case the power output employed was the same throughout. Hence the factor governing the hardening temperature was the heating-up time. The test specimens, 50 mm in diameter, were made from hardened and tempered steel with a hardness of ca 300 HB. Quenching took place in water directly when the specimens had reached the desired temperature.

Figs 24.11 and 24.12 show the results from tests using a frequency of 4,00,000 Hz. In this case the power output was regulated so that the heating-up times to the desired hardening temperature were different. Here too the specimens had a diameter of 50 mm and they were quenched in water directly as the current was cut off. Tests at temperatures of 875 and 900°C gave almost identical results. Use of the high frequency resulted in the depth of hardening being just about half the value obtained at 10000 Hz; compare Fig. 24.12 with Fig. 24.8.

All the experiments described above were made on steels that had been normalised or alternatively hardened and tempered. It was not possible to observe whether the mode of pre-treatment had made any difference. When hardening rolls, relatively deep hardening is required. This can be achieved by employing a low frequency, e.g., 500 Hz and by pre-heating the rolls. Fig. 24.13 shows the temperature as a function of the time during the heating-up and cooling of a pre-heated roll with a diameter of 300 mm. Fig. 24.14 shows the hardness distribution and the microstructure in the surface layer.

After being induction hardened the steel is usually tempered at 150–200°C, or higher if a lower hardness is desired. Parts that are to be ground should be tempered at least 160°C. If the parts are tempered at lower temperatures there is considerable risk of cracking during grinding. Fig. 24.15 and 24.16 illustrate how tempering influences hardness.

Tempering can also be performed by induction heating, this being particularly suitable for localised tempering. During tempering two axles are placed in a fixture with the flanges facing each other. The axles rotate during heating.

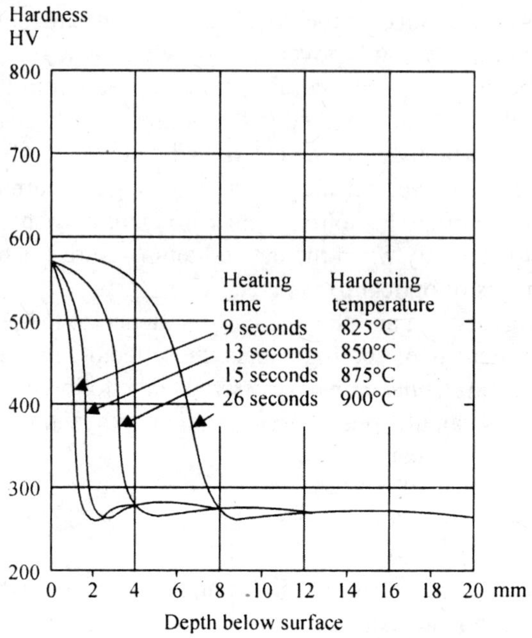

Fig. 24.7. SS 2225, AISI 4130 with 0.24 per cent C, 1.1 per cent Cr and 0.25 per cent Mo.

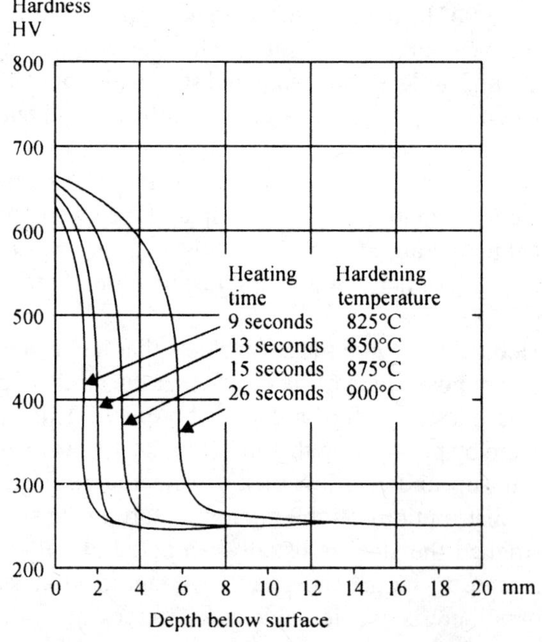

Fig. 24.8. SS 2234, AISI 4135 with 0.33 per cent C, 1.1 per cent Cr and 0.25 per cent Mo.

After tempering at 180°C the hardness values given in Table. 24.2 may be expected for the steels included in ISO 683/XII–1972.

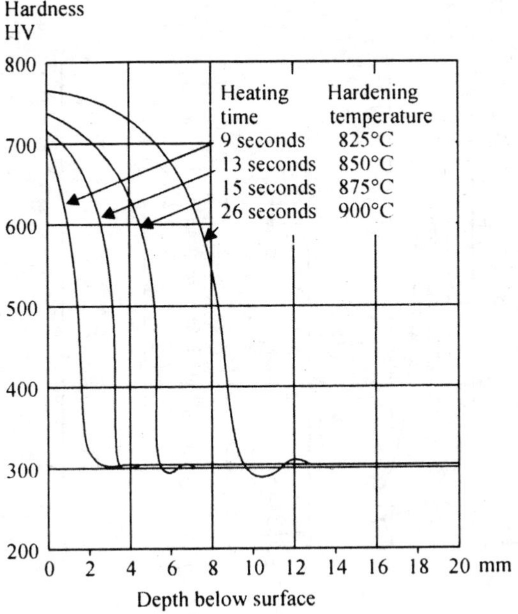

Fig. 24.9. 6.248 SS 2244, AISI 4142 with 0.40 per cent C, 1.1 per cent Cr and 0.25 per cent Mo.

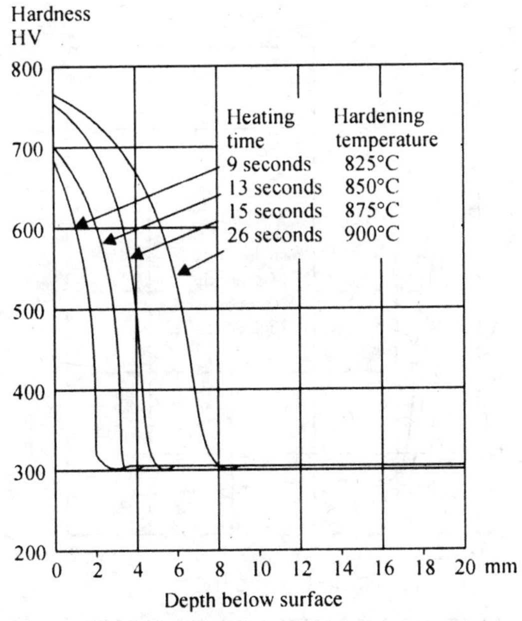

Fig. 24.10. 6.249 SS 2244, AISI 4142 with 0.45 per cent C, 1.1 per cent Cr and 0.25 per cent Mo.

Figs 24.7 to 24.10 Curves showing depth of hardening for induction-hardened Cr-Mo steels with different carbon contents. Frequency 10000 Hz.

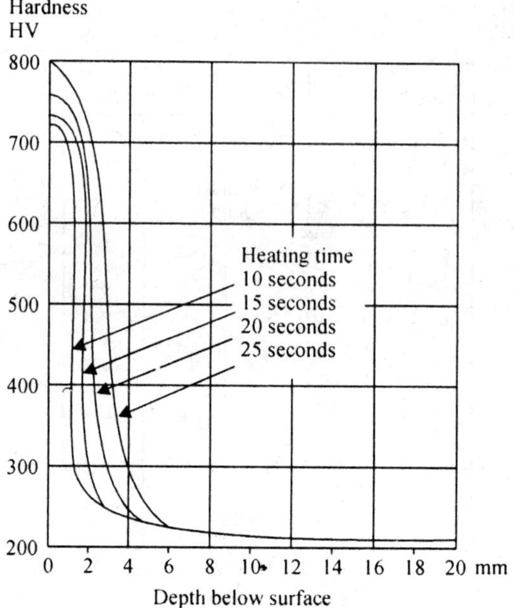

Fig. 24.11. C 50.

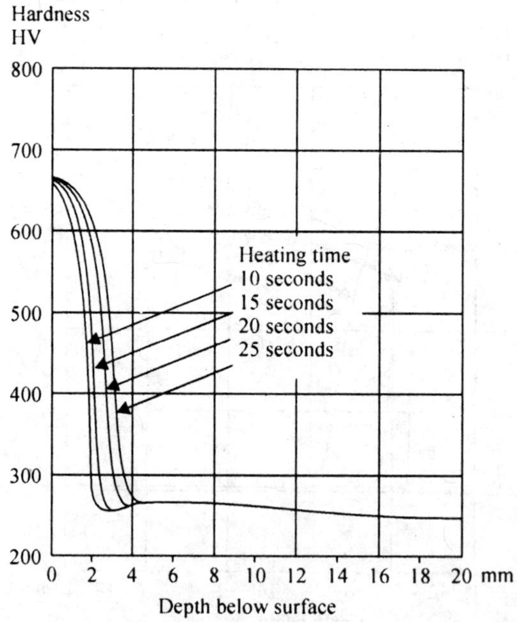

Fig. 24.12. SS 2234, AISI 4135, BS 708A37, En 19 B.

Figs 24.11 and 24.12 curves showing depth of hardening for induction-hardened test bars of carbon steel C 50 and SS 2234. Hardening temperature 900°C, frequency 4,00,000 Hz.

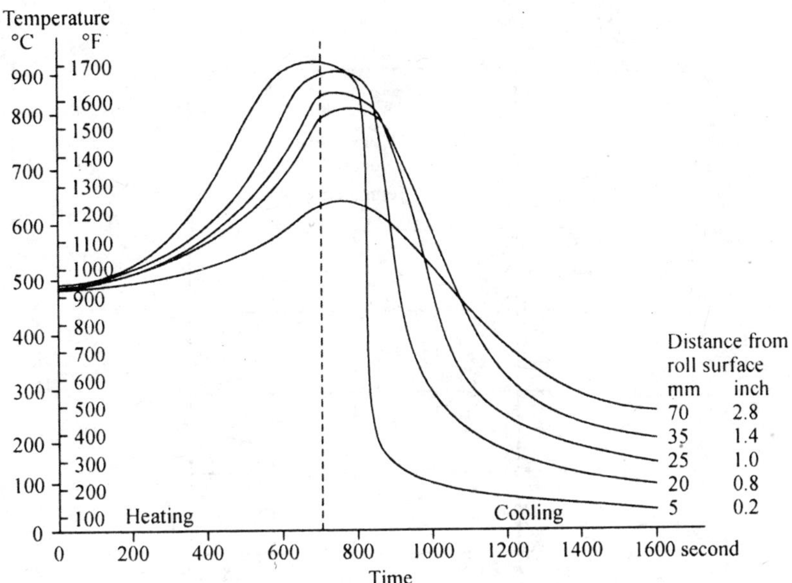

Fig. 24.13. Temperature as a function of time during heating and cooling of a roll for cold rolling, 300 mm diameter, pre-heated to 500°C.

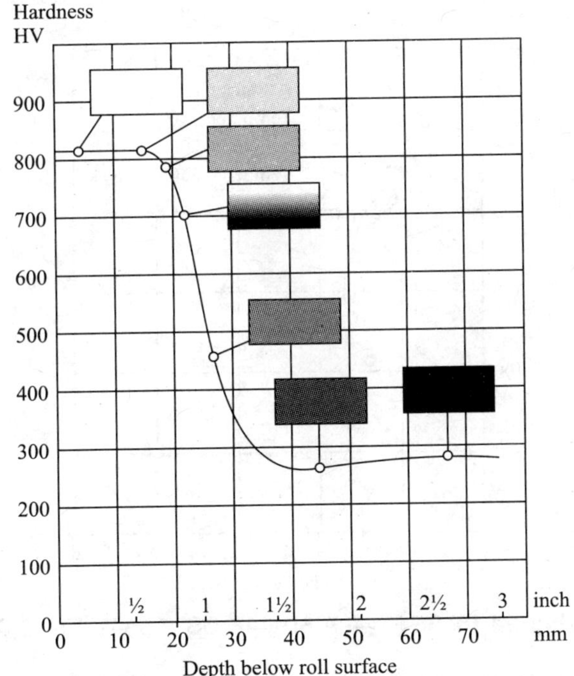

Fig. 24.14. Hardness and microstructure in the surface zone of induction-hardened roll for cold rolling .

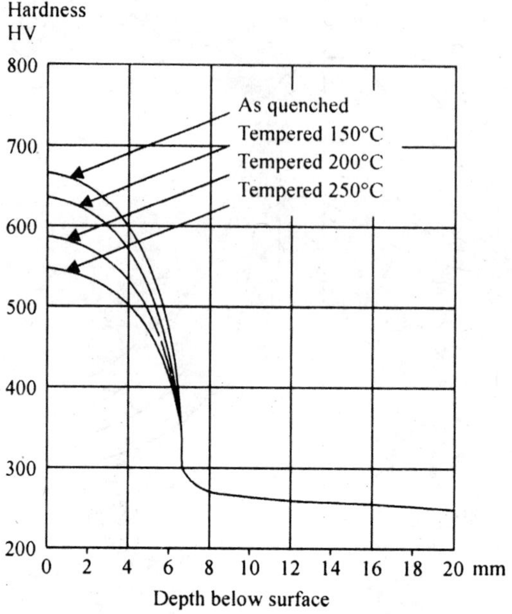

Fig. 24.15. SS 2234, AISI 4135, BS 708A37, En 19 B.

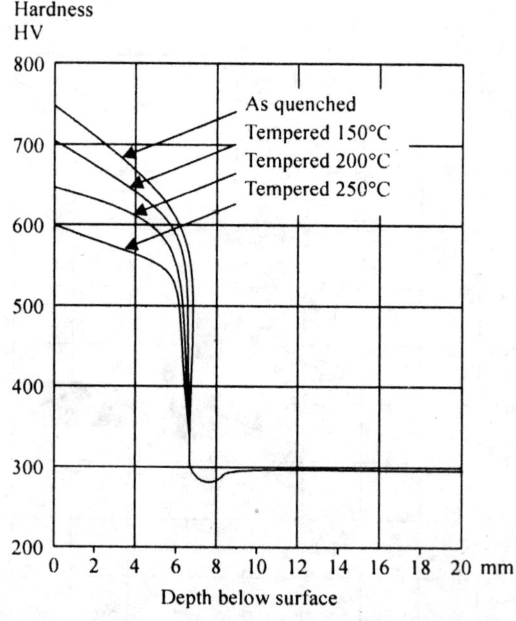

Fig. 24.16. SS 2244, AISI 4142, BS 708A42, En 19 C.

Figs 24.15 and 24.16 curves showing hardening for SS 2234 and SS 2244 after induction heating to 900°C, heating time 25 second, subsequent tempering at different temperatures. Frequency 10000 Hz.

Table 24.3. Hardness after induction heating and tempering 180°C.

Type of steel	*Hardness HRC*
1	50–55
2	53–57
3	56–61
4	57–62
5	58–63
6	56–61
7	52–57
8	54–58
9	55–60
10	55–60
11	55–60

EXAMPLES OF INDUCTION-HARDENED MACHINE COMPONENTS

Machine components which are induction-hardened are mainly bodies having rotational symmetry, such as gear wheels and shafts. Gears can be hardened in accordance with any of the methods described in Fig. 24.4 but the gear should preferably rotate. The power input and the temperature employed should be so matched that the central portions of the gear are not heated to temperatures beyond 200°C, which can be checked from the tempering colours. This is particularly important in the case of gears with finish-machined internal splines, which may be deformed during heating. Rotating the part during induction hardening, also called spin hardening, is generally suitable for gears with modules up to 5.

Acceptable contour hardening of a gear by means of spin hardening can be obtained if a high energy concentration, ca 2 kW/cm^2 and high frequency are employed. There is, however, a risk of overheating the tips of the gears and inadequate hardening at the tooth roots. When using intermediate frequencies the teeth first become hot at the roots, after which the heat diffuses towards the tops. In this case the teeth become through-hardened. More uniform contour hardening is achieved if heating is first started at an intermediate frequency until the tooth roots show a slight red heat, after which the tooth tips are heated by means of a high-frequency current for a short period. When hardening large gears, machines are employed which harden one tooth at a time by progressive heating.

In the following, sections summarise methods of induction hardening for gears are discussed. Figs 24.17a–d show the hardness zones for the differing methods.

Single-shot Hardening of Tooth Tips

The hardening is quite simple. The wear resistance of the teeth increases, but the strength remains unchanged. This method can be employed with high-frequency hardening up to module 3 and with intermediate-frequency hardening up to module 5.

Single-shot Hardening of Complete Teeth

Method (b) requires somewhat more experience than method (a). When using method (b) not only the wear resistance but also the bending strength of the teeth is increased. This method is recommended for modules up to 5.

Flank Hardening (Progressive Hardening)

This method is used for modules $\geqslant 2$ using high-frequency hardening and for modules $\geqslant 5$ using intermediate frequency hardening. In principle the properties obtained are the same as by method (a). The hardening process is relatively simple.

Tooth Gap Hardening (Progressive Hardening)

The sphere of application is the same as for method (b). A considerable, increase in bending and fatigue strength is achieved. This 'contour hardening' is actually the ideal method for gears, but it requires considerable experience. Trial hardening heat treatment and metallographical investigations are necessary before good results can be expected.

Induction hardening is very advantageous for long shafts on which only the bearing sections need hardening. Thanks to induction hardening we avoid deformation and distortion that could occur if the entire shaft were hardened. Shafts fitted with splines made from hardened and tempered steels often crack at the splines owing to overloading and fatigue. This susceptibility to cracking can easily be overcome by induction hardening. The depth of hardening below the base of the spline should at least be equal to height of the spline ribs.

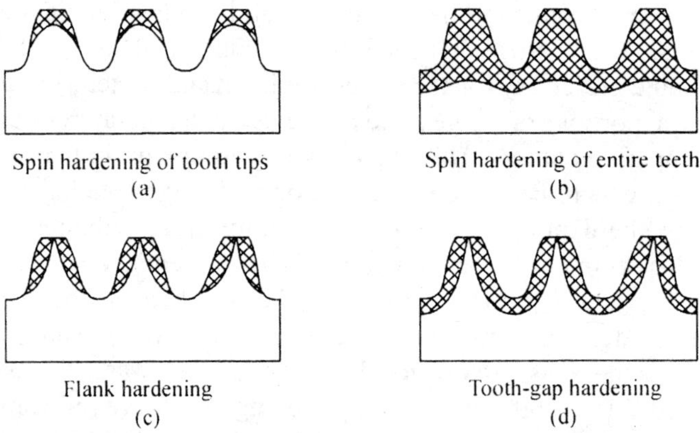

Fig. 24.17. Appearance of hardened zones after hardening by different methods.

ADVANTAGES AND DISADVANTAGES OF INDUCTION HARDENING

From the preceding section it is easy to understand that induction hardening can offer considerable advantages when compared with conventional methods of hardening.

Advantages

1. Restricted localised hardening.
2. Short heating-up periods.
3. Minimum surface decarburisation and oxidation.
4. Only slight deformation.
5. Increased fatigue strength.
6. Any straightening required can be carried out on the unhardened surfaces and—to a certain extent—on the hardened surfaces.

7. The process can be incorporated in a production line.
8. Low operating costs.

Disadvantages

1. The high capital costs necessitate high degree of equipment utilisation.
2. The method is restricted to components having a shape suitable for induction hardening.
3. Only a limited number of grades of steel can be induction hardened.

Induction heating can be used for heat-treatment processes other than those mentioned above, e.g., for continuous annealing or normalising of steel tubing or for brazing and welding.

ELECTROMAGNETIC INDUCTION

Electromagnetic induction is one method of generating heat within a part for hardening or tempering a steel or cast iron part. Any electrical conductor can be heated by electromagnetic induction. As alternating current from the converter flows through the inductor, or work coil, a highly concentrated, rapidly alternating magnetic field is established within the coil. The strength of this field depends primarily on the magnitude of the current flowing in the coil. The magnetic field thus established induces an electric potential in the part to be heated and because the part represents a closed circuit, the induced voltage causes the flow of current. The resistance of the part to the flow of the induced current causes heating.

The pattern of heating obtained by induction is determined by the: (i) shape of the induction coil producing the magnetic field; (ii) number of turns in the coil; (iii) operating frequency; (iv) alternating-current power input; and (v) nature of the workpiece.

The rate of heating obtained with induction coils depends on the strength of the magnetic field to which the part is exposed. In the workpiece, this becomes a function of the induced currents and of the resistance to their flow.

The depth of current penetration depends upon workpiece permeability, resistivity and the alternating current frequency. Since the first two factors vary comparatively little, the greatest variable is frequency. Depth of current penetration decreases as frequency increases. High-frequency current is generally used when shallow heating (thin case) is desired; intermediate and low frequencies are used in applications requiring deeper heating.

Most induction surface-hardening applications require comparatively high power densities and short heating cycles in order to restrict heating to the surface area. The principal metallurgical advantages that may be obtained by surface hardening with induction include increased wear resistance and improved fatigue strength.

HARDENING FOR WEAR RESISTANCE

A shallow hardened case, ranging in depth from 0.25 to 1.5 mm (0.010 to 0.060 inch), provides a part with good wear resistance in applications involving light to moderate loading. For shallow hardening, heating may be limited to the desired depths by use of equipment in the 10-kHz to 2-MHz frequency range. Rocker-arm shafts, sucker-rod couplings and pump shafts are typical parts requiring a shallow case depth. However, in applications where parts are subjected to heavy or impact-type loading, case depth must be increased over a range from 1.5 to 6.4 mm (0.060 to 0.250 inch) for adequate support and wear resistance. To obtain these case depths, a frequency range of from 1 to 10 kHz is recommended. Gears, track pins, heavy crankshaft bearings, camshafts and bearing races are typical parts requiring these greater case depths.

IMPROVING FATIGUE STRENGTH

The induction case hardening of bars and shafts to depths of 3.2 to 12.7 mm (1/8 to ½ inch) has resulted in greatly improved torsional and bending fatigue strength. The process has also made it possible, in some applications, to reduce cost by substituting carbon steels for deeper hardening alloy steels.

Long bars and shafts are usually passed through the inductor coil and are hardened by being progressively heated and quenched. These parts are usually rotated to obtain more uniform results in processing.

Truck, tractor and automobile axle shafts and hydraulic piston rods are typical of production parts that are processed in this manner. Equipment in the frequency range of 1 to 10 kHz is recommended for processing such parts.

Critically stressed areas of steering knuckles, flanged axle shafts and similar parts may be selectively case hardened by induction to improve torsional and bending fatigue properties. Depending on the specific application and desired case depth, frequencies ranging from 3 to 450 kHz have been effectively employed for selective hardening of these parts.

THROUGH HARDENING AND TEMPERING

Through hardening by induction heating requires low power densities and low frequencies, particularly when applied to heavy-wall tubing or solid bars. Progressive induction heating and quenching are common in processing long sections. Low frequencies (60 to 180 Hz) provide maximum depth of heating on large cross-sections.

Dual frequencies may be used to decrease the heating time when power requirements exceed 2000 kW. In dual-frequency heating, the higher-frequency power source may be operated at from 1 to 10 kHz and the low-frequency source at 60 to 180 Hz. The inductor coil must be proportioned to apply heat at a rate that does not exceed that at which it can be conducted inwardly.

Tempering of induction hardened surfaces may be done in conventional furnace equipment or in an induction coil at lower power densities.

CHARACTERISTICS OF COMMERCIAL EQUIPMENT

Induction hardening is generally done at frequencies of 1000 Hz or higher. The types of commercially available low-frequency and high-frequency equipment and pertinent data are shown in Table 24.4.

Line-Frequency Units

Line-frequency units consist essentially of a control circuit and power-factor correction capacitors, to which a transformer capable of matching low-impedance, low-voltage coils to a high-voltage power supply may be added.

When a low-voltage power supply is employed to heat ferrous alloys, a transformer is not required. In larger units, the power-factor correction capacitors are mounted on separate capacitor racks; the capacitors for smaller units are usually grouped with the work-handling and coil equipment.

Because these units do not convert frequency, power losses are negligible, but use of the equipment is limited to through-heating applications.

For these applications, line-frequency systems are capable of producing maximum temperature uniformity. Operating characteristics of line-frequency induction heating equipment are given in Table 24.4.

Table 24.4. Characteristics of line-frequency induction heating equipment.

60-Hz output, kW	*60-Hz input* Phase	*Voltage*
55	1	220/440
75	1	220/440
85	3	220/440
150	3	220/440
185	3	220/440
195	3	220/440
		2300/4160
300	3	220/440
		2300/4160
370	3	440/2400
		4160/4800
425	3	440/2400
		4160/4800
575	3	2300/4160/4800
650	3	2300/4160/4800
775	3	2300/4160/4800
875	3	2300/4160/4800

Static Frequency Converters

The most common types are the frequency tripler and the solid state converter. Basically, the tripler consists of an iron core transformer with three sets of windings, one for each input phase. Each phase voltage is applied to its own reactor and to the load in series. Operation of the tripler is based on the summation of pulses obtained from saturable reactors. Thus, by adding the pulses, a 60-Hz supply current is multiplied to 180 Hz. The operating characteristics of three-phase static frequency converters are shown in Table 24.5.

Table 24.5. Characteristics of static frequency converters for induction heating.

Output(a), kW	*Input(b) V*	*A*	*Cooling*
100	440 up	215	Water
200	440 up	400	Water
250	440 up	515	Water
300	440 up	585	Water
400	440 up	785	Water
500	440 up	975	Water
700	2300 up	255	Air/Water
1000	2300 up	350	Air

(Contd...)

Output(a), kW	*Input(b) V*	*A*	*Cooling*
1250	2300 up	440	Air
1500	4160 up	292	Air
2000	4160 up	395	Air

(a) 180 Hz, 750 V. (b) Three-phase.

A static frequency converter of the type described provides a highly efficient means for through heating steel bars of a given diameter, depending on output frequency. For example, at a frequency of 180 Hz, maximum heating efficiency can be achieved with diameters ranging from approximately 9 to 15 cm (3½ to 6 inch). Normally, diameters of 15 to 23 cm (6 to 9 inch) can be heated more efficiently at 60 Hz, whereas a diameter of 8 to 9 cm (3 to 3½ inch) can be heated more efficiently when output frequency is about 1 kHz.

Medium Frequency Motor-Generator Units

Motor-generator units consist of a high-frequency generator driven by a motor. Induction motors, which may be mounted integrally with the generator or separately on a common base, are used with 1, 3 and 10 kHz generators. The integrally mounted units can be either vertical or horizontal. The motors for generators of these frequencies may be of either the induction or synchronous type. Units using a synchronous motor are mounted horizontally and separately from the generator.

Medium Frequency Solid State Units

Medium frequency solid state units are used to produce frequencies from 180 Hz to 50 kHz. While several different types of power circuits are used, all convert the 60-Hz main-line frequency into single-phase high frequency at relatively high efficiency. Either fixed-frequency or variable-frequency output units are available. The fixed-frequency outputs operate load-match in a manner similar to that of motor generators, working at a fixed, specific frequency–for example, 10 kHz. Variable-frequency units operate over a range of frequencies for most effective load-matching. The power and tuning of variable-frequency unit can be changed by adjusting the frequency during operation.

Solid state units can be made to provide the same heating results as motor generator units. Selection of frequency, power and duration of power should be made independent of whether a motor generator or solid state induction heater is to be used.

Solid state units have the advantage of a conversion efficiency greater than 90 per cent, as compared with 75 to 80 per cent of a motor generator. Solid state units also do not have the starting in-rush of a motor generator. Finally, through built-in logic in the control circuits, the solid state units can match load impedances through frequency shift, providing better load input power without shifting capacitor or transformer taps during heating (Table 24.6).

Vacuum Tube Units

Electronic tube units consist of a power supply section and an oscillator section. The power section provides the high voltage for the oscillator tube after rectification to a pulsating direct current, usually by solid state diodes. The oscillator tube and a tank circuit consisting of a matched inductor coil and capacitor comprise the oscillator section. The oscillator tube controls the amount of electrical energy delivered to the tank circuit from which the energy is removed by the coupled load. A small and proportionate amount of the power in the tank circuit is fed back into the grid of the oscillator tube to control the current that is delivered to the tube and it in turn controls the amount of electrical energy entering the tank circuit. The frequency developed in the converter is determined by the inductance of the tank coil and the capacitor, which form a parallel tuned circuit. A load-matching network electrically coupled to the tank circuit is used to transmit tank circuit energy to the work.

Table 24.6. Characteristic of medium-frequency (1 to 50 kHz) solid state equipment for induction heating.

Output, kW	*Input, kV·A*	*Amperage at 480 V, input 30*
5	6	7
10	12	14
25	29	36
50	58	73
75	87	110
100	116	146
150	174	219
200	232	292
250	290	365
300	349	438
400	465	584
500	581	730
600	698	876
800	930	1168
1000	1163	1460
1500	1744	2190
2000	2326	2920

SELECTION OF FREQUENCY, POWER AND DURATION OF HEATING

The distribution of induced current in a part is maximum on the surface and decreases rapidly within the part; the effective penetration of current increases with a decrease in the frequency. The distribution of induced current is influenced also by the magnetic and electrical characteristics of the part being heated; and since these properties change with temperature, the current distribution will change as the work is heated.

Because the heat rapidly progresses to the interior by conduction as soon as the surface is heated, the actual depth of heating is determined by the duration of heating and the power density (kilowatts per square inch of surface exposed to the inductor), as well as by the frequency. Maximum power density, minimum duration of heating and high frequency produce a minimum depth of heating.

Selection of Frequency

In analysing the frequency and power required for a specific application, it is desirable to consider the frequency first. Primary considerations are the depth of heating and the size of the part. The lower frequencies are more suitable as the size of the part and the case depth increase. However, because power density (kilowatts per square inch of surface within the work coil or inductor) also has an important influence on the depth to which the part is heated, wide deviations from this chart may be made with successful results, because higher densities of power compensate for lower frequencies. This is particularly important where equipment of a given frequency already exists in the plant or where equipment is required for several different applications. In some instances, the determining factor in selecting the frequency is the power required to provide power density sufficient for successful hardening, as lower-frequency induction equipment is available with higher power ratings.

Efficiency

The use of wrong frequency will result in a decrease in electrical efficiency; sometimes in failure to maintain a minimum case depth where shallow cases are required; or in failure to heat uniformly throughout the piece where through hardening is required. Fig. 24.18 illustrates the decrease in transfer of energy or heating efficiency that could result if the incorrect frequency is selected. While the efficiency for all frequencies is similar with large bars, maximum efficiency is achieved with small bars by using higher frequencies. In parts of small diameter, suitable hardening temperatures cannot be obtained with the frequencies shown in Fig. 24.18. For such parts, MHz-frequencies are desirable, or the thin sections may sometimes be hardened successfully with normal frequencies by using a combination of induction heating and radiation from an induction heated liner that surrounds the part, except for a slit along its length to prevent electromagnetic shielding.

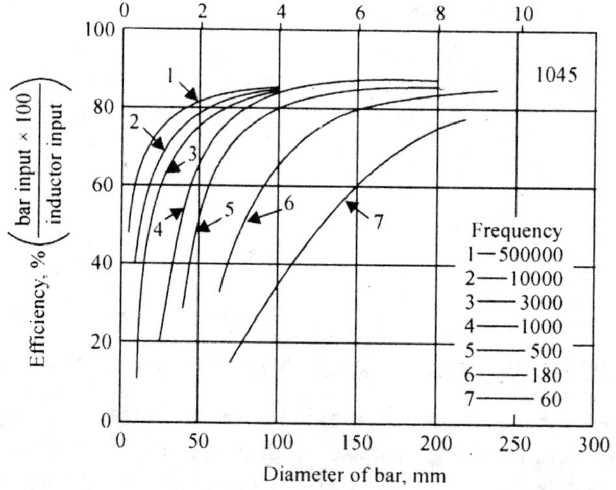

Bars are of various diameters heated to 1095°C (2000°F) with inside diameters of inductors 28.6 mm (1.125 inch) larger than outside diameters of bars.

Fig. 24.18. Efficiency of energy transfer at several frequencies to type 1045 steel bars.

Selection of Power

The size of the converter or the power required should be determined on the basis of power density, section size, heating method and production requirements, as indicated below.

Requirements for power density

Presented in kilowatts per square inch, this information may be available from previous experience, from experiment, or from data such as those in Tables 24.7 and 24.8.

Table 24.7. Power density required for surface hardening(a).

			Input					
				W/mm²			*kW/inch²(b)*	
Frequency, kHz	*Depth of hardening(c) mm*	*inch*	*Low (d)*	*Optimum (e)*	*High (f)*	*Low (d)*	*Optimum (e)*	*High (f)*
500	0.38–1.14	0.015–0.045	10.9	15.5	18.6	7	10	12
	1.14–2.29	0.045–0.090	4.7	7.8	12.4	3	5	8
10	1.52–2.29	0.060–0.090	12.4	15.5	24.8	8	10	16
	2.29–3.05	0.090–0.120	7.8	15.5	23.3	5	10	15
	3.05–4.06	0.120–0.160	7.8	15.5	21.7	5	10	14
3	2.29–3.05	0.090–0.120	15.5	23.3	26.35	10	15	17
	3.05–4.06	0.120–0.160	7.8	21.7	24.8	5	14	16
	4.06–5.08	0.160–0.200	7.8	15.5	21.7	5	10	14
1	5.08–7.11	0.200–0.280	7.8	15.5	18.6	5	10	12
	7.11–9.14	0.280–0.360	7.8	15.5	18.6	5	10	12

(a) This table is based on use of proper frequency and normal overall operating efficiency of equipment. Values in table may be used for static and progressive methods of heating; however, for some applications, higher inputs can be used when hardening progressively. (b) Kilowattage is read as maximum during heat cycle. (c) For greater depth of hardening, a lower kilowatt input is used. (d) Low kilowatt input may be used when generator capacity is limited. These kilowatt values may be used to calculate largest part hardened (single-shot method) with a given generator. (e) For best metallurgical results. (f) For higher production when generator capacity is available.

Size of section to be heated

In surface hardening, the area heated at one time, multiplied by power density, indicates the total power input (kilowatts). This area is obtained by multiplying the perimeter of the part by the length of the inductor. To calculate the power required for through heating, divide the desired production load by the values for pounds per kilowatt-hour given in Fig. 24.19.

Method of heating

When the static method of heating is used, the entire area to be hardened must be heated at one time. When the part is scanned, or progressively hardened, it is necessary to heat only a small segment or band at one time. Thus, a smaller converter can be used if the parts are progressively hardened.

Production requirements

Power densities, as indicated in Tables 24.7 and 24.8, may be adjusted within limits to meet certain production requirements. Beyond this, production rate may be increased by one or more of the following modifications. For hardening by the static method: (i) fixturing may be improved to reduce handling time; (ii) work may be quenched outside the inductor to permit immediate reloading; and (iii) two or more inductors may be used, if sufficient power is available.

For hardening by the scanning method, it may be possible to increase the inductor length or to use more than one inductor, depending on available power. If the production rate, as based on the size of converter, does not meet production requirements, two or more units of like capacity (or a single unit having two, three or four times the capacity determined) may be used to harden two, three or four pieces simultaneously in a multiple coil.

Table 24.8. Approximate power density required for through heating of steel(a).

Fre-quency, Hz(b)	Input(c) W/mm² 150–425°C 300–800°F	425–760°C 800–1400°F	760–980°C 1400–1800°F	980–1095°C 1800–2000°F	1095–1205°C 2000–2200°F	kW/inch² 150–425°C 300–800°F	425–760°C 800–1400°F	760–980°C 1400–1800°F	980–1095°C 1800–2000°F	1095–1205°C 2000–2200°F
60	0.09	0.23	(d)	(d)	(d)	0.06	0.15	(d)	(d)	(d)
180	0.08	0.22	(d)	(d)	(d)	0.05	0.14	(d)	(d)	(d)
1000	0.06	0.19	0.78	1.55	2.17	0.04	0.12	0.5	1.0	1.4
3000	0.05	0.16	0.62	0.85	1.09	0.03	0.10	0.4	0.55	0.7
10000	0.03	0.12	0.47	0.70	0.85	0.02	0.08	0.3	0.45	0.55

(a) For hardening, tempering or forming operations. (b) Table is based on use of proper frequency and normal over-all operating efficiency of equipment. (c) In general, these power densities are for section sizes of 12 to 50 mm (½ to 2 inch). Higher inputs can be used for smaller section sizes, and lower inputs may be required for larger section sizes. (d) Not recommended.

Selection of Duration of Heating

When the frequency and power density have been selected, the duration of the heating cycle becomes a fixed value for specific set of conditions. To calculate duration of heating for surface hardening by the static method, divide the value for kilowatt seconds per square inch by power density (kilowatts per square inch). The value of kilowatt seconds per square inch is affected by case depth requirements, type of steel and prior structure and may be derived by experiment, as were the data given in Fig. 24.20, or based on previous experience. To calculate heating time for surface hardening by the scanning method, divide kilowatt second per square inch by power density and inductor length.

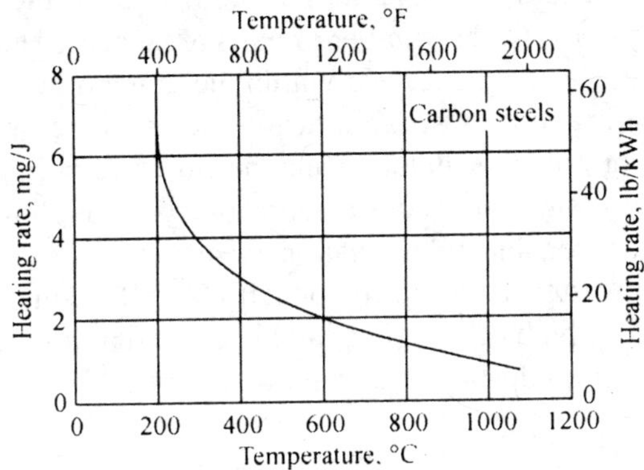

For converted frequencies, the total power transmitted by the inductor to the work is less than the power input to the machine, because of converter losses.

Fig. 24.19. Heating rate for through heating of carbon steels by induction.

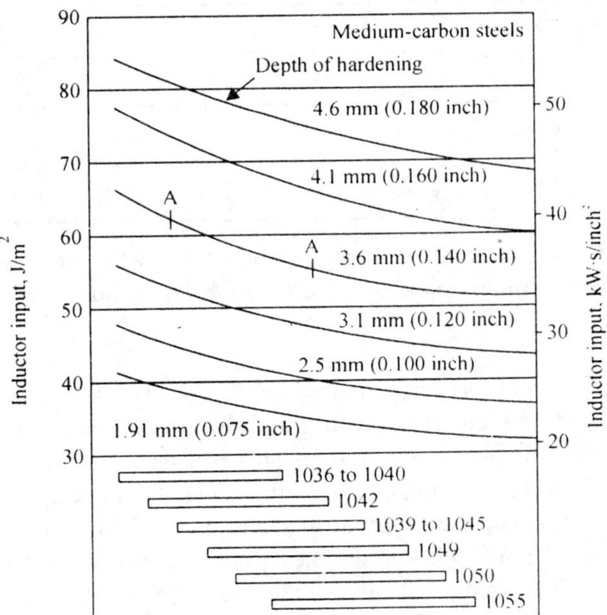

Fig. 24.20. Operating efficiency for surface hardening steels to various depths.

Data in Fig. 24.20 are for a single-turn inductor of 3000 or 10000-Hz frequency and a 1.5 to 2-mm (0.060 to 0.080-inch) air gap. Larger air gaps reduce efficiency, thus requiring more power. The values given are intended only as a guide, as different installations vary in efficiency, metering, wave form, voltage regulation and other electrical characteristics.

Selection of Coil Design

The success of many induction heating applications is related to the selection and design of the proper work coil, or inductor. This design is influenced by a number of factors, including the dimensions and configuration of the part to be heated, the heat pattern desired, whether the part is heated throughout its length at the same time or progressively, the number of parts to be heated at one time and the amount of power available. The strength of the magnetic field within the inductor, or work coil, is the basic factor that determines the rate of heating. For most rapid heating rates, therefore, inductors are designed to provide the maximum flow of current in the inductor and the closest coupling (distance between inductor inside diameter and the part) permissible, after consideration of work-handling features and arcing between the work and the coil. In practice, considerable variation exists in the design of coils for tube, spark-gap, solid state, static frequency converters and line frequencies. At all frequencies, however, the coils are generally of copper, because of its high conductivity and wide availability at moderate cost. The copper may be in the form of tubing or solid busbar, or a combination of both if required.

Design variables

The thickness of the copper conductor used with medium-frequency units is important. For efficient operation, the following minimum thicknesses of wall may be used as a guide in constructing multiturn coils:

Minimum wall thickness		*Frequency, Hz*
mm	*inch*	
3.0	0.120	1000
1.8	0.070	3000
1.0	0.040	10 000

Whenever possible in the construction of single-turn coils, with which current densities may be considerably higher, these minimums should be increased three or four times.

Cooling

Many surface hardening applications lend themselves to the use of single-turn inductors of busbar construction. Because of the high current densities and the extremely thin cross section in which the current confines itself, artificial cooling is required. This is usually accomplished by circulating water through channels provided for the purpose. These channels may be made by drilling connecting holes or milling out a path to make a completed loop around the bore of the inductor and then plugging the exposed ends of the holes or brazing a copper sheet over the milled passage to make a continuous watertight cooling channel. Cross-sectional areas of the cooling passages of 32.25 to 80.6 mm^2 (0.050 to 0.125 inch^2) provide adequate cooling at water pressures of 275 to 345 kPa (40 to 50 psi) for power inputs from 30 to 150 kW.

Quenching orifices

The single-turn inductor often can include another water chamber with a suitable pattern or series of orifices directed for spray quenching the heated area. Such a spray quench is effective for single-shot hardening of work that is stationary while being heated and quenched. For progressive hardening of a narrow band or area, the orifices must be directed at an angle so that the quenching medium will strike the heated work as it moves out of the inductor. A satisfactory angle, which eliminates backwash of the quench, is about 30° from the work.

Bore length

For hardening areas on cylindrical parts using a single-shot technique, the inductor bore should be longer than the area to be hardened, the actual overlap decreasing with an increase in power density. With intermediate power densities, the inductor bore should be approximately 9.5 mm (0.375 inch) longer than the area to be hardened if that area does not extend to the end of the part. If the area to be hardened is at the end of a shaft or protrudes from the shaft as on an eccentric, cam or gear, the inductor bore should be 6.4 mm (0.250 inch) longer than the area. For efficient transfer of energy the inductor bore may be 3.1 to 5.1 mm (0.120 to 0.200 inch) larger in diameter than the part. As the bore is increased, the work will approach a weaker portion of the magnetic field, heating time will be increased and the heating pattern will be deeper unless the power density is increased correspondingly.

Scanning

With the cylindrical areas described, single-shot inductors are generally operated at power densities of 7.75 to 23.25 W/mm^2 (5 to 15 kW/inch^2). When the length of the hardened area is such that the power density obtainable is lower than that required for a specified maximum case depth, it is logical to consider an inductor for progressive hardening, if the work is adaptable. The shorter inductor will usually permit the power density requirement to be satisfied. An inductor bore 25 mm (1 inch) long is satisfactory for shafts of 19 to 76-mm (3/4 to 3-inch) diam. Bore may be shorter if power is limited.

Irregular shapes

In constructing inductor coils for heating irregular shapes, it should be remembered that the portions of the work closest to the inductor will usually be in a stronger magnetic field and will heat more rapidly. If the irregularity is in the transverse plane of the part and does not have sharp corners, thin sections, or sharp re-entrant surfaces, a hardness pattern generally following the contour of the part may be expected with a circular inductor. If this is not satisfactory, an inductor should be constructed with a uniform air gap so that the bore will follow the irregularity. It may still be found necessary to relieve or increase the air gap around the section having least mass, in order to reduce the strength of the magnetic field and reduce the heating rate. This might be necessary for a cam or an automotive camshaft.

If the irregularity is in a longitudinal plane parallel to the axis of the inductor bore, as in a shaft with a bearing surface and an adjoining shoulder to be hardened, the inductor may again be machined initially to follow the shape of the part to produce a uniform air gap. The current density, however, will be strongest at the smallest bore dimension, because the current will follow the path of least impedance; therefore it will be found necessary to increase the dimension of the air gap at this point in order to decrease the heating rate.

The effect of the air gap on the hardness pattern produced for a number of irregular shapes is shown schematically in Fig. 24.21. In practice, it may be difficult to avoid overheating at sharp corners.

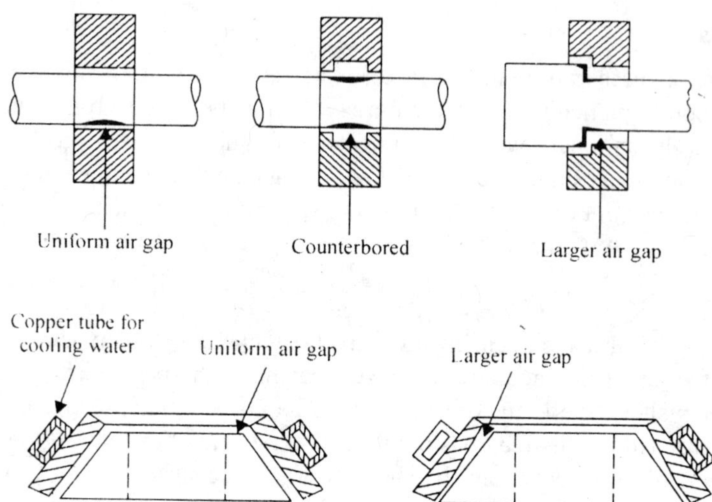

Fig. 24.21. Influence of air gap, or coupling, on hardness pattern in irregular shapes.

Transmission cables

Transmission line to be used from generator to station or between heating stations.

Coil coolants

Water is commonly used for cooling inductors, although in some applications oil, modified water, or a plastic quench may be employed to serve the dual purpose of cooling the inductor and quenching the workpiece in a continuous heating and quenching operation. Generally, the water should have a hardness of less than 10 grains/gal. If the water-cooling passages are small relative to the current load carried by the inductor, it may be necessary to use distilled or deionised water to avoid a deposit buildup that could eventually stop circulation. Preferably, the water should be filtered to remove foreign particles that might clog small passageways, especially when intricately designed inductors are being used. The water should have an inlet temperature below 35°C (95°F) and flow should be sufficient to prevent the outlet temperature from rising above 66°C (150°F). For high-voltage radio-frequency equipment and solid-state equipment, it is necessary to supply water that meets certain electrical-resistivity requirements, as specified by the equipment manufacturer. An electrical resistivity value greater that 50 Ω·m is a typical specification.

MATCHING OF IMPEDANCE

Efficient transfer of energy available from the high-frequency converter to the workpiece depends on good matching of impedance as well as on good coil design. This applies to all types of equipment for induction heating and is particularly important when work coils (inductors) have low impedance (one and two turns). Matching equipment consists of autotransformers, fixed-ratio step-down transformers, variable-ratio step-down transformers and capacitors.

Motor-Generator Units and Solid State Units

Motor-generator units are rated according to maximum power, amperage and voltage. The maximum rated power cannot be attained unless the maximum ratings of amperage and voltage are reached at the same time.

Fixed-ratio transformers

The simplest but least flexible method of matching impedance is that of the fixed-ratio step-down transformer with a capacitor bank between generator and inductor. This method is the least expensive and can be designed for a particular load, but only one type of load can be run at the rated power of the generator when properly matched. Other types of load may be run but at somewhat lower power. A more flexible method is to connect an autotransformer, capacitors and a fixed-ratio transformer between generator and inductor. This is often used when the generator voltage is high (800 V) to permit selection of a lower fixed-ratio step-down transformer.

Variable-ratio transformers

Two other methods are similar to those mentioned above except that they employ variable-ratio rather than fixed-ratio step-down transformers. In each of these methods considerably greater flexibility is attained.

Capacitors

Capacitors are required to effect proper matching of a circuit. These capacitors are connected across the transformer primary winding and are used for power factor correction. By adding the proper amount of capacitance to the circuit, the generator can be operated at close to unity power factor, thereby supplying more useful power to the load at rated output voltage. When a job is initially balanced out, it is best to start with a minimum amount of capacitance in the circuit. If the meters indicate a lagging power factor, the generator voltage should be removed and a small amount of capacitance added to the circuit. This procedure is followed until the meters indicate a kilovar reading of zero, or as close to zero as can be obtained. When multiturn coils are used, the coil for large parts can often be designed so that no transformer is needed and load matching is accomplished by the usual complement of capacitors alone. In tests of loaded coil impedance, it is usual to check the impedance matching with the generator in operation and the work cold. Starting with a low value of capacitance, capacitors are added to the circuit to secure unity power factor with the maximum values of voltage or amperage within the rating of the generator. If the voltage is low but the current is high, the load impedance presented to the generator is not great enough and coil turns must be increased or the coil more loosely coupled to the load. If the current is low but the voltage is high, excessive impedance is indicated and better matching may be accomplished by removing coil turns or coupling more closely to the work.

If changes in coil design and capacitance do not result in full loading of the generator (rated voltage and amperage at unity power factor), a transformer of the proper ratio is needed as with single-turn inductors.

Vacuum Tube Units

Vacuum tube units usually require an output transformer with a primary winding of high impedance to match the high impedance in the tube output circuit. Secondary windings are usually 1, 2, 3, 4 or 5 turns, depending on the size of the heating coil and the work load. Voltage output across the coil increases with the increase in secondary turns in the transformer. For certain applications, capacitors are sometimes used in the load circuits (primary or secondary) to afford some measure of correction of the power factor, but must be chosen with care because of the higher voltages encountered in these circuits.

When heating coils are used with an output transformer, the leads between the transformer and the coils must be kept as short and as close together as possible, so that inductance in the leads is of minimum value. Widely spaced or long leads have as much inductance as the heating coil itself and can result in an inability to draw full power from the equipment and also may cause radio and television interference.

Influence of Temperature

Because the magnetic permeability and electrical resistivity of ferrous loads change as heating occurs, the impedance of the work circuit will change as the work temperature increases and therefore a perfect impedance match throughout the entire heating cycle is impossible. If the impedance is matched while the work is cold, the demand for power will usually increase at first and then decrease as the work increases in temperature. This may be an advantage in avoiding overheating if sufficient power is available to continue heating at this stage of the heating cycle, but it may be a disadvantage if limited power is available and the change results in loss of efficiency at higher temperatures.

Matching the impedance with work that is hot will result in better transfer of energy at the higher temperatures but will cause slower initial heating of the cold work, or with tube converters, overloading. The change of impedance as the work becomes heated may be partially compensated for by the timed addition of capacitance or inductance to the output circuit.

SELECTION OF ACCESSORY EQUIPMENT

Integration of coil design, quench arrangement and fixtures and handling equipment frequently allows induction heat treating operations in high-speed production lines. On the other hand, manually operated or partially mechanised fixtures and work-handling arrangements are used successfully with minimum initial cost.

Work-Handling Equipment

The principles of machine design and the design of work-handling equipment can be readily adapted to induction heating accessory equipment.

The electric and electromagnetic characteristics of materials to be used in the immediate vicinity of the induction heating coil and quenching arrangements must be considered, as well as their resistance to corrosion and heat. Aluminium, brass and austenitic stainless steels are useful because they are less readily heated by induction. Stainless steel, chromium plated steel and brass resist corrosive quenching mediums such as water, caustic and brine. Nonmetallic materials such as ceramics, laminates, plastics, rubber and Transite are used for their corrosion resistance and electrical and thermal properties.

Typical arrangements

Work-handling equipment is available in a number of standard units. Typical arrangements include hopper and magazine feeding of small parts, or conveyer and rotary-table feeding of larger parts to the work coil and thence into the quench. Parts may be rotated to provide greater uniformity of temperature.

Scanning Devices

Vertical and horizontal scanning fixtures are often designed to provide rotation of the part and controlled movement of the coil and quench arrangement along the part. Such fixtures normally have arrangements for surface hardening either the whole length or desired sections of the part.

For selective hardening, parts are frequently heated in a multiple coil, which consists of two or more connected work coils. In such an arrangement, one multiple-coil unit may be reloaded, while a second unit goes through a hardening cycle. Upon completion of the heating cycle, parts may be dropped into the quench automatically, quenched in the heating position, rotated during quenching, or held and indexed into the quenching medium in either a horizontal or a vertical position.

Integration With Machine Tools

The automation of induction heating equipment is only the first step in combining automatic machining operations with automatic induction heat treating. The multispindle automatic screw machine, the rotating-dial type of fixture for multispindle machining, the modern process line or transfer line and the common conveyer all offer possibilities for including automatic induction heating.

Parts machined in stages on automatic machines can be hardened on the same equipment. After machining the surface to be hardened, the tooling is arranged to include an inductor on one station. It is preferable to use the inductor alone on this station, but it may be combined with machine tooling on adjacent areas, if necessary. It is often advantageous to integrate induction heating equipment with two or more other machines in order to achieve a continuous line operation. The manufacture of automotive shock absorber shafts provides a typical example of successful integration. In making these parts, bar stock is cut to length and hopper fed through a first centerless grinder for sizing to near finish tolerance. After the 'green grind', the shafts are continuously fed by spinfeed rollers through guide tubes, an induction coil and a quenching ring. The hardened shafts then proceed automatically through four tandem centerless grinders for final sizing and surface finishing.

QUENCHING

A great many induction hardening applications employ water as the quenching medium. Other media such as conventional quenching oil, water modified by organic polymer and compressed air are occasionally used. Water is easiest to handle, simple to install and maintain and generally less hazardous than other mediums. The modified-water compounds are compounds with organic polymers that are soluble in water. The temperature and concentration determine the quenching rate. The type of quench used will depend upon the metallurgical considerations.

Basic Systems for Quenching

Eleven basic arrangements for quenching induction hardened parts are shown schematically in Fig. 24.22, items: (i) through (ii). In correlation with the lettering there, these arrangements are briefly described as follows:

1. Heat in coil; manually lift part out of coil; submerge part in tank of agitated quench medium. Used where limited production does not warrant the cost of an automated quench.
2. Heat and quench in one position; quench by means of integral quench chamber in inductor. Called single-shot method.
3. Heat in coil with part stationary; quench ring moves in place. Single-shot adaptation of scanning method
4. Part is hydraulically lowered into quench tank after single-shot heating. Quench medium is agitated by submerged spray ring or propeller.
5. Vertical or horizontal scanning with integral spray quench. Single-turn inductor. Used for shallow hardening.
6. Vertical or horizontal scanning with multiturn coil and separate multirow quench ring. Used for deep-case or through hardening.
7. Coil scans and heats workpiece; self-quench or compressed air quench. Used in special applications with high-hardenability steels.
8. Horizontal cam-fed parts are pushed through coil, then dropped onto submerged quench conveyer.
9. Vertical scanning with single-turn inductor in combination with integral dual quench: one quench ring for scan hardening: the second for stationary quenching when the scanning travel stops.

Used for parts having a diameter or a flange section too large to travel through the inductor, wherein it is desired to harden up to the shoulder or flange.

10. Vertical scanning with single-turn inductor with integral spray quench and submerged quench in tank.
11. Split inductor and integral split quench ring. Used for hardening crankshaft bearing surfaces.

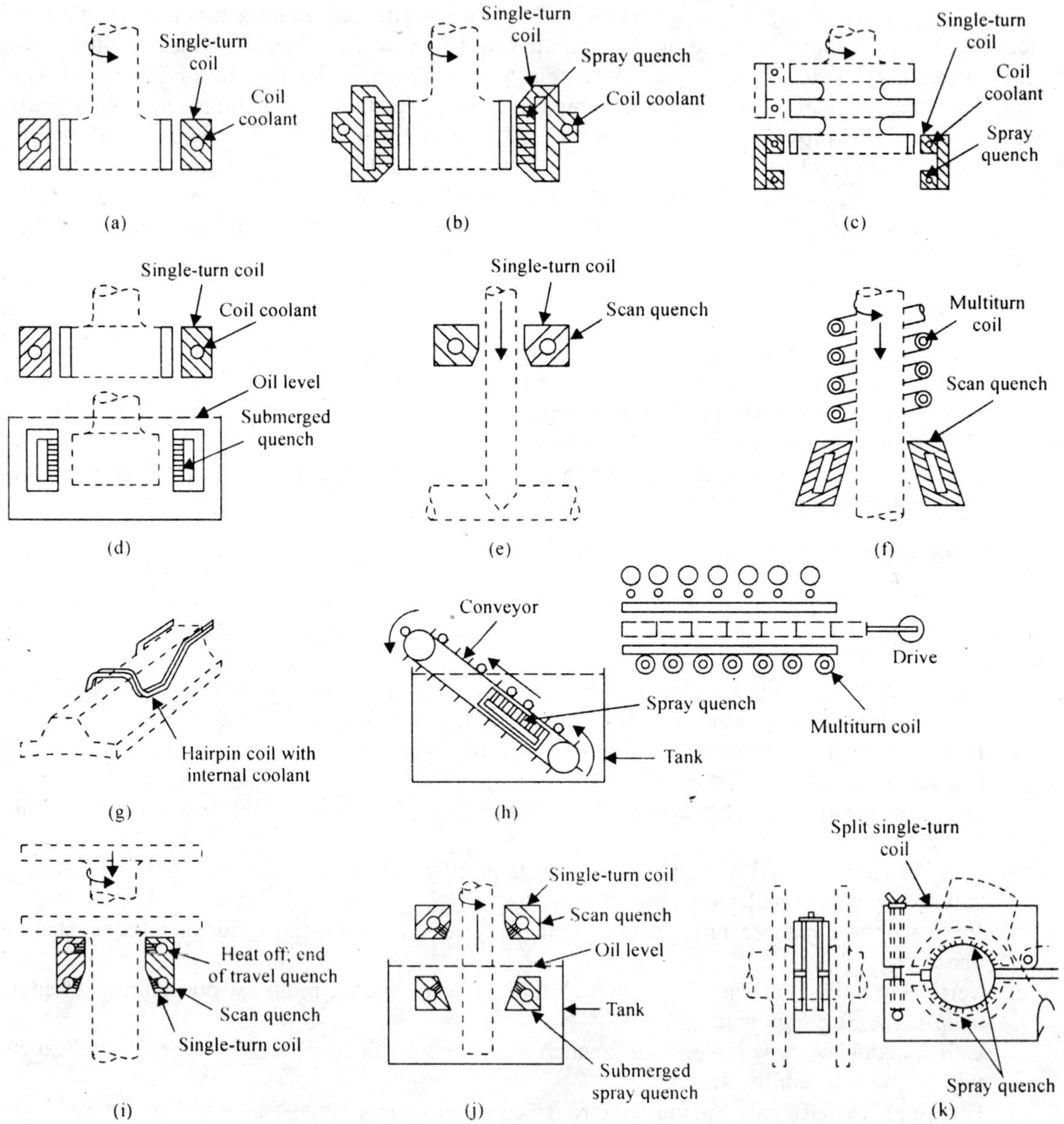

Fig. 24.22. Basic arrangements for quenching induction heated parts.

A water spray applied through a separate quench ring or from hollow inductors (for heating and quenching) has been used successfully in most applications involving plain carbon and low-alloy constructional steels; oil is specified for steels of higher hardenability and for parts with nonuniform sections when difficulty with cracking and distortion is anticipated. Quenching in water or oil may also be done by submersion in an agitated bath or by a combination of a spray quench ring and submersion in a tank on completion of the heating cycle.

Water Quenching

There are no stringent specifications for water used in quenching systems except that a reasonably clean and adequate supply of controlled-temperature water should be available. Extremely hard water may produce an excessive coating of deposits on quench orifice surfaces exposed to radiant heat.

Quench-water systems are usually separate from the supply of water for cooling the induction heating unit, because it is desirable to have a wider range of temperature control and variable pressure control for different quenching arrangements. Public water supplies and deep wells or,if necessary, water-recirculating systems with cooling towers are usually satisfactory. Recirculating systems consist of storage tanks, pumps and heat exchangers with adjustable controls; they are frequently protected with alarms and cutoffs for regulating pressure, temperature and flow. Pressures in water quenching vary from 205 to 585 kPa (30 to 85 psi); temperatures vary from 15 to 40°C (60 to 105°F).

Polyglycol Alcohol Solutions

An innovation in the field of water-base quenching media is the organic polymer solution in water which has been used with success on parts for which the steel selected has borderline hardenability—that is, where a spray oil quench does not provide a fast enough cooling rate to develop the hardness required and where a change to water would produce quench cracking.

The solution is made up by mixing from 1 to 20 per cent solutions of organic polymers with water. As diluted for quenching, the solution is nonflammable, does not produce objectionable fumes and does not cause skin irritation. System design is the same as for oil.

Oil Quenching

Despite the potential fire hazard that attends its use, oil spray quenching has a great many successful applications. A change from water to oil is made primarily to eliminate the quench cracking that often results in hardening irregular sections such as splines, gear teeth, shoulders, keyways and oil holes. Obviously, the change is also dependent upon whether or not the hardenability of the steel is sufficient to develop the required depth of hardening.

Typical specifications for oils used in quenching induction heated parts are given in Table 24.9.

Self-Quenching

A special quenching technique sometimes employed on parts with sufficient mass, such as large, coarse-pitch gears, is sometimes referred to as 'self-quenching', because most of the heat at the surface of the part is rapidly absorbed by the unheated mass of metal below the surface.

Forced-Air Quenching

A modification of self-quenching that has been successfully applied to smaller-pitch gears involves a progressive or scanning-type operation and uses a blast of compressed air fed through orifices in the inductor similar to those used with a liquid spray quench. This technique has been successfully applied to an external training gear made of 4340 steel, with an outside diameter of 1.8 m (72 inch), a face width of 76 mm (3 inch) and 144 teeth.

Table 24.9. Typical specifications for oils used in quenching induction heated parts.

General-purpose quenching oil (stock paraffin oil)	
Viscosity at 40°C (100°F)	70 to 85 SUS
Flash point	165°C (330°F) min.
Fire point	175°C (350°F) min.
Quenching temperature	50 to 60°C (120 to 140°F)
Fast quenching oil	
Viscosity at 40°C (100°F)	75 to 110 SUS
Flash point	175°C (350°F) min.
Fire point	200°C (390°F) min.
Quenching temperature	50 to 60°C (120 to 140°F)
Soluble-oil-water mixture	
Water-emulsifiable oil, 10 to 12% oil in water	

PROCESS CONTROL

Charts or graphs, representing the relationship between depth of hardening and heating time for various power densities and for various steels, can often be used to obtain a reasonably accurate estimate of the required cycle. Other estimates can be made from results for similar parts or by careful observation of the part itself as it is heating. However, the final cycle is usually determined by experimentation.

After a part has been hardened with one experimental arrangement, it is checked for hardness and examined for cracks. It is then sectioned with a rubber-bonded cutoff wheel, checked accurately for specified hardness at a given depth and prepared for a thorough examination of the microstructure. Should hardness, hardened depth or microstructure fail to meet desired standards, or if cracks are found, the cycle is adjusted accordingly. The procedure may be facilitated considerably with the aid of a high-speed recorder-controller together with either a thermocouple or special radiation devices that are available commercially. These units are designed to provide a minimum of inertia in response and therefore to record and control with suitable accuracy the temperature at the surface of parts heated at normal speeds.

Once the hardening cycle has been determined, production control of the process is achieved by repetition of the experimental cycle automatically, with respect to heating time or temperature, power density, delay between heating and quenching and the quenching cycle. In addition, parts are checked periodically by metallurgical examination for structure, case depth and hardness. The paragraphs under the four headings that follow include descriptions of equipment required to control each of these variables in a way that will ensure depth of hardening accurate to within ±10 per cent and suitable structure and hardness in a medium-carbon steel.

Heating Time

Heating time of parts heated in a fixed position relative to the inductor for 10 second or less should be controlled to within ±0.1 second. For heating times greater than 10 second but less than 60 second, this control should be within ±0.2 second and for heating times of 60 second or more, within ±1.0 second.

A multiple-circuit timer of the synchronous-motor-driven type, properly operated and maintained, will control the induction hardening process within the limits mentioned. A timer that is easily adjusted should be selected if the equipment is used for a variety of parts.

Power Density

Power density, as well as heating time, must be controlled carefully if heating is to be uniform from part to part. For a given part, power density will be affected by the electrical characteristics of the high-frequency converter, by variations in line voltage and in coupling and by part symmetry.

Delay Time

Delay time for single-shot operations is the interval between the time when the power is turned off and the time when the quenching medium contacts the workpiece. The delay is controlled easily by using an additional circuit of the same timer that controls the heating cycle.

Quenching Cycle

Control of quenching cycle involves accurate control of the quenching time, the temperature of the quenching medium, and, in spray quenching, the quenching pressure, velocity and direction. The quenching pressure should be measured as close to the discharge orifice as possible and a standard dial-type pressure gauge may be used for this purpose. A standard valve may be employed to regulate the pressure of the quenching medium.

Cooling of Equipment

The volume, pressure and temperature of water supplied for the cooling of tube, solid state or motor-generator units should be controlled within certain limits. Meter or gauges are used to indicate these limits and high- and low-pressure switches, flow switches and temperature-controlled switches act to stop operation of the induction heating unit if the preset limits are exceeded.

Salt or acid additives that decrease the electrical resistance of the water should be avoided.

MAINTENANCE OF EQUIPMENT

Induction hardening equipment is expensive and generally it is impossible to provide stand by equipment. A programme of preventive maintenance is therefore imperative and availability of certain critical parts for ready replacement is highly desirable.

Dust, dirt, moisture and high ambient temperature are the primary causes of failure in electrical equipment and one or all of these conditions are usually present in industrial locations where induction heating units are installed. In addition, cooling with air or water is necessary for almost all induction heating units and some problems develop in regard to suitable water-cooling systems.

PRECAUTIONS AND SAFE PRACTICES

Induction heating installations comprise various electrical and mechanical components, many of which are water-cooled and are equipped with protective devices. Instructions concerning safe practices and safety precautions are generally issued by the equipment manufacturer. These instructions should not be disregarded. Many potential hazards can be avoided by entrusting installations to qualified personnel who follow in detail the instructions provided. Installations should always conform to the safety practices and codes prevailing in a particular locality.

Devices for preventing overload and short circuits are usually built in as part of original equipment. In addition to these, there are ground detectors and surge detectors designed to protect motor-generator units from faulty coil or transformer setups at the heating stations, or from breakdown of insulation in the generator windings. Also, fire-prevention device are available to start or stop the quench in case of general power failure or malfunctioning of other protective devices. All protective devices should be checked periodically to make certain they are in good working condition.

SELECTION OF STEEL FOR INDUCTION HARDENING

Selection of steel should be based on the total of all processing factors as well as on the engineering requirements of service. Because the processing and engineering factors are interdependent, final selection is seldom determined by one item alone.

Selection of Prior Microstructure

The microstructure of the steel before induction hardening is important in selecting the heating cycle to be used. Steels most readily hardened have carbides that are small and uniformly dispersed, as obtained by quenching and tempering. Such structures are readily austenitised. Accordingly, minimum case depths can be developed with maximum surface hardness while using very rapid rates of heating.

Pearlite-ferrite structures, typical of normalised, hot rolled and annealed steels containing 0.40 to 0.50 per cent carbon, also respond successfully to induction hardening. Much of the induction hardened steel has such structures.

Spheroidised structures

Steels containing more than 0.50 per cent carbon are frequently spheroidised for improved machinability. Such spheroidised structures have the poorest response to induction hardening; the larger the carbide particles, the poorer the response. Coarsely spheroidised structures may require hardening temperatures 165°C (300°F) or more above the transformation temperature to redissolve the carbide and obtain uniform hardening results.

Nonhomogeneous structures

Any microstructure which is nonhomogeneous can cause difficulty in obtaining a completely austenitic structure. Examples of this are parts that have decarburised surfaces as found with cold-rolled material, some forgings, castings, or parts heat treated in decarburising atmospheres.

Selection of Hardening Temperature

Table 24.10 gives recommended hardening temperatures for a number of steels and irons commonly hardened by induction. For plain carbon steels and alloy steels not containing carbide-forming alloying elements, exhibiting a suitable prior structure, temperatures 25°C (50°F) higher than those used in furnace hardening are suitable, as indicated. This higher temperature compensates for the very short heating times.

Alloy steels

When carbide-forming elements (chromium, molybdenum, vanadium and tungsten) are present in constructional steels, it may be necessary to raise the hardening temperatures 55 to 110°C (100 to 200°F) above those shown in Table 24.10 for equivalent carbon contents. The exact increase in temperature depends on the length of the heating cycle, the specific alloying elements present, the specific effects desired from the alloying elements and the surface hardness required. These higher temperatures at the surface are used to obtain adequate solution of the carbon and alloying elements in the austenite with very short heating cycles. As the heating cycle is lengthened to produce deep cases or through heating, hardening temperatures should approach those given in the table.

Fully annealed structures may not provide full response with the short heating times used with induction heating. The use of steels with prior heat treating and tempering may be necessary.

Table 24.10. Recommended induction hardening temperatures and normal expected surface hardnesses for various metals(a).

Metal	Hardening temperature °C	°F	Quench(b)	Hardness(c), HRC
Carbon and alloy steel(d)				
0.30% C	900–925	1650–1700	Water	50
0.35	900	1650	Water	52
0.40	870–900	1600–1650	Water	55
0.45	870–900	1600–1650	Water	58
0.50	870	1600	Water	60
0.60	845–870	1550–1600	Water	64
			Oil	62
	815–845	1500–1550	Water	64
			Oil	62
Cast iron(e)				
Gray iron	870–925	1600–1700	Water	45
Pearlitic malleable	870–925	1600–1700	Water	48
Nodular	900–925	1650–1700	Water	50
Stainless steel(f)				
Type 420	1095–1150	2000–2100	Oil or air	50

(a) Metals listed in this table are typical of those successfully induction hardened and the listing is indicative rather than inclusive. (b) The quench selection depends upon the hardenability of the steel used, the diameter or cross section of the area heated, the case depth and hardness required, the need to minimise distortion and the propensity towards quench cracking. (c) Minimum surface hardness, Rock-well C. (d) Free-machining and alloy grades with equivalent carbon contents may be induction hardened. Alloy steels containing carbide-forming elements (chromium, molybdenum, vanadium or tungsten) should be heated 55 to 110°C (100 to 200°F) above the temperatures indicated. (e) Combined carbon should be 0.40 to 0.50 per cent min; hardness will vary with amount of combined carbon present. (f) Other martensitic grades of stainless steel, types 410, 416 and 440, have been induction hardened.

Grain Coarsening

With the usual induction hardening cycles, the hardening temperatures given in Table 24.10 will produce a fine austenitic grain size. This applies also to steels that contain carbide-forming elements and are heated 55 to 110°C (100 to 200°F) higher to provide alloy solution, since the alloy carbides inhibit austenitic grain growth as long as they remain undissolved. With longer heating cycles, there is danger of grain coarsening if the temperature is not reduced. An ASTM or fracture grain size of 6 or finer is specified for most parts hardened by induction heating.

In addition to grain coarsening, excessive hardening temperatures in induction heating may result in pronounced distortion and cracking. If specifications call for close limits of case depth and hardness pattern, over-heating may also cause failure to meet specifications, actual through hardening of thin sections, or possibly hardening into a critical area where machining or drilling is yet to be done. With pronounced overheating and long heating cycles, surfaces may scale.

Control of surface hardness

Values given in Table 24.10 are minimum values for the cited surface hardnesses, which are frequently specified in commercial induction hardening. The minimum values are given to avoid needlessly close hardness specifications and correspondingly close control and high cost. Considerable variation in the hardness of the cast irons may be expected, because of a variation in the combined carbon content.

As shown in Fig. 24.23, the hardness obtained on a given part may be several points Rockwell C higher than the minimum values recommended in Table 24.10. It is also apparent from Fig. 24.23 that the maximum hardness obtained in hardening by induction is usually higher than that obtained by hardening the same steel from a furnace. This higher hardness is most pronounced with the short induction heating cycles characteristic for parts that are surface hardened; it disappears in through hardened parts, parts that are surface hardened to produce a deep case, or parts that are tempered after induction hardening.

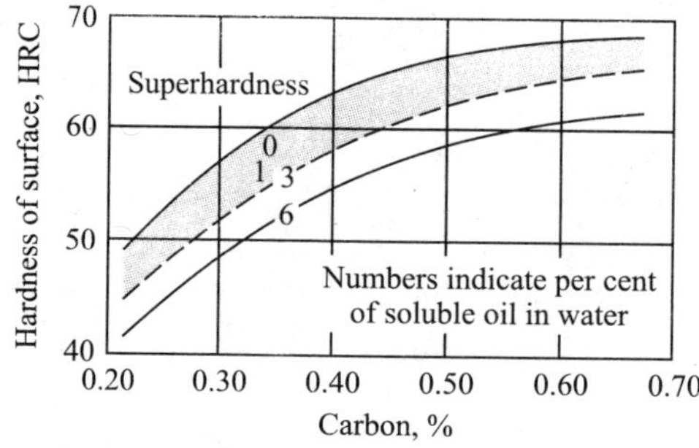

The two upper curves are for induction hardening; bottom, for furnace hardening;

Fig. 24.23. Relation between hardness and carbon content of water-quenched steels.

Low hardness

Low hardness at the surface after induction hardening may have a number of causes, a few of which are:

1. Lower carbon content than specified, possibly as a result of surface decarburisation before induction hardening. The usual induction hardening operations do not cause measurable decarburisation, because of the short time of heating characteristic of induction treatment.
2. Inadequate heating temperature or time, or both. Inadequate solution of carbon in the austenite before quenching will result in low hardness. Steels with unsuitable prior structures and alloy steels containing carbide-forming elements are most susceptible to low hardness from this source. Nonuniform performance of the induction heating unit and timer also may cause low surface hardness.
3. Unsatisfactory quenching conditions. Hot water or cold oil may result in low or spotty hardness, as will low pressures of either water or oil. Orifices in the quenching equipment that are plugged because of inadequate filtering may cause the same conditions.

4. Some hot-rolled steels can have different grain structures due to non-uniform deformation during hot rolling. Although these structures are rolled out linearly during hot rolling, they show up in cold drawn stock as spirals that are a result of the original hot coil being drawn and straightened into a bar. High and low hardnesses can be discovered by checking around the diameter. An examination along the diameter will show that these bands of hardness slowly spiral.

For a limited number of applications, surface hardnesses less than the minimums indicated in Table 24.10 are required, either to obtain greater notch toughness or to avoid quench cracking. These lower hardness values can be obtained by lessening the severity of quenching.

Control of Case Depth and Contour

The engineering specification of depth of hardening for a given part design is based on knowledge, or estimates, of the strength required to resist service loads. Frequently, as for shafts, the analysis involves considerations of requirements for fatigue strength.

Required case depth

The stress pattern is an important consideration in torsional fatigue. Radial stresses of both through hardened and case hardened shafts (induction or carburised) are always tensile and relatively unimportant. Residual circumferential and longitudinal stresses are usually tensile in through hardened parts and compressive in case hardened parts. The induction hardened shaft is therefore similar to a shaft made from shallow-hardening steel or to a carburised shaft, although the residual stresses are probably lower than in a carburised case. The induction hardened case must be deep enough to provide strength across the section to exceed the applied stress, but through hardening must be avoided.

Good Operating Practice

In surface hardening carbon and alloy steels by induction, if the hardened zone is too deep for the specific section thickness, high tensile stresses are established in the surface layers and cracking may result. Excessive depth of the hardened zone can result if the quench is too severe during the austenite-to-martensite transformation. Temperature that is too high can also cause excessive scale and pitting of the surface. The frequency, power density and time should be selected to provide the desired case depth. The quench should be selected to provide the cooling rate necessary to produce the desired hardness without cracking. The parts should be quenched to full hardness and then tempered to specified hardness (if lower than quenched). Stress relieving either through residual heat or furnace heat is always good practice where possible.

Control limits

Needlessly close control can result in added costs in steel, equipment and labour. The necessity for such control can start when the engineer specifies hardness depths and patterns that may be satisfactory for the design and service of the hardened part but which place unnecessary or even impracticable limitations on the size and frequency of the equipment, the tooling, fixtures and even the steel itself.

Hardness patterns

Hardness patterns on induction hardened parts are normally revealed by sectioning the piece with an abrasive wheel under water, rough grinding on a belt, polishing on fine emery paper and then etching with a 2 or 5 per cent nital solution until the pattern develops. The case depth is frequently reported as the depth from the surface at which the hardness drops below 50 HRC, although in initial studies for a given part, complete depth-hardness curves and microscopic analyses of the fully hardened and transition zones of the case are recommended. Fig. 24.24 shows depth-hardness curves for induction hardened parts made of steel and malleable iron.

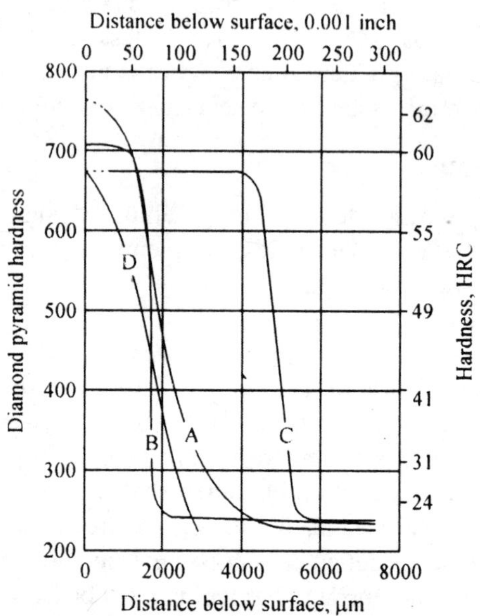

A is for a 47.6-mm-($1\frac{7}{8}$ inch) diam spindle made of 1045 steel, progressively hardened; 400 kHz, 50 kW, water spray quench. B is for normalised 5140 steel; 10 kHz, 24.8 W/mm^2 (16 kW/inch^2, 2 second. C is for normalised 5140 steel 10 kHz, 15.5 W/mm^2 (10kW/inch^2), 5 second. D is for pearlitic malleable iron; 10 kHz, 116.25 W/m^2 (75kW/inch^2, 1 ½ second, water spray quench.

Fig. 24.24. Depth-hardness curves for some parts surface hardened by induction.

Control factors

Basically, the depth of the hardened case obtained by induction heating followed by a suitable quench depends on the frequency, the power density, the heating time and the steel. In addition, the hardness pattern or contour depends on the shape of the part as related to coil design.

Effect of part shape

The shape of the part also may have a significant influence on the heat pattern. Parts with nonuniform cross section do not develop a uniform pattern, since edges and corners heat to a higher temperature.

In large gears or splines with a coarse pitch, unique hardness patterns are sometimes obtained not by controlling the heated contour but as a result of the combined influence of the hardenability of the steel and the severity of quench on the through heated gear teeth.

RESIDUAL STRESS, DISTORTION AND CRACKING

When a relatively shallow outside layer is selectively heated and quenched to martensite in induction surface hardening, the volume expansion associated with martensite formation gives rise to compressive stresses at the surface and tensile stresses in the interior. The tensile stresses usually start at the junction of case and core. The stresses are compressive at the surface, irrespective of the alloy content or the case depth. Oil quenching gives weaker stress patterns than water quenching.

Stress Distribution

The final stress distribution set up by quenching depends not only on the expansion accompanying martensite formation during the quench but also, to a lesser extent, on the localised plastic deformation resulting from nonuniform thermal expansion during rapid heating to the quenching temperature and nonuniform contraction during cooling to the temperature where martensite begins to form.

Thus, the actual stress distribution obtained depends on the rate of heating, the depth of hardening and the steel itself. One undesirable effect of increasing depth of hardening is an increasing depth of hardening is an increase in the magnitude of subsurface tensile stresses. Tempering reduces the stresses near the surface.

Compressive residual stress

Compressive residual stress at the surface of an induction hardened steering-knuckle pivot had a favourable influence on fatigue resistance. For this part, induction hardening to an equivalent case depth provided fatigue resistance superior to case hardening by carburising. Apparently, increased case depth further improved the fatigue resistance in this part. High tensile stresses, however, may exist at the juncture of case and core. If the heat pattern brings the high-tensile zone near the surface in an area such as a fillet that is subjected to repeated stresses in service, early fatigue failure may result. In addition, fatigue failure may also be initiated at the juncture of the case and the core.

Distortion

Steel parts that have been surface hardened by induction generally exhibit less total distortion or distortion more readily controllable than that for the same parts quenched from a furnace. The decrease in distortion is a result of the support given by the rigid, unheated core metal and of uniform, individual handling during heating and quenching. In scanning, distortion is controlled further by heating and quenching only a narrow band of the steel at one time. Unless a part through hardened by induction is scanned, the distortion encountered will approach the distortion that is experienced in furnace hardening.

Cracking

The forces that give rise to residual stresses and distortion will, in some instances, cause an induction heated part to crack during quenching.

Parts having severe discontinuities in configuration are particularly susceptible to cracking. The entire subject is so intimately associated with design that it can be discussed meaningfully only in relation to specific parts.

THROUGH HARDENING BY INDUCTION

The use of induction heating equipment for through hardening offers many of the same advantages over furnace hardening that it does for surface hardening. Some of these advantages are: little or no decarburisation, a minimum of scale, elimination of protective atmosphere in many applications, improved working conditions, improved quality of product, practical installation of induction equipment into the machine line and adaptability to automatic handling.

Methods

Parts can be through hardened by induction heating using the single-shot or scanning methods.

In the single-shot method, a part is positioned in the inductor coil, heated for a predetermined length of time and quenched either in the inductor or by removal to a separate system. Advantages of this method include simplicity of materials-handling equipment and ease of conversion to handle parts of different sizes. However, in the single-shot method, due to the 'Curie Effect', some generators do not operate continuously at rated output and consequently there is a loss in the over-all efficiency of the system. The Curie Effect refers to the change in permeability near the temperature at which steel becomes nonmagnetic, approximately 770°C (1420°F).

The scanning method permits the handling of larger parts with optimum efficiency. Using this method, bars or flats can be fed continuously through one or more inductor coils and into a quench. In some installations, the bar continues to advance through a tempering inductor after quenching, emerging from one integrated line completely hardened and tempered. Another advantage of the scanning method is that distortion is minimised and can normally be held to commercial mill tolerances for bars and flats.

Power Density and Frequency

Data for surface hardening by induction are not applicable to through hardening, because through hardening requires the heating of more metal and maintenance of a minimum temperature differential between the surface and the core. Thus, frequency and power density are required. The lower frequency heats to a greater depth; the lower power density prevents the surface from overheating while the heat flows into the core. Because the initial and operating costs of large power installations decrease with a decrease in frequency, it is advantageous to select the lowest frequency that will efficiently heat the specific parts.

Heating Rate

In a well-designed installation a rate of 1.3 mg/J (10 lb/kW·hr.) can be obtained when continuously heating bar stock or flats to hardening temperature. This rate cannot always be maintained in production when parts of various sizes are handled on the same line. Table 24.11 shows typical data for installations which have been in production service for many years.

Table 24.11. Typical operating conditions for progressive through hardening of steel parts by induction.

Section size					*Total heating time,*	*Scan time*		*Work temperature*			
								Entering coil		*Leaving coil*	
mm	*Inch*	*Material*	*Frequency, Hz*	*Power, kW(a)*	*s*	*s/mm*	*s/inch*	*°C*	*°F*	*°C*	*°F*
12.7 (round)	½ (round)	4130	180,	20	38	25	1	75	165	510	950
			9600(c)	21	17	25	1	510	950	925	1700
19 (round)	¾ (round)	1035 mod	180,	28.5	68.4	46	1.8	75	165	620	1150
			9600(c)	20.6	28.8	46	1.8	620	1150	955	1750
25 (round)	1 (round)	1041	180,	33	98.8	66	2.6	70	160	620	1150
			9600(c)	19.5	44.2	66	2.6	620	1150	955	1750
28.6 (round)	1⅛ (round)	1041	180,	36	114	76	3.0	75	170	620	1150
			9600(c)	19.1	51	76	3.0	620	1150	955	1750

(Contd...)

Section size mm	inch	Material	Frequency, Hz	Power, kW(a)	Total heating time, s	Scan time s/mm	s/inch	Work temperature Entering coil °C	°F	Leaving coil °C	°F
49.2 (round)	$1\frac{15}{16}$ (round)	14B35H	180,	35	260	178	7.0	75	165	635	1175
			9600(c)	32	119	178	7.0	635	1175	955	1750
15.9 (flat)	$\frac{5}{8}$ (flat)	1038	3000	300	11.3	38	1.5	20	70	870	1600
19 (flat)	$\frac{3}{4}$ (flat)	1038	3000	332	15	50	2.0	20	70	870	1600
22.2 (flat)	$\frac{7}{8}$ (flat)	1043	3000	336	28.5	97	3.8	20	70	870	1600
25 (flat)	1 (flat)	1036	3000	304	26.3	89	3.5	20	70	870	1600
28.6 (flat)	$1\frac{1}{8}$ (flat)	1036	3000	344	36.0	122	4.8	20	70	870	1600
17.5-33.3(d)	$\frac{11}{16}$-$1\frac{5}{16}$ (d)	1037 mod	3000	580	254	61	2.4	20	70	885	1625

Section size mm	inch	Material	Production kg/hr.	lb/hr.	Operating efficiency(b) mg/J	lb/kW·hr.	Inductor input(b) W/m²	kW/inch²
			91.6	202	1.27	10.1	0.67	0.43
12.7 (round)	½ (round)	4130	91.6	202	1.21	9.6	1.22	0.79
			113.4	250	1.10	8.77	0.62	0.40
19 (round)	$\frac{3}{4}$ (round)	1035 mod	113.4	250	1.53	12.14	0.85	0.55
			141.1	311	1.19	9.42	0.54	0.35
25 (round)	1 (round)	1041	141.1	311	2.00	15.9	0.57	0.37
			153.3	388	1.18	9.39	0.52	0.34
28.6 (round)	$1\frac{1}{8}$ (round)	1041	153.3	338	2.23	17.70	0.50	0.32
			194.6	429	1.55	12.3	0.29	0.19
49.2 (round)	$1\frac{15}{16}$ (round)	14B35H	194.6	429	1.69	13.4	0.48	0.31
15.9 (flat)	$\frac{5}{8}$ (flat)	1038	1448.9	3194	1.34	10.6	3.61	2.33
19 (flat)	$\frac{3}{4}$ (flat)	1038	1575.8	3474	1.32	10.5	3.19	2.06
22.2 (flat)	$\frac{7}{8}$ (flat)	1043	1609.4	3548	1.34	10.6	2.06	1.33
25 (flat)	1 (flat)	1036	1595.3	3517	1.46	11.6	2.25	1.45
28.6 (flat)	$1\frac{1}{8}$ (flat)	1036	1678.8	3701	1.36	10.8	2.08	1.34
17.5-33.3(d)	$\frac{11}{16}$-$1\frac{5}{16}$ (d)	1037 mod	2211.3	4875	1.06	8.4	0.40	0.26

(a) Power transmitted by the inductor at the operating frequency indicated. This power is approximately 25 per cent less than the power input to the machine, because of losses within the machine. (b) At the operating frequency of the inductor. (c) Dual frequencies are used. (d) Irregular shape.

Because of the relatively short heating times involved in induction heating, it is frequently necessary, in through hardening, to heat 25 to 55°C (50 to 100°F) above the recommended furnace temperature to form homogeneous austenite. Fine-grained material is required for best results.

APPLICATIONS AND COST RELATIONS

Hardening by induction can often contribute to cost savings by:

1. Substitution of carbon steel for alloy steel.

2. Increased production.
3. Reduction in prime labour personnel.
4. Elimination of a machining operation.
5. Elimination of a straightening operation.
6. In-line production setup.
7. Less handling of parts.
8. Automation of the hardening operation.
9. Less equipment maintenance.
10. Less floor space.
11. Less energy per piece.
12. Less finish grinding.
13. Less inspection.
14. Less scrap.

With respect to cost relations, the applications of induction hardening may be placed in two categories: (i) where induction hardening provides a unique solution to problems posed by special requirements and cost is secondary; and (ii) where induction hardening competes with other heat treating methods; cost is paramount.

High-Frequency Resistance Hardening

In conventional induction hardening, the coil is always separate and distinct from the workpiece and energy is transmitted from the coil to the work through an air gap, by means of induction. In high-frequency resistance hardening, a portion of the workpiece, in effect, becomes a part of the coil and energy is transmitted to the work by means of conduction. This technique may be employed when the use of a closed loop coil is not possible or is highly inefficient.

High-frequency resistance heating usually employs frequencies in the range of 10 to 500 kHz and is capable of producing higher current densities than can be obtained with induction. The use of high frequencies minimises problems normally encountered at the contact points in low-frequency resistance heating.

INDUCTION HARDENING OF CAST IRON

Cast irons may vary significantly in combined carbon content and results from induction hardening are nonuniform unless control is established over the combined carbon content. A minimum combined carbon content of 0.40 to 0.50 per cent carbon (as pearlite) is recommended for cast iron to be hardened by induction with the short heating cycles that are characteristic of the induction process. Heating castings with lower combined carbon content to high hardening temperatures for relatively long periods of time may dissolve some free graphite, but such a procedure is likely to coarsen the grain structure at the surface and will result in undesirably large amounts of retained austenite in the surface layers.

Flake-Graphite Gray Iron

The surface hardness attained from induction hardening of gray iron is influenced by the carbon equivalent (%C + 1/3% Si) when this hardness is measured by conventional Rockwell tests. The more graphite that is present in the microstructure, the lower the surface hardness will appear to be after hardening. Table 24.12 shows the surface hardness of induction hardened gray iron castings of various carbon equivalents from 3.63 to 4.23. The microstructure of these castings, which were cast in the same manner and cooled at similar rates, contained more and larger graphite flakes as the carbon equivalent increased. This resulted in lower apparent surface hardness after hardening, yet the hardened matrix was consistently 57 to 61 HRC (converted from microhardness).

Table 24.12. Effect of carbon equivalent on surface hardness of induction hardened gray irons.

Composition, %(a)			*Hardness, HRC*		
				Converted from	
C	*Si*	*Carbon equivalent(b)*	*As read*	*Rockwell 30-N*	*Micro-hardness*
3.13	1.50	3.63	50	50	61
3.14	1.68	3.70	49	50	57
3.19	1.64	3.74	48	50	61
3.34	1.59	3.87	47	49	58
3.42	1.80	4.02	46	47	61
3.46	2.00	4.13	43	45	59
3.52	2.14	4.23	36	38	61

(a) Each iron also contained: 0.50 to 0.90 Mn, 0.35 to 0.55 Ni, 0.08 to 0.15 Cr and 0.15 to 0.30 Mo. (b) Carbon equivalent = %C + $\frac{1}{3}$ % Si.

Pearlite

To induction harden gray iron within these limitations, it is necessary that the carbon be present in a form that will readily dissolve in the iron at the austenitising temperature. The combined carbon in pearlite is readily soluble and high percentages of pearlite in the matrix are desirable. Sufficient pearlite to give a combined carbon content of 0.40 to 0.50 per cent is satisfactory for induction hardening.

Ferrite

The prior structure and combined carbon content of as-cast gray iron are dependent on cooling rate (section size) and composition. Large sections, or slow cooling rates, may result in increased amounts of ferrite and decreased response to short heating cycles. Alloying elements, such as chromium, nickel, copper or molybdenum, will minimise the formation of ferrite on cooling.

Distortion

Induction hardening causes less distortion than would a similar quenching treatment from a furnace. Maximum warpage in the 55.9-cm (22-inch) length of a bar was found to be 0.03 mm (0.0015 inch) after induction hardening, compared with 0.17 to 0.25 mm (0.007 to 0.010 inch) for the same bars quenched from a furnace. For thin-wall cylinders, distortion is not a problem unless the thickness of the induction hardened layer exceeds 20 per cent of the wall thickness.

For selective hardening in place or by scanning, to depths up to 3.8 mm (0.150 inch), water quenching can be safely employed. However, the hardening of keyways, cross-drilled holes, or extremely thin walls may require the use of oil, to prevent excessive distortion or cracking.

Applications

A cast frame for a baler knotter had a nominal composition of 3.30 to 3.50 C, 0.50 to 0.70 Mn, 0.30 to 0.40 P, 0.15 max S and about 2.40 to 2.60 Si. A 10.9-mm (7/16 inch) section of the frame was heated for

12 second with a 14-kW, 10-kHz unit, using a modified hairpin-type inductor. After water quenching for 4 second, it developed an apparent hardness of approximately 45 HRC (not indicative of the true hardness of the matrix, which can harden to 61 HRC or higher).

Ductile Iron

The response of ductile iron to induction hardening is dependent on the amount of pearlite in the matrix of as-cast, normalised and tempered prior structures. In quenched and tempered iron, the secondary graphite nodules formed during tempering are close enough together to supply sufficient carbon to the matrix by re-solution during induction heating.

As cast

In the as-cast condition, the amount of pearlite can vary from 10 to 95 per cent, depending on cooling rate and silicon content. The pearlite content of the matrix increases as the cooling rate increases and as the amount of silicon decreases.

For induction heating cycles 3.5 seconds and longer and hardening temperatures of 955 to 980°C (1750 to 1800°F), 50 per cent pearlite is considered a minimum for satisfactory hardening. Structures containing less pearlite can be hardened by using higher temperatures, but at the risk of retaining austenite, forming ledeburite and damaging the surface. With more than 50 per cent pearlite, hardening temperatures may be reduced to within the range of 900 to 925°C (1650 to 1700°F).

Prior treatments

Because the amount of pearlite in as-cast ductile iron can fall below the minimum required for satisfactory response, a prior treatment to produce the desired microstructure is required, to ensure that all castings will respond to a given hardening cycle. Some castings, because of predominant section size or configuration, may satisfy the minimum microstructure in the area to be hardened. It is possible, to a great extent, to relate this microstructure to a Brinell hardness measurement. At 229 HB, for example, the microstructure can vary from 20 to 50 per cent ferrite (50 per cent pearlite). A hardness of 229 HB would therefore be the approximate minimum for induction hardening as-cast or normalised and tempered structures.

Normalised and tempered

Normalising, although it produces a uniform microstructure low in ferrite and with good induction hardening characteristics, usually produces a higher hardness than is desired for machinability and ductility. Therefore, the normalising treatment is usually followed by a tempering treatment.

To control the as-tempered hardness range within desired limits, it is necessary also to control the range of hardness developed in normalising. This can be accomplished by providing a sufficiently fast and uniform rate of air cooling from the normalising temperature, 900°C (1650°F). For heating cycles of 3.5 second and longer, at temperatures of 955 to 980°C (1750 to 1800°F), a prior structure containing 50 per cent pearlite would be considered a minimum. Examination of the normalised and tempered irons indicates that the poor response of structures of lower pearlite content is due to depletion of the matrix carbon. In the tempering operation, the carbon migrates from the pearlite matrix to the graphite nodules. In the heating cycle, carbon is reabsorbed in the matrix from the nodule; however, there is insufficient time for it to migrate throughout the ferritic areas. Another factor in the response of ductile iron is the graphite nodule count; the greater the number of nodules per unit area, the deeper the hardening for any given heat cycle. This effect is more evident as the percentage of ferrite increases (Fig. 24.25).

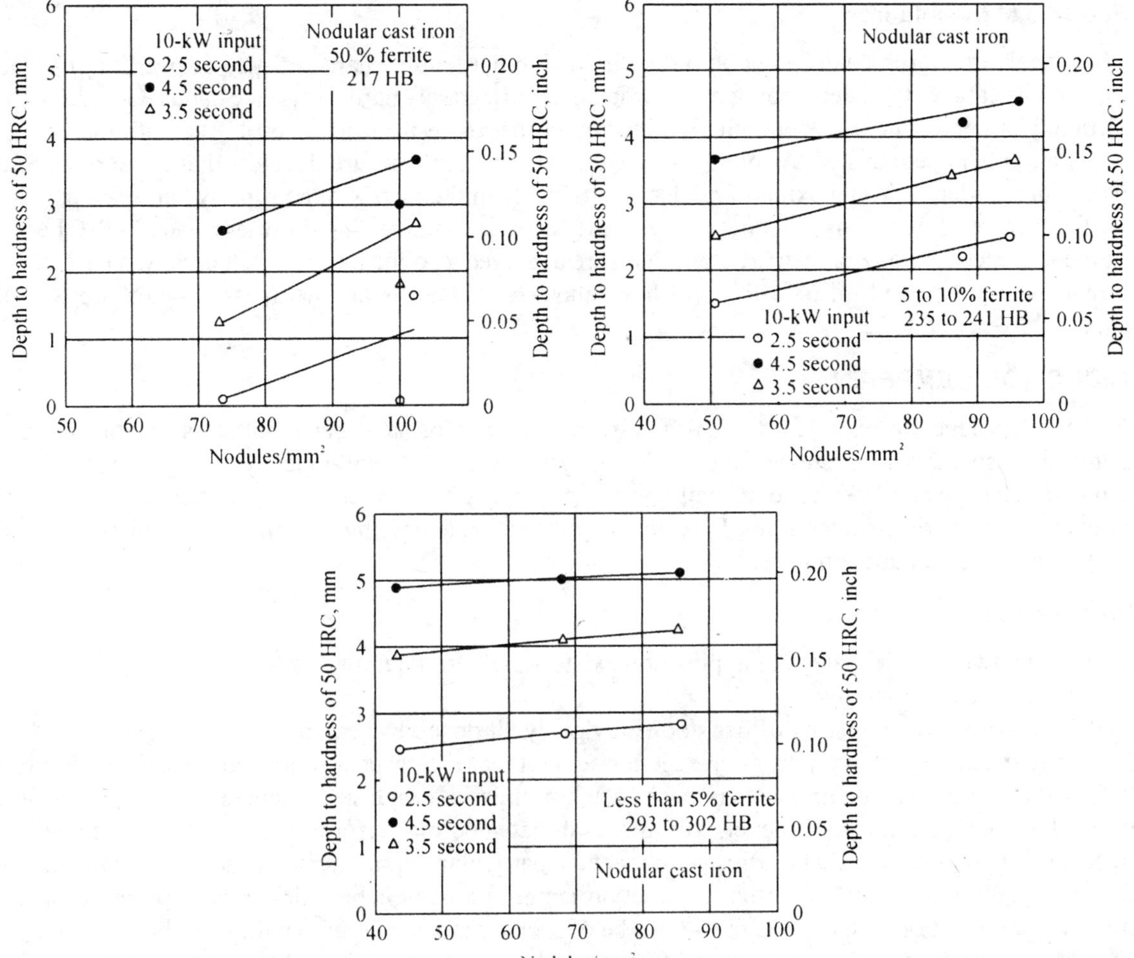

Fig. 24.25. Relation between depth of induction hardening to 50 HRC and graphite nodule count in normalised and tempered nodular iron.

Quenched and tempered

The response of quenched and tempered nodular iron to induction hardening is excellent over a wide range of microstructures containing up to 95 per cent ferrite. As a prior treatment, quenching and tempering has the advantage of permitting a lower prior hardness; there is a risk of distortion and quench cracking, however.

A quenched and tempered structure that provided good response to induction hardening was obtained by oil quenching from 900°C (1650°F) and tempering at 620°C (1150°F) for 1 hour. This treatment produced a hardness of 262 HB, which could have been lowered, if necessary, by increasing the tempering temperature to 675°C (1250°F). By induction heating to a depth of 4.7 mm (0.184 inch), a surface hardness of 54 to 56 HRC was developed and a depth of hardness to 50 HRC of 4.2 mm (0.164 inch) was obtained.

Pearlitic Malleable Iron

Considerable research has been done on the induction hardening characteristics of pearlitic malleable iron and its capability of developing high hardness over relatively narrow surface bands. Generally, little difficulty is encountered in obtaining hardnesses in the range from 55 to 60 HRC, with the depth of penetration being controlled by the rate of heating and the temperature developed at the surface of the part being hardened. The maximum hardness obtainable in the matrix of a properly hardened part is 67 HRC. However, conventional hardness measurements made on the casting show less than 67 HRC because of the presence of the graphite particles which are averaged into the hardness. Generally, a casting with a matrix microhardness of 67 HRC will have about 62 HRC average hardness, as measure with the standard Rockwell tester.

INDUCTION TEMPERING

Extensive production experience with thousands of tonnes of steel per month has demonstrated the commercial success of induction tempering for many applications. Metallurgically, the success of induction tempering has been related fundamentally to the possibility of compensating for short tempering times with higher tempering temperatures. Economically, induction tempering has proved particularly adaptable to automation in production lines.

Application

At present, two principal areas of application exist for induction tempering:

1. Selective tempering
2. Progressive tempering of bar stock previously hardened by scanning.

Many machine parts vary from one section to another with respect to load and wear requirements. Often, this variation in requirements is met by a compromise of properties obtained by uniform tempering to a single hardness level. However, it is apparent that superior performance might be expected if the mechanical properties could be adjusted to meet the particular requirements in each section by selective tempering. Within certain limitations, induction tempering is an economical means of accomplishing this. These limitations are that the parts must be of such a shape and size that they can be coupled by the inductor to heat uniformly to the desired temperature in critical sections. Although this is impossible or impracticable for some parts, many can be tempered selectively by induction to obtain different degrees of hardness in the same part, with a consequent improvement in quality.

Advantage

The principal advantage of induction tempering is the possibility of integration with machine lines to avoid excessive handling of work, thereby minimising labour cost. This is illustrated in the preparation of bar stock of specified mechanical properties before machining into cylinder-head studs and miscellaneous machine parts.

Selection of Equipment

Because tempering is performed below the lower transformation temperature 725°C (1333°F), lower-frequency induction tempering installations are generally used; such installations are necessary for tempering large sections, to minimise any temperature gradient from the surface to the interior. Table 24.13 presented as a guide for the selection of tempering equipment. It should be noted that line frequencies (60 Hz) may be used for tempering parts 25 to 50 mm (1 to 2 inch) in diameter or larger. Considerable overlapping occurs in suitable frequencies, so that equipment utilised for induction hardening can often be used for the tempering treatment as well.

Table 24.13. Selection of power source and frequency for various applications of induction tempering.

Size of section		*Maximum tempering temperature*		*Power lines, 50 or 60*	*Frequency converter,*		*Solid state or medium frequency*		*Vacuum tube, over*
mm	*inch*	*°C*	*°F*	*Hz*	*180 Hz*	*1000 Hz*	*3000 Hz*	*10000 Hz*	*200 kHz*
3.2–6.4	⅛–¼	705	1300	–	–	–	–	–	Good
6.4–12.7	¼–½	705	1300	–	–	–	–	Good	Good
12.7–25	½–1	425	800	–	Fair	Good	Good	Good	Fair
		705	1300	–	Poor	Fair	Good	Good	Fair
25–50	1–2	425	800	Fair	Fair	Good	Good	Fair	Poor
		705	1300	–	Fair	Good	Good	Fair	poor
50–152	2–6	425	800	Good	Good	Good	Fair	–	–
		705	1300	Good	Good	Good	Fair	–	–
Over 152	Over 6	705	1300	Good	Good	Good	Fair	–	–

Efficiency, capital cost and uniformity of heating are considered in the indicated selections. Indicated selections are defined as follows: *Good* indicates the optimum frequency that will heat 80 per cent or more of the pounds per kilowatt-hour indicated in Fig. 24.19. *Fair* indicates a frequency that will heat 60 to 80 per cent of the pounds per kilowatt-hour indicated in Fig. 24.19; or a frequency higher than optimum that increases capital cost and reduces uniformity of heating, thus requiring lower heat inputs. *Poor* indicates a frequency that will heat less than 60 per cent of the pounds per kilowatt-hour indicated in Fig. 24.19; or a frequency higher than optimum that increases capital cost and reduces uniformity of heating, thus requiring lower heat inputs.

Power density

Because the usual objective of induction tempering is to produce uniform hardness throughout the cross section, rather than to heat the surface, the power density is generally low from 0.08 to 0.8 W/mm^2 (0.05 to 0.5 kW/inch^2) of surface within the inductor. Power densities may be selected on the basis of experience, tests, or data such as those presented in Table 24.14.

Table 24.14. Approximate power density required for tempering.

	Input(b)			
	W/mm²		*kW/inch²*	
Frequency(a), Hz	*150–425°C (300–800°F)*	*425–705°C (800–1300°F)*	*150–425°C (300–800°F)*	*425–705°C (800–1300°F)*
60	0.09	0.23	0.06	0.15
180	0.08	0.22	0.05	0.14
1000	0.06	0.19	0.04	0.12
3000	0.05	0.16	0.03	0.10
10 000	0.03	0.12	0.02	0.08

(a) Table is based on use of proper frequency and normal over-all operating efficiency of equipment. (b) In general, these power, densities are for section sizes of 12 to 50 mm (½ to 2 inch). Higher inputs can be used for smaller section sizes and lower inputs may be required for larger section sizes.

Furthermore, the heating time is comparatively long, to help provide uniform heating throughout the part. To meet production requirements, length of the inductor can be increased, or more than one bar can be processed at a time.

AUXILIARY AND CONTROL ACCESSORIES

Because parts are cooled from the tempering temperature in air, no quenching facilities are required. However, equipment for handling is as important in induction tempering as in hardening; the fixtures already described for hardening are suitable. Frequently, the tempering operation is keyed to the hardening operation, or the same equipment may be used for both induction hardening and tempering either by merely changing the work coil or by reducing the power density and the heating time.

Radiation Pyrometer

In general, the control of induction tempering is achieved by selection of the power density and the rate of feed through the coil (scanning) on the basis of hardness tests of the tempered product. Automatic control may be obtained at tempering temperatures above 425°C (800°F) by use of a special radiation pyrometer and high-speed controller. This arrangement may be used to vary the speed of the scanning operation continuously or to control the power.

Voltage Regulator

With 60-Hz installations, an automatic voltage regulator in the power line ahead of the equipment is desirable if automatic control is not used. For adjustment, power input may be varied by one or more of these means:

1. A continuously variable transformer.
2. A tapped transformer.
3. A saturable reactor.
4. Taps on the coil itself.
5. An SCR (silicon-controlled rectifier) controller.

With motor-generator equipment employing automatic voltage regulation, or in scanning uniform work, control of the rectified field excitation has provided good results. Solid state equipment can use either constant-voltage or constant-power regulation, which is built into the unit. SCR controllers are used with vacuum-tube equipment to give uniform results in small sections.

SELECTION OF INDUCTION TEMPERING CYCLE

Fundamentally, the tempering cycle will be determined by the ultimate mechanical properties specified for the part. However, usual furnace tempering temperatures must be increased to compensate for the short heating times.

Fig. 24.26 shows the increase in tempering temperature required to produce a given hardness as the tempering time is decreased from 1 hour. (furnace tempering) to 60 second and 5 second (induction tempering), in 1050 steel quenched in brine from 855°C (1575°F). Pieces with small cross section may be air cooled immediately upon reaching the tempering temperature, while slower heating rates or short periods of time at temperature (5 to 60 second) before cooling are desirable for larger sections, to allow penetration of heat. In scanning, of course, the power density, the rate of travel and the length of the inductor will determine the time of tempering.

RESULTS OF INDUCTION TEMPERING

Selection of the proper induction tempering cycle and of suitable equipment results in uniform hardness from the surface to centre of the part. This is illustrated by Fig. 24.27, which shows hardness readings on the cross sections of a 19-mm ($\frac{3}{4}$-inch) diam bar made of 1037 steel, a 25-mm (1-inch) diam bar of 1041 steel and a 49-mm ($1\frac{15}{16}$-inch) diam bar of 14B35H steel. Fig. 24.27 shows the hardness values of these steels in both the as-quenched and induction tempered conditions.

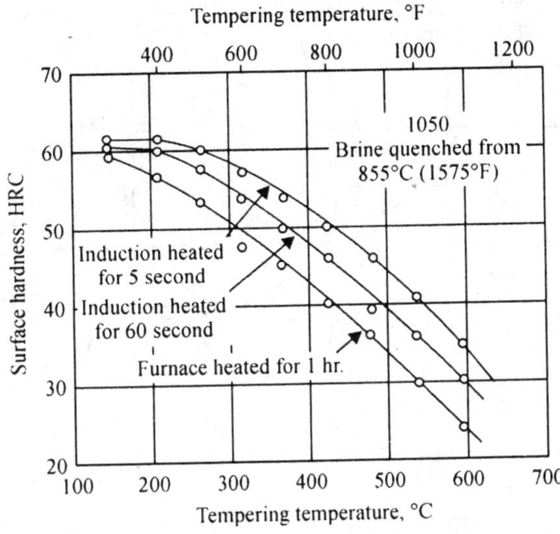

Fig. 24.26. Variation of hardness with tempering temperature for furnace and induction heating.

Statistical studies over several years on cross-sectional hardnesses for similar bar stock that was induction hardened and tempered by scanning reveal that 67 per cent of all average cross-sectional hardnesses resulted in a hardness spread of ±1.25 Rockwell C points and that approximately 95 per cent were within ±2.5 points. Variations in resultant hardness occur more often from differences in hardenability between respective heats of steel than from discrepancies in the induction tempering process. If each mill heat is run separately, it is possible to compensate for differences in hardenability.

Stress Distribution

Inspite of the short tempering times encountered in induction tempering, results indicate that the amount of stress relief is similar for parts uniformly tempered through the cross section if an equivalent hardness is obtained by higher tempering temperatures. After differential or selective tempering, the stress distribution may be substantially different from that achieved in furnace tempering; thus, analysis or a performance test should be made before adopting induction for selective tempering.

Induction Versus Electric-Furnace Tempering

Electric furnaces were used for tempering many years before induction heating was applied to tempering. For this reason, induction tempering is not normally used unless it proves advantageous over furnace tempering. The two examples above are not intended to imply that induction tempering is the most economical method for all parts. There are many parts which cannot be successfully induction tempered and in many instances furnace tempering is more economical.

Fig. 24.28 shows relative operating efficiency for electric tempering furnaces and induction tempering equipment. When making a graphical comparison between the two types of heating, allowances should be made for the somewhat higher temperatures that are required when tempering by induction to attain the same hardness range and microstructures.

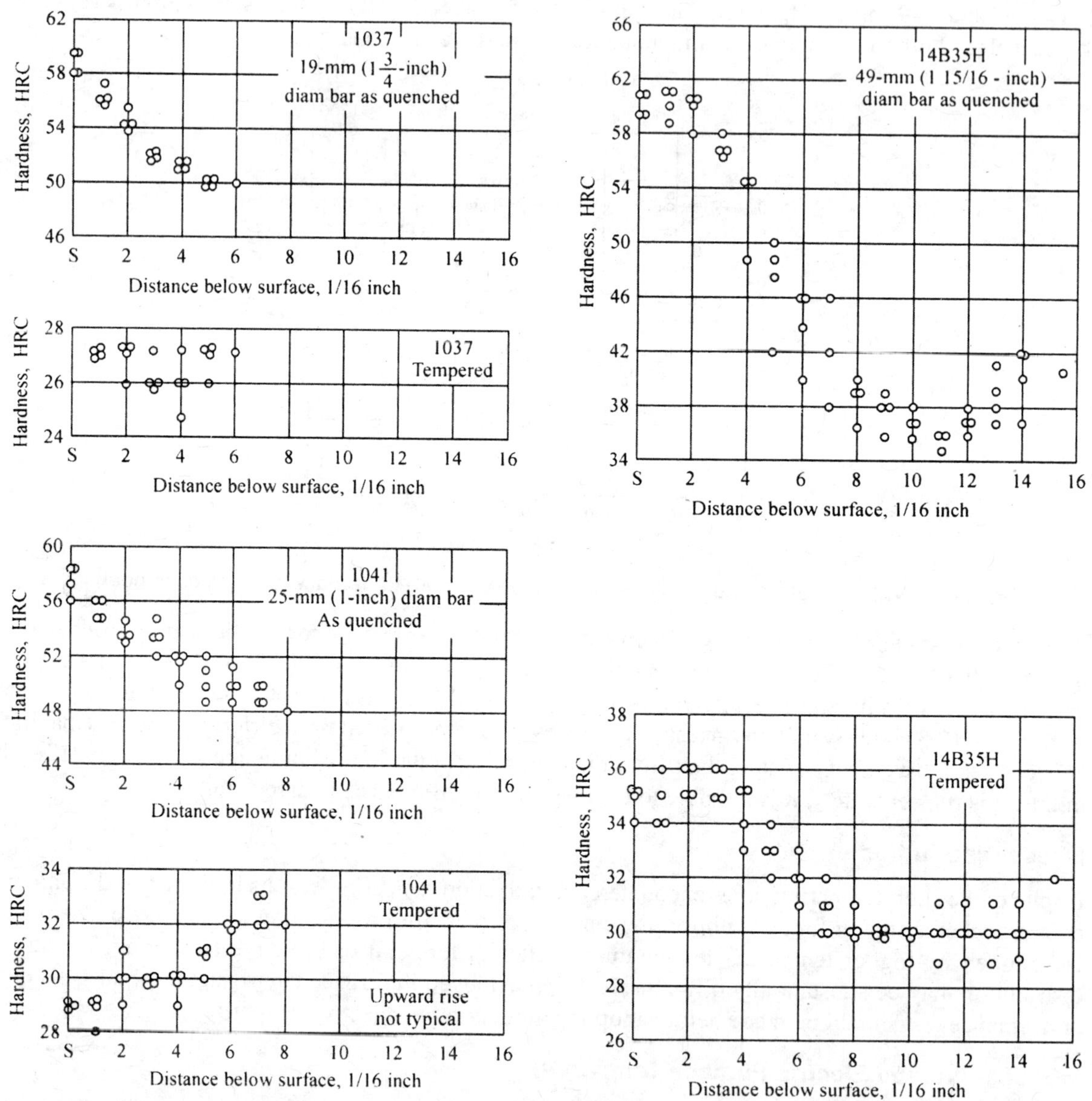

Fig. 24.27. Typical hardness survey on 19-mm (3/4-inch) diam 1037, 25-mm (1-inch) diam 1041 and 49-mm ($1\frac{15}{16}$-inch) diam 14B35H steel bars through hardened and tempered by induction.

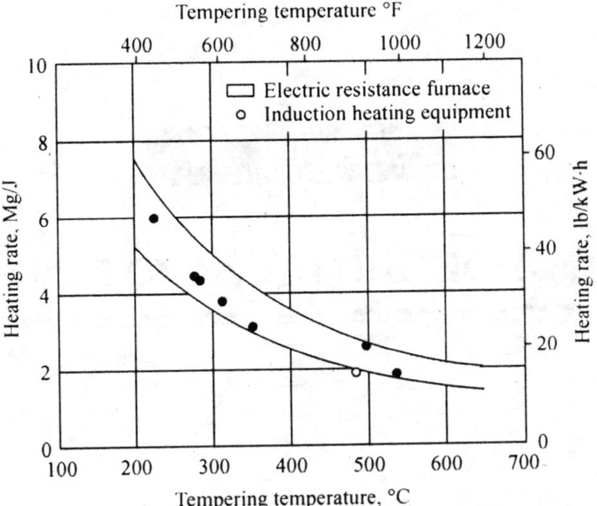

Curves represent upper and lower efficiencies expected with electric resistance furnace. Actual operating efficiencies of continuous induction tempering installations are represented by plotted points.

Fig. 24.28. Operating efficiencies for tempering by induction and electric furnace.

Chapter 25

Electron Beam Hardening

INTRODUCTION

Electron-beam heat treating is a selective hardening process in which the surface of a hardenable ferrous alloy is heated rapidly above the transformation temperature of the alloy by direct bombardment or impingement of an accelerated stream of electrons. At the end of a heating cycle of 0.5 to 2.5 seconds, the flow of electrons is stopped abruptly to allow the part or workpiece being processed to selfquench and to form a martensitic structure with a compressive stress on the surface of the hardened area. The electron-beam hardening process normally is applied to finish-machined or ground surfaces. Because the build-up of energy is rapid and well controlled, post-heat treatment operations such as grinding or straightening are not required in most instances.

APPLICATION CRITERIA

A part or workpiece to be heat treated is considered a suitable candidate for electron-beam heat treating if it meets the following criteria:

1. The material must contain adequate carbon to produce satisfactory case hardness. Fig. 25.1 presents relationships between carbon content and case hardness for plain carbon steels.
2. The stream of electrons must have line-of-sight access to the area requiring heat treatment and a beam-impingement angle of at least 25°. Guidelines for acceptable part configurations are shown in Fig. 25.2.
3. The component being heat treated may be processed in a vacuum envelope or chamber or at pressures up to 1 atm in air or inert gas. Vacuum chambers in high-production heat treating systems typically have interior volumes of 0.02 to 0.11 m^3 (0.8 to 4.0 ft^3).
4. The surface to be heat treated should be machined or ground to final dimensions. If grinding is required after heat treating, removal of 0.05 to 0.25 mm (0.002 to 0.010 inch) of stock normally is sufficient.
5. To prevent magnetic interaction or unintentional deflection of the electron beam, the component being heat treated must be demagnetised prior to hardening. Demagnetisation normally is required if the part has been fixtured with magnetic clamps or chucks in operations prior to heat treating.
6. The mass of the part must be sufficient to self-quench the heat treated area. The ability of a part to self-quench is determined by its composition as well as its configuration. Steels of high

hardenability (such as AISI 4150) can be through hardened in many instances. For plain carbon steels, however, five to eight unit of mass beneath the heated area are required for each unit of mass of hardened case. The self-quench phenomenon is described in detail later in this chapter.

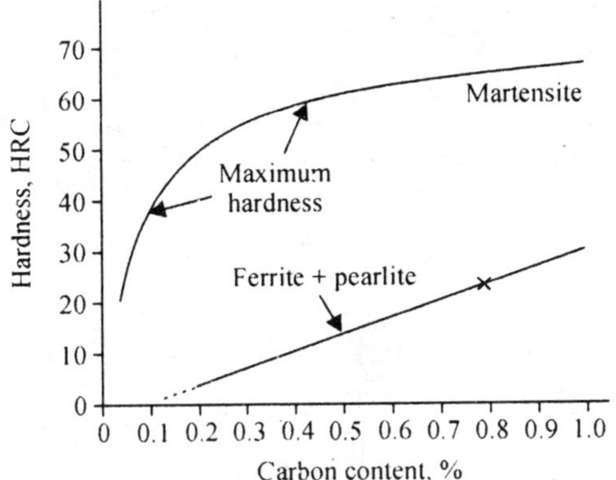

Fig. 25.1. Relationship between maximum hardness and carbon content for plain carbon steels.

ELECTRON-BEAM EQUIPMENT

In electron-beam heat treating, a highly concentrated beam of high-velocity electrons is used to heat selective surface areas. The electron beam is produced by an electron gun such as the one schematically represented in Fig. 25.3. Depicted is a 42 kW gun used for welding as well as for heat treating.

Beam Control

Free electrons escape from the filament when it is resistance heated to a very high temperature. These electrons are accelerated and collimated into a dense, extremely energetic beam by the accelerating potential between the cathode and the anode. The high-energy beam thus formed passes through a small-diameter hole in the anode. Because of the mutual repulsion among neighbouring electrons, the beam requires further collimation below the anode. This additional collimation is controlled with a focus coil that allows variation of the distance from the gun to the workpiece. A deflection coil deflects the reconverging beam to a designated location on the workpiece.

A high vacuum is needed in the region where the electrons are emitted and accelerated, both to protect the emitter from oxidation and to prevent interference with the electrons while they are still at low velocity. Therefore, the electron-gun housing is pumped and maintained at a vacuum of 10^{-5} torr. The workpieces are contained in an enclosure under a vacuum of approximately 5×10^{-2} torr. An intermediate vacuum level provides short evacuation times and higher production rates. Treating at one atmosphere does not require any evacuation time.

In electron-beam heat treating, the energy exchange is simply a matter of the electrons in the beam transferring their kinetic energy to the atomic structure of the target material in the form of heat. The electron beam, when sharply focused for welding, is capable of impingement power densities on the order of 10 MW/cm^2 (65 MW/inch^2). Because this powerful concentration of energy is easily controllable in power magnitude, power density and beam position, it is well suited for surface hardening as well. These power densities are much too high for non-destructive heat treating, however. Destructive heat treating in this context refers to controlled re-melting of ferrous and non-ferrous materials.

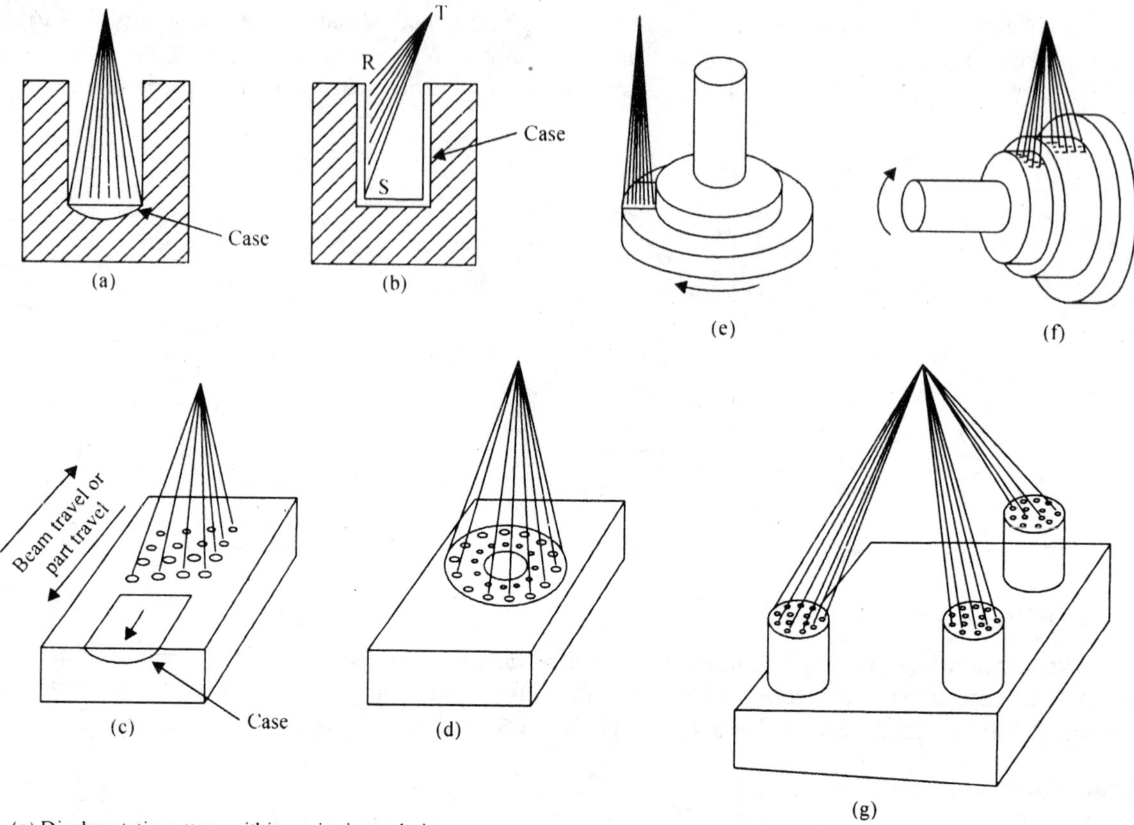

(a) Display static pattern within cavity in workpiece.
(b) Maintain angle of workpiece rotation, RST, at 25° minimum.
(c) Display static pattern and move the pattern or the workpiece to heat treat large areas.
(d) Display static pattern; this annular pattern has well-defined inside and outside diameters.
(e) Display static pattern and rotate workpiece.
(f) Display more than one pattern and rotate workpiece.
(g) Display multiple patterns on one workpiece or on a small group of workpieces for simultaneous hardening; patterns may be similar or dissimilar in geometric shape.

Fig. 25.2. Workpiece configurations and heating patterns for electron-beam heat treating.

An energy concentration of 3.1 kW/cm^2 (20 kW/inch^2) is more suitable for selective heat treating. To reduce the beam energy to this level, a single electron beam is programmed through a group of discrete beam positions referred to as a raster pattern.

Programmable Raster Pattern

The advancement that made electron-beam heat treating practical was the development of a programmable dedicated computer-control system coupled with electron-beam equipment. The control system generates a raster pattern by magnetically maneuvering the electron beam accurately from one programmed location to another while controlling energy input at each location, as shown in Fig. 25.4.

The power of the electron beam is distributed over the surface to be hardened at points spaced so that no 'hot spots' exist when a defocused beam is applied. These locations may be spaced as required to obtain the necessary power-density distribution. The electron beam is positioned successively at each specific location, dwelled for an adjustable time interval (20×10^{-6} second minimum) and then translated at high speed to the next co-ordinate point. Patterns that conform to the contours of each surface to be hardened are thus produced. These patterns are repeated throughout the entire heat treating cycle, typically at a rate of 500 times per second.

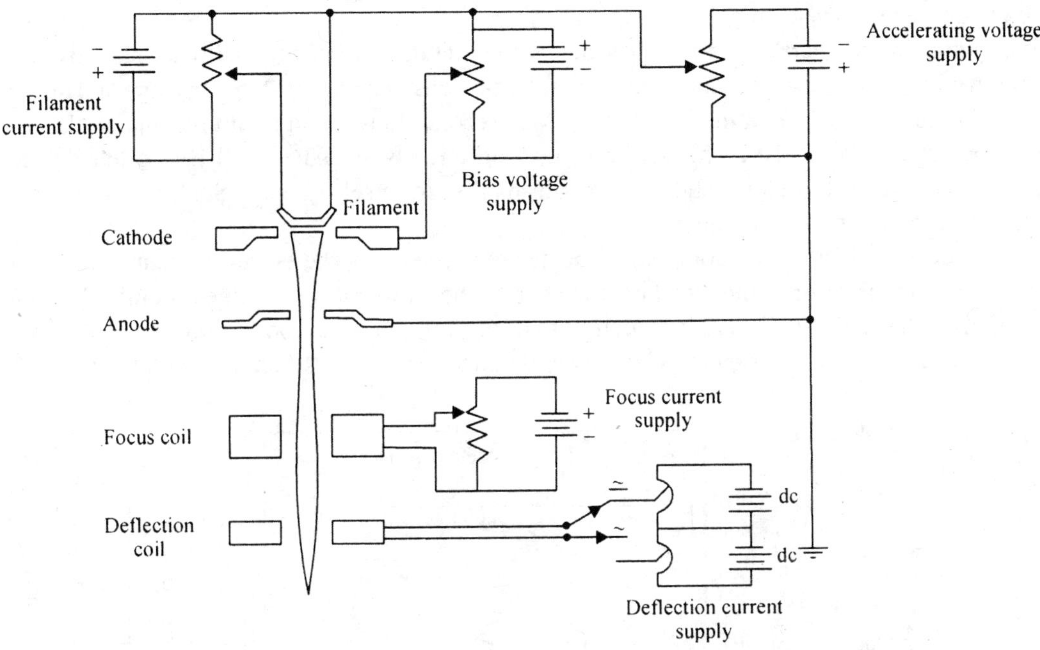

Fig. 25.3. Schematic representation of a 42 kW electron gun.

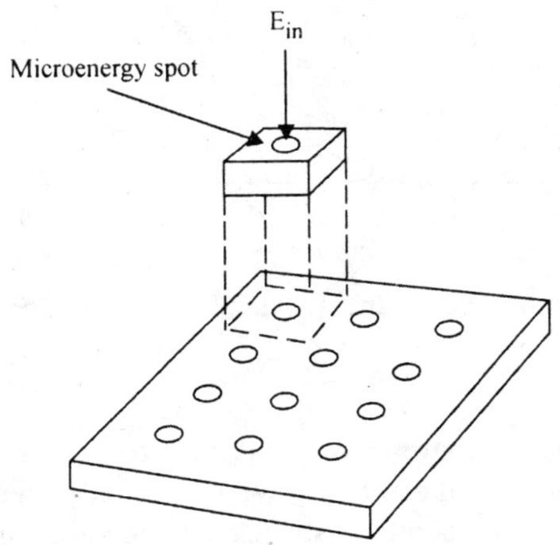

$E_{in} = \frac{P_i t_d}{F}$ (T), where E_{in} is energy input, P_i is instantaneous power, t_d is dwell time, T is total time and F is framing time.

Fig. 25.4. Programmed energy control.

Electron-beam power is also controlled as a function of time. Initial power density is sufficient to heat the surface rapidly to an austenitising temperature just below the melting point of the ferrous alloy being hardened. A thermal gradient is generated from the surface to the interior and the depth of the gradient increases as a function of time. As the thermal gradient gets deeper, the surface temperature tends to rise. To keep the surface temperature just below the melting point, the temperature rise is countered by decreasing the surface power density with respect to time, using the computer capability to adjust electron-beam power rapidly.

The austenitising temperature is held just below the melting point of the alloy to take advantage of rapid carbon migration. Typical minimum times required for austenitisation of the structure and dissolution of the iron carbide (Fe_3C) are on the order of ⅓ to ½ second. In this length of time, material to a depth of only about 0.25 mm (0.010 inch) can be heated sufficiently so that it will be hardened when the electron-beam energy is turned off and the workpiece allowed to self-quench. Soaking for longer times prepares deeper layers for hardening.

When the desired depth of carbon-bearing austenite is prepared, the electron-beam power is turned off. Heat flow to the interior of the metal continues until the temperature becomes equalised throughout the workpiece. The rate usually is high enough to exceed the critical cooling rate and to convert the austenite to hard martensite. A typical relationship among these various factors is shown in Fig. 25.5.

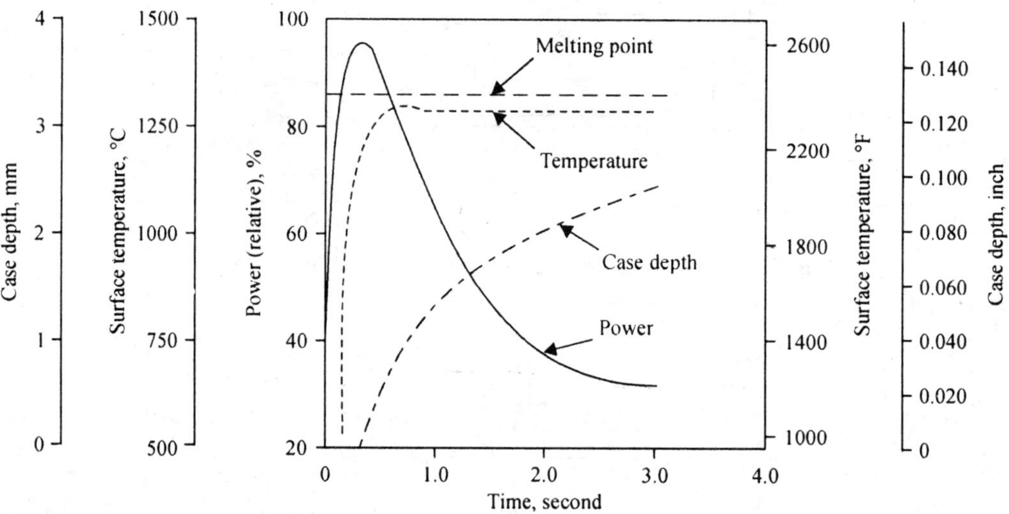

Fig. 25.5. Electron-beam energy input for heat treating.

Raster-Pattern Control

Although part configuration must be compatible with the electron-beam process, as indicated in Fig. 25.2, flexibility is possible through use of static or travelling raster patterns.

Static patterns are raster patterns whose geometric shapes may be varied infinitely. The patterns are displayed on the part with no relative movement between the part and the pattern.

Travelling patterns are used when heat treating of large areas is required. A static pattern of some geometric shape is displayed on the part and motion is applied either to the raster pattern or to the part.

CONTROL OF CASE DEPTH

Hardening depth is simply a function of time. Because the heat applied to the surface travels through the material by conduction, the longer the surface is maintained at a temperature above the critical value, the deeper the heat will penetrate and thus the greater will be the case depth. Moreover, if the temperature of the surface is raised fast enough and if the temperature is sufficiently high to enable the carbon to go quickly into solution, it is possible to obtain a reasonably deep case with very little heat-affected zone.

QUENCH MEDIA

To complete the surface-hardening process, the area heated must be quenched rapidly to develop full hardness. This limits the majority of electron-beam applications to those materials that are self-quenching. There must be sufficient mass beneath the surface being hardened to conduct the heat from the surface to the core of the workpiece rapidly. The transfer of heat must be rapid enough to prevent formation of pearlite and to reduce the surface temperature below the M_s (martensite-start) temperature before bainite transformation begins. For carbon steels, eight units of mass generally are required for each unit of case at the desired hardness.

In such a self-quenching process, quenching of the austenite occurs from the inside to the outside, which means that the inner layers harden first. Because transformation of ductile austenite to hard martensite causes expansion, the inner layer expands and plastically extends the surface layers while the surface is still austenitic. When the surface material finally transforms and expands, it generates moderate biaxial compressive forces in conjunction with the hard, virtually unyielding underlayer. This is an ideal residual-stress condition for fatigue resistance, similar to the benefits derived from controlled shot peening.

ADVANTAGES OF ELECTRON-BEAM HEAT TREATING

Electron-beam heat treating is a precise process that allows the heat treater control over every aspect of the flow of energy into a workpiece for selective surface hardening.

Low Energy Usage

Because the flow of energy is very rapid and the area being heated is held to a selective minimum, the core and the surface area not heat treated will not be adversely affected.

Minimal Part Distortion

When energy flow is controlled and rapid and is restricted to highly selective areas, minimum distortion may be realised.

Small Heat-Affected Zone

The combination of high energy, rapid flow of energy and rapid cooling generally yields very small heat-affected zones.

Flexible Surface Hardening

By moving the pattern over the part or moving the part beneath the pattern, large areas may be heat treated in short periods of time. Depending on case depth, these travelling speeds may be in excess of 150 cm/minute (60 inch/minute).

Compressive Surface Stress

Due to the self-quench phenomenon, ductile austenite forms hard martensite from the sub-surface to the surface. When the volume of the surface layer expands during formation of martensite, this expansion is resisted by the adjacent unyielding layer.

Maximum Hardness

Case hardness attained by means of the electron-beam process follows textbook relationships of hardness and carbon content, as illustrated in Fig. 25.1. The ultimate hardness achieved by this process may be slightly greater than that obtained by conventional heat treating methods because of the high efficiency of self-quenching, which is more effective and much faster than quenching with external quenchants.

Short Heat Treating Time

An accelerated electron beam delivers high-level energy to the surface being modified. Due to this high energy level and to the relatively low thermal conductivities of ferrous materials, heating times for electron-beam heat treating normally are on the order of 0.5 to 2.5 seconds for static patterns. For most applications, heating times range from 0.5 to 1.0 second.

Minimal Workpiece Oxidation

Carbon-bearing materials heat treated in an electron-beam system may be protected in a vacuum environment and thus may be treated without surface discolouration.

Chapter 26

Laser Surface Transformation Hardening

INTRODUCTION

Surface transformation hardening of ferrous materials is an established process widely used to enhance the mechanical properties of highly stressed machine parts, such as gears and bearings. Surface hardening increases the wear resistance of the material and, under favourable circumstances, increases the fatigue strength due to residual compressive stresses that are induced in the workpiece surface by the transformation hardening process. The surface hardening process is not fundamentally different from conventional through hardening of ferrous materials. In both processes, increased hardness and strength are obtained by quenching the material from the austenite region to form hard martensite. Surface hardening differs from conventional through hardening in that only a thin surface layer is heated to austenitisation temperatures prior to quenching, leaving the interior of the workpiece essentially unaffected.

Because ferrous materials are fairly good heat conductors, it is necessary to use very intense heat fluxes to heat the surface layer to austenitisation temperatures without unduly affecting the bulk temperature of the workpiece. This heat input is commonly obtained by the use of very hot flames or by high frequency induction heating. By selectively heating the workpiece surface to austenitisation temperatures, desired surface hardening is obtained by application of a quench medium to the hot surface or by self quenching.

Self-quenching occurs when the cold interior of the workpiece constitutes a sufficiently large heat sink to quench the hot surface by heat conduction to the interior at a rate high enough to allow martensite to form at the surface.

In recent years, industrial lasers have become available for metal-working uses, including surface hardening. A laser can generate very intense energy fluxes at the workpiece surface and the resulting temperature profiles in the workpiece usually can be made steep enough to negate the need for external quench media. The laser beam is a beam of light, which is essentially independent of the workpiece, easily controlled, requires no vacuum and generates no combustion products. It is ideally suited, therefore, for this purpose.

FUNDAMENTALS OF LASER SURFACE HARDENING

When a laser beam impinges on a surface, part of its energy is absorbed as heat at the surface. If the power density of the laser beam (usually given in watts per square centimetre) is sufficiently high, heat will be generated at the surface at a rate higher than heat conduction to the interior can remove it and the temperature in the surface layer will increase rapidly. In a very short time, a thin surface layer will have reached austenitising temperatures, whereas the interior of the workpiece is still cool. Even with a relatively moderate power density of 500 W/cm^2 (3300 W/inch^2), temperature gradients of 500°C/mm (25°F/mil) can be obtained. By moving the laser beam over the workpiece surface (Fig. 26.1), a point on the surface within the path of the beam is rapidly heated as the beam passes. This area is subsequently cooled rapidly by heat conduction to the interioı after the beam has passed. By selecting the correct power density and speed of the laser spot, the material will harden to the desired depth.

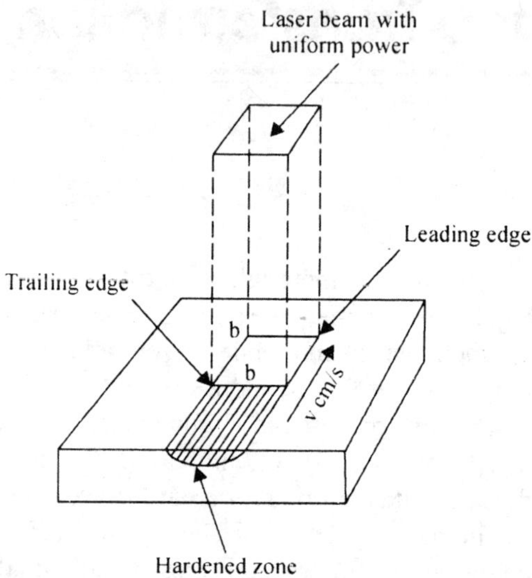

Fig. 26.1. Square laser beam with uniform power density on a flat plate.

A relatively broad area beam, usually in the shape of a square or a rectangle, is used in the laser hardening process. The power density of focused laser beam used for hardening is much lower than the power density of the small, intense focused spots used for welding and cutting. The power density is typically in the 1000 to 2000 W/cm^2 (6400 to 13000 W/inch^2) range, occasionally as high as 5000 or as low as 500 W/cm^2 (32000 to 3200 W/inch^2).

The resulting depth of case will depend on the hardening response of the material, but it will rarely be more than 2.5 mm (0.1 inch). For steel with low hardenability, such as plain carbon steel, the depth of case obtainable is much smaller, varying from perhaps 0.25 mm (0.01 inch) in mild steels to 1.3 mm (0.05 inch) in a medium carbon steel. Because of the very high heating and cooling rates obtainable, it is possible to harden steels not normally considered hardenable, such as SAE 1018. For the same reason, the hardness obtainable by the laser hardening process can, in some instances, be slightly higher than that considered possible with conventional methods.

Ferrous materials are not good absorbers of infrared and far-infrared electromatic radiation. For instance, polished pure iron has an absorptivity of about 4 per cent at room temperature. Grit-blasted cast iron has an absorptivity of 25 per cent at room temperature and this value increases to about 40 per cent at 800°C in an inert atmosphere. The formation of oxides on the ferrous surface can increase absorptivity beyond these values, but efficient use of the laser energy demands the introduction of a controlled high-value absorbing coating on the material surface. Chemical coatings, such as manganese phosphate and paints of graphite, silicon and carbon, have all been used successfully. Some of these coatings may burn off during the heating process and some may leave a residue which in itself can be an indicator of the maximum surface temperature reached. In any event, the absorptivity of these coatings at the beginning of the heating cycle is high (90 per cent or better) and continues to be higher than that of the bare material throughout the temperature excursion.

For shorter wavelength (1.06 μm near-infrared) radiation, as would be the case in (yttrium-aluminium-garnet) (YAG) laser transformation hardening, the room temperature absorptivity may be as high as 60 per cent, but an absorbing coating is still recommended for efficient use of the laser energy.

The major advantages of laser surface hardening include: close control of the power input with modern metal-working lasers; the laser can provide high power density, which in turn minimises the total energy input and, thereby, dimensional distortion and, the ability of the laser to reach normally inaccessible areas on the workpiece surface. Because no vacuum or protective atmosphere enclosure is needed and the distance from the workpiece to the last optical element of the laser system can be quite long, it is possible to process very large or irregular-shaped workpieces.

On the negative side, the depth of case obtainable is limited to about 2.5 mm (0.1 inch), usually less than half of this and the capital cost of the equipment may be high. Therefore, careful analysis of a potential application for laser hardening is needed to ascertain the cost-effectiveness of the process.

METALLURGY OF LASER HARDENING

As stated earlier, the surface hardening process is not fundamentally different from the conventional hardening processes because the same metallurgical reactions occur. In surface hardening, however, the heating portion of the process cycle must, out of necessity, be much shorter than that of conventional hardening. This is particularly true for laser hardening, where heating to the austenitising temperatures occurs within seconds or even fractions of a second. In fact, in laser surface hardening the heating period is frequently shorter than the cool-down time; hence, the expression 'up quench' is often used.

The response of ferrous materials to rapid cool-down from the austenite region has been studied in great detail for many decades and is well understood. The same cannot be said for the heating period of the process. Basically, the problems associated with these high heating rates are that the formation of austenite as well as the redistribution of carbon, necessary to form a homogeneous γFe-C solid solution, are processes that require small but finite time intervals. The kinetics of these processes under very high heating rates are still somewhat uncertain, making the design of a laser surface hardening process more difficult than that of a conventional through hardening procedure.

The basic reaction taking place during the heating period is the transformation of the bcc (α) iron into fcc (γ) iron. This occurs by nucleation and growth of the new phase in the matrix of the old phase. In slow heating, the process will start at A_1 (723°C or 1335°F) in a carbon steel and will be complete at the A_3 line. However, when the heating rate is high, the system is far from equilibrium conditions and the A_3 line will tend to be displaced upward to higher temperatures. Thus, although the temperature may be sufficiently high to form austenite under condition of slow heating, the same temperature level may be insufficient even to initiate austenitisation under high heating rates. Laser hardening parameters are, therefore, usually designed to give peak temperatures well above those employed in conventional hardening to ensure austenitisation. The equilibrium room temperature structure of iron and steel will contain carbon in the form of iron carbide or graphite as a separate phase. To bring about hardening upon quenching, this carbon must be uniformly dissolved in the austenite. To do so, the carbon must be redistributed by diffusion into areas that originated from practically carbon-free ferrite.

This is a time-dependent process even at the high temperatures used in laser hardening and, under certain conditions, laser processing may occur too rapidly to allow for complete diffusion. This problem is obviously most prevalent where the carbon distribution in the starting material is non-uniform, such as in coarse pearlitic structures, structures containing pro-eutectic cementite, spheroidised materials and cast irons, particularly cast iron with a high content of free ferrite.

The intrinsically high heating rates associated with laser hardening, combined with the need to allow sufficient time above the A_3 point to form homogeneous austenite, result in high peak temperatures. Under such conditions, there may be tendency to form a thin surface layer, 5 to 20 μm (195 to 785 micro-inch) thick, of a phase that has yet to be categorised accurately. Retained austenite is found in laser hardened materials as revealed by X-ray analysis, but it is not clear how variation in laser processing parameters influence the amount of retained austenite.

The upper limit for the surface temperature in laser processing is set by the melting point of the material, be cause surface melting is undesirable in most instances. On simple plane surfaces, this is easy to avoid, but on workpieces with more complex design the surface temperature may vary across the beam-impingement area, even if the power density over the beam cross-section is uniform. This can occur when the surface is curved and in particular, when the beam strikes an area with abrupt angles. The projected power density will then be non-uniform and the heat flow uneven, so that part of the illuminated surface may reach the melting point before the remaining area has reached sufficient temperatures to harden to the required depth. Sharp edges and corners can be troublesome, because they tend to concentrate the heat flow and lead to blunting of the edges by melting.

Another problem associated with some laser surface hardening applications is the necessity to overlap hardening passes. This may occur at the closure of a path around a cylindrical workpiece or in the overlap zone between parallel paths being processed sequentially. Because lateral heat flow from the moving laser spot is unavoidable, backtempering will take place by the heat flow into areas already hardened by previous passes or, in the case of closed paths, at the start of the pass. The lower the processing speed, the more pronounced the effect, because relatively more heat will have time to diffuse into the previously hardened area. Even at high speeds and with the application of external heat sinks or water cooling, this effect cannot be entirely eliminated. The use of special optics may eliminate the necessity of forming overlap areas. However, this will always entail illuminating the entire width of the work area with laser radiation and the available power of the laser limits the size of the workpiece that can be processed by the use of such optics. In many applications, the formation of a backtempered zone is not detrimental, provided that the processing is done in such a way that the backtempered area is located in a position of low service stresses or wear. Thus, for example, the interior walls of internal combustion engines may be hardened by straight or spiral laser passes, leaving unhardened material between the passes without harmful effects.

HEAT FLOW IN LASER HARDENING

In laser surface transformation hardening, thermal energy is generated by absorption of the laser radiation at the surface. The increase of temperature in the interior of the workpiece is by way of conduction only; no sources of thermal energy exist below the surface. Thus, if the rate of absorbed power and the thermal properties of the material are known, it is possible, at least in principle, to calculate the temperature distribution in the workpiece. This is of considerable value, because it is then possible to predict the results of laser processing in advance and to calculate the optimum processing parameters, such as power density, processing speed and spot size.

The long wavelength electromagnetic radiation (infrared) from a typical carbon dioxide laser is not efficiently absorbed by ferrous metals at room temperatures. It is, therefore, necessary to cost the workpiece with a substance that will aid in absorbing the laser energy. Commonly used coating materials are manganese phosphate, graphite or carbon-black paint. The paint, in the form of flat, black spray paint, is by far the most convenient coating to use. It is easy to apply and is fairly insensitive to variation in coating thickness.

When laser energy is absorbed at the surface at a rate of 500 W/cm^2 (3200 W/inch^2) or more, surface temperature rises very rapidly because the conduction of heat to the interior cannot keep up with the influx of energy to the surface. The higher the input flux, the more rapidly the temperature rises in the surface layer and as a consequence, the temperature gradient in the workpiece will be steeper. The maximum surface temperature allowable is the melting point of the material, although in practical applications, temperatures should be held well below this value. This clearly acts as a constraint on the depth of austenitisation that can be achieved. If lower laser power density is used and the processing speed is correspondingly decreased, the surface temperature will rise more slowly and the temperature gradient will be less steep. This allows austenitisation to a greater depth, as shown in Fig. 26.2. However, the rate of cooling by self quenching will be slower and it may be insufficient to allow the material to harden fully. Thus, the combined effects of melting temperature and hardenability act to impose a limit on the obtainable depth of case regardless of the power available for the processing. To obtain good self quenching, it is generally necessary to use high power density and high processing speed for steels of low hardenability. Steels of high hardenability can be processed to greater depth by relatively low power densities and low processing speed. For many such steels, the slow processing rate will be necessary to give the material time to form homogeneous austenite. Thus, deep case depth in plain carbon steels may be difficult to achieve. It is possible to use external quench procedures, thereby obtaining deeper case depths in steels of low hardenability. This is achieved at the cost of greater dimensional distortion, because the total power input increases with decreasing speed for constant maximum surface temperature. If the absorbed power density, laser spot dimensions, processing speed and thermal properties of the material are known, the temperature distribution in the workpiece can be calculated by means of several expressions found in the literature. The simplest expression is obtained if it is assumed that heat only flows normal to the workpiece surface, that is one-dimensional heat flow. If the workpiece is large in this direction, the temperature is given by:

$$T = T_o + 2Q/K\sqrt{\alpha t_D}\,\text{ierfc}\frac{\delta}{\sqrt{\alpha t_D}} \qquad \text{... (26.1)}$$

where, T is temperature in centigrade, T_o is room temperature in centigrade, Q is absorbed power in W/cm^2, K is thermal conductivity in W/cm C°, α is thermal diffusivity in cm^2/s, δ is depth below the surface in cm and t_D is dwell time in seconds.

Dwell time is the amount of time a given spot on the surface will be exposed to the laser beam. This is, therefore, equal to the length of the laser spot in the direction of travel divided by the speed of travel. The expression ierfcx is the integrated complementary error function defined by:

$$\text{ierfcx} = \int_X^{\infty} \text{erf} \times dy = \int_X^{\infty} (1 - \text{erfx})\, dy$$

where erfx is the error function:

$$\text{erfx} = \frac{2}{\sqrt{\pi}} \int_0^X e^{-Y^2} dy$$

These functions are tabulated in many sources and can easily be evaluated. Figure 26.3 shows a plot of ierfcx.

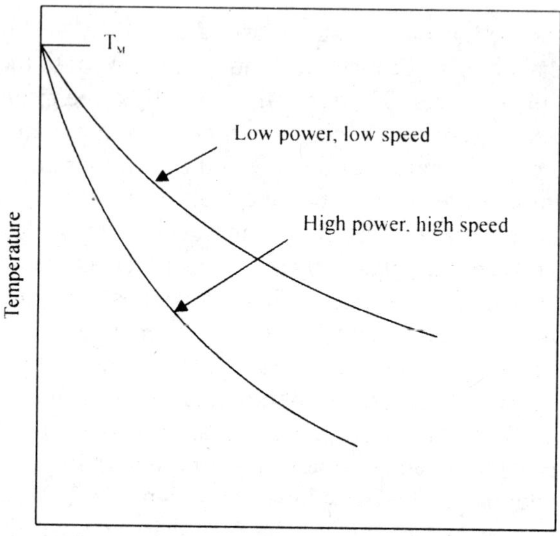

Fig. 26.2. Influence of processing parameters on heat penetration in laser surface transformation hardening.

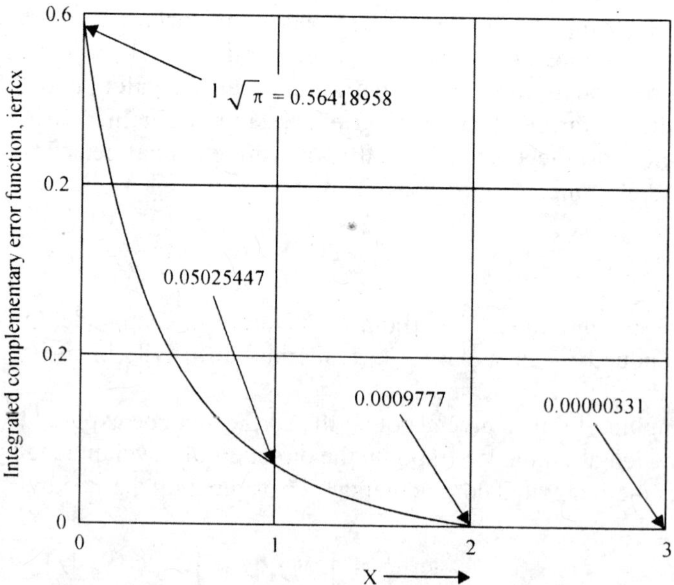

Fig. 26.3. The integrated complementary error function.

If we apply Eq. 26.1 to a moving laser spot on a plane surface, it becomes clear that as the leading edge of the spot reaches a given spot on a plane surface, the temperature will rapidly start to increase and the maximum temperature will be reached at the trailing edge of the spot, as shown in Fig. 26.4. A similar, but smaller, temperature increase will be experienced at points below the surface. If Eq. 26.1 is solved for various values of the depth δ under the trailing edge of the spot, a temperature profile can be constructed representing the maximum temperature conditions in the vicinity of the surface (Fig. 26.5). Knowing the transformation temperature A_3, it is then a simple matter to determine the largest obtainable depth of hardened case δ.

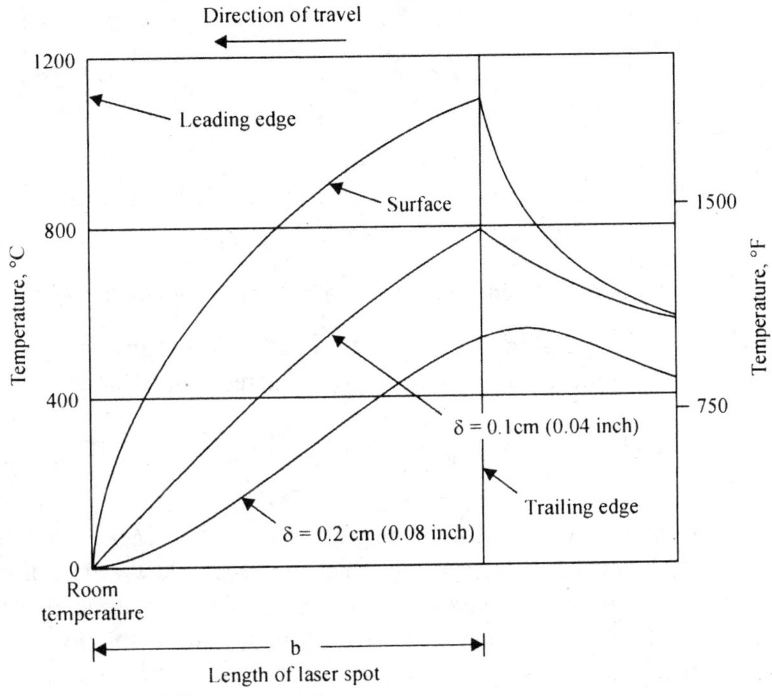

Power, 1200 W/cm^2 (7700 W/inch^2), Speed, 1cm/s (0.4 inch/s)

Fig. 26.4. Temperature distribution along the centre line of a 1 cm × 1 cm (0.4 inch × 0.4 inch) laser spot.

In reality, heat diffuses in all directions into the workpiece from the moving laser spot, not only normal to the surface. Furthermore, heat is lost by radiation from the surface and the thermal properties are not constants but are temperature dependent. Finally, in many instances, the workpiece is not large enough to be considered infinite, as Eq. 26.1 assumes. All of these factors can contribute to make the estimates obtained from Eq. 26.1 unreliable. Nevertheless, at high processing speeds (that is, low dwell time, t_D) and for reasonably thick specimens and large laser spots, Eq. 26.1 gives a good estimate for depth of obtainable case. For a laser spot with dimensions 1.27 by 1.2 cm (0.5 by 0.47 inch), Eq. 26.1 can be safely used down to a processing speed of 1 cm/s (0.4 inch/s) provided that the workpiece is at least 0.6 cm (0.2 inch) thick. It is possible to develop expressions that take into account the three-dimensional nature of the heat flow, but such expressions become very tedious and time consuming in use and it is often not worth the extra work.

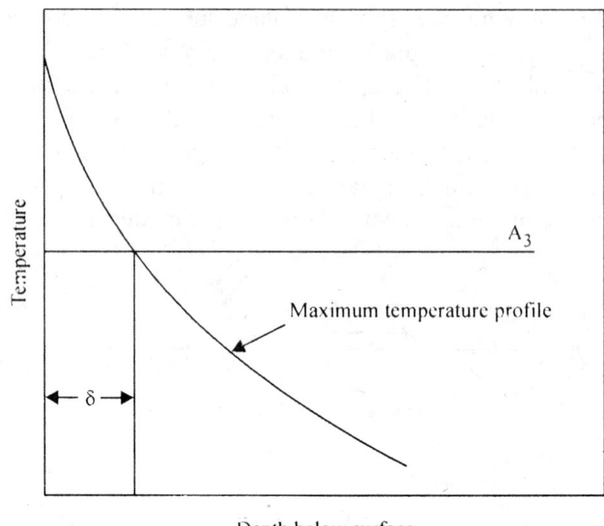

Fig. 26.5. Predicting depth of case in laser transformation hardening.

By a relatively simple expansion of Eq. 26.1, the cooling rate of any given point in the workpiece can be estimated if only one-dimensional heat flow is considered. This is done by means of the expression:

$$T = T_o + 2Q/K\left\{\sqrt{\alpha t}\ \text{ierfc}\frac{\delta}{2\sqrt{\alpha t}} - \sqrt{\alpha(t - t_D)}\ \text{ierfc}\frac{\delta}{2\sqrt{\alpha t - t_D}}\right\} \quad \text{... (26.2)}$$

In this equation, the symbols have the same meaning as in Eq. 26.1 except that t is the time elapsed since the leading edge passed over the spot for which we need to know the temperature; and $t - t_D$ is the time elapsed since the trailing edge passed over the spot, that is, the time elapsed after cooling by self quenching began. When using this equation, it is very important that the workpiece be large enough to provide an adequate heat sink.

Equation 26.2 can be used in conjunction with C-T diagrams to predict whether or not the cooling rate by self quenching will be adequate for hardening under a given set of circumstances. To aid in the use of Eqs. 26.1 and 26.2, Table 26.1 gives the value of the conductivity and the diffusivity for commonly used materials. The listed values are the average values from room temperature up to 1000°C (1830°F) and are given in units convenient for calculation of temperatures by Eqs. 26.1 and 26.2 when the power input (Q) is in W/cm^2. It should be noted that Q is not the applied power density but the estimated absorbed power in W/cm^2.

PROCESSING PARAMETERS

Many factors influence the results of laser surface transformation hardening including size of the laser spot, power density, processing speed, thermal properties of the material and laser hardenability of the material. The last factor encompasses the material's response to rapid heating and quenching and is, in part, dependent on the starting condition of the material, that is, whether the material is normalised, annealed, etc.

Table 26.1. Thermal properties of ferrous materials.

	Transformation temperature (e)		*Thermal diffusivity (a)*		*Thermal conductivity (a)*	
Material	*°C*	*°F*	*cm²/s*	*inch²/s*	*W/cm °C*	*W/inch² °F*
1025	850	1560	0.073	0.011	0.39	0.085
1045	800	1470	0.074	0.011	0.38	0.083
1078	725	1335	0.065	0.010	0.36	0.079
4140	790	1455	0.070	0.011	0.36	0.079
5130	780	1435	0.066	0.010	0.36	0.079
Gray cast iron	–	–	0.099 to 0.148	0.015 to 0.023	0.46 to 0.57	0.10 to 0.0125

(a) To obtain values in SI units, mm²/s and W/m K, multiply by 100.

The parameters of principal interest are power density and processing speed. It is obvious that processing speed should be as high as possible to attain high production rates. However, the speed at which the laser spot moves over the surface is not a good measure of the production rate by itself, because the dimensions of the spot normal to the direction of travel are equally important in determining the area coverage rate. Furthermore, another important factor in determining the results of processing is the time under the beam that a given spot on the surface experiences, that is, the spot dimension in the traveling direction divided by the speed, commonly referred to as dwell time. The relative dimensions of a rectangular spot do not influence the coverage rate as long as the power density stays constant. Therefore, the area of the laser spot that will give a specific result is limited by the available power. An exception to this is when the spot is very narrow in the direction of travel and/or when the speed is very low (or equivalent, the dwell time, t_D, very long). Under such conditions, lateral heat losses become large and the spot dimensions influence the results. The conditions under which this effect becomes noticeable is:

$$B = vb/4\alpha < 3.5$$

where b is the spot dimension in the direction of travel, v is the speed of the spot and α is the diffusivity, all in cgs units.

The following are general guidelines for choice of processing conditions:

1. Usable power densities in laser surface hardening are in the 500 to 5000 W/cm² (3200 to 32000 W/inch²) range. Corresponding dwell times are in the range 0.1 to 10 second. For carbon steels, the power density is usually from 1000 to 1500 W/cm² (6400 to 9600 W/inch²) and the dwell time 1 to 2 seconds.
2. Materials with high hardenability can be processed at low power density and high dwell time (low speed), whereas materials with low hardenability should be processed at high power density and low dwell times.
3. Rectangular or square laser spots with uniform power density are most suitable in obtaining uniform hardened case.
4. High power density and low dwell time give shallow case, but high cooling rates. The reverse is true for low power densities.
5. Maximum surface temperature is proportional to the square root of the speed. Hence, a doubling of the power density requires a quadrupling of the speed to obtain equivalent maximum surface temperatures.
6. Increasing the power density results in lower total energy input for the same maximum surface temperature.

7. Steel with normalised, annealed or spheroidised structures, steel with pro-eutectoid cementite, cast irons and steels with stable alloy carbides require longer dwell times than steels that have been hardened and tempered.
8. Small workpieces will require higher power densities and lower dwell times than large pieces, unless external quenching media are used.

METALWORKING LASERS

Several models of metalworking lasers of both domestic and foreign manufacture are commercially available. The majority of these are of either the neodymium, YAG solid state type or the carbon dioxide gas type. These lasers may have pulsed or continuous output power. Both types, whether pulsed or continuous wave, can be used for transformation surface hardening.

The power output of metalworking lasers at the present time is in the 50 W to 15 kW range and more powerful systems will be available in the future. Because the cost of a laser system is approximately proportional to its maximum power output, the minimum size of a laser system for a potential industrial application should be determined on the basis of such factors as required production rate, minimum practical spot size and power density necessary to achieve the desired results. It should be noted, however, that the laser is a versatile tool and can be used for a number of other metalworking processes such as welding, cutting and hard-facing. If a laser system is to be used in a variety of applications, it may be beneficial to have as much power available as possible.

Lasers emit electromagnetic radiation in the infrared portion of the spectrum and the laser beam from these machines is invisible. The carbon dioxide laser emits radiation with 10.6 μm wavelength and this radiation is easily absorbed in a variety of non-metallic substances. A thin sheet of lucite between the operator and the beam/workpiece interaction zone is sufficient to absorb potentially harmful stray radiation. The output of the YAG laser has a much shorter wavelength (1.064 μm) and therefore, the operator is required to wear special coloured protective eyeglasses. YAG lasers are limited to relatively low power levels. Therefore, in metalworking processes requiring more than 500 W of power, the carbon dioxide laser is usually employed because it can deliver much higher continuous output.

The primary output beam from the laser rarely is used in metalworking applications. Instead, the output beam is directed and shaped by optical systems to generate a laser spot of the desired size and shape on the workpiece surface. Such an arrangement allows substantial flexibility in the use of a laser system. Because the coherent radiation from a laser has low loss of power with distance, the laser itself can be situated at a considerable distance from the work area. Furthermore, different types of metalworking applications can be performed with the same laser by changing the optical system. Several jobs can be performed simultaneously with one laser by using several work stations, each with its own optical system designed for a specific application. Figure 26.6 shows a typical laser system arrangement. In this instance, each work station has its own control console, allowing manual or automatic control of beam power, duration of power delivery to the workpiece, rate of increase of power at the start of the run, rate of decrease at the end (ramp-up and ramp-down) and manipulation of the workpiece fixture protective gas flow. The laser power is automatically maintained at the desired level by a feedback control device in the laser. Positive feedback from temperature sensors, monitoring the temperature in the beam workpiece interaction zone, also can be used. The entire operation can be controlled by microprocessors. By directing the laser beam to the individual work stations in sequence and utilising the time when the power is used at another station for workpiece manipulation, maximum usage of the laser can be achieved.

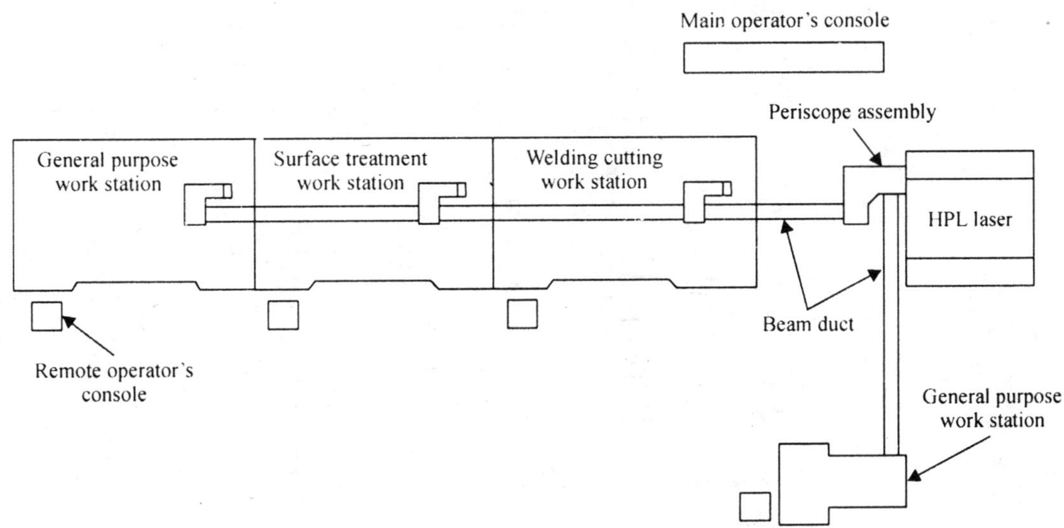

Fig. 26.6. Metalworking laser with multiple work stations.

OPTICAL SYSTEMS

In laser welding and cutting operations, a focused beam of intense power is used. For laser transformation hardening, this focused spot is replaced with a broad beam of much lower power density, partly because the hardening process requires lower power densities and partly to obtain reasonable rates of area coverage.

The simplest way to obtain such a spot is to position the workpiece surface in such a way that it intercepts the beam some distance from the focal plane of a converging beam. However, the power density distribution of spot obtained in this manner is rarely sufficiently uniform and at the same time large enough to give satisfactory results.

The most suitable laser spot for surface hardening is a square or rectangular spot with uniform power density, sometimes referred to as top hat profile (Fig. 26.1). Such a laser spot requires re-shaping of the output laser beam that is attained by the use of various optical systems.

Transmission optical elements, such as lenses and windows, can be used. However, because of the long wavelength of the radiation typical for CO_2 lasers, these elements must be made from special materials such as zinc selenide to avoid excessive absorption.

Reflective optical components are less fragile and better adapted to industrial environment. They consist of flat, spherical or parabolic mirrors made from copper or molybdenum which have excellent reflective characteristics to laser radiation. By the use of these mirrors, the beam can be re-shaped and re-directed to suit the particular application requirement.

The simplest way to form a square or rectangular beam is to use mirrors or lenses to form a focused spot and then to scan this spot rapidly back and forth in two perpendicular directions, as shown in Fig. 26.7. If scanning is done at a sufficient rate, the result will be equivalent to spreading the power of the focused spot over the area being scanned, because the thermal lag of the metal will act to even out the power input to a uniform power density. The scanning can be performed conveniently by electromechanical vibration of one or more of the mirrors in the optical system. In this way, it is possible to generate spots of desired size and shape on the workpiece surface.

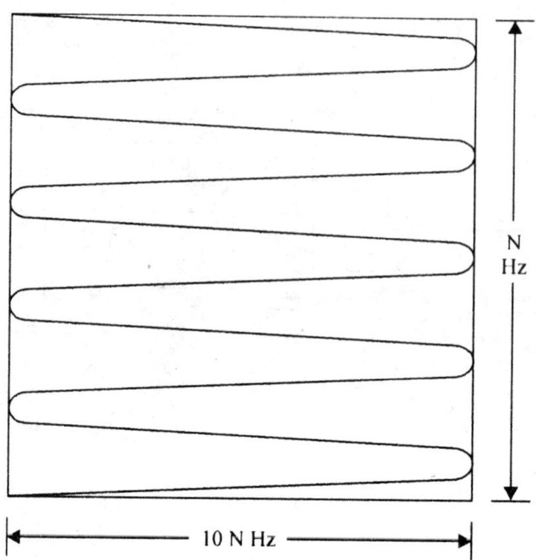

Fig. 26.7. Scanning a small focused laser spot to generate a square, broad area beam.

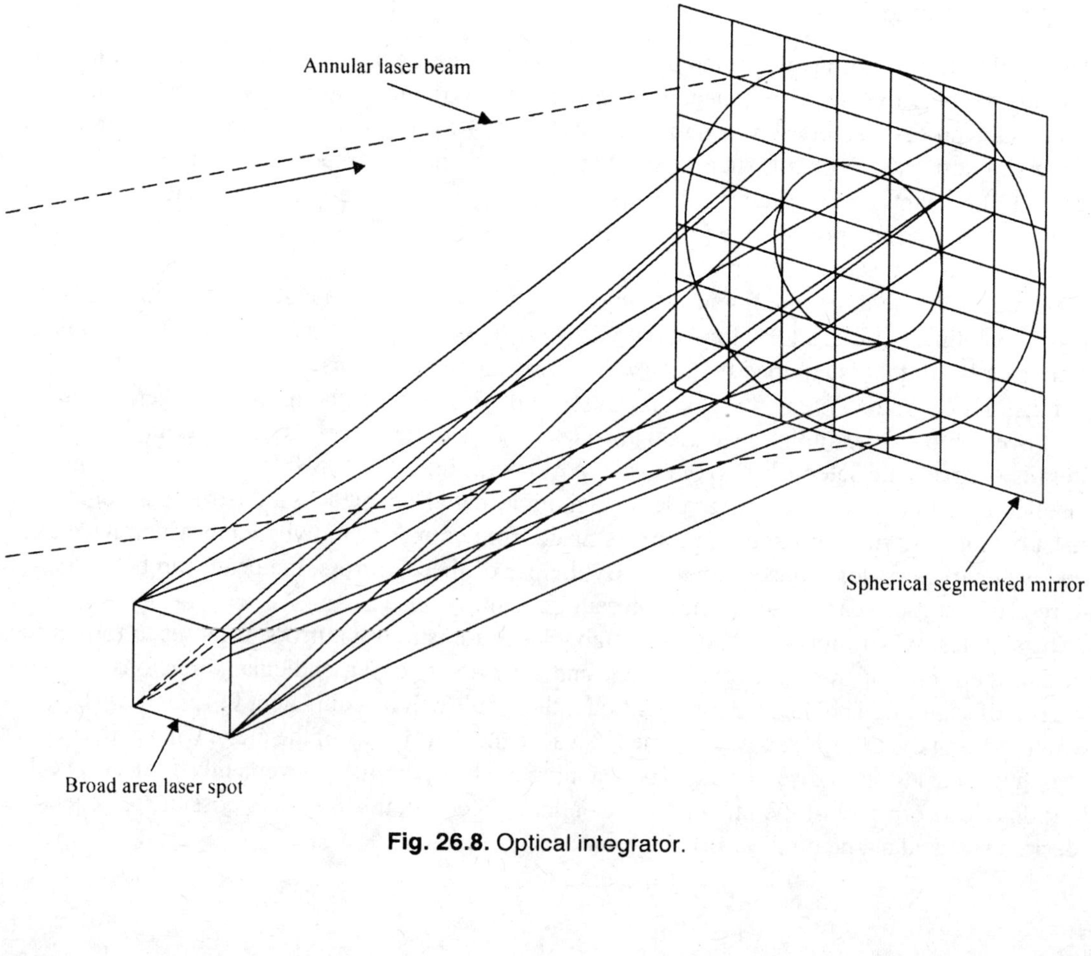

Fig. 26.8. Optical integrator.

Another method of forming a square or rectangular beam is to use a device known as an optical integrator. This device, shown in Fig. 26.8, is an array of flat mirrors mounted on a spherical surface placed to intercept the output laser beam. Each mirror forms an image of the part of the output beam that it intercepts and all cf the images from the array will be formed in the same position. In this way, the output beam is re-shaped to form the desired square or rectangular spot in the focal plane of the integrator, having uniform power density over the illuminated area on the workpiece surface.

Both scanning optics and integrator optics form beams that can be re-directed by flat mirrors or refocused by appropriately curved mirrors. In this way, the laser spot can be generated in the exact shape at desired location. Figure 26.9 shows how a spot can be projected on the inner wall of a tube.

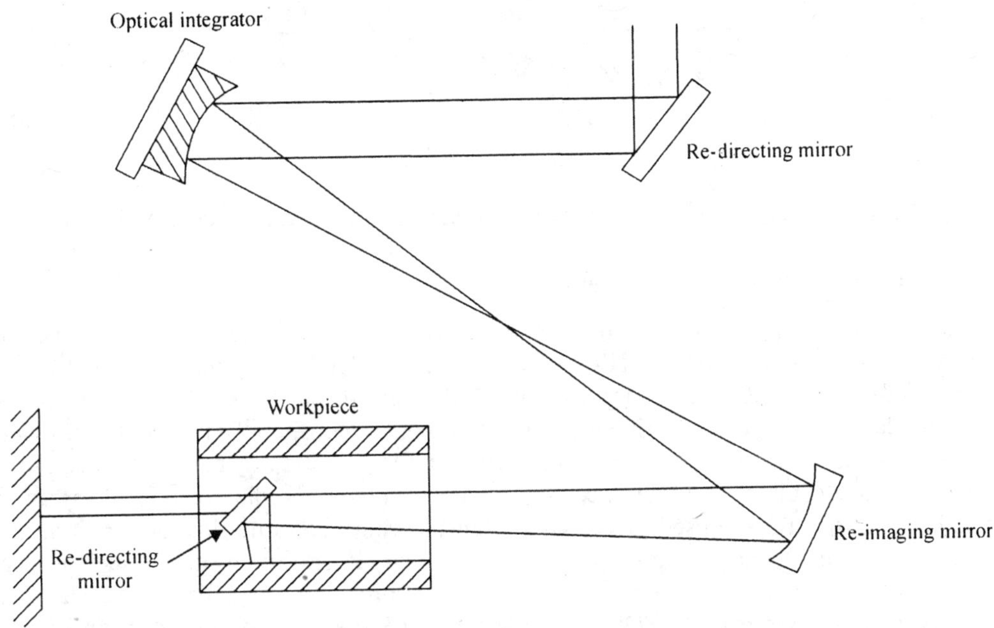

Fig. 26.9. Optical system for laser heat treatment of the inner surface of a cylinder.

For workpieces of cylindrical design, special optical components called toric mirrors can be used. These mirrors take advantage of the fact that the output beam from the laser can be made hollow or annular in cross-section. Directing the beam by suitable mirrors onto the workpiece causes a continuous band of laser irradiated surface to form around the periphery of the workpiece, either on the inner surface of a hollow cylinder or on the outer surface of a cylinder. The complete surface can be laser treated by moving the workpiece in the axial direction through the ring-shape laser spot. No backtempering is encountered because no start/stop or parallel spiral zones are generated. The relatively large area of the ring-shape laser spot limits the size of workpieces that can be processed with this technique to about 25 mm (1 inch) diameter for each 8 kW of power available.

Another technique that can be used for non-planar surfaces is beam splitting. The laser beam is split into two equal parts by a copper prism and the individual parts of the beam are directed at different angles to the workpiece surface by suitable reflective mirrors. Figure 26.10 shows the application of this technique to heat treating a wide semi-circular groove. If only a single beam, normal to the bottom of the groove, had been used, the angle of incidence of the laser beam at the left and right top of the groove would have been too shallow to generate sufficient heat. By using two beams as shown, this problem can be overcome.

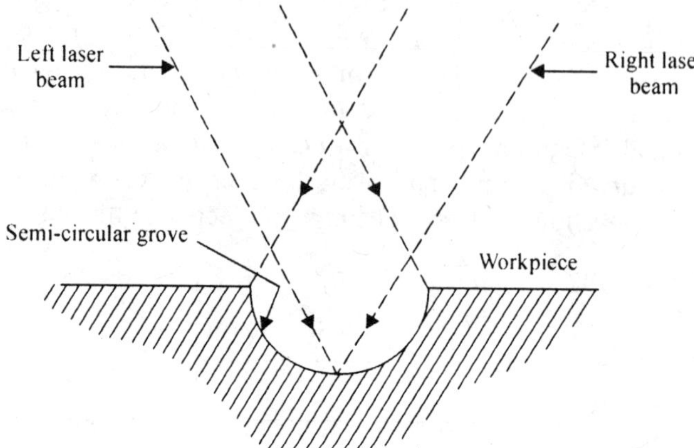

Fig. 26.10. Using two equal laser beams to surface harden a semi-circular groove.

Laser Surface Hardening of 1045 Steel Plate

The steel plate was 12.5 mm (0.5 inch) thick. Prior to laser processing, the material had been hardened and tempered to a hardness of 30 to 32 HRC. The desired case depth was 1 mm (0.04 inch); hardness at this depth should be 45 HRC. Single hardened band, 18 mm (0.7 inch) wide, was required. A 15 kW carbon dioxide laser was used, delivering a square beam with dimensions 18 mm by 18 mm (0.7 inch by 0.7 inch) to the workpiece surface by an optical integrator. Flat, black spray paint was used as the energy absorbing coating and applied to the workpiece surface after removal of surface oxide scale by sandblasting. No protective atmosphere was used, except for a fan blowing across the interaction zone to remove smoke. The workpiece was moved with respect to the stationary laser beam by mounting it on a controlled motion table in the work station. The approximate range of processing parameters was calculated from Eq. 26.1, using 0.38 W/cm°C (0.083 W/inch °F) for the thermal conductivity and 0.074 cm^2/s (0.01 $inch^2/s$) for the thermal diffusivity of 1045. Energy absorption was assumed to be 85 per cent of incident power. The room temperature was 20°C (68°F) and the transformation temperature of 1045 was taken to be 800°C (1470°F). The results obtained are given in Table 26.2. Hence, this workpiece could be laser processed to the required case depth at a speed of 762 mm/minute (30 inch/minute) (coverage rate 136 cm^2/minute or 21 $inch^2$/minute) at a power of 4.1 kW.

Table 26.2. Laser hardening of 1045 plate.

Speed		*Power, kw*	*Measured case*		*Calculated case*	
mm/minute	*inch/minute*		*mm*	*inch*	*mm*	*inch*
510	20	2500	0.52	0.020	0.04	0.002
510	20	3000	1.02	0.040	0.82	0.030
510	20	3600	1.37	0.055	1.56	0.060
760	30	3000	0.24	0.010	0	0
760	30	3600	0.66	0.025	0.60	0.024
760	30	4150	1.24	0.050	1.08	0.043

The relationship between claculated and measured case depth is shown in Fig. 26.11. As expected, using the simple one-dimensional Eq. 26.1, the correlation between claculated and observed results are better at the higher speed.

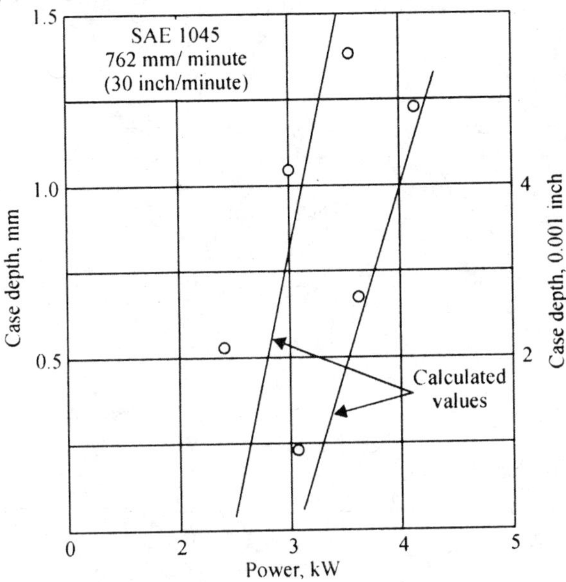

Fig. 26.11. Laser heat treating SAE 1045.

Laser Surface Hardening of 4140 Cylinder with Conical Top

The object was to surface harden the conical part of the workpiece shown in Fig. 26.12 to increase the wear resistance. The minimum required case depth was 1.1 mm (0.044 inch); the hardness at this depth should be 45 HRC. The workpiece had been hardened and tempered to 40 to 42 HRC prior to laser processing to obtain strength and ductility.

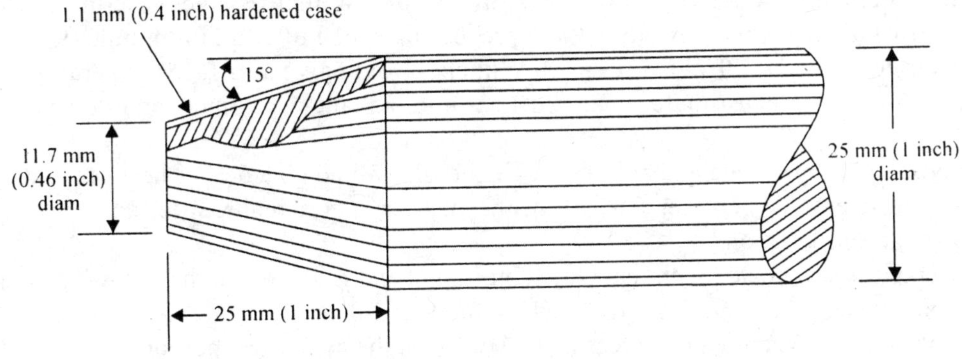

Fig. 26.12. Laser hardening of conical SAE 4140 workpiece.

Because the workpiece was cylindrical and had a maximum diameter of 25 mm (1 inch), a toric mirror could be used in conjunction with a 15 kW CO_2 laser. The processing arrangement is shown in

Fig. 26.13. In this fixture, the workpiece was rotated at 1300 rpm under the ring-shape laser spot formed by the toric mirror. This was done to ensure uniform power density around the periphery of the workpiece. The ring-shape spot had a width of 5 mm (0.2 inch) and hardening of the desired area could be obtained by moving the workpiece in the axial direction. In this way, the entire area could be covered without forming any overlap zones.

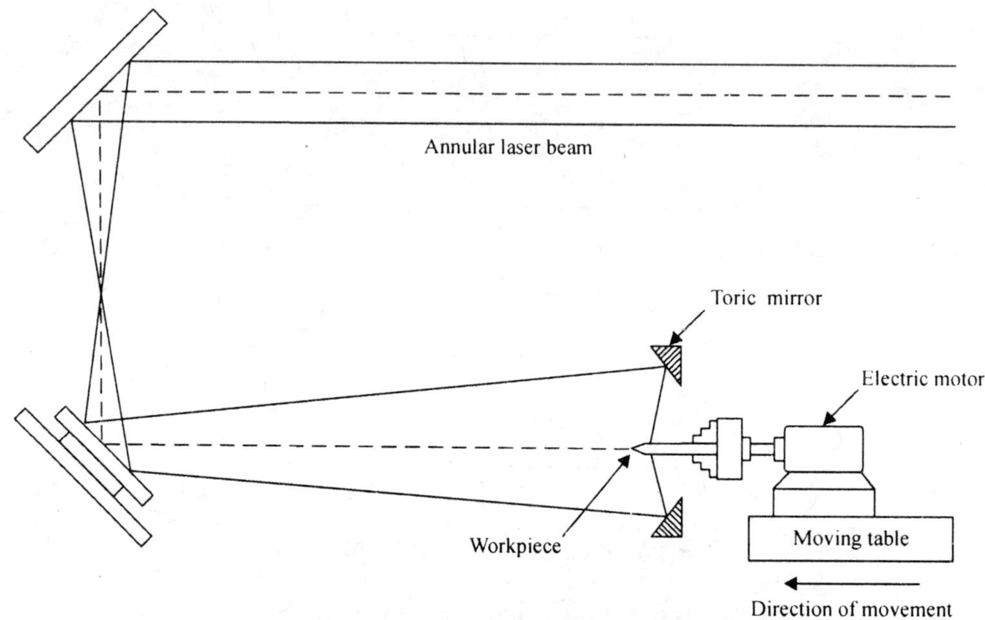

Fig. 26.13. Arrangement for laser heat treating SAE 4140 cylinder with conical tip.

Because the area under the ring-shape laser spot will increase as the beam sweeps from the tip of the cone to the cylindrical part of the workpiece, the power input had to be increased linearly with the distance in order to maintain constant power density. Theoretical calculations, using Eq. 26.1 with a thermal conductivity of 0.35 W/cm°C (0.083 W/inch°F) and a thermal diffusivity of 0.070 cm^2/s (0.01 $inch^2/s$), showed that at a processing speed of 229 mm/minute (9 inch/minute) and a power density of 1620 W/cm^2 (10 400 $W/inch^2$) would give a surface temperature of 1360°C (2480°F). This is only an estimate, as Eq. 26.1 assumes a plane workpiece rather than a cylindrical one.

To process at a slower rate and at lower power density would give deeper heat penetration, but the self-quenching rate would be lower. On the other hand, processing at higher speed and power would lead to shallower heat penetration.

The workpiece was coated with flat, black spray paint and processed at the following parameters:

1. Axial speed: 221 mm/minute (8.7 inch/minute); or 3.8 mm/s (0.15 inch/s).
2. Power: 3500 W at the tip, increasing to 7600 W at the cylindrical portion.
3. Rate of power increase: 620 W/s.

The hardened case obtained ranged from 1.65 mm (0.06 inch) at the tip to 2 mm (0.08 inch) at the cylindrical portion of the workpiece. The surface hardness was 58 to 59 HRC.

Laser Surface Hardening of Cast Iron Camshaft Lobes

The surface of the lobes of an automotive camshaft made from ductile cast iron (Fig. 26.14) was to be surface hardened to increase wear resistance. The desired case depth, defined as the depth where the hardness was 50 HRC, was 0.5 to 1.0 mm (0.02 to 0.04 inch).

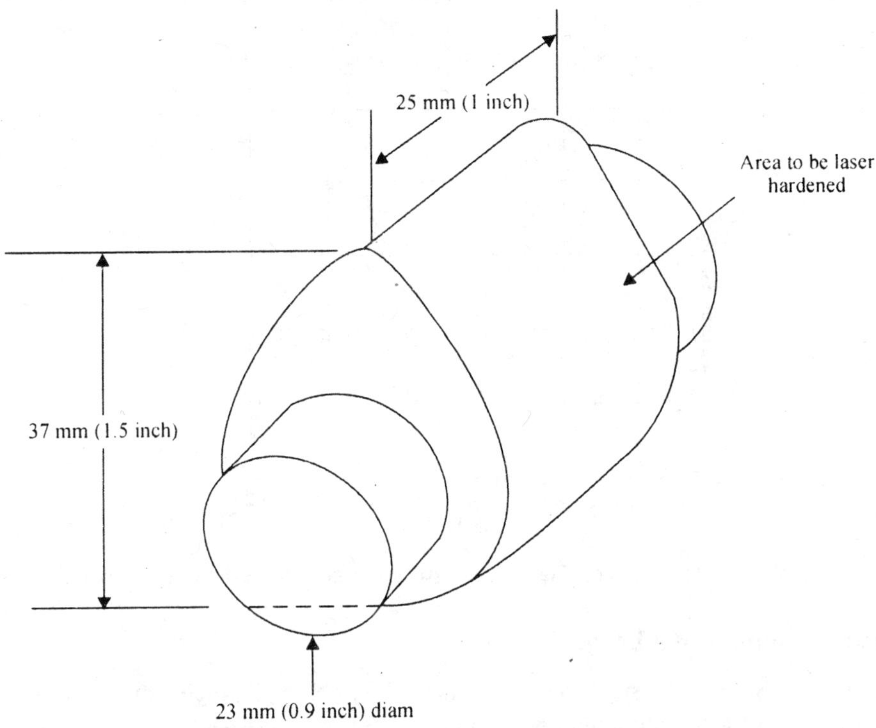

Fig. 26.14. Ductile cast iron cam from automotive camshaft.

A 15 kW CO_2 laser was used for the processing. The optical system delivered a focused spot, with a diameter of 10 mm (0.4 inch) to the workpiece. This spot was scanned over a distance of 22 mm (0.9 inch) normal to the direction of processing and 25 mm (1 inch) in the direction of processing. The frequency of scanning was 125 Hz in the normal direction and 700 Hz in the processing direction, forming a rectangular spot 22 mm by 25 mm (0.9 inch by 1.0 inch) on the camlobe surface.

To obtain an even hardened case around the periphery of the camlobe, it was necessary to vary the angular speed of rotation of the lobe under the laser beam. The reason is that the angle of incidence of the laser beam to the workpiece changed during rotation, from nearly normal incidence at the cylindrical portion of the lobe to a grazing incidence of only 20 to 30° at the flat portion. Furthermore, at constant rotational speed, the linear speed of processing would vary as the lobe rotated.

This was obtained by mounting the workpiece on a rotary table. The speed of rotation was varied by means of an electromechanical controller in a pre-determined manner.

The camlobe was laser hardened, using a manganese phosphate coating to increase energy absorption. Because of the design of the workpiece, it was difficult to predict the optimum processing parameters by calculations and the parameters were, therefore, evaluated by trial and error.

The results were:

Power input	9 kW
Power density	1600 W/cm^2 (10300 W/inch^2)
Linear speed of processing	760 mm/minute (30 inch/minute) at the cylindrical portion 180 mm/minute (7 inch/minute) at the flat portion
Depth of case	0.55 mm (0.022 inch)

The hardness profile of the surface layer of the camlobe is shown in Fig. 26.15.

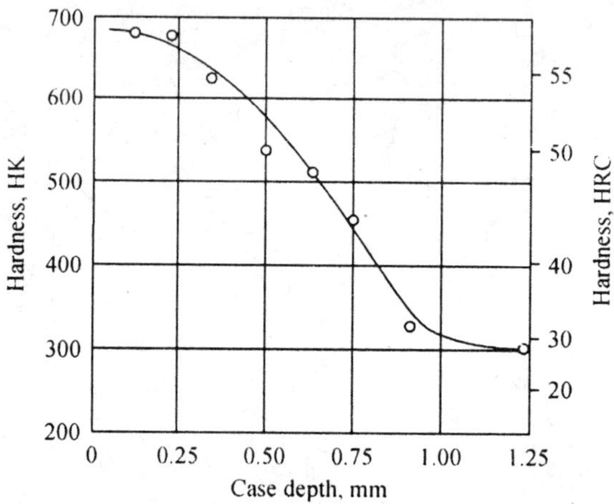

Fig. 26.15. Hardness profile of laser surface hardened cast iron camlobe.

Laser Surface Hardening of a Large Gear

The gear, made from SAE 1045 steel, had a diameter of 28 cm (11 inch) and a thickness of 10 cm (4 inch). The gear had 34 teeth and a diametral pitch of 3.35. To increase fatigue and wear properties, a hardened case of 1 mm (0.04 inch) was desired, extending in a continuous manner from the tip of one tooth to the tip of an adjacent tooth, as shown in Fig. 26.16.

In this application, two laser beams had to be directed toward the surface of the gear teeth at an angle of 15.5° from the normal to the root area, as shown in Fig. 26.16. These two beams must be abutting but not overlapping at the root between the teeth. A single beam with normal incidence to the root area could not be used because the angle of incidence at the adjacent fillets would be too shallow to generate a hardened case in this critical area, where fatigue cracks are likely to originate.

The output beam from a 15 kW CO_2 laser was split into two equal parts by a reflective copper wedge. Each beam was then directed to the workpiece surface in the form of a 12.5 mm by 12.5 mm (0.5 inch by 0.5 inch) spot by two optical integrators, as shown in Fig. 26.17. The projected area of each beam on the wokpiece surface was approximately 12.5 mm by 25.4 mm (0.5 inch by 1 inch); thus, the total irradiated area was 3.18 cm^2 (0.5 inch^2).

After surface preparation with flat, black spray paint, the gear was laser processed at a total power input of 8.8 kW and with a translation speed of 500 mm/inch (20 inch/minute). The calculated overall power density was 1380 W/cm^2 (8900 W/inch^2), but the actual power density varied somewhat over the interaction area because of the variation of incidence angle to the curved surface. Sequential runs were made on teeth 120° apart to minimise heat distortion. Water cooling was used on tooth flanks adjacent to the processing area to prevent backtempering of previously processed teeth. The obtained case depth (45 HRC) was 1.2 to 1.3 mm (0.046 to 0.052 inch) on mid flank and at the root and 1 mm (0.04 inch) at the fillets. Surface hardness was 59 to 60 HRC.

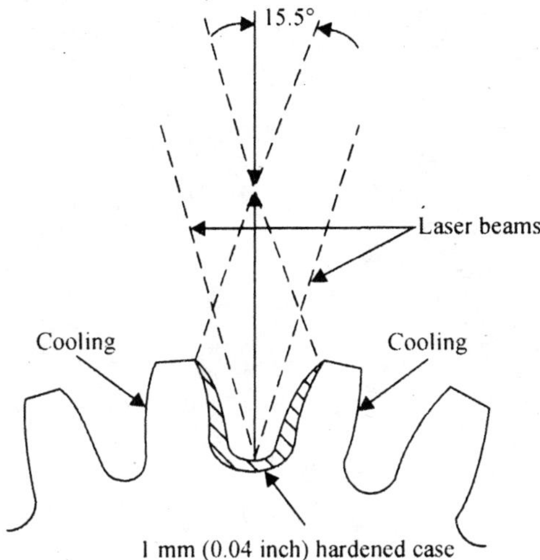

Fig. 26.16. Laser hardening of SAE 1045 gear.

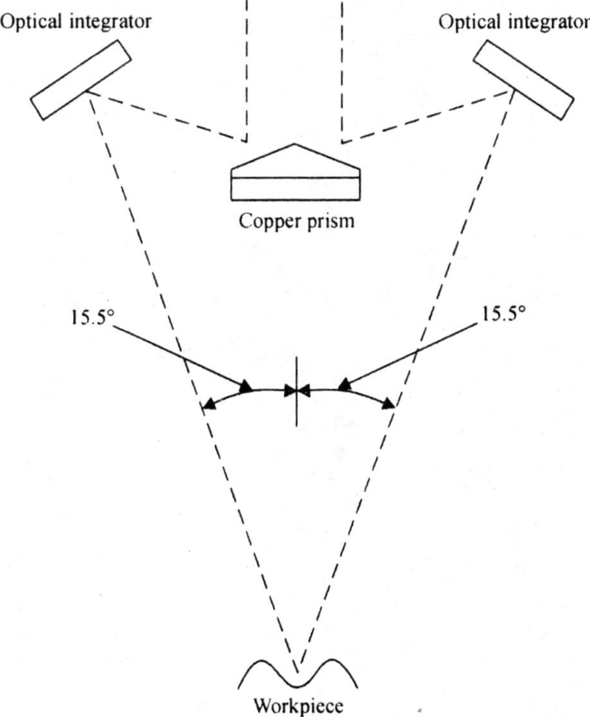

Fig. 26.17. Dual beam optics for laser gear heat treating.

References

Allen, L.K., *Heat Treatment of Metals*, Butterworths, London.

Alvarez, J.L., *Mechanical Properties of Steel*, John Wiley & Sons, New York.

Arceivala, K.J., *Case Hardening of Steel*, Marcel Dekker Inc., New York.

Barsom, K.G., *Heat Treatment of Metals*, Swedish Institute of Metal Research.

Batterman, S.A., *Diffusion in Solids*, McGraw-Hill, Tokyo.

Brown, M.H., *Alloying Elements in Steel*, Cambridge University Press, Cambridge.

Cambell, K.E. and Lemer, H.A., *Iron and Steel Alloys*, Academic Press, London.

Chang, J.C., *Fundamentals of Mettallography*, John Wiley & Sons, New York.

Coolingwood, R.W., *Temperability of Steel,* John Wiley & Sons, New York.

Cox, C.S., *Theory of Heat Treatment of Steel*, Marcel Dekker, New York.

Downe, S.A., *Hardenability concepts with Applications of Steel*, John Wiley & Sons, New York.

Friedel, J., *Tempering of Steel*, Van Nostrand, London.

Gould, G.W., *Physical Metallurgy Principles*, D. Van Nostrand, New York.

Hillert, M., *Decomposition of Austenite by Diffusional Processes, Intersciences*, New York, London.

Huff, C.B., *Cast Iron and Ductile Iron*, Elsevier Scientific Publishing Co., Amsterdam.

Jackwerth, E., *Heat Treatment Furnaces and Equipments*, Academic Press, London.

James, A. and Evison L., *Annealing of Steel*, John Wiley & Sons, New York.

Jencks, W.P., *Salt and Fluidised Bed Equipments*, John Wiley & Sons, New York.

Kiessling, R., *Heat Treatment of Non-ferrous Metals*, The Metals Society, London.

Kim, C.K., *Diffusio in Solids, Liquids and gases*, Marcel Dekker, New York.

Lechevallier, M.W., *Dimensional Stability of Steel*, Academic Press, London.

Lewis, R.L., *Heat Treatment Furnaces*, Chilton Book Co., USA.

Mason, C.F., *Testing of Metals*, Longman Group Ltd., London.

McCaull, J. and Crossland, J., *Hardening of Steel*, Harcourt Brace Jovanovich, New York.

Mitchell, R., *Temperature Control,* McGraw-Hill, New York.

Nyer, E.K., *Normalising of Steel*, Van Nostrand Reinhold, New York.

Phillips, D.J.H., *Testing of Metals*, Applied Science Publishers, London.

Reid, G.K., *Carburising and Nitriding*, Reinhold Publishing Corporation, New York.
Samson, R.A., *Physicalmetallurgy*, Pergamon Press, Oxford, London.
Schrowebel, J., *Induction Hardening of Steel*, Pergamon Press, New York.
Scott. H., *Influence of Tempering Temperature on Quenched Steel*, Progress Publishers, Moscow.
Snell, I.D. and Snell, C.T., *Chemical Heat Treatment*, D. Van Nostrand, New York.
Southwood, T.R.E., *Cyaniding and Carbonitriding,*, Swedish Institute of Metal Research, London.
Stephens, W.D., *Metallography*, John Wiley & Sons, New York.
Stumm, W. and Stumm-Zollineger, E., *Distortion in Tool Steel*, Wiley Interscience, New York.
Teal, J.M., *Nature of Metals and Alloys*, Pergamon Press, New York.
Thelning, K.E., *Steel and its Heat Treatments*, Butterworth, U.K.
Wyatt, G.M., *Non-metallic Inclusions in Steel*, Reston Publishing Co., Reston, Virginia.

Index

D

E

❑❑❑

READERS NOTES

READERS NOTES

READERS NOTES

READERS NOTES

READERS NOTES